Optical Fiber Communications Systems

Theory and Practice with MATLAB® and Simulink® Models

Le Nguyen Binh

CRC Press
Taylor & Francis Group
Boca Raton London New York

CRC Press is an imprint of the
Taylor & Francis Group, an **informa** business

CRC Press
Taylor & Francis Group
6000 Broken Sound Parkway NW, Suite 300
Boca Raton, FL 33487-2742

Printed in the United States of America on acid-free paper
10 9 8 7 6 5 4 3 2 1

International Standard Book Number: 978-1-4398-0620-3 (Hardback)

Library of Congress Cataloging-in-Publication Data

Binh, Le Nguyen.
Optical fiber communications systems : theory and practice with MATLAB and Simulink models / author, Le Nguyen Binh.
p. cm.
"A CRC title."
Includes bibliographical references and index.
ISBN 978-1-4398-0620-3 (alk. paper)
1. Optical fiber communication. 2. Optical fiber communication--Simulation methods. I. Title.

TK5103.592.F52.B56 2009
621.382'75--dc22 2009037730

Visit the Taylor & Francis Web site at
http://www.taylorandfrancis.com

and the CRC Press Web site at
http://www.crcpress.com

To the memory of my father

To my mother

To Phuong and Lam

Contents

Preface

This book covers the principles, practices, and modeling of optically amplified fiber communications systems using MATLAB® and Simulink® platforms. It is intended for use in university and professional training courses in the specialized field of optical communications based on single-mode optical fibers. It is based on lecture notes and should also appeal to undergraduate engineering and science students who have already completed courses in electromagnetic theory, signal processing, and digital communications, and as an introduction to modeling for optical engineers, designers, and practitioners in the industry.

The contents of this book have been developed as a set of lecture notes for senior students (Level 4) pursuing their bachelor degrees in computer systems engineering and master of telecommunications engineering at Monash University, Melbourne, Australia; thus the book is not a compendium of all the multifaceted aspects of lightwave optical fiber communications engineering.

Optical fiber communications technology has developed rapidly since the early 1970s and has, in combination with the advancement of digital processing technology, revolutionized global communications. Currently, the transmission of 40 Gb/s per channel in dense wavelength division multiplexed optical systems of 80 wavelength channels has been achieved, leading to the possibility of a transmission capacity of 3–10 Tb/s/single-mode fiber. The emerging technological development of 100 Gb/s Ethernet under either incoherent or coherent detection with the incorporation of electronic processing will further stretch the speed and capacity of optical fiber communications and networks in terrestrial and intercontinental information transport networking.

The applications of optical fibers and related optical technologies span across all optical components of optical communication engineering. The emphasis is on concepts and interpretation, mathematical procedures, and engineering applications. In this approach, the first two chapters present the groundwork in the propagation of lightwaves in planar slab optical waveguides and optical fibers. Single-mode fibers have been studied widely and thus only their principal parameters for operations and for the identification of structures are provided rather than a detailed analysis of the design of optical fibers.

MATLAB® software packages are now a common computing platform for students in university systems worldwide. It is thus sensible to make programs and simulation models available in MATLAB so that students and instructors can use these for laboratory experiments as well as for further research developments. Therefore, in this book we provide a detailed description of MATLAB and Simulink® models. Thus, the principles of operation of all optical components and optical systems are much more important than their detailed mathematical descriptions.

Chapter 1 gives an overview of the development of optical fiber communications technology over the last three decades of the twentieth century. Chapters 2 and 3 provide a fundamental understanding of lightwave transmission through optical fibers, and readers who are familiar with this can skip these chapters and move on to the other chapters on optical transmitters and receivers. Transmitters and receivers are treated independently, and they form the basic elements of optical communications systems.

Chapters 4 and 5 describe the optical transmitters for direct and external modulation techniques, respectively. A combination of coherent detection and digital signal processing will undoubtedly play a major role in next-generation ultra-high-speed optical transmission systems. Chapter 6 describes techniques to generate advance modulation formats such as discrete differential phase shift keying, continuous phase shift keying like minimum shift keying etc. Therefore, the detection of optical signals under direct coherent and incoherent receptions is described in Chapters 7 and 8. This is followed by two chapters on lumped Er:doped and distributed Raman optical amplifiers (Chapters 9 and 10) with extensive models for the amplification of signals and structuring the amplifiers on Simulink platforms.

Chapter 11 then outlines a design strategy for optically amplified transmission systems coupled with MATLAB and Simulink models, including dispersion and attenuation budget methodology and simulation techniques. Chapter 12 provides an introduction to advanced modulation formats for long-haul optical fiber transmission systems accompanied by Simulink models. Over the last decade, optical communications systems for extremely long and extremely high bit-rate transmission, which employ advanced modulation formats, have progressed rapidly. We thus present in Chapter 12 techniques for the generation of modulation formats and optical transmission. Chapters 11 and 12 deal with the advanced aspects of optical communications engineering for long-haul optical communications systems and intercontinental networks, with emphasis on the design and implementation of these optical communications beyond dispersion limits and networks.

Further emphasis has also been placed on wavelength division multiplexed optical fiber communications systems and networks, which provide an advanced perspective on current and future (beyond 2010) networking of multi-carrier optical multiplexed communications systems engineering.

The contents of this book have been taught to undergraduate students at Monash University for over a decade. Contributions and questions from many undergraduate and postgraduate students have enriched the content of this book. In particular, Prof. Ngo Quoc Nam, Dr. Nguyen Duc Nhan, Dr. Lam Quoc Huy, Calvin Li, K.-Y. Chin, Ho Siang Chiong, and Duc Lam have contributed to several software sections of the Monash Optical Communications Systems Simulators using both MATLAB and Simulink platforms as well as experimental platform setups. Furthermore, many challenging questions from students attending courses on this subject have made the author think and reflect on the solutions and the field of optical communications in general.

Throughout the last decade, the course developed at Monash University has undergone several changes, and advanced aspects of optical communications engineering have now been included to provide students at the honors level a deeper understanding of the future development of these optical systems and networks. Several fundamental issues involving coherent optical communications have been taught. However, we are now more certain of the development and deployment of optical systems and networks in the next few decades of the twenty-first century. These will be long-haul and wavelength multiplexed optical systems and distribution optical networks.

The contents of the chapters are thus focused to provide a practical understanding of the fundamental issues to students who may find this useful later in their engineering careers. Readers, especially lecturers, who may be interested in some sampled basic Simulink models described in this book can contact the author by e-mail.

There will no doubt be mistakes in the book and the author would appreciate constructive feedback from readers and scholars in order to improve the next edition of the book.

Last but not least, I would like to sincerely thank my wife Phuong and our son Lam for their understanding during the preparation of this book. My mother has always instilled strength and study passion in me. I therefore dedicate this book to her even though she would not understand a single word.

For MATLAB® and Simulink® product information, please contact:

The MathWorks, Inc.
3 Apple Hill Drive
Natick, MA 01760-2098 USA
Tel: 508 647 7000
Fax: 508-647-7001
E-mail: info@mathworks.com
Web: www.mathworks.com

Le Nguyen Binh, PhD
Glen Iris, Australia

Author

Le Nguyen Binh received his BE (Hons) and PhD in electronic engineering and integrated photonics in 1975 and 1980, respectively, both from the University of Western Australia, Nedlands, Western Australia, Australia. In 1981, he joined the Department of Electrical Engineering at Monash University, Clayton, Victoria, Australia, after three years with the Commonwealth Scientific and Industrial Research Organisation (CSIRO) Australia as a research scientist developing parallel microcomputing systems. He was appointed as a reader at Monash University in 1995. He has worked and led major research and development programs in the Department of Optical Communications at Siemens AG Central Research Laboratories in Munich, Germany, and the Advanced Technology Centre of Nortel Networks in Harlow, United Kingdom. He was a visiting professor in 2007–2008 and is currently an adjunct professor in the Faculty of Engineering at Christian Albretchs University of Kiel, Kiel, Germany. Dr. Binh has published more than 250 papers and technical reports in leading journals and international conferences. He has also written two other books: *Photonic Signal Processing* (2007) and *Digital Optical Communications* (2008) (both published by CRC Press, Boca Raton, Florida). He is also an associate editor of the *International Journal of Optics* and *Research Letters in Optics*, and the series editor of the Optics and Photonics series for CRC Press.

His current research interests include Tb/s optical transmission systems and networks, especially advanced modulation formats and associate electronic equalization methods, photonic signal processing, integrated photonics, and multi-bound soliton fiber lasers.

List of Abbreviations and Notations

A	Area, constant
A_{eff}	Effective area of fiber evaluated as the area of the guided mode
a	Fiber core radius—see also mode spot size and mode field diameter
a_k	Integer "1" or "0"
AM	Amplitude modulation/modulator
ASE	Amplified spontaneous emission
ASK	Amplitude shift keying
α_L	Attenuation (linear scale)
α_{dB}	Attenuation factor in dB
B_e	3 dB bandwidth—electrical
B_o	3 dB bandwidth—optical
b	Normalized propagation constant—see also β
b_0, b_1	Energy of a "0" or "1" transmitted and received at the front end of a photodetector
β	Propagation constant of fiber
β_1	First-order differentiation of the propagation with respect to the angular frequency–propagation delay
β_2	Second-order differentiation of the propagation constant–group velocity dispersion (GVD)
β_3	Third-order differentiation of the propagation constant–dispersion slope
BDPSK	Binary differential phase shift keying
C	Constant or capacitance
C_d	Junction capacitance of photodiode
C_f	Capacitance associated with feedback resistance—see also feedback impedance
C_T	Total capacitance
c	Velocity of light in vacuum—*note*: use exact value of c for calculation of wavelength grid for DWDM; $c = 2.998$ e8 m/s
CPFSK	Continuous phase frequency shift keying
CPM	Continuous phase modulation
CS-RZ	Carrier-suppressed return-to-zero format
CD	Chromatic dispersion
d	Fiber diameter ($= 2a$)
D	Dispersion factor (usual unit = ps/nm/km)
D_T	Total dispersion factor—chromatic dispersion
D_w	Waveguide dispersion factor (see also material dispersion factor M)
DBM	Duobinary modulation
DCF	Dispersion compensating fiber
DCM	Dispersion compensating module
DD	Direct detection
Δ	Relative refractive index difference
Demux	Demultiplexer
δ	Delta-factor—equivalent to the Q-factor optical carrier frequency (center)
DFB	Distributed feedback (laser)

DI	Delay interferometer
DPSK	Differential phase shift keying
DQPSK	Differential quadrature phase shift keying
DRA	Distributed Raman amplifier
DSF	Dispersion-shifted fiber
DSP	Digital signal processing (processor)
DuoB	Duobinary
EDFA	Erbium-doped fiber amplifier
ESI	Equivalent step index
η	Quantum efficiency
F	Fourier transform
F_n	Noise figure in linear scale (see also NF in log scale)
f	Frequency
f_o or ν	Optical frequency
G	Avalanche gain of APD
$\langle G^2 \rangle$	Average avalanche gain/multiplication factor of a photodetector (APD)
G^{2+x}	Avalanche gain/multiplication factor including material dependent factor x
G_o	Gain of optical amplifier (linear scale) or open loop gain of electronic amplifier
g_m	Transconductance—small signal model of an electronic amplifier
GVD	Group velocity dispersion
FWM	Four-wave mixing
h	Plank's constant $h = 6.624\text{e}{-34}$ J-s
I, i	Current large signal or bias current and small signal
i_d	Dark current of photodetector
i_{Neq}	Total noise equivalent current as seen from the input of an electronic amplifier
i_s	Signal current
IF	Intermediate frequency
IM/DD	Intensity modulation/direct detection
IM	Intensity modulation/modulator
I–Q	In-phase and quadrature
ITU	International Telecommunications Union
k_B	Boltzmann's constant 1.38e−23 J/K
L	Length of optical fiber
L_D	Dispersion length
LO	Local oscillator
MADPSK	Multilevel (M-ary) amplitude-differential phase shift keying
MI	Modulation instability
MMF	Multimode optical fibers
MSK	Minimum shift keying
MZDI	Mach–Zehnder delay interferometer
MZI	Mach–Zehnder interferometer
MZIM	Mach–Zehnder interferometer modulator or Mach–Zehnder interferometric intensity modulator
Mux	Multiplexers or multiplexing
NA	Numerical aperture

NF	Noise figure in dB
NLPN	Nonlinear phase noise
NLSE	Nonlinear Schröedinger equation
NRZ	Non-return-to-zero pulse shaping
NZDSF	Nonzero dispersion-shifted fiber (ITU-655)
OA	Optical amplifier (see also EDFA and ROA for lumped and distributed amplification)
O-DPSK	Offset differential phase shift keying
OFDM	Orthogonal frequency division multiplexing
OOK	On–off keying or amplitude shift keying (ASK)
OPLL	Optical phase locked loop
OSNR	Optical signal-to-noise ratio
P_o	Optical power out put
$p_s(t)$	Signal power
Pdf	Probability density function
PLL	Phase locked loop
PM	Phase modulator
PMD	Polarization mode dispersion in unit of $\text{ps}/\sqrt{\text{km}}$
PMF	Polarization maintaining fiber
q	Electronic charge $q = 1.6\text{e}{-16}$ C
QAM	Quadrature amplitude modulation
$\Re$	Responsivity
ROA	Raman optical amplifier
RS	Receiver sensitivity in dBm, calculated by referring to 1 mW of the minimum optical power at the front of the receiver and then taking log based 10 of the ratio
RZ	Return-to-zero pulse shaping
RZ33	RZ pulse of width of 33% of bit period format
RZ50	RZ pulse of width of 50% of bit period format
RZ67	RZ pulse of width of 67% of bit period format (normally CSRZ)
S_I	Spectral noise density of a bias current I
S_R	Spectral noise density of the thermal noise of a resistor R
SDH	Synchronous digital hierarchy
SMF	Single mode fiber
SPM	Self-phase modulation—nonlinear effects
SSMF	Standard single mode fiber (ITU-652)
Star-QAM	Star quadrature amplitude modulation, signal constellation distributed on circle like a star
T	Absolute temperature (K)
T_b	Bit period
T_r	Rise time
T_f	Fall time
TOD	Third-order dispersion
V	Normalized frequency parameter
V_e	Equivalent step index V-parameter
ν	Frequency of photon
XPM	Cross-phase modulation
Z_T	Transimpedance of an electronic amplifier

1

Introduction

Optical communication systems employ lightwaves to transmit information from one place to another separated by a few kilometers to thousands of kilometers for delivery to homes and from central exchanges between major cities. Furthermore, the reach can now be extended to transoceanic distances covering several thousands of kilometers as shown in Figure 1.1. Figure 1.2 depicts a map from KDD Submarine Cable Systems, Inc. that shows the submarine cable infrastructure in the Asian region in 1996. This reveals more details of the fiber cable networks in Southeast Asia and the Australia Oceania region. The connection and cable from Australia to America and Europe is the longest and most extensive layout of the optical transmission systems. The lightwave frequency is in the range of nearly 200 THz for 1550 nm wavelength and several wavelength channels can be multiplexed so as to take the total capacity over this spectral band to a few tens of terabytes per second. This band is only a very small part of the optical spectrum. Fortunately, this region is the lowest attenuation spectral window of silica fiber, which is the critical guiding medium with minimum broadening effects on the transmitted data pulse sequences. This electromagnetic spectrum for communications is shown in Figure 1.3. As observed, the spectrum of optical communication based on silica fiber occupies only a small fraction of the electromagnetic spectrum but extensive bandwidth and capacity can be made available for years to come.

The bit rate of the information can now reach several tens of gigabytes per second in the first decade of the twenty-first century. Ten gigabytes per second Ethernet has been standardized, and the introduction of 100 Gb/s Ethernet into the global fiber networks is inevitable. Similarly, for transmission rates under synchronous digital hierarchy (SDH), OC-192 and OC-768 for 10 and 40 Gb/s have been demonstrated over the last decade. Recently, the possibility of 1 Tb/s per wavelength channel has also been proposed but is yet to be demonstrated. The modulated lightwaves are guided through single-mode optical fibers and compensated over several spans that are made up by cascading dispersive and compensating fibers as well as optical amplifiers through which direct amplification of photons is achieved.

1.1 Historical Perspectives

Optical fiber communications have advanced at a tremendous pace since its inception in 1966. Its technological development has progressed through three principal phases. The multimode fiber era at the initial stage when silica fiber was first fabricated and manufactured in the early 1970s. Then at the end of the 1970s, single-mode fibers were manufactured, and the laser sources at a wavelength of 1300 nm were also available for research laboratories. At this wavelength, the fiber dispersion is almost zero, and the transmission system is limited by the attenuation of the lightwaves. Figure 1.4 shows the increase

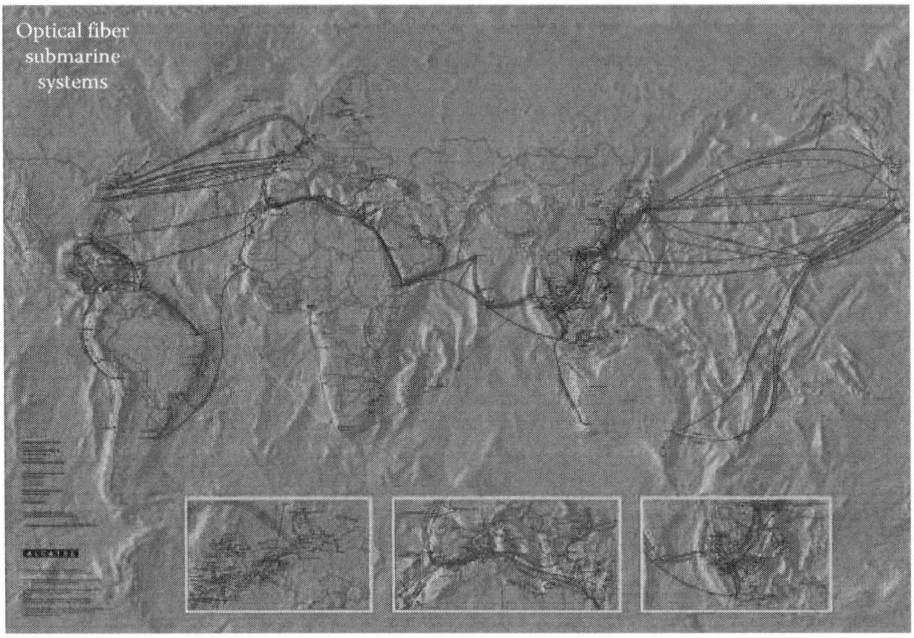

FIGURE 1.1
Alcatel map of submarine cable systems across the globe.

of the transmission capacity and distance as optical communications progressed during 1975–2010. A significant increase in the capacity–distance product after 1990 is due to the invention of in-line optical amplifiers using erbium-doped silica-based fibers. Prior to this, a certain improvement in the transmission capacity due to a change from the use of multimode to single-mode fibers allowed the distance between repeaters to increase to about 40 km.

Since then single-mode optical fibers with low loss at 1550 nm were used with sources in this region. The loss is nearly half of that at 1300 nm, so the repeater distance, in practice, was limited to 40 km. This scenario did not improve until the late 1980s when optical amplifiers were invented, in particular, the Er-doped fiber amplifier that offers significant optical gain in the 15,530–1,565 nm range. Amplification for the L-band would also be available using different doped agents in the silica fiber. However, Raman amplification can also be available, and the gain is distributed along the length of the propagation of the optical signals. These optical amplifiers have since then revolutionized the design and the implementation of optical transmission systems and networks in which the attenuation factor was no longer a major obstacle. This is why an entire chapter has been dedicated to both types of optical amplifiers and their Simulink® models.

The advancement of single-mode optical fibers of transmission and dispersion compensating types and single frequency source as well as wideband and low-noise optical receivers have permitted the transmission of high-quality signals over an extremely long haul (in the order of more than a few thousands) at bit rates reaching 40–100 Gb/s. Dispersion management techniques can be exploited to further extend the transmission distance.

Since the linewidth of laser sources can now be narrowed so as to be considered as single-frequency sources, the modulation by direct modulating the electron density in the

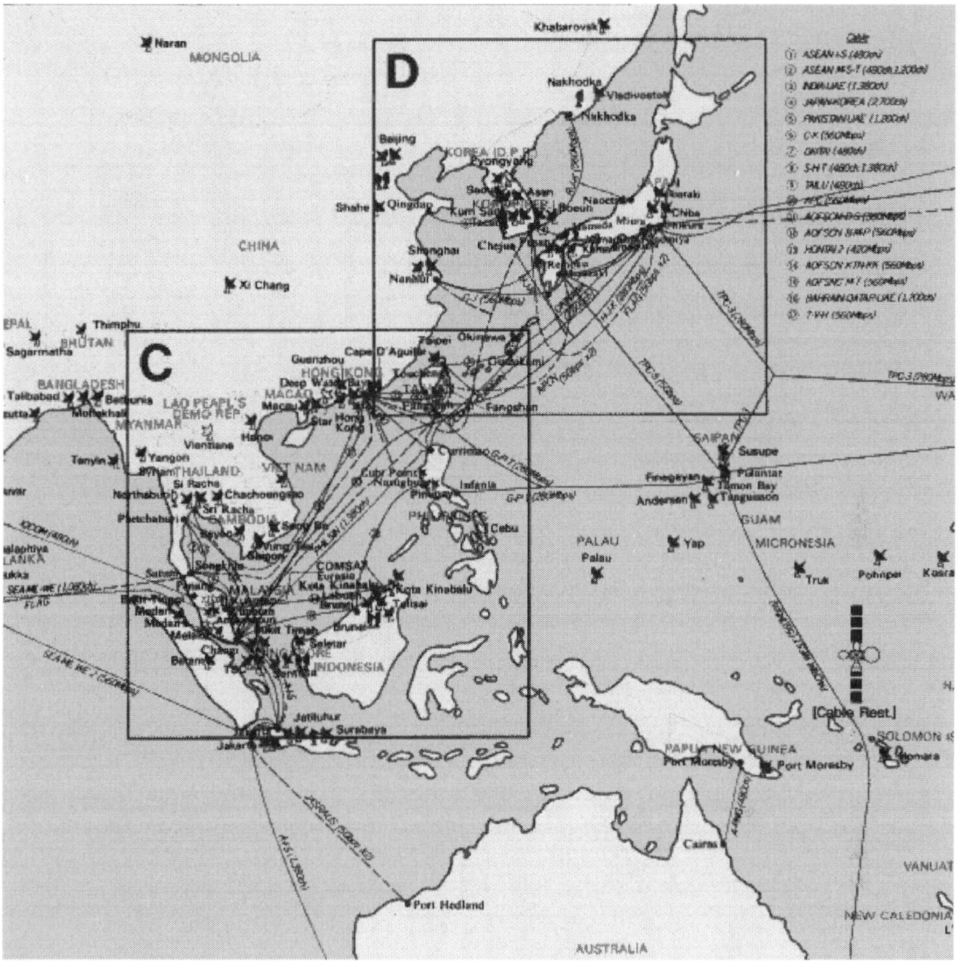

FIGURE 1.2
Optical fiber cable networks in Southeast Asia and the Australia Oceania region.

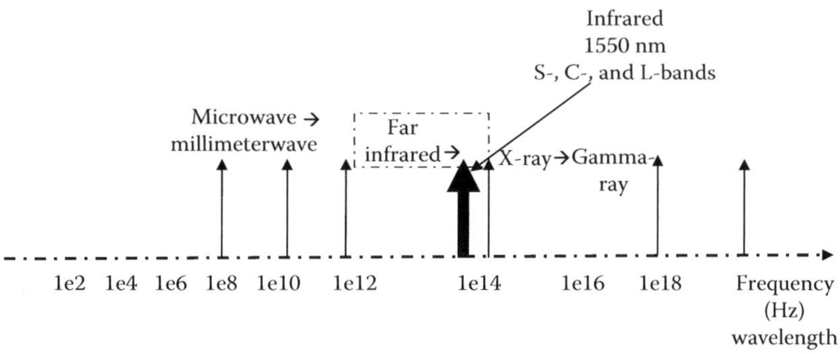

FIGURE 1.3
Electromagnetic spectrum of waves for communications and lightwaves region for silica-based fiber optical communications.

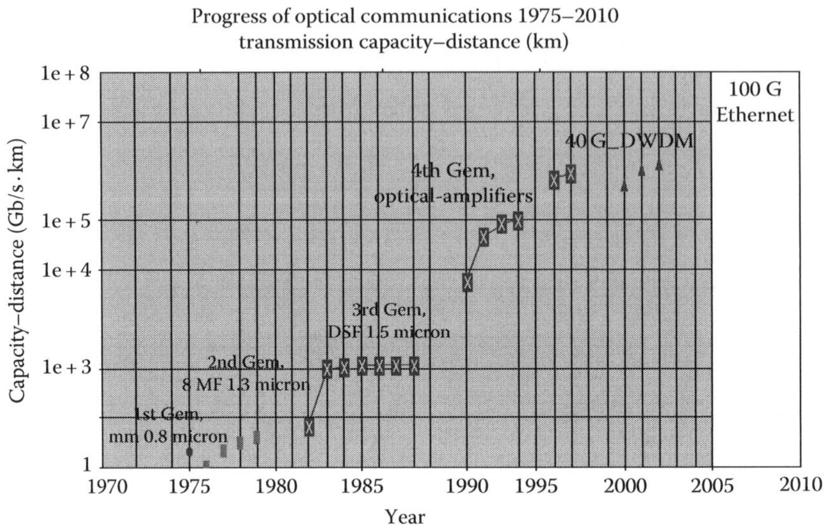

FIGURE 1.4
Progress of optical communications 1975–2010 with transmission capacity–distance product increase with year.

lasing cavity is seldom employed for a bit rate equal to or greater than 10 Gb/s. External modulation via the use of electro-optic effects and the interference of the continuously turned-on lightwaves is the technique that is commonly used currently. Thus, modulation formats have been used to achieve effective bandwidth in the optical passband and to combat effects of nonlinearity and dispersion. Thus, in this book we concentrate on models for external modulation of the continuous wave operation of the lasers and on advanced methods of detection and transmission of information over optically amplified multi-span single-mode optical fiber systems.

1.2 Digital Modulation for Advanced Optical Transmission Systems

In this book, we concentrate on the digital modulation format as a way of carrying information over long distance via the use of the optical carrier. The modulation of the lightwave carrier can be in the following forms:

The optical signal field has the ideal form during the duration of 1 bit period given as

$$E_s(t) = E_p(t)a(t)\cos\left[\omega(t)t + \theta(t)\right] \quad 0 \le t \le T \tag{1.1}$$

where

$E_s(t)$ is the optical signal field
$E_p(t)$ is the polarized field coefficient as a function of time
$a(t)$ is the amplitude variation
$\omega(t)$ is the optical frequency change with respect to time
$\theta(t)$ is the phase variation with respect to time

Depending on the modulation of the carrier by amplitude, frequency, or phase

- For amplitude shift keying (ASK), the amplitude $a(t)$ takes the value $a(t)>0$ for a "ONE" symbol and the value of 0 for a "ZERO" symbol. Other values such as the angular frequency and the phase parameter remain unchanged over a 1 bit period.
- For phase shift keying (PSK), the phase angle $\theta(t)$ takes a value of π rad for a "ONE" symbol and 0 rad for the symbol "ZERO" so that the distance between these symbols on the phase plane is at maximum and hence minimum interference or error can be obtained. These values are changed accordingly if the number of phase states is increased as shown in Figure 1.7. The values of $a(t)$, $\omega(t)$, and $E_p(t)$ remain unchanged.
- For frequency shift keying (FSK), the value of $\omega(t)$ takes the value ω_1 for the "ONE" symbol and ω_2 for the "ZERO" symbol. The values of $a(t)$, $\theta(t)$, and $E_p(t)$ remain unchanged. Indeed FSK is a form of phase modulation, provided that the phase is continuous. Sometimes, continuous phase modulation is also used as the term for FSK. In the case that the frequency spacing between ω_1 and ω_2 equals a quarter of the bit rate, the FSK is called minimum shift keying (MSK).
- For polarization shift keying (PolSK), we have $E_p(t)$ taking one direction for the "ONE" symbol and the other for the "ZERO" symbol. Sometimes, continuous polarization of lightwave is used to multiplex two optically modulated signal sequences to double the transmission capacity (Figures 1.5 and 1.6).

These four digital modulation formats form the basis of modulation formats in advanced optical fiber communications systems. Besides these formats, the pulse shaping also plays

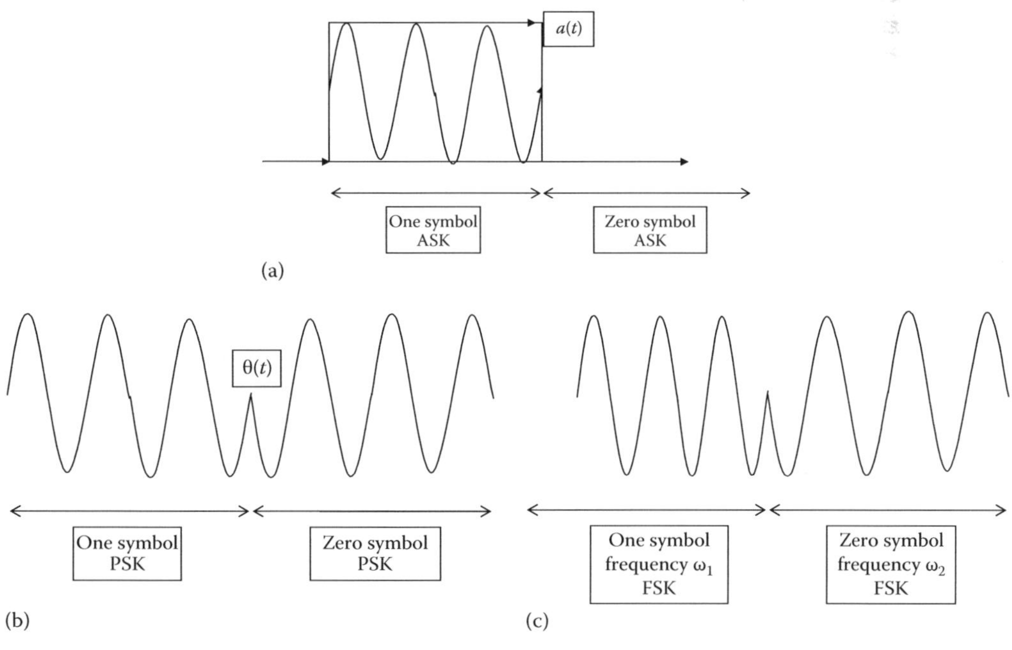

FIGURE 1.5
Illustration of ASK, PSK, and FSK with the symbol and the variation of the optical carrier: (a) amplitude, (b) phase, and (c) frequency.

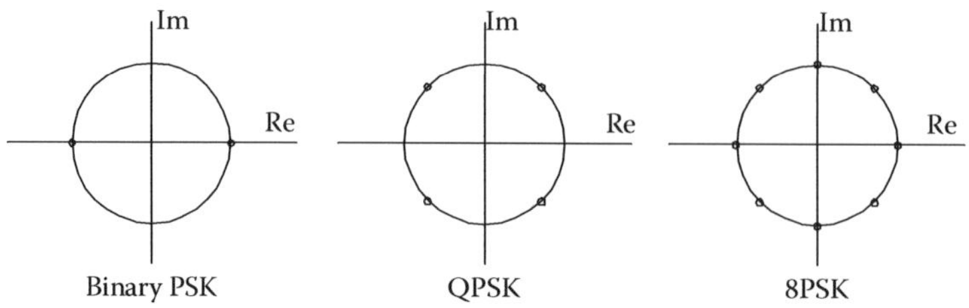

FIGURE 1.6
Phase of the carrier under modulation with π phase shift of the binary phase shift keying (BPSK) at the edge of the pulse period.

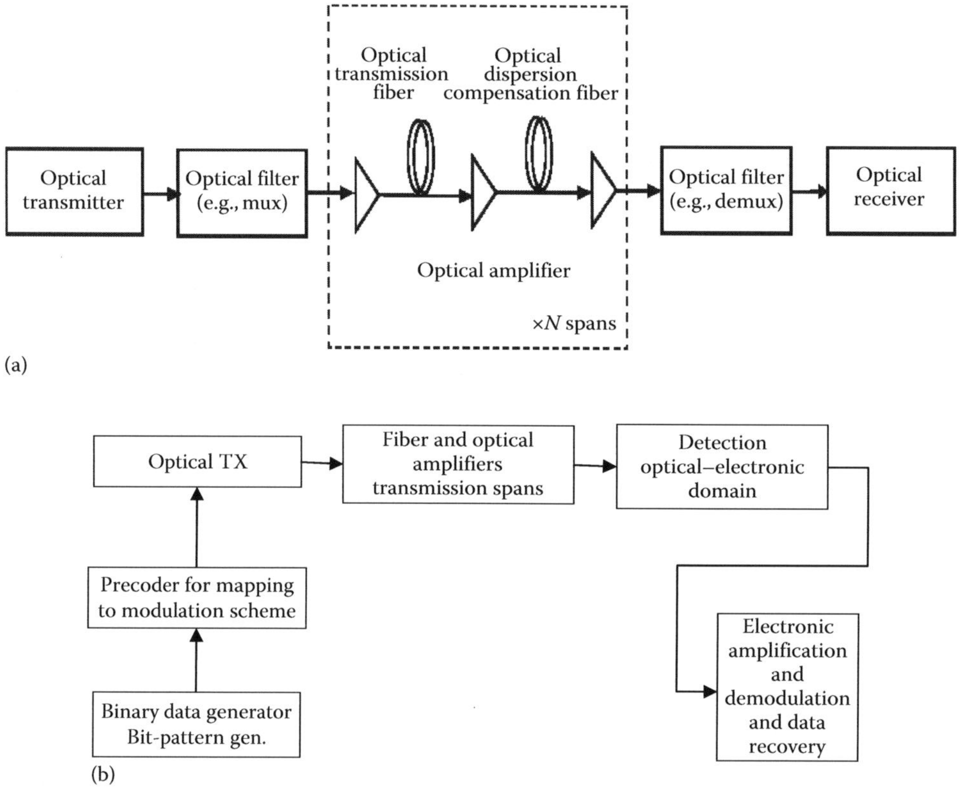

FIGURE 1.7
(a) Generalized diagram of optical transmission systems and (b) more details of the optical transmission system.

an important role in these advanced systems. They include non-return-to-zero (NRZ), return-to-zero (RZ), and duobinary (DuoB). RZ and NRZ are binary-level formats taking two levels "0 and 1" while DuoB is a tri-level shaping taking the values of "–1 0 1." The –1 in optical waves can be taken care of by an amplitude of "1" and a phase of π phase shift with respect to the "+1," which means a differential phase is used to distinguish between the +1 and –1 states.

The modulated lightwaves at the output of the optical transmitter are then fed into the transmission fibers and fiber spans, as shown in Figure 1.7.

1.3 Demodulation Techniques

The output transmitted signals, which are normally distorted, are then detected by a digital optical receiver. The main function of this optical receiver is to recognize whether the current received and thence the "bit symbol" voltage at the output of the amplifiers following the detector is a ONE or a ZERO. The modulation of the amplitude, the phase, or the frequency of the optical carrier requires an optical demodulation, that is, the demodulation of the optical carrier is implemented in the optical domain. This is necessary as due to the extremely high frequency of the optical carrier (in the order of nearly 200 THz for 1550 nm wavelength) it is impossible to demodulate in the electronic domain by direct detection using a single photodetector. Indeed it is quite straightforward to demodulate in the optical domain using optical interferometers to compare the phases of the carriers in two consecutive bits.

However, the phase and the frequency of the lightwave signals can be recovered via an intermediate step by mixing the optical signals with a local oscillator, a narrow linewidth laser, to beat it to the baseband or an intermediate frequency region. This is coherent detection technique. Figure 1.8 shows the schematics of optical receivers using direct detection and coherent detection.

The main difference between these detection systems and those presented in several textbooks is the electronic signal processing subsystem following the detection circuitry. In the first decade of the twenty-first century, we have witnessed tremendous progress in the speed of electronic ultra-large-scale integrated circuits, with the number of samples per second reaching a few tens of giga-samples. This has permitted considerations for applications of digital signal processing of distorted received optical signals in the electronic domain. Thus, flexibility in the equalization of signals in transmission systems and networks is very attractive.

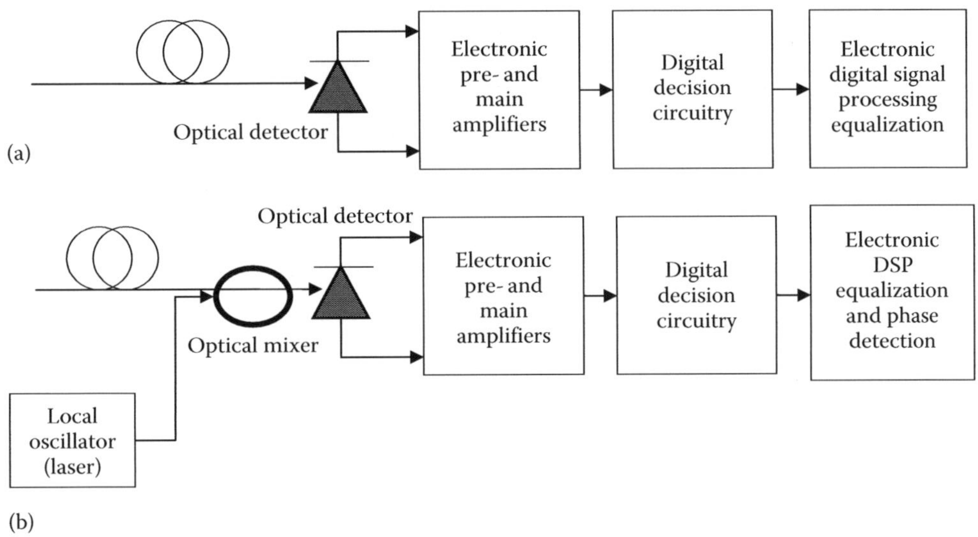

FIGURE 1.8
Schematics of optical receivers using (a) direct detection and (b) coherent detection.

1.4 MATLAB® and Simulink® Platforms

Why are MATLAB® and Simulink selected as a modeling and simulation platform for optical fiber transmission systems? Simulink is referred to as a separate software package within MATLAB. It is based on a number of block sets, thus the ease of use and the shortening of learning and development time. Furthermore, the use of MATLAB and Simulink would require users' understanding of principles and not a strong foundation in the mathematics of digital communications with various communications and mathematical blocks. There are no such optical communications block sets in MATLAB and Simulink, so one of the main objectives of this book is to provide the operational principles of optical communications blocks as examples for users who wish to model their systems. Last but not least, MATLAB packages have now been very popular in the global university computing environment. University students worldwide are now familiar with MATLAB, and Simulink is only an extension of MATLAB with several blocks of functions and monitoring equipment available to observations of signals and the behavior of the developed systems.

Therefore our secondary principal objective of this book is to describe several optical components and transmission systems in this Simulink platform so that senior and research students can adapt to their proposed transmission systems without resorting to expensive commercial packages such as VPI transmission system maker and Optiwave.

1.5 Organization of the Chapters in This Book

The presentation of this book follows the integration of optical components and modulation techniques in optical communications.

Chapters 2 and 3 give the fundamentals of waveguiding in optical fibers and essential parameters of such a waveguide that would influence the transmission and the propagation of optically modulated signals through the fibers. Naturally only single-mode optical fibers are treated for advanced optically amplified transmission systems. Thus, Chapter 2 gives the static parameters including the index profile distribution and the geometrical structure of the fiber. Mode spot size and mode field diameter of optical fibers are also given for the estimation of the nonlinear self-phase-modulation effects. Operational parameters such as the group velocity, the group velocity dispersion, the dispersion factor, and the dispersion slope of single-mode fiber as well as the attenuation factor are described in Chapter 3. The frequency responses including impulse and step responses of optical fibers are also given so that the chirping of an optically modulated signal can be understood from the point of view of phase evolution when propagating through an optical fiber, a quadratic phase-modulation medium. The propagation equation, the nonlinear Schröedinger equation (NLSE), that represents the propagation of the complex envelope of the optical signals is also described so that the modeling of the signal propagation can be related.

Chapter 4 gives a general outline of the modeling technique based on MATLAB and Simulink in which the basic operations of all subsystems of an optically amplified fiber transmission system are outlined. Basic Simulink models are also given.

Chapters 5 and 6 deal with optical transmitter configurations based on principles of direct modulation (Chapter 5) and external modulation (Chapter 6) based on the operation of interferometric effects for the generation of phase and frequency modulation, either continuous phase frequency shift keying (CPFSK) format or the inphase and quadrature phase (I–Q) structure of PSK format. In an optical transmitter, data modulation is implemented by using either external electro-optic phase modulators (EOPM) or Mach–Zehnder intensity modulators (MZIM). Phasor principles are extensively applied in this chapter to derive the modulation of the carrier phase and amplitude. A fast method for the evaluation of the statistical properties of the distribution of the received eye diagrams is described enabling the measurement of the bit error rate (BER) from the received eye diagram rather than resorting to the Monte Carlo method, which would consume considerable amount of time for computing the errors.

Chapters 7 and 8 give fundamental techniques of direct and coherent detection, respectively, for optical communication systems. Optical receivers and associated noises in such receiving systems are described. The principal motivation of the introduction of Chapter 7 on coherent detection is due to the emerging technological developments of photonic, optoelectronic components, and digital signal processors. The limitations and obstacles due to the linewidth of the laser source are no longer major hurdles. They are now used both in transmitters and as a local oscillator at the receiver. The high sampling speed of electronic digital processors make possible the estimation of the phase of the signal carrier after the signals beating with the local oscillator. Thus, it is necessary that coherent techniques be described for its applications in modern digital optical communication systems. These high-speed digital signal processors are also employed as electrical equalization systems to compensate for disturbance or residual dispersion in optical transmission systems. The equalization techniques in the electrical domain of digital optical communications are described in Chapter 8. In coherent receivers, the noise sources due to the beating between the oscillator and the amplified spontaneous emission (ASE) noise sources would dominate the electronic noises at the front end of the receiver. Analysis and simulation are described in this chapter.

Chapters 9 and 10 describe the optical amplification processes of lumped and distributed amplifiers based on erbium-doped fiber amplifier type and Raman-distributed amplifiers via pumping in the transmission fibers. Associate Simulink models are given and described in detail.

Chapter 11 gives detailed designs of optical transmission systems with an outline of design strategy using attenuation and dispersion budgets to ensure that there is a high probability of success in the transmission. This is followed by simulation and details of simulation model using Simulink. A few case studies with full details of Simulink models are given.

Chapter 12 presents the Simulink models and experimental demonstration for studying performance characteristics of optical transmission systems employing digital modulation schemes. Optical differential phase shift keying (DPSK) with the focus on its robustness to tight optical filtering is described in the first section that introduces the phase discrete modulation, especially the differential phase coding. It shows the feasibility of implementing cost-effective hybrid optical transmission systems in which 40 Gb/s DPSK channels can be co-transmitted simultaneously with the 10 Gb/s On–Off keying (OOK) channels over the existing 10 Gb/s network infrastructure. Thence, Section 5.3 describes the modulation of the phase of the carrier, but continuously rather than discrete as in Chapter 6, in which the MSK is described extensively as it is the most efficient continuous phase modulation due to the orthogonal properties of the spectra of the modulated signals at the two distinct carriers. Thence, it gives an introduction into multilevel amplitude and phase-modulation formats. The driving conditions for a dual drive MZIM for the generation of multilevel amplitude differential PSK are derived as well as the models in MATLAB and Simulink platforms. Dual-level MSK signals are also given. A 16-ADPSK and 16-Star QAM (quadrature amplitude modulation) are two typical case studies that would allow the reduction of the symbol rate from 100 to 25 Gb/s at which electronic processing would be able to assist in the detection using either direct or coherent detection. Practice problems are given wherever appropriate at the end of the chapters.

2

Optical Waveguides: Geometrical and Lightwave-Guiding Properties

This chapter describes the fundamental understanding of guiding lightwaves in planar and circular optical waveguides. A brief fundamental derivation of the wave equation from the Maxwell equations is given for symmetric planar waveguides leading to the phenomenon of guiding in optical fibers, circular optical waveguides.

Only geometrical parameters of optical fibers are presented here. The attenuation and dispersion effects of modulated lightwave signals are described in Chapter 3. However, nonlinear effects are given here so that the distortion impacts on signals can be treated in the next chapter.

2.1 Introduction

Optical fiber communications is no longer a term used in research centers or university research and teaching laboratories. It is now more than 35 years that optical fiber communications has progressed, and the technology has imposed a significant revolution to the information technologically oriented society beyond the twenty-first century. Indeed, it was a simple idea that was proposed in an article in the Proceedings of the Institution of Electrical Engineers of the United Kingdom by Charles Kao and G. Hockham. The idea was to guide optical waves in a wavelike light guide, and the term dielectric optical waveguide was first used in that article. It was thought as the most fundamental component of optical communications technology, guiding lightwaves whose frequency is extremely high and thus ultra-wideband regions can be available to accommodate information channels.

This chapter is dedicated to describe and give a fundamental understanding of optical waveguiding and the structures as well as the conditions under which the waveguide can be designed to guide lightwaves over an extremely long distance. The basic components for optical fiber communication networks are the optical guiding media for the transmission of optical signals in the form of lightwaves called optical waveguides or optical fibers. These lightwaves transmission media must satisfy the following conditions.

First, they must be able to guide lightwaves over a long distance without losing significantly the optical energy. Second, signal transmitting through them must not be distorted, and they must be structured in such a way that they can support a number of optical electromagnetic (EM) field modes.

The main objectives of this chapter are first to give an introduction and concepts of guided modes in optical waveguides in which lightwaves are confined to one dimension only, the planar optical waveguides. Second, the optical waveguides are extended to a circular geometrical structure, the optical fibers. The properties of the step-index optical fiber that supports only one mode, the single-mode optical fiber, are described and used as the fundamental elements for studying non-step-index fiber types and installed throughout global networks. The noncircular or non-step optical profile optical fibers can be transformed to a circular optical fiber by using an equivalent step-index (ESI) profile technique, which is given in the last section of this chapter.

Advanced design of dispersion-shifted, dispersion-flattened, and dispersion-compensated optical fibers are also briefly given. These fibers have gained much attention lately due to their potential applications in optical solution communication systems, dispersion-equalized optical fiber systems, etc.

We must note here that this chapter presents only the geometrical structures and the index distribution of circular optical waveguides and guiding conditions as well as some optical properties of the guided modes. In the next chapter, we shall describe the mechanism of distortion and the attenuation of optical signals when they are transmitted over a long length of optical fibers.

This chapter does not aim to give a full treatment of the theoretical development of optical waveguides but the essentials for students of physics and engineering that would allow them to apply to some applications such as the transmission of lightwave-modulated signals and possibly fiber sensors.

This chapter is organized as follows. Section 2.2 describes the fundamental aspects of planar optical waveguides with a derivation of the wave equation from the first principle via the formation of the Maxwell equation. Symmetric planar optical waveguides are given as a simple example of optical waveguiding so as to guide readers to the guided modes of both planar and circular optical waveguides. The V parameter is identical for both types of waveguides and hence it is easy for readers to appreciate the importance of this parameter of optical waveguides.

Section 2.3 describes the optical fibers or circular optical waveguides including their geometrical structures and conditions for supporting one and only one mode. An ESI profile method is also presented so that different index profile fibers can be designed. Gaussian approximation of the guided single-mode optical waveguide is given so that the operation parameters of optical fibers can be derived without resorting to mathematical complexity via the use of the wave equation. This is derived from practice; when measuring the mode field distribution across a single-mode optical fiber, one can find that its radial distribution is very close to that of a Gaussian function. Thus, all one can do is to substitute this function to the wave equation to obtain the propagation constant and hence other guided parameters.

Section 2.4 gives a brief introduction to the manufacturing of optical fibers so that readers can appreciate the technological development of optical fibers. At the end of this chapter, we include the technical data sheets of a number of optical fibers in an appendix: standard single-mode fibers (SSMF) ITU-G-652 standard (Corning® SMF-28™) and ITU-G655 nonzero dispersion-shifted fiber (Corning LEAF®—large effective area fiber). Three sample programs written in MATLAB® language are provided for readers who may be interested in the design of optical fibers.

2.2 Dielectric Slab Optical Waveguides

Dielectric slab waveguide is the simplest optical waveguiding structure that was first investigated in the early 1960s. It is from this structure that the proposal for an optical fiber for optical communications emerged.

The optical slab waveguide is now widely used as the basic structure for optical integrated circuits such as semiconductor lasers and optical modulators. This section gives the fundamental aspects of optical waveguiding with a symmetrical structure; that is, the optical guiding layer is sandwiched between two cladding regions of smaller dielectric constant or refractive index than that of the core region.

2.2.1 Structure

A dielectric slab waveguide consists of a slab of dielectric material of refractive index n_1 embedded in a material of index n_2, as shown in Figure 2.1.

Assuming that the structure is extended infinitely in the y and z directions, a guiding thickness of $2a$, and that the materials are isotopic and lossless (i.e., the permittivities are real and scalar) leads to a one-dimensional wave equation for the electric field of the lightwaves and hence a simplified mathematical solution. The main reason for this approximation is that a circular optical fiber can be easily analyzed by confining the optical waveguide in the y-direction into a circular structure. The wave equation for this type of structure involves the cylindrical coordinates and is too complicated to visualize analytically.

2.2.2 Numerical Aperture

If we assume at the moment that total internal reflections at the boundaries are required for guiding, what is the acceptance angle such that lightwaves can be launched? The ray path entering the optical fiber core for total internal reflection is shown in Figure 2.2.

$$n_0 \sin\theta_0 \le n_1 \cos\theta_c \tag{2.1}$$

By applying Snell's law at the air-core and core-cladding boundaries of the dielectric waveguide, the total internal reflection can take place only if

$$n_1 \sin\theta_c = n_2 \sin 90° = n_2 \tag{2.2}$$

where θ_c is the critical angle such that

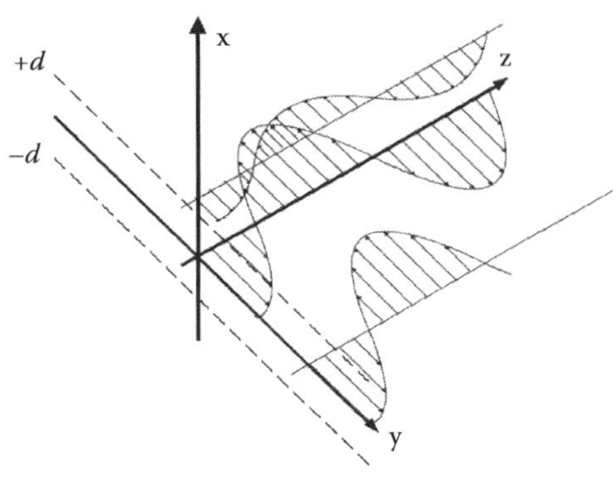

FIGURE 2.1
Real part of E-field in a slab waveguide ($t=0$).

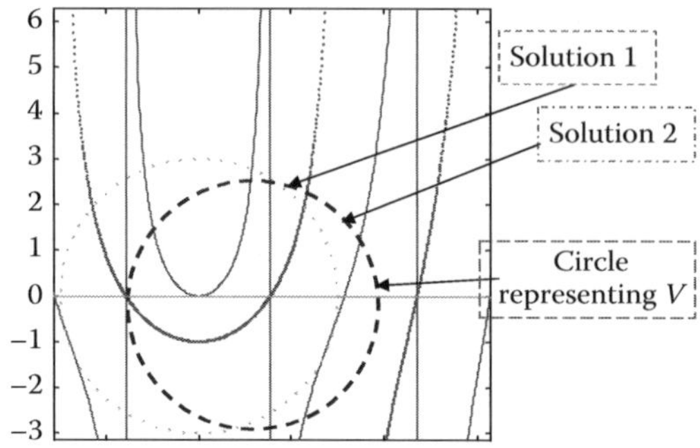

FIGURE 2.2
Graphics and solution of Equations 2.9, 2.10d, and 2.12. The curve indicated by _____ denotes the expression $v = u \tan u$; ... denotes $v = u/\tan u$; and ..._ denotes $V^2 = u^2 + u^2$.

$$\text{NA} = (\sin \theta_0)_{\max} = n_1 \cos \theta_c \tag{2.3}$$

Thus, the numerical aperture (NA), which is defined as the maximum value of sin θ_0, is given by (2.1)

$$\text{NA} = (n_1^2 - n_2^2)^{1/2} \tag{2.4}$$

then by substituting (2.2), we have

$$\text{NA} = (n_1^2 - n_2^2)^{1/2} \tag{2.5}$$

When the refractive indices n_1 and n_2 are different by a small index difference, the NA is approximated by

$$\text{NA} \simeq n_2(2\Delta)^{1/2} \tag{2.6}$$

where Δ is the refractive index difference between the refractive indices of the core and the cladding regions. Thus, the refractive index difference determines the magnitude of the NA and hence the acceptance angle at which an optical fiber can accept incident lightwaves. In practice, designers want to increase the NA to maximize the optical power that can be coupled to an optical fiber. However, it can be seen later that the larger the NA the larger the V parameter, and optical fibers can become multimode, thus suffering larger delay. So in designing a single-mode optical fiber, we have to consider several parameters that affect its performance from a communication systems point of view. Thus, we have to resort to the wave equation to understand the behavior of lightwaves in a single-mode optical fiber so that its transmission characteristics can be optimized.

2.2.3 Modes of the Symmetric Dielectric Slab Waveguides

The number of optical guide modes of lightwaves transmitted in an optical waveguide is very important because this indicates the concentration of energy of the lightwaves. It is similar to the elastic waves propagating in a string that is vibrated with the two open or closed ends. In order to find the number of optical guide modes, one has to find the conditions under which the lightwaves would be conformed. Normally, this condition is the eigenvalue equation derived from the wave equation that satisfies the boundary conditions.

Associated with each solution of the optical wave equation is the propagation wave number of each guided wave, which is derived from the eigenvalues of the equation. Thus, the propagation constants take discrete values.

2.2.3.1 Wave Equations

Assuming a monochromatic, single frequency or wavelength (we mean one and only one frequency component of lightwaves), wave propagates in the z-direction with its electric field component given by

$$E(x,y,z) = E(x)e^{(\omega t-\beta z)} \tag{2.7}$$

that is, field dependent on the x-direction and uniform along the y-direction, β is the propagation constant along the z propagating direction. Then, in the absence of charges and currents, Maxwell's equation reduces to

$$j\beta E_y = -j\omega\mu H_x \tag{2.8a}$$

$$0 = -j\omega\mu H_y \tag{2.8b}$$

$$\frac{dE_x}{dx} = -j\omega\mu H_z \tag{2.8c}$$

$$-j\beta H_x - \frac{dH_x}{dx} = j\omega\varepsilon E_y \tag{2.8d}$$

A visualization of the E-field guided in the slab waveguide is shown in Figure 2.1. By substituting H_x from (2.8a) and H_z from (2.8c) to (2.8d), we get

$$\frac{d^2E_y}{dx^2} + (\beta^2 - \omega^2\mu\varepsilon)E_y = 0 \tag{2.9}$$

where μ and ε are the permeability and permittivity of medium n_1 or n_2. Similarly, a wave equation involved H_y is given by

$$\frac{d^2H_y}{dx^2} - +(\beta^2 - \omega^2\mu\varepsilon)H_y = 0 \tag{2.10}$$

Equations 2.9 and 2.10 can be rewritten using $k=\omega/c$ and $c=(\mu_0\varepsilon_0)^{1/2}$, where c is the light velocity in vacuum and k is the wave number in vacuum, as follows:

$$\frac{d^2E_y}{dx^2} = -(k^2n_j^2 - \beta^2)E_y \tag{2.11}$$

$$\frac{d^2H_y}{dx^2} = -(k^2n_j^2 - \beta^2)H_y \tag{2.12}$$

where $n_j = n_1$ or n_2 ($j = 1, 2$)

From (2.11) and (2.12), we observe that the field variation along 0_x will always exhibit

- Sinusoidal behavior when $k^2n_j^2 > \beta^2$ oscillating: lightwaves are guided
- Exponential (decay) behavior when $k^2n_j^2 < \beta^2$ evanescent: lightwaves are radiating to the cladding region

In other words, the EM field is oscillating in regions where the longitudinal propagation constant is smaller than the plane-wave propagation constant in this region and evanescent with an exponential-like behavior elsewhere.

2.2.4 Optical Guided Modes

Optical waves are guided along the waveguide when their EM fields are oscillatory in the slab waveguide region and exponentially decay in the cladding region, that is

$$kn_2 < \beta < kn_1 \tag{2.13}$$

Now, we need to define a *transverse* propagation constant u/a and *transverse* decay constant v/a, as follows:

$$\frac{u^2}{a^2} = k^2n_1^2 - \beta^2 \tag{2.14}$$

$$\frac{v^2}{a^2} = -k^2n_2^2 + \beta^2 \tag{2.15}$$

Thus, we can observe from (2.7) that

$$\frac{u^2}{a^2} + \frac{v^2}{a^2} = k^2(n_1^2 - n_2^2) \tag{2.16}$$

or alternatively

$$V^2 = u^2 + v^2 = k^2a^2(n_1^2 - n_2^2) \tag{2.17}$$

in which the parameter V is defined as the normalized frequency (V), which is dependent only on the guide and light frequency. The V parameter and its expression is identical for both planar and optical fibers, which will be treated in the next section. The refractive index difference term is usually much <1. The V parameter can be normally approximated by a simpler expression.

Equations 2.11 and 2.17 show that the field E_y for TE modes and H_y for TM modes are a linear combination of cos(ux/a) and sin(ux/a) inside the core layer ($|x| \leq a$) and an exponential decay form on outside ($|x| > a$) with exp($-vx/a$) for $x>a$ and exp(vx/a) for $x<-a$. We therefore have two distinct cases: the even TE modes and the odd TE modes. Their solutions are given as follows.

2.2.4.1 Even TE Modes

For $|x| \leq a$: inside core region, oscillating waves, the field solution is

$$E_y \text{ or } H_x = A\cos\frac{ux}{a} \tag{2.18}$$

$$H_z = \frac{j}{\omega\mu_0}\frac{\delta E}{\delta x} = -\frac{j}{\omega\mu_0}\frac{u}{a}A\sin\frac{ux}{a} \tag{2.19}$$

- For $|x| > a$: outside core region, exponential decaying waves

$$E_y = Ce^{-\frac{V(x-a)}{a}} \tag{2.20}$$

other H_x and H_z can also be expressed in similar expressions. Applying the boundary condition (continuity of the E-field) at the interface between the cladding and the core, that is, at $x=a^+$ and $x=a^-$. Using Equations 2.18 through 2.20 we have

$$-j\frac{u}{\omega\mu}\sin u = -j\frac{v}{\omega\mu}\cos u \Rightarrow v = u\tan(u) \tag{2.21}$$

This is usually called the eigenvalue equation of the wave equation that dictates the number of guided (discrete) modes and the values of the propagation constants of these guided even TE modes.

2.2.4.2 Odd TE Modes

Similarly, the solutions for guided odd TE modes can be found as follows:

$$x > |a|$$

$$E_y \quad \text{or} \quad H_x = A\sin\frac{ux}{a} \tag{2.22a}$$

$$H_z = \frac{j}{\omega\mu_0}\frac{\delta E}{\delta x} = -\frac{j}{\omega\mu_0}\frac{u}{a}A\cos\frac{ux}{a} \tag{2.22b}$$

$$|x| \leq a$$

$$E_y = Ce^{-\frac{V(x-a)}{a}} \tag{2.22c}$$

and the eigenvalue equation can be found by applying the boundary conditions:

$$v = \frac{u}{\tan u} \tag{2.22d}$$

Equation 2.22 is obtained by using the continuity of the normal H field component at the boundaries. This is the eigenvalue equation for guided *odd* TE modes of a symmetric planar optical waveguide.

2.2.4.3 Graphical Solutions for Guided TE Modes (Even and Odd)

Combining Equations 2.17, 2.21, and 2.22d, we observe that the waveguides can support only discrete modes, and the propagation constant β related to u and v parameters can be found by solving graphically the intersection between circles of V and curves representing Equations 2.22d and 2.17. These solutions are illustrated in Figure 2.2. The field distribution of the even and the odd TE modes are represented in Figure 2.3.

2.2.5 Cutoff Properties

From Figure 2.2, we observe that

- For $V = 0$, that is, zero optical frequency or $\lambda = 0$, we observe that we always have at least one guided mode, TE_0.
- For $V < \pi/2$: one guided mode.
- For $V \rightarrow \pi/2$: odd TE_1 mode appears (second mode).
- For $V = \pi$: we have a third mode (TE_2).

That is, each time V reaches a multiple integer of $\pi/2$, a new mode reaches its cutoff that is corresponding to $v = 0$ and $\beta = kn_2$. We note that for TE-guided modes the optical waveguide can support at least one mode. However, for TM-guided modes, the waveguide can

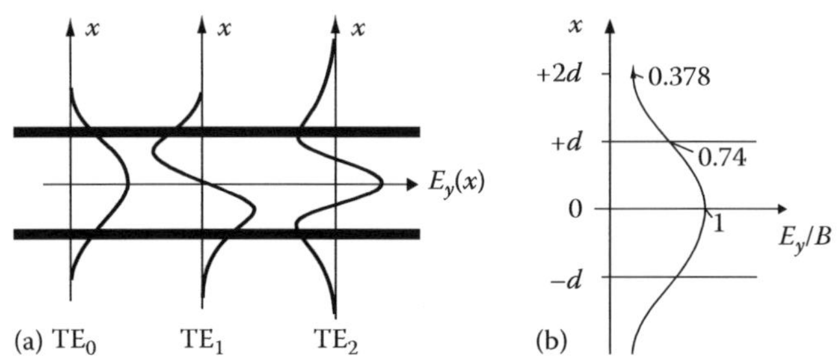

FIGURE 2.3
(a) Field distribution of TE-modes of order 0, 2 (even modes) and 1 (odd mode) and (b) field distribution of the fundamental mode.

support only if its thickness or the optical wavelength and the refractive index difference, v (lower case v), is at least $\pi/2$.

This is true for planar optical waveguides only. For circular optical waveguide such as in the case of circular optical fiber, the cutoff condition is different. We will deal with this type of optical waveguide in the next section.

It is noted that theoretical waveguide (planar type) always supports an optical guided mode. The cutoff is for cutting off the higher-order mode to make the waveguide support only one mode.

EXERCISE 2.1

A slab waveguide of width 10 μm has refractive indices of $n_1 = 1.505$ and $n_2 = 1.495$. Find its V parameter, the number of guided modes, and their propagation constants.

2.3 Optical Fiber: General Properties

2.3.1 Geometrical Structures and Index Profiles

An optical fiber consists of two concentric dielectric cylinders. The inner cylinder, or core, has a refractive index of $n(r)$ and radius a. The outer cylinder, or cladding, has index n_2 with $n(r) > n_2$ and a larger outer radius. A core of about 4–9 μm and a cladding diameter of 125 μm are the typical values for a silica-based, single-mode optical fiber. A schematic diagram of an optical fiber is shown in Figure 2.4.

The refractive index n of an optical waveguide is usually changed with radius r from the fiber axis ($r = 0$) and is expressed by

$$n^2(r) = n_2^2 + \mathrm{NA}^2 s\left(\frac{r}{a}\right) \tag{2.23}$$

where NA is the numerical aperture at the core axis, while $s(r/a)$ is the profile function that characterizes any profile shape ($s = 1$ at maximum) with a scaling parameter (usually the core radius).

2.3.1.1 Step-Index Profile

In a step-index profile, the refractive index remains constant in the core region; thus

$$s\left(\frac{r}{a}\right) = 1 \quad \text{for } r \le a \tag{2.24}$$

$$s\left(\frac{r}{a}\right) = 0 \quad \text{for } r > a \tag{2.25}$$

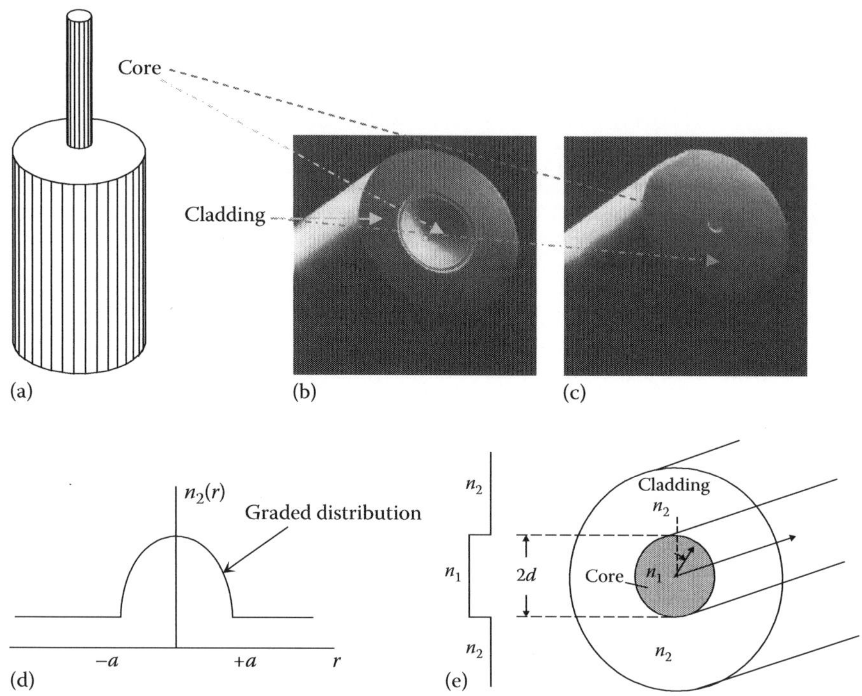

FIGURE 2.4
(a) Schematic diagram of the step-index fiber: coordinate system, structure. The refractive index of the core is uniform and slightly larger than that of the cladding. For silica glass, the refractive index of the core is about 1.478 and that of the cladding about 1.47 at 1550 nm wavelength region. (b) Cross section of an etched fiber—multimode type—50 μm diameter. (c) Single-mode optical fiber etched cross section. (d) Graded index profile. (e) Fiber cross section and step-index profile $d = a =$ radius of fiber.

so we have for a step-index profile

$$n^2(r) = n_1^2 \quad \text{for } r < a \tag{2.26}$$

and

$$n^2(r) = n_2^2 \quad \text{for } r > a \tag{2.27}$$

EXERCISE 2.2

Refer to the technical specification of the single-mode optical fiber Corning SMF-28. State whether the index profile of the fiber is a perfect step-index profile or graded index profile. Is it true that the profile is a perfect step-index distribution? If not, then what is the real manufactured profile?

2.3.1.2 Graded-Index Profile

We consider hereunder the two most common types of graded-index profiles: the power-law-index and the Gaussian profile.

2.3.1.3 Power Law–Index Profile

The core refractive index of an optical fiber is usually following a graded profile. In this case, the refractive index rises gradually from the value n_2 of the cladding glass to value n_1 at the fiber axis. Therefore, $s(r/a)$ can be expressed as

$$s\left(\frac{r}{a}\right) = \begin{cases} 1 - \left(\frac{r}{a}\right)^{\alpha} & \text{for } r \le a \\ 0 & \text{for } r < a \end{cases} \tag{2.28}$$

with α = power exponent. Thus, the index profile distribution $n(r)$ can be expressed in the usual way as [by using (2.28) and (2.24) and substituting $\mathrm{NA}^2 = n_1^2 - n_2^2$]

$$n^2(r) = \begin{cases} n_1^2\left[1 - 2\Delta\left(\frac{r}{a}\right)^{\alpha}\right] & \text{for } r \le a \\ n_2^2 & \text{for } r > a \end{cases} \tag{2.29}$$

where $\Delta = \mathrm{NA}^2/n_1^2$ is the relative refractive difference. Observing the equation describing the profile shape in (2.29), there are three special cases:

- $\alpha = 1$: the profile function $s(r/a)$ is linear and the profile is called a triangular profile.
- $\alpha = 2$: the profile is a quadratic function with respect to the radial distance and the profile is called the parabolic profile.
- $\alpha = \infty$: the profile is a step type profile.

2.3.1.4 Gaussian-Index Profile

In the Gaussian-index profile, the refractive index changes gradually from the core center to a distance very far away from it and $s(r)$ can be expressed as follows:

$$s\left(\frac{r}{a}\right) = e^{-\left(\frac{r}{a}\right)^2} \tag{2.30}$$

2.3.2 Fundamental Mode of Weakly Guiding Fibers

The electric and magnetic fields, $E(r, \phi, z)$ and $H(r, \phi, z)$, of the optical fibers in cylindrical coordinates can be found by solving Maxwell's equations. However, only lower-order modes of ideal step-index fibers are important to the present optical fiber communication systems. The fact is that $\Delta < 1\%$; thus, optical waves are weakly guided, and E and H are then approximate solutions of the scalar wave equation:

$$\left[\frac{\delta^2}{\delta r^2} + \frac{1}{r}\frac{\delta}{\delta r} + k^2 n_j^2\right]\varphi(r) = \beta^2\varphi(r) \tag{2.31}$$

where $n_j = n_1$, n_2, and $\psi(r)$ is the spatial field distribution of the nearly transverse EM waves

$$E_x = \psi(r)e^{-i\beta z}$$
$$H_y = \left(\frac{\varepsilon}{\mu}\right)^{1/2} E_x = \frac{n_2}{Z_0} E_x \tag{2.32}$$

where E_y, E_z, H_x, and H_z are negligible, $\varepsilon = n_2^2\varepsilon_0$, and $Z_0 = (\varepsilon\mu)^{1/2}$ is the vacuum impedance. That is, the waves can be seen as a plane wave traveling down along the fiber tube. These plane waves are reflected between the dielectric interfaces; in other words, it is trapped and guided in and along the core of the optical fiber.

2.3.2.1 Solutions of the Wave Equation for Step-Index Fiber

The field spatial function $\psi(r)$ would have the form of Bessel functions [from (2.31)] as follows:

$$\varphi(r) = A\frac{J_0(ur/a)}{J_0(u)} \quad \text{for } 0 < r < a \tag{2.33}$$

$$\varphi(r) = A\frac{K_0(vr/a)}{K_0(v)} \quad \text{for } r > a \tag{2.34}$$

where

J_0 and K_0 are Bessel functions of the first kind and modified of the second kind, respectively

u and v are defined similarly as given in Equations 2.14 and 2.15

Thus, following Maxwell's equations, we can find that E_z can take two possible solutions that are orthogonal:

$$E_z = -\frac{A}{kan_2}\begin{pmatrix}\sin\phi \\ \cos\phi\end{pmatrix}\begin{cases}\dfrac{uJ_1\left(u\dfrac{r}{a}\right)}{J_0(u)} & \text{for } 0 \le r < a \\[2ex] \dfrac{vK_1\left(\dfrac{vr}{a}\right)}{K_0(v)} & \text{for } r > a\end{cases} \tag{2.35}$$

The terms u and v must satisfy two equations simultaneously:

$$u^2 + v^2 = V^2 = ka(n_1^2 - n_2^2)^{1/2} = kan_2(2\Delta)^{1/2} \tag{2.36}$$

$$u\frac{J_1(u)}{J_0(u)} = v\frac{K_1(v)}{K_0(v)} \tag{2.37}$$

where (2.37) is obtained by applying the boundary conditions at the interface $r = a$ (E_z is the tangential component and must be continuous at this dielectric interface). Equation 2.37 is usually called the eigenvalue equation. The solution of this equation would give the values of β that would take discrete values and are the propagation constants of the guided lightwaves.

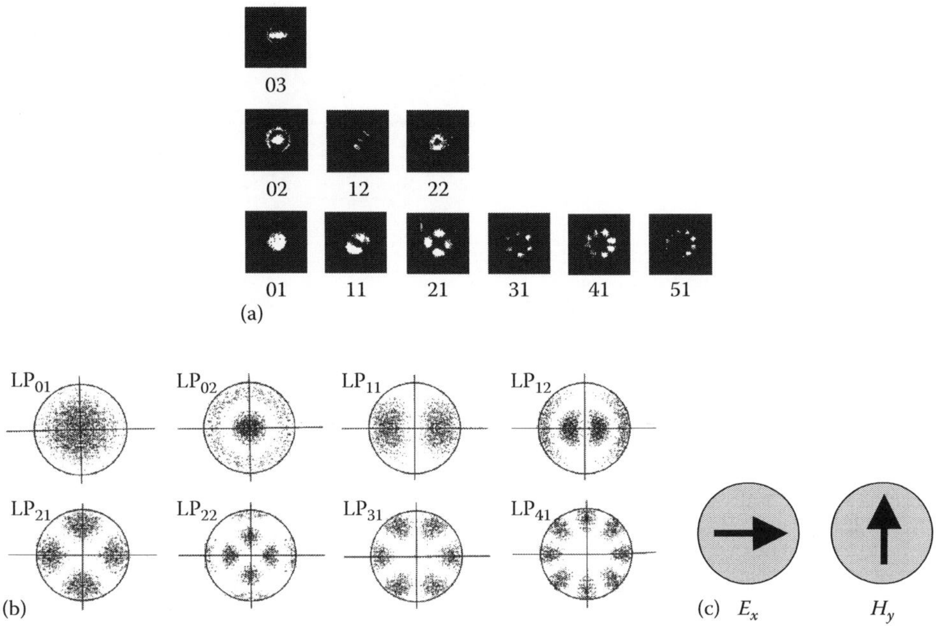

FIGURE 2.5
(a) Spectrum of guided modes in a multimode fiber, (b) calculated intensity distribution of LP guided modes in a step-index optical fiber with $V=7$, and (c) electric and magnetic field distribution of an LP01 mode polarized along O of the fundamental mode of a single-mode fiber.

Equation 2.21 shows that the longitudinal field is in the order of u/kan_2 with respect to the transverse component. In practice, $\Delta \ll 1$; by using (2.36), we observe that this longitudinal component is negligible compared to the transverse component. We thus consider the mode as *transversely polarized*. The fundamental mode is then usually denominated as LP_{01} mode (LP = linearly polarized) for which the field distribution is shown in Figure 2.5, and the graphical representation of the eigenvalue equation (2.37) calculated as the variation of $\beta/k=b$ as the normalized propagation constant and the V parameter is shown in Figure 2.6e.

2.3.2.1.1 Fundamental Mode Revisited

We note again that **E** and **H** are approximate solutions of the scalar wave equation and the main properties of the fundamental mode of weakly guiding fibers that can be observed as follows:

- The propagation constant β (in z-direction) of the fundamental mode must lie between the core and cladding wave numbers. This means that the effective refractive index of the guided mode lies within the range of the cladding and core refractive indices.
- Accordingly, the fundamental mode must be nearly a transverse electromagnetic wave as described by Equations 2.32.

$$\frac{2\pi n_2}{\lambda} < \beta < \frac{2\pi n_1}{\lambda} \tag{2.38}$$

- The spatial dependence $\psi(r)$ is a solution of the scalar wave equation (2.31).

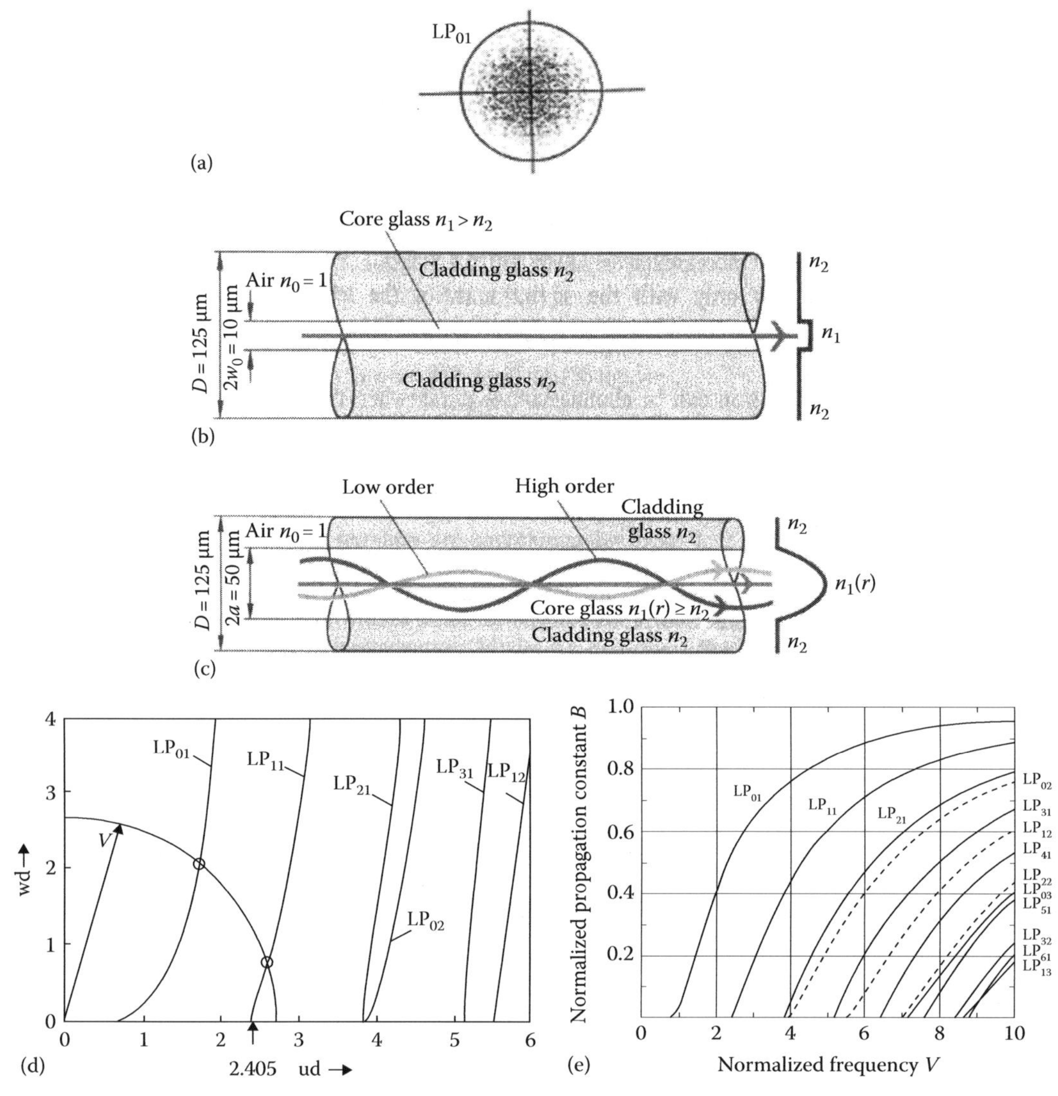

FIGURE 2.6
(a) Guided modes as seen in the transverse plane of a circular optical fiber, (b) "Ray" model of lightwave propagating in a single-mode fiber, (c) Ray model for multimode graded-index fiber, (d) graphical illustration of solution for eigenvalues (propagation constant—wave number of optical fibers), and (e) *b*–*V* characteristics of guided fibers.

2.3.2.1.2 *Gaussian Approximation*

The main objective is to find a good approximation for the field $\psi(r)$ and the propagation constant β. These can be found using the eigenvalue equation and the Bessel solutions as shown in the previous section. It is desirable if we can approximate the field to a good accuracy to obtain simple expressions to have a clearer understanding of light transmission on single-mode optical fiber without going through graphical or numerical methods. Furthermore, experimental measurements and numerical solution for step and power-law profiles show that $\psi(r)$ is approximately Gaussian in appearance. We thus approximate the field of the fundamental mode as follows:

$$\varphi(r) \cong Ae^{-\frac{1}{2}\left(\frac{r}{r_0}\right)^2} \tag{2.39}$$

where r_0 is defined as the spot size, that is, at which the intensity equals to e^{-1} of the maximum. Thus, if the wave equation (2.31) is multiplied by $r\psi(r)$ and by using the following identity

$$r\varphi\frac{\delta^2\varphi}{\delta r^2} + \varphi\frac{\delta\varphi}{\delta r} = \frac{\delta}{\delta r}\left(r\varphi\frac{\delta\varphi}{\delta r}\right) - r\left(\frac{\delta\varphi}{\delta r}\right)^2 \tag{2.40}$$

then by integrating from 0 to ∞ and using $\left[r\Psi\frac{d\varphi}{dr}\right]_0^\infty = 0$, we have

$$\beta^2 = \frac{\int_0^\infty\left[-\left(\frac{\delta\varphi}{\delta r}\right)^2 + k^2n^2(r)\varphi^2\right]r\delta r}{\int_0^\infty r\varphi^2\delta r} \tag{2.41}$$

The procedure to find the spot size is then followed by substituting $\psi(r)$ (Gaussian) in Equation 2.39 into Equation 2.41, then differentiating, and setting $\delta^2\beta/\delta r$ evaluated at r_0 to 0, that is, the propagation constant β of the fundamental mode must give the largest value of r_0.

Knowing r_0 and β, the fields E_x and H_y (Equation 2.32) are fully specified.

Case 1: Step-index fiber

Substituting the step-index profile given by Equation 2.32 and $\psi(r)$ in Equation 2.39 into Equation 2.41 leads to an expression for β in terms of r_0 given by

$$V = ka\text{NA} \tag{2.42}$$

The spot size is thus evaluated by setting

$$\frac{\delta^2\beta}{\delta r_0} = 0 \tag{2.43}$$

and r_0 is then given by

$$r_0^2 = \frac{a^2}{\ln V^2} \tag{2.44}$$

Substituting (2.44) into (2.42), we have

$$(a\beta)^2 = (akn_1)^2 - \ln V^2 - 1 \tag{2.45}$$

This expression is physically meaningful only when $V > 1$ (r_0 is positive)

Case 2: Gaussian-index profile fiber

Similarly, for the case of a Gaussian-index profile, by following the procedures for step-index profile fiber, we can obtain

$$(a\beta)^2 = (an_1k)^2 - \left(\frac{a}{r_0}\right)^2 + \frac{V^2}{\left(\frac{a}{r_0}+1\right)} \tag{2.46}$$

and

$$r_0^2 = \frac{a^2}{V-1} \quad \text{by using} \quad \frac{\delta^2\beta}{\delta r_0} = 0 \tag{2.47}$$

This is for maximizing the propagation constant of the guided waves. The propagation constant is at maximum when the "light ray" is very close to the horizontal direction. Substituting (2.47) into (2.46), we have

$$(a\beta)^2 = (akn_1)^2 - 2V + 1 \tag{2.48}$$

$$r_0 = \frac{a}{\ln V^2}$$

thus Equations 2.47 and 2.48 are physically meaningful only when $V > 1$ ($r_0 > 0$).

It is obvious from Equation 2.49 that the spot size of the optical fiber with a V parameter of 1 is extremely large. This is a very important point that one must not design the optical fiber with a near unit value of the V parameter. In practice, we observe that the spot size is large but finite (observable). In fact if V is smaller than 1.5, the spot size becomes large and this will be investigated in detail in Chapter 3.

2.3.3 Cutoff Properties

Similar to the case of planar dielectric waveguides, from Figure 2.6 we observe that when we have $V < 2.405$, only the fundamental LP_{01} exists (see Figure 2.7).

In fact the value 2.405 is the first zero of the Bessel function $J_0(u)$. In practice, one cannot really distinguish the V value between 2.3 and 3.0. Experimental observation shows that optical fiber can still support only one mode. Thus, designers do usually take the value of V as 3.0 or less to design a single-mode optical fiber.

The V parameter is inversely proportional to the optical wavelength. Thus, if an optical fiber is launched with lightwaves whose optical wavelength is smaller than the operating wavelength at which the optical fiber is single mode then the optical fiber is supporting more than one mode. The optical fiber is said to be operating in a multimode region.

Thus, one can define the cutoff wavelength for optical fibers as the wavelength (λ_c) above which only the fundamental mode is guided in the fiber. This cutoff wavelength can be found by using the V parameter as $V_c = V$ (at cutoff) $= 2.405$; thus,

$$\lambda_c = \frac{2\pi a \mathrm{NA}}{V_c} \tag{2.50}$$

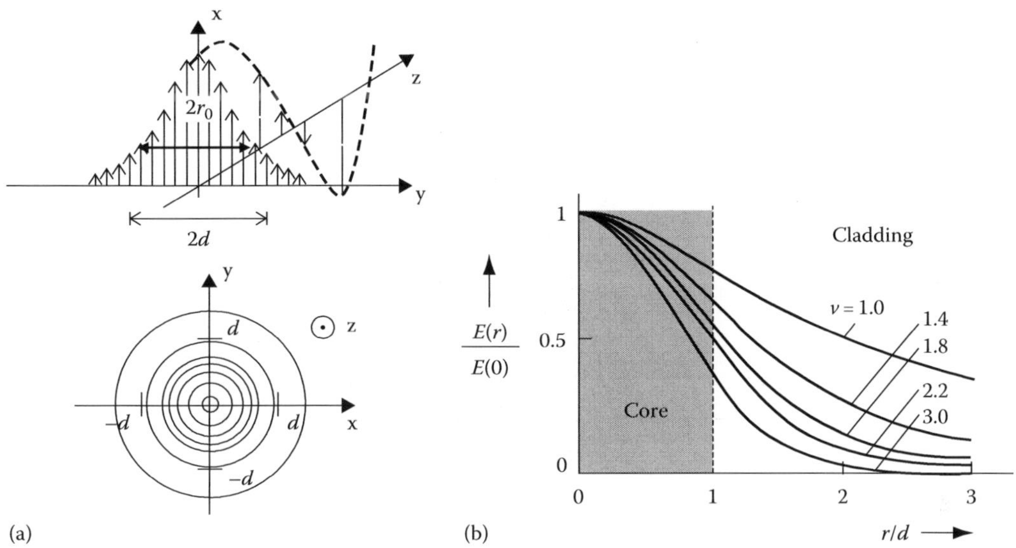

FIGURE 2.7
(a) Intensity distribution of the LP01 mode and (b) variation of the spot size—field distribution with radial distance r with V as a parameter. *Note*: for single-mode operation the V parameter must be less than or equal to 2.405. However in practice $V<3$ is acceptable.

EXERCISE 2.3

An optical fiber has the following parameters: a core refractive index of 1.46, a relative refractive index difference of 0.3%, a cladding diameter of 125 μm, and a core diameter of 8.0 μm. (a) Find the fiber NA and hence the fiber acceptance angle. (b) What is the cut-off wavelength of this fiber? (c) What is the number of optical guided modes that can be supported if the optical fiber is excited with lightwaves of a wavelength of 810 nm? (d) If the cladding diameter is reduced to 50 and 20 μm, comment on the field distribution of the guided single mode.

In practice, the fibers tend to be effectively single mode for larger values of V, say $V<3$, for the step profile, because the higher-order modes suffer radiation losses due to fiber imperfections. Thus, if $V=3$, from (2.36) we have $a<3\lambda/2$ NA; in the case that $\lambda=1\,\mu m$, the NA must be very small ($\ll 1$) for radius a to have some reasonable dimension. Usually Δ is about 1% or less for standard single-mode optical fibers in telecommunication systems.

2.4 Power Distribution

The axial power density or intensity profile $S(r)$, the z-component of the Poynting vector, is given by

$$S(r) = \frac{1}{2} E_x H_y^* \tag{2.51}$$

Substituting (2.2) into (2.15) and (2.32) into (2.51), we have

$$S(r) = \frac{1}{2}\left(\frac{\varepsilon}{\mu}\right)^{1/2} e^{-\left(\frac{r}{r_0}\right)^2} \tag{2.52}$$

The total power is then given by

$$P = 2\pi\int_0^\infty rS(r)dr = \frac{1}{2}\left(\frac{\varepsilon}{\mu}\right)^{1/2} r_0^2 \tag{2.53}$$

and hence the fraction of power $\eta(r)$ within 0 to r across the fiber cross section is given by

$$\eta(r) = \frac{\int_0^r rS(r)dr}{\int_0^\infty rS(r)dr} = 1 - e^{-\left(\frac{r^2}{r_0^2}\right)} \tag{2.54}$$

Table 2.1 gives the expressions for P and $\eta(r)$ of step-index and Gaussian profile fibers [by substituting the appropriate values of r_0 into (2.32) and (2.33)].

A rule of thumb and with experimental confirmation that an optical fiber is best for low dispersion propagation of the guided mode when its optical power contained in the core is about 70%–80% of the total power of the mode.

TABLE 2.1

Analytical Expressions for Total Optical Guided Power and Its Fractional Power Confined Inside the Core Region for Step-Index and Gaussian-Index Profiles

	Step Index	Gaussian
$S(r/a)$ for $V>1$	$\frac{1}{2}\left(\frac{\varepsilon}{\mu}\right)^{1/2} e^{-\left(\frac{r}{a}\right)^2 \ln V^2}$	$\frac{1}{2}\left(\frac{\varepsilon}{\mu}\right)^{1/2} e^{-\left(\frac{r}{a}\right)(V-1)}$
Power P for $V>1$	$\frac{1}{2}\left(\frac{\varepsilon}{\mu}\right)^{1/2} \frac{a^2}{\ln V^2}$	$\frac{1}{2}\left(\frac{\varepsilon}{\mu}\right)^{1/2} \frac{a^2}{V-1}$
	$\frac{1}{2}\left(\frac{\varepsilon}{\mu}\right)^{1/2} \frac{a^2}{\ln V^2}$	$\frac{1}{2}\left(\frac{\varepsilon}{\mu}\right)^{1/2} \frac{a^2}{V-1}$
$\eta(r)$ for $V>1$	$1-e^{-\left(\frac{r}{a}\right)^2 \ln V^2}$	$1-e^{-\left(\frac{r}{a}\right)^2 (V-1)}$
	$=1-\frac{1}{V^2}$ for $r=a$	

EXERCISE 2.4

Using Gaussian approximation for the intensity distribution of the fundamental mode of the single-mode optical fiber with $V=2$, find the fraction of power in the core region with $a=4\,\mu\text{m}$.

EXERCISE 2.5

Find the radius a for maximum confinement of light power, that is, maximum r_0, for step-index and parabolic profile optical fibers.

2.4.1 Approximation of Spot Size r_0 of a Step-Index Fiber

As stated earlier, spot size r_0 would play a major role in determining the performance of the single-mode fiber. It is useful if we can approximate the spot size as long as the fiber is operating over a certain wavelength. When a single-mode fiber is operating above its cutoff wavelength, a good approximation (greater than 96% accuracy) for r_0 is given by

$$\frac{r_0}{a}=0.65+1.619V^{-3/2}+2.879V^{-6}=0.65+0.434\left(\frac{\lambda}{\lambda_c}\right)^{+3/2}+0.0419\left(\frac{\lambda}{\lambda_c}\right)^{+6} \tag{2.55}$$

for $0.8\le\frac{\lambda}{\lambda_c}\le 2.0$ and for step-index fiber only.

EXERCISE 2.6

What is the equivalent range for the V parameter of Equation 2.34? Inspect the b–V and $V^2\frac{d^2(Vb)}{dV^2}$ versus V and b; if possible, do a curve fitting to obtain the approximate relationship for r/a and V (MATLAB procedure is recommended).

EXERCISE 2.7

Refer to the technical specification of Corning SMF-28 and LEAF

1. State the core diameter of the fibers, the spot size, or mode field diameters (MFDs) of the fibers.
2. Thence estimate the effective areas of these fibers.
3. What is the ration of the effective area and the physical area of the cores of the fibers?

2.4.2 Equivalent Step–Index Description

As we have seen in Section 2.2.2, there are two possible orthogonally polarized modes (E_x, H_y) and (E_y, H_x) that can be propagating *simultaneously*. These modes are usually approximated by a single LP mode. The properties of these modes are well known and well understood for step-index optical fibers, and analytical solutions are also readily available.

Unfortunately, practical SM optical fibers never have perfect step-index profile due to the variation of the dopant diffusion and polarization. These non-step-index fibers can be approximated, under some special conditions, by *an ESI* profile technique.

A number of index profiles of modern single-mode fibers, for example, nonzero dispersion-shifted fibers are shown in Figure 2.8. The ESI profile is determined by approximating the fundamental mode electric field spatial distribution $\psi(r)$ by a Gaussian function, as described in Section 2.3.1.1. The electric field can thus be totally specified by the e^{-1} width of this function or *mode spot size* (r_0). Alternatively, the term MFD is also used and equivalent to twice the size of the mode spot size r_0.

2.4.2.1 Definitions of ESI Parameters

The ESI description can be used to design SM fiber with graded index, W-, or segmented core profiles (under some limitations). These non-step-index profiles can be described by ESI parameters denoted as follows:

V_e = effective or equivalent V parameter
a_e = ESI core radius
λ_{ec} = ESI cutoff wavelength
Δ_e = equivalent relative index difference

These parameters are related to two moments M_0 and M_1 defined as follows:

$$M_n = \int_0^\infty \left[n^2(r) - n^2(a)\right] r^n dr \quad (2.56)$$

For $n = 1, 2$, the effective V_e parameter and effective core radius r_e are given by

$$V_e^2 = 2k^2 \int_0^\infty \left[n^2(r) - n^2(a)\right] r\, dr \quad (2.57)$$

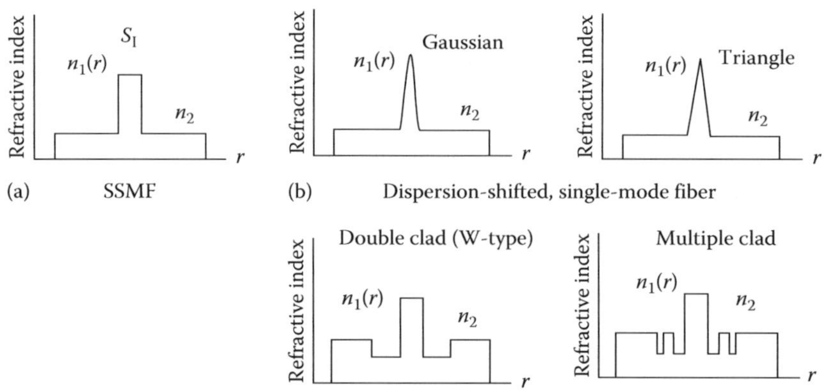

FIGURE 2.8
Index profiles of a number of modern fibers, for example, dispersion-shifted, single-mode fibers.

$$V_e^2 = 2k^2 M_1 \quad \text{and} \quad a_e = 2\frac{M_1}{M_0} \tag{2.58}$$

It is followed from (2.57) and (2.58), the parameters λ_{ec} and Δ_e by setting

$$V_e^2 = 2k^2 a_e^2 n_1^2 \Delta_e \tag{2.59}$$

and $V_e = 2.405$ (cutoff condition for step index). Therefore, the cutoff wavelength for an ESI profile fiber is

$$\lambda_{ec} = \frac{2\pi\sqrt{2M_1}}{2.405} \tag{2.60}$$

It is noteworthy that V_e as given in Equation 2.36 is equivalent to the mode *volume*. Physically, the significance of V_e can be compared to the *average density* of a disk with a local density equal to $[n^2(r) - n^2(a)]$.

2.4.2.2 Accuracy and Limits

The ESI approximation is generally accurate to within 2% at least over the wavelength range $0.8 < \lambda/\lambda_c < 1.5$. For most practical purposes, this range is the operating wavelength to minimize the dispersion property of SM optical fibers.

2.4.2.3 Examples on ESI Techniques

2.4.2.3.1 Graded-Index Fibers

These index profiles of graded fibers are given by Equation 2.29. We thus have

$$n^2(r) - n^2(a) = s(r/a) = 1 - \left(\frac{r}{a}\right)^{\alpha} \tag{2.61}$$

substituting (2.61) into (2.57) gives

$$\frac{V_e}{V} = \left(\frac{\alpha}{\alpha + 2}\right)^{1/2} \tag{2.62}$$

where $V^2 = k^2 a^2$ NA is the V parameter of a step-index fiber with the core index at the fiber axis of n_1. Hence, we have

$$\lambda_{ec} = \frac{V}{2.405}\left(\frac{\alpha}{\alpha + 2}\right)^{1/2} \tag{2.63}$$

EXERCISE 2.8

For a single-mode optical fiber with a triangular profile index distribution whose equivalent V parameter is equal to 2 at 1550 nm wavelength, what is the V parameter value at the center of the core of the fiber? If the diameters of the core of the two fibers are kept identical, what is the ration of the refractive indices at the core center of the fibers? Repeat for a parabolic profile.

2.4.2.3.2 Graded-Index Fiber with a Central Dip

The graded fiber index profile with a central dip gradually increases to the outer cladding, as shown in Figure 2.9.

Similar to (2.28) for graded index fiber with maximum index at the core axis, we have

$$S(r/a) = 1 - \gamma(1-x)^{\alpha} \quad \text{for } 0 < r < a \tag{2.64}$$

where γ is the depth and $0<\gamma<1$. When $r=0$, we have a step-index profile, and when $r=1$, we have the central axis refractive index equal to the cladding index.

Using (2.56) and (2.57), V_e can be easily found and given by

$$\frac{V_e^2}{V^2} = 1 - \frac{2\gamma}{(\alpha+1)(\alpha+2)} \tag{2.65}$$

2.4.2.4 General Method

The general technique to find the ESI parameters for optical fibers can be started by rising the stationary expression in (2.41) for expressing β of the actual fiber as compared to *its* equivalent propagation constant β_e as follows:

$$\beta^2 = \beta_e^2 + k^2 \frac{\int_0^{\infty} \left[n^2(r) - n_e^2(r)\right] r\psi^2(r)dr}{\int_0^{\infty} r\psi^2(r)dr} \tag{2.66}$$

where $n_e^2(r)$ is the equivalent counterpart of $n(r)$ when the fiber is expressed in its equivalent step form. The field expression $\psi(r)$ is assumed (in fact to be obtained) to be similar for the actual fiber and its step equivalence. Once the field $\psi(r)$ can be replaced by the approximate exact field shape, we can find V_e and a_e that minimize $\beta^2 - \beta_e^2$ in (2.66). Generally, these parameters are functions of both V and a; thus it is impossible to get one ESI technique applicable to a wide range of wavelength, and it is required to apply complicated numerical calculations.

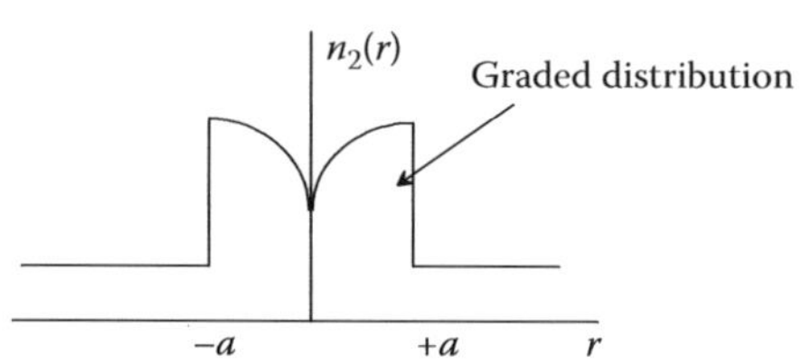

FIGURE 2.9
Refractive index profile of a graded index fiber with a central dip. This is a typical profile of the manufactured fiber if a good collapsing of the preform is not achieved.

2.5 Nonlinear Optical Effects

In this section, the nonlinear effects are described. These effects play important roles in the transmission of optical pulses along single-mode optical fibers. The nonlinear effects can be classified into three types: the effects that change the refractive index of the guided medium due to the intensity of the pulse, the self-phase modulation (SPM); the scattering of the lightwave to other frequency-shifted optical waves when the intensity reaches over a certain threshold, the Brillouin and Raman scattering phenomena; and the mixing of optical waves to generate a fourth wave, the degenerate four-wave mixing. Besides these nonlinear effects, there is also a photorefractive effect that is due to the change of refractive index of silica due to the intensity of ultraviolet optical waves. This phenomenon is used to fabricate grating whose spacing between dark and bright regions satisfy Bragg diffraction condition. These are fiber Bragg gratings and would be used as optical filters and dispersion compensator when the spacing varied or chirped.

2.5.1 Nonlinear Phase Modulation Effects

All optical transparent materials are subject to the change of the refractive index with the intensity of the optical waves, the optical Kerr effect. This physical phenomenon is originated from the anharmonic responses of electrons of optical fields leading to the change of the material susceptibility. The modified refractive index $n_{1,2}^K$ of the core and cladding regions of the silica-based material can be written as follows:

$$n_{1,2}^K = n_{1,2} + \bar{n}_2 \frac{P}{A_{\mathrm{eff}}} \tag{2.67}$$

where

n_2 is the nonlinear index coefficient of the guided medium, the average typical value of n_2 is about $2.6 \times 10^{-20}\ \mathrm{m^2/W}$

P is the average optical power of the pulse

A_{eff} is the effective area of the guided mode

The nonlinear index changes with the doping materials in the core. Not only the nonlinear index coefficient but also the effective area is very small, about 50–70 μm^2, and the length of the fiber under the propagation of optical signals is very long and the accumulated phase change is quite substantial. This leads to the SPM and cross-phase modulation (XPM) effects in the optical channels.

2.5.1.1 Self-Phase Modulation

Under a linear approximation, we can write the modified propagation constant of the guided LP mode in a single-mode optical fiber as follows:

$$\beta^K = \beta + k_0 \bar{n}_2 \frac{P}{A_{\mathrm{eff}}} = \beta + \gamma P$$

where

$$\gamma = \frac{2\pi \bar{n}_2}{\lambda A_{\text{eff}}} \tag{2.68}$$

is an important nonlinear parameter of the guided medium taking an effective value from 1 to 5 (km W)$^{-1}$ depending on the effective area of the guided mode and the operating wavelength. Thus, the smaller the mode spot size or MFD the larger the nonlinear SPM effect. For dispersion-compensating fiber, the effective area is about 15 μm^2 while for SSMF and NZ_DSF the effective area ranges from 50 to 80 μm^2. Thus, the nonlinear threshold power of dispersion compensating fiber (DCF) is much lower than that of SSMF and NZ-DSF. We would see later that the maximum launched power into DCF would be limited at about 0 dBm or 1 mW in order to avoid nonlinear distortion effect while it is about 5 dBm for SSMF.

The accumulated nonlinear phase changes due to the nonlinear Kerr effect over the propagation length L is given by

$$\phi_{\text{NL}} = \int_0^L (\beta^K - \beta)dz = \int_0^L \gamma P(z)dz = \gamma P_{\text{in}} L_{\text{eff}} \quad \text{with} \quad P(z) = P_{\text{in}} e^{-\alpha z} \tag{2.69}$$

This is the representation of the attenuation of the optical signals along the propagation direction z. In order to consider that the nonlinear SPM effect is small compared to the linear chromatic dispersion effect, one can set $\phi_{\text{NL}} \ll 1$ or $\phi_{\text{NL}} = 0.1\,\text{rad}$ and the effective length of the propagating fiber is set at $L_{\text{eff}} = 1/\alpha$ with optical losses equalized by cascaded optical amplification subsystems. Then the maximum input power to be launched into the fiber can be set at

$$P_{\text{in}} < \frac{0.1\alpha}{\gamma N_A} \tag{2.70}$$

For $\gamma = 2$ (W km)$^{-1}$ and $N_A = 10$, $\alpha = 0.2$ dB/km (or $0.0434 \times 0.2\,\text{km}^{-1}$) $P_{\text{in}} < 2.2$ mW or about 3 dBm and accordingly 1 mW for DCF. In practice, due the randomness of the arrival "1" and "0", this nonlinear threshold input power can be set at about 10 dBm as the total average power of all wavelength-multiplexed optical channels launched into the propagation fiber.

2.5.1.2 Cross-Phase Modulation

The change of the refractive index of the guided medium as a function of the intensity of the optical signals can also lead to the phase of optical channels in different spectral regions close to that of the original channel. This is the XPM effect. This is critical in wavelength division multiplexed (WDM) channels, and even more critical in dense WDM when the frequency spacing between channels is 50 GHz or even narrower. In such systems, the nonlinear phase shift of a particular channel depends not only on its power but also on that of other channels. The phase shift of the ith channel can be written as follows [1]:

$$\phi_{NL}^{i} = \gamma L_{eff}\left(P_{in}^{i} + 2\sum_{j\neq i}^{M} P_j\right) \quad \text{with } M = \text{number of multiplexed channels} \tag{2.71}$$

The factor 2 in (2.71) is due to the bipolar effects of the susceptibility of silica materials. The XPM thus depends on the bit pattern and the randomness of the synchronous arrival of the "1." It is hard to estimate; so the numerical simulation would normally be employed to obtain the XPM distortion effects by numerical simulation using the wave propagation of the signal envelop via the nonlinear Schrödinger equation. The evolution of slowly varying complex envelopes $A(z,t)$ of optical pulses along a single-mode optical fiber is governed by the nonlinear Schrödinger equation (NLSE) [1–3]:

$$\frac{\partial A(z,t)}{\partial z} + \frac{\alpha}{2}A(z,t) + \beta_1\frac{\partial A(z,t)}{\partial t} + \frac{j}{2}\beta_2\frac{\partial^2 A(z,t)}{\partial t^2} - \frac{1}{6}\beta_3\frac{\partial^3 A(z,t)}{\partial t^3} = -j\gamma\left|A(z,t)\right|^2 A(z,t) \tag{2.72}$$

where

z is the spatial longitudinal coordinate
α accounts for fiber attenuation
β_1 indicates differential group delay (DGD)
β_2 and β_3 represent second- and third-order factors of fiber CD
γ is the nonlinear coefficient

This equation will be described in detail in Chapter 3. The phase modulation due to nonlinear phase effects is then converted to amplitude modulation and thence the cross talk to other adjacent channels. This is shown in Figure 2.10.

2.5.2 Stimulated Scattering Effects

The scattering of lightwave by the impurities can happen due to the absorption and the vibration of the electrons and the dislocation of molecules in silica-based materials. The backscattering and absorption is commonly known as Raleigh scattering losses in fiber propagation in which phenomena the frequency of the optical carrier does not change.

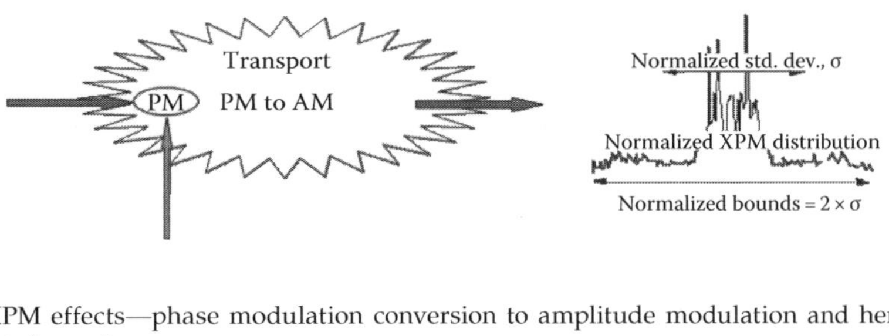

FIGURE 2.10
Illustration of XPM effects—phase modulation conversion to amplitude modulation and hence interference between adjacent channels.

Other scattering processes in which the frequency of the lightwave carrier is shifted to other frequency regions are commonly known as inelastic scattering: Raman scattering and Brillouin scattering. In both cases, the scattering of photons to a lower-energy-level photon with energy difference between these levels is fallen with the energy of phonons. Optical phonons are resulted from the electronic vibration for Raman scattering while acoustic phonons or mechanical vibration of the linkage between molecules lead to Brillouin scattering. At high power, when the intensity reaches over a certain threshold, the number of scattered photons is exponentially grown and the phenomenon is a simulated process. Thus, the phenomena can be called as stimulated Brillouin scattering (SBS) and stimulated Raman scattering (SRS). SRS and SBS were first observed in the 1970s [9–11].

2.5.2.1 Stimulated Brillouin Scattering

Brillouin scattering comes from the compression of the silica materials in the presence of an electric field, the electrostriction effect. Under the pumping of an oscillating electric field of frequency f_p, an acoustic wave of frequency F_a is generated. Spontaneous scattering is energy transfer from the pump wave to the acoustic wave and then a phase matching to transfer a frequency-shifted optical wave of frequency as a sum of the optical signal waves and the acoustic wave. This acoustic wave frequency shift is around 11 GHz with a bandwidth of around 50–100 MHz (due to the gain coefficient of the SBS) and a beating envelope would be modulating the optical signals. Thus, jittering of the received signals at the receiver would be formed, hence the closure of the eye diagram in the time domain.

Once the acoustics wave is generated, it beats with the signal waves to generate the side band components. This beating beam acts as a source and further transfer the signal beam energy into the acoustic wave energy and further amplify this wave to generate additional jittering effects. The Brillouin scattering process can be expressed by the following coupled equations [1, p. 60]:

$$\begin{aligned} \frac{dI_p}{dz} &= -g_B I_p I_s - \alpha_p I_p \\ -\frac{dI_s}{dz} &= +g_B I_p I_s - \alpha_s I_s \end{aligned} \tag{2.73}$$

where
- I_p is the intensity of pump beam
- I_s is the intensity of signal beam
- g_B is the Brillouin scattering gain coefficient
- α_s, α_p are the losses of signal and pump waves

The SBS gain g_B is frequency dependent with a gain bandwidth of around 50–100 MHz for pump wavelength at around 1550 nm. For silica fiber, g_B is about 5e–11 mW^{-1}. The threshold power for the generation of SBS can be estimated (using Equation 2.73) as follows:

$$\begin{aligned} g_B P_{th\_SBS} \frac{L_{eff}}{A_{eff}} &\approx 21 \\ \text{with the\_effective\_length } L_{eff} &= \frac{1-e^{-\alpha L}}{\alpha} \end{aligned} \tag{2.74}$$

For SSMF, this SBS power threshold is about 1.0 mW. Once the launched power exceeds this power threshold level, the beam energy is reflected back. Thus, the average launched power is usually limited to a few decibels relative to 1 mW of power due to this low threshold power level.

2.5.2.2 Stimulated Raman Scattering

Stimulated Raman scattering occurs in silica-based fiber when a pump laser source is launched into the guided medium, the scattering light from the molecules and dopants in the core region would be shifted to higher energy level and then jump down to lower energy level; hence, the amplification of photons in this level. Thus, a transfer of energy from different frequencies and energy levels of photons occurs. The stimulated emission happens when the pump energy level reaches above the threshold level. The pump intensity and signal beam intensity are coupled via the coupled equations:

$$\begin{aligned} \frac{dI_p}{dz} &= -g_R I_p I_s - \alpha_p I_p \\ -\frac{dI_s}{dz} &= +g_R I_p I_s - \alpha_s I_s \end{aligned} \tag{2.75}$$

where

I_p is the intensity of pump beam
I_s is the intensity of signal beam
g_R is the Raman scattering gain coefficient
α_s, α_p are the losses of signal and pump waves

The spectrum of the Raman gain depends on the decay lifetime of the excited electronic vibration state. The decay time is in the range of 1 ns and Raman-gain-bandwidth is about 1 GHz. In single-mode optical fibers, the bandwidth of the Raman gain is about 10 THz. The pump beam wavelength is usually about 100 nm below the amplification wavelength region. Thus, in order to extend the gain spectra, a number of pump sources of different wavelengths are used. Polarization multiplexing of these beams is also used to reduce the effective power launched in the fiber so as to avoid the damage of the fiber. The threshold for stimulated Raman gain is given by

$$\begin{aligned} g_R P_{th\_SRS} \frac{L_{eff}}{A_{eff}} &\approx 16 \\ \text{with the\_effective\_length } L_{eff} &= \frac{1-e^{-\alpha L}}{\alpha} \quad \text{or} \quad \approx 1/\alpha \text{ for long\_length} \end{aligned} \tag{2.76}$$

For fiber (SSMF) with an effective area of 50 μm^2, g_R~1e–13 m/W then the threshold power is about 570 mW near the C-band spectral region. This would require at least two pump laser sources that should be polarization multiplexed. The SRS is used frequently in modern optical communication systems, especially when no undersea optical amplification is required, the distributed amplification of SRS offers significant advantages as compared to lumped amplifiers such as erbium-doped fiber amplifiers (EDFA). The broadband gain and low gain ripple of SRS is also another advantage for dense wavelength division multiplexing (DWDM) transmission.

2.5.3 Four-Wave Mixing

Four wave mixing (FWM) is considered as a scattering process in which three photons are mixed to generate the fourth wave. This happens when the momentum of the four waves satisfy a phase matching condition. That is the condition of maximum power transfer. Figure 2.11 illustrates the mixing of different wavelength channels to generate interchannel cross talk. The phase matching can be represented by a relationship between the propagation constant along the z-direction in a single-mode optical fiber as follows:

$$\beta(\omega_1)+\beta(\omega_2)-\beta(\omega_3)-\beta(\omega_4)=\Delta(\omega \tag{2.77}$$

where

ω_1, ω_2, ω_3, and ω_4 are the frequencies of the first to the fourth waves

Δ is the phase mismatching parameter

In the case that the channels are equally spaced with a frequency spacing of Ω as in DWDM optical transmission, we have $\omega_1=\omega_2$; $\omega_3=\omega_1+\Omega$; and $\omega_4=\omega_1-\Omega$. One can use Taylor's series expansion around the propagation constant at the center frequency of the guide carrier β_0; we can then obtain [1, p. 67]

$$\Delta(\omega)=\beta_2\Omega^2 \tag{2.78}$$

The phase matching is thus optimized when β_2 is 0, which means that in the region where there is no dispersion, FWM is maximum and thus there is biggest interchannel cross talk. This is the reason why dispersion-shifted fiber is not commonly used when the zero dispersion wavelength is fallen in the spectral region of the operation of channel. In modern transmission fiber, the zero dispersion wavelength is shifted to outside the C-band, say 1510 nm, so that there is a small dispersion factor at 1550 nm and the C-band ranging from 2 to 6 ps/nm km, for example, Corning LEAF or nonzero dispersion-shifted fibers (NZ-DSF). This small amount of dispersion is sufficient to avoid the FWM with a channel spacing of 100 or 50 GHz.

The XPM signal is proportional to instantaneous signal power. Its distribution is bounded <5 channels and otherwise effectively unbounded. Thus, the link budgets include XPM evaluated at maximum outer bounds.

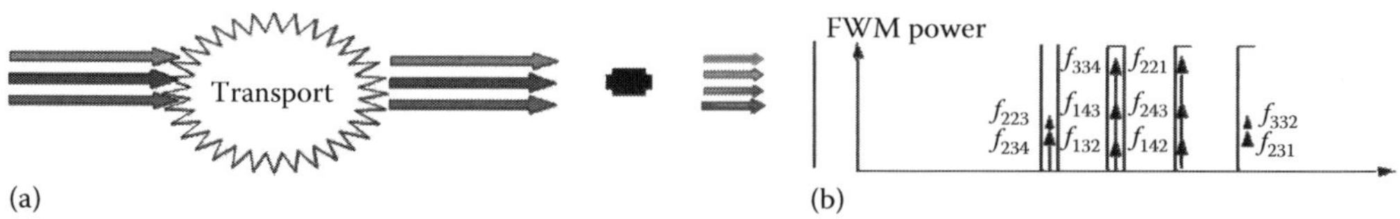

FIGURE 2.11
Illustration of four-wave mixing of optical channels: (a) momentum vectors of channels and (b) frequencies resulted from mixing of different channels.

2.6 Optical Fiber Manufacturing and Cabling

This section provides a brief description of the manufacturing of optical fibers and the cabling of several fibers for optical communication systems. The manufacturing techniques and cabling process affect the transmission and physical properties of the fibers. We focus on the aspects for a general understanding of optical transmission systems [2,4,–8].

As we have described in previous sections, the standard single-mode optical fiber structure is a cylindrical core with a refractive index slightly higher than that of the cladding region. For optical communications operating in the 1300 and 1700 nm wavelength regions, the silica material is the base material. A "pure" silica tube is the starting structure and a combination of silica, germanium oxide GeO_2, and P_2O_5 are then deposited inside the tube. Other dopants such as B_2O_3 and fluoride can also be used to reduce the refractive index of some small regions of the core. These are the segmented core and W-type fibers, which would be described in Chapter 3.

Once the deposition of the impurities are done (see Figure 2.14), the tube is collapsed to produce a preform, as shown in Figure 2.12. Also shown in this figure is a schematic of the fiber drawing machine and fiber drawing tower. The refractive index of the preform is also shown in this figure as noted in its caption and its detailed area is shown in Figure 2.13.

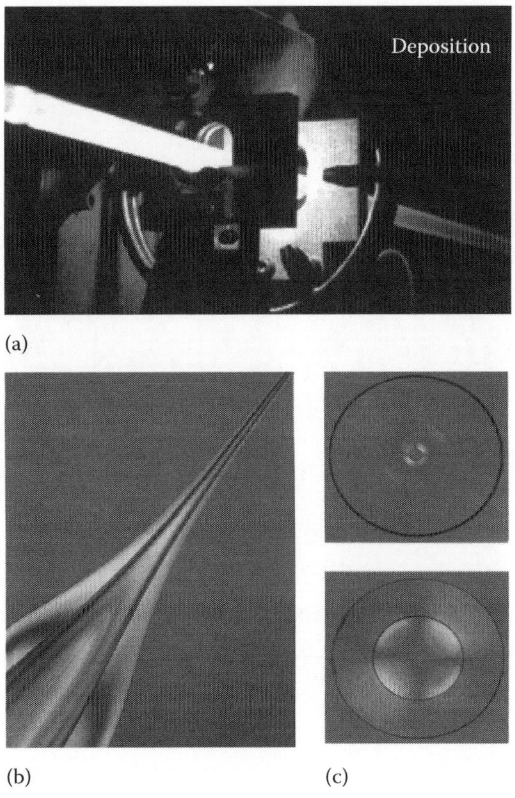

FIGURE 2.12
(a) Schematic of a fiber deposition and the fabrication of fiber preform, the deposition of core material, and collapsing; (b) fiber preform, before drawing into fiber stands; and (c) cross section of fiber preform with refractive index profile exactly the same as the fiber index profile of single mode (upper) and multimode (lower) types.

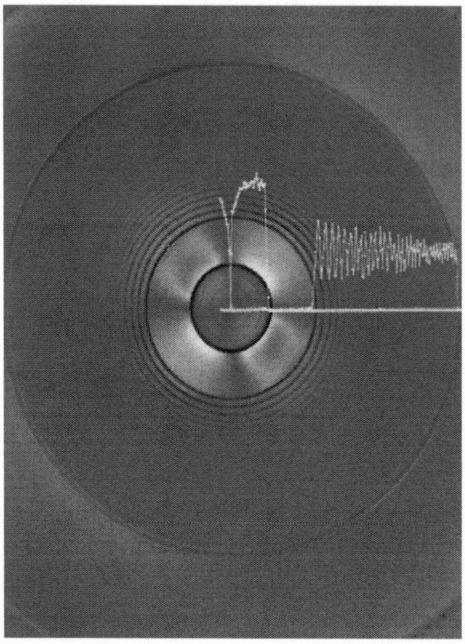

FIGURE 2.13
Real index profile across an SM fiber preform. *Note*: non-step-like profile—so why modeled as step-index structure?

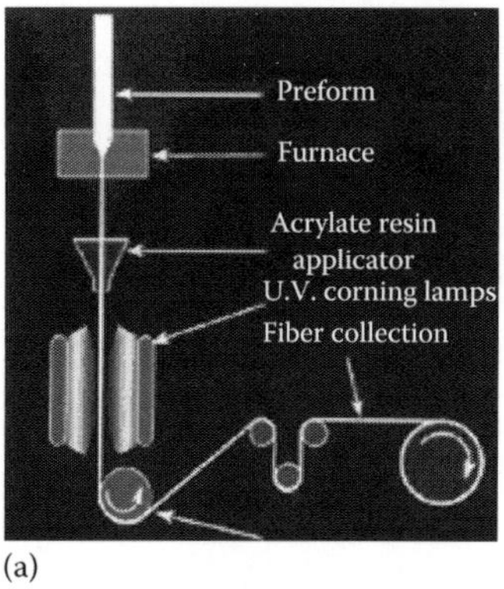

(a)

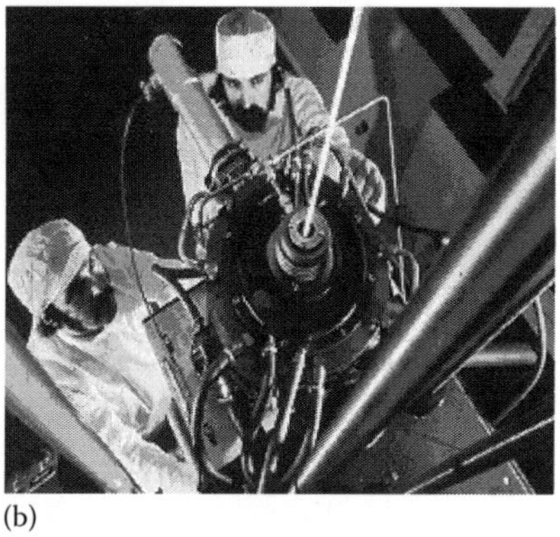

(b)

FIGURE 2.14
(a) Schematic of a fiber drawing machine and (b) picture of a fiber microwave furnace and diameter monitoring and feedback control.

Figure 2.15a–c shows the installation of optical fiber cables in different scenario by aerial cabling, buried ploughing, submarine laying. Figure 2.15d shows the fixed connection between two silica fiber ends.

2.7 Concluding Remarks

This chapter introduced the fundamental concepts of optical slab waveguide and then the circular optical waveguide or optical fibers. The basic properties of the fiber structures,

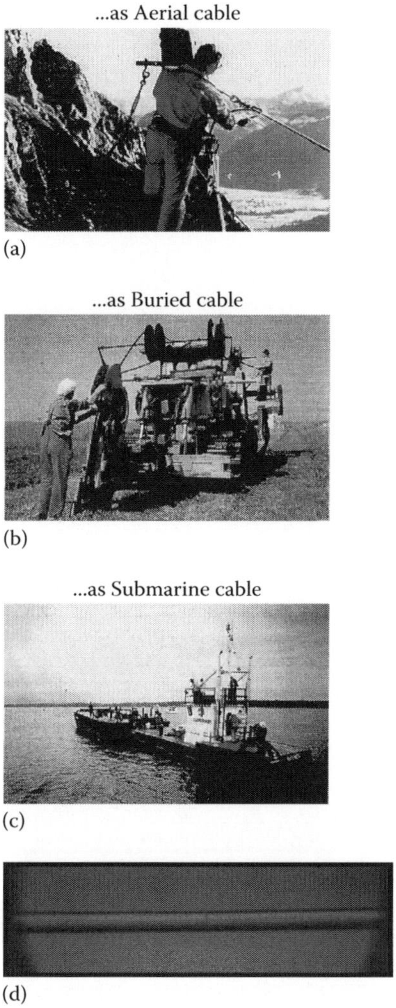

FIGURE 2.15
Installation of fiber cables: (a) installation of fiber cables by hanging, (b) installation of fiber cables by ploughing, (c) installation of undersea fiber cables, and (d) splicing two optical fibers.

its profile, the spot size, the cutoff wavelength, and the Gaussian approximation were described. The Gaussian approximation makes the understanding of the optical guided mode simple. It also allows us to directly obtain the optical mode distribution and thus several other approximations required to obtain the simplest form of important parameters of single-mode optical fibers.

Once the basic properties of a single-mode optical fiber are found, they form the basic set of parameters so that optical fibers whose effective index profiles are non-step can be found based on the ESI technique that would convert the parameters to an equivalent step-like profile and hence other optical properties.

Only structural and wave properties of lightwave signals traveling in optical fibers are presented here. As an optical communication systems engineer, one has to understand and develop techniques for analyzing and identifying the transmission of digital and analog signals through optical fibers and the attenuation and broadening of optical signals after transmitting through the transmission medium, namely, attenuation and broadening via the dispersion of lightwave pulses. These topics will be treated in the next chapter.

Furthermore, the design examples given in Chapter 11 and Chapter 12 employing special designed fibers whose design methods are described in this chapter and Chapter 3 for dispersion flattening and compensating, especially when they are used for Raman optical amplification medium in distributed or lumped configurations.

2.8 Problems

PROBLEM 2.1

An optical fiber with a step-index profile with a core diameter of 62.5 μm, a NA of 0.2 at a wavelength of 1300 nm is used for signal distribution and transmission in a local area network.

(a) What is the V parameter of this optical fiber?
(b) How many guided modes would it support? Can you comment on this number regarding the velocities of lightwaves?
(c) Select a cladding diameter. Give reasons for your selection.
(d) Find the maximum acceptance angle of this fiber. Hence, estimate the coupling loss of a laser source with a uniform radiation cone of 30°.

PROBLEM 2.2A

An optical fiber has the following parameters: index profile = steplike; core diameter = 9.0 μm; NA = 0.11; and a cladding refractive index = 1.48.

(a) Find the normalized frequency of the fiber at 1550 nm wavelength.
(b) Is the fiber operating in the single-mode or multimode region at 1550 nm? If it is in the single-mode region, estimate its mode filed diameter and its spot size. Sketch its field and intensity distribution across the fiber cross section.
(c) Find the cutoff wavelength of the fiber. If lightwaves of wavelength smaller than this cutoff wavelength are launched into the fiber, will the fiber still operate in single-mode region?

PROBLEM 2.2B

A single-mode step-index optical fiber has the following parameters: core diameter = 8.0 μm, cladding diameter = 0.125 mm, core refractive index = 1.460, and relative index difference = 0.2% at 1550 nm.

(a) Confirm that the fiber can be operating in the single-mode regime at 1550 nm wavelength.
(b) Find the fiber cutoff wavelength.
(c) What is the fiber MFD if it is operating at 1550 nm wavelength?

PROBLEM 2.3

For the optical fiber in Problem 2.2, if the refractive index profile is parabolic (α = 2) or triangular (α = 1) with the above NA at the central axis, repeat (a), (b), and (c).

PROBLEM 2.4

Single-mode optical fibers produced by Corning (and Optical Waveguides Pty. Ltd., Noble Park, Victoria, Australia) have typical characteristics as per technical data sheet attached at the end of this chapter.

(a) Using the physical characteristics of the fiber and technical data on its NA, confirm the fiber's functional characteristics such as the cutoff wavelength range.
(b) If this fiber is launched with an 850 nm laser, how many modes would it support? Sketch the mode fields for LP_{01} and LP_{11} modes.
(c) If lightwaves at 1300 nm travel over 10 km of this fiber, calculate the travel time of these waves.
(d) Estimate the fiber core diameter at 1300 nm wavelength.
(e) If the same spot size of (d) is required for the fiber to operate at 1550 nm, can you advise the manufacturer on any change in the physical parameters of the fiber.

PROBLEM 2.5

(a) The optical fiber in Problem 2.3 is used in an optical fiber transmission system with a laser source operating at 1310 nm having an output power of 1.0 mW and an optical receiver that can detect an average optical power of 0.1 mW. The fiber length is 50 km. Is it possible to detect the optical power at the end of the fiber length?
(b) Referring to the technical data of the standard optical fiber estimate the spreading of optical pulse after transmitting through the 50 km length fiber if the source has an optical line width of 2.0 nm.

PROBLEM 2.6

A step-index optical fiber used for an optical communication system operating at 1310 nm has a core radius of 25 μm and refractive indices in the core and cladding regions of 1.460 and 1.4550, respectively.

(a) What is the NA of the fiber?
(b) Estimate the number of guided modes.

PROBLEM 2.7

(a) Show that for a graded index fiber having a core refractive index $n^2(r) = n_2^2\left[1 + 2\Delta s\left(\frac{r}{a}\right)\right]$ with $s(r/a) = 1 - (r/a)^\alpha$ the acceptance angle $\alpha(r)$ is given by

$$\sin\alpha(r) = \left[n^2(r) - n_2^2\right]^{1/2}$$

(b) If the optical fiber has a parabolic profile shape, show that $\sin\alpha(r) = \mathrm{NA}\sqrt{1-\left(\frac{r}{a}\right)^{1/2}}$ where $\mathrm{NA} = n_2(2\Delta)^{1/2}(1+\Delta)$.

(c) A parabolic graded index silica optical fiber has a cladding refractive index of 1.460 and a relative index difference at the core axis of 1%. Find the maximum acceptance angle at the core axis of the fiber. Plot $\sin\alpha(r)$ as a function of r. What is the acceptance angle of the fiber at the core and cladding interface. Comment on the launch of a laser source into this fiber.

PROBLEM 2.8

(a) For a single-mode optical fiber having a graded index central dip, that is, $s(r/a) = 1 - (1 - r/a)^{\alpha}$, the ESI parameters of V and radius are given by

$$\frac{V_e}{V} = \left[1 - \frac{2\gamma}{(\alpha+1)(\alpha+2)}\right]^{1/2} \quad \text{and} \quad \frac{a_e}{a} = \frac{(\alpha+1)(\alpha+2)(\alpha+3) - 6\gamma}{(\alpha+1)(\alpha+2)(\alpha+3) - 2\gamma}$$

where $V = ka(2\Delta)^{1/2}$.

(b) The fiber has a physical core radius of 8.0 μm, a maximum relative index difference of 0.3%, and a cladding refractive index of 1.460. Find its ESI parameters for the normalized frequency and radius at 1550 nm wavelength. Also, find its ESI cutoff wavelength and its MFD at this wavelength.

Appendix: Technical Data of Single-Mode Optical Fibers

Corning SMF-28 Optical Fiber

Product Information

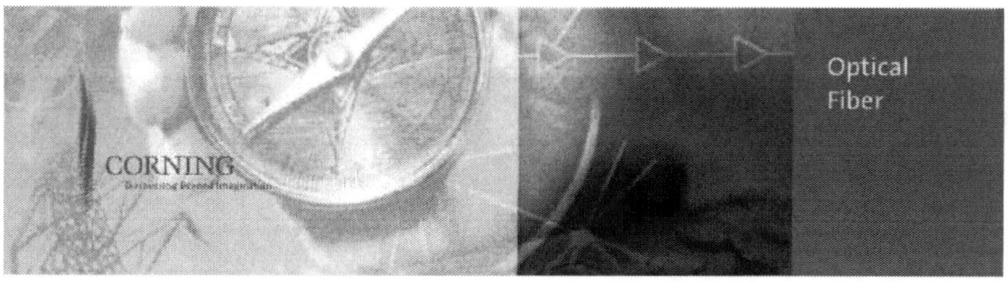

PI1036
Issued: April 2001
Supercedes: March 2001
ISO 9001 Registered

Corning Single-Mode Optical Fiber

The Standard for Performance

Corning SMF-28 single-mode optical fiber has set the standard for value and performance for telephony, cable television, submarine, and utility network applications. Widely used in the transmission of voice, data, and/or video services, SMF-28 fiber is manufactured to the most demanding specifications in the industry. SMF-28 fiber meets or exceeds ITU-T Recommendation G.652, TIA/EIA-492CAAA, IEC Publication 60793-2, and GR-20-CORE requirements.

Taking advantage of today's high-capacity, low-cost transmission components developed for the 1310 nm window, SMF-28 fiber features low dispersion and is optimized for use in the 1310 nm wavelength region. SMF-28 fiber can also be used effectively with TDM and WDM systems operating in the 1550 nm wavelength region.

Features and Benefits

- Versatility in 1310 and 1550 nm applications.
- Outstanding geometrical properties for low-splice loss and high-splice yields.
- OVD manufacturing reliability and product consistency.
- Optimized for use in loose tube, ribbon, and other common cable designs.

The Sales Leader

Corning SMF-28 fiber is the world's best-selling fiber. In 2000, SMF-28 fiber was deployed in over 45 countries around the world. All types of network providers count on this fiber to support network expansion into the twenty-first century.

Protection and Versatility

SMF-28 fiber is protected for long-term performance and reliability by the CPC™ coating system. Corning's enhanced, dual acrylate CPC coatings provide excellent fiber protection and are easy to work with. CPC coatings are designed to be mechanically stripped and have an outside diameter of 245 μm. They are optimized for use in many single- and multi-fiber cable designs including loose tube, ribbon, slotted core, and tight buffer cables.

Patented Quality Process

SMF-28 fiber is manufactured using the Outside Vapor Deposition (OVD) process, which produces a totally synthetic ultra-pure fiber. As a result, Corning SMF-28 fiber has consistent geometric properties, high strength, and low attenuation. Corning SMF-28 fiber can be counted on to deliver excellent performance and high reliability, reel after reel. Measurement methods comply with ITU recommendations G.650, IEC 60793-1, and Bellcore GR-20-CORE.

Optical Specifications

Attenuation

Standard Attenuation Cells

Wavelength (nm)	Attenuation Cells (dB/km)	
	Premium[a]	Standard
1310	≤0.35	≤0.40
1550	≤0.25	≤0.30

[a] Lower attenuation available in limited quantities.

Point Discontinuity

No point discontinuity greater than 0.10 dB at either 1310 or 1550 nm.

Attenuation at the Water Peak

The attenuation at 1383 ± 3 nm shall not exceed 2.1 dB/km.

Attenuation vs. Wavelength

Range (nm)	Ref. λ (nm)	Max. α Difference (dB/km)
1285–1330	1310	0.05
1525–1575	1550	0.05

The attenuation in a given wavelength range does not exceed the attenuation of the reference wavelength (λ) by more than the value α.

Attenuation with Bending

Mandrel Diameter (mm)	Number of Turns	Wavelength (nm)	Induced Attenuation[a] (dB)
32	1	1550	≤0.50
50	100	1310	≤0.05
50	100	1550	≤0.10

[a] The induced attenuation due to fiber wrapped around a mandrel of a specified diameter.

Cable Cutoff Wavelength (λ_{ccf})

$\lambda_{ccf} \leq 1260\,\text{nm}$

Mode-Field Diameter

9.2 ± 0.4 μm at 1310 nm
10.4 ± 0.8 μm at 1550 nm

Dispersion

Zero dispersion wavelength (λ_0):
$1302\,\text{nm} \leq \lambda_0 \leq 1322\,\text{nm}$

Zero dispersion slope (S_0):
$\leq 0.092\,\text{ps/(nm}^2\cdot\text{km)}$

$$\text{Dispersion} = D(\lambda): \approx \frac{S_0}{4}\left[\lambda - \frac{\lambda_0^4}{\lambda^3}\right] \text{ps/(nm}\cdot\text{km)},$$

$$\text{for } 1200\text{ nm} \leq \lambda \leq 1600\text{ nm}$$

λ = Operating wavelength

Polarization Mode Dispersion

Fiber Polarization Mode Dispersion (PMD)	
PMD link value	≤0.1[a]
Maximum individual fiber	≤0.2

[a] Complies with IECSC 86A/WGI, Method 1, September 1997.

The PMD link value is a term used to describe the PMD of concatenated lengths of fiber (also known as the link quadrature average). This value is used to determine a statistical upper limit for system PMD performance.

Individual PMD values may change when cabled. Corning's fiber specification supports network design requirements for a $0.5\,\text{ps}/\sqrt{\text{km}}$ maximum PMD.

Environmental Specifications

	Induced Attenuation (dB/km)	
Environmental Test Condition	**1310 nm**	**1550 nm**
Temperature dependence: −60°C to +85°C[a]	≤0.05	≤0.05
Temperature–humidity cycling −10°C to +85°C[a], up to 98% RH	≤0.05	≤0.05
Water immersion, 23°C ± 2°C[a]	≤0.05	≤0.05
Heat aging, 85°C ± 2°C[a]	≤0.05	≤0.05

[a] Reference temperature = +23°C.

Operating Temperature Range

−60°C to +85°C

Dimensional Specifications

***Standard Length (km/reel)*:** 2.2–50.4[a]

[a]Longer spliced lengths available at a premium.

Glass Geometry

Fiber curl: ≥4.0 m radius of curvature
Cladding diameter: 125.0 ± 1.0 μm
Core-clad concentricity: ≤0.5 μm
Cladding non-circularity: ≤1.0%

$$\text{Defined as:}\left[1-\frac{\text{Min. cladding diameter}}{\text{Max. cladding diameter}}\right]\times 100$$

Coating Geometry

Coating diameter: 245 ± 5 μm
Coating-cladding concentricity: <12 μm

Mechanical Specifications

Proof Test

The entire fiber length is subjected to a tensile proof stress ≥100 kpsi (0.7 GN/m^2).[a]

[a]Higher proof test levels available at a premium.

Performance Characterizations

Characterized parameters are typical values.

Core Diameter**:** 8.2 µm

Numerical Aperture**:** 0.14

NA is measured at the 1% power level of a one-dimensional far-field scan at 1310 nm.

***Zero Dispersion Wavelength* (λ_0):** 1313 nm

***Zero Dispersion Slope* (S_0):** 0.086 ps/(nm^2·km)

Refractive Index Difference**:** 0.36%

Effective Group Index of Refraction**, (N_{eff}© nominal MFD):**

1.4677 at 1310 nm
1.4682 at 1550 nm

***Fatigue Resistance Parameter* (n_d):** 20

Coating Strip Force:

Dry: 0.6 lbs. (3 N)
Wet, 14-day room temperature: 0.6 lbs. (3 N)

***Rayleigh Backscatter Coefficient* (for 1 ns pulse width):**

1310 nm: −77 dB
1550 nm: −82 dB

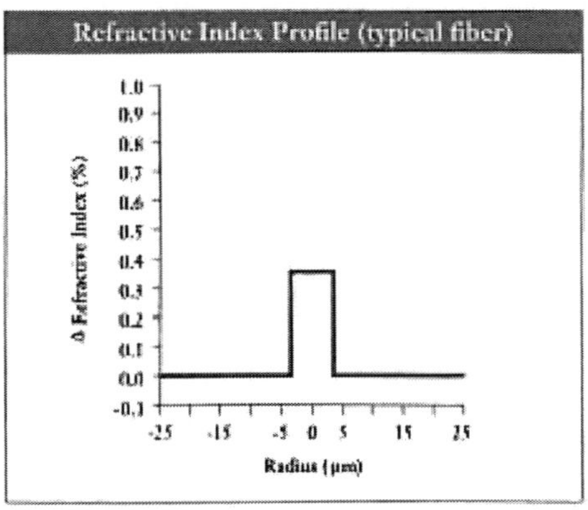

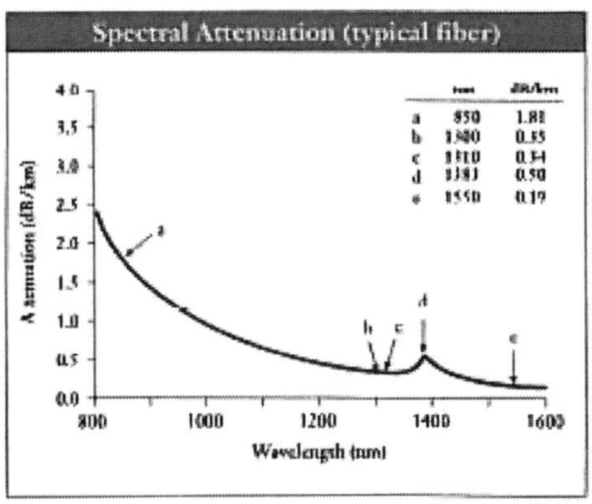

Corning SMF-28e Optical Fiber

Product Information

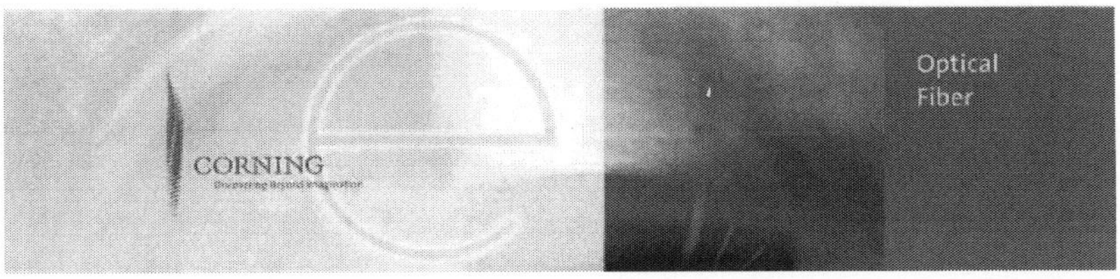

PI1344
Issued: April 2001
ISO 9001 Registered

Corning Single-Mode Optical Fiber

Introducing SMF-28e Optical Fiber

Corning SMF-28e single-mode optical fiber continues Corning's long history and performance as a premium fiber supplier. While allowing low water peak attenuation for system installers seeking a low water peak fiber, it meets all requirements for standard single-mode fiber.

Reduced Attenuation

SMF-28e fiber is designed with low water peak attenuation and complies with the requirements of newly adopted U.S. standard, TIA/EIA 492-CAAB, and the international standard, ITU G.652.C. These requirements define standard single-mode fibers for use across a broad wavelength range including the Extended Band (1360–1460 nm).

SMF-28e fiber provides superior attenuation performance throughout the 1260–1625 nm wavelength range, including a specified attenuation at 1383 nm ≤ 0.31 dB/km.

The Impact of Low Water Peak

Typical optical fiber displays an attenuation increase at or about 1383 nm. This "water peak" region (1375–1400 nm) is where light is strongly absorbed by hydroxyl (–OH) ions that are present in the glass core leading to increased attenuation over this wavelength range. SMF-28e fiber has had these ions removed during manufacture, thus reducing the attenuation and leading to a smooth curve across the Extended Band.

SMF-28e fiber boasts superior low water peak performance, providing excellent first-day attenuation as well as unparalleled lifetime performance.

Features and Benefits

- Versatility in the 1310, 1383, and 1550 nm windows
- Low attenuation throughout the Extended Band
- Characterized out to 1625 nm
- Longer transmission distance
- Enables emerging technologies like CWDM and SOA
- More options for network management
- Lengths up to 50.4 km/spool
- Protected by CPC coating system

Optical Specifications

Premium Attenuation

Wavelength (nm)	Attenuation (dB/km)
1310	≤0.35
1383	≤0.31[a]
1550	≤0.25

[a] Attenuation increases due to hydrogen aging at this wavelength will be ≤0.01 dB/km and evaluated in accordance with the IEC 60793-2 test procedure.

Point Discontinuity

No point discontinuity greater than 0.10 dB at either 1310 or 1550 nm.

Attenuation vs. Wavelength

Range (nm)	Ref. λ (nm)	Max. α Difference (dB/km)
1285–1330	1310	0.05
1525–1575	1550	0.05

The attenuation in a given wavelength range does not exceed the attenuation of the reference wavelength (λ) by more than the value α.

Attenuation with Bending

Mandrel Diameter (mm)	Number of Turns	Wavelength (nm)	Induced Attenuation[a] (dB)
32	1	1550	≤0.50
50	100	1310	≤0.05
50	100	1550	≤0.10
75	100	1625	≤0.50

[a] The induced attenuation due to fiber wrapped around a mandrel of a specified diameter.

Cable Cutoff Wavelength (λ_{ccf})

$\lambda_{ccf} \leq 1260\,\text{nm}$

Mode-Field Diameter

9.2 ± 0.4 μm at 1310 nm
10.4 ± 0.8 μm at 1550 nm

Dispersion

Zero dispersion wavelength (λ_0)
$1302\,\text{nm} \leq \lambda_0 \leq 1322\,\text{nm}$

Zero dispersion slope (S_0):
$\leq 0.092\,\text{ps/(nm}^2\cdot\text{km)}$

$$\text{Dispersion} = D(\lambda): \approx \frac{S_0}{4}\left[\lambda - \frac{\lambda_0^{\,4}}{\lambda^3}\right] \text{ps/(nm}\cdot\text{km)},$$

$$\text{for } 1200\text{ nm} \le \lambda \le 1625\text{ nm}$$

λ = Operating wavelength

Polarization Mode Dispersion

Fiber Polarization Mode Dispersion (PMD)

	Value $(\text{ps}/\sqrt{\text{km}})$
PMD link value	≤0.1[a]
Maximum individual fiber	≤0.2

[a] Complies with IEC SC 86A/WGI, Method 1, September 1997.

The PMD link value is a term used to describe the PMD of concatenated lengths of fiber (also known as the link quadrature average). This value is used to determine a statistical upper limit for system PMD performance.

Individual PMD values may change when cabled. Corning's fiber specification supports network design requirements for a $0.5\,\text{ps}/\sqrt{\text{km}}$ maximum PMD.

Environmental Specifications

	Induced Attenuation (dB/km)	
Environmental Test Condition	**1310 nm**	**1550 nm**
Temperature dependence: −60°C to +85°C[a]	≤0.05	≤0.05
Temperature–humidity cycling −10°C to +85°C[a], up to 98% RH	≤0.05	≤0.05
Water immersion, 23°C ± 2°C	≤0.05	≤0.05
Heat aging, 85°C ± 2°C[a]	≤0.05	≤0.05

[a] Reference temperature = +23°C.

Operating Temperature Range

−60°C to +85°C

Dimensional Specifications

Standard Length (km/spool): 2.2–50.4

Glass Geometry

Fiber curl: ≥4.0 m radius of curvature
Cladding diameter: 125.0 ± 1.0 μm
Core-clad concentricity: ≤0.5 μm
Cladding non-circularity: ≤1.0%

$$\text{Defined as:}\left[1-\frac{\text{Min. cladding diameter}}{\text{Max. cladding diameter}}\right]\times 100$$

Coating Geometry

Coating diameter: $245 \pm 5\,\mu m$
Coating-cladding concentricity: $<12\,\mu m$

Mechanical Specifications

Proof Test

The entire fiber length is subjected to a tensile stress ≥100 kpsi (0.7 GN/m^2)[a].
[a]Higher proof test levels available at a premium.

Performance Characterizations

Characterized parameters are typical values.

***Core Diameter*:** 8.2 μm

***Numerical Aperture*:** 0.14

NA is measured at the 1% power level of a one-dimensional far-field scan at 1310 nm.

***Zero Dispersion Wavelength* (λ_0):** 1313 nm

***Zero Dispersion Slope* (S_0):** 0.086 ps/(nm^2·km)

***Refractive Index Difference*:** 0.36%

***Effective Group Index of Refraction* (N_{eff}):**

1.4677 at 1310 nm
1.4682 at 1550 nm

***Fatigue Resistance Parameter* (n_d):** 20

***Coating Strip Force*:**

Dry: 0.6 lbs. (3 N)
Wet, 14 day room temperature: 0.6 lbs. (3 N)

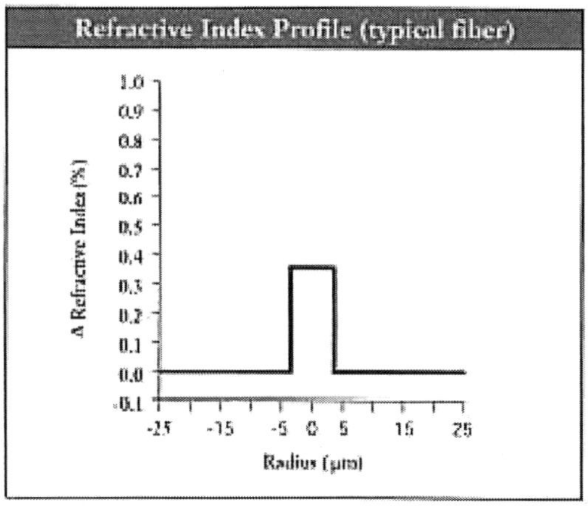

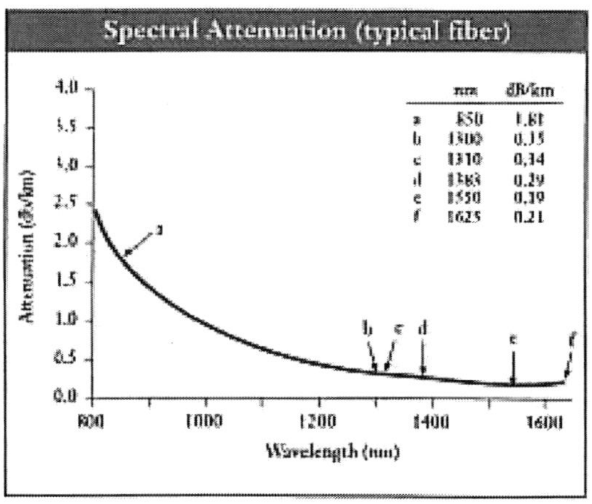

Corning LEAF Optical Fiber

Product Information

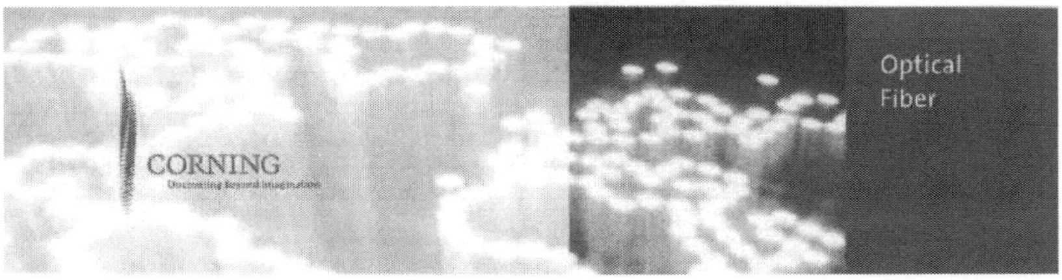

PI1107

Issued: May 2001
Supercedes: April 2001
ISO 9001 Registered

A Powerful Network Needs:

Backbone by LEAF Fiber

With the ever-accelerating race for bandwidth, network designers are challenged to build a network for the present that will also maximize future technologies. Deploy the fiber that revolutionized network technology and gives you room to move. Break the bandwidth barrier with a fiber so technologically advanced it gives you the optical backbone you need for today's and tomorrow's networks—Corning LEAF optical fiber.

Find out what the world's most powerful networks have in common: Backbone by LEAF fiber.

The Large Effective Area Advantage

LEAF fiber's large effective area (A_{eff}) offers higher power-handling capability, improved optical signal-to-noise ratio, longer amplifier spacing, and maximum dense wavelength division multiplexing (DWDM) channel plan flexibility compared with other nonzero dispersion-shifted fibers (NZ-DSFs). Fiber with a large A_{eff} also provides a critical performance advantage—the ability to uniformly reduce all nonlinear effects (Figure 2.16). Nonlinear effects represent the greatest performance limitations in today's multichannel DWDM systems.

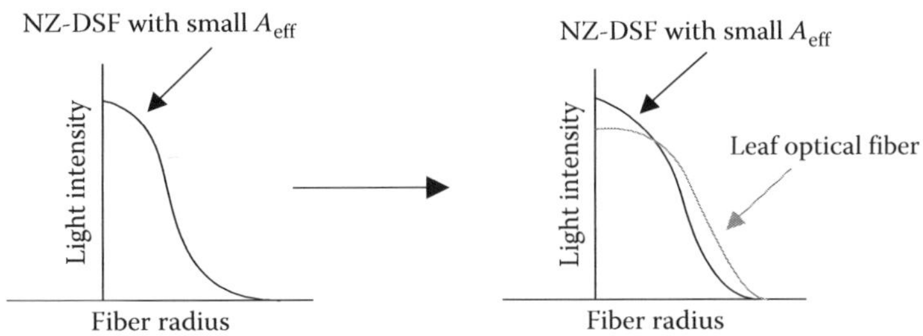

FIGURE 2.16
LEAF fiber's larger A_{eff} increases the area where the light can propagate, thereby reducing nonlinear effects.

The Next Generation

In addition to outperforming other NZ-DSFs in the conventional band (C-Band: 1530–1565nm), LEAF fiber facilitates the next technological development in fiber-optic networks—the migration to the long band (L-Band: 1565–1625nm). In both C-Band and L-Band operation, LEAF fiber has demonstrated greater ability to handle more channels by reducing nonlinear effects such as four-wave mixing, self-phase modulation, and cross-phase modulation in multichannel DWDM transmission.

Reduce Network Costs

With its increased optical reach advantage, LEAF fiber requires fewer amplifiers and regenerators, and therefore provides immediate and long-term cost savings. LEAF fiber is also compatible with installed base fibers and photonic components. In fact, LEAF fiber's slightly larger mode-field diameter improves its splicing performance, especially when connecting to standard single-mode fiber such as Corning SMF-28 fiber. And, as with all Corning optical fiber, LEAF fiber's geometry package is the best in the industry. With LEAF fiber, it is easy and economical to increase the information-carrying capacity of your network.

Fiber for Today and Tomorrow

While LEAF fiber is exceptionally suited to operate with already-installed 2.5Gb/s systems, it is techno-economically optimized for today's high-channel-count 10Gb/s systems, and provides the ability to upgrade in the future to tomorrow's high-bit systems. Additionally, LEAF fiber's unparalleled specifications on polarization mode dispersion (PMD) allow fiber installed today to operate at data rates higher than 10Gb/s. The combination of LEAF fiber's large A_{eff} and its demonstrated Raman upgradeability allows transmission engineers to design and build networks advantaged over other fiber plants. As the world's most advanced NZ-DSF, LEAF fiber is ready for future technology when your network is.

LEAF Fiber—All about Value

With LEAF fiber's proven large A_{eff} advantage, the industry's best geometry package, and inherent future-proof design, LEAF fiber continues to be the fiber of choice for today's high-capacity and tomorrow's all-optical networks. Network providers on the cutting edge have embraced large A_{eff} technology as the fiber "backbone" for high-data-rate networks now and in the future.

Technology Awards

Corning Incorporated has received multiple industry awards for its patented LEAF optical fiber. Independent panels of experts have chosen LEAF fiber based on its technical merits for the following awards:

"Annual Technology Award" from *Fiberoptic Product News*

"Commercial Technology Achievement Award for Fiber-Optics" from *Laser Focus World Magazine*

"Circle of Excellence Award" from *Photonics Spectra Magazine*

"R&D 100 Award" from *R&D Magazine*

Coating

Corning fiber is protected for long-term performance and reliability by the CPC coating system. Corning's enhanced, dual acrylate CPC coatings provide excellent fiber protection and are easy to work with. CPC coatings are designed to be mechanically stripped and have an outside diameter of 245 μm. CPC coatings are optimized for use in many single- and multi-fiber cable designs, including loose tube, ribbon, slotted core, and tight buffer cables.

Optical Specifications

Attenuation

≤0.25 dB/km at 1550 nm
≤0.25 dB/km at 1625 nm
No point discontinuity greater than 0.10 dB at 1550 nm
Attenuation at 1383 ± 3 nm shall not exceed 1.0 dB/km

Attenuation vs. Wavelength

Range (nm)	Ref, λ (nm)	Max, Increase α (dB/km)
1525–1575	1550	0.05
1625	1550	0.05

The attenuation in a given wavelength range does not exceed the attenuation of the reference wavelength (λ) by more than the value α. In all cases, a maximum attenuation of ≤0.25 dB/km applies at 1550 and 1625 nm.

Attenuation with Bending

Mandrel Diameter (mm)	Number of Turns	Wavelength (nm)	Induced Attenuation[a] (dB)
32	1	1550 and 1625	≤0.50
75	100	1550 and 1625	≤0.50

The induced attenuation due to fiber wrapped around a mandrel of a specified diameter.

Mode-Field Diameter

9.20–10.00 μm at 1550 nm

Dispersion

Total dispersion: 2.0–6.0 ps/(nm · km) over the range 1530–1565 nm
4.5 to 11.2 psec/(nm · km) over the range of 1565 to 1625 nm

Fiber Polarization Mode Dispersion (PMD)

	Value $\left(ps/\sqrt{km}\right)$
PMD link value	≤0.4[a]
Maximum individual fiber	≤0.1

[a] Complies with IEC SC 86A/WGI, Method 1, September 1997 (n-24, Q-0.1%).

The PMD link design value is a term used to describe the PMD of concatenated lengths of fiber (also known as PMD_Q). This value represents a statistical upper limit for total link PMD.

PMD values may change when fiber is cabled. Corning's fiber specification supports emerging network design requirements for high-data-rate systems operating at 10 Gb/s (TDM) rates and higher.

Environmental Specifications

Environmental Test Condition	Induced Attenuation (dB/km) 1550 nm
Temperature dependence: −60°C to +85°C[a]	≤0.05
Temperature–humidity cycling −10°C to +85°C[a], up to 98% RH	≤0.05
Water immersion, 23°C	≤0.05
Heat aging, 85°C[a]	≤0.05
Operating temperature range: −60°C to +85°C.	

[a] Reference temperature = +23°C.

Dimensional Specifications

Standard Length (km/reel): 4.4–25.2[a]

[a]Longer spliced lengths available at a premium.

Glass Geometry

Fiber curl: ≥4.0 m radius of curvature
Cladding diameter: 125.0 ± 1.0 μm
Core/clad concentricity: ≤0.5 μm
Cladding non-circularity: ≤1.0%

$$\text{Defined as:}\left[1-\frac{\text{Min. cladding diameter}}{\text{Max. cladding diameter}}\right]\times 100$$

Coating Geometry

Coating diameter: 245 ± 5 μm
Coating/cladding concentricity: <12 μm

Mechanical Specifications

Proof Test

The entire fiber length is subjected to a tensile stress ≥100 kpsi (0.7 GN/m^2)[a]

[a]Higher proof test available at a premium.

Performance Characterizations

Characterized parameters are typical values.

Effective Area (A_{eff})

72 μm²

Effective Group Index of Refraction (N_{eff})

1.469 at 1550 nm

Fatigue Resistance Parameter (n_s)

20

Coating Strip Force

Dry, 2.8 N (0.6 lbs)
Wet, 14 days room temperature:
2.7 N (0.6 lbs)

Consistency with Global Standards

The values in this product information sheet demonstrate Corning LEAF fiber's conformity with ITU-T Recommendation G.655, IEC 60793-2 for B4 class fibers and Belcore/Telcordia GR-20-CORE.

Dispersion Calculation

$$\text{Dispersion} = D(\lambda) = \left(\frac{D(1565\text{ nm}) - D(1530\text{ nm})}{35} * (\lambda - 1565)\right) + D(1565\text{ nm})$$

λ-Operating wavelength up to 1565

$$\text{Dispersion} = D(\lambda) = \left(\frac{D(1625\text{ nm}) - D(1565\text{ nm})}{60} * (\lambda - 1625)\right) + D(1625\text{ nm})$$

λ-Operating wavelength up to 1565–1625

Special selections of LEAF fiber attributes are available upon request.

References

1. G.P. Agrawal, *Fiber Optic Communications Systems*, 3rd edn. New York: Academic Press, 2002, Chapter 2, pp. 37–50.
2. L.B. Jeunhomme, *Single Mode Fiber Optics*. New York: Marcel Dekker, 1983.
3. G. Mahlkc and P. Gossing, *Fiber Optic Cables*. Chichester, U.K.: Siemens A.G. and J. Wiley, 1987.
4. D. Jones, *Optical Fiber Communications Systems*. New York: Holt, Rinhart and Winston, 1988.

5. A.W. Snyder, Understanding monomode optical fibers, *Proceedings of the IEEE*, 69(1): 6–13, January 1981.
6. D. Keck, Single mode fibers outperform multimode cables, *IEEE Spectrum*, 20: 30–37, March 1983.
7. D. Marcuse, Loss-analysis of single-mode fiber splices, *Bell System Technical Journal*, 56(5): 703–718, 1977.
8. P.S. Henry, Lightwave primer, *IEEE Journal Quantum Electronics*, QE-21(12): 1862–1879, December 1985.
9. R.H. Stolen, E.P. Ippen, and A.R. Tynes, Raman oscillation in glass optical waveguide, *Applied Physics Letters*, 20: 62, 1972.
10. E.P. Ippen and R.H. Stolen, Stimulated Brillouin scattering in optical fibers, *Applied Physics Letters*, 21: 539, 1972.
11. R.G. Smith, Optical power handling capacity of low loss optical fibers as determined by Stimulated Raman and Brillouin Scattering, *Applied Optics*, 11: 2489, 1972.

3

Optical Fibers: Signal Attenuation and Dispersion

This chapter describes the mechanism and properties of lightwave-modulated signals when propagating in optical fibers, only in single-mode optical fibers (SMF), in particular, standard SMF (SSMF), nonzero dispersion-shifted fibers (NZDSF), and dispersion-compensating fibers. Attenuation and dispersion effects, the two principal phenomena in the design of transmission systems, are also described in detail.

3.1 Introduction

In Chapter 2, the basic structure and fundamental aspects of lightwaves propagating in optical fibers were treated, in particular, SMF, which form the basic structure for standard communication transmission systems. This chapter deals with the transmission of optical signals over optical fibers, mainly the loss and spreading of optical signals transmitted through optical fibers, namely, attenuation and dispersion effects.

Attenuation and dispersion are the two most important effects that play a major role in optical fiber transmission systems. The attenuation of an optical signal would limit the availability of optical power along the transmission path, and for very low attenuation, dispersion limits the repeater spacing below what would be possible from the attenuation factor.

The fiber loss has been reduced from 100 dB/km (i.e., transmission possible over only a few meters) at 1300 nm in 1970 to about 0.25 and 0.15 dB/km, which are very close to the theoretically possible transparent limits and the transmission over several hundreds of kilometers of fibers, at 1300 and 1550 nm wavelength regions, respectively, in 1980.

The dispersion and pulse broadening of optical fibers have been also reduced due to the "smart design" of optical fiber structures. In the early 1970s, we saw a remarkable development of theories for understanding lightwave guiding in optical fibers of multimode types. The breakthrough in the reduction of loss in optical fibers and the ability to manufacture optical fibers with a very small core diameter led to the design of SMF. The remarkable theoretical development of optical waveguiding in a "weakly guided" (i.e., a very small difference between the core and cladding regions) fiber structure led to a plane-wavelike transmission of lightwaves. Further, the availability of narrow-linewidth lasers allowed systems engineers to design and implement several high-speed, long-distance fiber-optic communication systems.

The attenuation that arises from intrinsic material properties and from waveguide properties is described, and a general attenuation coefficient is derived. The chromatic dispersion for an SMF in linear limit, implying the optical power launched into the fiber to be less than the threshold for nonlinear effects, is then treated. The effects of optical waveguide parameters on the dispersion factors are analyzed. The balance of the opposite-signed dispersion factors between materials and waveguides is analyzed so that a minimum dispersion factor can be designed for optical fibers with dispersion-compensated or dispersion-shifted characteristics.

In the following chapters, we will describe optical amplifiers and techniques to compensate for the broadening of optical pulses so that an ultra-long, ultra-high-speed optical fiber communication system can be designed and implemented. Readers can refer to [1–4] for more information.

3.2 Signal Attenuation in Optical Fibers

Optical loss in optical fibers is one of the two fundamental limiting factors, as it reduces the average optical power reaching the receiver. The optical loss is the sum of three major components: intrinsic loss, microbending loss, and splicing loss.

3.2.1 Intrinsic or Material Attenuation

Referring to Figure 3.1, intrinsic loss mainly consists of the absorption loss due to OH impurities and the Rayleigh scattering loss. The intrinsic loss is a function of λ^{-6}. Thus, the longer the operating wavelength, the lower the loss. However, it also depends on the transparency of the optical materials that are used to form the optical fibers. For silica fibers, the optical material loss is low over the wavelength range 0.8–1.8 μm. Over this wavelength range, there are three optical windows for which optical communication is used. The first window, 810 nm, has a bandwidth range of about 20.0 nm over the central wavelength. The second and third windows, most commonly used in present optical communications, are over 1300 and 1550 nm having ranges of 80 and 40 nm, respectively. The intrinsic losses are about 0.3 and 0.15 dB/km at 1550 and 1300 nm regions, respectively.

This improvement is a few hundred thousand times the original transmission of signals over 5.0 m with a loss of about 60 dB/km. Most communication fiber systems operate at

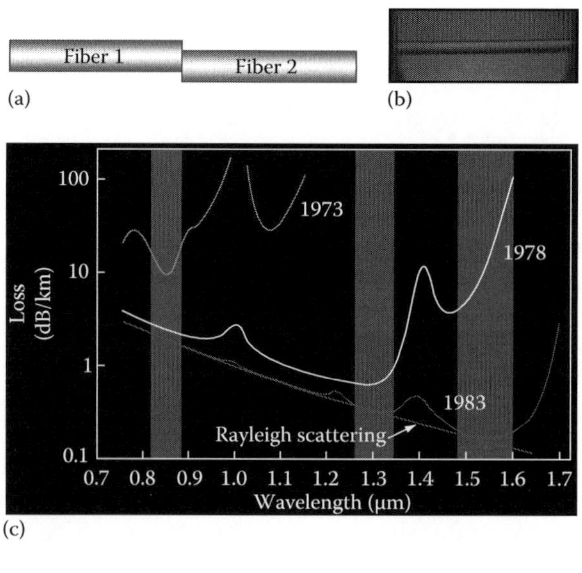

FIGURE 3.1
(a) Misalignment in splicing two optical fibers generating losses, (b) aligned spliced fibers, and (c) spectral attenuation of silica fiber over the years.

1300 nm due to their minimum dispersion at this range. For "power-hungry" systems, optical or extra-long systems should operate at 1550 nm.

3.2.2 Absorption

The absorption loss in silica glass is composed mainly of ultraviolet (UV) and infrared (IR) absorption tails of pure silica. The IR absorption tail of pure silica formed due to the vibration of the basic tetrahedron is shown in Figure 3.1, which causes strong resonances around 8–13 μm with a loss of about 10^{10} dB/km. This loss is shown by curve IR. Overtones and combinations of these vibrations lead to various absorption peaks in the low-wavelength range, as shown by curve UV.

Various impurities that also lead to spurious absorption effects in the wavelength range of interest (1.2–1.6 μm) are transition metal ions and water in the form of OH ions. These sources of absorption have been practically reduced in recent years.

3.2.3 Rayleigh Scattering

The Rayleigh scattering loss, L_R, which is due to microscopic inhomogeneities of the material, shows a λ^{-4} dependence and is given by

$$L_R = (0.75 + 4.5\Delta)\lambda^{-4} \text{ dB/km} \tag{3.1}$$

where

Δ is the relative index difference, as defined in Chapter 1
λ is the wavelength in μm

Thus, to minimize the loss, Δ should be made as low as possible.

3.2.4 Waveguide Loss

The losses due to waveguide structures arise from power leakage, bending and microbending of the fiber axis, and defects and joints between fibers. Power leakage is significant only for depressed cladding fibers.

3.2.5 Bending Loss

When a fiber is bent, the plane wave front associated with the guided mode is pivoted at the center of curvature and its longitudinal velocity along the fiber axis increases with the distance from the center of curvature. As the fiber is bent further over a critical curve, its phase velocity exceeds that of the plane wave in the cladding, and radiation occurs.

The bend loss, L_B, for a radius R (radius of curvature) is given by

$$L_B = -10\,\text{Log}_{10}\left(1 - 890\frac{r_0^6}{\lambda^4 R^2}\right) \quad \text{for silica} \tag{3.2}$$

3.2.6 Microbending Loss

Microbending loss results due to power coupling from the guided fundamental mode of a fiber to radiation modes. This coupling takes place when the fiber axis is bent randomly in a high spatial frequency. Such bending can occur during the packing of the fiber in the cabling process, as shown in Figure 3.2.

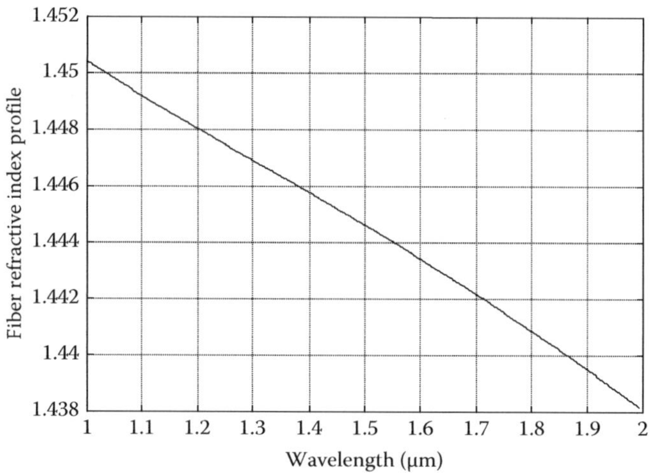

FIGURE 3.2
Variation in the refractive index as a function of the optical wavelength of silica.

The microbending loss of an SMF is a function of the fundamental mode spot size, r_0. Fibers with large spot sizes are extremely sensitive to microbending. It is therefore desirable to design the fiber to have as small a spot size as possible to minimize the bending loss. The microbending loss can be expressed by the relation

$$L_{\mathrm{m}} = 2.15\times10^{-4} r_0^6 \lambda^{-4} L_{\mathrm{mm}} \ \mathrm{dB/km} \tag{3.3}$$

where L_{mm} is the microbending loss of a 50 μm core multimode fiber having an NA of 0.2.

3.2.7 Joint or Splice Loss

Ultimately, the fibers will have to be spliced together to form the final transmission link. With a fiber cable that averages 0.4–0.6 dB/km, a splice loss in excess of 0.2 dB/splice drastically reduces the unrepeated distance that can be achieved. It is, therefore, extremely important that the fiber be designed in such a way that the splice loss is minimized.

Splice loss is mainly due to the axial misalignment of the fiber core, as shown in Figure 3.1.

Splicing techniques, which rely on the alignment of the outside surface of the fibers, require extremely close tolerances on core-to-outside surface concentricity. Offsets of the order of 1 μm can produce significant splice loss. This loss is given by

$$L_{\mathrm{s}} = \frac{10}{\ln 10}\left(\frac{d}{r_0}\right)^2 \ \mathrm{dB} \tag{3.4}$$

where d is the *axial* misalignment of the fiber cores. It is obvious that minimizing the optical loss requires trade-offs between the different sources of loss. It is advantageous to have a large spot size to minimize both Rayleigh and splicing losses, whereas minimizing bending and microbending losses requires a small spot size. In addition, as will be described in the next section, the spot size plays a significant role in the chromatic dispersion properties of SMF.

3.2.8 Attenuation Coefficient

Under general conditions of power attenuation inside an optical fiber, the attenuation coefficient of the optical power, P, can be expressed as

$$\frac{dP}{dz} = -\alpha P \tag{3.5}$$

where α is the attenuation coefficient. This attenuation coefficient can include all effects of power loss when signals are transmitted through the optical fibers.

Considering optical signals with an average optical power, P_{in}, entering at the input of the fiber length, L, and the output optical power, P_{out}, in relation to the attenuation coefficient, α, we have

$$P_{out} = P_{in}\, e^{(-\alpha L)} \tag{3.6}$$

It is customary to express α in dB/km using the relation

$$\alpha(\text{dB/km}) = -\frac{10}{L}\log_{10}\left(\frac{P_{out}}{P_{in}}\right) = 4.343\alpha \tag{3.7}$$

Standard optical fibers with a small Δ would exhibit a loss of about 0.2 dB/km, that is, the purity of silica is very high. Such purity of a bar of silica would allow us to see through a 1 km glass bar, without distortion, a person standing at the other end! The attenuation curve for silica glass is shown in Figure 3.1.

3.3 Signal Distortion in Optical Fibers

3.3.1 Basics on Group Velocity

Consider a monochromatic field given by

$$E_x = A\cos(\omega t - \beta z) \tag{3.8}$$

where

A is the wave amplitude

ω is the radial frequency

β is the propagation constant along the z-direction

Setting $(\omega\tau - \beta z)$ constant, the wave phase velocity is given by

$$v_p = \frac{dz}{dt} = \frac{\omega}{\beta} \tag{3.9}$$

Now we consider that the propagating wave consists of two monochromatic fields of frequencies, $\omega + \delta\omega$ and $\omega - \delta\omega$, of

$$E_{x1} = A\cos\left[(\omega + \delta\omega)t - (\beta + \delta\beta)z)\right] \tag{3.10}$$

$$E_{x2} = A\cos\left[(\omega - \delta\omega)t - (\beta - \delta\beta)z)\right] \tag{3.11}$$

The total field is then given by

$$E_x = E_{x1} + E_{x2} = 2A\cos(\omega t - \beta z)\cos(\delta\omega t - \delta\beta z) \tag{3.12}$$

If $\omega \gg \delta\omega$, then $\cos(\omega t - \beta z)$ varies much faster than $\cos(\delta\omega t - \delta\beta z)$ as shown in Figure 3.3. The phasor representation is given in Figure 3.4. Now, by setting $(d\omega t - d\beta z)$ constant, the group velocity is defined as

$$v_g = \frac{d\omega}{d\beta} \rightarrow v_g^{-1} = \frac{d\beta}{d\omega} \tag{3.13}$$

The group delay, t_g, per unit length (setting L at 1.0 km) is thus given as

$$t_g = \frac{L(1\,\text{km})}{v_g} = \frac{d\beta}{d\omega} \tag{3.14}$$

The pulse spread, $\Delta\tau$, per unit length due to group delay of light sources of spectral width, σ_λ (i.e., the full-width half-maximum [FWHM] of the optical spectrum of the light source), is

$$\Delta\tau = \frac{dt_g}{d\lambda}\sigma_\lambda \tag{3.15}$$

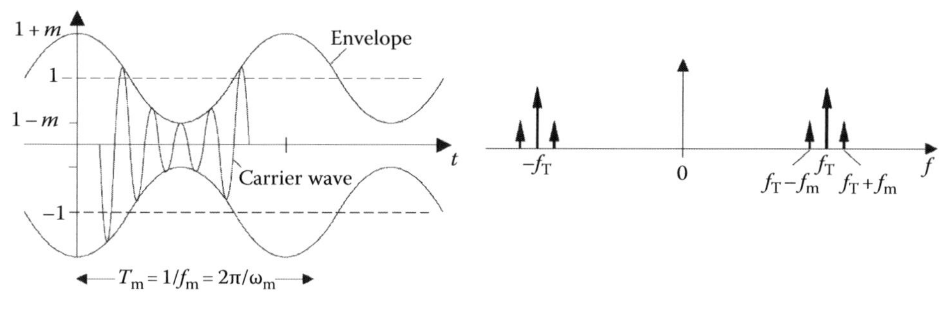

FIGURE 3.3
Time signal and spectrum.

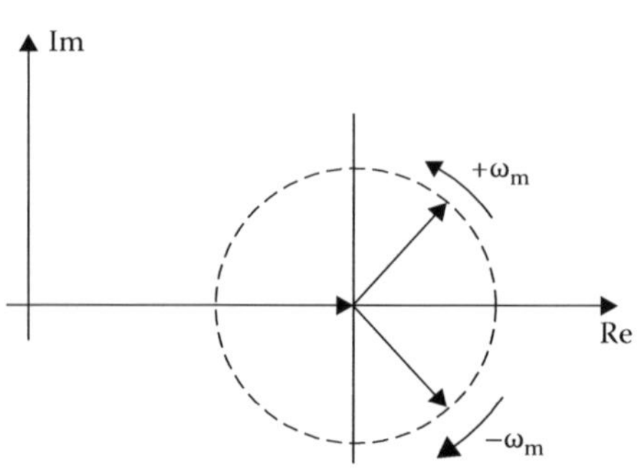

FIGURE 3.4
Vector phasor diagram of the complex envelope.

i.e., the spread of the group delay due to the spread of the source wavelength in ps/km. Thus, the linewidth of the light source makes a great difference in the distortion of the optical signal transmitted through the optical fiber. The narrower the source linewidth, the less dispersed are the optical pulses. The typical linewidth of Fabry–Perot semiconductor lasers is about 1–2.0 nm, while the DFB (distributed feedback) laser exhibits a linewidth of 100 MHz. (How many nm is this 100 MHz optical frequency equivalent to?)

An optical signal traveling along a fiber becomes increasingly distorted. This distortion is a consequence of intermodal delay effects and intramodal dispersion. Intermodal delay effects are significant in multimode optical fibers since each mode has a different value of group velocity at a specific frequency. Although intramodal dispersion is pulse spreading that occurs within a single mode, it is the result of the group velocity that is a function of the wavelength, λ, and is therefore referred to as chromatic dispersion.

The two main causes of intramodal dispersion are

- Material dispersion, which arises from the variation of the refractive index, $n(\lambda)$, as a function of the wavelength. This causes a wavelength dependence of the group velocity of any given mode.
- Waveguide dispersion, which occurs because the mode propagation constant, $\beta(\lambda)$, is a function of the wavelength, λ; the core radius, a; and the refractive index difference.

The group velocity associated with the fundamental mode is frequency dependent because of chromatic dispersion. As a result, different spectral components of the light pulse travel at different group velocities, a phenomenon referred to as *group velocity dispersion* (GVD), intramodal dispersion, or as material dispersion and waveguide dispersion. The movement of the phasors is illustrated in Figure 3.5.

3.3.2 Group Velocity Dispersion

3.3.2.1 Material Dispersion

The refractive index of silica as a function of wavelength is shown in Figure 3.2. The refractive index is plotted over the wavelength region of 1.0–2.0 μm, which is the most important range for silica-based optical communication systems, as the loss is lowest at 1300 and 1550 nm windows.

The propagation constant, β, of the fundamental mode guided in the optical fiber can be written as

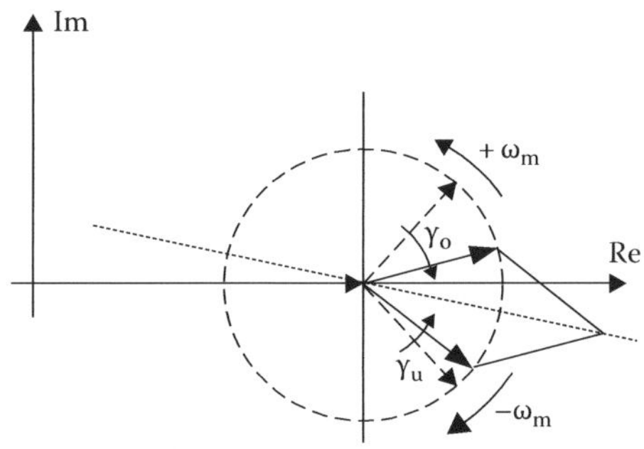

FIGURE 3.5
Magnitude of the complex envelope when not sinusoidal; the envelope is subject to nonlinear distortions.

$$\beta(\lambda) = \frac{2\pi n(\lambda)}{\lambda} \tag{3.16}$$

The group delay, t_{gm}, per unit length of (3.9) can be obtained from

$$t_{gm} = \frac{d\beta}{d\omega} \tag{3.17}$$

where we can use

$$d\omega = d\left(\frac{2\pi c}{\lambda}\right) = -\frac{2\pi c}{\lambda^2} d\lambda \tag{3.18}$$

Thus,

$$t_{gm} = -\frac{\lambda^2}{2\pi c} \frac{d\beta}{d\lambda} \tag{3.19}$$

Substituting (3.16) in (3.19), we have

$$t_{gm} = \frac{1}{c}\left[n(\lambda) - \frac{\lambda\, dn(\lambda)}{d\lambda}\right] \tag{3.20}$$

Thus, the pulse dispersion per unit length, $\Delta\tau_m/\Delta\lambda$, due to the material (using Equation 3.20) for a source having an RMS spectral width, σ_λ, is

$$\Delta\tau_m = -\frac{\lambda}{c} \frac{d^2 n}{d\lambda^2} \sigma_\lambda \tag{3.21}$$

If $\Delta\tau_m = M(\lambda)\sigma_\lambda$, then

$$M(\lambda) = -\frac{\lambda}{c} \frac{d^2 n}{d\lambda^2} \tag{3.22}$$

where $M(\lambda)$ is the *material dispersion factor or* the *material dispersion parameter*; its unit is commonly expressed in ps/(nm · km).

Thus, if the refractive index can be expressed as a function of the optical wavelength, then the material dispersion can be calculated. In fact, in practice, optical material engineers have to characterize all optical properties of new materials. The refractive index, $n(\lambda)$, can usually be expressed in terms of Sellmeier's dispersion formula as

$$n^2(\lambda) = 1 + \sum_k \frac{G_k \lambda^2}{(\lambda^2 - \lambda_k^2)} \tag{3.23}$$

where

G_k are Sellmeier's constants

k is an integer and normally takes a value in the range of $k = 1$–3

TABLE 3.1

Sellmeier's Coefficients for Several Optical Fiber Silica-Based Materials with Germanium Doped in the Core Region

Sellmeier's Constants	Germanium Concentration, C (mol%)			
	0 (Pure Silica)	3.1	5.8	7.9
G_1	0.6961663	0.7028554	0.7088876	0.7136824
G_2	0.4079426	0.4146307	0.4206803	0.4254807
G_3	0.8974794	0.8974540	0.8956551	0.8964226
λ_1	0.0684043	0.0727723	0.0609053	0.0617167
λ_2	0.1162414	0.1143085	0.1254514	0.1270814
λ_3	9.896161	9.896161	9.896162	9.896161

Sources: After Kobayashi, S. et al., *Rev. Electrical Comm. Lab.*, 26, 453, 1978; Kobayashi, S. et al., Refractive-index dispersion of doped fused silica, *International Conference on Integrated Optics and Optical Fibre Communication*, Tokyo, Japan, 1977, pp. 9–312; Gloge, D., *Appl. Opt.*, 10(11), 2442, November 1971.

In the late 1970s, several silica-based glass materials were manufactured and their properties were determined. The refractive indices are usually expressed using Sellmeier's coefficients. These coefficients are given in Table 3.1 for several optical fiber materials.

By using curve fitting, the refractive index of pure silica, $n(\lambda)$, can be expressed as

$$n(\lambda) = c_1 + c_2\lambda^2 + c_3\lambda^{-2} \tag{3.24}$$

where

$c_1 = 1.45084$
$c_2 = -0.00343\,\mu\text{m}^{-2}$
$c_3 = 0.00292\,\mu\text{m}^2$

Thus, from Table 3.1 and Equation 3.24, we can use Equation 3.22 to determine the material dispersion factor for a certain wavelength range.

For the doped core of the optical fiber, Sellmeier's expression (3.16) can be approximated to the form in (3.17) by using the curve-fitting technique. The material dispersion factor, $M(\lambda)$, becomes zero at wavelengths around 1350 nm and about −10 ps/(nm · km) at 1550 nm. However, the attenuation at 1350 nm is about 0.4 dB/km compared with 0.2 dB/km at 1550 nm, as given in Table 3.1.

3.3.2.2 Waveguide Dispersion

The effect of waveguide dispersion can be approximated by assuming that the refractive index of the material is independent of the wavelength. Let us now consider the group delay, i.e., the time required for a mode to travel along a fiber of length L. This kind of dispersion depends strongly on Δ and V parameters. To estimate the results of fiber parameters, we define a *normalized propagation constant*, b, as

$$b = \frac{\dfrac{\beta^2}{k^2} - n_2^2}{n_1^2 - n_2^2} \tag{3.25}$$

for a small Δ. We note that β/k is, in fact, the *effective refractive index* of the guided optical mode propagating along the optical fiber, i.e., the guided waves traveling along the axial direction of the fiber "see" it as a medium with a refractive index of an equivalent effective index.

In case the fiber is a weakly guided waveguide with the effective refractive index taking a value significantly close to that of the core or cladding index, (3.25) can then be approximated by

$$b \cong \frac{\frac{\beta}{k} - n_2}{n_1 - n_2} \tag{3.26}$$

Solving (3.26) for β, we have

$$\beta = n_2 k(b\Delta + 1)$$

The group delay for waveguide dispersion is then given by (per unit length)

$$t_{\text{wg}} = \frac{d\beta}{d\omega} = \frac{1}{c}\frac{d\beta}{dk} \tag{3.27}$$

$$t_{\text{wg}} = \frac{1}{c}\left[n_1 + n_2\Delta\frac{d(bk)}{dk}\right] = \frac{1}{c}\left[n_1 + n_2\Delta\frac{d(bV)}{dV}\right] \tag{3.28}$$

Equation 3.28 can be obtained from (3.27) by using the expression of V. Thus, the pulse spreading, $\Delta\tau_\omega$, due to the waveguide dispersion per unit length by a source having an optical bandwidth (or linewidth, σ_λ) is given by

$$\Delta\tau_\omega = \frac{dt_{\text{gw}}}{d\lambda}\sigma_\lambda = -\frac{n_2\Delta}{c\lambda}V\frac{d^2(Vb)}{dV^2}\sigma_\lambda \tag{3.29}$$

and the *waveguide dispersion factor or the waveguide dispersion parameter* (similar to the material dispersion factor) is then defined as

$$D(\lambda) = -\frac{n_2(\lambda)\Delta}{c\lambda}V\frac{d^2(Vb)}{dV^2} \tag{3.30}$$

in ps/(nm · km). In the range of $0.9 < \lambda/\lambda_c < 2.6$, the factor $V\frac{d^2(Vb)}{dV^2}$ can be approximated (to <5% error) by

$$V\frac{d^2(Vb)}{dV^2} \cong 0.080 + 0.549(2.834 - V)^2 \tag{3.31}$$

or, alternatively, using the definition of the cutoff wavelength and the expression of the V parameters, we obtain

$$V\frac{d^2(Vb)}{dV^2} \cong 0.080 + 3.175\left(1.178 - \frac{\lambda_c}{\lambda}\right)^2 \tag{3.32}$$

Exercise to readers/students: Prove the equivalent equations of (3.27) to (3.32). The sign assignment of the material and waveguide dispersion factors must be the same. Otherwise, a negative and a positive of these dispersion factors would create confusion. Can you explain what would happen to the pulse if it is transmitted through an optical fiber having a total negative dispersion factor?

Thus, from Equations 3.32 and 3.31, we can calculate the waveguide dispersion factor and, hence, the pulse dispersion factor for a particular source spectral width, σ_λ. It is noted that the dispersion considered in this chapter is for a step-index fiber only. For a grade-index fiber, ESI parameters must be found, and the chromatic dispersion can then be calculated.

3.3.2.2.1 Alternative Expression for the Waveguide Dispersion Parameter

Alternatively, the waveguide dispersion parameter can be expressed as a function of the propagation constant, β, by using $\omega = 2\pi c/\lambda$ and Equation 3.32; then the waveguide dispersion factor can be written as

$$D(\lambda) = -\frac{2\pi c}{\lambda^2}\beta_2 = -\frac{2\pi c}{\lambda^2}\frac{d\beta^2}{d\omega^2} \tag{3.26}$$

Thus, the waveguide dispersion factor is directly related to the second-order derivative of the propagation constant with respect to the optical radial frequency.

An example of a design of an optical fiber operating in the single-mode region is given in Figure 3.6. The cladding material is pure silica. Shown in this figure are the curves of the material dispersion, the waveguide dispersion, and the total dispersion for an SMF with a nonuniform refractive index profile in the core.

A system for measurement of the relative group delay and chromatic dispersion of single mode optical fibers is shown in Figure 3.7.

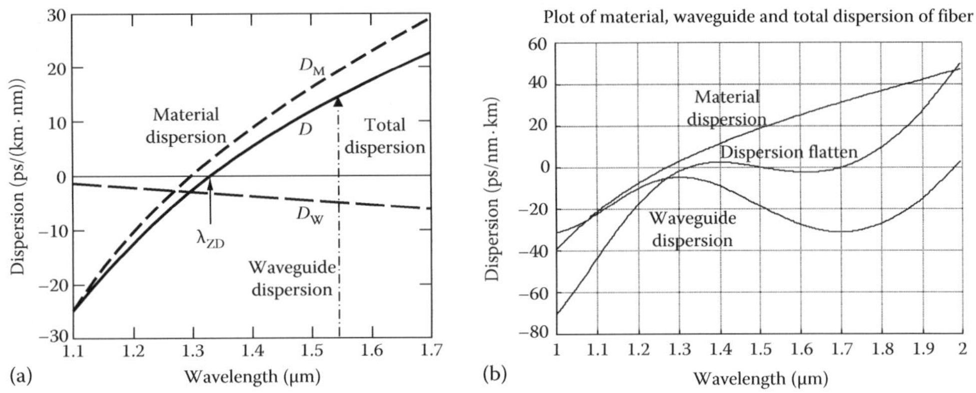

FIGURE 3.6
Chromatic dispersion factors of (a) an SSMF and (b) a dispersion-flattened fiber. Plotted curves representing the material dispersion factor as a function of the optical wavelength for a silica-based optical fiber with a zero-dispersion wavelength at 1290 nm. This curve is generated as an example. For SSMF that are currently installed throughout the world, the total dispersion is around +17 ps/(nm · km) at 1550 nm and almost zero at 1310 nm. Can we estimate the waveguide dispersion curve for the SSMF at around 1300 and 1550 nm windows?

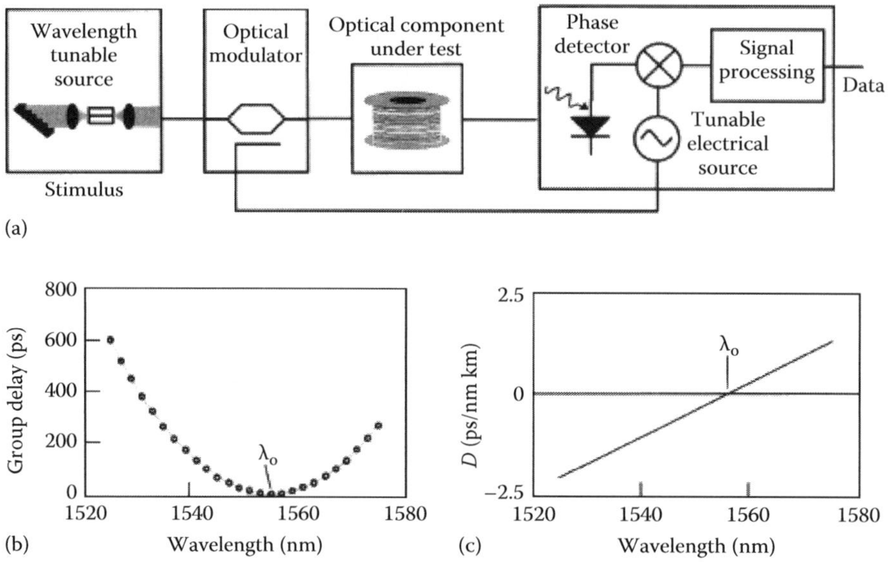

FIGURE 3.7
Chromatic dispersion measurement of a two-port optical device. (b) Relative group delay versus wavelength and (c) dispersion parameter versus wavelength.

3.4 Transfer Function of Single-Mode Fibers

A linear time-variant system, such as an SMF, would have a transfer function of

$$H(f) = |H(f)|e^{-j\alpha(f)} \tag{3.33}$$

The property of the frequency response of the fiber can be converted from the passband to the baseband as shown in Figure 3.8. Figure 3.9a shows the chirp of optical carrier under the pulse over a long length of single mode optical fiber. The chirp occurs significantly near the edge of the pulse as observed in Figure 3.9b when it is in the near and much less in the far field. In the far field region the pulse becomes Gaussian like as shown in Figures 3.10 and 3.11.

3.4.1 Higher-Order Dispersion

We also observe from Figure 3.6 that the bandwidth–length product of the optical fiber can be extended to infinity if the system is operating at a wavelength at which the total dispersion is zero. However, the dispersive effects do not disappear completely at this zero-dispersion wavelength. Optical pulses still experience broadening because of higher-order dispersion effects. It is easy to understand that the total dispersion factor cannot be zero "flattened" over the optical spectrum. This is higher-order dispersion, which is governed

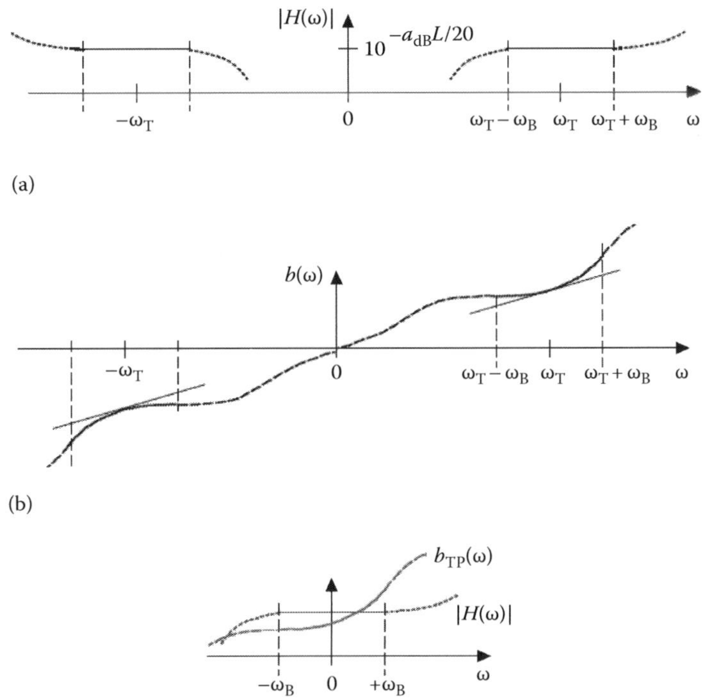

$$H_{LP}(f) = \text{const.}\cdot e^{-jb_{LP}(f)} = \text{const.}\cdot e^{+j\pi\frac{\lambda_T^2}{c}D_1 L \cdot f^2}$$

(c) Band-pass to low-pass transformation

FIGURE 3.8
Frequency response of an SMF: (a) Magnitude response in the band-pass regime and (b) phase response in the band-pass regime. (c) Baseband equivalence.

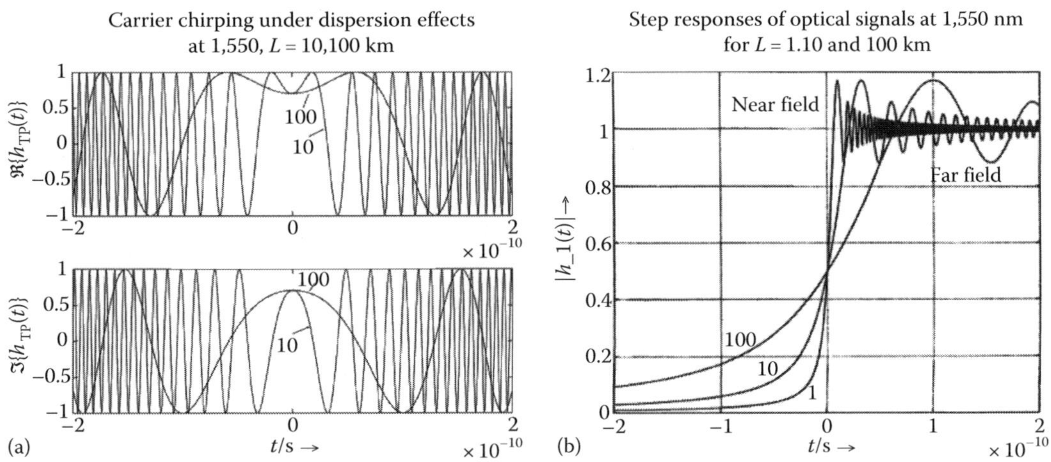

FIGURE 3.9
(a) Carrier-chirping effects and (b) step response of an SMF of $L = 1$, 10, and 100 km.

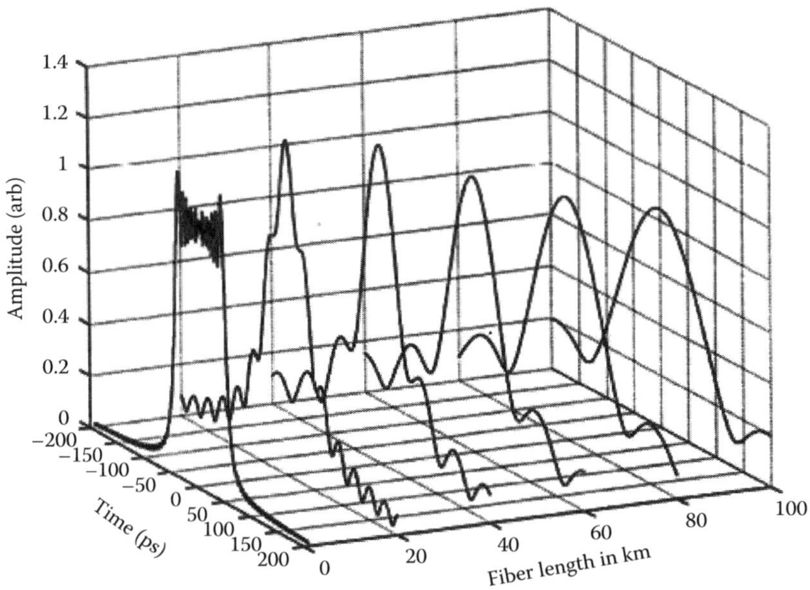

FIGURE 3.10
Pulse response from near field (~<2 km) to far field (>80 km).

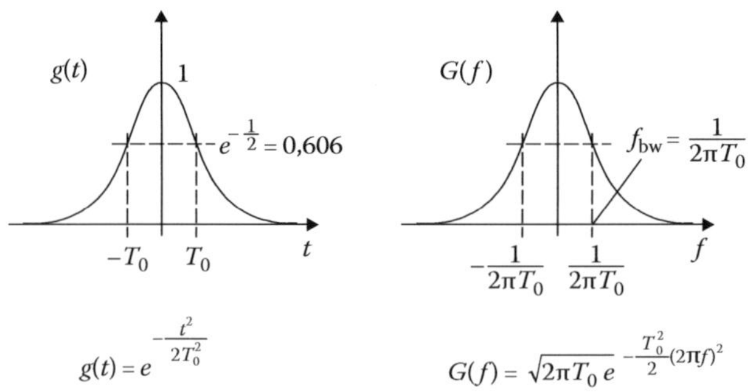

FIGURE 3.11
Fiber response to Gaussian pulse. Gaussian → Gaussian!

by the slope of the total dispersion curve, called the dispersion slope, $S = d(D+M)/d\tilde{\lambda}$; $S(\lambda)$ can thus be expressed as

$$S(\lambda) = \left(\frac{2\pi c}{\lambda^2}\right)^2 \frac{d^3\beta}{d\lambda^3} + \left(\frac{4\pi c}{\lambda^3}\right) \frac{d^2\beta}{d\lambda^2} \tag{3.34}$$

S is also known as the differential dispersion parameter.

3.4.2 Transmission Bit Rate and Dispersion Factor

The effect of dispersion on the system bit rate, B_r, is obvious and can be estimated by using the criterion

$$B_r \cdot \Delta\tau < 1 \tag{3.35}$$

where $\Delta\tau$ is the total pulse broadening. With the fiber length; the total dispersion, D_T $(= M(\lambda) + D(\lambda))$; and a source linewidth, σ_λ; the criterion becomes

$$B_r \cdot L \cdot |D_T| \sigma_\lambda \leq 1 \tag{3.36}$$

For a total dispersion factor of 1 ps/(nm · km) and a semiconductor laser of linewidth 2–4 nm, the bit rate–length product cannot exceed 100 Gb/s · km. That is, if a 100 km transmission distance is used, then the bit rate cannot be higher than 1.0 Gb/s.

A block diagram of an optical fiber communication system from the transmitter to the fiber to the optical receiver is shown in Figure 3.12. Note that the photodetector is represented as the square of the magnitude of the optical field that is to indicate the conversion of the optical power to the electronic current flowing out of the detector. Figure 3.13 shows the evolution of the sequence of pulses (right column) and the eye diagram (left column) at various distance from the launched input.

3.4.3 Polarization Mode Dispersion

Figure 3.14 illustrates the propagation of the polarization components of the guide mode of the single mode fiber via the random variation of the fiber birefringence axis through a long length of fiber. Figure 3.15 shows the transmission effects of the polarization mode dispersion, that is the degradation of the eye diagram due to differential delay between the polarized modes.

The delay between two principal states of polarization (PSPs) is normally negligibly small at 10 Gb/s. However, at a high bit rate and in ultra-long-haul transmission, polarization mode dispersion (PMD) severely degrades the system performance [7–10]. The instantaneous value of differential group delay (DGD), $\Delta\tau$, varies along the fiber and follows a Maxwellian distribution [8,11,12] (see Figure 3.16).

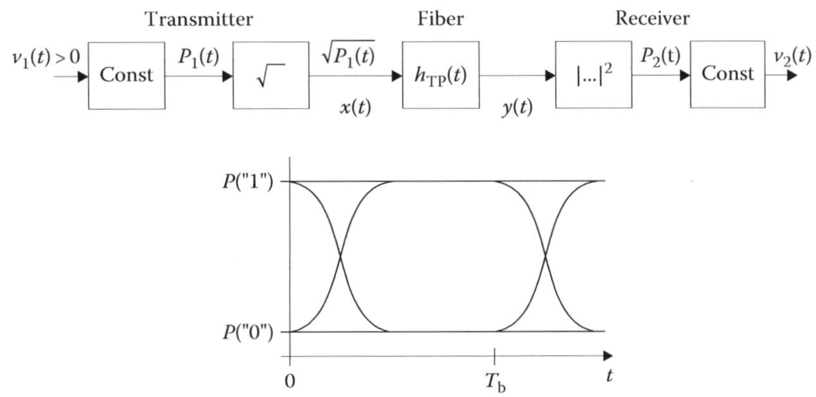

FIGURE 3.12
Schematic of an optical transmission system and its equivalent transfer function.

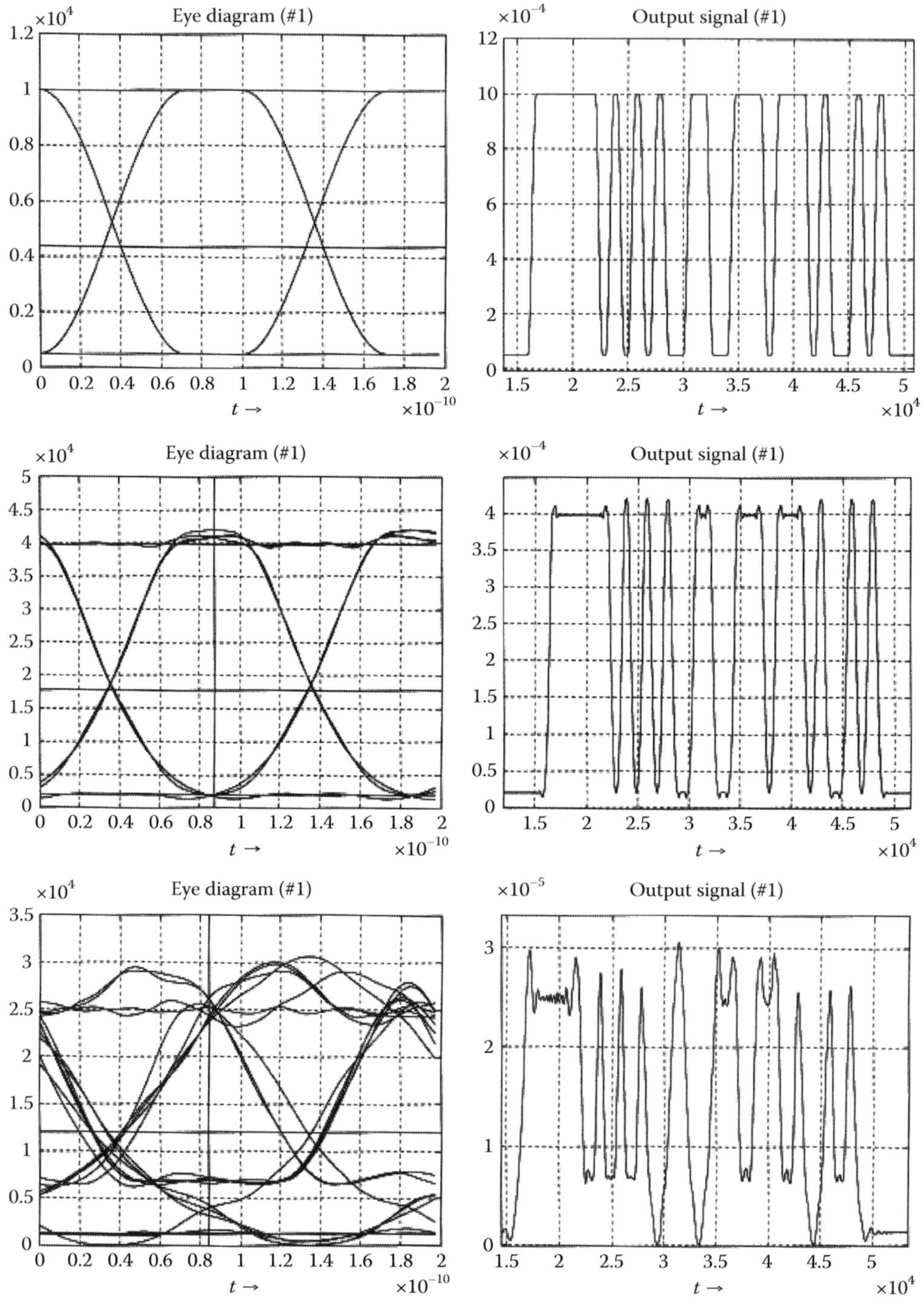

FIGURE 3.13
Eye diagram of time signals at 10 Gb/s transmission over an SSMF for 0, 20, and 80 km lengths.

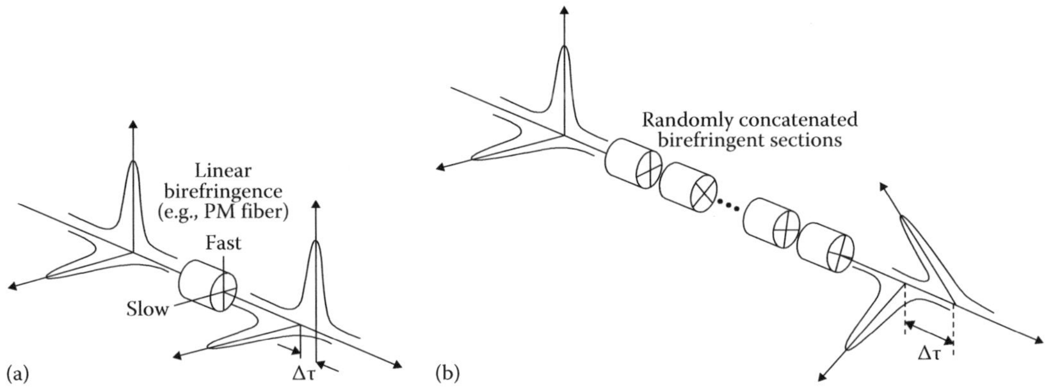

FIGURE 3.14
Conceptual model of PMD. (a) Simple birefringence device and (b) randomly concatenated birefringence.

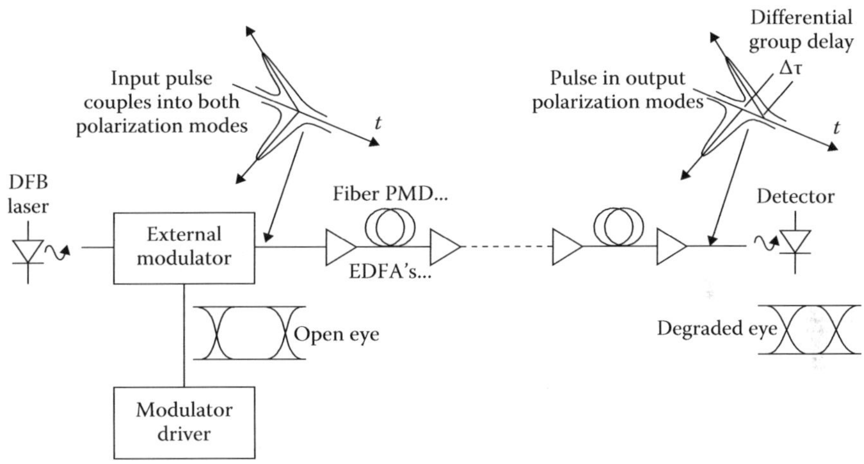

FIGURE 3.15
Effect of PMD in a digital optical communication system—degradation of the received eye diagram.

The Maxwellian distribution is governed by the following expression:

$$f(\Delta\tau)=\frac{32(\Delta\tau)^2}{\pi^2\langle\Delta\tau\rangle^3}\exp\left\{-\frac{4(\Delta\tau)^2}{\pi\langle\Delta\tau\rangle^2}\right\} \quad \Delta\tau \geq 0 \tag{3.37}$$

The mean DGD value, $\langle\Delta\tau\rangle$, is commonly termed "fiber PMD" and provided in the fiber specifications. The following expression gives an estimate of the maximum transmission limit, L_{max}, due to the PMD effect [8–10]:

$$L_{\max} = \frac{0.02}{\langle\Delta\tau\rangle^2 \cdot R^2} \tag{3.38}$$

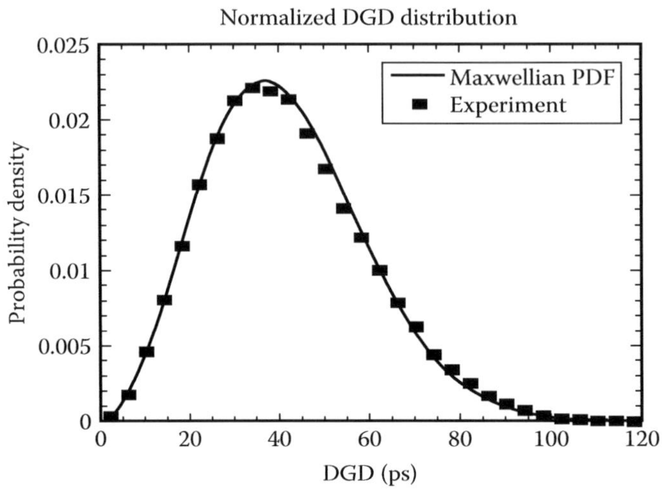

FIGURE 3.16
Maxwellian distribution of a PMD random process.

where R is the bit rate. Based on (3.38), L_{max} for both old fiber vintages and contemporary fibers is obtained as follows:

- $\langle\Delta\tau\rangle = 1$ ps/km (old fiber vintages): For a bit rate of $R = 40$ Gb/s, the maximum distance $L_{max} = 12.5$ km; for $R = 10$ Gb/s, $L_{max} = 200$ km.
- $\langle\Delta\tau\rangle = 0.1$ ps/km (contemporary fibers for modern optical systems): For a bit rate of $R = 40$ Gb/s, the maximum transmission distance is $L_{max} = 1{,}250$ km; for $R = 10$ Gb/s, $L_{max} = 20{,}000$ km.

EXERCISE 3.1

Inspect the technical specifications of Corning fibers given in the Appendix of Chapter 2 and extract the values of the PMD. Explain the difference between the values of the fibers. What is the standard value allowable for the PMD in modern fibers?

3.4.4 Fiber Nonlinearity

The nonlinear effects in optical fibers were described in Chapter 2. This section revisits these effects and their influence on the propagation of optical signals over long lengths of fibers. The nonlinearity and linear effects in optical fibers can be classified as shown in Figure 3.17.

Fiber RI is dependent on both operating wavelengths and lightwave intensity. This intensity-dependent phenomenon is known as the Kerr effect and is the cause of fiber nonlinear effects. The power dependence of RI is expressed as [1]

$$n' = n + \bar{n}_2 (P/A_{eff}) \tag{3.39}$$

where

P is the average optical power of the guided mode
$\bar{n}_2$ is the fiber nonlinear coefficient
A_{eff} is the effective area of the fiber

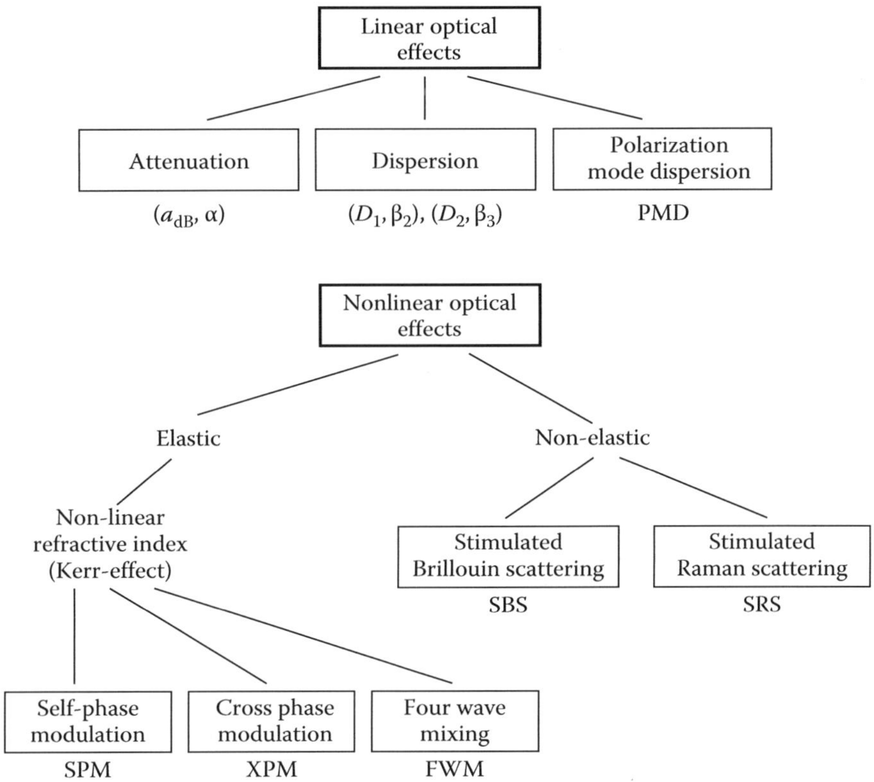

FIGURE 3.17
Linear and nonlinear fiber properties of SMF.

Fiber nonlinear effects include intrachannel SPM, interchannel XPM, FWM, stimulated Raman scattering (SRS), and stimulated Brillouin scattering (SBS). SRS and SBS are not the main degrading factors as their effects are only getting noticeably large with very high optical power. On the other hand, FWM severely degrades the performance of an optical system with the generation of ghost pulses only if the phases of optical signals are matched with each other. However, with high local dispersions, such as in SSMF, the effects of FWM become negligible [1]. In terms of XPM, its effects can be considered to be negligible in a DWDM system in the following scenarios: (1) highly locally dispersive system and (2) large channel spacing. However, XPM should be taken into account for optical transmission systems deploying NZDSF where local dispersion values are small. Thus, SPM is usually the dominant nonlinear effect for systems employing transmission fibers with high local dispersions, e.g., SSMF and DCFs. The effect of SPM is normally coupled with the nonlinear phase shift, ϕ_{NL}, defined as [1]

$$\phi_{NL} = \int_0^L \gamma P(z)dz = \gamma L_{eff} P$$

$$\gamma = \omega_c \bar{n}_2 / (A_{eff} c)$$

$$L_{eff} = (1 - e^{-\alpha L})/\alpha \qquad (3.40)$$

where

ω_c is the lightwave carrier

L_{eff} is the effective transmission length

α is the fiber attenuation factor, which normally takes a value between 0.17 and 0.2 dB/km in the 1550 nm spectral window

The temporal variation of the nonlinear phase shift, ϕ_{NL}, results in the generation of new spectral components far apart from the lightwave carrier, ω_c, indicating the broadening of the signal spectrum. This spectral broadening, $\delta\omega$, can be obtained from the time dependence of the nonlinear phase shift as follows:

$$\delta\omega = -\frac{\partial\phi_{NL}}{\partial T} = -\gamma\frac{\partial P}{\partial T}L_{eff} \tag{3.41}$$

Equation 3.41 indicates that $\delta\omega$ is proportional to the time derivative of the average signal power, P. Additionally, the generation of new spectral components occurs mainly at the rising and falling edges of optical pulses, i.e., the amount of generated chirps are larger for an increased steepness of the pulse edges.

The wave propagation equation can be represented as

$$\begin{aligned}&\frac{\partial A(z,t)}{\partial z} + \frac{\alpha}{2}A(z,t) + \beta_1\frac{\partial A(z,t)}{\partial t} + \frac{j}{2}\beta_2\frac{\partial^2 A(z,t)}{\partial t^2} - \frac{1}{6}\beta_3\frac{\partial^3 A(z,t)}{\partial t^3}\\ &= -j\gamma\left|A(z,t)\right|^2 A(z,t) - \frac{1}{\omega_0}\frac{\delta}{\delta t}\left(\left|A\right|^2 A\right) - T_R A\frac{\delta\left(\left|A\right|^2\right)}{\delta t}\end{aligned}$$

in which we have ignored the pure delay factor involving β_1. The last term on the RHS represents the Raman scattering effects. The mutual effect of the nonlinear dispersion effects and the nonlinear effects can lead to the modulation of the lightwave pulses and, thus, unstable states of the optical pulses. This phenomenon is usually called modulation instability, normally observed in soliton lasers.

The gain spectrum of modulation instability is shown in Figure 3.18 [13].

3.5 Advanced Optical Fibers: Dispersion Shifted, Dispersion Flattened, and Dispersion Compensated

In the beginning of the 1980s, there was great interest to reduce the total dispersion $[M(\lambda) + D(\lambda)]$ of SMF at 1550 nm, where the loss is the lowest for silica fibers. There are two significant trends: one is to reduce the linewidth and to stabilize the laser center wavelength, and the other is to reduce the dispersion at this wavelength. The fibers designed for long-haul transmission systems usually exhibit a near-zero dispersion at a certain spectral window. These are dispersion-shifted fibers, i.e., at this wavelength we prefer to have the total dispersion $= [M(\lambda) + D(\lambda)] \sim 0$. The material dispersion factor, $M(\lambda)$, is natural but slightly affected by the variation of the doping material and concentration. However,

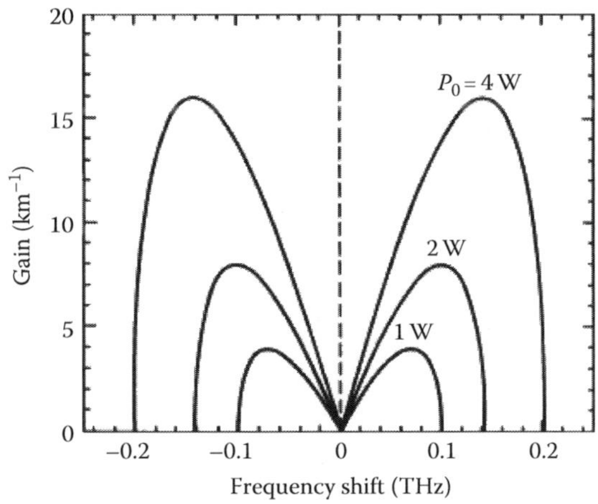

FIGURE 3.18
Spectrum of the optical gain due to modulation instability at three different average power levels in an optical fiber with $\beta_2 = 20\,\text{ps}^2/\text{km}$ and $\gamma = 2\,\text{W/km}$.

$D(\lambda)$ can be tailored, by designing appropriate refractive index profiles and geometrical structure, to balance the material dispersion effects. Note that the dispersion factors due to the material and the waveguide take algebraic values; thus, they can be designed to take opposite values to cancel each other.

Advanced optical fiber design techniques can offer the design of dispersion-flattened fibers, for which the dispersion factor is flat over the wavelength range from 1300 to 1600 nm, by tailoring the refractive index profiles of the core of optical fibers in distributions such as the W profile, the segmented profile, and the multilayer core structure.

QUESTION

What is the principal phenomenon for an optical fiber so that the dispersion characteristic is flattened over the wavelength range of 1300–1550 nm?

Another type of optical fiber would be required for compensating the dispersion effect of the optical signal after being transmitted over a length of the optical fiber, the dispersion-compensated fiber, whose dispersion factor is many times larger than that of the standard communication fiber with an opposite sign. This can be designed by setting the total dispersion to the required compensated dispersion, and thus the waveguide dispersion can be found over the required operating range. Thence, optical fiber structures can be tailored.

3.6 Effects of Mode Hopping

Until now we have assumed that the source center emission wavelength was unaffected by the modulation. In fact, when a short current pulse is applied to a semiconductor laser, its center emission wavelength may hop from one mode to its neighbor, a longer wavelength. In the case where a multi-longitudinal-mode laser is used, this hopping effect is negligible; however, it is very significant for a single longitudinal-mode laser.

3.7 Numerical Solution: Split-Step Fourier Method

3.7.1 Symmetrical Split-Step Fourier Method

The evolution of slowly varying complex envelopes, $A(z, t)$, of optical pulses along an SMF is governed by the nonlinear Schröedinger equation (NLSE) [3]:

$$\frac{\partial A(z,t)}{\partial z}+\frac{\alpha}{2}A(z,t)+\beta_1\frac{\partial A(z,t)}{\partial t}+\frac{j}{2}\beta_2\frac{\partial^2 A(z,t)}{\partial t^2}-\frac{1}{6}\beta_3\frac{\partial^3 A(z,t)}{\partial t^3}=-j\gamma\left|A(z,t)\right|^2 A(z,t) \quad (3.42)$$

where
- z is the spatial longitudinal coordinate
- α accounts for fiber attenuation
- β_1 indicates DGD
- β_2 and β_3 represent second- and third-order dispersion factors of fiber CD
- γ is the nonlinear coefficient

In a single-channel transmission, Equation 3.42 includes the following effects: fiber attenuation, fiber CD and PMD, dispersion slope, and SPM nonlinearity. The fluctuation of optical intensity caused by the Gordon–Mollenauer effect [16] is also included in this equation.

The solution of NLSE and, hence, the modeling of pulse propagation along an SMF are obtained numerically by using the split-step Fourier method (SSFM) [3]. In the SSFM, fiber length is divided into a large number of small segments, δz. In practice, fiber dispersion and nonlinearity are mutually interactive at any distance along the fiber. However, these mutual effects are small within δz, and thus the effects of fiber dispersion and fiber nonlinearity over δz are assumed to be statistically independent of each other. As a result, the SSFM can separately define two operators: (1) the linear operator that involves fiber attenuation and fiber dispersion effects and (2) the nonlinearity operator that takes into account fiber nonlinearities. These linear and nonlinear operators are formulated as follows:

$$\begin{aligned}\hat{D}&=-\frac{i\beta_2}{2}\frac{\partial^2}{\partial T^2}+\frac{\beta_3}{6}\frac{\partial^3}{\partial T^3}-\frac{\alpha}{2}\\ \hat{N}&=i\gamma\,|A|^2\end{aligned} \quad (3.43)$$

where
- A replaces $A(z,t)$ for simpler notation
- $T=t-z/v_g$ is the reference time frame moving at group velocity

Equation 3.43 can be rewritten in a shorter form given by

$$\frac{\partial A}{\partial z}=(\hat{D}+\hat{N})A \quad (3.44)$$

and the complex amplitudes of optical pulses propagating from z to $z+\delta z$ are calculated using the approximation below:

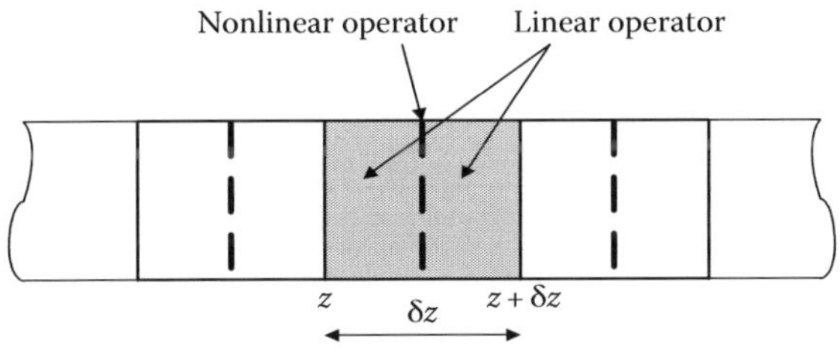

FIGURE 3.19
Schematic illustration of the symmetrical SSFM.

$$A(z+h,T) \approx \exp(h\hat{D})\exp(h\hat{N})A(z,T) \tag{3.45}$$

Equation 3.45 is accurate to the second order of the step size, δz [3]. The accuracy of the SSFM can be improved by including the effect of fiber nonlinearity in the middle of the segment rather than at the segment boundary (see Figure 3.19). This modified SSFM is known as the symmetrical SSFM.

Equation 3.45 can now be modified as

$$A(z+\delta z,T) \approx \exp\left(\frac{\delta z}{2}\hat{D}\right)\exp\left(\int_{z}^{z+\delta z}\hat{N}(z')dz'\right)\exp\left(\frac{\delta z}{2}\hat{D}\right)A(z,T) \tag{3.46}$$

This method is accurate to the third order of the step size, δz. In the symmetrical SSFM, the optical pulse propagates along a fiber segment, δz, in two stages. First, the optical pulse propagates through the linear operator that has a step of δz/2 in which the fiber attenuation and dispersion effects are taken into account. Then, the fiber nonlinearity is calculated in the middle of the segment. After that, the pulse propagates through the second half of the linear operator. The process continues repetitively in consecutive segments of size δz until the end of the fiber. It should be highlighted that the linear operator is computed in the frequency domain while the nonlinear operator is calculated in the time domain.

3.7.2 MATLAB® Program and MATLAB® and Simulink® Models of SSFM

A MATLAB® program is given below. This program performs the propagation of the optical signals along the optical fiber transmission distance, as shown in Figure 3.20. This program must be included in the folder that stores the MATLAB and Simulink® models. In this folder, an initialization program (Appendix 3.A.5) must also be included to set the data and parameters required for the Simulink model and subroutines. Furthermore, the SSMF including Raman gain amplification effects is given in Appendix 3.A.4.

3.7.2.1 MATLAB® Program

```
function output=ssprop_matlabfunction_modified(input)
nt=input(1);
u0=input(2:nt+1);
dt=input(nt+2);
dz=input(nt+3);
nz=input(nt+4);
```

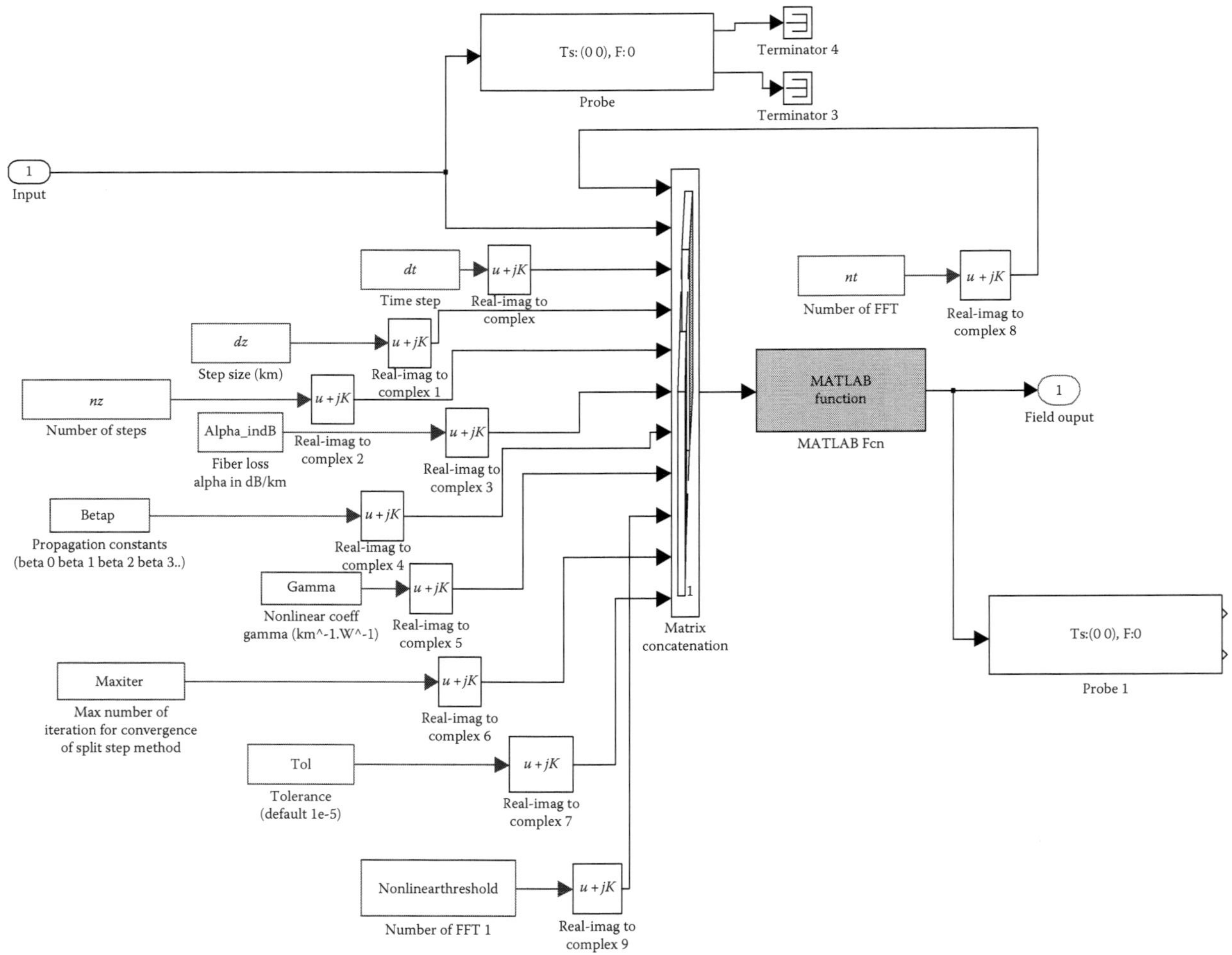

FIGURE 3.20
MATLAB and Simulink models of the SSMF. MATLAB function includes the MATLAB program; other inputs are data to pass to the MATLAB program SSMF, usually defined in the initialization file (given in Appendix 3.A.4).

```
alpha_indB=input(nt+5);
betap=input(nt+6:nt+9);
gamma=input(nt+10);
P_non_thres=input(nt+11)
maxiter=input(nt+12);
tol=input(nt+13);
  tic;
%tmp=cputime;
% This section solves the NLSE for pulse propagation in an optical fiber
using the SSF method
% The following effects are included: group velocity dispersion
% (GVD),higher order dispersion, loss, and self-phase modulation (gamma).
%
% USAGE
%
% u1=ssprop(u0,dt,dz,nz,alpha,betap,gamma);
% u1=ssprop(u0,dt,dz,nz,alpha,betap,gamma,maxiter);
% u1=ssprop(u0,dt,dz,nz,alpha,betap,gamma,maxiter,tol);
%
% INPUT
%
% u0-starting field amplitude (vector)
% dt-time step-[in ps]
% dz-propagation stepsize-[in km]
% nz-number of steps to take, ie, ztotal=dz*nz
% alpha-power loss coefficient [in dB/km], need to convert to linear to
% %have P=P0*exp(-alpha*z)
% betap-dispersion polynomial coefs, [beta_0 … beta_m] [in ps^(m-1)/km]
% gamma-nonlinearity coefficient [in (km^-1.W^-1)]
% maxiter-max number of iterations (default=4)
% tol-convergence tolerance (default=1e-5)
%% OUTPUT
%% u1-field at the output
%
% Convert alpha_indB to alpha in linear domain
%------
alpha=log(10)*alpha_indB/10;     % alpha (1/km)
%------
ntt=length(u0);
w=2*pi*[(0:ntt/2-1),(-ntt/2:-1)]'/(dt*nt);
%w=2*pi*[(ntt/2:ntt-1),(1:ntt/2)]'/(dt*ntt);
clear halfstep
halfstep=-alpha/2;
for ii=0:length(betap)-1;
  halfstep=halfstep-j*betap(ii+1)*(w.^ii)/factorial(ii);
end
clear LinearOperator
% Linear Operator in Split Step method
LinearOperator=halfstep;
% pause
halfstep=exp(halfstep*dz/2);
u1=u0;
ufft=fft(u0);
% Nonlinear operator will be added if the peak power is greater than the
```

```
% Nonlinear threshold
iz=0;
while (iz<nz) & (max((abs(u1).^2+abs(u0).^2))>P_non_thres)
  iz=iz+1;
  uhalf=ifft(halfstep.*ufft);
  for ii=1:maxiter,
    uv=uhalf .* exp(-j*gamma*(abs(u1).^2+abs(u0).^2)*dz/2);
    ufft=halfstep.*fft(uv);
    uv=ifft(ufft);
    %fprintf('You are using SSFM\n');
    if (max(uv-u1)/max(u1)<tol)
      u1=uv;
      break;
    else
      u1=uv;
    end
  end
  if (ii==maxiter)
    warning(sprintf('Failed to converge to %f in %d iterations',...
tol,maxiter));
  end
  u0=u1;
end
if (iz<nz) & (max((abs(u1).^2+abs(u0).^2))<P_non_thres)
% u1=u1.*rectwin(ntt);
  ufft == fft(u1);
  ufft=ufft.*exp(LinearOperator*(nz-iz)*dz);
  u1=ifft(ufft);
%fprintf('Implementing Linear Transfer Function of the Fibre
Propagation');
end
toc;
output=u1;
```

3.7.2.2 MATLAB® and Simulink® Models

The MATLAB program (subroutine) is incorporated in the MATLAB and Simulink models for signal propagation. Under the mask of the block MATLAB function `[ssprop _ matlabfunction _ modified(input)]` is the inclusion of the MATLAB program subroutine, as shown in Figure 3.21.

3.7.3 Modeling of Polarization Mode Dispersion

The first-order PMD can be implemented by modeling the optical fiber as two separate paths representing the propagation of two PSPs. The symmetrical SSFM can be implemented on each polarized transmission path, and then their outputs are superimposed to form the output optical field of the propagated signals. The transfer function to represent the first-order PMD is given by

$$H(f) = H^{+}(f) + H^{-}(f) \tag{3.47}$$

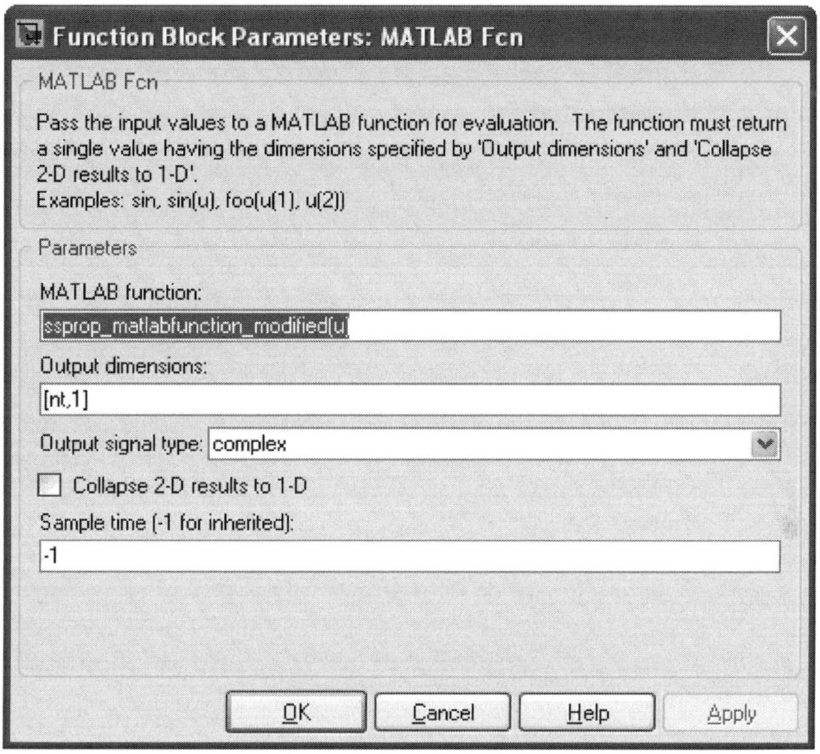

FIGURE 3.21
Screen of the MATLAB and Simulink models under the mask of the MATLAB function of the SSMF algorithm.

where

$$H^{+}(f) = \sqrt{k}\exp\left[j2\pi f\left(-\frac{\Delta\tau}{2}\right)\right] \tag{3.48}$$

and

$$H^{-}(f) = \sqrt{k}\exp\left[j2\pi f\left(-\frac{\Delta\tau}{2}\right)\right] \tag{3.49}$$

in which

k is the power-splitting ratio; $k = 1/2$ when a 3 dB or 50:50 optical coupler/splitter is used
$\Delta\tau$ is the instantaneous DGD value following a Maxwell distribution (refer to Equation 3.37) [14,15]

3.7.4 Optimization of Symmetrical SSFM

3.7.4.1 Optimization of Computational Time

A huge amount of time is spent in performing the symmetrical SSFM for FFT and IFFT operations, in particular, when fiber nonlinear effects are involved. In practice, when optical pulses propagate toward the end of a fiber span, the pulse intensity is greatly attenuated due to fiber attenuation. As a result, fiber nonlinear effects become negligible for the rest of that fiber span and, hence, the transmission takes place in a linear domain in this range. In this research, a technique to configure the symmetrical SSFM is proposed in order to reduce the computational time. If the peak power of an optical pulse is lower than the nonlinear threshold of the transmission fiber, for example, around −4 dBm, the symmetrical SSFM is switched to a linear-mode operation. This linear mode involves only fiber dispersions and fiber attenuation, and its low-pass equivalent transfer function for the optical fiber is

$$H(\varpi) = \exp\left\{-j\left[\left(\frac{1}{2}\right)\beta_2\varpi^2 + \left(\frac{1}{6}\right)\beta_3\varpi^3\right]\right. \tag{3.50}$$

If β_3 is not considered in this transfer function for the fiber, which is normally the case due to its negligible effects on 40 Gb/s and lower-bit rate transmission systems, the transfer function has a parabolic phase profile [14,15].

3.7.4.2 Mitigation of Windowing Effect and Waveform Discontinuity

In the symmetrical SSFM, mathematical operations of FFT and IFFT play very significant roles. However, due to a finite window length required for FFT and IFFT operations, these operations normally introduce overshooting at two boundary regions of the FFT window, commonly known as the windowing effect of FFT. In addition, since the FFT operation is a block-based process, there exists the issue of waveform discontinuity, i.e., the rightmost sample of the current output block does not start at the same position of the leftmost sample of the previous output block. The windowing effect and the waveform discontinuity problems are resolved with the solutions shown in Figure 3.22.

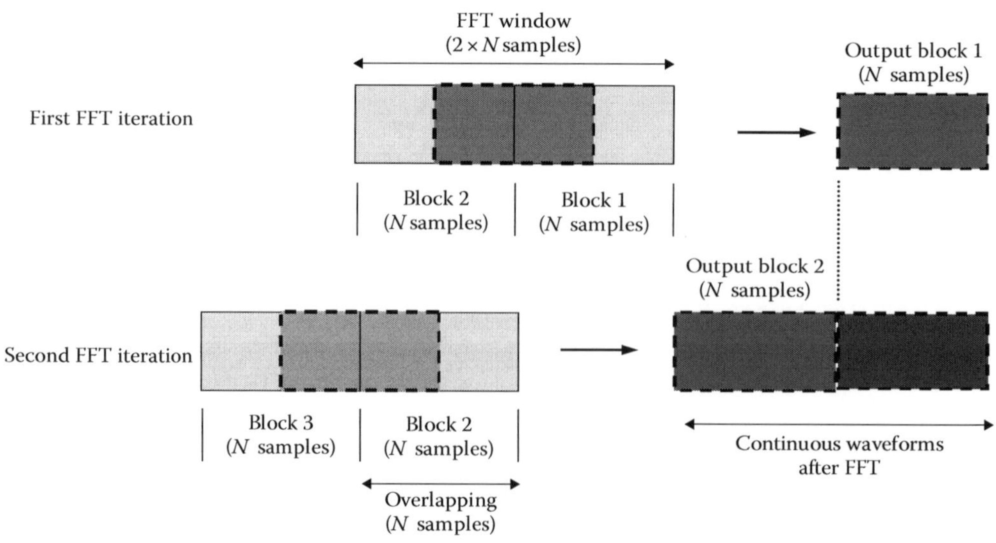

FIGURE 3.22
Proposed solution for mitigating windowing effect and waveform discontinuity caused by FFT and IFFT operations.

The actual window length for FFT/IFFT operations consists of two blocks of samples (2N sample length). The output, however, is a truncated version with the length of one block (N samples), and output samples are taken in the middle of the two input blocks. The next FFT window overlaps the previous one by one block of N samples.

3.8 Concluding Remarks

The attenuation and dispersion of optical signals transmitted through silica optical fibers are described in this chapter. Attenuation can be reduced by using the optical wavelength in the longer wavelength range. For example, for silica fibers, the preferred wavelength is 1.55 μm. However, due to natural forces, a dispersion factor of about 18 ps/(nm · km) generates pulse broadening for a signal transmitted at this wavelength in a circular fiber.

Longer wavelength carriers can be used in the mid-IR range of about 2.5–5 μm. At this wavelength range, different kinds of glasses must be used, such as chalcogenite-type or fluoride-type glasses. Another technique presently used to compensate for the dispersion effect is to reduce the linewidth of the lasers by equalizing techniques, such as spectrum inversion at the transmitter ends or at the center of the transmission length by optical filtering at the receiving end. Alternatively, the optical fiber can be tailored to achieve dispersion-shifted or dispersion-flattened characteristics.

3.9 Problems

PROBLEM 3.1

What are the wavelength ranges of IR light, UV light, and far-IR light? What are the approximate wavelengths of the colors in the color band of resistors? Do they correspond to the color of the rainbow?

PROBLEM 3.2

A GeO_2-doped, silica-based optical fiber has the following parameters: (1) step-index profile, (2) refractive index difference of 0.5% at the core, and (3) core diameter of 9.0 μm.

(a) Calculate the refractive index of the fiber core and cladding at 1.310 and 1.55 μm wavelengths.
(b) What is the estimated loss of this fiber at these wavelengths?
(c) What are the V parameters of this fiber at these wavelengths?
(d) What are the material dispersion and waveguide dispersion factors at these wavelengths? Hence, what are the total dispersion factors?
(e) This fiber is to be used in optical systems having a bit rate of 2.2 Gb/s. What is the maximum fiber length at which the signal can be transmitted without suffering the allowable signal degradation?

PROBLEM 3.3

(a) Give a brief account of the pros and cons for optical fiber communication systems operating at 810, 1300, and 1550 nm wavelength regions.
(b) Why do silica optical fibers become very lossy at a 1400 nm wavelength region?
(c) What are the typical optical fiber losses at the above wavelength regions and, hence, what are the typical cable losses?

PROBLEM 3.4

(a) Show that the material dispersion factor is zero at the wavelength given by

$$\lambda^4 = -\frac{3c_3}{c_2}$$

(b) The coefficients c_1, c_2, and c_3 for pure and GeO_2-doped silica fibers are as follows:

Coefficients (c's)	Pure Silica	7.9% GeO_2-Doped Silica
c_1	1.45084	1.46286
c_2 (in μm^{-2})	−0.00334	−0.00331
c_3 (in μm^2)	0.00292	0.00320

Find the zero-dispersion wavelengths due to the material of these fibers.
(c) Derive an expression for the group delay per kilometer unit length. Plot this group delay versus wavelength for part (b).
(d) Hence, find the transit time difference of lightwaves propagating through the fiber emitted by light sources centered at 810 and 1550 nm with a linewidth of 10 nm.

PROBLEM 3.5A

Using the approximate expression for the normalized propagation constant, *b*, as a function of *V*, derive the group velocity delay due to the waveguide and, hence, the dispersion factor due to the waveguide as a function of *V*.

PROBLEM 3.5B

Using the data of the SMF manufactured by Corning, calculate

(a) The material dispersion factor
(b) The waveguide dispersion factor

at 1330 and 1550 nm.

PROBLEM 3.6

Design a dispersion-shifted SMF at 1550 nm. Each group of three to four students is requested to select a particular group of doped materials with different Sellmeier's coefficients for the design.

PROBLEM 3.7
Mini-Project: Design of Single-Mode Optical Fibers

Project objectives

Design

Design the geometrical and index profile of silica optical fibers to meet the required dispersion properties. Further, a set of fiber performance with dispersion as the main factor must be investigated as a function of the fiber core radius and the relative refractive index difference.

System applications

Your designed fiber must be incorporated with a dispersion-compensated fiber, in this case, the SSMF, whose specifications are given in lectures. If the total length of transmission is 10,000 km, specify the length of your designed fiber and the SMF so that the average fiber dispersion is 0.01 ps/(nm · km). It can be assumed that the spacing of optical amplifiers is 100 km.

Design specifications

A number of types of SMF are required to be designed for optical communication systems in long- and short-haul transmissions. Table 3.2 gives the characteristics of the required fibers. Tables 3.3 and 3.4 give Sellmeier's coefficients for different materials and doping concentrations so that different refractive indices can be achieved.

TABLE 3.2
Index Distribution Profile and Requirements for Dispersion of the Fibers to be Designed

Fiber No.	Optical Fibers' Profile	Core Material Type	System Requirement	Other Requirements
		See Tables 3.3 and 3.4 for Sellmeier's Constants		**Maximum Dispersion in the Wavelength Range 1510–1590 nm in ps/(nm · km)**
1	Triangular	A–J	Dispersion-shifted wavelength at 1550 nm	<1.5
2	Parabolic	A–J	Dispersion-shifted wavelength at 1520 nm	<1.5
3	Triangular	A–J	Dispersion-shifted wavelength at 1560 nm	<2
4	Parabolic	A–J	Dispersion-shifted wavelength at 1530 nm	<3
5	Triangular	A–J	Dispersion-compensated wavelength at 1550 nm	>0.5
6	Parabolic	A–J	Dispersion-compensated wavelength at 1540 nm	>1.0
9	Triple clad	A–J	Dispersion-flattened wavelength over 1500–1590 nm	Not more than 0.2 ps/(nm · km)
10	Triangular in core and segmented in cladding	A–J	Dispersion-flattened wavelength over 1500–1590 nm	Not more than 0.2 ps/(nm · km)

TABLE 3.3

Sellmeier's Coefficients for Several Optical Fiber Silica-Based Materials with Germanium Doped in the Core Region

Sellmeier's Constants	Germanium Concentration, *C* (mol%)			
Types	**A**	**B**	**C**	**D**
	0	**3.1%**	**5.8%**	**7.9%**
G_1	0.6961663	0.7028554	0.7088876	0.7136824
G_2	0.4079426	0.4146307	0.4206803	0.4254807
G_3	0.8974794	0.8974540	0.8956551	0.8964226
λ_1	0.6840432	0.0727723	0.0609053	0.0617167
λ_2	0.1162414	0.1143085	0.1254514	0.1270814
λ_3	9.896161	9.896161	9.896162	9.896161

Source: After Kobayashi, S. et al., 1978.
Note: Wavelength in μm and G_k in μm^{-2}.

TABLE 3.4

Sellmeier's Coefficients for Several Optical Fiber Silica-Based Materials with Germanium Doped in the Core Region

Sellmeier's Constants	Concentration Composition			
Types	**E**	**F**	**G**	**H**
	Quenched SiO_2	**13.5 GeO_2:86.5 SiO_2**	**9.1 P_2O_5:90.0 SiO_2**	**13.3B_2O_3:86.7 SiO_2**
G_1	0.696750	0.711040	0.695790	0.690618
G_2	0.408218	0.408218	0.452497	0.401996
G_3	0.890815	0.704048	0.712513	0.898817
λ_1	0.069066	0.064270	0.061568	0.061900
λ_2	0.115662	0.129408	0.119921	0.123662
λ_3	9.900559	9.425478	8.656641	9.098960
	I	**J**	**K**	**L**
	1.0 F:99.0 SiO_2	**16.2 Na_2O:32.5 B_2O_3: 50.6 SiO_2**		
G_1	0.691116	0.796468		
G_2	0.399166	0.497614		
G_3	0.890423	0.358924		
λ_1	0.068227	0.094359		
λ_2	0.116460	0.093386		
λ_3	9.993707	5.999652		

Source: After Fleming, J.W., *Electron. Lett.*, 14, 326, 1978.
Note: Wavelength in μm and G_k in μm^{-2}.

Mini-project: Design of optical fibers

Practical limits for the fiber core radius and the relative index difference must be taken into account. A set of curves must be obtained with the core radius or the relative index difference as a parameter. Make sure that the material dispersion factors are correctly modeled.

Software environment

The preferred package is MATLAB or MATLAB 7.0 and above for Windows.

Assessment of design assignments

A major design assignment on optical fibers for communication systems is counted for 10% of the total marks allocated for the optical systems part. A 20% would be awarded to design groups selecting the index profile type 9 or 10.

The design assignment is specified for different groups of students. The maximum number of students in each group is two. A group with a higher number of students can only be accepted in exceptional circumstances and, in this case, the complexity of the design assignment is increased accordingly.

Appendix 3.A.1: MATLAB® Program of the Design of Optical Fibers—A Solution to the Mini-Project Design

```
%----OPTICAL COMMUNICATIONS LABORATORY 1----%
% Copyright CRC Press
% clear after each run.
c=2.997925e8;
G1=0.711040;
G2=0.408218;
G3=0.704048;
lambda1=0.064270e-6;
lambda2=0.129408e-6;
lambda3=9.425478e-6;
a=4.1e-6;
delta=0.0025;
start=input('Enter lambda start point (nm) -: ');
finish=input('Enter lambda end point (nm) --: ');
resolution=input('Enter lambda resolution (nm) --: ');
disp('');
lambda=start*1e-9;
lambdavector(1,1)=lambda;
for(i=1:(((finish-start)/resolution)+1))
  n1squared=1+((G1*power(lambda,2))/
(power(lambda,2)-power(lambda1,2)))+((G2*power(lambda,2))/
(power(lambda,2)-power(lambda2,2)))+((G3*power(lambda,2))/(power(lambda,2)
-power(lambda3,2)));
  n1=sqrt(n1squared);
  n1vector(1,i)=n1;
  n2=n1*(1+delta);
  n2vector(1,i)=n2;
  V=(2*pi/lambda)*a*n1*sqrt(2*delta);
  Vvector(1,i)=V;
  dy1dx=(-2*G1*power(lambda1,2)*lambda)/(power(power(lambda,2)-power
(lambda1,2),2));
  dy2dx=(-2*G2*power(lambda2,2)*lambda)/(power(power(lambda,2)-power
(lambda2,2),2));
  dy3dx=(-2*G3*power(lambda3,2)*lambda)/(power(power(lambda,2)-power
(lambda3,2),2));
```

```
d2y1dx2=(2*G1*power(lambda1,2)*(3*power(lambda,2)+power(lambda1,2)))/
(power(power(lambda,2)-power(lambda1,2),3));
d2y2dx2=(2*G2*power(lambda2,2)*(3*power(lambda,2)+power(lambda2,2)))/
(power(power(lambda,2)-power(lambda2,2),3));
d2y3dx2=(2*G3*power(lambda3,2)*(3*power(lambda,2)+power(lambda3,2)))/
(power(power(lambda,2)-power(lambda3,2),3));
  d2ndx2=0.5*(((d2y1dx2+d2y2dx2+d2y3dx2)*power(n1,2)-0.5*(power
(dy1dx+dy2dx+dy3dx,2)))/power(n1,3));
  M=(-d2ndx2/c)*lambda;
  Mvector(1,i)=M; %row vector
  Dw=(-n2*delta)/c*(0.080+0.549*power(2.834-V,2))*(1/lambda);
  Dwvector(1,i)=Dw;
  if(i<(((finish-start)/resolution)+1))
    lambdavector(1,i+1)=lambdavector(1,i)+(resolution*1e-9);
    lambda=lambdavector(1,i+1);
  end
end
plot(lambdavector,Mvector,lambdavector,Dwvector,lambdavector,Mvecto
r+Dwvector);
grid;
```

Appendix 3.A.2: Program Listings of the Design of Standard Single-Mode Fibers

```
% totdisp_SMF.m
%
% MatLab script for calculating of total dispersion for
% Non-Zero Dispersion Shifted Fiber. The script plots
% the material dispersion, the waveguide dispersion, and
% the total dispersion for the designed fiber.
%
% Optical Fiber Design
%
lambda=[1.1:0.01:1.700]*1e-6;
G1=0.7028554;               %Sellmeier's coefficients for germanium
G2=0.4146307;               %doped silica (concentration B in table)
G3=0.8974540;
lambda1=0.0727723e-6;       %Wavelengths for germanium doped silica
lambda2=0.1143085e-6;
lambda3=9.896161e-6;
c=299792458;                %Speed of light
pi=3.1415926;               %Greek letter pi
a=4.1e-6;                   %Core radius
delta=0.003;                %Greek letter delta (ref. index difference
                            %between core and cladding)
% Calculating the refractive index
% ------------
npow2oflambda=1+(G1.*lambda.^2./(lambda.^2.-lambda1*lambda1)) ...
  + (G2.*lambda.^2./(lambda.^2.-lambda2*lambda2)) ...
  + (G3.*lambda.^2./(lambda.^2.-lambda3*lambda3));
noflambda=sqrt(npow2oflambda);
```

```
pointer=find(lambda==1.550e-6);
n1=noflambda(pointer)      %Refractive index in the core
% Calculating the material dispersion
% ------------
t1=diff(noflambda);
t2=diff(lambda);
t3=t1./t2;
t4=diff(t3);
t5=diff(lambda);
t5=adjmat(t5);
lambda=adjmat(lambda);
lambda=adjmat(lambda);
%Material dispersion
Matdisp=- (lambda./c) .* (t4./t5);
% Converting to ps/nm.km
Matdisp=Matdisp.*1e6;
figure(1)
clf
hold
xlabel('nm')
ylabel('ps/nm.km')
title('Standard Single Mode Fiber')
plot(lambda, Matdisp, '.-')
grid on
% Calculating waveguide dispersion
% -----------
V=(2 * pi * a * n1 * sqrt(2 * delta)) ./(lambda);
Dlambda1=- (n1 * delta)./(c * lambda);
%plot(lambda,Dlambda2)
Dlambda2=0.080+0.549 * (2.834-V).^2;
%plot(lambda,Dlambda2)
Dlambda=Dlambda1 .* Dlambda2;
% Converting to ps/nm.km
Dlambda=Dlambda.*1e6;
plot(lambda,Dlambda, '-')
% Calculating total dispersion
% ------------
TotDisp=Matdisp+Dlambda;
plot(lambda, TotDisp, ':')
legend('Material Dispersion','Waveguide Dispersion','Total Dispersion',0)
% Finding the dispersion at 1460, 1550 and 1625nm
% -----------------
pointer=find(lambda==1.460e-6);
Disp1460=TotDisp(pointer)
pointer=find(lambda==1.550e-6);
Disp1550=TotDisp(pointer)
pointer=find(lambda==1.6250e-6);
Disp1625=TotDisp(pointer)
% refind_SMF.m
%
% MatLab script for calculating of possible values for
% refractive index in the core, n1, and cladding, n2,
% its corresponding relative refractive index.
%
```

```
% PROJECT DESIGN: Optical Fiber Design
%
lambda=[1.1:0.01:1.700]*1e-6;
n1=[1.0487:0.001:1.8587]; %Refractive index of the core
n2=[1.0435:0.001:1.8535]; %Refractive index of the cladding
% Calculating the refractive index for the index profile
% ------------------
delta=(n1-n2) ./ n1;
deltap=delta * 100;
plot(n1, deltap)
grid on
xlabel('n1')
ylabel('Delta-Relative Refractive Index(%)')
title('Refractive Index')
```

Appendix 3.A.3: Program Listings of the Design of Nonzero Dispersion-Shifted Fibers

```
% totdisp_NZDSF.m
%
% MatLab script for calculating of total dispersion for
% Non-Zero Dispersion Shifted Fiber. The script plots
% the material dispersion, the waveguide dispersion and
% the total dispersion of the designed fiber.
%
% PROJECT 1: Optical Fiber Design
%
lambda=[1.1:0.001:1.700]*1e-6;
G1=0.7028554;             %Sellmeier's coefficients for germanium
G2=0.4146307;             %doped silica (concentration B in table)
G3=0.8974540;
lambda1=0.0727723e-6;     %Wavelengths for germanium doped silica
lambda2=0.1143085e-6;
lambda3=9.896161e-6;
c=299792458;              %Speed of light
pi=3.1415926;             %Greek letter pi
a=2.4e-6;                 %Core radius
delta=0.0043;             %Greek letter delta (ref. index difference
                          %between core and cladding)
% Calculating the refractive index
% -----------
npow2oflambda=1+(G1.*lambda.^2./(lambda.^2.-lambda1*lambda1)) ...
  +(G2.*lambda.^2./(lambda.^2.-lambda2*lambda2)) ...
  +(G3.*lambda.^2./(lambda.^2.-lambda3*lambda3));
noflambda=sqrt(npow2oflambda);
pointer=find(lambda==1.550e-6);
n1=noflambda(pointer)     %Refractive index in the core @1550nm
% Calculating the material dispersion
% -----------
t1=diff(noflambda);
t2=diff(lambda);
```

```
t3=t1./t2;
t4=diff(t3);
t5=diff(lambda);
t5=adjmat(t5);
lambda=adjmat(lambda);
lambda=adjmat(lambda);
%Material dispersion
Matdisp=- (lambda./c) .* (t4./t5);
% Converting to ps/nm.km
Matdisp=Matdisp.*1e6;
figure(1)
clf
hold
xlabel('nm')
ylabel('ps/nm.km')
title('Non-Zero Dispersion Shifted Fiber')
plot(lambda, Matdisp, '.')
grid on
% Calculating waveguide dispersion
% ------------
V=(2*pi*a*n1*sqrt(2*delta))./(lambda);
pointer=find(lambda==1.550e-6);
V1550=V(pointer)
Dlambda1=- (n1 * delta)./(c * lambda);
Dlambda2=0.080+0.549 * (2.834-V).^2;
Dlambda=Dlambda1 .* Dlambda2;
% Converting to ps/nm.km
Dlambda=Dlambda.*1e6;
plot(lambda,Dlambda, '-')
% Calculating total dispersion
% ------------
TotDisp=Matdisp+Dlambda;
plot(lambda, TotDisp, '+')
legend('Material Dispersion', 'Waveguide Dispersion', 'Total Dispersion', 0)
% Finding the dispersion at 1460, 1550 and 1625nm
% -----------------
pointer=find(lambda==1.460e-6);
Disp1460=TotDisp(pointer)
pointer=find(lambda==1.550e-6);
Disp1550=TotDisp(pointer)
pointer=find(lambda==1.6250e-6);
Disp1625=TotDisp(pointer)
% refine_NZDSF.m
%
% MatLab script for calculating of possible values for
% refractive index in the core, n1, and cladding, n2,
% its corresponding relative refractive index.
%
lambda=[1.1:0.01:1.700]*1e-6;
n1=[1.0487:0.001:1.8587]; %Refractive index of the core
n2=[1.0324:0.001:1.8424]; %Refractive index of the cladding
% Calculating the refractive index for the index profile
% -----------------
delta=(n1-n2) ./ n1;
```

```
deltap=delta * 100;
plot(n1, deltap)
grid on
xlabel('n1')
ylabel('Delta-Relative Refractive Index(%)')
title('Refractive Index')
```

Appendix 3.A.4: Program Listings of the Split-Step Fourier Method with SPM and Raman Gain Distribution

```
function output=ssprop_matlabfunction_raman(input)
nt=input(1);
u0=input(2:nt+1);
dt=input(nt+2);
dz=input(nt+3);
nz=input(nt+4);
alpha_indB=input(nt+5);
betap=input(nt+6:nt+9);
gamma=input(nt+10);
P_non_thres=input(nt+11);
maxiter=input(nt+12);
tol=input(nt+13);
%Ld=input(nt+14);
%Aeff=input(nt+15);
%Leff=input(nt+16);
tic;
%tmp=cputime;
%--------------------
% Original author: Thomas E. Murphy (tem@alum.mit.edu)
% Adapted and modified by Thanh Liem Huynh (thanh.huynh@eng.monash.edu.au)
%--------------------
% This function ssolves the nonlinear Schrodinger equation for
% pulse propagation in an optical fiber using the split-step
% Fourier method described in:
%
% Agrawal, Govind. Nonlinear Fiber Optics, 2nd ed. Academic
% Press, 1995, Chapter 2
%
% The following effects are included in the model: group velocity
% dispersion (GVD), higher order dispersion, loss, and self-phase
% modulation (gamma). Raman gain as a distributed amplification
%
% USAGE
%
% u1=ssprop(u0,dt,dz,nz,alpha,betap,gamma);
% u1=ssprop(u0,dt,dz,nz,alpha,betap,gamma,maxiter);
% u1=ssprop(u0,dt,dz,nz,alpha,betap,gamma,maxiter,tol);
%
% INPUT
%
```

```
% u0-starting field amplitude (vector)
% dt-time step-[in ps]
% dz-propagation stepsize-[in km]
% nz-number of steps to take, ie, ztotal=dz*nz
% alpha-power loss coefficient [in dB/km], need to convert to linear to
have P=P0*exp(-alpha*z)
% betap-dispersion polynomial coefs, [beta_0 ... beta_m] [in ps^(m-1)/km]
% gamma-nonlinearity coefficient [in (km^-1.W^-1)]
% maxiter-max number of iterations (default=4)
% tol-convergence tolerance (default=1e-5)
%
% OUTPUT
%
% u1-field at the output
%-----
% Convert alpha_indB to alpha in linear domain
%-----
alpha=1e-3*log(10)*alpha_indB/10; % alpha (1/km)-see Agrawal p57
%-----
%P_non_thres=0.0000005;
ntt=length(u0);
w=2*pi*[(0:ntt/2-1),(-ntt/2:-1)]'/(dt*nt);
%t=((1:nt)'-(nt+1)/2)*dt;
gain=numerical_gain_hybrid(dz,nz);
for array_counter=2:nz+1
    grad_gain(1)=gain(1)/dz;
    grad_gain(array_counter)=(gain(array_counter)-gain(array_counter-1))/dz;
end
gain_lin=log(10)*grad_gain/(10*2);
clear halfstep
  halfstep=-alpha/2;
    for ii=0:length(betap)-1;
      halfstep=halfstep-j*betap(ii+1)*(w.^ii)/factorial(ii);
    end
    square_mat=repmat(halfstep, 1, nz+1);
    square_mat2=repmat(gain_lin, ntt, 1);
    size(square_mat);
    size(square_mat2);
    total=square_mat+square_mat2;
clear LinearOperator
    % Linear Operator in Split Step method
    LinearOperator=halfstep;
    halfstep=exp(total*dz/2);
u1=u0;
ufft=fft(u0);
% Nonlinear operator will be added if the peak power is greater than the
% Nonlinear threshold
iz=0;
while (iz<nz) && (max((gamma*abs(u1).^2+gamma*abs(u0).^2))>P_non_thres)
  iz=iz+1;
  uhalf=ifft(halfstep(:,iz).*ufft);
  for ii=1:maxiter,
    uv=uhalf .* exp((-j*(gamma)*abs(u1).^2+(gamma)*abs(u0).^2)*dz/2);
    ufft=halfstep(:,iz).*fft(uv);
```

```
    uv=ifft(ufft);
    if (max(uv-u1)/max(u1)<tol)
      u1=uv;
      break;
    else
      u1=uv;
    end
end
% fprintf('You are using SSFM\n');
if (ii==maxiter)
    fprintf('Failed to converge to %f in %d iterations',tol,maxiter);
    end
    u0=u1;
end
if (iz<nz) && (max((gamma*abs(u1).^2+gamma*abs(u0).^2))<P_non_thres)
% u1=u1.*rectwin(ntt);
    ufft=fft(u1);
    ufft=ufft.*exp(LinearOperator*(nz-iz)*dz);
    u1=ifft(ufft);
    %fprintf('Implementing Linear Transfer Function of the Fibre
Propagation');
end
%toc;
output=u1;
```

Appendix 3.A.5: Program Listings of the Initialization File

```
% This file initialization file - delaring all parameters and data
required for SImulink model and Split Step Fourier
clear all
close all
% CONSTANTS
c=299792458;                % speed of light (m/s)
% NUMERICAL PARAMETERS
numbitspersymbol=1
P0=0.003;                   % peak power (W)
FWHM=25                     % pulse width FWHM (ps)
%halfwidth=FWHM/1.6651      % for Gaussian pulse
halfwidth=FWHM              % for square pulse
bitrate=1/halfwidth;        % THz
baudrate=bitrate/numbitspersymbol;
signalbandwidth=baudrate;
%%%%%%%%%%%%%%%%%%%%%%%%%%%%%%%
% for DPSK
Vpi=5;
halfVpi=Vpi/2;
twoVpi=Vpi*2;
% nt=2^8;     % number of points in FFT
PRBSlength=2^5;
% Make sure : FFT time window (=nt*dt)=PRBSlength * FWHM...
```

```
% FFTlength nt=PRBSlength/block * numbersamples/bit=PRBSlength * (FWHM/dt)
% num_samplesperbit=FWHM/dt should be about 8-16 samples/bit
num_samplesperbit=32; % should be 2^n
dt=FWHM/num_samplesperbit ;      % sampling time(ps); % time step (ps)
nt=PRBSlength*num_samplesperbit; % FFT length
% nt=2^9;
% nt=num_samplesperbit;
dz=0.2;       % distance stepsize (km)
nz=500;
%melbourne to gippsland
%170km two spans
nz_MelbToGipps=500;
%undersea link
%290km
nz_Raman=250;
nz_undersea=950;
nz_DCF=145;
%George Town to Hobart
nz_GtownToHobart=500;
% number of z-steps
maxiter=10;  % max # of iterations
tol=1e-5;    % error tolerance
% OPTICAL PARAMETERS
nonlinearthreshold=0.010; % 10mW -- % Nonlinear Threshold Peak Power
lambda=1550; % wavelength (nm)
optical_carrier=c/(lambda*1e-9);
%dBperkm=0.2;        % loss (dB/km)
alpha_indB=0.17;    % loss (dB/km)
D=18.5;       % GVD (ps/nm.km); if anomalous dispersion(for
              compensation),D is negative
beta3=0.06;   % GVD slope (ps^3/km)
ng=1.46;      % group index
n2=2.6e-20;   % nonlinear index (m^2/W)
Aeff=76;      % effective area (um^2)
% CALCULATED QUANTITIES
T=nt*dt;      % FFT window size (ps) -Agrawal: should be about 10-20 times
              % of the pulse width
alpha_loss=log(10)*alpha_indB/10; % alpha (1/km)
beta2=-1000*D*lambda^2/(2*pi*c);   % beta2 (ps^2/km);
%-----------------
% beta 3 can be calculated from the Slope Dispersion (S) as follows:]
% Slope Dispersion
% S=0.092;   % ps/(nm^2.km)
% beta31=(S-(4*pi*c./lambda.^3))./(2*pi*c./lambda.^2)
%-----------------
gamma=2e24*pi*n2/(lambda*Aeff);   % nonlinearity coef (km^-1.W^-1)
t=((1:nt)'-(nt+1)/2)*dt;   % vector of t values (ps)
t1=[(-nt/2+1:0)]'*dt;      % vector of t values (ps)
t2=[(1:nt/2)]'*dt;  % vector of t values (ps)
w=2*pi*[(0:nt/2-1),(-nt/2:-1)]'/T;      % vector of w values (rad/ps)
v=1000*[(0:nt/2-1),(-nt/2:-1)]'/T;      % vector of v values (GHz)
vs=fftshift(v);     % swap halves for plotting
v_tmp=1000*[(-nt/2:nt/2-1)]'/T;
% STARTING FIELD
```

```
% P0=0.001     % peak power (W)
% FWHM=20      % pulse width FWHM (ps)
%halfwidth=FWHM/1.6651     % for Gaussian pulse
%For square wave input, the FWHM=Half Width
%halfwidth=FWHM;
L=nz*dz
Lnl=1/(P0*gamma)     % nonlinear length (km)
Ld=halfwidth^2/abs(beta2)  % dispersion length (km)
N=sqrt(abs(Ld./Lnl))       % governing the which one is dominating:
                           % dispersion or Non-linearities
ratio_LandLd=L/Ld   % if L ≪ Ld -> NO Dispersion Effect
ratio_LandLnl=L/Lnl % if L ≪ Lnl -> NO Nonlinear Effect
% Monitor the broadening of the pulse with relative the Dispersion
% Length
% Calculate the expected pulsewidth of the output pulse
% Eq 3.2.10 in Agrawal "Nonlinear Fiber Optics" 2001 pp67
FWHM_new=FWHM*sqrt(1+(L/Ld)^2)
% N≪1 -> GVD ; N≫1 -> SPM
Leff=(1-exp(-alpha_loss*L))/alpha_loss
expected_normPout=exp(-alpha_loss*2*L)
NlnPhaseshiftmax=gamma*P0*Leff
betap=[0 0 beta2 beta3]';
% Constants for ASE of EDFA
% PSD of ASE: N(at carrier freq)=2*h*fc*nsp*(G-1) with nsp=Noise
% Figure/2 (assume saturated gain)
%*************** Standdard Constant ******************************
h=6.626068e-34;     %Plank's Constant
%*****************************************
```

References

1. A.P. Agrawal, *Fiber Optic Communications Systems*, 2nd edn. New York: Academic Press, 1997.
2. L.B. Jeunhomme, *Single Mode Fiber Optics*. New York: Marcel Dekker, 1983.
3. G. Mahlkc and P. Gossing, *Fiber Optic Cables*. Chichester, U.K.: Siemens A.G. and John Wiley, 1987.
4. W.D. Jones, *Optical Fiber Communications Systems*. New York: Holt, Rinhart and Winston, 1988.
5. S. Kobayashi, S. Shibata, N. Shibata, and T. lzawa, Refractive-index dispersion of doped fused si1ica, in *International Conference on Integrated Optics and Optical Fibre Communication*, Tokyo, Japan, 1977, pp. 9–312.
6. D. Gloge, Dispersion in weakly guiding fibres, *Applied Optics*, 10(11): 2442–2445, November 1971.
7. J.P. Gordon and H. Kogelnik, PMD fundamentals: Polarization mode dispersion in optical fibers, *Proceedings of the National Academy of Sciences*, 97(9): 4541–4550, April 2000.
8. Corning. Inc, An introduction to the fundamentals of PMD in fibers, White Paper, July 2006.
9. A. Galtarossa and L. Palmieri, Relationship between pulse broadening due to polarisation mode dispersion and differential group delay in long singlemode fiber, *Electronics Letters*, 34(5): 492–493, March 1998.
10. J.M. Fini and H.A. Haus, Accumulation of polarization-mode dispersion in cascades of compensated optical fibers, *IEEE Photonics Technology Letters*, 13(2): 124–126, February 2001.

11. A. Carena, V. Curri, R. Gaudino, P. Poggiolini, and S. Benedetto, A time-domain optical transmission system simulation package accounting for nonlinear and polarization-related effects in fiber, *IEEE Journal on Selected Areas in Communications*, 15(4): 751–765, 1997.
12. S.A. Jacobs, J.J. Refi, and R.E. Fangmann, Statistical estimation of PMD coefficients for system design, *Electronics Letters*, 33(7): 619–621, March 1997.
13. G.P. Agrawal, *Nonlinear Fiber Optics*, 3rd edn. San Diego, CA: Academic Press, 2001.
14. A.F. Elrefaie and R.E. Wagner, Chromatic dispersion limitations for FSK and DPSK systems with direct detection receivers, *IEEE Photonics Technology Letters*, 3(1): 71–73, 1991.
15. A.F. Elrefaie, R.E. Wagner, D.A. Atlas, and A.D. Daut, Chromatic dispersion limitation in coherent lightwave systems, *IEEE Journal of Lightwave Technology*, 6(5): 704–710, 1988.
16. J.P. Gordon and L.F. Mollenauer, Phase noise in photonic communications systems using linear amplifiers, *Optics Letters*, 15: 1351–1353, December 1990.
17. S. Kobayashi, N. Shibata, S. Shibata, and T. Izawa, Characteristics of optical fibers in infrared wavelength region, *Rev. Electrical Comm. Lab.* 26: 453–467, 1978.
18. J.W. Fleming, Material dispersion in lightguide glasses, *Electron Lett.*, 14: 326–328, 1978.

4

Overview of Modeling Techniques for Optical Transmission Systems Using MATLAB® and Simulink®

This chapter describes general aspects for modeling of optical transmission systems. MATLAB® and Simulink® are the modeling platforms, and a number of models are introduced. Generic models are described from the transmitters to receivers and signal processing to evaluate the transmission performances.

4.1 Overview

A digital photonic transmission system can be divided into three main subsystems: optical transmitter, optical fiber channel, and optical receiver, as demonstrated in Figure 4.1.

The first key subsystem is the optical transmitter. The main function of an optical transmitter is to generate lightwaves carrying a particular modulation format. Modulation formats are classified into three groups depending on whether amplitude, phase, or frequency component of the lightwave carrier is modulated. In modern photonic communications, the modulation process is implemented by using external optical modulators that can be categorized into phase and intensity modulators. Advantages of external data modulation over direct data modulation on semiconductor lasers and operational descriptions of external optical modulators are presented in Section 4.2.

The second subsystem is the optical fiber channel. The classification of optical fibers is based on their dispersion characteristics, and some notable types are standard single-mode fiber (SSMF), commonly known as ITU-G.652 type, and nonzero dispersion-shifted fiber (NZ-DSF), ITU-G.655 type. SSMF has a chromatic dispersion (CD) factor of about ±17 ps/(nm·km) at 1550 nm wavelength compared to small values of ±(2–6) ps/(nm·km) for NZ-DSF. The dispersion induced from these optical transmission fibers is compensated by a dispersion-compensating fiber (DCF) that has negative CD factors. In addition, recent progress in fiber design has also introduced several new types of fibers, such as dispersion-flattened fibers (DFF) and Corning–Vascade fibers [1,2].

Optical fibers consist of several impairments that cause severe degradations to the system performance. These impairments are grouped into fiber dispersions and fiber nonlinearities. Fiber dispersion includes second-order CD dispersion, the third-order dispersion slope, and polarization mode dispersion (PMD). On the other hand, fiber nonlinearities, which are power-dependent impairments, contain a number of effects, including intrachannel self-phase modulation (SPM), interchannel cross-phase modulation (XPM), and four-wave mixing (FWM). Dispersion and nonlinearity impairments are embedded into nonlinear Schrödinger equation (NLSE) that governs signal propagation along the optical fiber. The most popular method to solve NLSE numerically is the symmetric split-step Fourier method (SSFM) [3]. This method, however, encounters a number of issues,

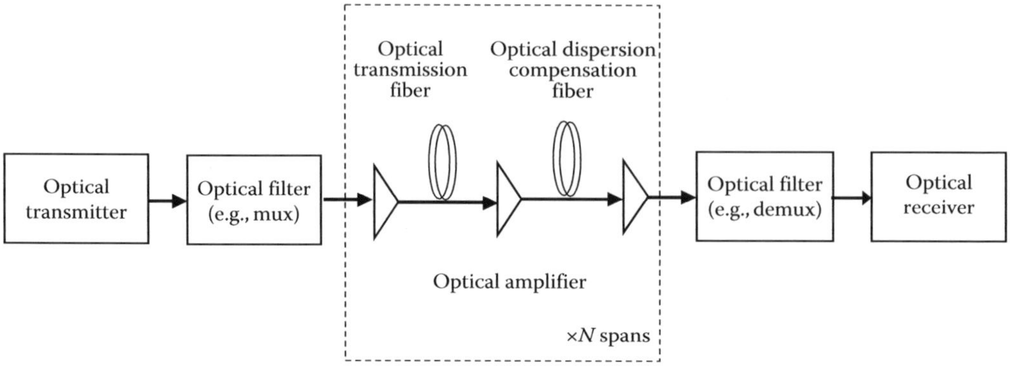

FIGURE 4.1
Generalized diagram of optical transmission systems.

such as long computation time and artificial errors caused by the windowing effect of fast Fourier transform (FFT) and inverse FFT (IFFT) operations. Detailed descriptions on fiber impairments, the symmetric SSFM, and techniques to overcome the modeling limiting factors are provided in Sections 4.3 and 4.4.

Optical signals are attenuated when propagating along the optical fiber channel, thus necessitating signal amplification. This amplification is carried out in the photonic domain by using erbium-doped fiber amplifiers (EDFA). Figure 4.1 illustrates the conventional configuration in which DCFs are normally accompanied by two EDFAs. The first EDFA compensates the attenuation of the preceding SSMF span while the other EDFA boosts optical intensity to a designated level before launching into the next transmission span. From the system point of view, there are two key parameters modeling an EDFA: amplified spontaneous emission (ASE) noise and noise figure (NF). These parameters are formulated in Section 4.3. Unless specifically stated, it is assumed in this chapter that optical amplifiers are operating in their saturation modes. Figure 4.1 also shows optical filters whose bandwidths have a significant impact on system performance. The modeling of these optical filters as well as electrical filters is presented in Section 4.5.3.

The last key subsystem is the direct detection optical receiver, which can be classified into two types: single-photodiode receivers and Mach–Zehnder delay interferometer (MZDI) balanced receivers, for incoherent detection of optical on–off keying (OOK) and differential phase shift keying (DPSK) signals, respectively [4,5]. The performance of these receivers is influenced by several noise sources whose formulations are provided in Section 4.5.

The evaluation of system performance is critical in fiber-optic communications for both modeling/simulation and practical experiments. The performance of optical transmission systems can be characterized by several measures. A quick measure is based on simple metrics such as eye opening (EO) and eye opening penalty (EOP) of detected signals. However, the most popular measure is bit error rate (BER) as a function of either optical signal-to-noise ratios (OSNR), average received powers, or average input powers. Moreover, power or OSNR penalties can be inferred from the obtained BER curves, at a particular BER level. Conventional methods for calculating BER to evaluate system performance are presented in Section 4.6, while Section 4.7 provides thorough descriptions on statistical methods for novel applications in fiber-optic communications. Finally, Section 4.8 discusses the advantages of the developed MATLAB and Simulink modeling platforms

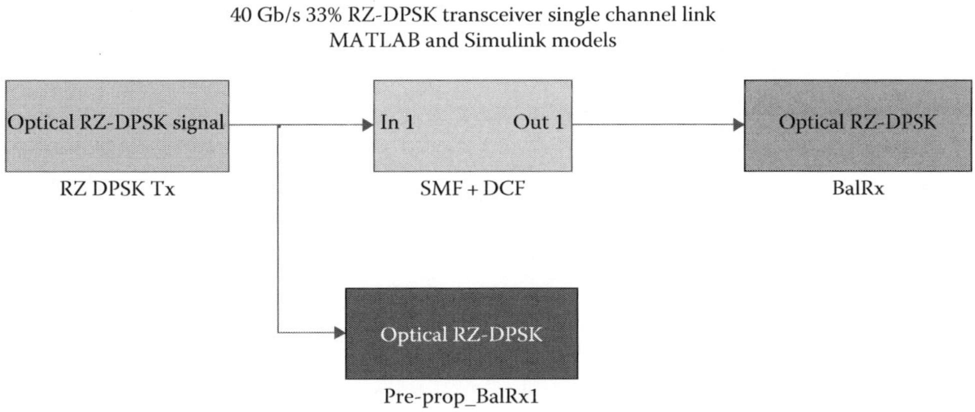

FIGURE 4.2
Generic overview of an optical transmission model on MATLAB and Simulink platforms for generation, propagation, and detection of 40 Gb/s DPSK RZ pulse shaping with 33% pulse with two-bit period ratio.

before a summary of this chapter is provided. A generic model of an optical transmission system developed on the Simulink platform is shown in Figure 4.2.

4.2 Optical Transmitter

An optical transmitter normally consists of a narrow linewidth laser source, external optical modulators, a bit pattern data generator, and, optionally, an electrical precoder or an electrical shaping filter (see Figure 4.3).

The narrow linewidth laser source is normally a distributed feedback (DFB) laser [6,7] with generated wavelengths complying with the ITU-grid standard. The laser can be biased at a constant current to provide a continuous-wave (CW) laser whose optical carrier passes through an external optical modulator for the data modulation process. In simulation, all signals are to be sampled due to the fact that the simulation model is normally implemented in digital computers and sampling is essential. If an optical carrier is used with frequency in the range of 200 THz, then the sampling rate would be in the order of a few hundred tera-samples per second. At this rate, the computer may run out of memory, and the execution time is very long. Thus, normally, the amplitude of the data sequence is represented as complex amplitude whose imaginary part and hence the phase angle of the carrier can be incorporated without using the real carrier.

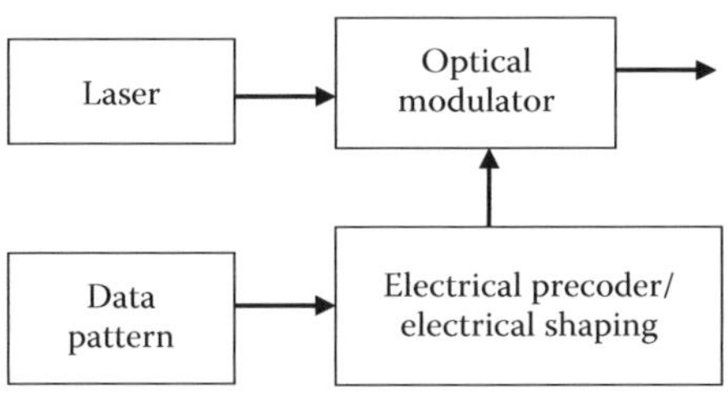

FIGURE 4.3
Generic block diagram of an optical transmitter.

4.2.1 Background of External Optical Modulators

In the 1980s and the early 1990s, direct modulation of semiconductor lasers was the main modulation technique. However, this technique faces several limiting factors [8,9]. Direct modulation induces unwanted chirps, resulting in signal spectral broadening and, hence, causing severe dispersion penalties. Directly modulated optical signals experience fluctuation in the intensity; this arises from the relative intensity noise (RIN) of the semiconductor laser. Laser phase noise induced from the nonzero linewidth of laser sources also limits the application of the direct modulation technique in high-speed transmission systems.

External data modulation has thus been the preferred technique over direct modulation as it minimizes the above-mentioned problems for digital photonic transmission systems. External data modulation can be implemented using either electro-absorption modulators (EAM) or electro-optic modulators (EOM). EOMs have been preferable due to the advantages of electro-optic materials, such as linear response characteristic, high extinction ratio, and, particularly, capability to control either phase, frequency, or amplitude of the lightwave carrier [10,11]. The operation of EOM is based on principles of Pockels electro-optic effects of solid-state, polymeric, or semiconductor materials [8,10]. Over the years, waveguides of EOMs are mainly integrated on lithium niobate ($LiNbO_3$) materials. These have prominent properties such as high electro-optic coefficients, low attenuation, and the possibility of generating chirp-free signals [10,12]. $LiNbO_3$ EOMs have been developed since the early 1980s, but were not popular until the advent of EDFA in the late 1980s [10,13]. They were employed in coherent optical communications to mitigate the problems of broad linewidth and RIN of the laser source in direct modulation, as mentioned earlier. The knowledge of using these external modulators has recently been revisited for the generation of advanced modulation formats in incoherent transmission systems. Data modulation is conducted in the photonic domain by using either an optical phase modulator or an intensity modulator.

4.2.2 Optical Phase Modulator

An electro-optic phase modulator (EOPM) employs a single electrode, as shown in Figure 4.4. Its operation is based on the Pockels electro-optic effect, that is, when a driving voltage is applied to the electrode, the refractive index (RI) of the electro-optic waveguide changes accordingly, thus slowing down the lightwaves and hence inducing a delay on the optical waves. Since the delays correspond to phase changes, the EOPM is able to manipulate the phase of the lightwave carrier.

The induced phase variation corresponding to a particular radio frequency (RF) electrical driving voltage, $V(t)$, is given as

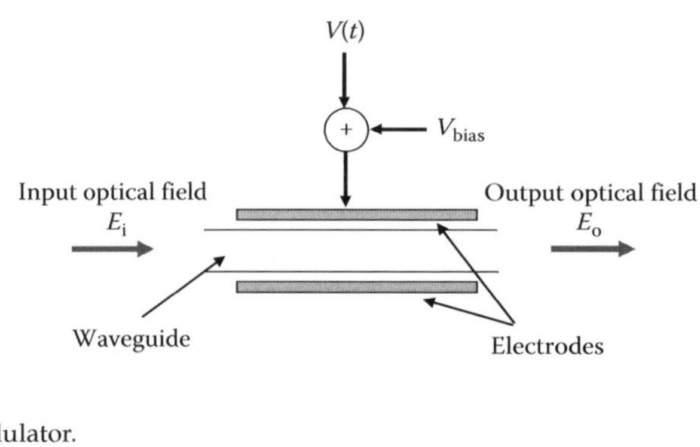

FIGURE 4.4
Electro-optic phase modulator.

$$\varphi(t) = \pi \frac{(V(t) + V_{\text{bias}})}{V_\pi} \tag{4.1}$$

where

V_π is the driving voltage required to create a π phase shift on the lightwave carrier and has typical values within a range of 3–6 V [12,14]

$V(t)$ is a time-varying driving signal voltage

V_{bias} is a DC bias voltage that is normally coupled with the signal source using a T-bias device

The complex-envelope representation of the optical field E_o at the output of EOPM is expressed as

$$E_o(t) = E_i(t)e^{j\varphi(t)} \tag{4.2}$$

EOPMs operating at high speed, for example, 40 Gb/s, and using resonant-type electrodes have been recently reported [15,16]. In addition, due to the linear response characteristic of electro-optical materials [reflected in (4.1)], it is possible to modulate the EOPM, and hence the optical frequency, by manipulating the slope of the time-varying RF driving signal.

4.2.3 Optical Intensity Modulator

An intensity modulator is composed of two EOPMs in a parallel structure to form a Mach–Zehnder interferometer that is commonly known as the Mach–Zehnder intensity modulator (MZIM) (see Figure 4.5).

The field of the input optical waves is split equally when entering two arms of MZIM. Each arm of the MZIM is actually an EOPM for modulating the phase of the optical carrier. At the output of MZIM, the optical fields from the two arms are coupled and interfered with each other either constructively or destructively. This enables the "ON–OFF" modulation of the lightwave intensity. Figure 4.6 shows a sample of a packaged MZIM and its modulation transfer characteristics [17] whose 3 dB bandwidth reaches 26 GHz.

MZIM can be classified into single-drive and dual-drive types, which are described in the following sections.

4.2.3.1 Single-Drive MZIM

A single-drive MZIM has only one single driving voltage applied to either arm of MZIM. For instance, it is assumed that there is no voltage driving onto arm 1 while a voltage $V(t)$

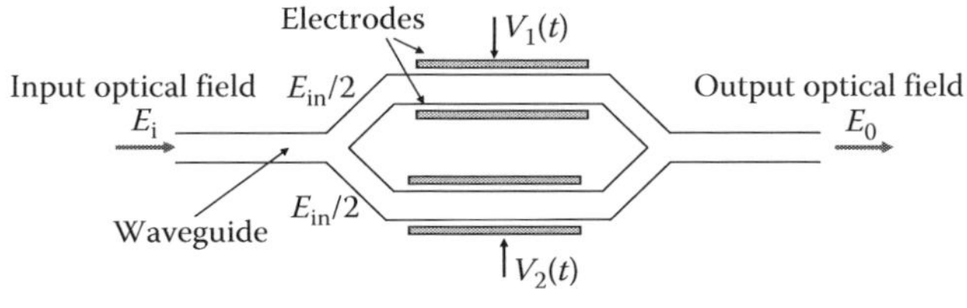

FIGURE 4.5
Optical intensity modulator with a Mach–Zehnder interferometric structure.

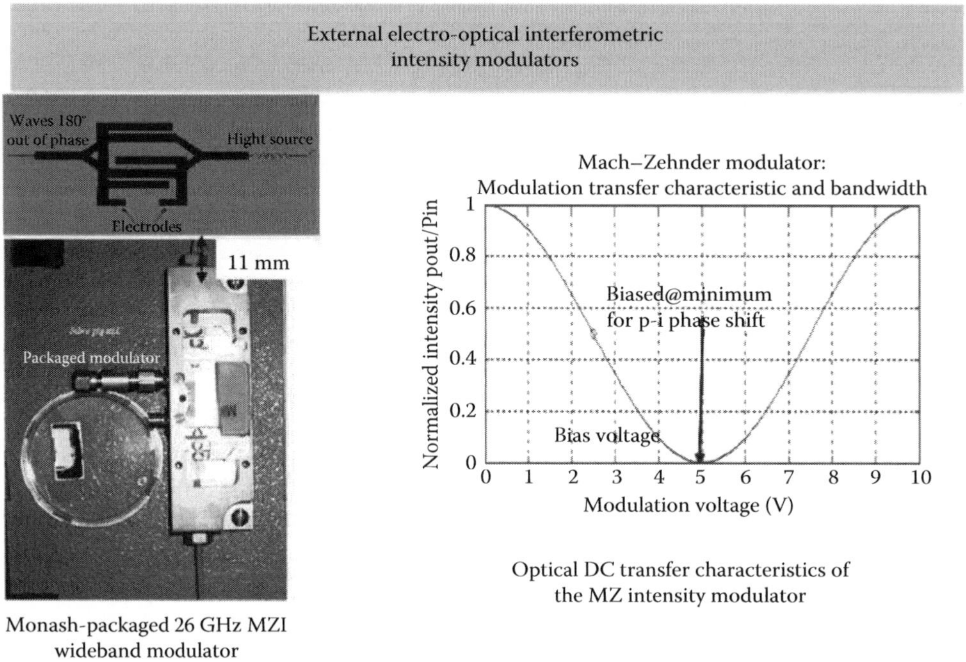

FIGURE 4.6
A Monash-packaged MZIM and its operating the transfer curve.

is applied on arm 2 (refer to Figure 4.4). The transmitted optical field $E_o(t)$ at the output of a single-drive MZIM is a function of $V(t)$ and a bias DC voltage V_{bias}. Written in the low-pass equivalent format, the expression of $E_o(t)$ is given by

$$E_o(t) = \frac{E_i(t)}{2}\left[1 + e^{j\pi\frac{(V(t)+V_{bias})}{V_\pi}}\right] = E_i \cos\left[\frac{\pi}{2}\frac{(V(t)+V_{bias})}{V_\pi}\right] e^{-j\left[\frac{\pi}{2}\frac{(V(t)+V_{bias})}{V_\pi}\right]} \tag{4.3}$$

The phase term in (4.3) implies the existence of the phase modulation of the optical carrier, that is, the chirping effect. Thus, by using a single-drive MZIM, generated optical signals are theoretically not chirp-free, particularly in the case of using a z-cut $LiNbO_3$ MZIM that has an asymmetrical structure of field distributions [12]. However, an x-cut MZIM can provide a modest amount of chirping thanks to its symmetrical distribution of the electrical fields [12]. In practice, a small amount of chirp might be useful for transmission [8].

4.2.3.2 Dual-Drive MZIM

A dual-drive MZIM has a push–pull arrangement for dual driving voltages, that is, ($V_1(t)$ and $V_2(t)$ are inverse to each other ($V_2(t) = -V_1(t)$) and, hence, can completely eliminate the chirping effect in the modulation of the lightwave carrier. The transmitted optical field $E_o(t)$ in (4.3) can be rewritten as

$$E_o(t) = \frac{E_i(t)}{2}\left[e^{j\pi\frac{(V(t)+V_{bias})}{V_\pi}} + e^{j\pi\frac{-(V(t)+V_{bias})}{V_\pi}}\right] = E_i \cos\left[\frac{\pi}{2}\frac{(V(t)+V_{bias})}{V_\pi}\right] \tag{4.4}$$

In (4.4), the phase term no longer exists, indicating that the chirping effect is totally eliminated.

4.3 Impairments of Optical Fiber

4.3.1 Chromatic Dispersion

This section presents key properties of the CD impairment in a single-mode fiber. The initial point when mentioning fiber CD is the expansion of the mode propagation constant or "wave number" parameter β using the Taylor series [8]

$$\beta(\omega)=\frac{\omega n(\omega)}{c}=\beta_0+\beta_1\Delta\omega+\frac{1}{2}\beta_2\Delta\omega^2+\frac{1}{6}\beta_3\Delta\omega^3+\cdots+\frac{1}{n!}\beta_n\Delta\omega^n \tag{4.5}$$

where
 ω is the angular optical frequency
 $n(\omega)$ is the fiber RI

Parameters β_n represent the nth derivative of β and their meanings are described as follows:

- β_0 involves the phase velocity, v_p, of the optical carrier, ω_0, and v_p is defined as

$$v_p=\frac{\omega_0}{\beta_0}=\frac{c}{n(\omega_0)} \tag{4.6}$$

The subscript 0 of the optical carrier frequency indicates the central frequency component of the carrier. The frequency difference Δω indicates the difference of the sideband component and the central carrier.

- β_1 determines the group velocity v_g that is related to β of the guided mode by

$$v_g=\frac{1}{\beta_1}=\left(\frac{d\beta}{d\omega}\bigg|_{\omega=\omega_0}\right)^{-1} \tag{4.7}$$

- β_2 is the derivative of the group velocity v_g with respect to frequency and, hence, β_2 clearly shows the frequency dependence of the group velocity. This means that different frequency components of an optical pulse propagate along the optical fiber at different velocities, thus leading to the spreading of the pulse, that is, the dispersion. The parameter β_2 is commonly known as the group velocity dispersion (GVD). The optical fiber exhibits normal dispersions for $\beta_2>0$ or anomalous dispersions for $\beta_2<0$. A pulse having the spectral width of Δω and traveling through a length L of fiber is broadened by an amount of time ΔT given by $\Delta T=\beta_2 L\Delta\omega$ [3]. In practice, a more common factor to represent fiber CD of a single-mode optical fiber is the dispersion factor D with the unit of ps/(nm · km). D is closely related to β_2 by

$$D = -\left(\frac{2\pi c}{\lambda^2}\right)\beta_2 \tag{4.8}$$

where λ is the operating wavelength.

- β_3 is the derivative of β_2 and contributes to the dispersion slope $S(\lambda)$ as follows:

$$S = \frac{dD}{d\lambda} = \left(\frac{2\pi c}{\lambda^2}\right)\beta_3 + \left(\frac{4\pi c}{\lambda^3}\right)\beta_2 \tag{4.9}$$

4.3.2 Chromatic Dispersion as a Total of Material Dispersion and Waveguide Dispersion

From the viewpoint of fiber design [8,18], D is a sum of material dispersion (D_M) and waveguide dispersion (D_W):

$$D = -\left(\frac{2\pi c}{\lambda^2}\right)\beta_2 \equiv D_M + D_W \tag{4.10}$$

The following equations describe how D_M and D_W are obtained. In this case, a step-index optical fiber with a core radius a is considered and the RIs of the core and cladding of the SSMF are denoted as n_1 and n_2, respectively. The significant transverse propagation constants of guided lightwaves u and v in the core and cladding regions are formulated as

$$u = a\sqrt{k^2 n_1^2 - \beta^2} \tag{4.11}$$

$$v = a\sqrt{\beta^2 - k^2 n_2^2} \tag{4.12}$$

where $k^2 n_1^2$ and $k^2 n_2^2$ are the plane-wave propagation constants in the core and the cladding, respectively. The guided wave number β is calculated as

$$\beta = \sqrt{k^2 (b(n_1^2 - n_2^2) + n_2^2)} \tag{4.13}$$

where b is the normalized propagation constant whose values for guided modes fall within the range of [0,1] and b is calculated as

$$b = \frac{\frac{\beta}{k} - n_2}{n_1 - n_2} \tag{4.14}$$

The normalized frequency V is expressed as

$$V = ak\sqrt{n_1^2 - n_2^2} \tag{4.15}$$

The waveguide dispersion D_W can be calculated using the following equation [8,18–20]:

$$D_W = -\left(\frac{n_1 - n_2}{c\lambda}\right) V \frac{d^2(Vb)}{dV^2} \tag{4.16}$$

where $V\ (d^2(Vb)/dV^2)$ is defined as the normalized waveguide dispersion parameter. An effective approximation based on polynomial interpolation has been developed to calculate the waveguide dispersion parameter of a multi-cladding DCF [18].

The material dispersion of an optical fiber is due to the wavelength dependence of RI of the core and the cladding. The RI $n(\lambda)$ is estimated by the Sellmeier equation [20]:

$$n^2(\lambda) = 1 + \sum_{i=1}^{M} \frac{B_i \lambda^2}{(\lambda^2 - \lambda_i^2)} \tag{4.17}$$

where
- λ_i indicates the ith resonance wavelength
- B_i is its corresponding oscillator strength
- n stands for n_1 or n_2 for either the core or the cladding RI, respectively

The material dispersion factor D_M is then obtained by

$$D_M = -\frac{\lambda}{c}\left(\frac{d^2 n(\lambda)}{d\lambda^2}\right) \tag{4.18}$$

where c is the light velocity in vacuum. For pure silica and over the wavelength of 1.25–1.66 μm, D_M can also be approximated by an empirical relation [8,20]:

$$D_M = 122\left(1 - \frac{\lambda_{ZD}}{\lambda}\right) \tag{4.19}$$

where λ_{ZD} is the zero material dispersion wavelength, which is defined as the wavelength at $D_M(\lambda) = 0$, for instance, $\lambda_{ZD} = 1.276\,\mu m$ for pure silica. λ_{ZD} can vary according to various doping concentrations in the core and the cladding of different materials, such as germanium (Ge) or fluorine (F).

With the demand of reducing effects of fiber CD, several types of fibers, including DSF and NZ-DSF, were proposed. The latter type requires a nonzero local dispersion value in order to avoid phase matching between the wavelengths in a dense wavelength division multiplexing (DWDM) transmission system, that is, to avoid the FWM effect. Values of D for various types of fibers within C-band wavelengths of the ITU-grid standard are demonstrated in Figure 4.7, and the circle values are located at 1550 nm wavelength.

The optical transmission medium also involves DCF and several other types of fibers, such as DFF and Corning–Vascade fibers. DCF is usually used inline with SSMF in a dispersion-managed optical system, so that fiber CD is fully compensated within a span. The dispersion factor D of a DCF has negative values. On the other hand, DFF is manufactured for the specific purpose of flattening dispersion factors over a wide range of wavelengths in order to reduce fiber CH effects. However, these wavelengths travel at nearly the

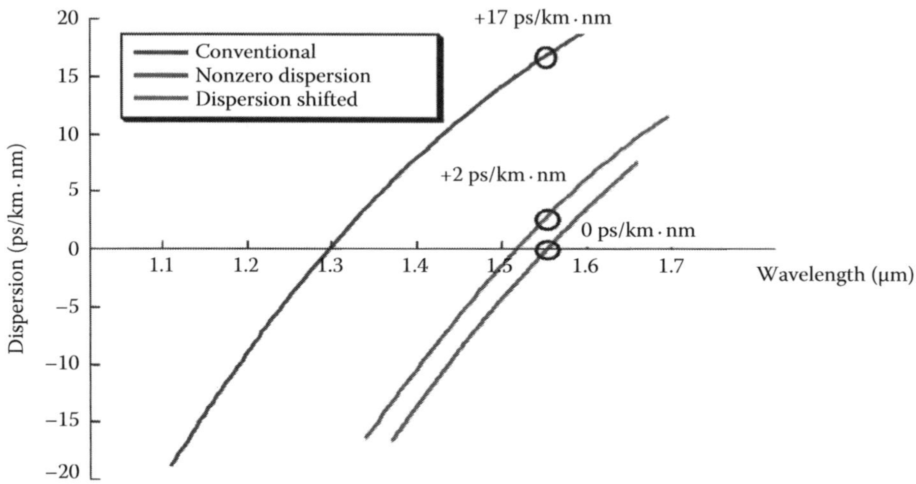

FIGURE 4.7
Typical values of fiber CD for different types of fibers.

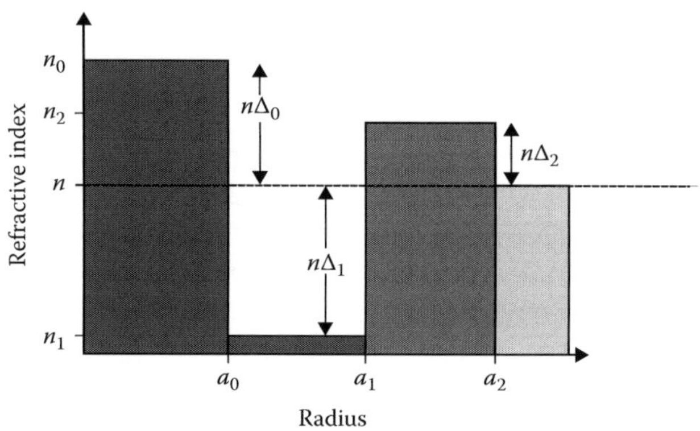

FIGURE 4.8
Index profile of a triple-cladding fiber for a design of DCF or DFF.

same velocities and, thus, the phase-matching conditions are easily satisfied. This results in FWM effects with the introduction of a ghost pulse. However, DFF has an important application in the parametric amplification utilizing FWM nonlinearity [3]. Furthermore, Corning–Vascade fibers are designed for ultra-long-haul and transoceanic optical transmission systems, and they provide a complete built-in CD compensation.

The designs of DCF and DFF are more complicated than that of SSMF, as it involves a multi-cladding/core structure (see Figure 4.8) compared to the simple step-index profile of SSMF [18].

4.3.3 Dispersion Length

An important parameter to govern the effects of fiber CD on optical pulses is the dispersion length L_D. This length corresponds to the distance at which a pulse has broadened by one-bit interval [8,21]. For high-capacity and long-haul transmission employing external modulation, L_D can be estimated as [21]

$$L_D = \frac{10^5}{D \cdot R^2} \tag{4.20}$$

where
- R is the bit rate (Gb/s)
- D is in ps/(nm · km)
- L_D is in kilometers

Equation 4.20 provides a reasonable approximation even though the accurate computation of L_D depends on a number of factors: the modulation format, pulse shaping, and the optical receiver performance. It is found that this transmission limit is inversely proportional to the square of the bit rate. Thus, for 10 Gb/s OC-192 OOK systems ($D = \pm 17$ ps/(nm · km)), the dispersion length L_D has a value of approximately 60 km SSMF. This SSMF length corresponds to a residual dispersion of about ±1000 ps/nm. In the case of 40 Gb/s OC-768 systems, L_D is about 4 km or equivalent to ±60 ps/nm.

4.3.4 Polarization Mode Dispersion

PMD represents another cause of pulse spreading and relates to differential group delay (DGD) between two orthogonal principal states of polarization (PSPs) of the propagating optical field, as illustrated in Figure 4.9.

Fiber PMD is caused by either the asymmetry of the fiber core or the deformation of optical fibers. These are consequences of defects in the manufacturing process, the external stress to the fiber, the aging problem, or the variation of temperature over time.

The delay between two PSPs is normally negligibly small at 10 Gb/s. However, at a high bit rate and in ultra-long-haul transmission, PMD severely degrades the system performance [22–25]. The instantaneous value of DGD ($\Delta\tau$) varies along the fiber and follows a Maxwellian distribution [23,26,27] (see Figure 4.10).

The Maxwellian distribution is governed by the following expression:

$$f(\Delta\tau) = \frac{32(\Delta\tau)^2}{\pi^2 \langle \Delta\tau \rangle^3} \exp\left\{ -\frac{4(\Delta\tau)^2}{\pi \langle \Delta\tau \rangle^2} \right\} \quad \Delta\tau \geq 0 \tag{4.21}$$

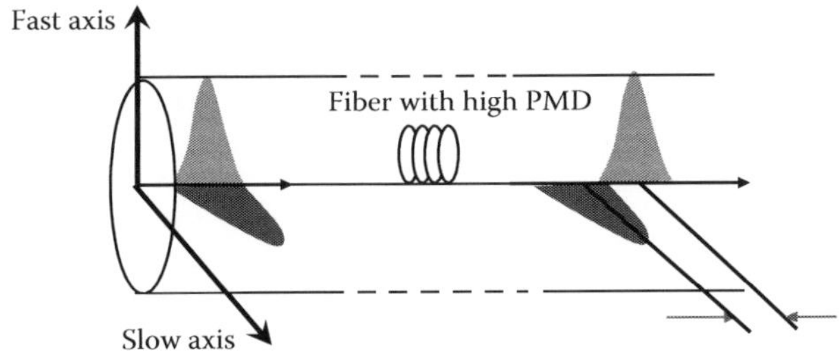

FIGURE 4.9
Delay caused by DGD of two PSPs along the fiber propagation.

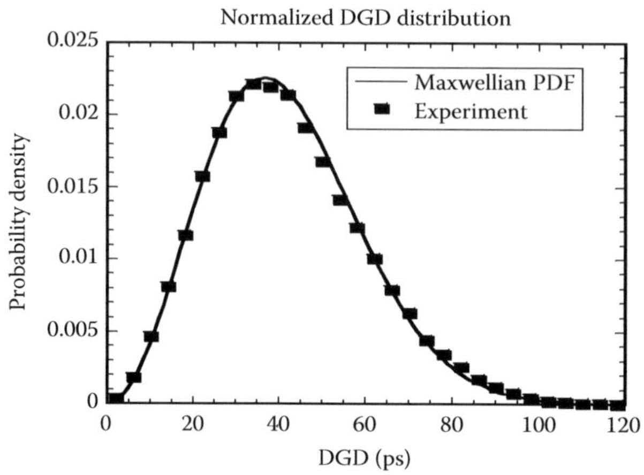

FIGURE 4.10
Maxwellian distribution of PMD random process.

The mean DGD value $\langle\Delta\tau\rangle$ is commonly termed "fiber PMD" and is provided in the fiber specifications. The following expression gives an estimate of the maximum transmission limit L_{max} due to the PMD effect [28]:

$$L_{max} = \frac{0.02}{\langle\Delta\tau\rangle^2 \cdot R^2} \tag{4.22}$$

where R is the bit rate. Based on (4.22), L_{max} for both old fiber vintages and contemporary fibers are obtained as follows:

- $\langle\Delta\tau\rangle = 1$ ps/km (old installed fibers)
  - $R = 40$ Gb/s; $L_{max} = 12.5$ km
  - $R = 10$ Gb/s; $L_{max} = 200$ km
- $\langle\Delta\tau\rangle = 0.1$ ps/km (contemporary modern fibers)
  - $R = 40$ Gb/s; $L_{max} = 1{,}250$ km
  - $R = 10$ Gb/s; $L_{max} = 20{,}000$ km

4.3.5 Fiber Nonlinearity

Fiber RI is dependent on both operating wavelengths and the intensity of the guided mode distributed across the area of the fiber core and cladding. This intensity-dependent phenomenon is known as the Kerr effect and is the cause of fiber nonlinear effects. The power dependence of RI is expressed as [3]

$$n' = n + \bar{n}_2\left(\frac{P}{A_{eff}}\right) \tag{4.23}$$

where

P is the average optical intensity inside the fiber
$\bar{n}_2$ is the fiber nonlinear coefficient
A_{eff} is the effective area of the fiber

Fiber nonlinear effects include intrachannel SPM, interchannel XPM, FWM, stimulated Raman scattering (SRS), and stimulated Brillouin scattering (SBS). SRS and SBS are not the main degrading factors as their effects are only getting noticeably large with very high optical power. On the other hand, FWM degrades severely the performance of an optical system with the generation of ghost pulses only if the phases of optical signals are matched with each other. However, with high local dispersions such as in SSMF, effects of FWM become negligible [3,29]. In terms of XPM, its effects can be considered to be negligible in a DWDM system in the following scenarios [30–35]: (1) highly locally dispersive system and (2) large channel spacing. However, XPM should be taken into account for optical transmission systems deploying the NZ-DSF fiber where local dispersion values are small. Thus, SPM is usually the dominant nonlinear effect for systems employing a transmission fiber with high local dispersions, for example, SSMF and DCF. The effect of SPM is normally coupled with the nonlinear phase shift ϕ_{NL} defined as [3]

$$\phi_{NL} = \int_0^L \gamma P(z)dz = \gamma L_{eff} P$$

$$\gamma = \frac{\omega_c \bar{n}_2}{(A_{eff} c)}$$

$$L_{eff} = \frac{(1 - e^{-\alpha L})}{\alpha} \tag{4.24}$$

where

ω_c is the lightwave carrier

L_{eff} is the effective transmission length

α is the fiber attenuation factor that normally has a value of 0.17–0.2 dB/km for the current 1550 nm window of operating wavelengths

The temporal variation of the nonlinear phase ϕ_{NL} results in the generation of new spectral components far apart from the lightwave carrier ω_c, indicating the broadening of the signal spectrum. This spectral broadening $\delta\omega$ can be obtained from the time dependence of the nonlinear phase shift as follows:

$$\delta\omega = -\frac{\partial \phi_{NL}}{\partial T} = -\gamma \frac{\partial P}{\partial T} L_{eff} \tag{4.25}$$

Equation 4.25 indicates that $\delta\omega$ is proportional to the time derivative of the average signal power P. Additionally, the generation of new spectral components occurs mainly at the rising and falling edges of optical pulses, that is, the amount of generated chirps is larger for an increased steepness of the pulse edges.

4.4 Modeling of Fiber Propagation

4.4.1 Symmetrical Split-Step Fourier Method

The evolution of slowly varying complex envelopes $A(z, t)$ of optical pulses along a single-mode optical fiber is governed by the NLSE [3]:

$$\frac{\partial A(z,t)}{\partial z} + \frac{\alpha}{2} A(z,t) + \beta_1 \frac{\partial A(z,t)}{\partial t}$$
$$+ \frac{j}{2}\beta_2 \frac{\partial^2 A(z,t)}{\partial t^2} - \frac{1}{6}\beta_3 \frac{\partial^3 A(z,t)}{\partial t^3} = -j\gamma |A(z,t)|^2 A(z,t) \tag{4.26}$$

where
 z is the spatial longitudinal coordinate
 α accounts for fiber attenuation
 β_1 indicates DGD
 β_2 and β_3 represent second- and third-order factors of fiber CD
 γ is the nonlinear coefficient

In a single-channel transmission, (4.26) includes the following effects: fiber attenuation, fiber CD and PMD, dispersion slope, and SPM nonlinearity. The fluctuation of optical intensity caused by the Gordon–Mollenauer effect [36] is also included in this equation.

The solution of NLSE and, hence, the modeling of pulse propagation along a single-mode optical fiber is solved numerically by using SSFM [3]. In SSFM, fiber length is divided into a large number of small segments δz. In practice, fiber dispersion and nonlinearity are mutually interactive at any distance along the fiber. However, these mutual effects are small within δz, and thus effects of fiber dispersion and fiber nonlinearity over δz are assumed to be statistically independent of each other. As a result, SSFM can separately define two operators: (1) the linear operator that involves fiber attenuation and fiber dispersion effects and (2) the nonlinearity operator that takes into account fiber nonlinearities. These linear and nonlinear operators are formulated as follows:

$$\hat{D} = -\frac{i\beta_2}{2}\frac{\partial^2}{\partial T^2} + \frac{\beta_3}{6}\frac{\partial^3}{\partial T^3} - \frac{\alpha}{2}$$
$$\hat{N} = i\gamma |A|^2 \tag{4.27}$$

where
 A replaces $A(z,t)$ for simpler notation
 $T = t - z/v_g$ is the reference time frame moving at the group velocity

Equation 4.26 can be rewritten in a shorter form, given by

$$\frac{\partial A}{\partial z} = (\hat{D} + \hat{N})A \tag{4.28}$$

and the complex amplitudes of optical pulses propagating from z to $z+\delta z$ are calculated using the approximation that follows:

$$A(z+h,T) \approx \exp(h\hat{D})\exp(h\hat{N})A(z,T) \tag{4.29}$$

Equation 4.29 is accurate to the second order of the step size δz [3]. The accuracy of SSFM can be improved by including the effect of fiber nonlinearity in the middle of the segment

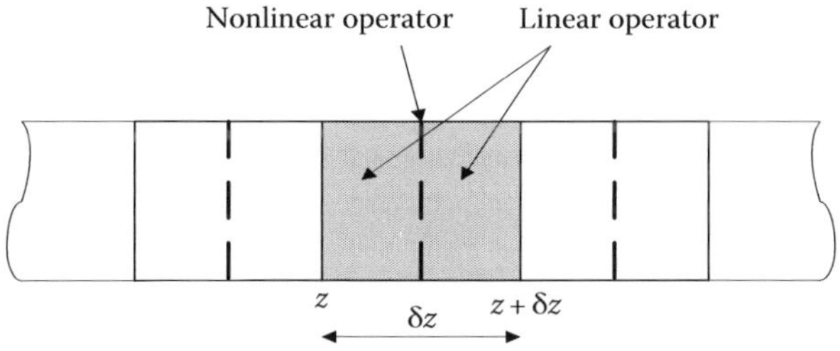

FIGURE 4.11
Schematic illustration of symmetric SSFM.

rather than at the segment boundary (see Figure 4.11). This modified SSFM is known as the symmetric SSFM.

Equation 4.29 can now be modified as

$$A(z+\delta z, T) \approx \exp\left(\frac{\delta z}{2}\hat{D}\right)\exp\left(\int_{z}^{z+\delta z}\hat{N}(z')dz'\right)\exp\left(\frac{\delta z}{2}\hat{D}\right)A(z,T) \tag{4.30}$$

This method is accurate to the third order of the step size δz. In symmetric SSFM, the optical pulse propagates along a fiber segment δz in two stages. First, the optical pulse propagates through the linear operator that has a step of δz/2 that takes into account fiber attenuation and dispersion effects. Then, the fiber nonlinearity is calculated in the middle of the segment. After that, the pulse propagates through the second half of the linear operator. The process continues repetitively in consecutive segments of size δz until the end of the fiber. It should be highlighted that the linear operator is computed in the frequency domain while the nonlinear operator is calculated in the time domain.

4.4.2 Modeling of Polarization Mode Dispersion

The first-order PMD can be implemented by modeling the optical fiber as two separate paths representing the propagation of two PSPs. Symmetrical SSFM is carried out on each transmission path before the outputs of these two paths are superimposed to calculate the output optical field. The transfer function to represent the first-order PMD is given by

$$H(f) = H^{+}(f) + H^{-}(f) \tag{4.31}$$

where

$$H^{+}(f) = \sqrt{k}\exp\left[j2\pi f\left(-\frac{\Delta\tau}{2}\right)\right] \quad \text{and} \quad H^{-}(f) = \sqrt{k}\exp\left[j2\pi f\left(-\frac{\Delta\tau}{2}\right)\right],$$

in which k is the power-splitting ratio and $k=1/2$ when using a 3 dB or 50:50 optical coupler/splitter, and Δτ is the instantaneous DGD value following a Maxwell distribution [refer to (4.21)].

4.4.3 Optimization of Symmetrical SSFM

4.4.3.1 Optimization of Computational Time

A huge amount of time is spent in symmetric SSFM for FFT and IFFT operations, in particular, when fiber nonlinear effects are involved. In practice, when optical pulses propagate toward the end of a fiber span, the pulse intensity is greatly attenuated due to the fiber attenuation. As a result, fiber nonlinear effects become negligible for the rest of that fiber span and, hence, the transmission operates in a linear domain in this range. In this research, a technique to configure symmetric SSFM is proposed in order to reduce the computational time. If the peak power of an optical pulse is lower than the nonlinear threshold of the transmission fiber, for example, around −4 dBm, symmetrical SSFM is switched to a linear mode operation. This linear mode involves only fiber dispersions and fiber attenuation, and its low-pass equivalent transfer function for the optical fiber is

$$H(\varpi) = \exp\left\{-j\left[\left(\frac{1}{2}\right)\beta_2\varpi^2 + \left(\frac{1}{6}\right)\beta_3\varpi^3\right]\right\} \tag{4.32}$$

If β_3 is not considered in this fiber transfer function, which is normally the case due to its negligible effects on 40 Gb/s and lower bit rate transmission systems, the transfer function given earlier has a parabolic phase profile [37,38].

4.4.3.2 Mitigation of Windowing Effect and Waveform Discontinuity

In symmetric SSFM, mathematical operations of FFT and IFFT play very significant roles. However, due to a finite window length required for FFT and IFFT operations, these operations normally introduce overshooting at two boundary regions of the FFT window, commonly known as the windowing effect of FFT. In addition, since the FFT operation is a block-based process, there exists the issue of waveform discontinuity, that is, the rightmost sample of the current output block does not start at the same position of the leftmost sample of the previous output block. The windowing effect and the waveform discontinuity problems are resolved with the following solutions (see Figure 4.12):

- The actual window length for FFT/IFFT operations consists of two blocks of samples ($2N$ sample length). The output, however, is a truncated version with the length of one block (N samples), and output samples are taken in the middle of the two input blocks.
- The next FFT window overlaps the previous one by one block of N sample.

4.5 Optical Amplifiers

In the modeling of an EDFA, two key parameters are ASE noise and NF.

4.5.1 ASE Noise

The following formulation accounts for the average power of ASE noise (N_{ASE}) [8]:

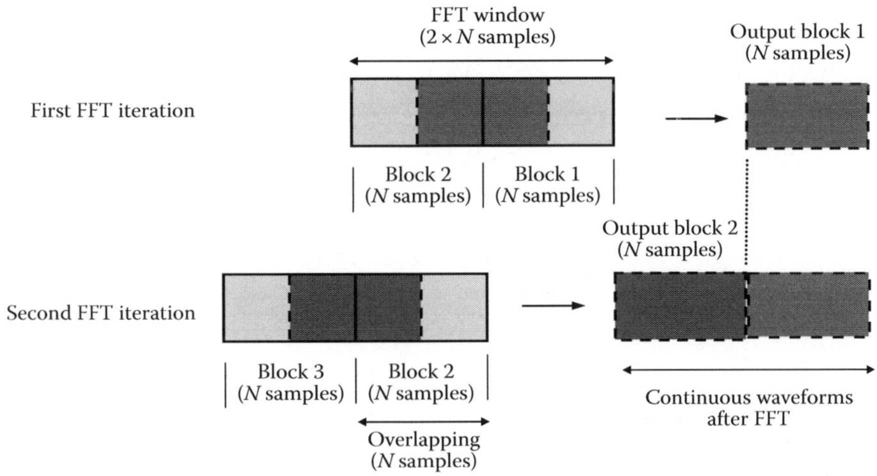

FIGURE 4.12
Proposed solution for mitigating windowing effect and waveform discontinuity caused by FFT/IFFT operations.

$$N_{\text{ASE}} = mn_{\text{sp}}hf(G-1)B_{\text{o}} \tag{4.33}$$

where
- f is the optical operating frequency
- G is the amplifier optical gain
- n_{sp} is the spontaneous emission factor
- h is the Planck constant
- m is the number of polarization modes (usually $m=2$ for systems not using polarization multiplexing)
- B_{o} is the bandwidth of an optical filter

4.5.2 Noise Figure

NF of an EDFA is defined as the ratio of the output OSNR to the OSNR at the input of the EDFA, given by

$$\text{NF} = \frac{\text{OSNR}_{\text{in}}}{\text{OSNR}_{\text{out}}} \approx 2n_{\text{sp}} \quad \text{for } G \gg 1 \tag{4.34}$$

Assuming that EDFAs are operating in the saturation regions and by substituting NF from (4.34) into (4.33), the N_{ASE} can be related to NF as follows:

$$N_{\text{ASE}} = (\text{NF}\cdot G - 1)hfB_{\text{o}} \tag{4.35}$$

4.5.3 Optical and Electrical Filters

In this research, filtering of noise-corrupted optical signals is conducted with a Gaussian-type optical filter whose impulse response is governed by the following expression:

$$h_{\text{Gauss}}(t) = \frac{1}{\sqrt{2\pi}\xi} e^{\left(\frac{-t^2}{2\xi^2}\right)} \tag{4.36}$$

where $\xi = \frac{\sqrt{\ln(2)}}{2\pi BT}$, in which B is one-sided 3dB bandwidth of the Gaussian filter and T is the bit period. Variations of these two parameters are reflected by the change of the BT product.

The modeling of an electrical filter can also be implemented with a Gaussian filter having an impulse response similar to (4.36). Alternatively, a fifth-order Bessel filter is used, which can be easily designed using the filter design toolbox in MATLAB. The MATLAB pseudo-codes for designing a fifth-order Bessel filter are shown as follows:

```
[b,a]=besself(5thorder,2*pi*BT/sampling_factor)
[bz,az]=impinvar(b,a,1);
[hf t1]=impz(bz,az,2*delay* sampling_factor +1, sampling_factor)
```

where the BT product is defined similarly as to that of the Gaussian filter.

4.6 Optical Receiver

The original message is recovered in the electrical domain; thus, a conversion of lightwaves to electrical signals is required. In optical communications, this process is widely implemented with a P-type-intrinsic-N-type (PIN) photodiode for either coherent or incoherent detection. The first type requires a local oscillator to coherently convert modulated lightwaves from the optical frequency range down to the intermediate frequency (IF) range. On the other hand, incoherent detection, which has been preferable since the last decade, is based on the square-law envelope detection of optical signals. Incoherent detection however still requires the recovery of the clock timing. In the modeling of digital photonic transmission systems in this research, the ideal clock timing is assumed. Another key parameter of optical receivers is the responsivity parameter of the PIN photodiode, which is a measure for the efficiency of the photonic–electronic conversion. Photodiodes with a high responsivity of around 0.8–0.9 A/W are commercially available. In this research, it is assumed that the responsivity is equal to 1 A/W. Moreover, the induced electrical current is usually amplified with a trans-impedance amplifier, and then passed through an electrical filter. At this stage, electrical eye diagrams are observed, and the sampling process of electrically filtered received signals is carried out for recovering the data information.

Received signals are corrupted by noise from several sources. These include the shot noise (σ^2_{shot}), the electronic noise (σ^2_{elec}), the dark current noise (σ^2_{dark}), and the interactions between signals and the ASE noise ($\sigma^2_{\text{signal,ASE}}$) as well as between the ASE noise itself ($\sigma^2_{\text{ASE,ASE}}$). The summation of these noise sources reflects the total receiver noise:

$$\sigma^2_{\text{total}} = \sigma^2_{\text{shot}} + \sigma^2_{\text{elec}} + \sigma^2_{\text{dark}} + \sigma^2_{\text{signal,ASE}} + \sigma^2_{\text{ASE,ASE}} \tag{4.37}$$

These noise sources are modeled with normal distributions whose variances represent the noise power, and they are described as follows:

- *Shot noise* (σ^2_{shot}) is caused by the intrinsic optoelectronic phenomenon of the semiconductor photodiode. A random number of electron–hole pairs is generated with the receipt of photons, causing the randomness of the induced photocurrent. The shot noise is given by

$$\sigma^2_{\text{shot}} = 2 \cdot q \langle i_s \rangle B_e \tag{4.38}$$

  where
  B_e is the 3 dB bandwidth of the electrical filter
  $<i_s>$ is the average signal current (with the unit of A/Hz)

- *Electronic noise* source σ^2_{elec} is injected from the trans-impedance amplifier. It is modeled by an equivalent noise current i_{Neq} over the bandwidth B_e of the electrical filter. The unit of i_{Neq} is $\text{A}/\sqrt{\text{Hz}}$, and the value of σ^2_{elec} is obtained from

$$\sigma^2_{\text{elec}} = (i_{\text{Neq}})^2 B_e \tag{4.39}$$

- *Dark current* i_{dark} is normally specified for a photodiode and has the unit of A/Hz. Hence, the noise power σ^2_{dark} is calculated as

$$\sigma^2_{\text{dark}} = 2 \cdot q i_{\text{dark}} B_e \tag{4.40}$$

- The variances of amplitude fluctuations due to the beating of signal and ASE noise and between ASE noise itself are given in the following expressions:

$$\sigma^2_{\text{signal,ASE}} = 4 \cdot i_S i_N \frac{B_e}{B_{\text{opt}}} \tag{4.41}$$

$$\sigma^2_{\text{ASE,ASE}} = i_N^2 \frac{B_e}{B_{\text{opt}}^2} (2 \cdot B_{\text{opt}} - B_e) \tag{4.42}$$

  where
  B_{opt} is the 3 dB bandwidth of the optical filter
  i_N is the noise-induced photocurrent

In practice, the value of $\sigma^2_{\text{ASE,ASE}}$ is normally negligible compared to the value of $\sigma^2_{\text{signal,ASE}}$ and can be ignored without affecting the performance of the receiver. In addition, in an optically preamplified receiver, in which, the optical signal is amplified at a stage before the photodetector, $\sigma^2_{\text{signal,ASE}}$ is the dominant factor compared to other noise sources.

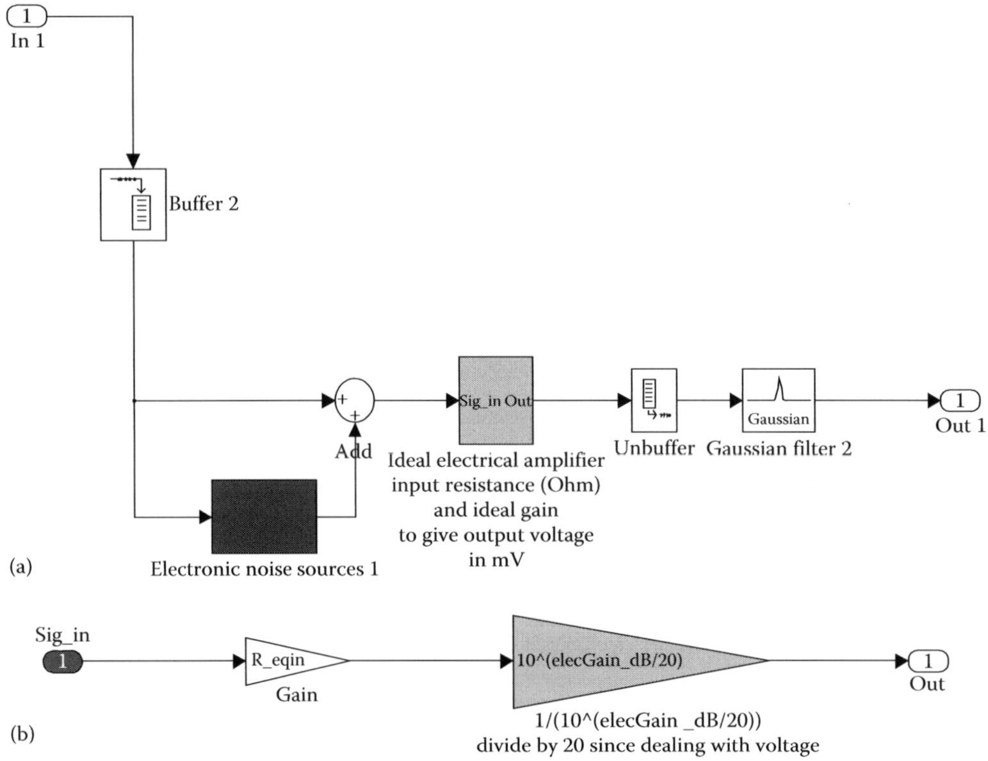

FIGURE 4.13
Simulink model of optical receiver. (a) Noise and gain. (b) Gain section.

A simplified Simulink model is shown in Figure 4.13a and b. The noise-equivalent current, as seen from the input of the electronic preamplifier, is modeled as a random noise source superimposed on the signal input to the receiver; relevant quantum shot noise is also calculated by the model at each sampled numerical instant.

4.7 Performance Evaluation

The performance of an optical transmission system can be evaluated by using conventional techniques, such as the Monte Carlo method and the single Gaussian distribution method. However, these conventional techniques have several limiting factors. The main limitation in the Monte Carlo method is the large amount of time needed for a simulation experiment, whereas the single Gaussian method does not take into account distortions caused by dynamic effects of the optical fibers. To cope with these issues, two statistical methods are presented in this research for novel applications in optical fiber communication, and they offer flexible and fast-processing methods to obtain the BER performance. Several major characteristics of these methods are as follows.

- The first method implements the expected maximization (EM) theorem, in which the probability density function (pdf) of received electrical signals is estimated as a mixture of multiple Gaussian distributions (MGD). Although the application of this method in optical communications has recently been reported [39,40], the guidelines for optimizing the accuracy of this method are yet to be presented.

- The second method, which is based on the generalized extreme values (GEV) theorem, is the generalized Pareto distribution (GPD) method. This method predicts the probability of the occurrence of extreme values that occur within the long tail of the signal pdf. Although the GPD method is popularly used in several fields, such as finance [41], meteorology [42], and climate forecasting [43], it has not yet been applied in the field of optical communications.

BER values calculated from all of the above-described evaluation methods are normally plotted as a function of OSNR, which is discussed in the next section.

4.7.1 Optical Signal-to-Noise Ratio

OSNR is a metric for the quality assessment of received signals that are corrupted by ASE noise of EDFAs. OSNR is defined as the ratio of the average optical signal power to the average optical noise power. For a single EDFA with the output power P_{out} and the noise power N_{ASE}, OSNR is computed as [8]

$$\mathrm{OSNR} = \frac{P_{\mathrm{out}}}{N_{\mathrm{ASE}}} = \frac{P_{\mathrm{out}}}{(\mathrm{NF} \cdot G - 1)hf\,\Delta f} \tag{4.43}$$

where

NF is the noise figure

G is the amplifier gain

hf is the photon energy

Δf is the optical measurement bandwidth

When addressing an OSNR value, it is important to define an optical reference bandwidth for the calculation of OSNR. A bandwidth Δf of 12.5 GHz (or $\Delta\lambda = 0.1$ nm) is the typical reference bandwidth for calculating OSNR values.

4.7.2 OSNR Penalty

OSNR penalty is obtained from the BER curves and determined at a particular BER. A value of the OSNR penalty is obtained by comparing the values of OSNR before and after the change of the parameters that are under test, and given by

$$\mathrm{OSNR\_Penalty} = 10 \log\left(\frac{\mathrm{OSNR}_{\mathrm{before}}}{\mathrm{OSNR}_{\mathrm{after}}}\right) \tag{4.44}$$

4.7.3 Eye Opening

OSNR is a time-averaged indicator for the ratio of average optical signal power to average optical noise power. Hence, OSNR is used most effectively when noise is the main degrading factor to the system performance. However, the OSNR metric is becoming less accurate, and the system degradation is mainly due to waveform distortions. These waveform distortions can originate either from the ISI problem caused by fiber CD and PMD, from fiber nonlinearities, or from effects of narrowband optical/electrical filtering. In contrast, waveform distortions are taken into account by using the EO metric. The EO is determined from the difference between the "mark" and "space" levels. In addition, EOP is the penalty of an EO when compared to a reference EO. This reference EO is usually obtained from a back-to-back configuration when signal waveforms are not distorted at all. The EOP is normally in log scale (dB) and given by

$$\mathrm{EOP} = 10\log\left(\frac{\mathrm{EO}_{\mathrm{ref}}}{\mathrm{EO}_{\mathrm{received}}}\right) \tag{4.45}$$

EO and EOP metrics are useful for noise-free systems, as they provide a good measure for pulse distortions. If noise is present, calculations of EO and EOP become less precise. In addition, the accuracy of EO and EOP calculations relies on the sampling instance. Detected pulses are usually sampled at the middle of the eye diagrams where the EO is widest.

4.7.4 Conventional Evaluation Methods

4.7.4.1 Monte Carlo Method

BER in simulation experiments is computed as the ratio of the number of error occurrences (N_{error}) to the total number of transmitted data bits (N_{tot}):

$$\mathrm{BER} = \frac{N_{\mathrm{error}}}{N_{\mathrm{trans}}} \tag{4.46}$$

The Monte Carlo method provides a precise BER calculation as it takes into account effects arisen from all the fiber impairments. However, the Monte Carlo method requires a huge number of transmitted information bits to obtain low values of BER, thus leading to excessive computational time. A BER of 1e–9, which is considered as "error free" in fiber-optic communications, requires at least 1e10 transmitted bits. Furthermore, a BER of up to 1e–12 is now becoming more common for modern high-capacity and high-speed digital photonic transmission systems. These huge numbers are not feasible for simulation experiments conducted by a personal computer at present. Additionally, time-consuming operations, such as FFT/IFFT, while implementing the symmetrical SSFM also significantly increase the computational time.

However, the Monte Carlo method is still used effectively in simulation experiments, because commercial optical transmission systems usually employ forward error correction (FEC) coding schemes to enhance significantly the system performance. Pre-FEC BERs can thus be as low as 2e–3 (the FEC limit), provided that no sign of long-burst errors is observed [21]. Figure 4.14 demonstrates significant improvements on the system performance by using FEC schemes: 7% single-stage and 23% concatenated Reed–Solomon code [21].

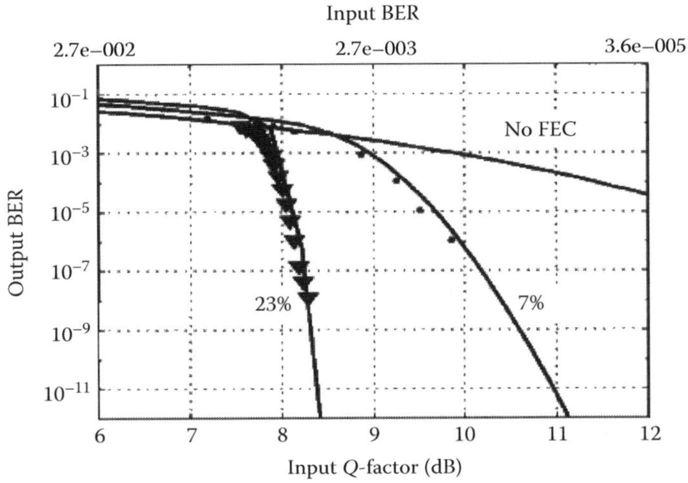

FIGURE 4.14
Demonstration of effectiveness of FEC schemes in improving the system performance.

With the current processing speed of contemporary personal computers, a BER of up to 1e–6 is of interest for simulation experiments. BERs calculated from the Monte Carlo method are usually taken as benchmarks for BER values computed by other methods.

4.7.4.2 Single Gaussian Statistical Method

This method implements a statistical process to calculate BER. It should be noted that signals are normally in voltage since photodiode-induced currents are amplified by a trans-impedance electrical amplifier. A particular voltage that serves as a reference for distinguishing "0" and "1" levels is known as the threshold voltage (V_{th}). In addition, the received signals are sampled at a particular instance. As a result, based on this V_{th}, normalized histograms of received signals for "0" and "1" are obtained, thus leading to the achievement of the pdf. The pdf normally complies to a distribution such as Gaussian and Chi-square. The BER is calculated as [44]

$$\text{BER} = P(\text{"1"})P(\text{"0"} \mid \text{"1"}) + P(\text{"0"})P(\text{"1"} \mid \text{"0"}) \tag{4.47}$$

where
$P(\text{"1"})$ is the probability that a bit "1" is sent
$P(\text{"0"}|\text{"1"})$ is the probability of error due to receiving "0" where actually a "1" is sent
$P(\text{"0"})$ is the probability that a "0" is sent
$P(\text{"1"}|\text{"0"})$ is the probability of error due to receiving "1" where actually a "0" is sent

In the case of binary digital transmission, the probability of transmitting a "0" or a "1" is equal, that is, $P(\text{"1"}) = P(\text{"0"}) = 1/2$. $P(\text{"0"}|\text{"1"})$ and $P(\text{"1"}|\text{"0"})$ are calculated from integrating the overlap region of the pdf, which exceeds the threshold voltage.

A popular assumption is that the pdfs of received electrical signals follow a Gaussian distribution. This enables a fast calculation for BER values by using the complementary error functions (erfc) [44]:

$$\text{BER} = \frac{1}{2}\left[\text{erfc}\left(\frac{|\mu_1 - V_{\text{th}}|}{\sqrt{2}\sigma_1}\right) + \text{erfc}\left(\frac{|\mu_0 - V_{\text{th}}|}{\sqrt{2}\sigma_0}\right)\right] \tag{4.48}$$

where μ_0, μ_1 and σ_0, σ_1 are means and variances of the Gaussian pdfs of "0" and "1" received signals, respectively.

Apart from BER, the quality (Q) factor is another common measure metric to assess the system performance. Q values are calculated from μ_0, μ_1 and σ_0, σ_1 as follows [44]:

$$Q = \frac{\mu_1 - \mu_0}{\sigma_1 - \sigma_0} \tag{4.49}$$

and Q factors are either in a linear scale or a dB scale. The BER can be obtained from the Q factor by [44]

$$\text{BER} = \frac{1}{2}\text{erfc}\left(\frac{Q}{\sqrt{2}}\right) \tag{4.50}$$

4.7.4.2.1 Improving Accuracy of Histograms

The pdfs of "0" and "1"' received signals are determined from normalized histograms. The estimation of these normalized histograms thus affects considerably the BER's accuracy. Thus, a proper estimation of histogram values is very important. A histogram is normally divided into a number of bins with the same binwidth, with a sufficiently large number of transmitted bits (N_0); a good estimate for the width (W_{bin}) of each equally spaced histogram bin is given by $W_{\text{bin}} = \sqrt{N_0}$ [45].

4.7.5 Novel Statistical Methods

The single Gaussian distribution (the Q factor) method considers only effects of noise corruption on detected signals while ignoring waveform distortions caused by fiber dispersion and fiber nonlinear effects. These distortions result in multi-peak and non-Gaussian pdfs (see Figure 4.15). Such pdfs cannot be correctly estimated by the conventional single Gaussian distribution method.

The issue discussed earlier is resolved with the following two statistical methods: the MGD method based on the expectation maximization (EM) theorem and the GPD method based on the GEV theorem. The implementation of these methods is conducted in MATLAB. It should be noted that the conventional single Gaussian distribution method is a particular case of the MGD method.

4.7.5.1 Multiple Gaussian Distributions Method

One of the most popular applications of the EM theorem is to obtain parameters of mixed probability densities. This theorem is based on the fact that most of the deterministic distributions can be considered as a superposition of multiple distributions. The pdf $p(x|\Theta)$ for a set of received data can be expressed as the mixture of M different distributions [46]:

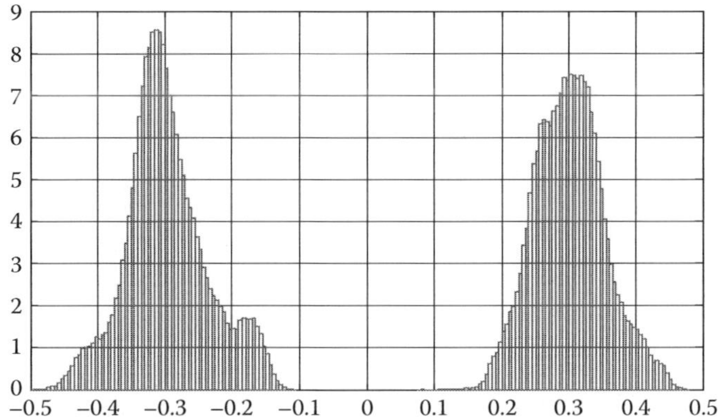

FIGURE 4.15
Demonstration of multi-peak and non-Gaussian distributions of received signals.

$$p(x|\Theta)=\sum_{i=1}^{M} w_i p_i(x|\theta_i); \quad \Theta=(w_1,\ldots,w_M,\theta_1,\ldots,\theta_M) \tag{4.51}$$

where $p_i(x|\theta_i)$ represents the pdf of each distribution in the mixture and each pdf has a weight w_i such that $\sum_{i=1}^{M} w_i = 1$. This weight indicates the probability of each pdf. When adopted for optical communications, the EM algorithm is implemented with a mixture of MGD. A critical stage in the MGD method that affects the accuracy of BER calculations is to estimate the number of Gaussian distributions for use in the mixture.

4.7.5.1.1 Selection of Number of Gaussian Distributions in MGD Method

The number of Gaussian distributions to be used is estimated by the number of peak and valley pairs in the first and second derivatives of the original data set. This is illustrated in Figure 4.16. This figure is based on the "Heming Lake Pike" example [48,49]. In this example, the data of five age-groups give the lengths of 523 pikes, and they were sampled in 1965 from Heming Lake, Manitoba, Canada. The components are heavily overlapped, and the resultant pdf is obtained with a mixture of five Gaussian distributions (refer to the Figure 4.16a). The number of Gaussian distributions is then estimated from the number of peak and valley pairs in the first and second derivative curves of the original data set. As seen from Figure 4.16b, the first derivative of the mixed pdf shows clearly four pairs of peaks (lightly filled circles) and valleys (darker circles), suggesting that there should be at least four Gaussian distributions contributing to the original pdf. However, by taking the second derivative (Figure 4.16c), it is realized that there are actually up to five contributed Gaussian distributions.

4.7.5.1.2 Steps for Implementing MGD Method to Obtain BERs

- Obtaining the pdf by the normalized histogram of received electrical signals.
- Estimating the number of Gaussian distributions (N_{Gaus}) used for fitting the pdf of the original data set. This selection is based on the guidelines explained earlier.

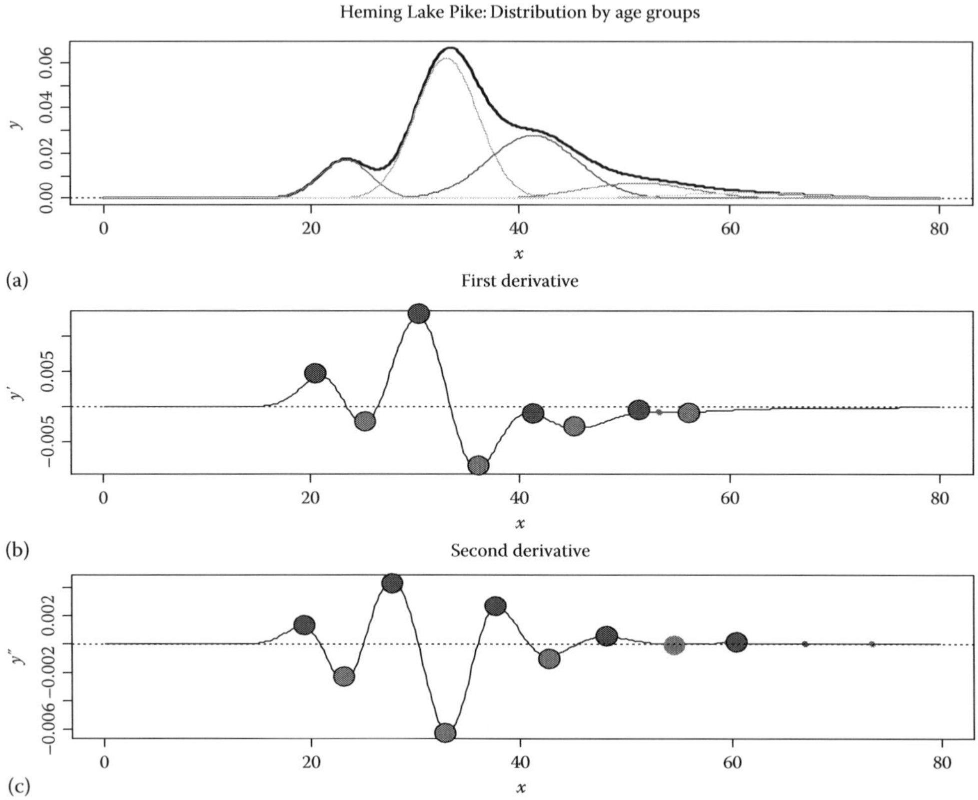

FIGURE 4.16
(a) Process of estimating the number of Gaussian pdfs for MGD fitting based on the number of pairs of inflection peaks (b) and valleys in the first and second derivative curves (c). (Courtesy of E.F. Glynn, Mixtures of Gaussians, Stowers Institute for Medical Research, Kansas City, MO, February 2007.)

- Implementing the EM algorithm in MATLAB with the mixture of N_{Gaus} Gaussian distributions to calculate values of mean, variance, and weight for each distribution.
- Calculating BERs by integrating the tails of these Gaussian distributions when these tails cross the threshold value.

4.7.5.2 Generalized Pareto Distribution Method

GPD is a subset of the GEV theorem that consists of GEV and GPD distributions. Both methods can be used to determine the probabilities of extreme events occurring in the tails of the data pdf. However, the key difference between these two methods is that GEV requires the whole signal pdf whereas GPD only needs the tail regions of the signal pdf. There is only one report on the application of the GEV theorem in fiber-optic communications [50]. However, this chapter studies the GEV distribution for an OOK optical system and involves noise effects while neglecting effects of fiber impairments. Moreover, similar to the single Gaussian distribution, GEV distribution fails to estimate precisely the multi-peak pdf of received signals, which are mainly caused by waveform distortions.

When nonlinearity is the dominant shortcoming to the performance of optical transmission systems, sampled received signals usually introduce a long-tailed pdf. This differs from the Gaussian pdf that has slow roll-off tails. As a result, the conventional BER based on the assumption of the Gaussian pdf is no longer valid, and often underestimates the BER.

A wide range of analytical techniques have recently been studied for optical communications such as importance sampling, multi-canonical, and covariance matrix methods [51–54]. Although these techniques provide precise BERs, they are quite complicated. In contrast, the GPD method has been widely used in various fields [41,42], and it has become available in recent MATLAB versions (since MATLAB version 7.1). Thus, GPD provides the potential of a fast and convenient method for evaluating the system performance in both practical and simulation scenarios.

The pdf for the GPD fitting function is defined as follows [55]:

$$y = f(x|k,\sigma,\theta) = \left(\frac{1}{\sigma}\right)\left(1 + k\frac{(x-\theta)}{\sigma}\right)^{-1-\frac{1}{k}}$$

$$\text{for } \theta < x \text{ when } k > 0 \quad \text{or} \quad \text{for } \theta < x < \frac{-\sigma}{k} \; k < 0 \tag{4.52}$$

where

k is the shape parameter ($k \neq 0$)
σ is the scale parameter
θ is the threshold parameter

Key constraints for (4.52) are described here:

- When $k>0$ and $\theta<x$, there is no upper bound for x.
- When $k<0$ and $\theta<x<-\sigma/k$, zero probability for the case $x>-\sigma/k$.
- When $k=0$, (4.52) changes to

$$y = f(x|0,\sigma,\theta) = \left(\frac{1}{\sigma}\right)e^{-\frac{(x-\theta)}{\sigma}} \quad \text{for } \theta < x \tag{4.53}$$

- When $k=0$ and $\theta=0$, GPD is equivalent to the exponential distribution.
- When $k>0$ and $\theta=\sigma$, GPD is equivalent to the Pareto distribution.

Accordingly, the GPD method has three basic classes of the underlying distributions:

- Distributions whose tails decrease exponentially, such as the normal distribution, have shape parameters equal to zero.
- Distributions with tails decreasing as a polynomial, such as Student's t distribution, lead to a positive shape parameter.
- Distributions having finite tails, such as the beta distribution, have negative shape parameters.

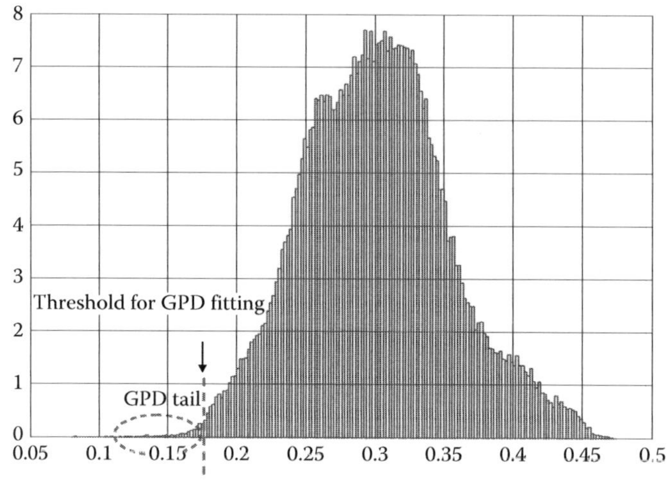

FIGURE 4.17
Selection of threshold for GPD fitting.

The first step and also the most critical step affecting the accuracy of BER calculations is to find the precise threshold for the GPD fitting function.

4.7.5.2.1 Selection of GPD Threshold

The threshold value (V_{GPD}) used for the GPD fitting function indicates the start of the GPD. There have been several suggested guidelines to aid the selection of the GPD threshold [56–58]. However, they are either too complicated or not applicable in optical communications.

In this research, simple guidelines to determine the GPD threshold value are proposed. These guidelines are based on the observation that extreme values in the long-tail region normally comply with a slow exponential slope compared to a faster decaying slope when they are close to the peak of the distribution. The interception region of these two slopes gives a good estimated region for the GPD threshold value (see Figure 4.17).

The accuracy in the selection of V_{GPD} is evaluated by using the cumulative density function (cdf) and the quantile–quantile plot (Q–Q plot). If there is a high correlation between the distribution of the tail of the original data set and the implemented GPD function, there should be a good fit between the empirical cdf of the original data set and the GPD-fitted cdf (see Figure 4.18). Furthermore, this high correlation is also reflected in a linear trend of Q–Q plot, as observed in Figure 4.19.

Figures 4.20 and 4.21 demonstrate the inaccuracy of the GPD method caused by an improper selection of V_{GPD}. Figure 4.20 shows the discrepancy between the GPD-fitted cdf and the empirical cdf of data, while Figure 4.21 clearly displays a nonlinear trend of the Q–Q plot instead of a linear trend.

4.8 MATLAB® and Simulink® Modeling Platforms

Simulation package can be developed using MATLAB and Simulink platforms for modeling advanced digital optical transmission systems. The modeling platform developed in

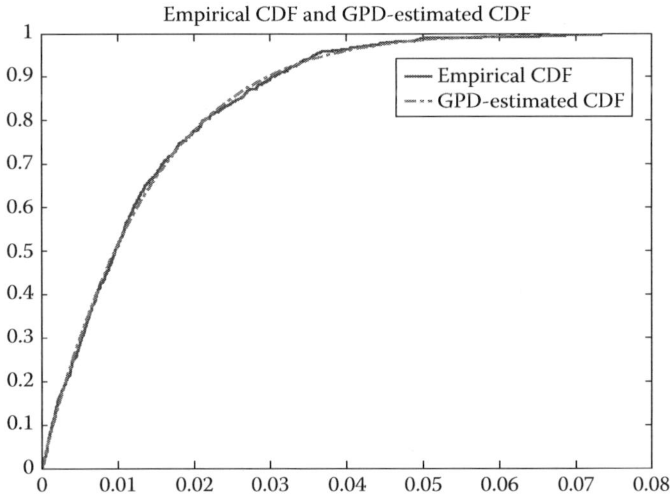

FIGURE 4.18
Demonstration of high correlation between GPD-fitted and empirical cdf.

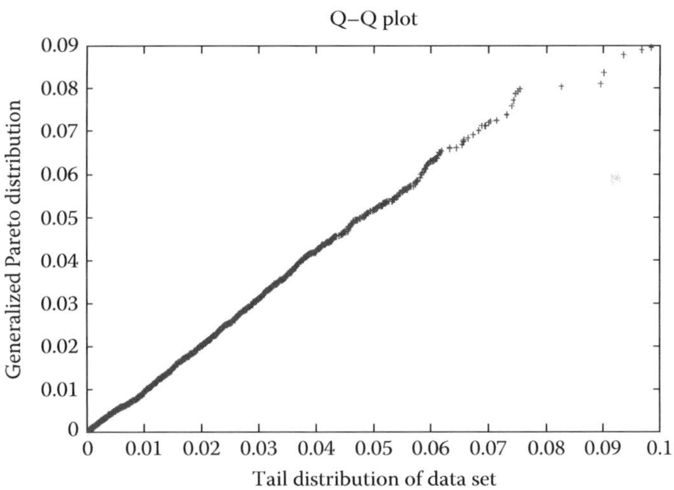

FIGURE 4.19
Q–Q plot of a high correlation GPD fitting function.

this research mainly aims to investigate and verify benefits and shortcomings of advanced modulation formats used in fiber-optic communications. Thus, single-channel optical transmission systems are of main interest in this research. The MATLAB and Simulink modeling platforms have several advantages as follows:

- The simulator provides toolboxes and block-sets for setting up any complicated transmission configurations. In addition, the initialization process for all key parameters of subsystem components can be automatically conducted at the start of any simulation. Furthermore, the initialization file is written in a separate MATLAB file; hence, the simulation parameters can be modified easily.

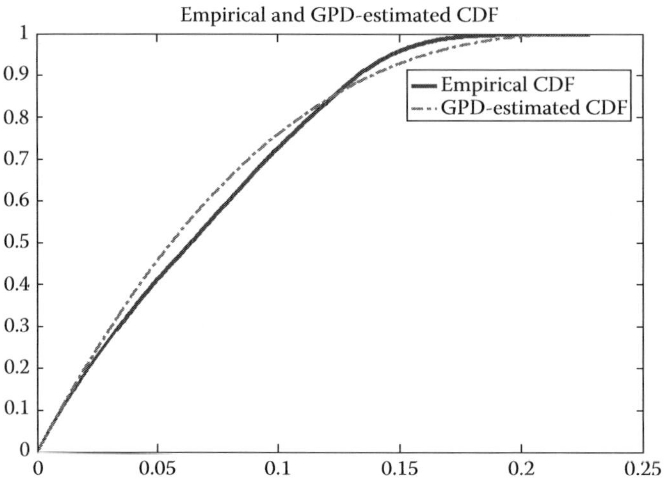

FIGURE 4.20
Demonstration of incompliance between the empirical cdf and the GPD-fitted cdf due to improper selection of V_{GPD}.

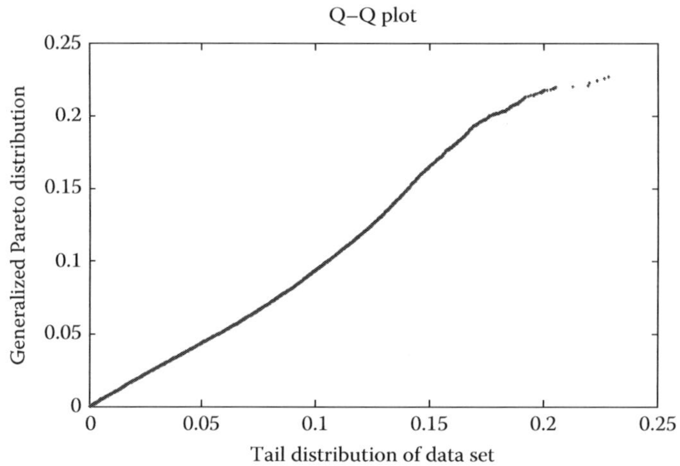

FIGURE 4.21
Nonlinear trend in the Q–Q plot caused by an improper selection of V_{GPD}.

- Signal monitoring can be carried out easily at any point along the propagation path in a simulation with simple plug-and-see monitoring scopes provided by Simulink block-sets.
- Numerical data can be easily stored for post-processing in MATLAB. This offers a complete package for generating and processing numerical data for the achievement of BER.

A generic arrangement of an optical fiber transmission system is shown in Figure 4.22. Each major block in this figure represents a subsystem of the transmission model including an optical transmitter, fiber spans and a propagation model, and an optical

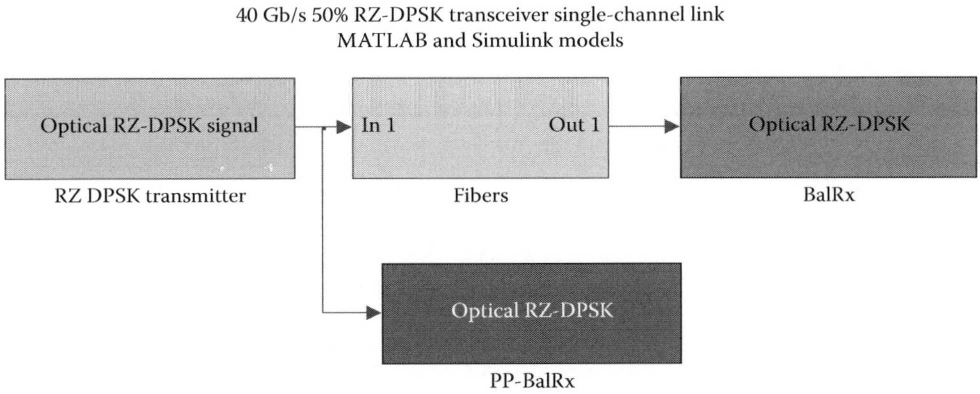

FIGURE 4.22
Overview of an optical fiber transmission system composed of an optical transmitter generating RZ-DQPSK modulation format, fiber propagation model including optical amplifiers and DCFs, and optical receiver at the transmission end and monitoring at the output of the transmitter.

receiver and signal monitoring, as shown in Figures 4.23 through 4.25. The reverse is true. That means when the model of each subsystem development is complete, one can combine and form a subsystem using "create subsystem" command in Simulink. The data inputs can be loaded to the Simulink model by using an initial initialization file, as shown in Figure 4.26. This initialization can be included in the Simulink model using the "model property" and "call back" in the PreLoadFcn, PostLoadFcn and InitFcn, as shown in Figure 4.27. SSFM files or any MATLAB files required for the split-step Fourier propagation along fibers or for any blocks of the Simulink model can also be included here.

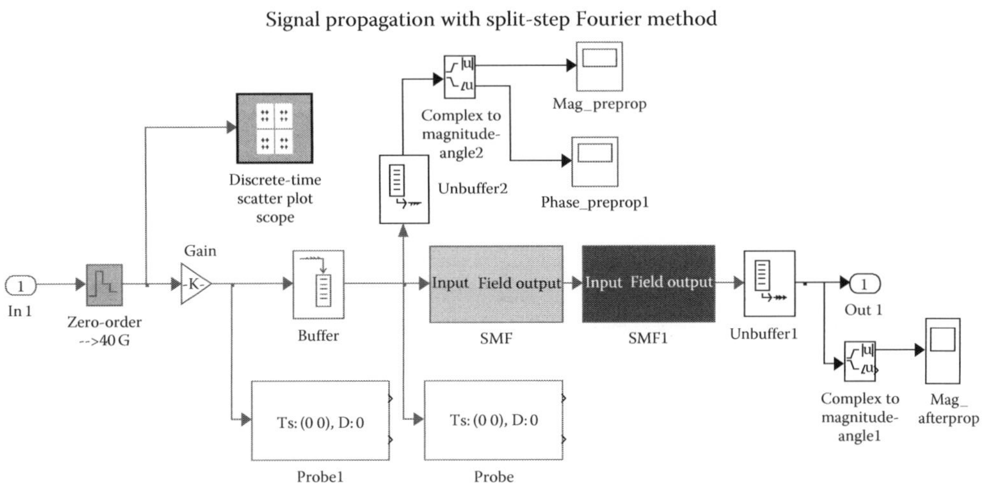

FIGURE 4.23
Fiber propagation model with transmission and dispersion-compensating fiber without optical amplifiers.

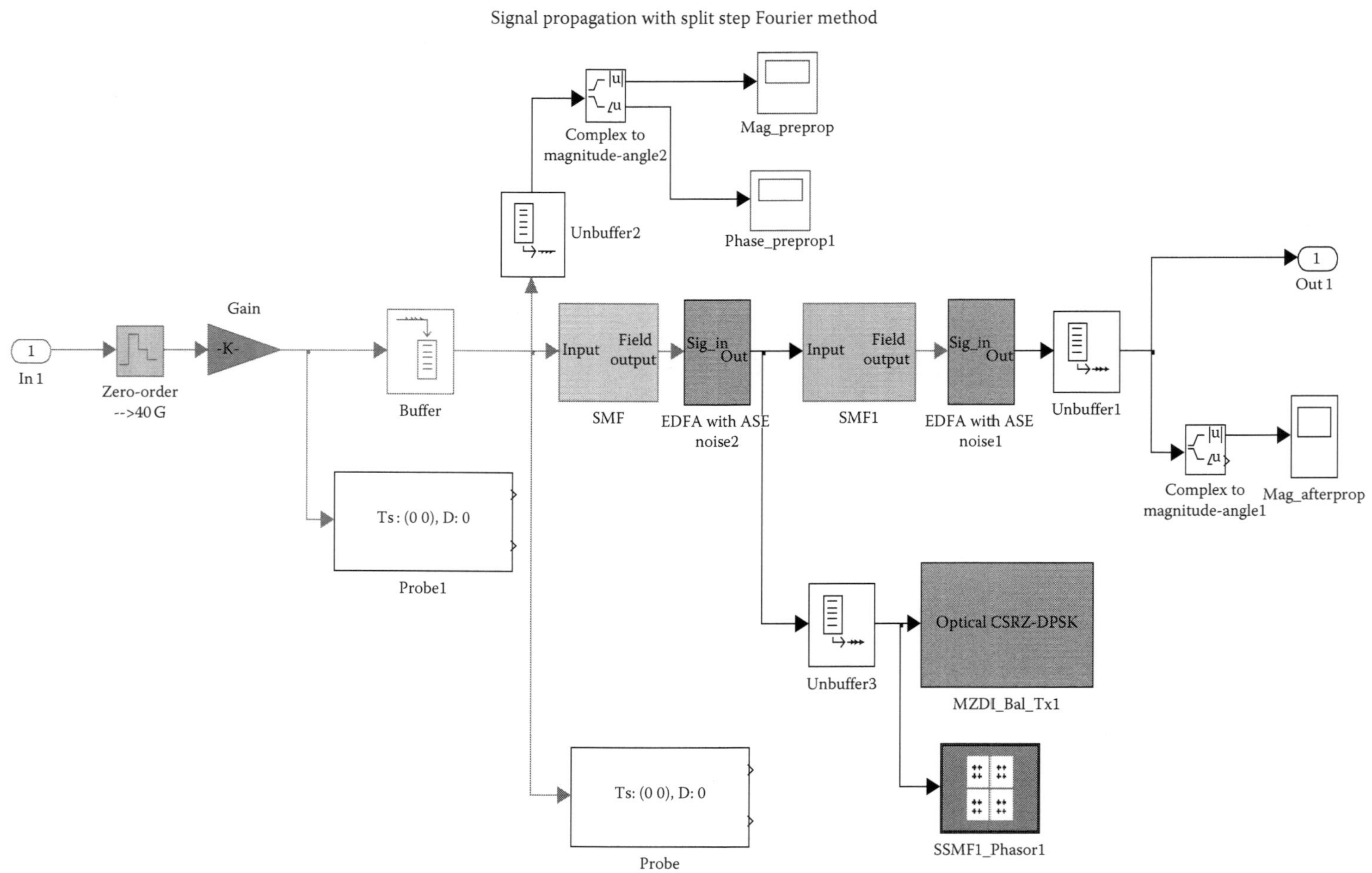

FIGURE 4.24
Fiber propagation model with transmission and dispersion-compensating fiber with optical amplifiers integrated after both transmission and dispersion-compensating fibers.

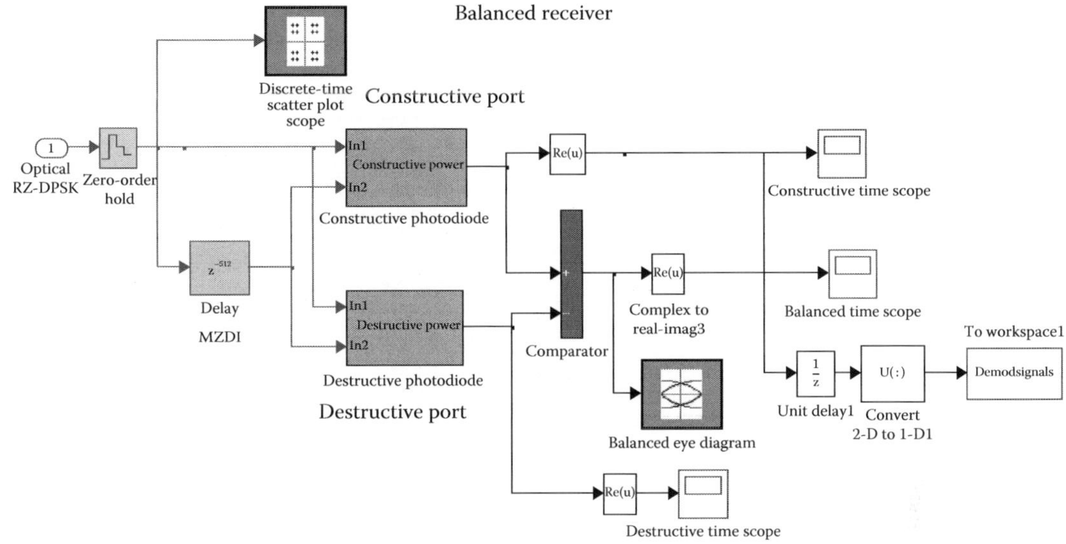

FIGURE 4.25
Balanced optical receiver with a phase comparison at the input to obtained constructive and destructive interfered modes for identification of the "1" and "0" of the phase difference of consecutive bits in differential coded sequence.

```
% Initialization file for data and parameters transfer to Simulink model
% This file also includes parameters required for solving the NLSE for
% pulse propagation in an optical fiber using the split-step
% Fourier method as given in: GP. Agrawal,"Nonlinear Fiber Optics", 2nd ed., AP
% Agrawal, 1995, Chapter 2 - see Ref [3]
%
% The following effects are included in the model: group
% velocity dispersion (GVD), GVD-slope / third-order
% dispersion, loss, and self-phase modulation (n2). The core
% routine is implementing the split-step propagation
clear all
close all
% CONSTANTS
c=3e8;                          % speed of light (m/s)
% NUMERICAL PARAMETERS
numbitspersymbol=1;
P0=0.003162;                    % peak power (W)
FWHM=25                         % pulse width FWHM (ps)
%halfwidth=FWHM/1.6651          % for Gaussian pulse
halfwidth=FWHM                  % for square pulse
bitrate=1/halfwidth;            % THz
baudrate=bitrate/numbitspersymbol;
signalbandwidth=baudrate;
%%%%%%%%%%%%%%%%%%%%%%%%%%%%%%%%
%%%%%%%%%%%%%%%%%%%%%%%%%%%%%%%%
  PRBSlength=2^5;
% Make sure : FFT time window (=nt*dt)=PRBSlength * FWHM...
% FFTlength nt=PRBSlength/block * numbersamples/bit=PRBSlength * (FWHM/dt)
```

```
% num_samplesperbit=FWHM/dt should be about 8 - 16 samples/bit
num_samplesperbit=64;                  % should be 2^n
dt=FWHM/num_samplesperbit ;            % sampling time(ps); % time step (ps)
nt=PRBSlength*num_samplesperbit;       % FFT length
dz=0.1;                                % distance step-size (km)
nz=10;                                 % number of z-steps
maxiter=20;                            % max # of iterations
tol=1e-5;                              % error tolerance
% OPTICAL PARAMETERS
nonlinearthreshold=0.005;      % 5mW - % Nonlinear Threshold Peak Power for silica
                               % core fibre
lambda=1550;
lambda_carrier=14648.4375;     % wavelength (nm)with Level 4 group the carrier freq
                               % is scaled down to 500G or 200GHz
optical_carrier=c/(lambda_carrier*1e-9);     %carrier freq
%dBperkm=0.2;                                % loss (dB/km)
alpha_indB=0.001;                            % loss (dB/km)
D=17;                                        % NZDSF GVD (ps/nm.km); if anomalous
                                             % dispersion(for compensation),D is
                                             % negative
%D=17;                                       % SSMF GVD (ps/nm.km); if anomalous
                                             % dispersion(for compensation),D is
                                             % negative
beta3=0.3;                                   % GVD slope (ps^3/km)
ng=1.46;                                     % group index
n2=2.6e-20;                                  % nonlinear index (m^2/W)
Aeff=47;                                     % effective area (um^2)
% CALCULATED QUANTITIES
T=nt*dt;                                     % FFT window size (ps) -Agrawal:
                                             % should be about 10-20 times of the
                                             % pulse width
alpha_loss=log(10)*alpha_indB/10;            % alpha (1/km)
beta2=-1000*D*lambda^2/(2*pi*c);             % beta2 (ps^2/km);
%------------------
% beta 3 can be calculated from the Slope Dispersion (S) as follows:]
% Slope Dispersion S=0.092;                  % ps/(nm^2.km)
% beta31=(S - (4*pi*c./lambda.^3))./(2*pi*c./lambda.^2)
%------------------
gamma=2e24*pi*n2/(lambda*Aeff);              % nonlinearity coef (km^-1.W^-1)
t=((1:nt)'-(nt+1)/2)*dt;                     % vector of t values (ps)
t1=[(-nt/2+1:0)]'*dt;                        % vector of t values (ps)
t2=[(1:nt/2)]'*dt;                           % vector of t values (ps)
w=2*pi*[(0:nt/2-1),(-nt/2:-1)]'/T;           % vector of w values (rad/ps)
v=1000*[(0:nt/2-1),(-nt/2:-1)]'/T;           % vector of v values (GHz)
vs=fftshift(v);                              % swap halves for plotting
v_tmp=1000*[(-nt/2:nt/2-1)]'/T;
% STARTING FIELD
L=nz*dz
Lnl=1/(P0*gamma)                             % nonlinear length (km)
Ld=halfwidth^2/abs(beta2)                    % dispersion length (km)
N=sqrt(abs(Ld./Lnl))                         % governing the which one is dominat-
ing: dispersion or Non-linearities
ratio_LandLd=L/Ld                            % if L«Ld -> NO Dispersion Effect
ratio_LandLnl=L/Lnl                          % if L«Lnl -> NO Nonlinear Effect
% Monitor the broadening of the pulse with relative the Dispersion Length
% Calculate the expected pulsewidth of the output pulse
% Eq 3.2.10 in Agrawal "Nonlinear Fiber Optics" 2001 pp67
```

```
FWHM_new=FWHM*sqrt(1+(L/Ld)^2);
% N«1 -> GVD ; N»1 —> SPM
Leff=(1 - exp(-alpha_loss*L))/alpha_loss;
expected_normPout=exp(-alpha_loss*2*L);
NlnPhaseshiftmax=gamma*P0*Leff ;
betap=[0 0 beta2 beta3]';
% Constants for ASE of EDFA
% PSD of ASE: N(at carrier freq)=2*h*fc*nsp*(G-1) with nsp=Noise
% Figure/2 (assume saturated gain)
%*************** Standdard Constant ******************************
h=6.626068e-34;        %Plank's Constant
%******************************************
```

FIGURE 4.26
A typical initialization file.

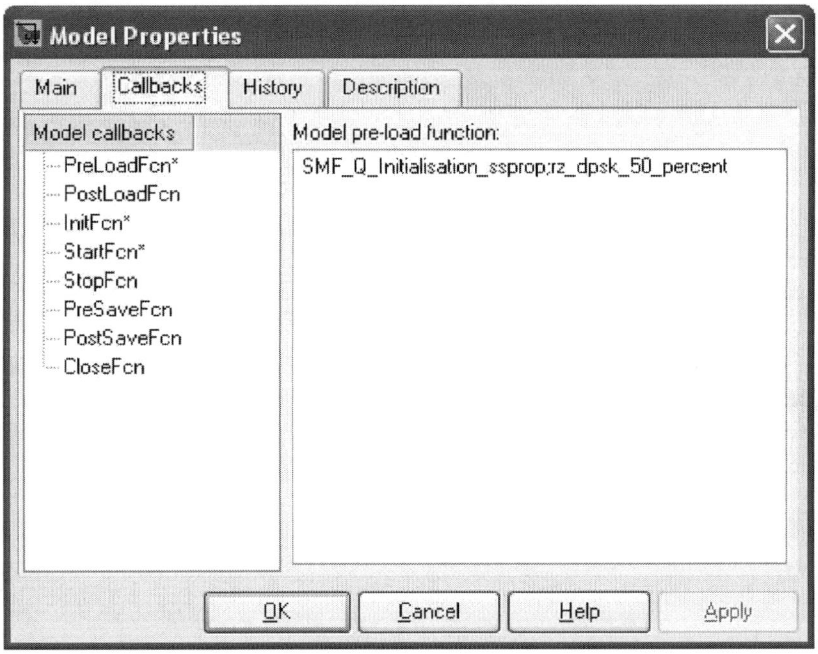

FIGURE 4.27
Model property setup for automatic transfer of parameters in initialization file to Simulink model.

4.9 Concluding Remarks

This chapter presented an overview of modeling of critical subsystem components of a digital photonic transmission system, starting from the optical transmitter, onto the generic properties of optical fibers as a transmission medium, and to the optical receiver. A detailed description of the fiber impairments, including CD, PMD, and Kerr-effect nonlinearities, are provided. Moreover, the symmetrical SSFM is optimized to reduce the computational time and to mitigate artificial errors induced from the windowing effect of FFT/IFFT. The modeling of the ASE noise of optical amplifiers and electronic noises of the receiver are also discussed.

Apart from the conventional evaluation methods, such as Monte Carlo and single Gaussian distribution methods, two statistical methods have been adopted for novel application in optical communications. These are the MGD method implementing the EM theorem and the GPD method based on the GEV theorem. The guidelines for optimizing the accuracy of these two methods were provided. These methods offer fast processing techniques to obtain BER values. Last but not least, MATLAB and Simulink models have been developed for advanced digital photonic transmission systems. These models are provided in the chapter. The Simulink platform takes advantages of the user-friendly MATLAB and Simulink, thus making it easy for further development.

References

1. S. Ten—Corning Inc., Advanced fibers for submarine and long-haul applications, in *Proceedings of IEEE LEOS'04*, Rio Grande, Puerto Rico, 2004, Paper WJ2.
2. S. Ten—Corning Inc., Advanced fibers for submarine networks, in *Proceedings of SubOptic'07* (invited paper), Baltimore, MD, 2007.
3. G.P. Agrawal, *Nonlinear Fiber Optics*, 3rd edn. San Diego, CA: Academic Press, 2001.
4. K.P. Ho, *Phase-Modulated Optical Communication Systems*. New York: Springer, 2005.
5. A.H. Gnauck and P.J. Winzer, Optical phase-shift-keyed transmission, *IEEE Journal of Lightwave Technology*, 23(1): 115–130, 2005.
6. H. Ghafouri-Shiraz, *Distributed Feedback Laser Diodes*. New York: Wiley, 1995.
7. G. Morthier and P. Vankwikelberge, *Handbook of Distributed Feedback Laser Diodes*. Norwood, MA: Artech House, 1995.
8. G.P. Agrawal, *Fiber-Optic Communication Systems*, 3rd edn. New York: Wiley, 2002.
9. M. Lax, Rate equations and amplitude noise, *IEEE Journal of Quantum Electronics*, 3(2): 37–46, 1967.
10. I.P. Kaminow and T. Li, *Optical Fiber Communications*, vol. IVA. San Diego, CA/New York: Academic Press/Elsevier Science, 2002, Chapter 6.
11. T. Kawanishi, T. Sakamoto, and M. Izutsu, High-speed control of lightwave amplitude, phase, and frequency by use of electrooptic effect, *IEEE Journal of Selected Topcis in Quantum Electronics*, 13(1): 92–103, 2007.
12. G.P. Agrawal, *Fiber-Optic Communication Systems*, 3rd edn. New York: Wiley, 2002, Chapter 3.
13. L. Thylen, Integrated optics in LiNbO3: Recent developments in devices for telecommunications, *IEEE Journal of Lightwave Technology*, 6: 847–861, June 1988.
14. SHF Communication Technologies AG, Operating Manual, DPSK Optical Transmitter—SHF 5003, SHF Communication Technologies AG, Germany.
15. T. Kawanishi, S. Shinada, T. Sakamoto, S. Oikawa, K. Yoshiara, and M. Izutsu, Reciprocating optical modulator with resonant modulating electrode, *Electronics Letters*, 41(5): 271–272, 2005.
16. R. Krahenbuhl, J.H. Cole, R.P. Moeller, and M.M. Howerton, High-speed optical modulator in $LiNbO_3$ with cascaded resonant-type electrodes, *IEEE Journal of Lightwave Technology*, 24(5): 2184–2189, 2006.
17. L.N. Binh, Tutorial Part I on optical systems design, in *Proceedings of ICOCN 2002*, Singapore, November 2002.
18. L.N. Binh, T.L. Huynh, K.Y. Chin, and D. Sharma, Design of dispersion flattened and compensating fibers for dispersion-managed optical communications systems, *International Journal of Wireless and Optical Communications*, 2(1): 63–82, June, 2004.
19. J.B. Jeunhomme, *Single Mode Fibre Optics, Principles and Applications*, 2nd edn. New York: Marcel Dekker Pub., 1990.
20. J.A. Buck, *Fundamentals of Optical Fibers*. New York: Wiley, 1995.

21. I.P. Kaminow and T. Li, *Optical Fiber Communications*, Vol. IVB. New York: Elsevier Science, 2002, Chapter 5.
22. J.P. Gordon and H. Kogelnik, PMD fundamentals: Polarization mode dispersion in optical fibers, *PNAS*, 97(9): 4541–4550, April 2000.
23. Corning. Inc., An introduction to the fundamentals of PMD in fibers, White Paper, July 2006.
24. A. Galtarossa and L. Palmieri, Relationship between pulse broadening due to polarisation mode dispersion and differential group delay in long singlemode fiber, *Electronics Letters*, 34(5): 492–493, March 1998.
25. J.M. Fini and H.A. Haus, Accumulation of polarization-mode dispersion in cascades of compensated optical fibers, *IEEE Photonics Technology Letters*, 13(2): 124–126, February 2001.
26. A. Carena, V. Curri, R. Gaudino, P. Poggiolini, and S. Benedetto, A time-domain optical transmission system simulation package accounting for nonlinear and polarization-related effects in fiber, *IEEE Journal on Selected Areas in Communications*, 15(4): 751–765, 1997.
27. S.A. Jacobs, J.J. Refi, and R.E. Fangmann, Statistical estimation of PMD coefficients for system design, *Electronics Letters*, 33(7): 619–621, March 1997.
28. I. Kaminow and T. Koch, *Optical Fiber Communications IIIA*. San Diego, CA: Academic Press, 1997.
29. J. Leibrich and W. Rosenkranz, Efficient numerical simulation of multichannel WDM transmission systems limited by XPM, *IEEE Photonics Technology Letters*, 15(3): 395–397, 2003.
30. D. Marcuse, A.R. Chraplyvy, and R.W. Tkach, Dependence of cross-phase modulation on channel number in fiber WDM systems, *IEEE Journal of Lightwave Technology*, 12(5): 885–890, 1994.
31. T. Mizuochi, K. Ishida, T. Kobayashi, J. Abe, K. Kinjo, K. Motoshima, and K. Kasahara, A comparative study of DPSK and OOK WDM transmission over transoceanic distances and their performance degradations due to nonlinear phase noise, *IEEE Journal of Lightwave Technology*, 21(9): 1933–1943, 2003.
32. K. Hoon, Cross-phase-modulation-induced nonlinear phase noise in WDM direct-detection DPSK systems, *IEEE Journal of Lightwave Technology*, 21(8): 1770–1774, 2003.
33. S. Bigo, G. Bellotti, and M.W. Chbat, Investigation of cross-phase modulation limitation over various types of fiber infrastructures, *IEEE Photonics Technology Letters*, 11(5): 605–607, 1999.
34. C. Furst, J.P. Elbers, C. Scheerer, and C. Glingener, Limitations of dispersion-managed DWDM systems due to cross-phase modulation, in *Proceedings of Annual Meeting LEOS'00*, vol. 1, Rio Grande, Puerto Rico, 2000, pp. 23–24.
35. H.J. Thiele, R.I. Killey, and P. Bayvel, Influence of transmission distance on XPM-induced intensity distortion in dispersion-managed, amplified fibre links, *Electronics Letters*, 35(5): 408–409, 1999.
36. J.P. Gordon and L.F. Mollenauer, Phase noise in photonic communications systems using linear amplifiers, *Optics Letters*, 15(23): 1351–1353, December 1990.
37. A.F. Elrefaie and R.E. Wagner, Chromatic dispersion limitations for FSK and DPSK systems with direct detection receivers, *IEEE Photonics Technology Letters*, 3(1): 71–73, 1991.
38. A.F. Elrefaie, R.E. Wagner, D.A. Atlas, and A.D. Daut, Chromatic dispersion limitation in coherent lightwave systems, *IEEE Journal of Lightwave Technology*, 6(5): 704–710, 1988.
39. D. Ye and W.D. Zhong, Improved BER monitoring based on amplitude histogram and multi-Gaussian curve fitting, *Journal of Optical Networking*, 6(6): 584–598, 2007.
40. L. Ding, W.-D. Zhong, C. Lu, and Y. Wang, New bit-error-rate monitoring technique based on histograms and curve fitting, *Optics Express*, 12(11): 2507–2511, 2004.
41. E.J. Bomhoff, *Financial Forecasting for Business and Economics*, Spiral edn. San Diego, CA: Academic Press, 1995.
42. B.B. Brabson and J.P. Palutikof, Test of the generalized Pareto distribution for predicting extreme wind speed, *Journal of Applied Meteorology*, 39: 1627–1640, 2000.
43. T. Schneider, Analysis of incomplete climate data: Estimation of mean values and covariance matrices and imputation of missing values, *Journal of Climate*, 14: 853–871, March 2000.
44. J.G. Proakis, *Digital Communications*, 4th edn. New York: McGraw-Hill, 2001.
45. W.H. Tranter, K.S. Shanmugan, T.S. Rappaport, and K.L. Kosbar, *Principles of Communication Systems Simulation with Wireless Applications*. Upper Saddle River, NJ: Prentice Hall, 2004.

46. A.P. Dempster, N.M. Laird, and D.B. Rubin, Maximum-likelihood from the incomplete data via the EM algorithm, *Journal of Royal Statistics Society*, 39(1): 1–38, 1977.
47. E.F. Glynn, Mixtures of Gaussians, Stowers Institute for Medical Research, Kansas City, MO, February 2007.
48. P.D.M. Macdonald, Analysis of length-frequency distributions, in R.C. Summerfelt and G.E. Hall (eds.) *Age and Growth of Fish*. Ames, IA: Iowa State University Press, 1987, pp. 371–384.
49. P.D.M. Macdonald and T.J. Pitcher, Age-groups from size-frequency data: A versatile and efficient method of analysing distribution mixtures, *Journal of the Fisheries Research Board of Canada*, 36: 987–1001, 1979.
50. Y. Kopsinis, J. Thompson, and B. Mulgrew, Performance evaluation of optical communication systems using extreme value theory, in *Proceedings of IEE Seminar on Optical Fibre Communications and Electronic Signal Processing*, London, U.K., December 2005.
51. Y. Yadin, M. Shtaif, and M. Orenstein, Bit-error rate of optical DPSK in fiber systems by multicanonical Monte Carlo Simulations, *IEEE Photonics Technology Letters*, 17(6): 1355–1357, 2005.
52. W. Pellegrini, J. Zweck, C.R. Menyuk, and R. Holzlohner, Computation of bit error ratios for a dense WDM system using the noise covariance matrix and multicanonical Monte Carlo methods, *IEEE Photonics Technology Letters*, 17(8): 1644–1646, 2005.
53. D. Yevick, Multicanonical communication system modeling-application to PMD statistics, *IEEE Photonics Technology Letters*, 14(11): 1512–1514, 2002.
54. N.B. Mandayam and B. Aazhang, Importance sampling for analysis of direct detection optical communication systems, *IEEE Transactions on Communications*, 43(234): 229–239, 1995.
55. Matlab Helpdesk, Statistical Toolbox, Generalized Pareto Distribution, http://www.mathworks.com/access/helpdesk/help/toolbox/stats
56. R.L. Smith, Extreme value theory, in W. Ledermann (ed.) *Handbook of Applicable Mathematics*. Chichester, U.K.: Wiley, 1989.
57. A.C. Davidson, Models for exceedances over high thresholds, *Journal of Royal Statistics Society*, B52: 393–442, 1990.
58. R.L. Smith and I. Weissman, Estimating the extremal index, *Journal of Royal Statistics Society*, B56: 515–528, 1994.

5

Optical Modulation

This chapter deals with the modulation of lightwave sources by either direct or external modulation. For direct modulation, the generation of the lightwave within the cavity of the laser is manipulated by the injection or withdrawal of electrons, and hence it is called the stimulated emission process. On the other hand, in the case of external modulation, the laser is turned on at all times, and the generated optical continuous waves are then modulated, frequency, phase, or amplitude, through an optical modulator. The properties of modulated signals are described. The uses of these transmitters in optical communication transmission systems are given, especially those for long-haul transmission at a very high bit rate. MATLAB® and Simulink® models are also described.

5.1 Introduction

A photonic transmitter would consist of single or multiple lightwave sources that can be modulated either directly by manipulating the driving current of the laser diode or externally via an integrated optical modulator. These techniques are known as direct and external modulation techniques.

This chapter presents the techniques for the generation of lightwaves and modulation techniques of lightwaves, either directly or externally. Direct modulation is the technique that directly manipulates the stimulated emission from inside the laser cavity via the use of electro-optic effects. In external modulation, the laser is turned on at all times, then the generated lightwaves are coupled to an integrated optical modulator through which the electro-optic effect is used with the electrical traveling waves (signals), and the amplitude or phase of the lightwaves are modulated. Advanced modulation formats have attracted much attention for the enhancement of the transmission efficiency for coherent optical communications since the mid-1980s. Hence, the preservation of the narrow linewidth of the laser source is critical for operation bit rates in the range of several tens of gigabits per second. Thus, external modulation is essential.

For direct modulation, fundamentally, electrical signals inject electronic carriers into the lasing cavity and then convert them to optically modulated lightwave signals, which would then be transmitted through the optical fiber transmission line, the optical fibers.

In this module, a number of practical lasers, such as special lasers, constricted mesa lasers, distributed-feedback (DFB) lasers, and Fabry–Perot (FP) lasers, are modeled for the analysis and simulation of the ultra-long, high-speed optical fiber communication system.

A complete analysis and study of the effect of physical parameters of the laser rate equation on its dynamic behavior are described. The simulated results of the laser output are confirmed with the experimental and analytical results published in various literatures.

For external modulation, three typical types of optical modulators are briefly presented in this chapter, including the modulation of lightwaves using lithium niobate ($LiNbO_3$)

electro-optic modulators. Their operating principles, the physical structures of the devices, the parameters and applications of the devices, and the driving condition for the generation of different modulation formats as well as their impacts on system performance are described. Further details of the modulation of lightwaves to generate different modulation formats, such as phase and frequency shift keying schemes, will be treated in Chapter 6.

5.2 Direct Modulation

5.2.1 General Introduction

The principal optical component of an optical fiber communications transmitter is the optical source. The crucial role of the optical transmitter is to convert an electrical information input signal into its corresponding optical domain, which would then be launched into the optical communication channel, the single-mode optical fiber (SMF). Most long-haul optical communication systems use semiconductor lasers as optical sources by direct modulation or are incorporated with an external modulator for modulation without switching the laser on and off. In a similar manner, our analysis for ultra-long, high-speed optical communication systems examined in this chapter employs semiconductor lasers with direct-intensity modulation.

Laser is a coherent source generated by a stimulated emission process and, therefore, produces a relatively narrow spectral width of emitted light that allows ultra-long, high-speed optical fiber communication systems to operate. Lightwaves emitted from the semiconductor laser can be modulated either directly by controlling the diode-injected current or externally coupled to an optical modulator; normally an electro-optic integrated device is used, through which guided lightwaves are electro-optically modulated by applying a traveling electric wave. The phase-modulated optical lightwaves are combined with non-modulated phase paths resulting in an amplitude-modulated lightwave output [1–5].

Optical systems operating at the gigabit per second range (up to 100 Gb/s) can be directly modulated without using an external modulator [6], while external modulators, such as $LiNbO_3$ Mach–Zehnder modulator and electro-absorption (EA) modulators, can operate up to several tens of gigahertz.

Several designs of semiconductor lasers have been produced to cater for ultra-long, high-speed optical transmission systems, for example, *FP* laser, *DFB* laser, and constricted mesa laser. These are capable of handling approximately 200 Mb/s, 600 Mb/s, and 15 Gb/s system bit rates, respectively [1]. These laser sources are used for modeling the optical transmitter except for the "special" laser source that is modeled in a separate section.

Further understanding of the semiconductor laser acting as an optical transmitter is presented in Sections 5.2.2.3 to 5.2.2.8. Three laser rate equations, namely, the photon rate equation, the carrier rate equation, and the optical phase rate equation, are described for generating temporally coherent lightwaves. The effects of each physical parameter of the rate equation on the laser response are studied and analyzed in Section 5.2.2.5.

For ultra-long, high-speed optical transmission systems, the dynamic response of a laser source is critical for the output characteristic of an optical source. Relaxation oscillation (RO) and switch-on delay are two examples of dynamic response, covered in Section 5.2.2.9. These effects are simulated and discussed in Section 5.4.3.

In order to model a real practical system, a Langevin force (a noise term) is introduced into each rate equation, and noise considerations are given in Section 5.2.2.11. In Appendix, eye diagrams of laser output lightwaves are generated and compared to those obtained in practical systems. Further research and development on the optical semiconductor laser done by various research centers and institutions are summarized in Section 5.7.

5.2.2 Physics of Semiconductor Lasers

Under normal conditions, all semiconductor materials absorb, rather than emit, light. The absorption process can be understood by referring to Figure 5.1, where the energy levels E_1 and E_2 correspond to the ground state and the excited state of atoms, respectively. If the photon energy $h\nu$ of the incident light of frequency υ is about the same as the energy difference, $E_g = E_2 - E_1$, the photon is absorbed by the atom, which ends up in the excited state.

Excited atoms eventually return to their normal "ground" state and emit light in the process. Light emission can occur through two fundamental processes: spontaneous emission and stimulated emission (see Figure 5.1). In a spontaneous emission process, photons are emitted in random directions with no phase relationship, whereas in stimulated emission, the process is initiated by an existing photon, and the emitted photons match the original photon in energy (or frequency), phase, and the direction of propagation. All lasers, including semiconductor lasers, emit light through the process of stimulated emission and are said to emit coherent light. Therefore, over a large number of stimulated emission events, the laser source acts like a linear amplifier [5].

At room temperature, a spontaneous emission always dominates a stimulated emission in thermal equilibrium. Thus, it can never emit coherent light. Therefore, lasers should necessarily operate away from thermal equilibrium by pumping lasers with an external energy source. Even for an atomic system pumped externally, stimulated emission may not be the dominant process, since it has to compete with the absorption process. Thus, population inversion is a prerequisite for laser operation where the atomic density in excited states must be relatively greater than that at the ground level [3].

5.2.2.1 The Semiconductor p–n Junction for Lasing Lightwaves

The p–n junction, as shown in Figure 5.2, is a homojunction type, because the same type of semiconductor material is fabricated on both sides of the junction. The disadvantage of a homojunction is that the electron–hole recombination can occur over a relatively wide

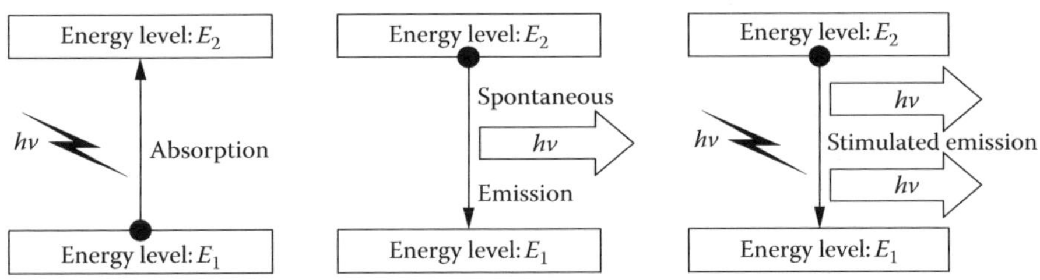

FIGURE 5.1
Three fundamental processes in a semiconductor laser—absorption, spontaneous emission, and stimulated emission—occurring between the two energy states of an atom where E_1 and E_2 are valance band and conduction band, respectively, and $E_2 > E_1$.

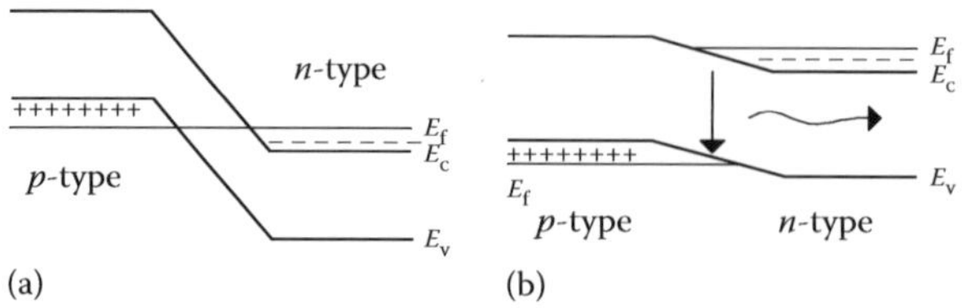

FIGURE 5.2
Energy-band diagram of a p–n homojunction of a semiconductor laser: (a) thermal equilibrium and (b) under forward bias.

region (10 μm), which is determined by the diffusion length of charged carriers (electron and holes). However, the homojunction structure leads to the spatial dispersion of charged carriers and thus non-confinement to the immediate vicinity of the junction; therefore, it is difficult to realize high carrier densities.

The heterojunction in which the carrier confinement occurs as a result of bandgap discontinuity at the junction between two different layers of semiconductors that have the same crystalline structure but different bandgaps is shown in Figure 5.3. The carrier confinement problem for homojunction can thus be resolved by sandwiching this semiconductor material between the p-type and n-type layers. This is shown as the lasing active region. The bandgap of the active region is smaller than the layers surrounding it. Therefore, the electron–hole recombination process only occurs in a relatively narrow region (0.1 μm), and high carrier densities can be realized at a given injection current, which is far better than that for the case of homojunction.

In addition, the refractive index of the active layer is slightly larger than the surrounding layers. As a result of the refractive index difference, the active layer acts as a dielectric waveguide. The main function of this waveguide and active layer is to confine the generated optical energy in the active layer so that resonance can occur to effectively generate lightwaves in the cavity. Thus, the heterostructure p–n junction semiconductor laser is a very efficient device where the optical waves are mostly confined to the active region, and similarly for the carriers, thus maximizing stimulated emission and hence the laser optical gain.

5.2.2.2 Optical Gain Spectrum

Optical gain occurs in the active region when the injected carrier density in the active layer exceeds a certain limit known as the transparency value where population inversion is realized. Spectra distribution of emitted light affects the performance of optical

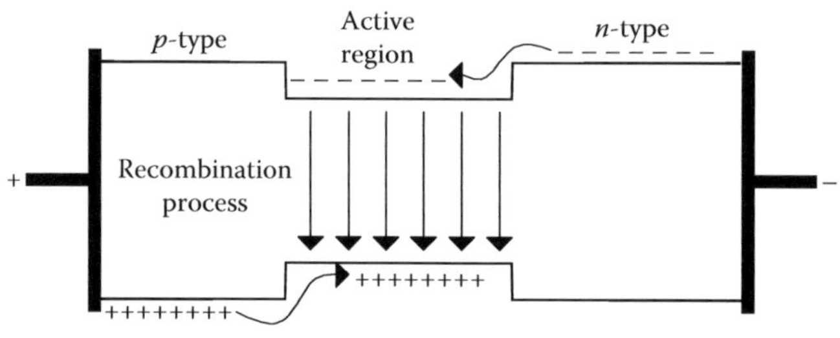

FIGURE 5.3
Energy-band diagram of a heterojunction.

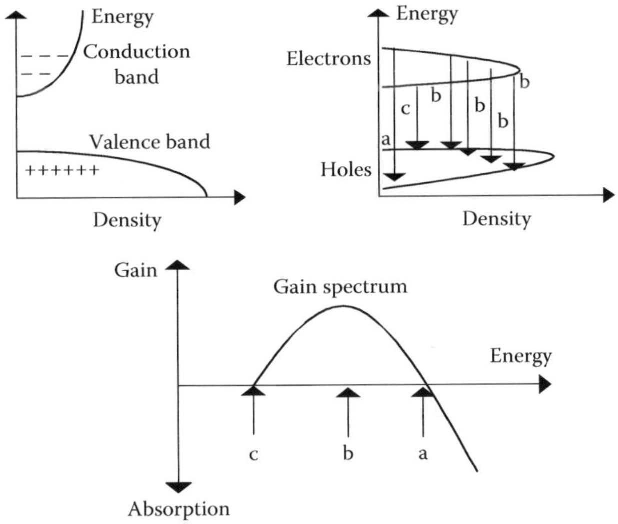

FIGURE 5.4
Optical gain and absorption as a function of bandgap energy.

communication systems through fiber dispersion. The optical gain spectrum is found by considering all possible transitions from conduction to valance band, as shown in Figure 5.4. The width of the gain spectrum ranges from 30 to 100 nm. It can be further reduced by the laser resonance cavity that consists of two partially reflected mirrors placed at the ends of a very short lasing length (in the order of 10–40 ppm).

5.2.2.3 Types of Semiconductor Lasers

Semiconductor lasers are mainly classified by their structure. In order to achieve high bit-rate transmission, one has to narrow the laser spectral width and improve the optical pulse response by minimizing the rise-time constant. Some common types of semiconductor lasers, which are commercially available, include *FP* lasers, *DFB* lasers, distributed Bragg reflector (*DBR*) lasers, single quantum well (*SQW*) lasers, multi-quantum well (*MQW*) lasers, constricted mesa lasers, and buried heterostructure (*BH*) lasers.

5.2.2.4 Fabry–Perot Heterojunction Semiconductor Laser

The optical gain alone is essential but not sufficient for laser operation. The other necessary ingredients are the optical feedback mechanism and the optical guiding in the active layer.

Similar to the resonance in second- and third-order electrical circuitry, the optical energy feedback phenomenon from the two end mirrors located at the end of the guiding cavity, as indicated in Figure 5.5a, is a positive feedback. This would thus stimulate optical resonance and oscillation. This cavity is commonly known as the *FP* cavity. Each modal resonance frequency in and along the cavity is called a longitudinal mode. Since a standing wave is formed along the cavity, it is bound to have several longitudinal modes (due to multiple resonance modes) to exist within the gain spectrum of the device (Figure 5.5). However, improved longitudinal mode selectivity can be achieved using structures that give adequate loss discrimination between the desired mode and all of the unwanted

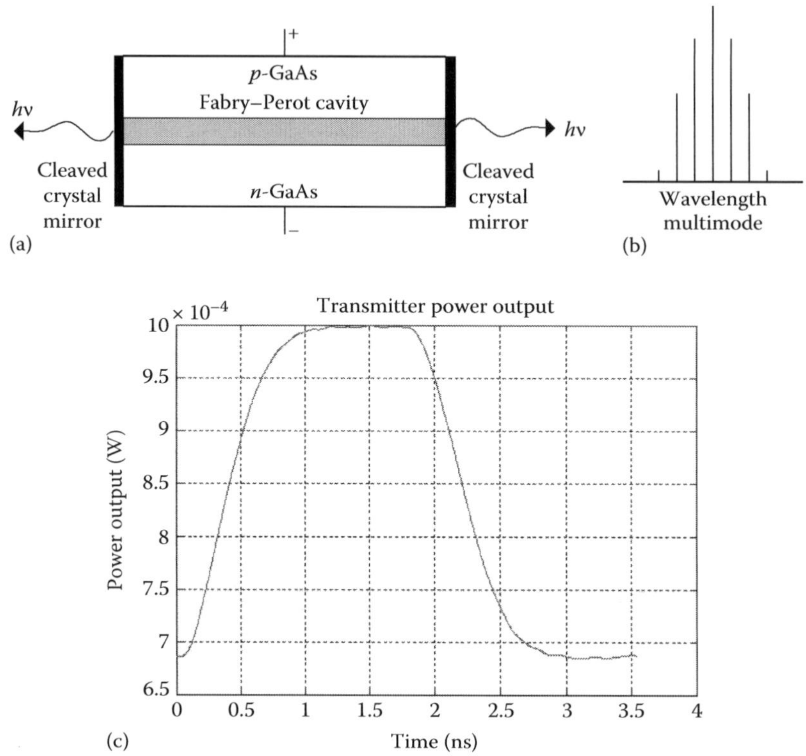

FIGURE 5.5
(a) Schematic diagram of *FP* laser; (b) *FP* laser output spectral; and (c) *FP* laser output response (565MHz) obtained by simulation.

modes of the laser resonator, for example, by using short cavity resonator, coupled cavity resonators [3], and DFB. In Figure 5.5, the broad spectral width (≈5nm) limits the *FP* lasers transmission rate to about 565Mb/s for *NRZ* line coding.

Lightwaves reflected and traveling back and forth and confined in the active layer that is designed and acts as an optical waveguide with the refractive index in the active layer higher than that in the cladding layers. This confinement would give better power conservation and field matching to the fiber pigtail.

5.2.2.5 Distributed-Feedback Semiconductor Laser

In a *DFB* laser, as the name implies, optical energy feedback mechanism is not localized at the facets but is distributed throughout the length of the cavity. This is achieved through an internal built-in corrugated grating that leads to a periodic perturbation of the refractive index (Figure 5.6a). Feedback occurs by means of Bragg diffraction that couples the waves propagating in the forward and backward directions. We can now refer to Figure 5.6b, where each impedance boundary causes a reflection, and the grating period (Λ) must satisfy the Bragg condition:

$$\Lambda = m\left(\frac{\lambda_B}{2n_{eq}}\right) \tag{5.1}$$

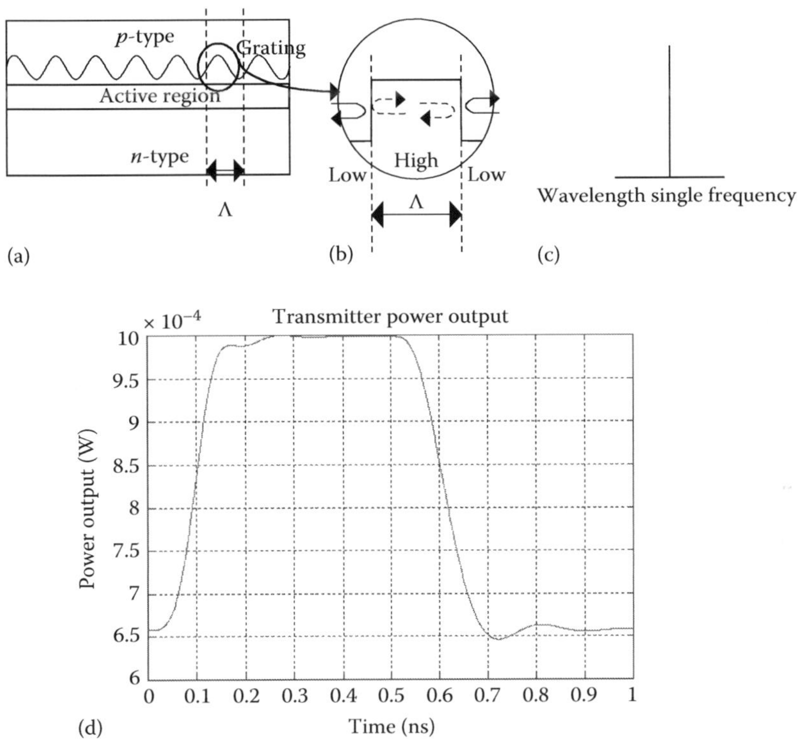

FIGURE 5.6
(a) Schematic diagram of DFB laser; (b) each impedance boundary causes a reflection; (c) DFB laser output spectral; and (d) DFB laser output response (2 GHz) simulated by MOCSS©.

where

n_{eq} denotes the equivalent (or effective, normally derived from the eigenvalue of the wave equation) refractive index of the optically guided waves in the active waveguide cavity

λ_B denotes the operating wavelength

the integer m represents the order of Bragg diffraction

The region of periodically varying refractive index serves to couple two counter-propagating traveling waves. The coupling is a maximum for wavelengths close to λ_B. First-order gratings provide the strongest coupling, but second-order gratings are sometimes used because they are easier to fabricate, with their larger spatial period (Λ). Thus, *DFB* lasers by means of Bragg diffraction could produce single-frequency modes with a very narrow spectral width (≈a few hundred megahertz). The modulation rate of *DFB* lasers can be about 600 Mb/s to 2 Gb/s for *NRZ* line coding without broadening its linewidth.

5.2.2.6 Constricted Mesa Semiconductor Laser

Constricted mesa laser is a component of the family of index-guided lasers. The index-guided laser structure is modified from the gain-guided laser [1] that eliminates the problems of having "kinks," astigmatism, unstable far-field patterns, and self-pulsations [1]. In other words, this is to linearize the optical power-driven current (*P–I*) characteristic of a laser, and hence reduce the distortion of the emitted laser pulse.

Real refractive index variation is introduced into the lateral structure of the laser to achieve this purpose. In a high-speed modulation laser, parasitic capacitance could be the limiting factor in achieving excellent high-speed transmission. In a laser with significant parasitic capacitance, the electrical pulse reaching the laser is broadened, and the capacitance provides a source of current during the time when the photon density is high. Consequently, the laser output may consist of two or more pulses as the electron density is repetitively built up and extinguished (ROs in Section 5.2.2.9). Thus, constricted mesa lasers are fabricated with current-blocking layers by the regrowth of semi-insulating materials, SiO_2 or SiN_x, as shown in Figure 5.7. Hence, the modulation speed of 20 GHz for the *NRZ* line-coding format is achieved in modeling the simulation package.

5.2.2.7 Special Semiconductor Laser Source

In addition to the above-mentioned lasers available for MOCSS simulation, a "special" semiconductor laser has been modeled, which is tailored for the purpose of analysis of the high-speed, ultra-long optical fiber communication systems. An optical confinement factor of 0.9 (which is remarkably higher than that of a normal laser) and a narrow spectral linewidth of 0.05 nm enable it to modulate very high-speed signals up to 20 GHz without any observable defects. At around 2.5 GHz, it provides a smooth running response with slight RO (in Section 5.2.2.9) and overshooting (Figure 5.8). Thus, it produces a reasonable optical pulse shape for us to analyze, especially at the fiber output and the receiver terminal.

5.2.2.8 Single-Mode Optical Laser Rate Equations

The operating characteristics of semiconductor lasers are well described by a set of rate equations that govern the interaction of photons and electrons inside the active region. A rigorous derivation of these rate equations generally starts from Maxwell's equations together with a quantum-mechanical approach for the induced polarization [10]. However, the rate equations can also be obtained heuristically by considering the physical phenomena

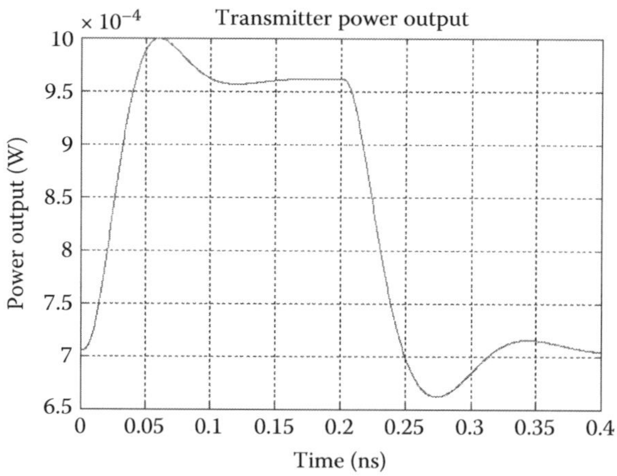

FIGURE 5.7
Constricted mesa laser output response (5 GHz) from MOCSS* simulation.

* MOCSS = Monash Optical Communication Systems Simulator.

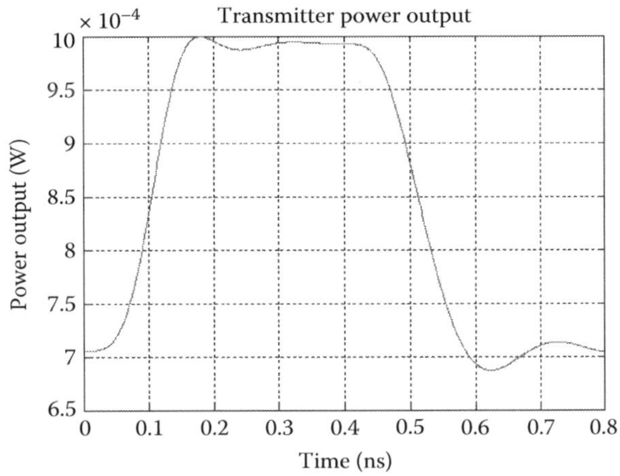

FIGURE 5.8
Special laser output response (2.5 GHz) by simulation.

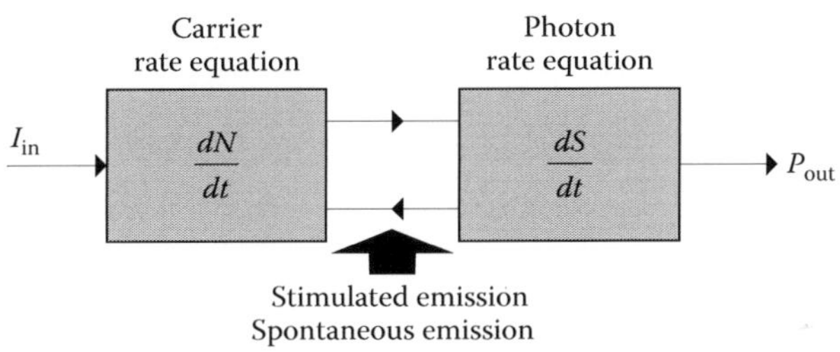

FIGURE 5.9
Single-mode laser model as expressed by the rate equations (Equations 5.2 through 5.4).

through which the number of photons S and the number of electrons N change with time inside the active region, as illustrated in Figure 5.9.

For a single-mode laser, three rate equations are given in Equations 5.2 through 5.4. These rate equations can be used for the computer simulation of the frequency chirp and output power waveform:

$$\frac{dN(t)}{dt} = \frac{I(t)}{qV_a} - \frac{N(t)}{\tau_n} - v_g a_0 \frac{N(t) - N_0}{1 + \varepsilon S(t)} S(t) \tag{5.2}$$

$$\frac{dS(t)}{dt} = \left(\Gamma a_0 v_g \frac{N(t) - N_0}{1 + \varepsilon S(t)} - \frac{1}{\tau_p} \right) S(t) + \frac{\beta \Gamma N(t)}{\tau_n} \tag{5.3}$$

$$\frac{d\phi_m(t)}{dt} = \frac{\alpha}{2} \left(\Gamma v_g a_0 \left(N(t) - N_0 \right) - \frac{1}{\tau_p} \right) \tag{5.4}$$

The parameters of these rate equations are as follows:

Γ is the optical confinement factor
v_g is the guided lightwave group velocity

a_0 is the gain coefficient
N_0 is the carrier density at transparency
ε is the gain compression factor
τ_p is the photon lifetime
β is the fraction of spontaneous emission coupled to the lasing mode
τ_n is the electronic carrier lifetime
q is the electronic charge
V_a is the active (lasing) layer volume
α is the optical linewidth enhancement factor
ϕ_m is the optical phase

In Equation 5.2, the electron density, $N(t)$, increases due to the injection of a current, $I(t)$, into the active layer volume, V_a, and decreases due to stimulated and spontaneous emission of photon density, $S(t)$. Similarly, in Equation 5.3, the photon density $S(t)$, is increased by stimulated and spontaneous emission, $S(t)$, and decreased by internal and mirror losses with a photon lifetime, τ_p, given by

$$\tau_p = \frac{1}{v_g(\alpha_i + \alpha_m)} \tag{5.5}$$

where α_i and α_m denote waveguide loss and mirror loss, respectively. The carrier lifetime, τ_n, is related to the loss of electrons due to both spontaneous emission and non-radiative recombination [3], given by

$$\tau_n = \frac{N(t)}{R_{sp} + R_{nr}} \tag{5.6}$$

where
R_{sp} denotes the rate of spontaneous emission
R_{nr} denotes non-radiative recombination

The time variations of the optical power output [11] is given by

$$m(t) = \frac{S(t)V_a\eta_0 h\nu}{2\Gamma\tau_p} \tag{5.7}$$

where
$h\nu$ denotes the photon energy
η_0 denotes the total differential quantum efficiency

5.2.2.9 Dynamic Response of Laser Source

The dynamic behavior of the injection laser is critical, especially when it is used in ultra-long, high-speed optical fiber communication systems. The application of a step current to the device results in a switch-on delay, often followed by high-frequency damped oscillations (≈10 GHz) known as *ROs*. *RO* occurs when the electron (as the carrier) and photon populations within the structure come into equilibrium, as shown in Figure 5.10. The input

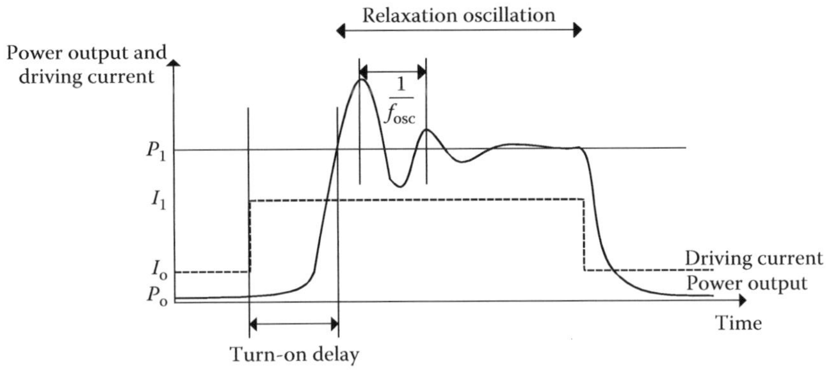

FIGURE 5.10
Typical response to a rectangular drive pulse.

electrical pulse causes the electron density to rise to a maximum, which is maintained during a turn-on delay until a large photon density builds up and depletes the carriers. This behavior is easily seen from the carrier rate equations (Equations 5.2 through 5.4) and is explained in Section 5.3.

In addition, when a current pulse reaches a laser that has significant parasitic capacitance after the initial delay time, the pulse will be broadened because the capacitance provides a source of current over the period during which the photon density is high. The turn-on delay is caused by the initial buildup of the photon density resulting from stimulated emission. It is related to the minority carrier lifetime and the current through the device. It can be reduced by biasing the laser current near the threshold current level (pre-biasing). However, any further increase in laser current will decrease the extinction ratio (P_1/P_0). The resonant or oscillating frequency, f_{osc}, can be increased by increasing I_1. The overshooting can be reduced by shaping the electrical input pulse [8].

5.2.2.10 Frequency Chirp

The direct current modulation of a single longitudinal-mode semiconductor laser can cause a dynamic shift of the peak wavelength emitted from the device [7]. This phenomenon, which results in dynamic linewidth broadening under direct modulation of injection current, is referred to as frequency chirping. It arises from gain-induced variations in the laser refractive index due to the strong coupling between the free carrier density and the index of refraction, which is present in any semiconductor structure [11]. Hence, even small changes in carrier density will result in a phase shift of the optical field, giving an associated change in the resonance frequency within both *FP* and *DFB* laser structures:

$$\Delta \nu(t) = \frac{1}{2\pi}\left(\frac{d\phi_m(t)}{dt}\right) = \frac{1}{2\pi}\left[\frac{\alpha}{2}\left(\Gamma v_g a_0 (N(t) - N_0) - \frac{1}{\tau_p}\right)\right] \tag{5.8}$$

where

$\Delta\nu$ denotes the frequency chirp
$N(t)$ denotes the carrier density
$d\phi_m/dt$ denotes the optical phase rate equation (Equation 5.4)

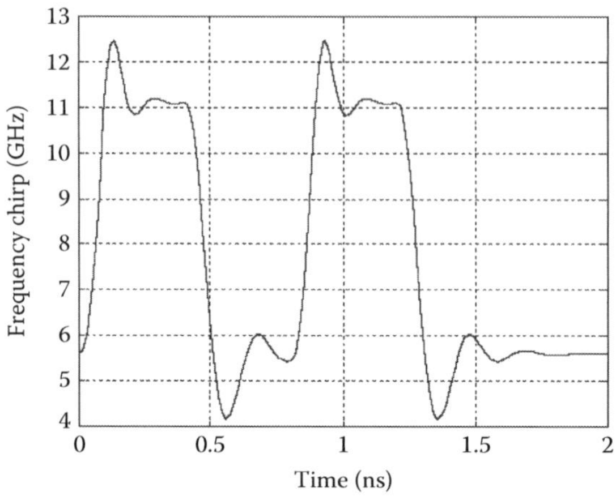

FIGURE 5.11
Frequency chirp effects on directly modulated optical pulses.

The effects of the chirp on optical signals generated from the laser can be observed as shown in Figure 5.11. Overshooting oscillation of the pulse at low and high levels would contribute to higher error rate.

A number of techniques can be employed to reduce the frequency chirp effects by

1. Biasing the laser sufficiently above threshold so that the modulation current does not drive the device below the threshold where the rate of change of optical output power varies rapidly with time. However, this method gives an extinction ratio penalty of the order of several decibels at the receiver.
2. Damping the ROs by shaping the electrical drive pulses that would result in small fluctuations of optical output power [8].
3. Using quantum well lasers, Bragg wavelength detuned *DFB* lasers, and multi-electrode *DFB* lasers, which provides an improvement in frequency chirp. However, it requires a complex fabrication process.
4. Using an external modulator, such as $LiNbO_3$ Mach–Zehnder interferometric amplitude-type modulator, with the laser source emitting continuous lightwaves of appropriate wavelengths or frequencies [9].

5.2.2.11 Laser Noises

The output of a semiconductor laser exhibits fluctuations in its intensity, phase, and frequency chirp (as described in Section 5.2.2.10) even when the laser is biased at a constant current with negligible current fluctuations. This is due to two fundamental noise mechanisms: the spontaneous emission and the electron–hole recombination (shot noises). Usually noises generated in semiconductor lasers are dominated by photons randomly generated by spontaneous emission. The intrinsic intensity fluctuation in a semiconductor laser is a noise source in directly modulated optical communication systems and, in some specific applications, may significantly reduce the signal-to-noise ratio (*SNR*), whereas phase fluctuation leads to a limited spectral linewidth when semiconductor lasers are operated continuously at a constant current. The rate equations can be used to study laser noises by adding an extra noise term, known as the Langevin force, to each of them. Then Equations 5.2 through 5.4 become

$$\frac{dS(t)}{dt} = \left(\Gamma a_0 v_g \frac{N(t) - N_0}{1 + \varepsilon S(t)} - \frac{1}{\tau_p}\right) S(t) + \frac{\beta \Gamma N(t)}{\tau_n} + F_S(t) \tag{5.9}$$

$$\frac{dN(t)}{dt} = \frac{I(t)}{qV_a} - \frac{N(t)}{\tau_n} - v_g a_0 \frac{N(t) - N_0}{1 + \varepsilon S(t)} S(t) + F_N(t) \tag{5.10}$$

$$\frac{d\phi_m(t)}{dt} = \frac{\alpha}{2}\left(\Gamma v_g a_0 \left(N(t) - N_0\right) - \frac{1}{\tau_p}\right) + F_\phi(t) \tag{5.11}$$

where $F_S(t)$, $F_N(t)$, and $F_\phi(t)$ denote Langevin noise sources due to the spontaneous emission, the carrier generation recombination process in photon number, and the generated phase, respectively [12]. These are assumed to be Gaussian random processes with zero mean value under the Markovian assumption (system has no memory), and the correlation function of the form by Markovian approximation is

$$\langle F_i(t) F_j(t') \rangle = 2D_{ij}\delta(t - t') \tag{5.12}$$

where
- $i, j = S, N$, or ϕ
- angular brackets denote the ensemble average
- D_{ij} is called the diffusion coefficient [18] and is listed as follows:

$$D_{SS} = \frac{\beta V_a N_{sd} \left(V_a S_{sd} + 1\right)^3}{\tau_n} \tag{5.13}$$

$$D_{NN} = \frac{V_a N_{sd}}{\tau_n}\left[\beta V_a S_{sd} + 1\right] \tag{5.14}$$

$$D_{\phi\phi} = \frac{R_{sp}}{4S} \tag{5.15}$$

where
- N_{sd} and S_{sd} represent the steady-state average values of the carrier and photon populations, respectively (Equations 5.16 and 5.17)
- R_{sp} denotes the rate of spontaneous emission (Equation 5.18)

$$N_{sd} = \frac{1}{\Gamma a_0 v_g \tau_p} + N_0 \tag{5.16}$$

$$S_{sd} = \frac{\tau_p}{\tau_n} N_{sd}\left(\frac{I}{I_b} - 1\right) \tag{5.17}$$

$$R_{sp} = 2\Gamma v_g \sigma_g \left(N(t) - N_0\right) \tag{5.18}$$

where
I_b denotes bias current
σ_g denotes the gain cross section, which is normally $2 \times 10^{-20}\,\text{m}^2$

By making use of the Wiener–Kinchen theorem, which states that the Fourier transform of the autocorrelation function of a process is equal to the power spectral density of that process, we take the Fourier transform of a delta function as simply a white Gaussian noise. For ergodic random process and since $\overline{F_i}^2 = 0$, the variance of F_i or VAR(F_i) = $2D_{ij}$, (refer to Equation 5.12). Thus, the Langevin force is white Gaussian with a mean of zero and standard deviation of $\sqrt{2D_{ij}}$ given as follows:

$$F_S(t) = N(0,1)\sqrt{2D_{SS}} \tag{5.19}$$

$$F_N(t) = N(0,1)\sqrt{2D_{NN}} \tag{5.20}$$

$$F_\phi(t) = N(0,1)\sqrt{2D_{\phi\phi}} \tag{5.21}$$

where $N(0, 1)$ denotes the Gaussian distributed random process with mean zero and standard deviation of 1, and these three Langevin forces are to be fitted in (5.19) through (5.21). The noise power spectrum $s(f)$ of the photon density as a function of the Fourier frequency, f, is written using the rate equation as follows:

$$s(f) = \frac{\tau_p f_r^4 \left\langle F_N^2 \right\rangle + \Psi^2 \dfrac{\left\langle F_S^2 \right\rangle}{4\pi^2} + \tau_p f_r^4 \Psi \dfrac{\left\langle F_S F_N \right\rangle}{\pi} + \dfrac{\left\langle F_S^2 \right\rangle}{4\pi^2} f^2}{\left(f^2 - f_r^2\right) + \Psi^2 f^2} \tag{5.22}$$

and relative intensity noise (RIN) of laser source is given by

$$\text{RIN} = 10\,\text{Log}_{10} \frac{s(f)^2}{S_{sd}^2} \tag{5.23}$$

where f_r denotes the resonant frequency given by

$$f_r = \frac{1}{2\pi}\sqrt{\frac{1}{\tau_p \tau_n}\left(1 + \Gamma a_0 v_g N_0 \tau_p\right)\left(\frac{I}{I_b} - 1\right)} \tag{5.24}$$

and $\langle F_S F_N \rangle$ denotes the cross-correlation given by

$$\left\langle F_S F_N \right\rangle = -\frac{\beta V_a N_{sd}(V_a S_{sd} + 1)}{\tau_n} + \frac{V_a S_{sd}}{\tau_p} \tag{5.25}$$

and constant Ψ denotes

$$\Psi = \frac{1}{2\pi}\left(\Gamma a_0 v_g S_{sd} + \frac{1}{\tau_n}\right) \tag{5.26}$$

The relative intensity noise is calculated and shown in Figure 5.12a as a function of the modulating frequency. It is observed that there is a resonant peak due to the cavity resonance of the laser waveguide cavity at the modulating frequency/time of the traveling back and forth of the photons in the cavity. Figure 5.12b shows the variation of this resonance with respect to the change of temperature.

Figure 5.13 shows some measurements of the generated laser intensity of a laser with respect to the change of the driving current. There is a threshold current level above which lasing happens. This threshold current also varies when the temperature is increased. This is main reason why one has to control the laser temperature so that the intensity generated from the laser remains stable. Figure 5.14 shows the inside structure of a packaged semiconductor laser. Figure 5.15 shows the circuitry of the control of the driving current to the laser in which a Peltier cooler is included to cool the laser and keep the temperature of the laser at certain level.

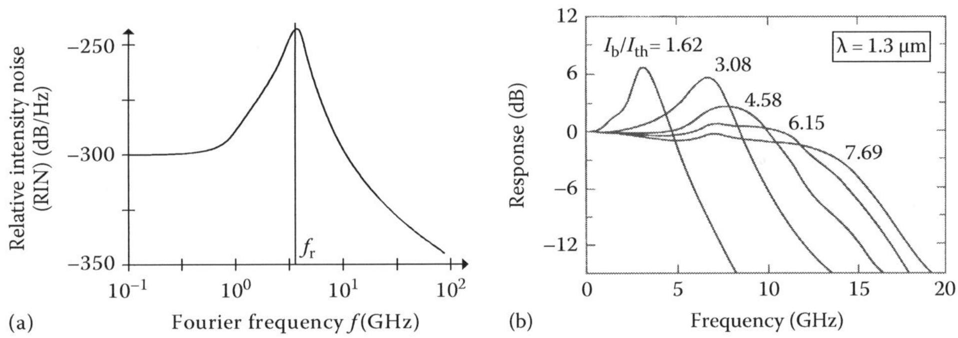

FIGURE 5.12
Relative intensity noise for intensity modulation: (a) modeling using rate equations and (b) measured frequency response.

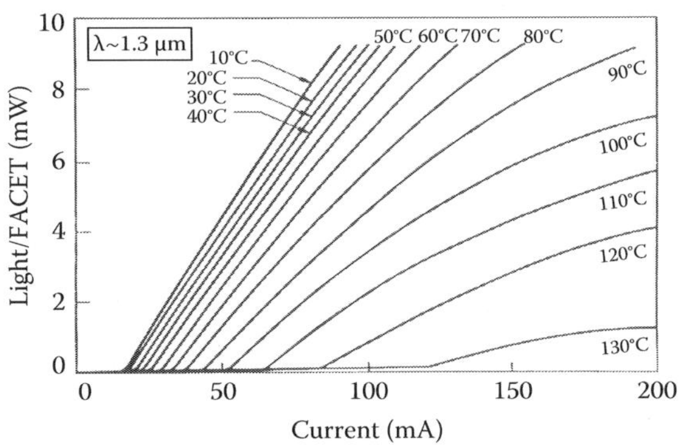

FIGURE 5.13
Laser P–I characteristics with temperature as a parameter.

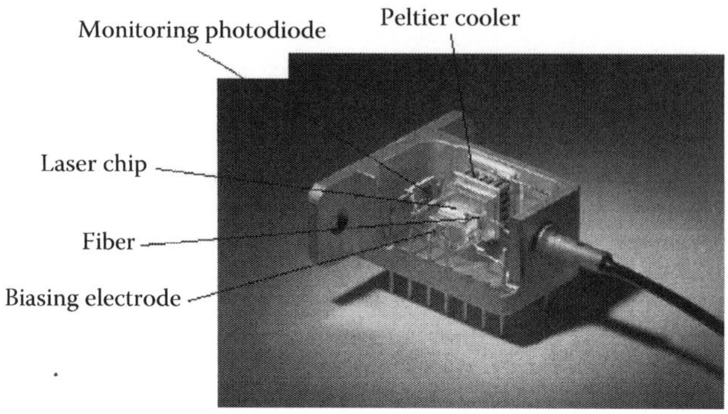

FIGURE 5.14
Packaging of a semiconductor laser.

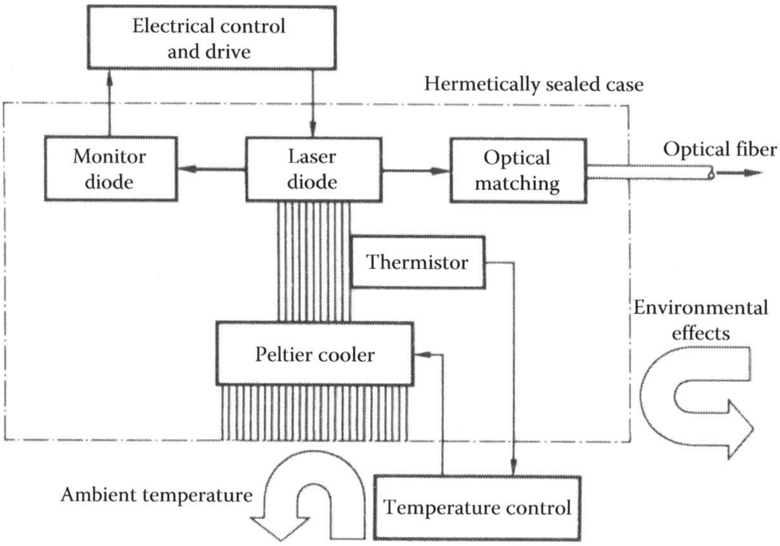

FIGURE 5.15
Block diagram of driving circuitry for a semiconductor laser.

5.3 External Modulation Using Optical Modulators

The modulation of lightwaves via an external optical modulator can be classified into three types depending on the special effects that alter the lightwaves properties, especially the intensity or the phase of the lightwave carrier. In an external modulator, the intensity is normally manipulated by manipulating the phase of the carrier lightwaves guided in one path of an interferometer. The Mach–Zehnder interferometric structure is the most common type [21,25].

An EA modulator employs the Franz and Keldysh effect, which is observed as the lengthening of the wavelength of the absorption edge of a semiconductor medium under the influence of an electric field [22–25]. In quantum structures, such as the MQW structure, this effect is called the Stark effect, or the EA effect. The EA modulator can be integrated with a laser structure on the same integrated circuit chip. For the $LiNbO_3$ modulator, the device is externally connected to a laser source via an optical fiber.

The total insertion loss of a semiconductor intensity modulator is about 8–10 dB including the fiber-waveguide coupling loss, which is rather high. However, this loss can be compensated by a semiconductor optical amplifier (SOA) that can be integrated on the same circuit. Compared with $LiNbO_3$, its total insertion loss is about 3–4 dB, which can be affordable as the Er-doped fiber amplifier (EDFA—see Chapter 9) is now readily available.

The driving voltage for an EA modulator is usually lower than that required for $LiNbO_3$. However, the extension ratio is not as high as that of the $LiNbO_3$-type modulator, which is about 25 dB as compared to 10 dB for the EA modulator. This feature contrasts the operating characteristics of the $LiNbO_3$ and EA modulators. Although the driving voltage for EA modulators is about 4–3 V and that for $LiNbO_3$ modulators is about 5–7 V, the former type would be preferred for the intensity modulation or the phase modulation format due to this extinction ratio that offers much lower "zero" noise level and hence high quality factor.

5.3.1 Phase Modulators

Figure 5.16 shows the schematic of an optical phase modulator using integrated waveguide on lithium niobate (LN).

The phase modulator is a device that manipulates the "phase" of optical carrier signals under the influence of an electric field created by an applied voltage. When a voltage is not applied to the RF electrode, the number of periods of the lightwaves, n, exists in a certain path length. When a voltage is applied to the RF electrode, one or a fraction of one period of the wave is added, which now means that $(n+1)$ waves exist in the same length. In this case, the phase has been changed by 2π, and the half voltage of this is called the driving voltage. In the case of long-distance optical transmission, waveform is susceptible to degradation due to nonlinear effects such as self-phase modulation. A phase modulator can be used to alter the phase of the carrier to compensate for this degradation. The magnitude of the change of the phase depends on the change of the refractive index created via the electro-optic effect that in turn depends on the orientation of the crystal axis with respect to the direction of the electric field established by the applied signal voltage.

An integrated optical phase modulator operates in a similar manner except that the lightwave carrier is guided via an optical waveguide, which is a diffused or ion-exchanged confined region for $LiNbO_3$ modulators and a rib-waveguide structure for semiconductor-type modulators. Two electrodes are deposited so that an electric field can be established across the waveguiding cross section in order that there is a change of the refractive index via the electro-optic or the EA effect, as shown in Figure 5.16. For ultrafast operation, one of the electrodes is a traveling-wave-type or hot electrode, and the other is a ground electrode. The traveling-wave electrode must be terminated with matching impedance at the end so as to avoid wave reflection. Usually, a quarter wavelength impedance is used to match the impedance of the traveling-wave electrode to that of the 50 Ω transmission line.

A phasor representation of a phase-modulated lightwave can be achieved by circular rotation at a radial speed of ω_c. Thus, the vector with an angle ϕ represents the magnitude and the phase of the lightwave.

5.3.2 Intensity Modulators

Figure 5.16b shows the structure of an integrated intensity modulator in which a traveling-wave electrode is placed across the surface of an optical waveguide to create a change in

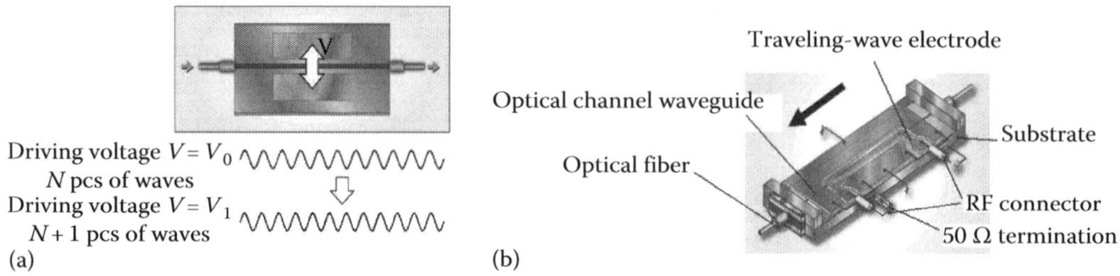

FIGURE 5.16
Electro-optic phase modulation in an integrated modulator using $LiNbO_3$. Electrode impedance matching is not shown: (a) schematic diagram and (b) integrated optic structure.

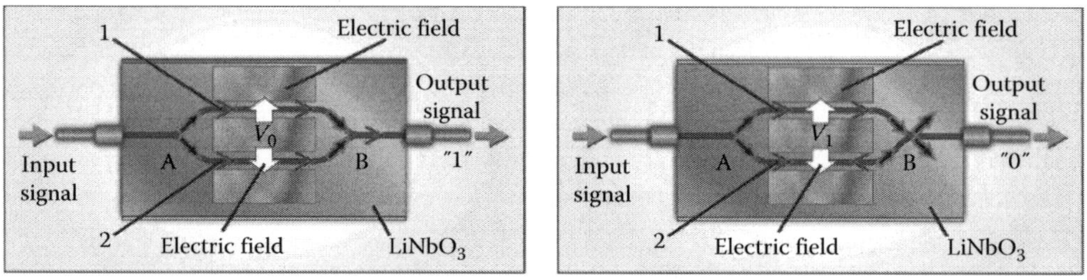

FIGURE 5.17
Intensity modulation using interferometric principles in waveguide structures in $LiNbO_3$: (a) ON—constructive interference mode and (b) OFF—destructive interference mode. Optically guided wave paths 1 and 2. Electric field is established across the optical waveguide.

the refractive index of the waveguide, thence a phase change on the guided lightwave. Detailed of modulation to generate the "ON" and "OFF" at the output of the intensity modulator are illustrated in Figure 5.17.

A basic, structured LN modulator comprises (1) two waveguides, (2) two Y junctions, and (3) RF/DC traveling-wave electrodes. The optical signal coming from the lightwave source is launched into the LN modulator through the polarization-maintaining fiber; it is then equally split into two branches at the first Y junction on the substrate. When no voltage is applied to the RF electrodes, the two signals are recombined constructively at the second Y junction and coupled into a single output. In this case, the output signal from the LN modulator is recognized as "ONE." When a voltage is applied to the RF electrode, due to the electro-optic effects of the LN crystal substrate, the waveguide refractive index is changed, and hence the carrier phase is advanced in one arm but retarded in the other. Thence, the two signals are recombined destructively at the second Y junction; they are transformed into a higher-order mode and radiated at the junction. If the phase retarding is in multiple odd factors of π, the two signals are completely out of phase, the combined signals are radiated into the substrate, and the output signal from the LN modulator is recognized as a "ZERO." The voltage difference that induces this "ZERO" and "ONE" is called the driving voltage of the modulator, and is one of the important parameters in deciding the modulator's performance.

5.3.3 Phasor Representation and Transfer Characteristics

Consider an interferometric intensity modulator consisting of an input waveguide that splits into two branches and then recombines to a single output waveguide. If the two electrodes are initially biased with voltages V_{b1} and V_{b2}, then the initial phases exerted on the lightwaves would be $\phi_1 = \pi V_{b1}/V_\pi = -\phi_2$, which are indicated by the bias vectors shown in Figure 5.18b. From these positions, the phasors are swinging according to the magnitude and the sign of the pulse voltages applied to the electrodes. They can be switched to the two positions that can be constructive or destructive. The output field of the lightwave carrier can be represented by

$$E_0 = \frac{1}{2} E_{iRMS} e^{j\omega_c t} \left(e^{j\phi_1(t)} + e^{j\phi_2(t)}\right) \tag{5.27}$$

where
- ω_c is the carrier radial frequency
- E_{iRMS} is the root mean square value of the magnitude of the carrier
- $\phi_1(t)$ and $\phi_2(t)$ are the temporal phases generated by the two time-dependent pulse sequences applied to the two electrodes

With the voltage levels varying according to the magnitude of the pulse sequence, one can obtain the transfer curve as shown in Figure 5.18a. This phasor representation can be used to determine exactly the biasing conditions and magnitude of the RF or digital signals required for driving the optical modulators to achieve 50%, 33%, or 67% bit period pulse shapes.

Figure 5.18a shows the P-V characteristic curve of a Mach–Zehnder intensity modulator (MZIM) between the output optical power and the driving voltage applied to the electrode. If the input voltage is a sequence of pulse then the output optical pulse sequence would follow a cosine square shape. If the MZIM is biased at the quadrature point and the amplitude of the driving voltage is within the linear range of the transfer curve then the output signal would be linear. Figure 5.18b shows the representation in phasor form of the fields of the lightwaves passing through the two paths of the modulator at the state of

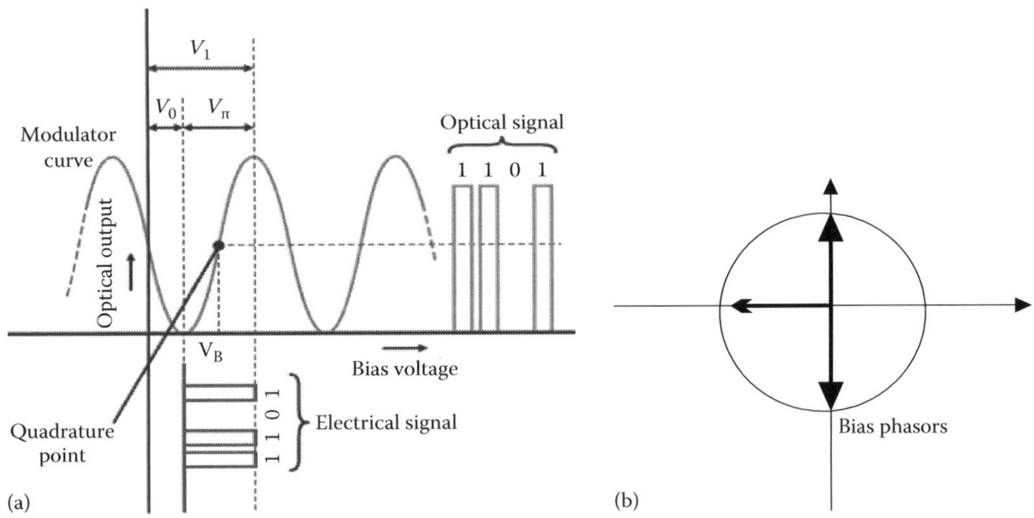

FIGURE 5.18
Electrical to optical transfer curve of an interferometric intensity modulator.

biasing and the $\pi/2$ phase shift changes applied to the optical fields. The power transfer function of the Mach–Zehnder modulator is expressed as follows*:

$$P_0(t) = \alpha P_i \cos^2 \frac{\pi V(t)}{V_\pi} \tag{5.28}$$

where

- $P_0(t)$ is the transmitted output power
- α is the modulator total insertion loss
- P_i is the input power (usually from the laser diode)
- $V(t)$ is the time-dependent signal applied voltage
- V_π is the driving voltage so that a π phase shift is exerted on the lightwave carrier

It is necessary to set the static bias on the transmission curve through the bias electrode. It is a common practice to set the bias point at a 50% transmission point or a $\pi/2$ phase difference between the two optical waveguide branches, the quadrature bias point. As shown in Figure 5.18, electrical digital signals are transformed into optical digital signals by switching the voltage to both ends of quadrature points in the positive and negative directions.

5.4 Bias Control

One factor that affects the modulator performance is the drift of the bias voltage. For the Mach–Zehnder interferometric modulator (MZIM), it is very critical when it is required to bias at the quadrature point or at minimum or maximum locations on the transfer curve. DC drift is the phenomenon occurring in $LiNbO_3$ due to the buildup of charges on the surface of the crystal substrate. Under this drift, the transmission curve gradually shifts in the long term [19,20]. In the case of the $LiNbO_3$ modulator, the bias point control is vital as the bias point will shift in the long term. To compensate for the drift, it is necessary to monitor the output signals and feed them back into the bias control circuits to adjust the DC voltage so that operating points stay at the same point as shown in Figure 5.19, for example, the quadrature point. It is the manufacturer's responsibility to reduce the DC drift so that DC voltage is not beyond the limit throughout the lifetime of the device.

5.4.1 Chirp-Free Optical Modulators

Due to the symmetry of the crystal refractive index of the uniaxial anisotropy of the class *m* of $LiNbO_3$, the crystal cut and the propagation direction of the electric field affect both modulator efficiency, denoted as driving voltage, and modulator chirp. The uniaxial property of $LiNbO_3$ is shown in Figure 5.20. As depicted in Figure 5.21, in the case of the Z-cut structure, as a hot electrode is placed on top of the waveguide, RF field flux is

* Note that this equation is representing single-drive MZIM—it is the same for dual-drive MZIM provided that the bias voltages applied to the two electrodes are equal and opposite in signs. The transfer curve of the field representation would have half the periodic frequency of the transmission curve shown in Figure 5.18.

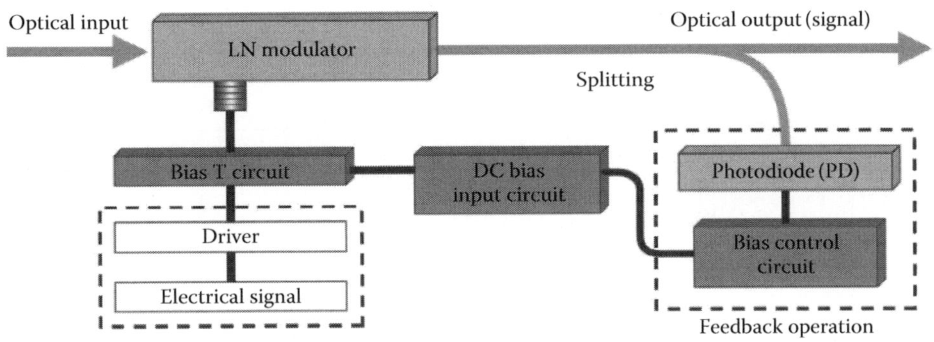

FIGURE 5.19
Arrangement of bias control of integrated optical modulators.

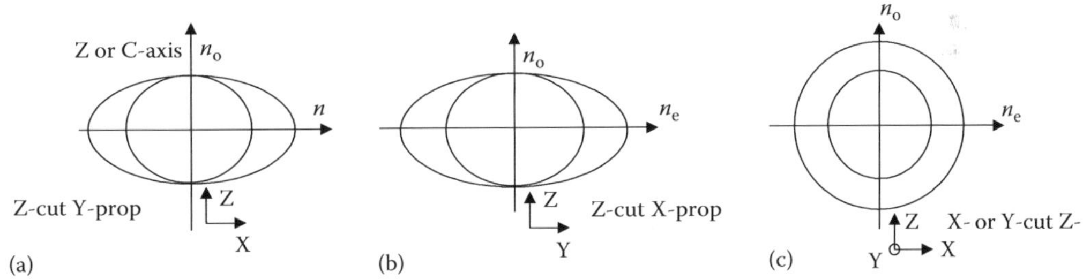

FIGURE 5.20
Refractive index contours of $LiNbO_3$ uniaxial crystal with Z or C denoting the principal axis: (a) lightwave propagation in the Z-axis polarizes along Y-cut $LiNbO_3$, (b) propagation direction Z-axis and X-cut crystal, and (c) Y-prop and Y-cut crystal.

more concentrated, and this results in the improvement of overlap between RF and optical field. However, overlap between RF in ground electrode and waveguide is reduced in the Z-cut structure so that the overall improvement of driving voltage for Z-cut structure compared to X-cut is approximately 20%. The different overlapping area for the Z-cut structure results in a chirp parameter of 0.7, whereas X-cut and Z-propagation has almost-zero chirp due to its symmetric structure. A number of commonly arranged electrode and waveguide structures are shown in Figure 5.22 to maximize the interaction between the traveling electric field and the optically guided waves. Furthermore, a buffer layer, normally SiO_2, is used to match the velocities between these waves so as to optimize the optical modulation bandwidth.

5.4.1.1 Structures of Photonic Modulators

Figure 5.23a and b shows the structure of an MZ intensity modulator using single- and dual-electrode configurations, respectively. The thin-line electrode is called the "hot" electrode or the traveling-wave electrode. RF connectors are required for launching the RF data signals to establish the electric field required for electro-optic effects. Impedance termination is also required. Optical fiber pigtails are also attached to the end faces of the diffused waveguide. The mode spot size of the diffused waveguide is

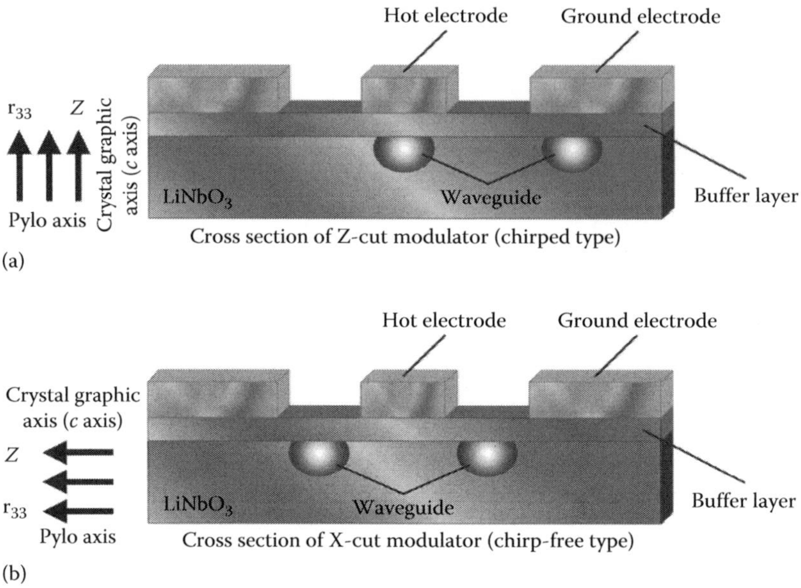

FIGURE 5.21
Different crystal cuts of $LiNbO_3$-integrated structures: The integration of electrodes and optical waveguides in (a) Z-cut and (b) X-cut.

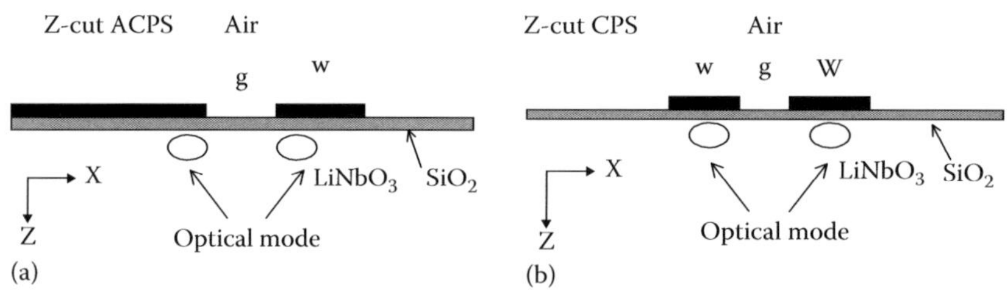

FIGURE 5.22
Commonly used electrode structure and crystal orientation for interferometric modulation to maximize the use of the overlap integral between the optically guided mode and the electric field distribution for largest electro-optic coefficients.

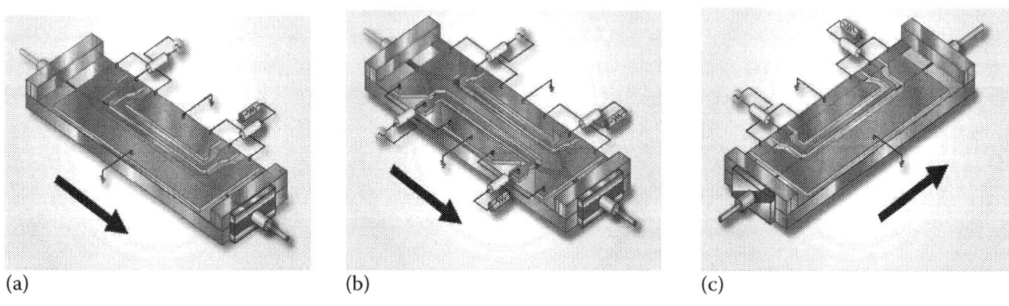

FIGURE 5.23
Intensity modulators using $LiNbO_3$: (a) single-drive electrode, (b) dual electrode structure, and (c) electro-optic polarization scrambler using $LiNbO_3$.

not symmetric, and hence some diffusion parameters are controlled so that maximizing the coupling between the fiber and the diffused or rib waveguide can be achieved. Due to this mismatch between the mode spot sizes of the circular and diffused optical waveguides, there occurs coupling loss. Furthermore, the difference between the refractive indices of the fiber and $LiNbO_3$ is quite substantial, and thus the Fresnel reflection loss would also incur.

Figure 5.23c shows the structure of a polarization modulator that is essential for multiplexing of two polarized data sequences so as to double the transmission capacity, for example, 40–80 Gb/s. Furthermore, this type of polarization modulator can be used as a polarization rotator in a polarization dispersion-compensating subsystem [16].

5.4.2 Typical Operational Parameters of Optical Intensity Modulators

Table 5.1 shows the operational parameters of external $LiNbO_3$ optical intensity modulators. The bandwidth is the dB bandwidth of the modulator in the optical domain, normally measured by scanning the driving voltage at different frequency sinusoidal wave and detected by a very wide-band optical receiver. The voltage at which complete depletion of the output intensity is called Vpi. The polarization extinction ratio indicates the polarization property of the waveguide of the modulator and only either TE or TM guided mode is modulated.

5.4.3 MATLAB® and Simulink® Models of External Optical Modulators

5.4.3.1 Phase Modulation Model and Intensity Modulation

In MATLAB and Simulink, there are a number of blocksets, such as Simulink common blockset, communications blockset, control system blockset, and signal-processing blockset. A phase modulation section of an optical waveguide can be implemented by using the phase shift block given in "utility blocks" of the communication blockset, as shown in Figure 5.24.

TABLE 5.1

Typical Operational Parameters of Optical Intensity Modulators

Parameters	Typical Values	Definition/Comments
Modulation speed	10 Gb/s, 40 Gb/s	Capability to transmit digital signals
Insertion loss	Max 5 dB	Defined as the optical power loss within the modulator including coupling loss between fiber and channel waveguides and propagation in the structure
Driving voltage	Max 4 V	The RF voltage required to have a full modulation
V_π		Voltage required for generating a phase difference of π between the two arms of the interferometer; that is the minimum transmission point
Optical bandwidth	Min 8 GHz	3 dB roll-off in efficiency at the highest frequency in the modulated signal spectrum
ON/OFF extinction ratio	Min 20 dB	The ratio of maximum optical power (ON) and minimum optical power (OFF)
Polarization extinction ratio	Min 20 dB	The ratio of two polarization states (TM- and TE-guided modes) at the output

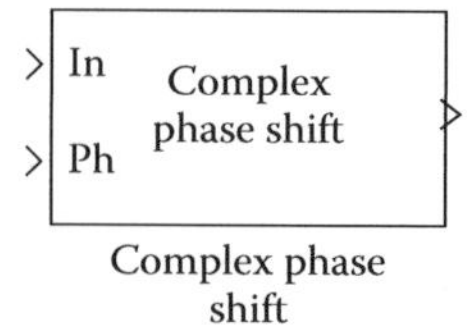

FIGURE 5.24
Phase shift block to be used for phase modulation in optical phase modulators or intensity modulators.

The integration of phase shift blocks to form an intensity modulator is shown in Figure 5.25.

This data modulator can be integrated with another optical modulator, the pulse carver, to generate the RZ pulse shaping with or without carrier suppression. This is implemented by setting an appropriate biasing of the pulse carver as shown in Figure 5.26. Note that the sinusoidal signal generators fed into the two-phase shifters of the modulator complement each other, and the voltage peak-to-peak magnitude must be two V_p in order to create periodic sequence with the suppression of carrier. The output pulse period is half that of the input sinusoidal wave; hence a frequency-doubling effect due to the nonlinear property of the voltage-output intensity transfer characteristics of the intensity modulator. If a dual-drive modulator is used, then the biasing voltage must be at $\pi/2$ so that there is always a π phase difference between the two arms of the Mach–Zehnder interferometer. The overall transmitter is shown in Figure 5.27. In addition to the optical modulation, for differential phase shift keying (DPSK), there must be a differential coder shown in Figure 5.28 for coding the different features of the bit sequence to map two bits into the port representing the in-phase and quadrature components to be fed into the optical modulator.

5.4.3.2 Dense Wavelength Division Multiplexing Optical Multiplexers and Modulators

In dense wavelength division multiplexing (DWDM) transmission systems, the optical transmitters are multiplexed before the conduction of the transmission via optical transmission spans. The Simulink model is shown in Figure 5.29. Different modulation

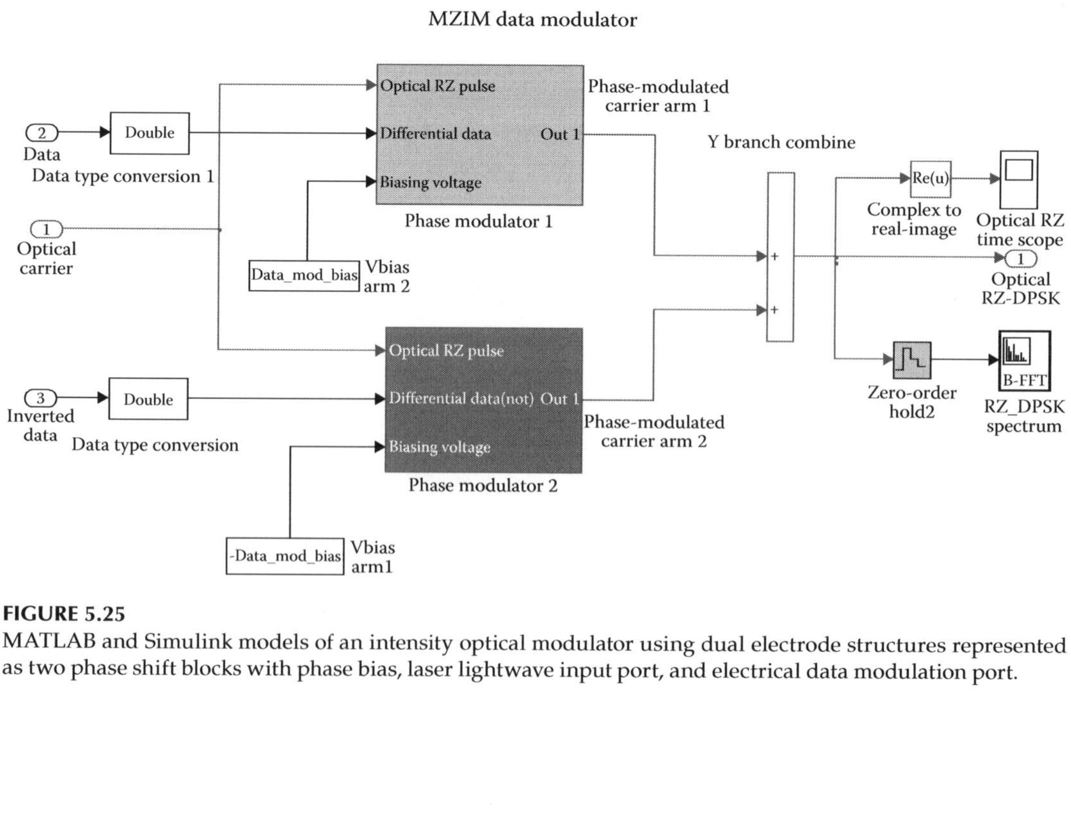

FIGURE 5.25
MATLAB and Simulink models of an intensity optical modulator using dual electrode structures represented as two phase shift blocks with phase bias, laser lightwave input port, and electrical data modulation port.

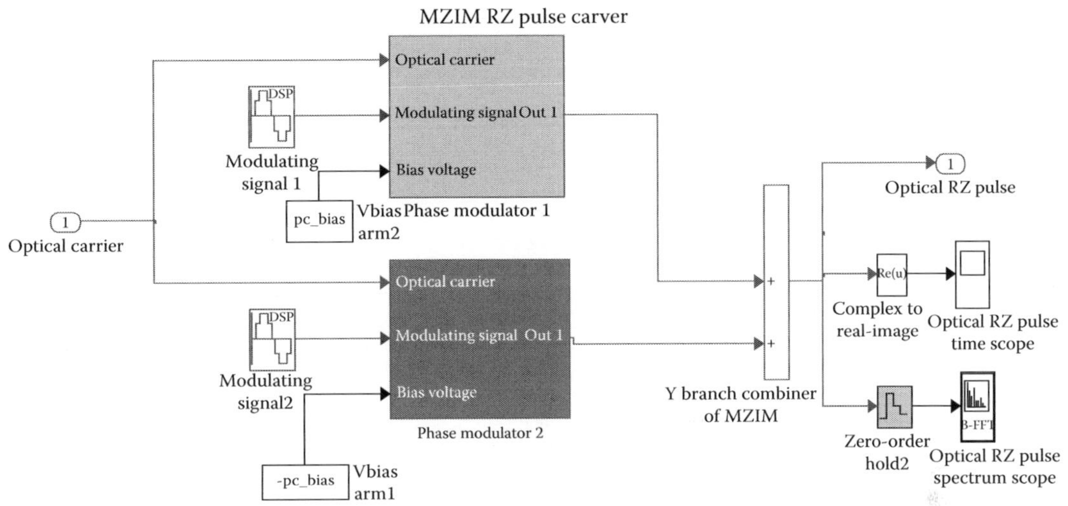

FIGURE 5.26
MATLAB and Simulink models of an optical pulse carver for the generation of RZ pulse shaping.

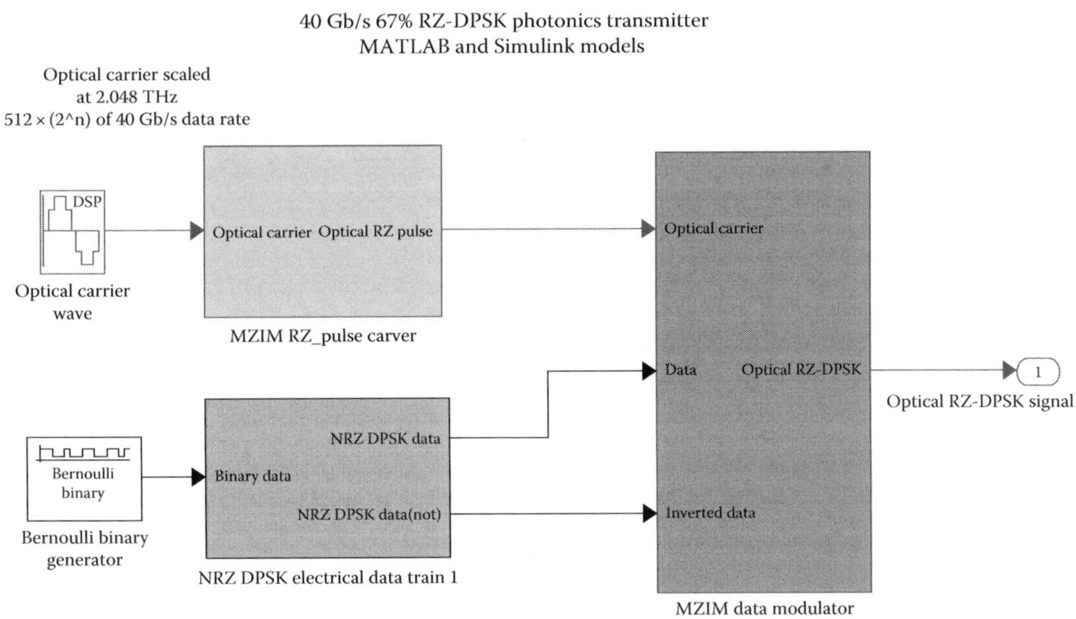

FIGURE 5.27
Photonic transmitters for carrier-suppressed RZ and data modulation using DPSK modulation scheme.

formats, the minimum shift keying and DPSK transmitters, are employed in this model. It is noted that the optical spectrum analyzer shown in this figure should be set sufficiently wide to accommodate the spectrum of all multiplexed channels. The 1/z block is required for sampling the waveform for appropriate setting for observation in the spectrum analyzer.

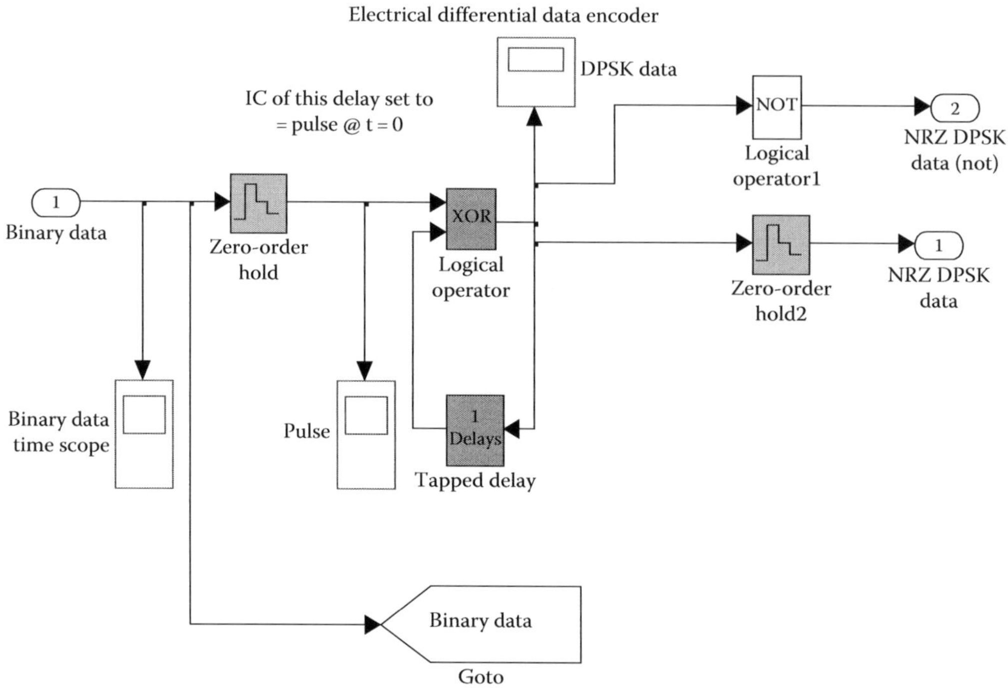

FIGURE 5.28
Electrical differential encoder for DPSK modulation scheme.

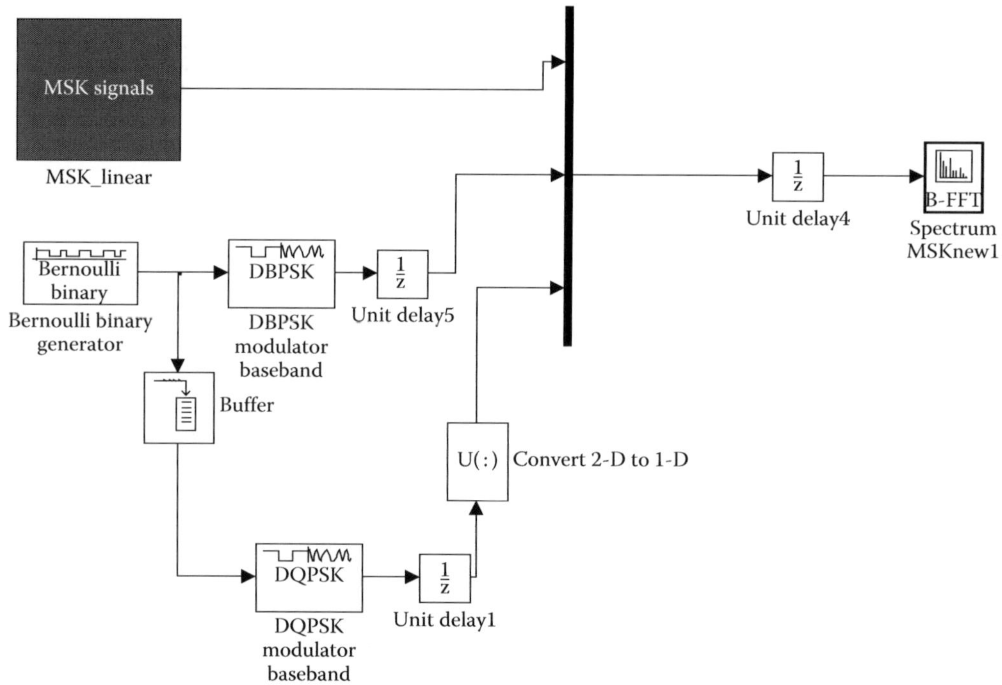

FIGURE 5.29
Simulink model for multiplexing of different modulation formats.

5.5 Concluding Remarks

Direct modulation of a laser source is described with modeling and practical considerations in this chapter. These lightwave sources dominate the optical communications in the bit rate up to 2.5 Gb/s but are no longer capable to operate in the higher bit rates, especially at rates greater than 10 Gb/s. In this case, external modulators are employed. These modulators have been described and coupled with advanced modulation techniques, as given in the rest of this chapter.

Since the proposed of dielectric waveguide and then the advent of optical circular waveguide, the employment of modulation techniques is only extensively exploited recently since the availability of optical amplifiers. The modulation formats allow the transmission efficiency, and hence the economy of ultrahigh-capacity information telecommunications. Optical communications have evolved significantly through several phases from single-mode systems to coherent detection and modulation, which was developed with the main aim of improving the optical power. The optical amplifiers defeated this main objective of modulation formats and allowed the possibility of incoherent detection and all possible formats employing the modulation of the amplitude, the phase, and frequency of the lightwave carrier.

Currently, photonic transmitters play a principal role in the extension of the modulation speed into several gigahertz range and make possible the modulation of the amplitude, the phase, and frequency of the optical carriers and their multiplexing. Photonic transmitters using $LiNbO_3$ have been proven in laboratory and installed systems. The principal optical modulator is the MZIM, which can be a single or a combined set of these modulators whereby to form binary or multilevel amplitude or phase modulation, and is even more effective for discrete or continuous phase shift keying techniques.

Optical modulators employed for the generation of optical signals using advanced modulation techniques at ultrahigh speed will be presented in the next chapter. The effects of the modulation on transmission performance will be given. MATLAB and Simulink assist the modeling of photonic transmitters. Optical modulation can be implemented using phase shift blocks. These techniques will be further developed in Chapter 6 for the implementation of models for advanced modulation formats and pulse shaping.

Appendix: MATLAB® Program for Solving the Laser Rate Equation

```
%%%%%%%%%%%%%%%%%%%%%%%%%%%%%%%%%%%%%%%%%%%%%%%%%%%%%%%%%%%%%%%%%%
%%%%%%%%%%%%%%%%%%%%%%%%%%%
% D F B   L A S E R   D I O D E   P A R A M E T E R S % 1.55micron ;
InGaAsP/InP ; 2Gbit/s ; 40Km %%%%%%%%%%%%%%%%%%%%%%%%%%%%%%%%%%
%%%%%%%%%%%%%%%%%%%%%%%%%%%%%%%%%%%%%%%%%%%%%%%%%%%%%%%%%
global q
global Vg
global Beta
global Imax
global Mune
global Gamma
global Anull
```

```
global Nnull
global Epsil
global Tphot
global Tcarr
global Vactv
global Alpha
global Trise
global Ibias
global MidLamda
global SpecWidth
Gamma=0.8 ;        % OPTICAL CONFINEMENT FACTOR 0.8 %
Anull=3.2e-20;     % GAIN COEFFICIENT %
Vg=7.5e+7 ;        % GROUP VELOCITY %
Nnull=1.0e+12;     % CARRIER DENSITY AT TRANSPARENCY %
Epsil=2.5e-23;     % GAIN COMPRESSION FACTOR %
Tphot=3.0e-12;     % PHOTON LIFE-TIME %
Beta=3.0e-5 ;      % FRACTION OF SPONTANEOUS EMISSION %
Tcarr=0.3e-9 ;     % CARRIER LIFE-TIME %
q=1.6e-19;         % ELECTRON CHARGE %
Vactv=1.5e-16;     % ACTIVE VOLUME LAYER %
Alpha=5 ;          % LINE WIDTH ENHANCEMENT FACTOR %
Trise=1e-12;       % RISE TIME - 1ps version oc2000_1
Ith=q/Tcarr*Vactv*(Nnull+1/Gamma/Anull/Vg/Tphot); % Threshold current
Ibias=7.1*Ith ;    % BIAS CURRENT
Imax=9.0*Ith ;     % Maximum Input Current
Mune=0.042 ;       % TOTAL DIFFERENTIAL QUANTUM EFFICIENCY %
% if Channel ==1
% MidLamda=1553.3*1e-9; % Operating Wavelength of Laser Source=1553.8nm
to match with centre of AWG see mulplex %
% change from 1553.8nm to 1553.3nm to match with ITU-grid AC and GA
% else
   MidLamda = OpLamda(Channel);
% end
SpecWidth = 40e-15;    % Laser Source Spectral Line Width=10MHz linewidth
set %
%%%%%%%%%%%%%%%%%%%%%%%%%%%%%%%%%%%%%%%%%%%%%%%%%%%
%%%%%%%%%%%% Solving The Rate equations
%%%%%%%%%%%%%%%%%%%%%%%%%%%%%%%%%%%%%%%%%%%%%%%%%%%
%%%%%%%%%%
function rate=rate_equ(t,y)
global q
global Vg
global Nsd
global Ssd
global Beta
global Gamma
global Anull
global Nnull
global Epsil
global Tphot
global Tcarr
global Vactv
global Alpha
global Trise
```

```
global Ibias
global Sigma
global Imax
global dummy
global Initial
global InPulse
global BitPeriod
Nsd=1/(Gamma*Anull*Vg*Tphot)+Nnull ;
Ssd=(Tphot/Tcarr)*Nsd*(Imax/Ibias-1) ;
Sigma=2e-20 ;
%---- Computing Rate Equations ------------
rate=zeros(3,1);
rate(1)=(Gamma*Anull*Vg*((y(2)-Nnull)/...
       (1+Epsil*y(1)))-1/Tphot)*y(1)+Beta*Gamma*y(2)/Tcarr+...
randn*sqrt(2*Beta*Vactv*Nsd*(Vactv*Ssd+1)∧3/Tcarr) ;
rate(2)=dgcoding(t,InPulse)/(q*Vactv)-y(2)/...
       Tcarr-Vg*Anull*y(1)*(y(2)-Nnull)/(1+Epsil*y(1))+...
randn*sqrt(2*Vactv*Nsd/Tcarr*(Beta*Vactv*Ssd+1)) ;
rate(3)=Alpha*(Gamma*Vg*Anull*(y(2)-Nnull)-1/Tphot)/2 +...
       randn*sqrt(Gamma*Vg*Sigma*(y(2)-Nnull)/y(1)) ;
  if t<0
    rate=Initial;
  end
  dummy=dummy+1;
  if dummy>=50
% waitbar(t/(BitPeriod*size(InPulse,2))) ;
    dummy=0 ;
  end
```

References

1. (a) R.C. Alferness, Optical guided-wave devices, *Science*, 234(4778): 825–829, November 14, 1986; (b) M. Rizzi and B. Castagnolo, Electro-optic intensity modulator for broadband optical communications, *Fiber and Integrated Optics*, 21: 243–251, 2002; (c) H. Takara, High-speed optical time-division-multiplexed signal generation, *Optical and Quantum Electronics*, 33(7–10): 795–810, July 10, 2001; E.L. Wooten, K.M. Kissa, A. Yi-Yan, E.J. Murphy, D.A. Lafaw, P.F. Hallemeier, D. Maack, D.V. Attanasio, D.J. Fritz, G.J. McBrien, and D.E. Bossi, A review of lithium niobate modulators for fiber-optic communications systems, *IEEE Journal of Selected Topics in Quantum Electronics*, 6(1): 69–80, January/February 2000; (d) K. Noguchi, O. Mitomi, H. Miyazawa, and S. Seki, A broadband Ti: $LiNbO_3$ optical modulator with a ridge structure, *Journal of Lightwave Technology*, 13(6): 1164–1168, June 1995.
2. J. Nodam, Electro-optic modulation method and device using the low-energy oblique transition of a highly coupled super-grid, *IEEE Journal of Lightwave Technology*, LT-4: 1445–1453, 1986.
3. M. Suzuki, Y. Noda, H. Tanaka, S. Akiba, Y. Kuahiro, and H. Isshiki, Monolithic integration of InGaAsP/InP distributed feedback laser and electroabsorption modulator by vapor phase epitaxy, *IEEE Journal of Lightwave Technology*, LT-5(9): 127, September 1987.
4. P.K. Tien, Integrated optics and new wave phenomena in optical waveguides, *Reviews of Modern Physics*, 49: 361–420, 1977.
5. A. Yariv, C.A. Mead, and J.V. Parker, *IEEE Journal of Quantum Electronics*, QE-2: 243, 1966.
6. (a) Stillman 1976; (b) Dutta 1984; (c) Noda 1985.

7. H. Nagata, Y. Li, W.R. Bosenberg, and G.L. Reiff, DC drift of X-cut $LiNbO_3$ modulators, *IEEE Photonics Technology Letters*, 16(10): 2233–2335, October 2004.
8. H. Nagata, DC drift failure rate estimation on 10 Gb/s X-cut lithium niobate modulators, *IEEE Photonics Technology Letters*, 12(11): 1477–1479, November 2000.
9. R.E. Epworth, K.S. Farley, and D. Watley, Polarization mode dispersion compensation, U.S. Patent 398/152, 398/202, 398/65.
10. C. Langton, Basic concept of modulation, www.complextoreal.com (access date: December 2007).
11. K.K. Pang, *Digital Transmission*. Melbourne, Australia: Mi-Tec Publishing, 2002, p. 58.
12. R. Redner and H. Walker, Mixture densities, maximum likelihood and the EM algorithm, *SIAM Review*, 26(2): 195–239, 1984.
13. T. Kawanishi, S. Shinada, T. Sakamoto, S. Oikawa, K. Yoshiara, and M. Izutsu, Reciprocating optical modulator with resonant modulating electrode, *Electronics Letters*, 41(5): 271–272, 2005.
14. R. Krahenbuhl, J.H. Cole, R.P. Moeller, and M.M. Howerton, High-speed optical modulator in $LiNbO_3$ with cascaded resonant-type electrodes, *Journal of Lightwave Technology*, 24(5): 2184–2189, 2006.
15. P. Kaminow and T. Li, *Optical Fiber Communications*, vol. IVA. New York: Elsevier Science, 2002, Chapter 16.
16. F. Amoroso, Pulse and spectrum manipulation in the minimum frequency shift keying (MSK) format, *IEEE Transactions on Communications*, 24: 381–384, March 1976.
17. E. Lach and K. Schuh, Recent advances in ultrahigh bit rate ETDM transmission systems, *IEEE Journal of Lightwave Technology*, 24(12): 4455–4467, 2006.
18. W.S. Lee, 80+Gb/s ETDM systems implementation: An overview of current technology, in *Proceedings of OFC 2006*, Anaheim, CA, 2006, Paper no OTuB3.
19. I.B. Djordjevic and B. Vasic, 100-Gb/s transmission using orthogonal frequency-division multiplexing, *IEEE Photonics Technology Letters*, 18(15): 1576–1578, August 1, 2006.
20. A.J. Lowery, L. Du, and J. Armstrong, Orthogonal frequency division multiplexing for adaptive dispersion compensation in long haul WDM systems, OFC 2006 Postdeadline Sessions, Anaheim, CA, March 9, 2006, Paper PDP39.
21. A.H. Gnauck, X. Liu, X. Wei, D.M. Gill, and E.C. Burrows, Comparison of modulation formats for 42.7-gb/s single-channel transmission through 1980 km of SSMF, *IEEE Photonics Technology Letters*, 16(3): 909–911, 2004.
22. G.P. Agrawal, *Fiber-Optic Communication Systems*, 3rd edn. New York: Wiley, 2002.
23. A. Hirano, Y. Miyamoto, and S. Kuwahara, Performances of CSRZ-DPSK and RZ-DPSK in 43-Gbit/s/ch DWDM G.652 single-mode-fiber transmission, in *Proceedings of OFC'03*, vol. 2, Atlanta, GA, 2003, pp. 454–456.
24. A.H. Gnauck, G. Raybon, P.G. Bernasconi, J. Leuthold, C.R. Doerr, and L.W. Stulz, 1-Tb/s (6/spl times/170.6 Gb/s) transmission over 2000-km NZDF using OTDM and RZ-DPSK format, *IEEE Photonics Technology Letters*, 15(11): 1618–1620, 2003.
25. Y. Yamada, H. Taga, and K. Goto, Comparison between VSB, CS-RZ and NRZ format in a conventional DSF based long haul DWDM system, in *Proceedings of ECOC'02*, vol. 4, Copenhagen, Denmark, 2002, pp. 1–2.
26. O. Painter, P.C. Sercel, K.J. Vahala, D.W. Vernooy, and G.H. Hunziker, Resonant optical modulators, U.S. Patent WO/2002/050575, 27.06.2002.
27. C. Schubert, R.H. Derksen, M. Möller, R. Ludwig, C.-J. Weiske, J. Lutz, S. Ferber, A. Kirstädter, G. Lehmann, and C. Schmidt-Langhorst, Integrated 100-Gb/s ETDM receiver, *IEEE Journal of Lightwave Technology*, 25(1): 122–130, January 2007.

6

Optical Transmitters for Advanced Modulation Format

Following the presentation of optical modulation by direct and external modulation techniques in Chapter 5, this chapter deals with the advanced modulation formats of lightwave sources by external modulators. This is a broad extension of the modulation employing optical modulators connected at the output of a continuous-wave (CW) laser source. The laser is thus turned on at all times, and the generated optical CWs are then modulated, frequency, phase, or amplitude, through an external optical modulator. The uses of these transmitters in optical communication transmission systems are given, especially those for long-haul transmission at very high bit rate. MATLAB® and Simulink® models are described for the generation of advanced modulation format in the optical domain.

6.1 Introduction

A photonic transmitter would consist of single or multiple lightwave sources that can be modulated either directly by manipulating the driving current of the laser diode or externally via an integrated optical modulator [1–12,14]. These are called direct and external modulation techniques.

This chapter presents the techniques for the generation of lightwaves and modulation techniques of lightwaves, either directly or externally. Direct modulation is the technique that directly manipulates the stimulated emission from inside the laser cavity, via the use of electro-optic effects. In external modulation, the laser is turned on at all times, then the generated lightwaves are coupled to an integrated optical modulator through which the electro-optic effect is used with the electrical traveling waves that incorporates the coded information signals, and the amplitude or phase of the lightwaves are modulated. Advanced modulation formats have attracted much attention for the enhancement of the transmission efficiency for coherent optical communications since the mid-1980s. Hence, the preservation of the narrow linewidth of the laser source is critical for operation bit rates in the range of several tens of gigabytes per second. Thus external modulation is essential.

This chapter describes the modulation techniques for optical communication systems [15–23], namely, the modulation of the phase and the amplitude of lightwaves and pulse-shaping non-return to zero (NRZ) or return to zero (RZ); schemes such as amplitude shift keying (ASK), differential phase shift keying (DPSK), minimum shift keying (MSK), and frequency shift keying (FSK); multilevel amplitude and phase modulation techniques such as quadrature amplitude modulation (QAM) [24–28]; and multi-carrier modulation techniques such as orthogonal frequency division modulation (OFDM). Appropriate Simulink models are given for different modulation schemes.

6.2 External Modulation and Advanced Modulation Formats

Comparing with LN-based MZIM, the total insertion loss of semiconductor intensity modulator is about 8–10 dB including fiber-waveguide coupling loss, which is rather high. However, this loss can be compensated by a semiconductor optical amplifier (SOA) that can be integrated on the same circuit. Compared with $LiNbO_3$, its total insertion loss is about 3–4 dB, which can be affordable as the Er-doped fiber amplifier (EDFA—see Chapter 9) is now readily available.

The driving voltage for the EA modulator is usually lower than that required for $LiNbO_3$. However, the extension ratio is not as high as that of the $LiNbO_3$-type modulator, which is about 25 dB as compared to 10 dB for the EA modulator. This feature contrasts the operating characteristics of the $LiNbO_3$ and EA modulators. Although the driving voltage for the EA modulator is about 3–4 V and that for the $LiNbO_3$ modulator is 5–7 V, the former type would be preferred for intensity- or phase-modulation formats due to this extinction ratio that offers much lower "zero" noise level and hence high quality factor.

The phase modulator is a device that manipulates the "phase" of optical carrier signals under the influence of an electric field created by an applied voltage. When voltage is not applied to the RF electrode, the number of periods of the lightwaves, n, exists in a certain path length. When voltage is applied to the RF electrode, one or a fraction of one period of the wave is added, which now means $(n+1)$ waves exist in the same length. In this case, the phase has been changed by 2π and the half voltage of this is called the driving voltage. In the case of long-distance optical transmission, the waveform is susceptible to degradation due to nonlinear effects such as self-phase modulation. A phase modulator can be used to alter the phase of the carrier to compensate for this degradation. The magnitude of the change of the phase depends on the change of the refractive index created via the electro-optic effect that in turn depends on the orientation of the crystal axis with respect to the direction of the established electric field by the applied signal voltage.

An integrated optical phase modulator operates in a similar manner except that the lightwave carrier is guided via an optical waveguide, which is a diffused or ion-exchanged confined region for $LiNbO_3$ modulators and a rib-waveguide structure for semiconductor-type modulators. Two electrodes are deposited so that an electric field can be established across the waveguiding cross section so that there is a change of the refractive index via the electro-optic or EA effect, as shown in Figure 6.1. For ultrafast operations, one of the electrodes is a traveling wave type or hot electrode and the other is a ground electrode.

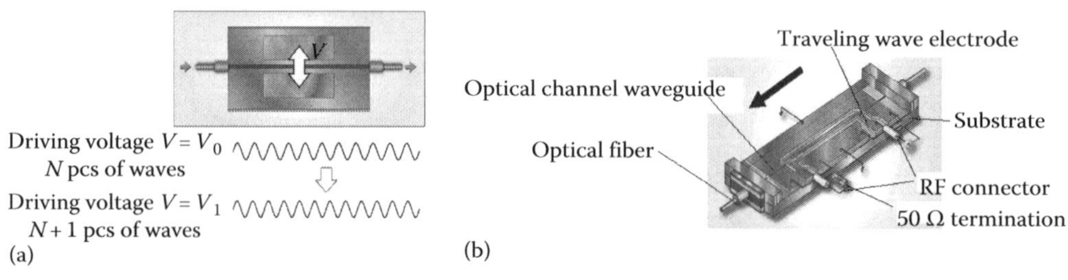

FIGURE 6.1
Electro-optic phase modulation in an integrated modulator using $LiNbO_3$. Electrode impedance matching is not shown: (a) schematic diagram and (b) integrated optic structure.

The traveling wave electrode must be terminated with matching impedance at the end so as to avoid wave reflection. Usually a quarter wavelength impedance is used to match the impedance of the traveling wave electrode to that of the 50 Ω transmission line.

A phasor representation of a phase-modulated lightwave can be achieved by the circular rotation at a radial speed of ω_c. Thus, the vector with an angle ϕ represents the magnitude and the phase of the lightwave.

Basic-structured LN modulator comprises (1) two waveguides, (2) two Y-junctions, and (3) RF/DC traveling wave electrodes. Optical signals coming from the lightwave source is launched into the LN modulator through the polarization maintaining fiber; it is then equally split into two branches at the first Y-junction on the substrate. When no voltage is applied to the RF electrodes, the two signals are recombined constructively at the second Y-junction and coupled into a single output. In this case, output signals from the LN modulator is recognized as "ONE." When voltage is applied to the RF electrode, due to the electro-optic effects of LN crystal substrate, the waveguide refractive index is changed, and hence the carrier phase is advanced in one arm and retarded in the other arm. Thence, the two signals are recombined destructively at the second Y-junction; they are transformed into higher-order mode and radiated at the junction. If the phase retarding is in multiple odd factor of π, the two signals are completely out of phase, the combined signals are radiated into the substrate, and the output signal from the LN modulator is recognized as a "ZERO." The voltage difference that induces this "ZERO" and "ONE" is called the driving voltage of the modulator, and is one of the important parameters in deciding the modulator's performance.

Consider an interferometric intensity modulator consisting of an input waveguide that splits into two branches and then recombines to a single output waveguide. If the two electrodes are initially biased with voltages V_{b1} and V_{b2}, then the initial phases exerted on the lightwaves would be $\phi_1 = \pi V_{b1}/V_\pi = -\phi_2$, which are indicated by the bias vectors shown in Figure 6.3b. From these positions, the phasors swing according to the magnitude and the sign of the pulse voltages applied to the electrodes. They can be switched to the two positions that can be constructive or destructive. The output field of the lightwave carrier can be represented by

$$E_0 = \frac{1}{2} E_{iRMS} e^{j\omega_c t} \left(e^{j\phi_1(t)} + e^{j\phi_2(t)} \right) \tag{6.1}$$

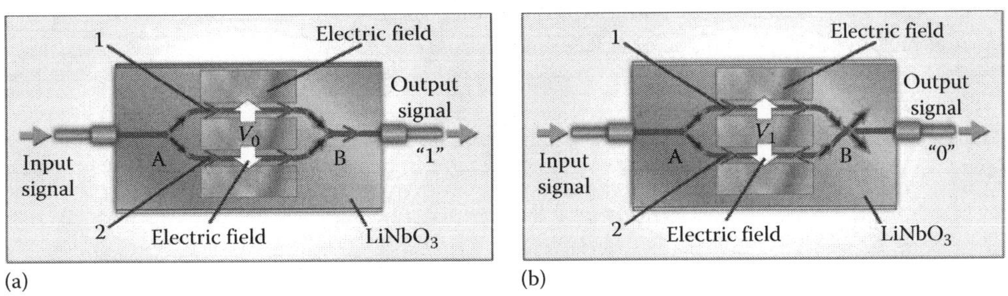

FIGURE 6.2
Intensity modulation using interferometric principles in waveguide structures in $LiNbO_3$: (a) ON—constructive interference mode and (b) destructive interference mode—OFF. Optically guided wave paths 1 and 2. Electric field is established across the optical waveguide.

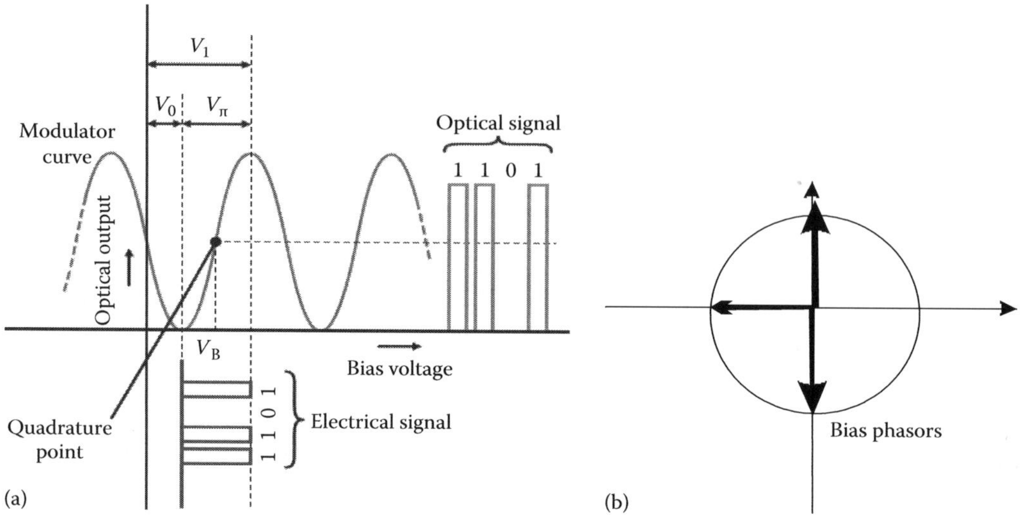

FIGURE 6.3
Electrical to optical transfer curve of an interferometric intensity modulator.

where
- ω_c is the carrier radial frequency
- E_{iRMS} is the root mean square value of the magnitude of the carrier
- $\phi_1(t)$ and $\phi_2(t)$ are the temporal phase generated by the two time-dependent pulse sequences applied to the two electrodes

With the voltage levels varying according to the magnitude of the pulse sequence, one can obtain the transfer curve, as shown in Figure 6.3a. This phasor representation can be used to determine exactly the biasing conditions and magnitude of the RF or digital signals required for driving the optical modulators to achieve 50%, 33%, or 67% bit period pulse shapes.

The power transfer function of Mach–Zehnder modulator is expressed as follows*:

$$P_0(t) = \alpha P_i \cos^2 \frac{\pi V(t)}{V_\pi} \tag{6.2}$$

where
- $P_0(t)$ is the output transmitted power
- α is the modulator total insertion loss
- P_i is the input power (usually from the laser diode)
- $V(t)$ is the time-dependent signal applied voltage
- V_π is the driving voltage so that a π phase shift is exerted on the lightwave carrier

* Note that this equation is representing single-drive MZIM—it is the same for dual-drive MZIM provided that the bias voltages applied to the two electrodes are equal and opposite in signs. The transfer curve of the field representation would have half the periodic frequency of the transmission curve shown in Figure 6.3.

It is necessary to set the static bias on the transmission curve through bias electrode. It is common practice to set bias point at 50% transmission point or a $\pi/2$ phase difference between the two optical waveguide branches, the quadrature bias point. As shown in Figure 6.3, electrical digital signals are transformed into optical digital signals by switching voltage to both ends of quadrature points in the positive and negative directions.

One factor that affects the modulator performance is the drift of the bias voltage. For the Mach–Zehnder interferometric modulator (MZIM), it is very critical when it is required to bias at the quadrature point or at minimum or maximum locations on the transfer curve. DC drift is the phenomena occurring in $LiNbO_3$ due to the buildup of charges on the surface of the crystal substrate. Under this drift, the transmission curve gradually shifts in the long term [19,20]. In the case of the $LiNbO_3$ modulator, the bias point control is vital as the bias point will shift in the long term. To compensate for the drift, it is necessary to monitor the output signals and feed it back into the bias control circuits to adjust the DC voltage so that operating points stay at the same point, for example, the quadrature point (see Figure 6.4). It is the manufacturer's responsibility to reduce DC drift so that DC voltage is not beyond the limit throughout the lifetime of the device.

Due to the symmetry of the crystal refractive index of the uniaxial anisotropy of the class *m* of $LiNbO_3$, the crystal cut and the propagation direction of the electric field affect both modulator efficiency, denoted as driving voltage, and modulator chirp. The uniaxial property of $LiNbO_3$ is shown in Figure 6.5. As shown in Figure 6.6, in the case of the Z-cut structure, when a hot electrode is placed on top of the waveguide, RF field flux is

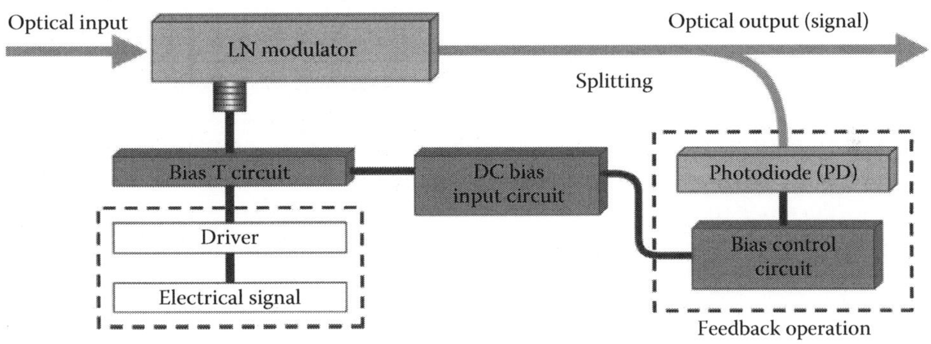

FIGURE 6.4
Arrangement of bias control of integrated optical modulators.

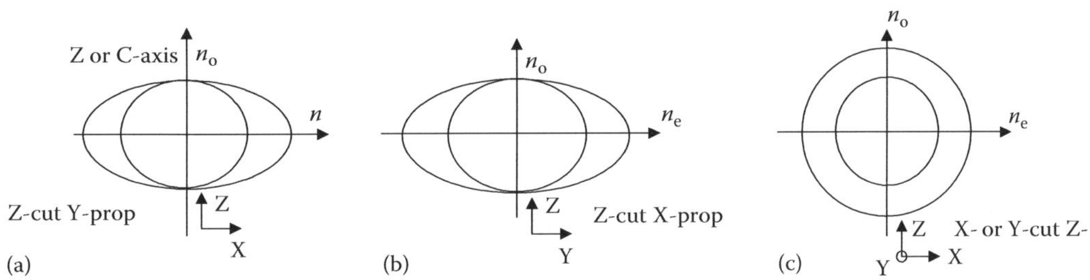

FIGURE 6.5
Refractive index contours of $LiNbO_3$ uniaxial crystal with Z or C denoting the principal axis: (a) lightwave propagation in the Z-axis polarize along Y-cut $LiNbO_3$, (b) propagation direction Z-axis and X-cut crystal, and (c) Y-prop and Y-cut crystal.

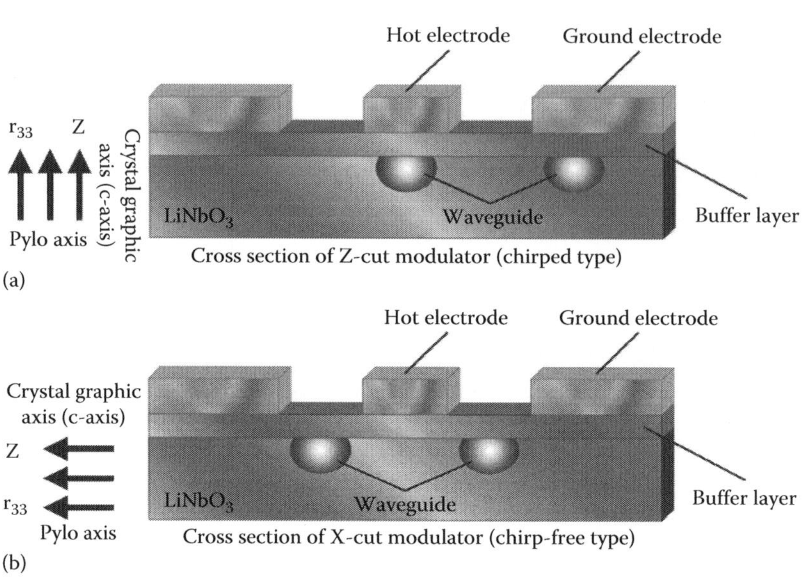

FIGURE 6.6
Different crystal cuts of $LiNbO_3$ integrated structures: (a) integration of electrodes and optical waveguides in Z-cut and (b) X-cut.

more concentrated, and this results in the improvement of overlap between RF and optical field. However, overlap between RF in ground electrode and waveguide is reduced in the Z-cut structure so that the overall improvement of driving voltage for Z-cut structure compared to X-cut is approximately 20%. The different overlapping area for the Z-cut structure results in a chirp parameter of 0.7 whereas X-cut and Z-propagation has almost zero-chirp due to its symmetric structure. A number of commonly arranged electrode and waveguide structures are shown in Figure 6.7 to maximize the interaction between the traveling electric field and the optically guided waves. Furthermore, a buffer layer, normally SiO_2, is used to match the velocities between these waves so as to optimize the optical modulation bandwidth.

Figure 6.8a and b shows the structure of an MZ intensity modulator using a single- and dual-electrode configurations, respectively. The thin-line electrode is called the "hot" electrode or the traveling wave electrode. RF connectors are required for launching the

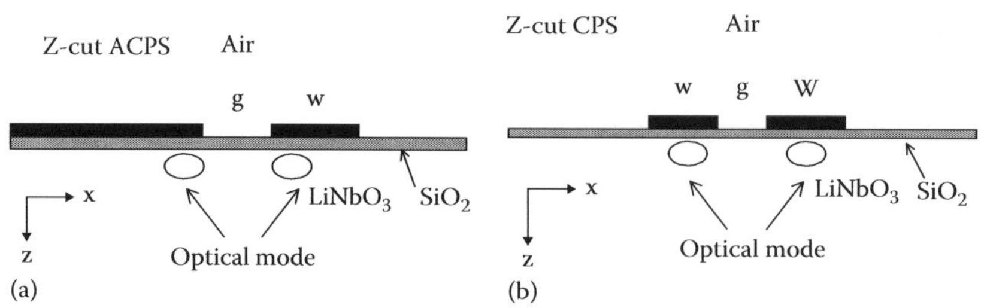

FIGURE 6.7
Commonly used electrode structure and crystal orientation for interferometric modulation to maximize the use of the overlap integral between the optically guided mode and the electric field distribution for largest electro-optic coefficients.

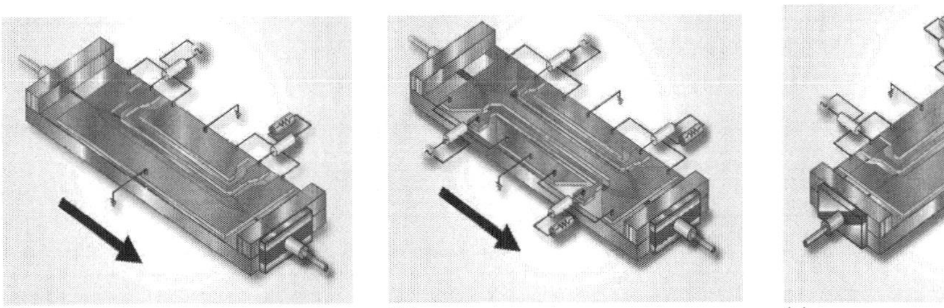

(a) (b) (c)

FIGURE 6.8
Intensity modulators using $LiNbO_3$: (a) single-drive electrode, (b) dual electrode structure, and (c) electro-optic polarization scrambler using $LiNbO_3$.

RF data signals to establish the electric field required for electro-optic effects. Impedance termination is also required. Optical fibers pigtails are also attached to the end faces of the diffused waveguide. The mode spot size of the diffused waveguide is not symmetric, and hence some diffusion parameters are controlled so that maximizing the coupling between the fiber and the diffused or rib waveguide can be achieved. Coupling loss occurs due to this mismatch between the mode spot sizes of the circular and diffused optical waveguides. Furthermore, the difference between the refractive indices of the fiber and $LiNbO_3$ is quite substantial and thus Fresnel reflection loss would also incur.

Figure 6.8c shows the structure of a polarization modulator that is essential for multiplexing of two polarized data sequences so as to double the transmission capacity, for example, 40–80 Gb/s. Furthermore, this type of polarization modulator can be used as a polarization rotator in a polarization dispersion-compensating subsystem [15].

Typical operational parameters of optical intensity modulators are given in Table 6.1.

6.2.1 Modulation Formats and Pulse Shaping

6.2.1.1 Generation

Figure 6.9 shows the conventional structure of an RZ-ASK transmitter in which two external $LiNbO_3$ MZIMs can be used. The MZIM shown in this transmitter can be either single- or dual-drive (push–pull) type. Operational principles of the MZIM were presented in

TABLE 6.1
Typical Operational Parameters of Optical Intensity Modulators

Parameters	Typical Values	Definition/Comments
Modulation speed	10 Gb/s	Capability to transmit digital signals
Insertion loss	Max 5 dB	Defined as the optical power loss within the modulator
Driving voltage	Max 4 V	The RF voltage required to have a full modulation
Optical bandwidth	Min 8 GHz	3 dB roll-off in efficiency at the highest frequency in the modulated signal spectrum
ON/OFF extinction ratio	Min 20 dB	The ratio of maximum optical power (ON) and minimum optical power (OFF)
Polarization extinction ratio	Min 20 dB	The ratio of two polarization states (TM- and TE-guided modes) at the output

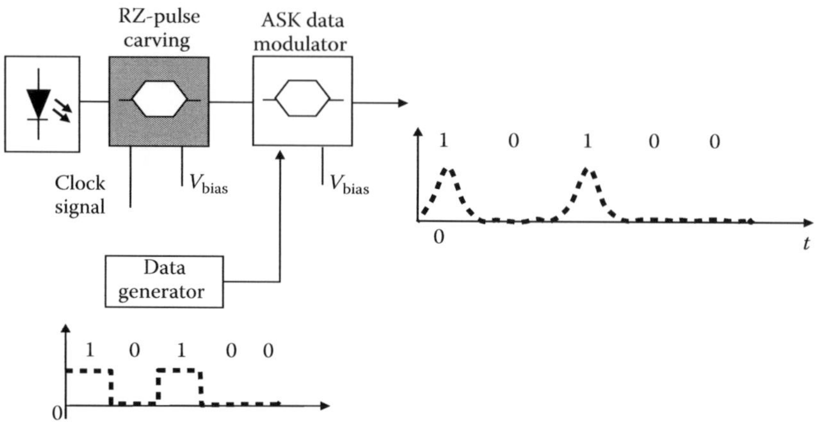

FIGURE 6.9
Conventional structure of an OOK optical transmitter utilizing two MZIMs.

Section 5.2. The optical ON/OFF keying (OOK) transmitter would normally consist of a narrow-linewidth laser source to generate lightwaves whose wavelength satisfies the ITU grid standard.

The first MZIM, commonly known as the pulse carver, is used to generate the periodic pulse trains with a required RZ format. The suppression of the lightwave carrier, commonly known as the carrier-suppressed RZ (CSRZ), can also be carried out at this stage if necessary. Compared to other RZ types, CSRZ pulse shape is found to have attractive attributes for long-haul wavelength division multiplexed (WDM) transmissions including the π phase difference of adjacent modulated bits, the suppression of the optical carrier component in optical spectrum, and narrower spectral width.

Different types of RZ pulses can be generated depending on the driving amplitude of the RF voltage and the biasing schemes of the MZIM. The equations governing the RZ pulse electric field waveforms are as follows:

$$E(t)=\begin{cases}\sqrt{\dfrac{E_b}{T}}\sin\left[\dfrac{\pi}{2}\cos\left(\dfrac{\pi t}{T}\right)\right] & \text{67\% duty-ratio RZ pulses or CSRZ}\\[2ex] \sqrt{\dfrac{E_b}{T}}\sin\left[\dfrac{\pi}{2}\left(1+\sin\left(\dfrac{\pi t}{T}\right)\right)\right] & \text{33\% duty-ratio RZ pulses or RZ33}\end{cases} \tag{6.3}$$

where

E_b is the pulse energy per transmitted bit
T is the one bit period

The 33% duty-ratio RZ pulse are denoted as RZ33 pulse whereas the 67% duty cycle RZ pulse is known as the CSRZ type. The art in the generation of these two RZ pulse types stays at the difference of biasing point on the transfer curve of an MZIM.

The bias voltage conditions and the pulse shape of these two RZ types, the carrier suppression and non-suppression of maximum carrier, can be implemented with the

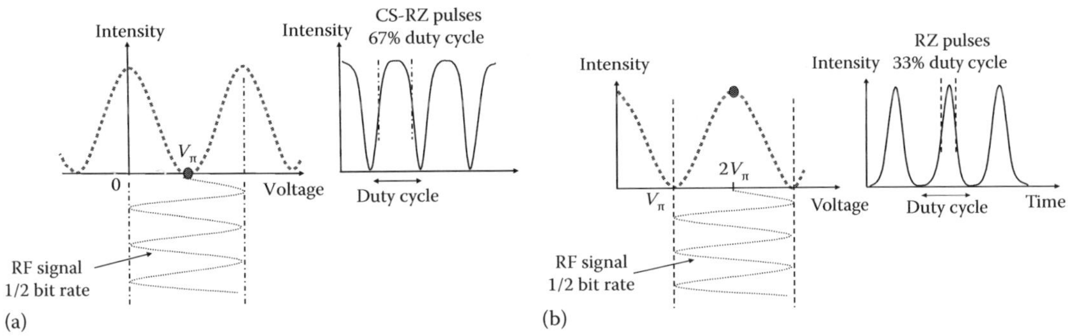

FIGURE 6.10
Bias point and RF driving signals for the generation of (a) CSRZ and (b) RZ33 pulses.

biasing points at the minimum and maximum transmission point of the transmittance characteristics of the MZIM, respectively. The peak-to-peak amplitude of the RF driving voltage is $2V_\pi$, where V_π is the required driving voltage to obtain a π phase shift of the lightwave carrier. Another important point is that the RF signal is operating at only a half of the transmission bit rate. Hence, pulse carving is actually implementing the frequency doubling. The generations of RZ33 and CSRZ pulse train are demonstrated in Figure 6.10a and b.

The pulse carver can also utilize a dual-drive MZIM that is driven by two complementary sinusoidal RF signals. This pulse carver is biased at $-V_{\pi/2}$ and $+V_{\pi/2}$ with the peak-to-peak amplitude of $V_{\pi/2}$. Thus, a π phase shift is created between the state "1" and "0" of the pulse sequence and hence the RZ with alternating phase 0 and π. If the carrier suppression is required, then the two electrodes are applied with voltages V_π and swing voltage amplitude of V_π.

Although RZ modulation offers improved performance, RZ optical systems usually require more complex transmitters than those in the NRZ ones. Compared to only one stage for modulating data on the NRZ optical signals, two modulation stages are required for the generation of RZ optical pulses.

6.2.1.2 Phasor Representation

Recalling (6.3) we have

$$E_o = \frac{E_i}{2}\left[e^{j\varphi_1(t)} + e^{j\varphi_2(t)}\right] = \frac{E_i}{2}\left[e^{j\pi v_1(t)/V_\pi} + e^{j\pi v_2(t)/V_\pi}\right] \tag{6.4}$$

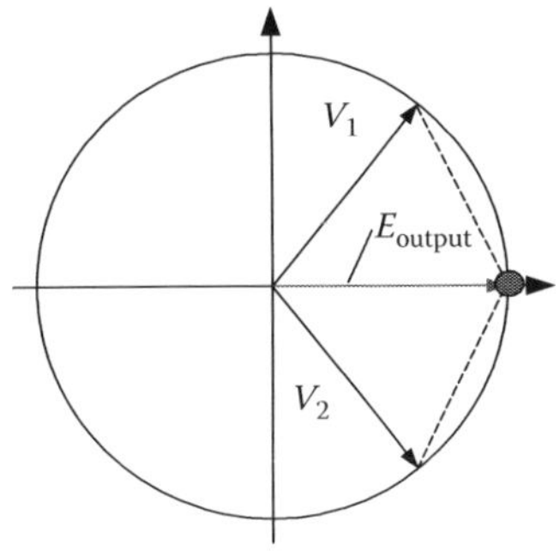

FIGURE 6.11
Phasor representation for the generation of output field in dual-drive MZIM.

It can be seen that the modulating process for the generation of RZ pulses can be represented by a phasor diagram, as shown in Figure 6.11. This technique gives a clear understanding of the superposition of the fields at the coupling output of two arms of the MZIM. Here, a dual-drive MZIM is used, that is, the data driving signals $[V_1(t)]$ and inverse data ($\overline{\text{data}}: V_2(t) = -V_1(t)$ are applied to each arm of the MZIM, respectively, and the RF voltages swing in inverse directions. Applying the phasor representation, vector addition, and simple trigonometric calculus, the process of generation RZ33 and CSRZ is explained in detail and verified.

The width of these pulses are commonly measured at the position of full-width half maximum (FWHM). It is noted that the measured pulses are intensity pulses whereas we are considering the addition of the fields in the MZIM. Thus, the normalized E_o field vector has the value of $\pm 1/\sqrt{2}$ at the FWHM intensity pulse positions, and the time interval between these points gives the FWHM values.

6.2.1.3 Phasor Representation of CSRZ Pulses

Key parameters including the V_{bias}, the amplitude of the RF driving signal, are shown in Figure 6.12a. Accordingly, its initialized phasor representation is demonstrated in Figure 6.12b.

The values of the key parameters are outlined as follows: (1) V_{bias} is $\pm V_\pi/2$; (2) swing voltage of driving RF signal on each arm has the amplitude of $V_\pi/2$ (i.e., $V_{p-p} = V_\pi$); (3) RF signal operates at half of bit rate ($B_R/2$); and (4) at the FWHM position of the optical pulse, the $E_{out} = \pm 1/\sqrt{2}$ and the component vectors V_1 and V_2 form a phase of $\pi/4$ with vertical axis as shown in Figure 6.13.

Considering the scenario for the generation of 40 Gb/s CSRZ optical signal, the modulating frequency is f_m ($f_m = 20\,\text{GHz} = B_R/2$). At the FWHM positions of the optical pulse, the phase is given by the following expressions:

$$\frac{\pi}{2}\sin(2\pi f_m) = \frac{\pi}{4} \Rightarrow \sin 2\pi f_m = \frac{1}{2} \Rightarrow 2\pi f_m = \left(\frac{\pi}{6}, \frac{5\pi}{6}\right) + 2n\pi \tag{6.5}$$

Thus, the calculation of TFWHM can be carried out and hence, the duty cycle of the RZ optical pulse can be obtained as given in the following expressions:

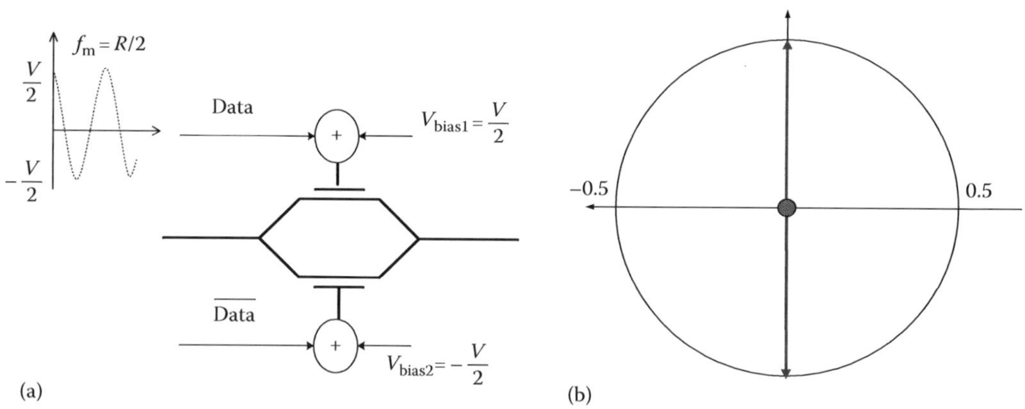

FIGURE 6.12
Initialized stage for the generation of CSRZ pulse: (a) RF driving signal and the bias voltages and (b) initial phasor representation.

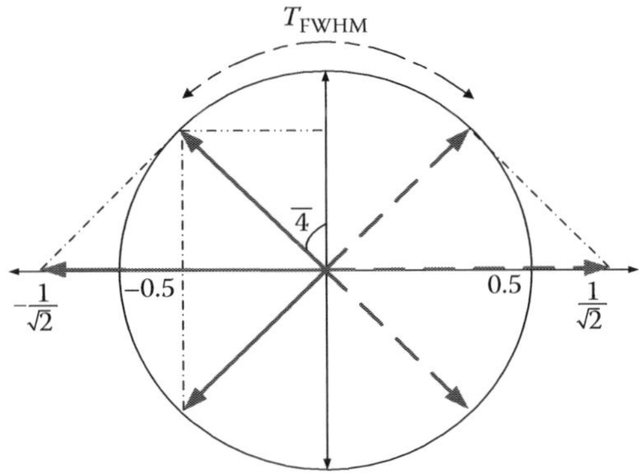

FIGURE 6.13
Phasor representation of CSRZ pulse generation using dual-drive MZIM.

$$T_{FWHM} = \left(\frac{5\pi}{6} - \frac{\pi}{6}\right)\frac{1}{R2\pi} = \frac{1}{3}\pi \times \frac{1}{R} \Rightarrow \frac{T_{FWHM}}{T_{BIT}} = \frac{1.66\times10^{-4}}{2.5\times10^{-11}} = 66.67\% \tag{6.6}$$

The result obtained in (6.6) clearly verifies the generation of CSRZ optical pulses from the phasor representation.

6.2.1.4 Phasor Representation of RZ33 Pulses

Key parameters including the V_{bias}, the amplitude of driving voltage, and its correspondent initialized phasor representation are shown in Figure 6.14a and b, respectively.

The values of the key parameters are (1) V_{bias} is V_π for both arms; (2) swing voltage of driving RF signal on each arm has the amplitude of $V_\pi/2$ (i.e., $V_{p-p} = V_\pi$); and (3) RF signal operates at half of bit rate ($B_R/2$).

At the FWHM positions of the optical pulse, the $E_{output} = \pm 1/\sqrt{2}$ and the component vectors V_1 and V_2 form a phase of $\pi/4$ with horizontal axis as shown in Figure 6.15.

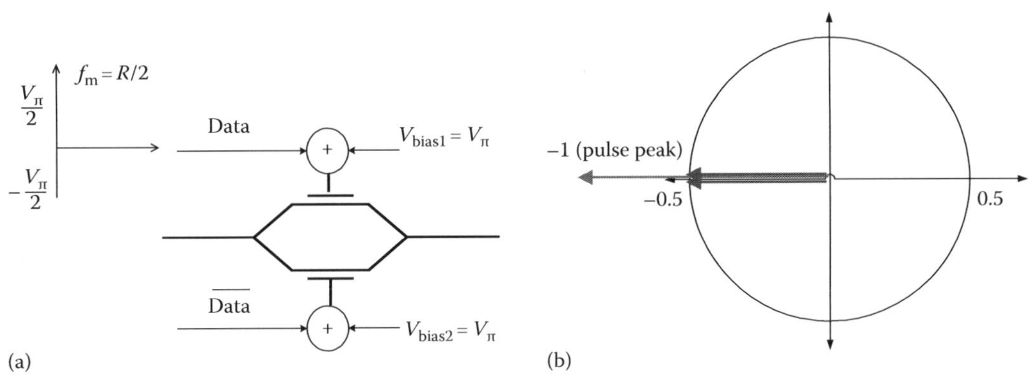

FIGURE 6.14
Initialized stage for the generation of RZ33 pulse: (a) RF driving signal and the bias voltage and (b) initial phasor representation.

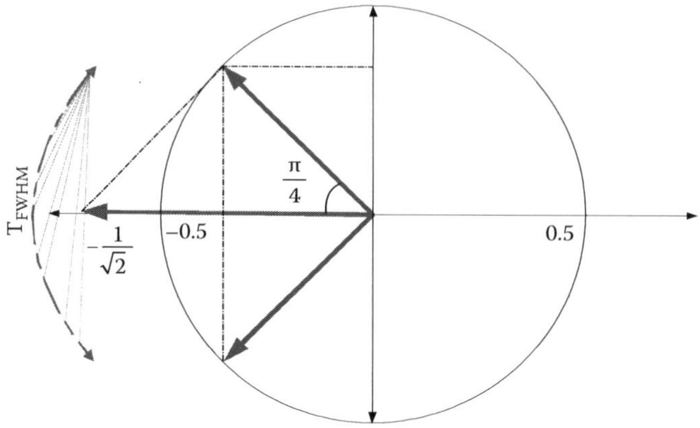

FIGURE 6.15
Phasor representation of RZ33 pulse generation using dual-drive MZIM.

Considering the scenario for the generation of 40 Gb/s CSRZ optical signal, the modulating frequency is f_m ($f_m = 20\,\text{GHz} = B_R/2$). At the FWHM positions of the optical pulse, the phase is given by the following expressions:

$$\frac{\pi}{2}\cos(2\pi f_m t) = \frac{\pi}{4} \Rightarrow t_1 = \frac{1}{6f_m} \tag{6.7}$$

$$\frac{\pi}{2}\cos(2\pi f_m t) = -\frac{\pi}{4} \Rightarrow t_2 = \frac{1}{3f_m} \tag{6.8}$$

$$T_{FWHM} = \frac{1}{3f_m} - \frac{1}{6f_m} = \frac{1}{6f_m} \quad \therefore \frac{T_{FWHM}}{T_b} = \frac{1/6f_m}{1/2f_m} = 33\% \tag{6.9}$$

Thus, the calculation of TFWHM can be carried out, and hence the duty cycle of the RZ optical pulse can be obtained. The result obtained in (6.9) clearly verifies the generation of RZ33 optical pulses from the phasor representation.

6.2.2 Differential Phase Shift Keying

6.2.2.1 Background

Digital encoding of data information by modulating the phase of the lightwave carrier is referred to as optical PSK. In early days, optical PSK was studied extensively for coherent photonic transmission systems. This technique requires the manipulation of the absolute phase of the lightwave carrier. Thus, precise alignment of the transmitter and demodulator center frequencies for the coherent detection is required. These coherent optical PSK systems face severe obstacles such as broad linewidth and chirping problems of the laser source. Meanwhile, the DPSK scheme overcomes those problems, since the DPSK optically modulated signals can be detected incoherently. This technique only requires

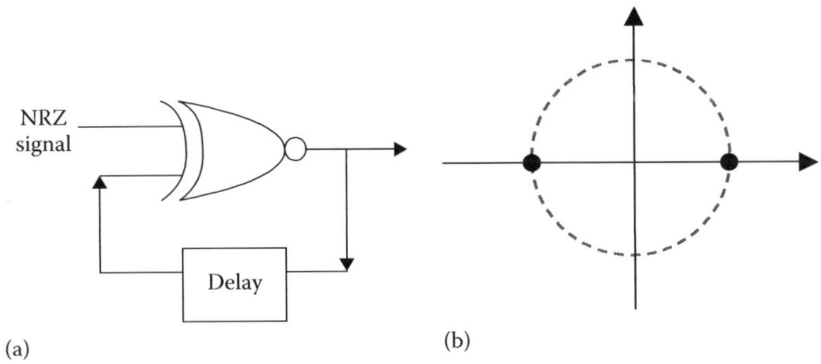

FIGURE 6.16
(a) DPSK precoder and (b) signal constellation diagram of DPSK.

the coherence of the lightwave carriers over a one-bit period for the comparison of the differentially coded phases of the consecutive optical pulses.

A binary "1" is encoded if the present input bit and the past encoded bit are of opposite logic whereas a binary 0 is encoded if the logics are similar. This operation is equivalent to an XOR logic operation. Hence, an XOR gate is employed as a differential encoder. NOR can also be used to replace XOR operation in differential encoding as shown in Figure 6.16a. In DPSK, the electrical data "1" indicates a π phase change between the consecutive data bits in the optical carrier, while the binary "0" is encoded if there is no phase change between the consecutive data bits. Hence, this encoding scheme gives rise to two points located exactly at π phase difference with respect to each other in signal constellation diagram. For continuous PSK, such as the MSK, the phase evolves continuously over a quarter of the section, thus a phase change of π/2 between one phase state to the other. This is indicated by the filled circle, as shown in Figure 6.16b.

6.2.2.2 Optical DPSK Transmitter

Figure 6.17 shows the structure of a 40 Gb/s DPSK transmitter in which two external $LiNbO_3$ MZIMs are used. Operational principles of an MZIM were presented earlier. The MZIMs shown in Figure 6.17 can be of either single- or dual-drive type. The optical DPSK transmitter also consists of a narrow linewidth laser to generate a lightwave whose wavelength conforms to the ITU grid.

The RZ optical pulses are then fed into the second MZIM through which the RZ pulses are modulated by the precoded binary data to generate RZ-DPSK optical signals. Electrical data pulses are differentially precoded in a precoder using the XOR coding scheme. Without pulse carver, the structure shown in Figure 6.17 is an optical NRZ-DPSK transmitter. In data modulation for DPSK format, the second MZIM is biased at the minimum transmission point. The precoded electrical data has peak-to-peak amplitude equal to $2V_\pi$ and operates at the transmission bit rate. The modulation principles for the generation of optical DPSK signals are demonstrated in Figure 6.18.

The electro-optic phase modulator (E-OPM) might also be used for the generation of DPSK signals instead of MZIM. Using optical phase modulator, the transmitted optical signal is chirped whereas using MZIM, especially the X-cut type with Z-propagation, chirp-free signals can be produced. However, in practice, a small amount of chirp might be useful for transmission [26].

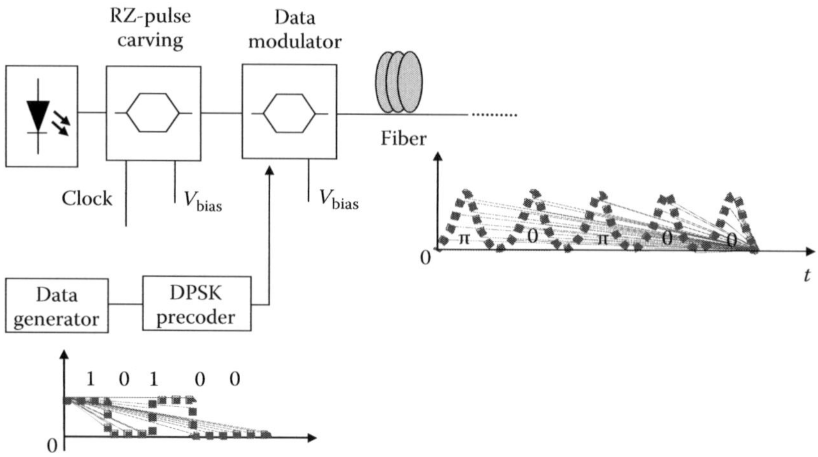

FIGURE 6.17
DPSK optical transmitter with RZ pulse carver.

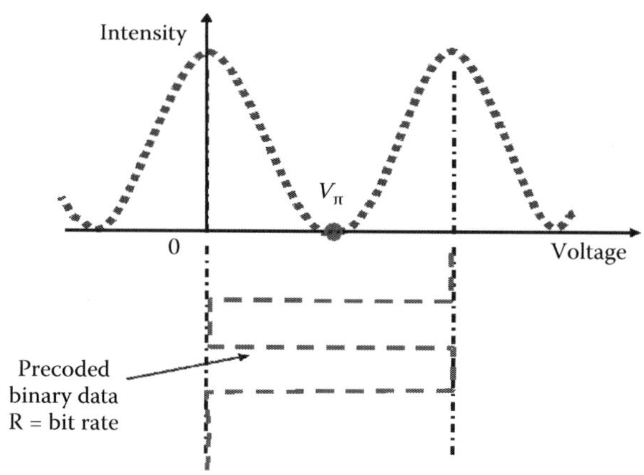

FIGURE 6.18
Bias point and RF driving signals for the generation of optical DPSK format.

6.3 Generation of Modulation Formats

6.3.1 Introductory Remarks

Modulation is the process facilitating the transfer of information over a medium, for example, a wireless or optical environment. Three basic types of modulation techniques are based on the manipulation of a parameter of the optical carrier to represent the information digital data. These are ASK, PSK, and FSK. In addition to the manipulation of the carrier, the occupation of the data pulse over a single period would also determine the amount of energy concentrates and the speed of the system required for transmission. The pulse can remain constant over a bit period or RZ level within a portion of the period.

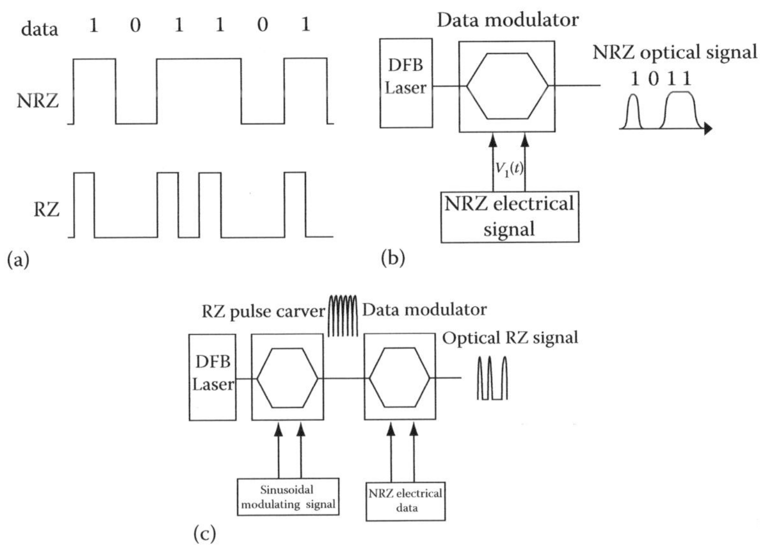

FIGURE 6.19
(a) Baseband NRZ and RZ line coding for 101101 data sequence, (b) block diagram of NRZ photonics transmitter, and (c) RZ photonics transmitter incorporating a pulse carver.

These formats would be named NRZ or RZ. They are combined with the modulation of the carrier to form various modulations formats, which are presented in this section.

Figure 6.19 shows the baseband signals of the NRZ and RZ formats and its corresponding block diagram of a photonic transmitter.

6.3.2 Amplitude-Modulation ASK-NRZ and ASK-RZ

6.3.2.1 ASK-NRZ and ASK-RZ Pulse Shaping

There are a number of formats used in advanced optical communications; based on the intensity of the pulse they may include NRZ, RZ, and duo-binary. These ASK formats can also be integrated with the phase modulation to generate discrete- or continuous-phase NRZ or RZ formats. Currently, the majority of 10 Gb/s installed optical communications system have been developed with NRZ due to its simple transmitter design and bandwidth-efficient characteristics. However, RZ format has higher robustness to fiber nonlinearity and polarization mode dispersion (PMD). In this section, the RZ pulse is generated by MZIM commonly known as *pulse carver*, as shown in Figure 6.20. There are a number of variations in RZ format based on the biasing point in transmission curve shown in Table 6.2. The phasor representation of the biasing and driving signals can also be observed in this table.

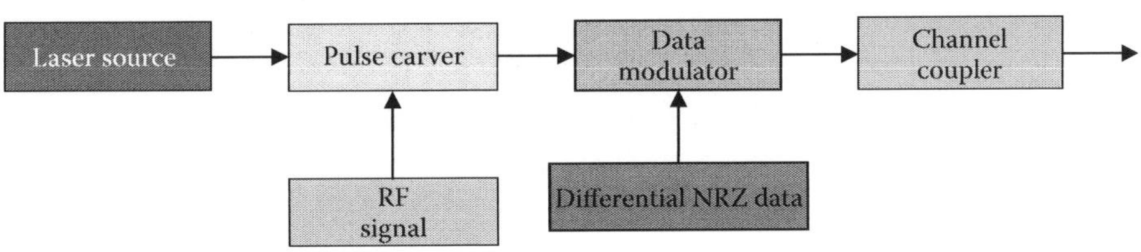

FIGURE 6.20
Block diagrams of RZ-DPSK transmitter.

TABLE 6.2

Summary of RZ Format Generation and Characteristics of Single-Drive MZIM Based on Biasing Point, Drive Signal Amplitude, and Frequency

Biasing Point in the cos² Transmission Transfer Curve	RZ Generation Characteristics	Phasor Representation
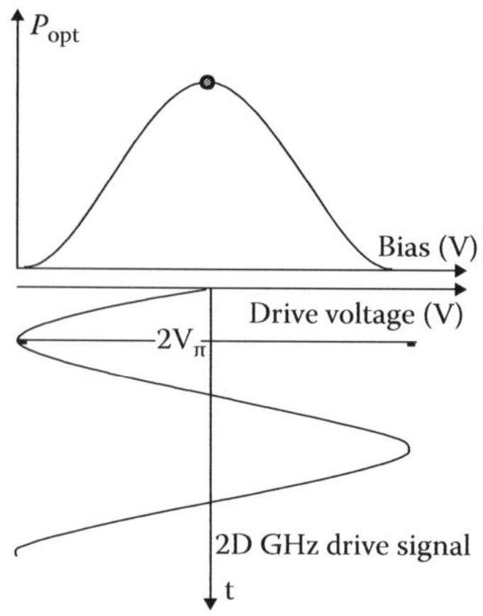	(a) Biasing point: maximum, $V_{2\pi}$ (b) Drive signal amplitude: $2V_\pi$ (c) Drive sinusoidal signal frequency: 20 GHz (d) Pulse width: 9–10 ps about 70% of half the data rate (e) RZ pulse frequency: 40 GHz frequency doubling effect	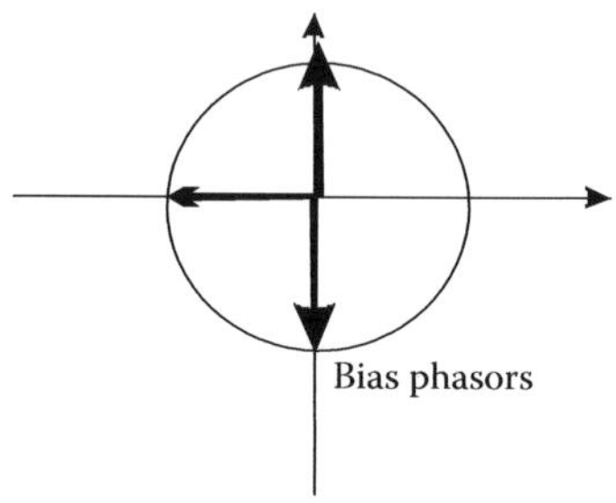
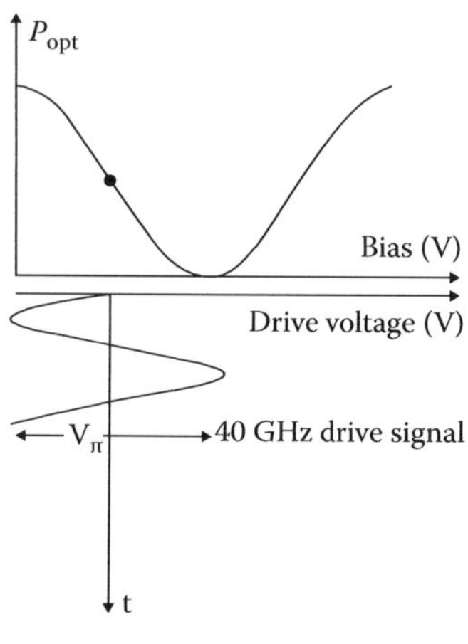	(a) Biasing point: linear region, $V_{\pi/2}$ (b) Drive signal amplitude: V_π (c) Drive sinusoidal signal frequency: 20 GHz (d) Pulse width: 9–10 ps or about 70% of half the data rate (e) RZ pulse frequency: 20 GHz	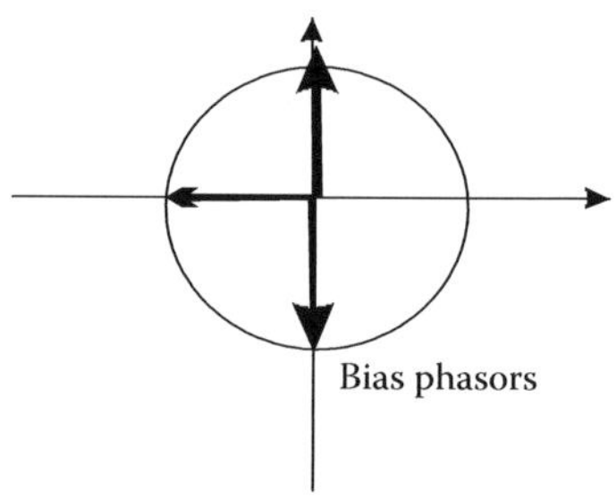
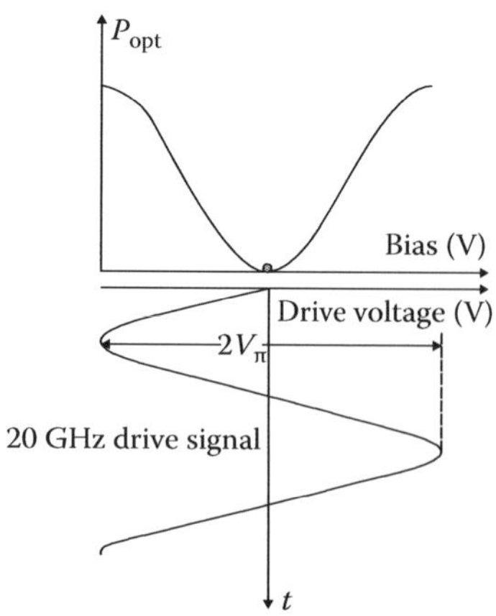	(a) Biasing point: minimum, V_π (b) Drive signal amplitude: $2V_\pi$ (c) Drive sinusoidal signal frequency: 20 GHz (d) Pulse width: 9–10 ps about 70% of half the data rate (e) RZ pulse frequency: 40 GHz frequency doubling effect. (CSRZ is generated using this scheme)	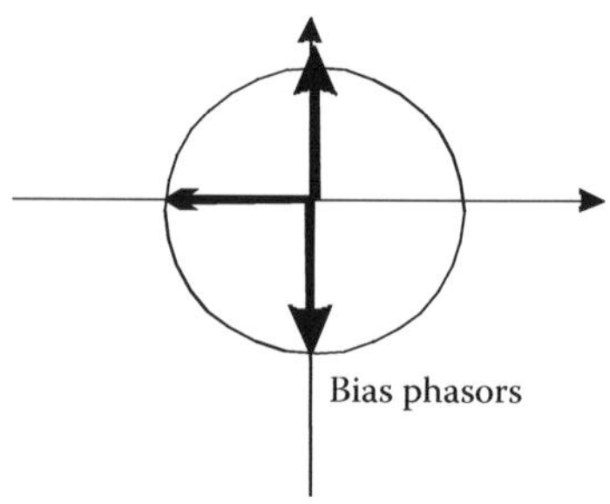

CSRZ has been found to have more attractive attributes in long-haul WDM transmissions compared to conventional RZ format due the possibility of reducing the upper level of the power contained in the carrier that serve no purpose in the transmission but only increase the total energy level, thus approaching the nonlinear threshold level faster. CSRZ pulse has an optical phase difference of π in adjacent bits, removing the optical carrier component in optical spectrum and reducing the spectral width. This offers an advantage in compact WDM channel spacing.

6.3.2.2 *Amplitude-Modulation Carrier-Suppressed RZ Formats*

The suppression of the carrier can be implemented by biasing the MZ interferometer in such a way that there is a π phase shift between the two arms of the interferometer. The magnitude of the sinusoidal signals applied to an arm or both arms would determine the width of the optical output pulse sequence. The driving conditions and phasor representation are shown in Table 6.2.

6.3.3 Discrete Phase-Modulation NRZ Formats

The term discrete phase modulation is referred to as DPSK, whether DPSK or quadrature DPSK (DQPSK), to indicate that the states of the phases of the lightwave carrier are switched from one distinct location on the phasor diagram to the other state, or, for example, from 0 to π or −π/2 to −π/2 for binary PSK (BPSK), or even more evenly spaced PSK levels as in the case of M-ary PSK.

6.3.3.1 *Differential Phase Shift Keying*

Information encoded in the phase of an optical carrier is commonly referred to as optical PSK. In early days, PSK required precise alignment of the transmitter and demodulator center frequencies. Hence, PSK system was not widely deployed. With the introduction of the DPSK scheme, coherent detection is not critical since DPSK detection only requires source coherence over a one-bit period by comparison of tow consecutive pulses.

A binary "1" is encoded if the present input bit and the past encoded bit are of opposite logic and a binary 0 is encoded if the logic is similar. This operation is equivalent to an XOR logic operation. Hence, an XOR gate is usually employed in a differential encoder. NOR can also be used to replace XOR operation in differential encoding as shown in Figure 6.21.

In optical application, electrical data "1" is represented by a π phase change between the consecutive data bits in the optical carrier, while state "0" is encoded with no phase change between the consecutive data bits. Hence, this encoding scheme gives rise to two points located exactly at π phase difference with respect to each other in signal constellation diagram as indicated in Figure 6.21b.

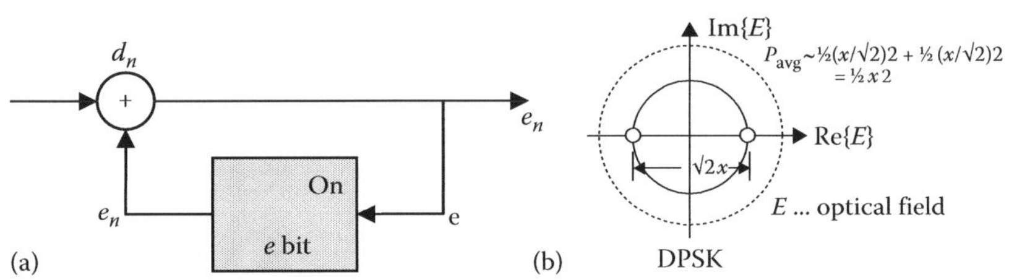

FIGURE 6.21
(a) The encoded differential data are generated by $e_n = d_n \oplus e_{n-1}$; and (b) signal constellation diagram of DPSK.

An RZ-DPSK transmitter consists of an optical source, a pulse carver, a data modulator, a differential data encoder, and a channel coupler. Channel coupler model is not developed in simulation by assuming no coupling losses when optical RZ-DPSK-modulated signal is launched into the optical fiber. This modulation scheme has combined the functionality of dual-drive MZIM modulator of pulse carving and phase modulation.

The pulse carver, usually an MZ interferometric intensity modulator, is driven by a sinusoidal RF signal for single-drive MZIM and two complementary electrical RF signals for dual-drive MZIM, to carve pulses out from optical carrier signal forming RZ pulses. These optical RZ pulses are fed into the second MZ intensity modulator where RZ pulses are modulated by differential NRZ electrical data to generate RZ-DPSK. This data phase modulation can be performed using straight-line phase modulator but the MZ waveguide structure has several advantages over phase modulator due to its chirpless property. Electrical data pulses are differentially precoded in a differential precoder, as shown in Figure 6.21a. Without pulse carver and sinusoidal RF signal, the output pulse sequence follows NRZ-DPSK format, that is, the pulse would occupy 100% of the pulse period, and there is no transition between the consecutive "1s."

6.3.3.2 Differential Quadrature Phase Shift Keying

This differential coding is similar to DPSK except that each symbol consists of two bits that are represented by the two orthogonal axial discrete phase at (0, π) and (−π/2, +π/2) as shown in Figure 6.22, or two additional orthogonal phase positions are located on the imaginary axis of Figure 6.21b.

6.3.3.3 NRZ-DPSK

Figure 6.23 shows the block diagram of a typical NRZ-DPSK transmitter. Differential precoder of electrical data is implemented using the logic explained in the previous section. In modulating the phase of an optical carrier, MZ modulator known as data phase modulator is biased at minimum point and driven by data swing of $2V_\pi$. The modulator showed an excellent behavior so that the phase of the optical carrier will be altered by π exactly when the signal transits the minimum point of the transfer characteristics.

6.3.3.4 RZ-DPSK

The arrangement of RZ-DPSK transmitter is essentially similar to RZ-ASK as shown in Figure 6.24 with the data intensity modulator replaced with data phase modulator. The

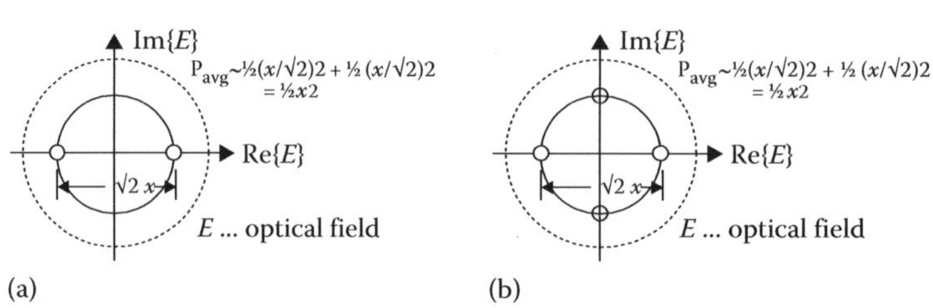

FIGURE 6.22
Signal constellation diagram of (a) DPSK: two bold dots are orthogonal to the DPSK constellation and (b) DQPSK.

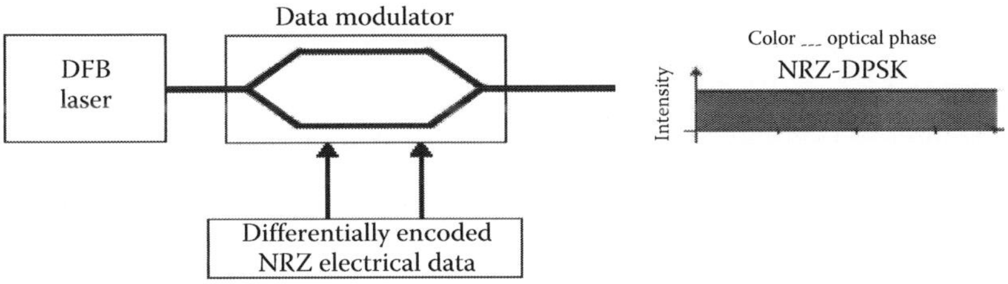

FIGURE 6.23
Block diagram of NRZ-DPSK photonics transmitter.

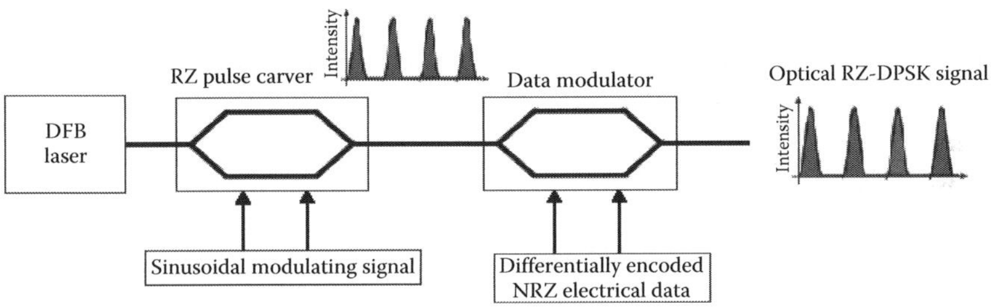

FIGURE 6.24
Block diagram of RZ-DPSK photonics transmitter.

difference between them is the biasing point and the electrical voltage swing. Different RZ formats can also be generated.

6.3.3.5 Generation of M-Ary Amplitude Differential Phase Shift Keying Using One MZIM

As an example, a 16-ary M-ary amplitude differential phase shift keying (MADPSK) signal can be represented by a constellation shown in Figure 6.25. It is, indeed, a combination of a 4-ary ASK and a DQPSK scheme. At the transmitting end, each group of four

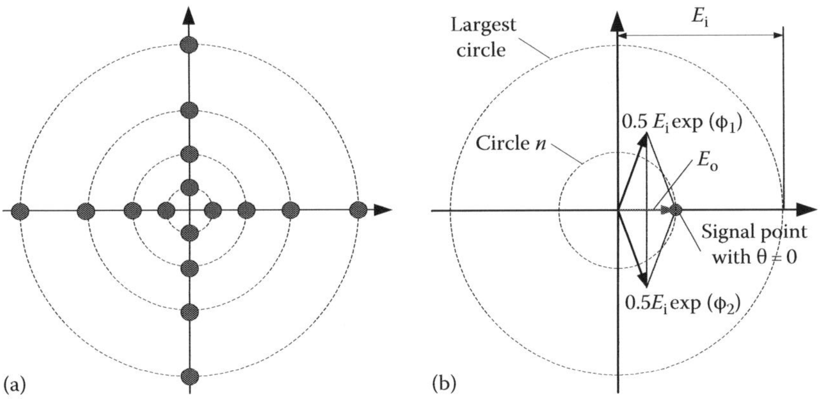

FIGURE 6.25
Signal constellation of 4-ary ADPSK format and phasor representation of a point on the constellation point for driving voltages applied to dual-drive MZIM.

bits $[D_3D_2D_1D_0]$ of user data are encoded into a symbol, among them two least significant bits $[D_1D_0]$ are encoded into four phase states $[0, \pi/2, \pi, 3\pi/2]$ and the other two most significant bits, $[D_3D_2]$, are encoded into four amplitude levels. At the receiving end, as MZ delay interferometers (MZDI) are used for phase comparison and detection, a diagonal arrangement of the signal constellation shown in Figure 6.25a is preferred. This simplifies the design of transmitter and receiver, minimizes the number of phase detectors, hence leading to high receiver sensitivity.

In order to balance the bit error rate (BER) between ASK and DQPSK components, the signal levels corresponding to four circles of the signal space should be adjusted to a reasonable ratio that depends on the noise power at the receiver. As an example, if this ratio is set to $[I_0/I_1/I_2/I_3]=[1/1.4/2/2.5]$, where I_0, I_1, I_2, and I_3 are the intensity of the optical signals corresponding to circle 0, circle 1, circle 2, and circle 3, respectively, then by selecting E_i equal to the amplitude of the circle 3 and V_π equal to 1, the driving voltages should have the values given in Table 6.3. Inversely speaking, one can set the outermost level such that its peak value is below the nonlinear SPM threshold; the voltage level of the outermost circle would be determined. The inner most circle is limited to the condition that the largest signal-to-noise ratio (SNR) should be achieved. This is related to the optical SNR (OSNR) required for a certain BER. Thus from the largest amplitude level and smallest amplitude level we can then design the other points of the constellation.

Furthermore, to minimize the effect of inter-symbol interference, 66%-RZ and 50%-RZ pulse formats are also used as alternatives to the traditional NRZ counterpart. These RZ pulse formats can be created by a pulse carver that precedes or follows the dual-drive MZIM modulator. Mathematically, waveforms of NRZ and RZ pulses can be represented by the following equations, where E_{on}, $n=0$, 1, 2, and 3 are peak amplitudes of the signals in the circle 0, circle 1, circle 2, and circle 3 of the constellation, respectively:

$$p(t)=\begin{cases} E_{on} & \text{for NRZ} \\ E_{on}\cos\left(\dfrac{\pi}{2}\cos^2\left(\dfrac{1.5\pi t}{T_s}\right)\right) & \text{for 66\%-RZ} \\ E_{on}\cos\left(\dfrac{\pi}{2}\cos^2\left(\dfrac{2\pi t}{T_s}\right)\right) & \text{for 50\%-RZ} \end{cases} \tag{6.10}$$

A typical arrangement of the signals of the precoder and driving signals for the MZIM is shown in Figure 6.26.

TABLE 6.3

Driving Voltages for 16-Ary MADPSK Signal Constellation

Circle 0			Circle 1			Circle 2			Circle 3		
Phase	**$V_1(t)$ V**	**$V_2(t)$ V**	**Phase**	**$V_1(t)$ V**	**$V_2(t)$ V**	**Phase**	**$V_1(t)$ V**	**$V_2(t)$ V**	**Phase**	**$V_1(t)$ V**	**$V_2(t)$ V**
0	0.38	−0.38	0	0.30	−0.30	0	0.21	−0.21	0	0.0	0.0
$\pi/2$	0.88	0.12	$\pi/2$	0.80	0.20	$\pi/2$	0.71	0.29	$\pi/2$	0.5	0.5
π	−0.62	0.62	π	−0.7	0.70	π	−0.79	0.79	π	−1.0	1.0
$3\pi/2$	−0.12	−0.88	$3\pi/2$	−0.20	−0.8	$3\pi/2$	−0.29	−0.71	$3\pi/2$	−0.5	−0.5

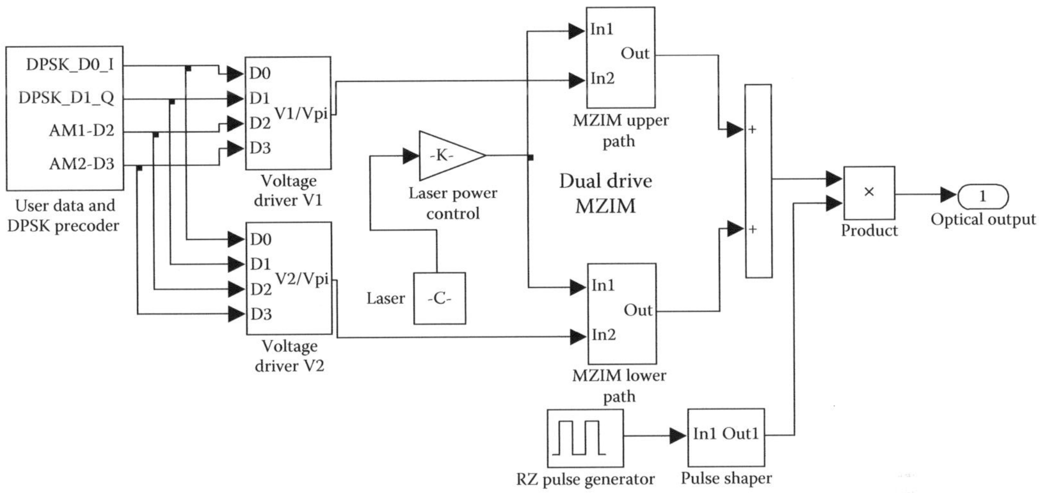

FIGURE 6.26
MATLAB and Simulink models of an MADPSK photonic transmitter. The MZIM is represented by two phase shifter blocks.

6.3.3.6 Continuous Phase Modulation PM-NRZ Formats

In the previous section, the optical transmitters for discrete PSK modulation formats have been described. Obviously, the phase of the carrier has been used to indicate the digital states of the bits or symbols. These phases are allocated in a noncontinuous manner around a circle corresponding to the magnitude of the wave. Alternatively, the phase of the carrier can be continuously modulated and the total phase changes at the transition instants, usually at the end of the bit period, would be the same as those for discrete cases. Since the phase of the carrier continuously varies during the bit period, this can be considered as an FSK modulation technique, except that the transition of the phase at the end of one bit and the beginning of next bit would be continuous. The continuity of the carrier phase at these transitions would reduce the signal bandwidth and hence it is more tolerable to dispersion effects and has a higher energy concentration for effective transmission over the optically guided medium. One of the examples of the reduction of the phase at the transition is the offset DQPSK, which is a minor but important variation on the QPSK or DQPSK. In OQPSK, the Q channel is shifted by half a symbol period so that the transition instants of I and Q channel signals do not happen at the same time. The result of this simple change is that the phase shifts at any one time are limited, and hence the offset QPSK is a more constant envelope than the normal QPSK.

The enhancement of the efficiency of the bandwidth of the signals can be further improved if the phase changes at these transitions are continuous. In this case, the change of the phase during the symbol period is continuously changed by using a half-cycle sinusoidal driving signal with the total phase transition over a symbol period as a fraction of π, depending on the levels of this PSK modulation. If the change is $\pi/2$ then we have an MSK scheme. The orthogonality of the I and Q channels will also reduce the bandwidth of the carrier-modulated signals further. In this section, the basic principles of optical MSK and the photonic transmitters for these modulation formats are described. Ideally, the driving signal to the phase modulator should be a triangular wave so that a linear phase variation of the carrier in the symbol period is linear. However, when a sinusoidal function is used, there are some nonlinear variations; we thus term this type of MSK a nonlinear MSK format. This nonlinearity contributes to some penalty in the optical SNR (OSNR), which will be explained in a later chapter. Furthermore, the MSK as a special form of ODQPSK is also described for optical systems (Chapter 11, Section 11.3).

6.3.3.7 Linear and Nonlinear MSK

6.3.3.7.1 Signals and Precoding

MSK is a special form of continuous-phase FSK (CPFSK) signal in which the two frequencies are spaced in such a way that they are orthogonal and hence minimum spacing between them, defined by

$$s(t)=\sqrt{\frac{2E_b}{T_b}}\cos\left[2\pi f_1 t+\theta(0)\right] \quad \text{for symbol 1} \tag{6.11}$$

$$s(t)=\sqrt{\frac{2E_b}{T_b}}\cos\left[2\pi f_2 t+\theta(0)\right] \quad \text{for symbol 0} \tag{6.12}$$

As shown by these equations, the signal frequency change corresponds to higher frequency for data-1 and lower frequency for data-0. Both frequencies, f_1 and f_2, are defined by

$$f_1 = f_c + \frac{1}{4T_b} \quad \text{and} \quad f_2 = f_c - \frac{1}{4T_b} \tag{6.13}$$

Depending on the binary data, the phase of signal changes; data-1 increases the phase by $\pi/2$, while data-0 decreases the phase by $\pi/2$. The variation of phase follows paths of sequence of straight lines in phase trellis (Figure 6.27), in which the slopes represent frequency changes. The change in carrier frequency from data-0 to data-1, or vice versa, is equal to half the bit rate of incoming data [6]. This is the minimum frequency spacing that allows the two FSK signals representing symbols 1 and 0, to be coherently orthogonal in the sense that they do not interfere with one another in the process of detection.

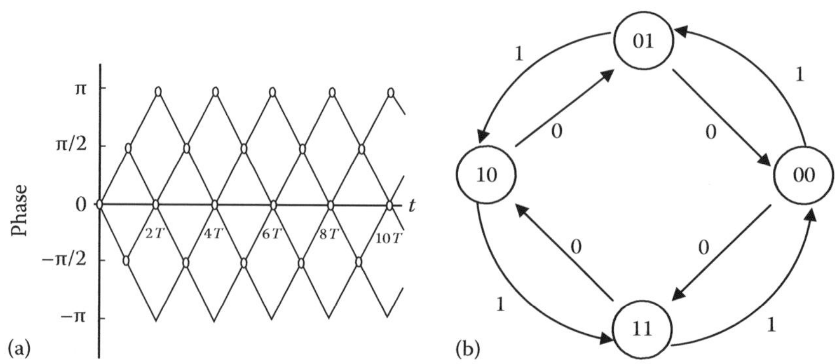

FIGURE 6.27
(a) Phase trellis for MSK and (b) state diagram for MSK.

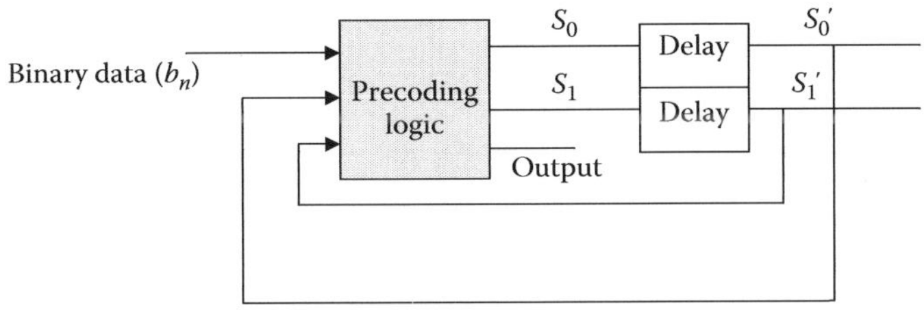

FIGURE 6.28
Combinational logic, the basis of the logic for constructing the precoder.

An MSK signal consists of both I and Q components, which can be written as follows:

$$s(t)=\sqrt{\frac{2E_b}{T_b}}\cos\left[\theta(t)\right]\cos\left[2\pi f_c t\right]-\sqrt{\frac{2E_b}{T_b}}\sin\left[\theta(t)\right]\sin\left[2\pi f_c t\right] \tag{6.14}$$

The in-phase component consists of half-cycle cosine pulse defined by

$$s_I(t)=\pm\sqrt{\frac{2E_b}{T_b}}\cos\left(\frac{\pi t}{2T_b}\right) \quad -T_b \le t \le T_b, \tag{6.15}$$

while the quadrature component would take the following form:

$$s_Q(t)=\pm\sqrt{\frac{2E_b}{T_b}}\sin\left(\frac{\pi t}{2T_b}\right) \quad 0 \le t \le 2T_b, \tag{6.16}$$

During even bit interval, the I component consists of positive cosine waveform for a phase of 0, while negative cosine waveform for a phase of π; during odd bit interval, the Q component consists of positive sine waveform for a phase of π/2, while negative sine waveform for a phase of −π/2 (see Figure 6.27). Any of four states can arise: 0, π/2, −π/2, and π. However, only state 0 or π can occur during any even bit interval and only π/2 or −π/2 can occur during any odd bit interval. The transmitted signal is the sum of I and Q components and its phase is continuous with time.

Two important characteristics of MSK are each data bit is held for a two-bit period, meaning that the symbol period is equal to a two-bit period ($h=½$), and the I and Q components are interleaved. I and Q components are delayed by a one-bit period with respect to each other. Therefore, only I or Q component can change at a time (when one is at zero-crossing, the other is at maximum peak). The precoder can be a combinational logic as shown in Figure 6.28.

A truth table can be constructed based on the logic state diagram and combinational logic diagram shown in Figure 6.28 (Table 6.4). For positive half-cycle cosine wave and positive half-cycle sine wave, the output is 1; for negative half-cycle cosine

TABLE 6.4
Truth Table Based on MSK State Diagram

$b_nS_0'S_1'$	$S_0'S_1'$	Output
100	01	1
001	00	1
100	01	1
101	10	0
010	01	1
101	10	0
110	11	0
111	00	1
000	11	0
011	10	0

wave and negative half-cycle sine wave, the output is 0. Then, a K-map can be constructed to derive the logic gates of the precoder, based on the truth table. The following three precoding logic equations are derived:

$$S_0 = \overline{b_n}\,\overline{S_0'}\,\overline{S_1'} + b_n \overline{S_0'}\, S_1' + \overline{b_n} S_0' S_1' + b_n S_0' \overline{S_1'} \tag{6.17}$$

$$S_1 = \overline{S_1'} = \overline{b_n}\,\overline{S_1'} + b_n \overline{S_1'} \tag{6.18}$$

$$\text{Output} = \overline{S_0} \tag{6.19}$$

The resultant logic gates construction for the precoder is depicted in Figure 6.28.

6.3.3.7.2 MSK as a Special Case of Continuous-Phase FSK

CPFSK signals are modulated in upper and lower sideband frequency carriers f_1 and f_2, which can be expressed as follows:

$$s(t) = \sqrt{\frac{2E_b}{T_b}} \cos\left[2\pi f_1 t + \theta(0)\right] \quad \text{symbol 1} \tag{6.20}$$

$$s(t) = \sqrt{\frac{2E_b}{T_b}} \cos\left[2\pi f_2 t + \theta(0)\right] \quad \text{symbol 0} \tag{6.21}$$

where

$$f_1 = f_c + \frac{1}{4T_b}$$

$$f_2 = f_c - \frac{1}{4T_b}$$

where T_b is the bit period.

The phase slope of lightwave carrier changes linearly or nonlinearly with the modulating binary data. In the case of linear MSK, the carrier phase linearly changes by $\pi/2$ at the end of the bit slot with data "1," while it linearly decreases by $\pi/2$ with data "0." The variation of phase follows paths of well-defined phase trellis in which the slopes represent frequency changes. The change in carrier frequency from data-0 to data-1, or vice versa, equals half the bit rate of incoming data [14]. This is the minimum frequency spacing that allows the two FSK signals representing symbols 1 and 0, to be coherently orthogonal in the sense that they do not interfere with one another in the process of detection. MSK carrier phase is always continuous at bit transitions. The MSK signal in 6.20 and 6.21 can be simplified as follows:

$$s(t) = \sqrt{\frac{2E_b}{T_b}} \cos[2\pi f_c t + d_k \frac{\pi t}{2T_b} + \Phi_k], \quad kT_b \le t \le (k+1)T_b \tag{6.22}$$

and the baseband equivalent optical MSK signal is represented as follows:

$$\tilde{s}(t) = \sqrt{\frac{2E_b}{T_b}} \exp\left\{ j\left[d_k \frac{\pi t}{2T} + \Phi(t,k) \right] \right\}, \quad kT \le t \le (k+1)T$$

$$= \sqrt{\frac{2E_b}{T_b}} \exp\left\{ j\left[d_k 2\pi f_d t + \Phi(t,k) \right] \right\} \tag{6.23}$$

where

$d_k = \pm 1$ are the logic levels

f_d is the frequency deviation from the optical carrier frequency

$h = 2f_d T$ is defined as the frequency modulation index

In the case of optical MSK, $h=1/2$ or $f_d = 1/(4T_b)$.

6.3.3.7.3 MSK as Offset Differential Quadrature Phase Shift Keying

Equation 6.16 can be rewritten to express MSK signals in the form of I–Q components as follows:

$$s(t) = \pm\sqrt{\frac{2E_b}{T_b}} \cos\left(\frac{\pi t}{2T_b}\right) \cos\left[2\pi f_c t\right] \pm \sqrt{\frac{2E_b}{T_b}} \sin\left(\frac{\pi t}{2T_b}\right) \sin\left[2\pi f_c t\right] \tag{6.24}$$

The I and Q components consist of half-cycle sine and cosine pulses defined by

$$s_I(t) = \pm\sqrt{\frac{2E_b}{T_b}} \cos\left(\frac{\pi t}{2T_b}\right) \quad -T_b < t < T_b \tag{6.25}$$

$$s_Q(t) = \pm\sqrt{\frac{2E_b}{T_b}} \sin\left(\frac{\pi t}{2T_b}\right) \quad 0 < t < 2T_b \tag{6.26}$$

During even bit intervals, the in-phase component consists of positive cosine waveform for a phase of 0, while negative cosine waveform for a phase of π; during odd bit interval, the Q component consists of positive sine waveform for a phase of π/2, while negative sine waveform for a phase of −π/2. Any of four states can arise: 0, π/2, −π/2, and π. However, only state 0 or π can occur during any even bit interval and only π/2 or −π/2 can occur during any odd bit interval. The transmitted signal is the sum of I and Q components and its phase is continuous with time.

Two important characteristics of MSK are the following: each data bit is held for a two-bit period, meaning the symbol period is equal to a two-bit period ($h = 1/2$), and the I and Q components are interleaved. I and Q components are delayed by a one-bit period with respect to each other. Therefore, only I or Q component can change at a time (when one is at zero-crossing, the other is at maximum peak).

6.3.4 Photonic MSK Transmitter Using Two Cascaded Electro-Optic Phase Modulators

E-OPMs and interferometers operating at high frequency using resonant-type electrodes have been studied and proposed in Ref. [29]. In addition, high-speed electronic driving circuits evolved with the ASIC technology using 0.1 μm GaAs P-HEMT or InP HEMTs

[15] enables the feasibility in the realization of the optical MSK transmitter structure. The baseband equivalent optical MSK signal is represented in (6.25) and (6.26).

The first E-OPM enables the frequency modulation of data logics into upper sidebands (USB) and lower sidebands (LSB) of the optical carrier with a frequency deviation of f_d. Differential phase precoding is not necessary in this configuration due to the nature of the continuity of the differential phase trellis. By alternating the driving sources $V_d(t)$ to sinusoidal waveforms for simple implementation or combination of sinusoidal and periodic ramp signals, which was first proposed by Amoroso in 1976 [16], different schemes of linear and nonlinear phase shaping MSK transmitted sequences can be generated whose spectra are shown in Figure 6.40.

The second E-OPM enforces the phase continuity of the lightwave carrier at every bit transition. The delay control between the E-OPMs is usually implemented by the phase shifter shown in Figure 6.29. The driving voltage of the second E-OPM is precoded to fully compensate the transitional phase jump at the output $E_{01}(t)$ of the first E-OPM. Phase continuity characteristics of the optical MSK signals is determined by the algorithm in (6.25) and (6.26).

$$\Phi(t,k) = \frac{\pi}{2}\left(\sum_{j=0}^{k-1} a_j - a_k I_k \sum_{j=0}^{k-1} I_j\right) \tag{6.27}$$

where

$a_k = \pm 1$ are the logic levels

$I_k = \pm 1$ is a clock pulse whose duty cycle is equal to the period of the driving signal $V_d(t)$

f_d is the frequency deviation from the optical carrier frequency

$h = 2f_dT$ is previously defined as the frequency modulation index

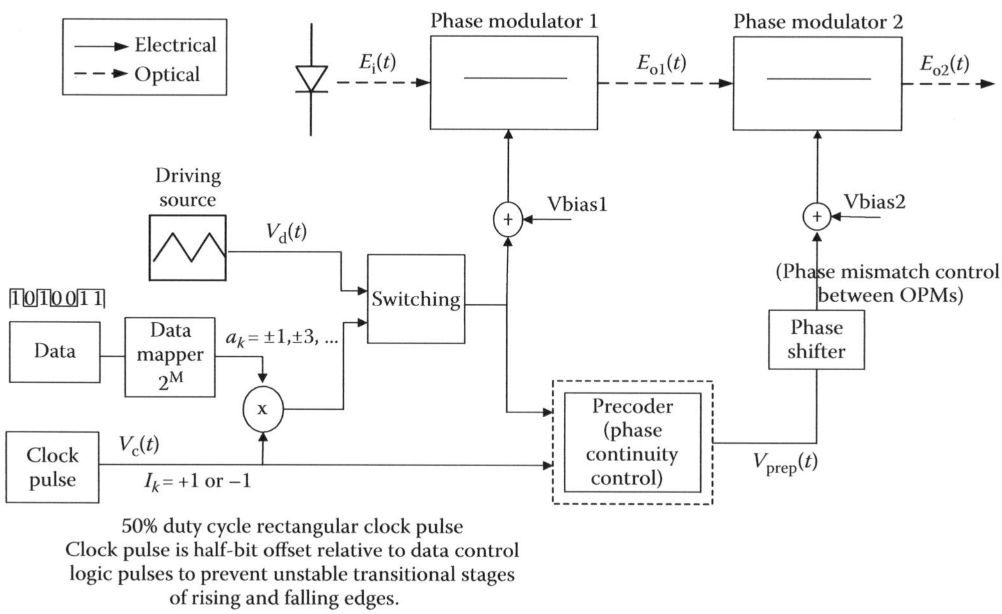

FIGURE 6.29
Block diagram of optical MSK transmitter employing two cascaded optical phase modulators.

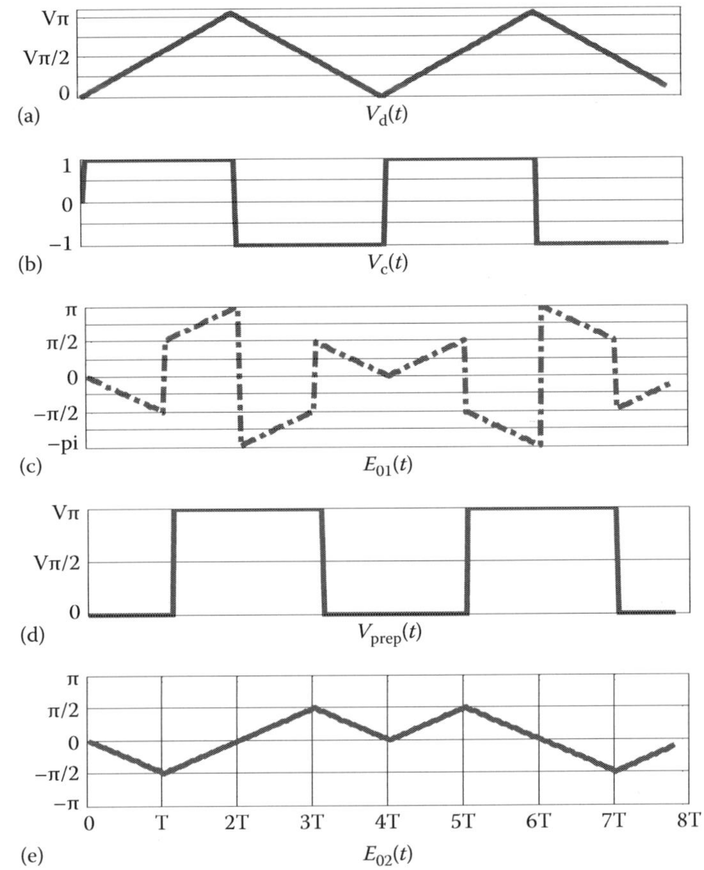

FIGURE 6.30
Evolution of time-domain phase trellis of optical MSK signal sequence [−1 1 1 −1 1 −1 1 1] as inputs and outputs at different stages of the optical MSK transmitter The notation is denoted in Figure 6.29 accordingly: (a) $V_d(t)$: periodic triangular driving signal for optical MSK signals with duty cycle of four-bit period, (b) $V_c(t)$: the clock pulse with duty cycle of $4T$, (c) $E_{01}(t)$: phase output of oPM1, (d) $V_{prep}(t)$: precomputed phase compensation driving voltage of oPM2, and (e) $E_{02}(t)$: phase trellis of a transmitted optical MSK sequences at output of oPM2.

In the case of optical MSK, $h=1/2$ or $f_d=1/(4T)$. The phase evolution of the continuous-phase optical MSK signals is explained in Figure 6.29. In order to mitigate the effects of unstable stages of rising and falling edges of the electronic circuits, the clock pulse $V_c(t)$ is offset with the driving voltages $V_d(t)$.

6.3.4.1 Configuration of Optical MSK Transmitter Using Mach–Zehnder Intensity Modulators: I–Q Approach

The conceptual block diagram of optical MSK transmitter is shown in Figure 6.31. The transmitter consists of two dual-drive electro-optic Mach–Zehnder modulators (MZMs) generating chirpless I and Q components of MSK-modulated signals, which is considered as a special case of staggered or offset QPSK. The binary logic data is precoded and de-interleaved into even- and odd-bit streams that are interleaved with each other by one bit duration offset.

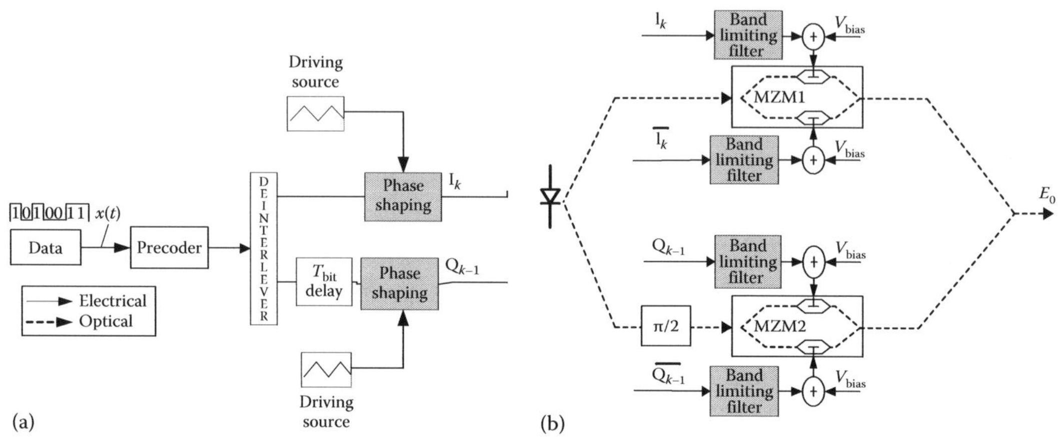

FIGURE 6.31
Block diagram configurations of band-limited phase-shaped optical MSK: (a) encoding circuit and pulse shaping and (b) parallel MZIM optical modulators.

Two arms of the dual-drive MZM modulator are biased at $V_\pi/2$ and $-V_\pi/2$ and driven with data and $\overline{\text{data}}$. Phase-shaping driving sources can be a periodic triangular voltage source in the case of linear MSK generation or simply a sinusoidal source for generating a nonlinear MSK-like signal, which also shows linear phase trellis property but with small ripples introduced in the magnitude. The magnitude fluctuation level depends on the magnitude of the phase-shaping driving source. High spectral efficiency can be achieved with tight filtering of the driving signals before modulating the electro-optic MZMs. Three types of pulse-shaping filters are investigated including Gaussian, raised cosine, and squared-root raised cosine filters. The optical carrier phase trellis of linear and nonlinear optical MSK signals are shown in Figure 6.32.

6.3.4.2 Single Sideband Optical Modulators

A single sideband (SSB) modulator can be formed using a primary interferometer with two secondary MZM structures, the optical Ti-diffused waveguide paths that form a nested primary MZ structure as shown in Figure 6.33. Each of the two primary arms contains an MZ structure. Two RF ports are there for RF modulation and three DC ports are there for biasing the two secondary MZMs and one primary MZM. The modulator consists of an X-cut Y-propagation $LiNbO_3$ crystal, where you can produce an SSB modulation just by driving each MZ. DC voltage is supplied to produce the π phase shift between upper and lower arms. DC bias voltages are also supplied from DC_B to produce the phase shift between third and fourth arms. A DC bias voltage is supplied from DC_C to achieve a π/2 phase shift between $MZIM_A$ and $MZIM_B$. The RF voltage applied $\Phi_1(t) = \Phi \cos \omega_m t$ and $\Phi_2(t) = \Phi \sin \omega_m t$ are inserted from RF_A and RF_B, respectively, by using a wideband π/2 phase shifter. Φ is the modulation level and ω_m is the RF angular frequency.

6.3.4.3 Optical RZ-MSK

The RZ format of the optical MSK modulation scheme can also be generated. A bit is used to generate the ASK-like feature of the bit. A Simulink structure of such a transmitter

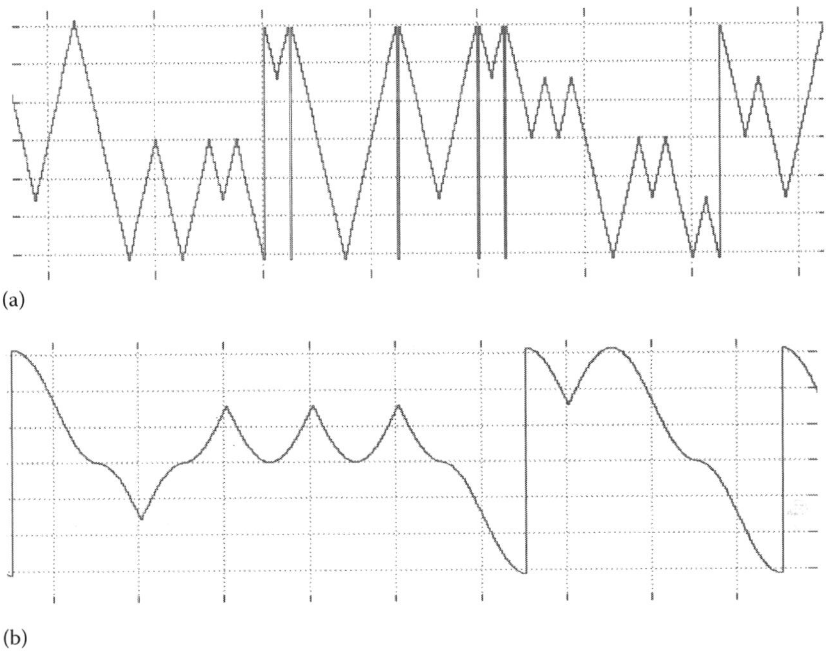

FIGURE 6.32
Phase trellis of linear and nonlinear MSK transmitted signals.

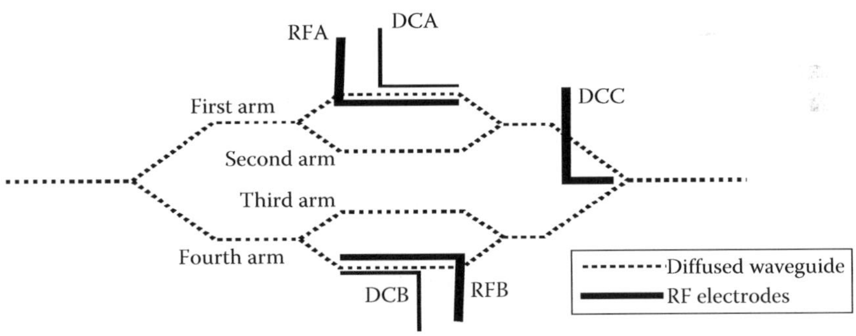

FIGURE 6.33
Schematic diagram (not to scale) of a SSB FSK optical modulator formed by nested MZ modulators.

is shown in Figure 6.34. The encoder as shown in the far left of the model provides two outputs, one for MSK signal generation and the other for amplitude modulation for the generation of the RZ or NRZ format. The amplitude and the phase of the RZ signals at the receivers are shown in Figure 6.35 after a 100 km span transmission and full compensation. The phase of the RZ MSK must be nonzero so as they satisfy the continuity of the phase between one state to the other.

6.3.5 Multi-Carrier Multiplexing Optical Modulators

Another modulation format that can offer much higher single-channel capacity and flexibility in dispersion and nonlinear impairment mitigation is the employment of

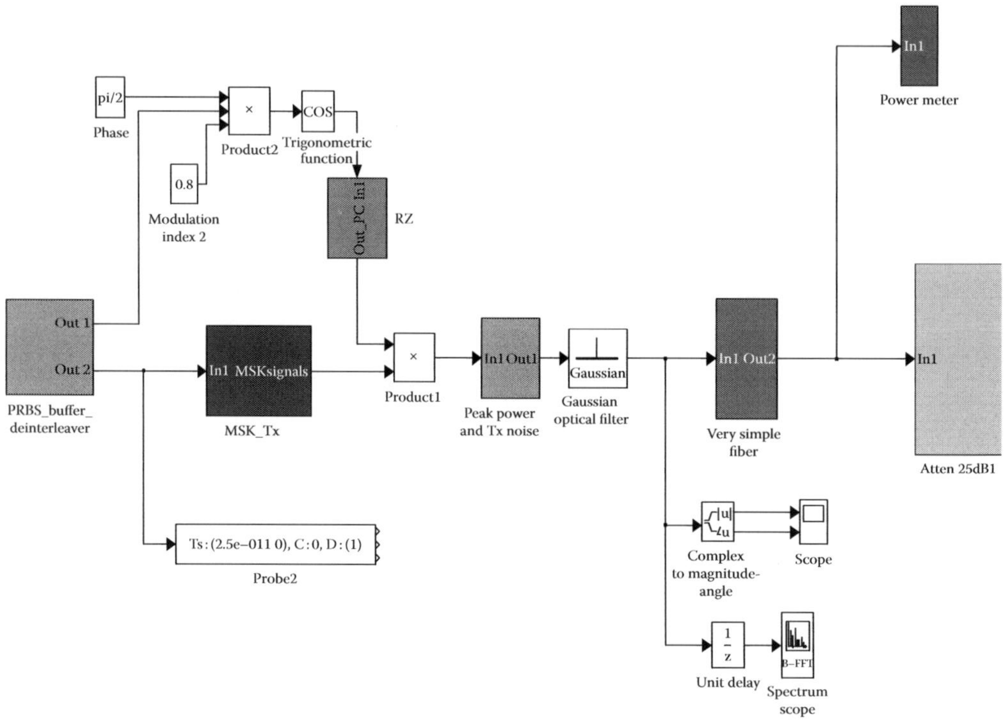

FIGURE 6.34
MATLAB and Simulink models of an RZ-MSK optical transmission system.

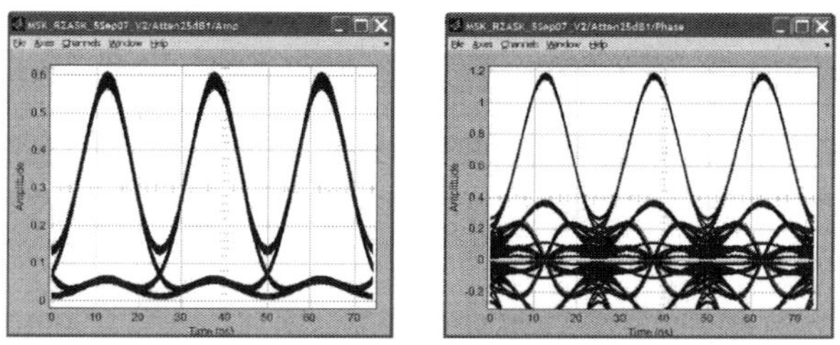

(a) (b)

FIGURE 6.35
RZ eye diagram at outputs of the (a) amplitude receiver and (b) phase detection.

multi-carrier multiplexing. When these sub-carrier channels are orthogonal, hence the term orthogonal frequency division multiplexing (OFDM) is used.

Our motivation in the introduction of OFDM is due to its potential as an ultrahigh capacity channel for the next generation Ethernet, the optical Internet. The network interface cards for 1 and 10 Gb/s Ethernet are readily commercially available. Traditionally, the Ethernet data rates have grown in 10 Gb/s increments, so the data rate of 100 Gb/s can be expected as the speed of the next generation of Ethernet. The 100 Gb/s all-electrically time-division-multiplexed (ETDM) transponders are becoming increasingly important because they are viewed as a promising technology that may be able to meet speed

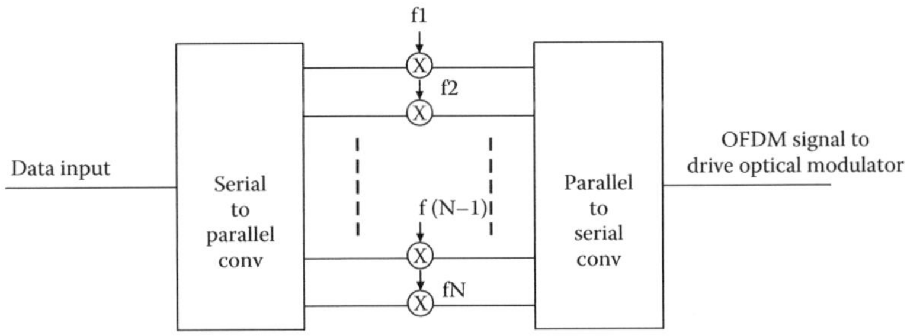

FIGURE 6.36
Multi-carrier FDM signal arrangement. The middle section is the RF converter as shown in Figure 6.26.

requirements of the new generation of Ethernet. Despite the recent progress in high-speed electronics, ETDM [19] modulators and photodetectors are still not widely available, so that alternative approaches to achieving a 100 Gb/s transmission using commercially available components and QPSK is very attractive.

OFDM is a combination of multiplexing and modulation. The data signal is first split into independent subsets and then modulated with independent sub-carriers. These sub-channels are then multiplexed for OFDM signals. OFDM is thus a special case of FDM but instead of one stream, it is a combination of several small streams into one bundle.

A schematic signal flow diagram of a multi-carrier multiplexing (MCM) is shown in Figure 6.36. The basic OFDM transmitter and receiver configurations are given in Figure 6.37a and b, respectively. Data streams (e.g., 1 Gb/s) are mapped into a two-dimensional signal point from a point signal constellation such as QAM. The complex-valued signal points from all sub-channels are considered as the values of the discrete Fourier transform (DFT) of a multi-carrier OFDM signal. The symbol interval length in an OFDM system is the number of sub-carriers as desired, where is the symbol interval length in a single-carrier system. By selecting the number of sub-channels, sufficiently large, the OFDM symbol interval can be made much larger than the dispersed pulse width in a single-carrier system, resulting in an arbitrary small intersymbol interference (ISI). The OFDM symbol, shown in Figure 6.38, is generated under software processing as follows: input QAM symbols are zero-padded to obtain input samples for inverse fast Fourier transform (IFFT), the samples are inserted to create the guard band, and the OFDM symbol is multiplied by the window function (raised cosine function can be used). The purpose of cyclic extension is to preserve the orthogonality among sub-carriers even when the neighboring OFDM symbols partially overlap due to dispersion.

A system arrangement of the OFDM for optical transmission in laboratory demonstration is shown in Figure 6.39 [13]. Each individual channel at the input would carry the same data rate sequence. These sequences can be generated from an arbitrary waveform generator. The multiplexed channels are combined and converted to time domain using the IFFT module and then converted to the analog version via the two digital-to-analog converters. These orthogonal data sequences are then used to modulate I and Q optical waveguide sections of the electro-optical modulator to generate the orthogonal channels in the optical domain. Similar decoding of I and Q channels are performed in the electronic domain after the optical transmission and optical-electronic conversion via the photodetector and electronic amplifier.

In OFDM, the serial data sequence, with a symbol period of Ts and a symbol rate of 1/Ts is split up into N-parallel sub-streams (sub-channels).

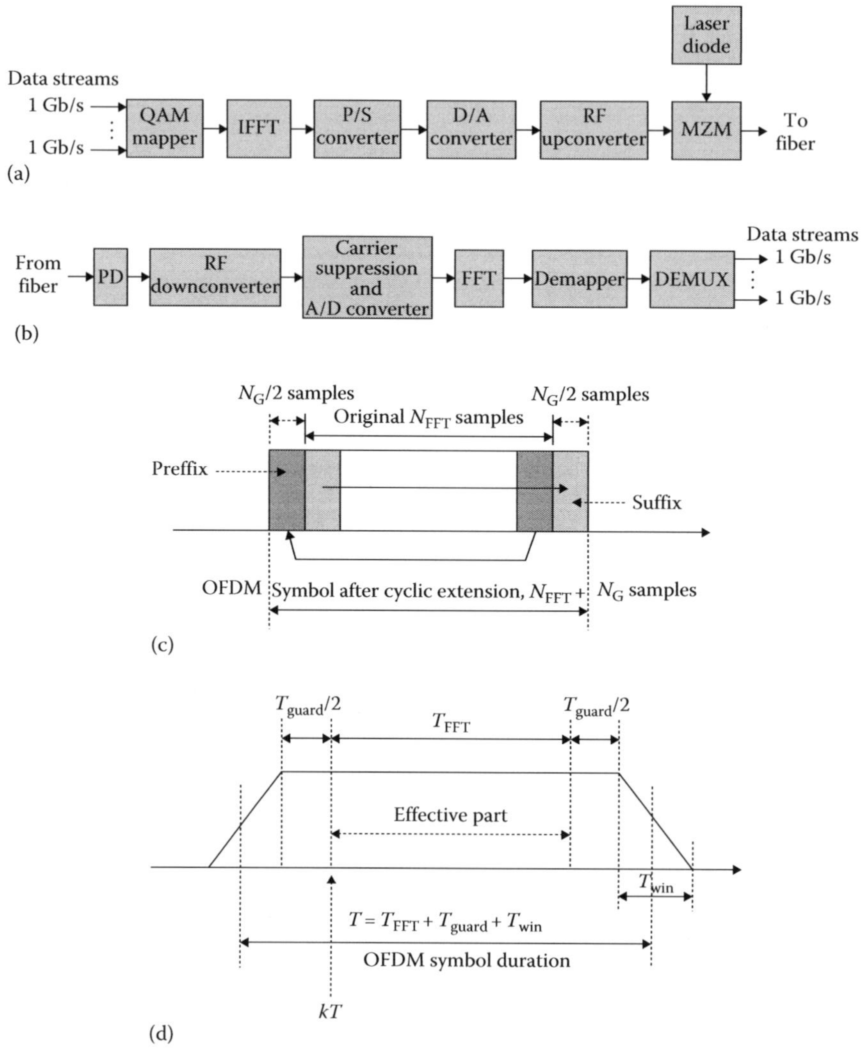

FIGURE 6.37
Schematic diagram of the principles of generation and recovery of OFDM signals: (a) electronic processing and optical transmitter; (b) optoelectronic and receiver configurations; (c) OFDM symbol cyclic extension; and (d) OFDM symbol after windowing. (Extracted from Krahenbuhl, R., *J. Lightwave Technol.*, 24(5), 2184, 2006.)

6.3.6 Spectra of Modulation Formats

Utilizing this double-phase modulation configuration, different types of linear and nonlinear continuous phase modulation phase shaping signals including MSK, weakly nonlinear MSK, and linear-sinusoidal MSK can be generated. The third scheme was introduced by Amoroso [16] and its side lobes decay with a factor of 8 compared to 4 of MSK. The simulated optical spectra of DBPSK and MSK schemes at 40 Gb/s are contrasted in Figure 6.40. Table 6.5 outlines the characteristics and spectra of all different modulation schemes.

Figure 6.42 shows the power spectra of the DPSK-modulated optical signals with various types of pulse shapes including NRZ, RZ33, and CSRZ. For the convenience of the

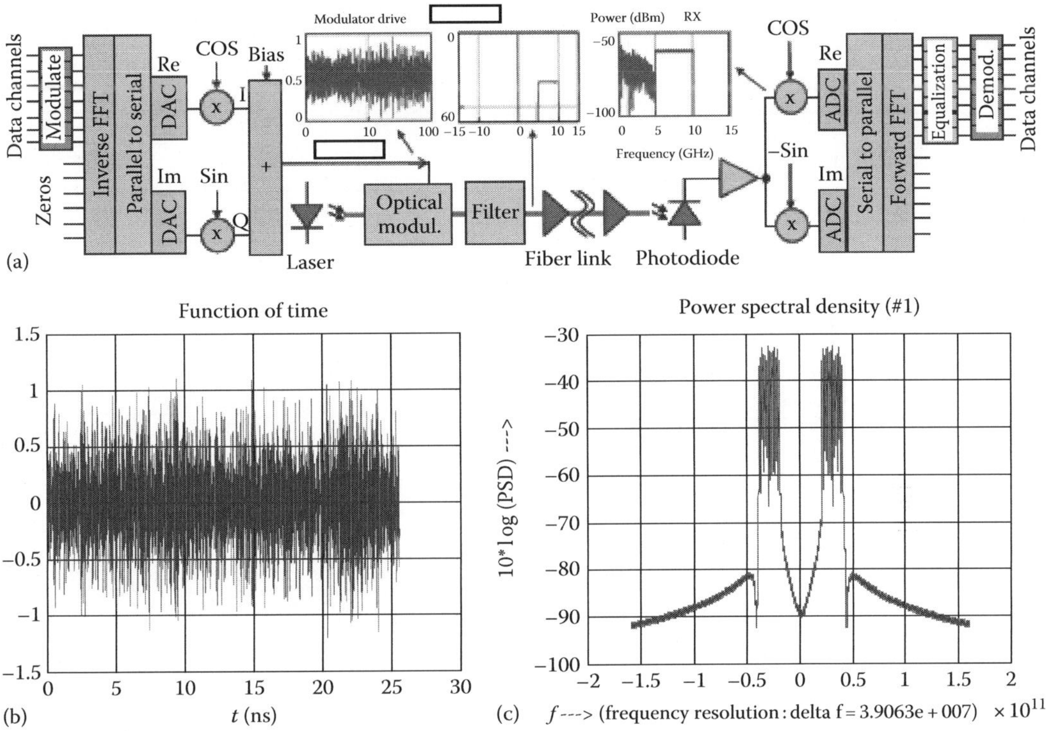

FIGURE 6.38
An optical FFT/IFFT-based (a) OFDM system including representative waveforms and spectra. (Extracted from Djordjevic, I.B. and Vasic, B., *IEEE Photonics Technol. Lett.*, 18(15), 1576, August 1, 2006.), (b) typical time domain OFDM signals, and (c) power spectral density of OFDM signal with 512 sub-carriers, with a shift of 30 GHz for the line rate of 40 Gb/s and QPSK modulation.

comparison, the optical power spectra of the RZ OOK counterparts are also shown in Figure 6.40.

Several key notes observed from Figures 6.41 and 6.42 are outlined as follows. (1) The optical power spectrum of the OOK format has high power spikes at the carrier frequency or at signal modulation frequencies, which contribute significantly to the severe penalties caused by the nonlinear effects. Meanwhile, the DPSK optical power spectra do not contain these high-power frequency components; (2) RZ pulses are more sensitive to the fiber dispersion due to their broader spectra. In particular, RZ33 pulse type has the broadest spectrum at the point of −20 dB down from the peak. This property of the RZ pulses thus leads to faster spreading of the pulse when propagating along the fiber. Thus, the peak values of the optical power of these CSRZ or RZ33 pulses decrease much more quickly than the NRZ counterparts. As a result, the peak optical power quickly turns to be lower than the nonlinear threshold of the fiber, which means that the effects of fiber nonlinearity are significantly reduced; and (3) however, NRZ optical pulses have the narrowest spectrum. Thus, they are expected to be most robust to the fiber dispersion. As a result, there is a trade-off between RZ and NRZ pulse types. RZ pulses are much more robust to nonlinearity but less tolerant to the fiber dispersion. The RZ33/CSRZ–DPSK optical pulses are proven to be more robust against impairments especially self-phase modulation and PMD compared to the NRZ-DPSK and the CSRZ/RZ33-OOK counterparts.

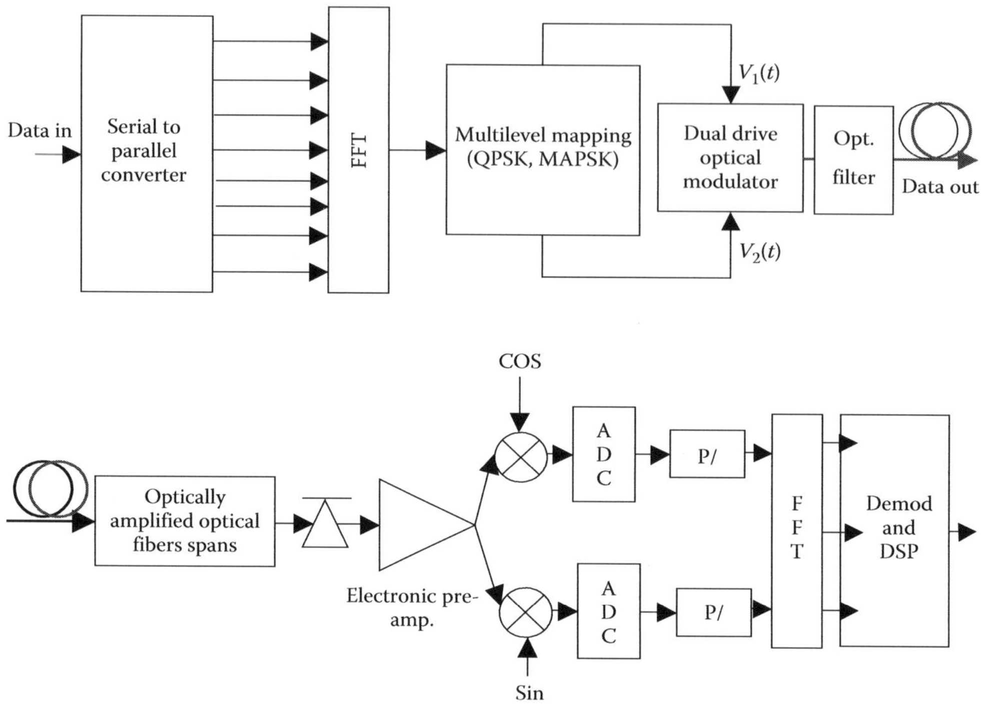

FIGURE 6.39
Schematic diagram of an optical FFT/IFFT-based OFDM system. S/P and P/S ~ serial to parallel conversion and vice versa.

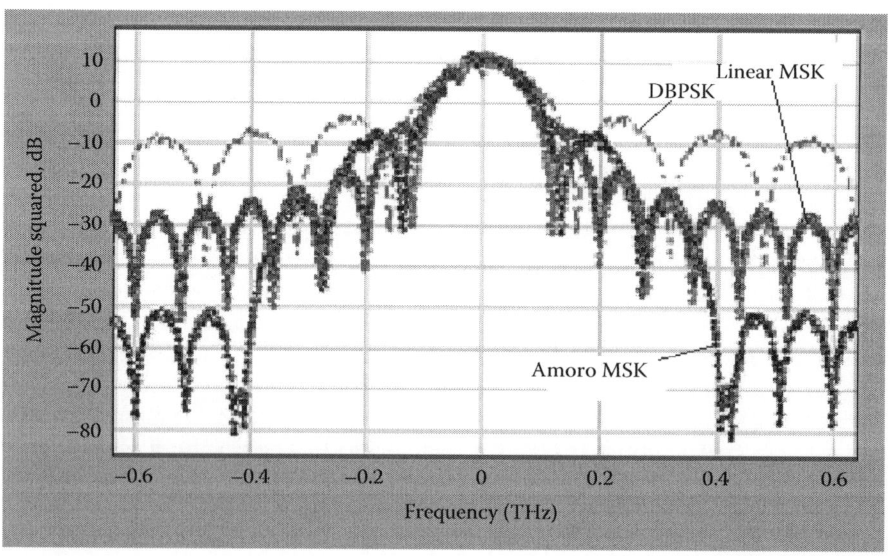

FIGURE 6.40
Spectra of 40 Gb/s DBPSK, and linear and nonlinear MSKs.

TABLE 6.5

Typical Parameters of Optical Intensity Modulators for the Generation of Modulation Formats

Modulation Techniques	Spectra	Formats	Definition/Comments
Amplitude modulation—ASK-NRZ	DSB + carrier	ASK-NRZ	Biased at quadrature point or offset for pre-chirp
AM—ASK-RZ	DSB + carrier	ASK-RZ	Two MZIMs required—one for RZ pulse sequence and other for data switching
ASK-RZ-carrier suppressed	DSB-CSRZ	ASK-RZCS	Carrier-suppressed, biased at π phase difference for the two electrodes
SSB	SSB + carrier	SSB NRZ	Signals applied to MZIM are in phase quadrature to suppress one sideband. Alternatively an optical filter can be used
CSRZ DSB	DSB-carrier	CSRZ-ASK	RZ pulse carver is biased such that a π phase shift between the two arms of the MZM to suppress the carrier and then switch on and off or phase modulation via a data modulator.
DPSK-NRZ DPSK-RZ, CSRZ-DPSK		Differential BPSK RZ or NRZ/RZ-carrier suppressed	
DQPSK		DQPSK-RZ or NRZ	Two bits per symbol
MSK	SSB equivalent	Continuous-phase modulation with orthogonality	Two bits per symbol and efficient bandwidth with high side-lobe suppression.
Offset-DQPSK		Oriented π/4 as compared to DQPSK constellation	Two bits per symbol
MCM (e.g., OFDM)	Multiplexed bandwidth – base rate per sub-carrier		
Duo-binary	Effective SSB		Electrical low-pass filter required at the driving signal to the MZM
FSK	Two distinct frequency peaks		
Continuous-phase FSK	Two distinct frequency peaks with phase continuity at bit transition	MSK	When the frequency difference equals to a quarter of the bit rate, the signals for "1" and "0" are orthogonal and thus called MSK.
PM	Chirped carrier phase		Chirpless MZM should be used to avoid inherent crystal effects, hence carrier chirp.

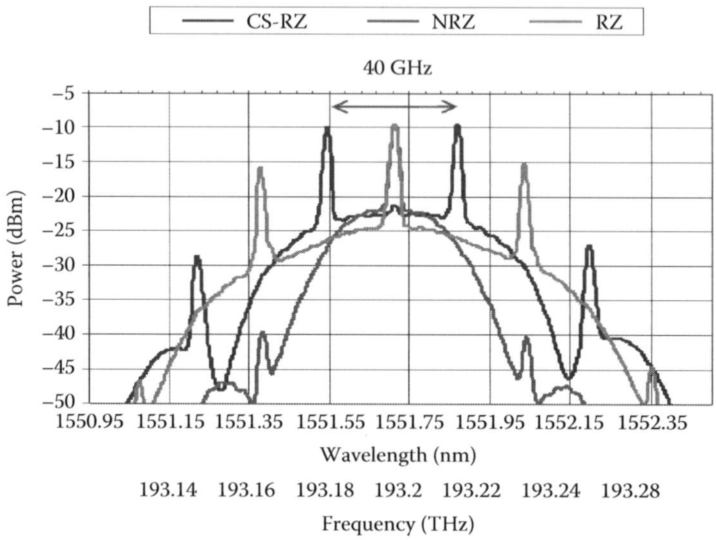

FIGURE 6.41
Spectra of CSRZ/RZ/NRZ-OOK-modulated optical signals.

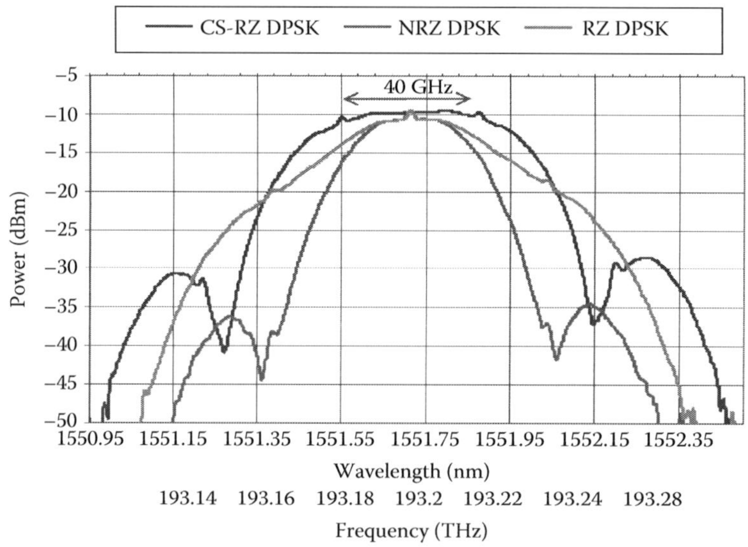

FIGURE 6.42
Spectra of CSRZ/RZ33/NRZ—DPSK-modulated optical signals.

Optical power spectra of three I–Q optical MSK modulation formats that are of linear, weakly nonlinear, and strongly nonlinear types are shown in Figure 6.43. It can be observed that there are no significant distinctions of the spectral characteristics between these three schemes. However, the strongly nonlinear optical MSK format does not highly suppress the side lobe as compared to the linear MSK type. All three formats offer better spectral efficiency compared to the DPSK counterpart as shown in Figure 6.44. This figure compares the power spectra of three modulation formats: 80 Gb/s dual-level MSK,

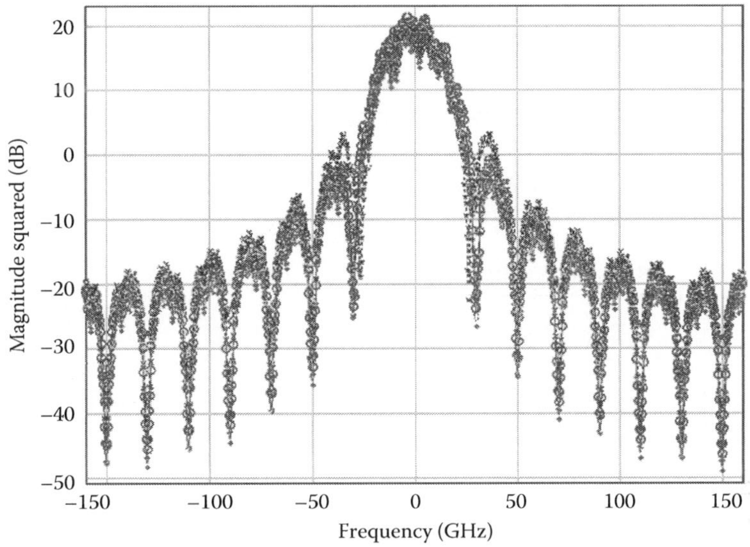

FIGURE 6.43
Optical power spectra of three types of I–Q optical MSK formats: linear, weakly nonlinear, and strongly nonlinear which are very close to each other.

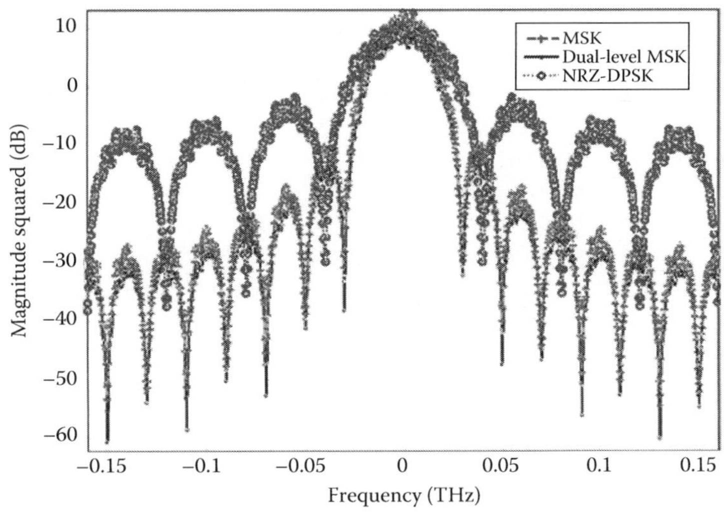

FIGURE 6.44
Spectral properties of three modulation formats: MSK, dual-level MSK, and DPSK.

40 Gb/s MSK, and NRZ-DPSK optical signals. The normalized amplitude levels into the two optical MSK transmitters comply with the ratio of 0.75/0.25.

Several key notes can be observed from Figure 6.44 and are outlined as follows. (1) The power spectrum of the optical dual-level MSK format has identical characteristics to that of the MSK format. The spectral width of the main lobe is narrower than that of the DPSK. The base-width takes a value of approximately ±32 GHz on either side compared to

±40 GHz in the case of the DPSK format. Hence, the tolerance to the fiber dispersion effects is improved; (2) high suppression of the side lobes with a value of approximately greater than 20 dB in the case of 80 Gb/s dual-level MSK and 40 Gb/s optical MSK power spectra; thus, more robustness to interchannel cross talk between DWDM channels; and (3) the confinement of signal energy in the main lobe of spectrum leads to a better SNR. Thus, the sensitivity to optical filtering can be significantly reduced [5.4–5.6]. A summary of the spectra of different modulation formats is given in Table 6.5.

6.4 Concluding Remarks

Since the proposal of dielectric waveguide and then the advent of optical circular waveguide, the employment of modulation techniques is only extensively exploited recently since the availability of optical amplifiers. The modulation formats allow the transmission efficiency, and hence the economy of ultra-high-capacity information telecommunications. Optical communications have evolved significantly through several phases from single-mode systems to coherent detection and modulation, which was developed with the main aim of improving the optical power. The optical amplifiers defeated that main objective of modulation formats and allowed the possibility of incoherent and all possible formats employing the modulation of the amplitude, the phase and frequency of the lightwave carrier.

Currently, photonic transmitters play a principal part in the extension of the modulation speed into several gigahertz range and make possible the modulation of the amplitude, the phase, and the frequency of the optical carriers and their multiplexing. Photonic transmitters using $LiNbO_3$ have been proven in laboratory and installed systems. The principal optical modulator is the MZIM, which can be a single or a combined set of these modulators whereby to form binary or multilevel amplitude or phase modulation and even more effective for discrete or continuous PSK techniques. The effects of the modulation on transmission performance will be given in the following chapters.

Spectral properties of the optical 80 Gb/s dual-level MSK, 40 Gb/s MSK, and 40 Gb/s DPSK with various RZ pulse shapes are compared. The spectral properties of the first two formats are similar. Compared to the optical DPSK, the power spectra of optical MSK and dual-level MSK modulation formats have more attractive characteristics. These include the high spectral efficiency for transmission, higher energy concentration in the main spectral lobe, and more robustness to interchannel cross talk in DWDM due to greater suppression of the side lobes. In addition, the optical MSK offers the orthogonal property, which may offer a great potential in coherent detection, in which the phase information is reserved via I and Q components of the transmitted optical signals. In addition, the multilevel formats would permit the lowering of the bit rate and hence substantial reduction of the signal effective bandwidth and the possibility of reaching the highest speed limit of the electronic signal processing, the digital signal processing, for equalization and compensation of distortion effects. The demonstration of the ETDM receiver at 80 G and higher speed [18] would allow the applications of these modulation formatted scheme very potential in ultra-high-speed transmission.

6.5 Problems

PROBLEM 6.1

Refer to the *CSRZ* optical transmitter of Section 6.3.2.2.

1. Sketch the time-domain pulse sequence over a 10-bit period for a bit rate of 40 Gb/s at the output of the pulse pattern generator.
2. Give a brief description of the principles of the suppression of the carrier. Which component of the transmitter would implement the suppression?
3. What are the functions of the laser, the modulator, and the push-pull modulator. If the V_π for the two modulators are 5 V, sketch the transfer characteristics of the modulators—that is, the output power versus the input driving voltage. Make sure that you set appropriate biasing voltages for the modulators. The output power of the laser is 10 dBm and the total insertion loss for each modulator is 4 dB. For the pulse pattern generator, the output power at the output port data is 10 dBm and that at the clock output port is $2V_{p-p}$. All line impedances are 50 Ω.
4. Is it necessary to use a booster optical amplifier to increase to total average power launching into an optical fiber for transmission. If it does then what is the gain and noise of the optical amplifier. Note that the nonlinear limit of an SSMF is around 5 dBm.
5. Sketch the spectra at the outputs of the laser, the modulator, and the push-pull modulator.

PROBLEM 6.2

Optical modulator and phasor diagram—dual-drive MZM.

1. Repeat Question 3 with a bias voltage of $V_b = V_\pi/2$ and V_π and a time-varying signal of $v_s(t) = V_{\pi/2} \cos\omega_s t$ with $\omega_s = 2\pi f_s$ and $f_s = 20$ GHz.
2. Now the modulator is a dual-drive MZM, repeat Question 1 with a bias voltage of $V_b = V_\pi/2$ and V_π and a time-varying signal of $v_s(t) = V_{\pi/2} \cos\omega_s t$ with $\omega_s = 2\pi f_s$ and $f_s = 20$ GHz.

PROBLEM 6.3

An optical fiber communication system consists of (1) an optical transmitter using a 1550 nm DFB laser with a linewidth of 10 pm (picometers), an external optical modulator whose bandwidth is 20 GHz, and a total insertion loss of 5 dB. (2) The modulator is driven with a bit-pattern-signal generator with a 10 dBm electrical power output into a 50 Ω line. A microwave amplifier is used to boost the electrical data pulse to an appropriate level for driving the optical modulator. The data bit rate is 10 Gb/s and its format is NRZ. (3) An 80 km SSMF is used for the transmission of the modulated signals.

Sketch the block diagram of the transmission system.

If the V_π of the external modulator is 5 V. What is the gain of the microwave amplifier so that an extension ration of 20 dB can be achieved for the output pulses of "1" and "0" at the output of the modulator. Make sure that you sketch the amplitude and power output of the modulator versus the driving voltage into the modulator. What type of connector would you use to connect the microwave amplifier to the modulator and to the bit-pattern-generator.

If the DFB laser emits 0 dBm optical power at its pigtail output then what is the average of optical power contained in the signal spectrum. You may assume that the pulse sequence generated at the output of the bit-pattern-generator is a perfect rectangular shape. What is the effective 3 dB bandwidth of the signal power spectrum? Thence estimate the total pulse broadening of the pulse sequence at the end of the 80 km fiber length. Similarly, estimate the pulse sequence if the bit rate is 40 Gb/s.

Now if a dispersion-compensating fiber of 20 km is used to compensate for the signal distortion in the 80 km fiber, what is the required dispersion factor of this fiber so that there would be no distortion? If the loss of the dispersion-compensating fiber is 1.0 dB/km at 1550 nm, estimate the average optical power of the signal at the output of the dispersion-compensating fiber.

Based on the dispersion limit given in the following equation, plot the dispersion length as a function of the bit rate for NRZ format.

Note: The dispersion limit, under linear regime operation, can be estimated in the following equation:

$$L_D = \frac{c}{\lambda}\frac{\rho}{B_R^2 D}$$

where

B_R is the bit rate

D is the dispersion factor (s/m^2)

ρ is the duty cycle ration, that is the ration between the "ON" and "OFF" in a bit period

L_D is in meters

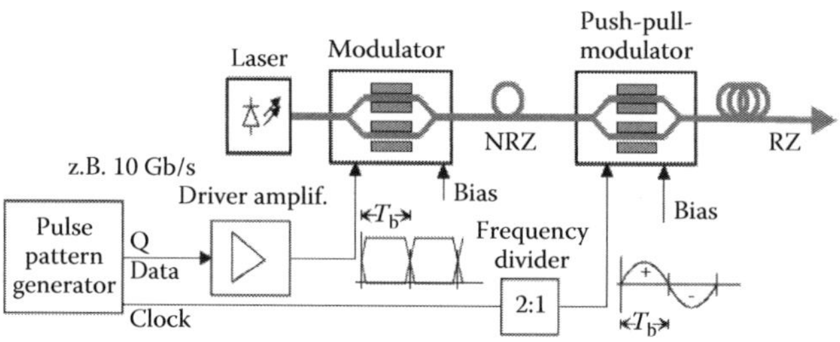

PROBLEM 6.4

Repeat Question 6 for RZ format and ASK modulation. Sketch the structure of the RZ optical transmitter; note that an extra optical modulator must be used and coupled with the data modulator of Question 1, the optical pulse carver. Give details of the pulse carver including driving voltage, driving signal, and synchronization with the data generator.

PROBLEM 6.5

Spectral Efficiency

1. A DWDM optical transmission system that can transmit optical channels whose channel spacing is 100 GHz. What is the spectral efficiency if the bit rate of each channel is 40 Gb/s and the modulation is NRZ-ASK?
2. Repeat 1 for RZ-ASK modulation format.
3. Repeat 1 and 2 for the channel spacing of 50 GHz.

PROBLEM 6.6

Give a structure of an optical transmitter for the generation of RZ-ASK modulation format. Make sure that you assign the optical power of lightwaves generated from the light source, and that at the output of the optical modulators so that a maximum of

10 dBm of optical power is launched into the SSMF so that it is below the nonlinear SPM effect limit.

Describe the operation of the optical modulator, the pulse carver, so that it can generate periodic pulse sequence before feeding into the data generator. Make sure that you provide the amplitude and intensity versus the driving signal voltage levels that are used to drive the optical modulators.

PROBLEM 6.7

Nonlinear SPM Effect

The nonlinear refractive index coefficient of silica-based SSMF is $n_2 = 2.5 \times 10^{-20}$ m^2/W.

What is the effective area of the SSMF? You can refer to the technical specification of the Corning SMF-28 and its mode field diameter to estimate this area.

Estimate the change of the refractive index as a function of the average optical power. Hence estimate the total phase change due to this nonlinear effect after propagating though a length L (in kilometers) of this fiber.

Hence estimate the maximum length L of the SSMF that the lightwaves can travel so that not higher than 0.1 rad of the phase change on this lightwave carrier would be suffered.

Show how you can generate an RZ format and a suppression of the lightwave carrier. Show that the width of the RZ pulse on this case is 67% of the bit period. *Hint*: you may represent the lightwaves in the path of the optical modulator, an optical interferometer, by using phasors. First sketch the phasor of the input lightwave. Then those of the two paths and then the phase applied onto these paths. Then sum up at the output to give the resultant output. For the pulse width you can estimate the width over which the amplitude falls to $1/\sqrt{2}$ of its maximum.

Now show that you can generate an RZ pulse sequence with 50% and 33% pulse width of the bit period.

PROBLEM 6.8

Balanced Receiver

Sketch the schematic diagram of an optical balanced receiver—a balanced receiver would consist of a delay interferometer and a back-to-back connected pair of photodetector with its output connected to the input of an optical pre-amplifier.

What is the functionality of the delay interferometer. What is the temporal length of the delay unit?

What are the roles of the two optical couplers and their ideal coupling coefficients?

What is the relationship between the two output ports of the delay interferometer?

Suppose that a sequence of four bits of a DPSK 10 Gb/s data channel is presented at the input of a balanced receiver. The phases of the lightwave carrier contained within these four bits are π π o π at the transition of the bit period.

Sketch the carrier wave and the pulse envelop. The lightwave has a wavelength of 1550 nm—however to illustrate the wave you are expected to sketch only a few periods of the waves contained within the bit period at the input of the receiver.

Sketch the electrical signal at the output of the electronic preamplifier not including noises.

Now assume that an optical amplifier is used as an optical preamplifier and is placed at the input of the balanced receiver that would give an optical signal power of −10 dBm for the "0" and "1" of the DPSK sequence. The responsibility of the photodetector is 0.9, and the electronic preamplifier has a trans-impedance of 150 Ω, a total equivalent noise current spectral density of 2 pA/(Hz)$^{1/2}$, and a bandwidth of 15 GHz. Sketch the signal waveform at the output of the electronic preamplifier.

PROBLEM 6.9

Duo-Binary Modulation Format

Design a block diagram of a precoder that would generate tri-level modified duo-binary format signals. Make sure that the coefficients of the filters are specified. *Hint*: you may refer to Section 6.3.3. Hence, derive the spectrum of the signals after the precoder of the modified duo-binary.

If possible obtain the precoders for AMI and duo-binary and their frequency responses. Compare the frequency responses of the three modulation schemes.

Sketch the structure of the tri-level duo-binary precoder with its output of −1, 0, and +1.

Now show how to use the coded signals to drive a dual-drive MZIM to generate optical duo-binary signals.

PROBLEM 6.10

Duo-Binary Modulation Format

A modulation format that would allow the detection of the modulated signals is duo-binary, which is a special case of partial response coding (see any textbook on digital communications).

Give a brief account of the principles of operation of this line code. A duo-binary coder using a delay and add coding structure is shown below. If three-level duo-binary coded signals are required, design the precoder for this type of modulation.

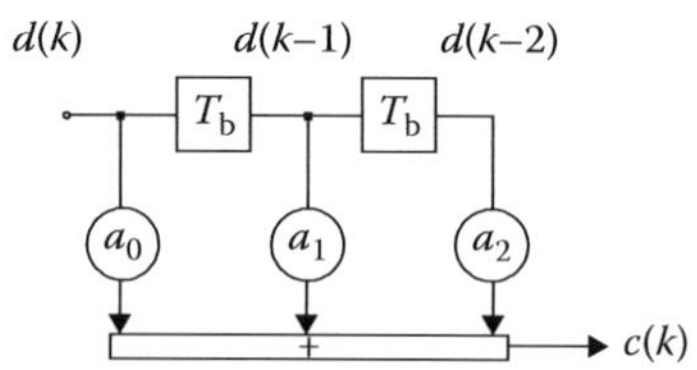

Now if we are setting the delay time Tb as a bit period, transform the structure into the z-transform diagram and hence obtain the transfer function of the coder in the z-domain, thence the frequency response of this coder. Plot the frequency response of the transfer function of the filter in continuous domain.

Find the impulse response of the coder, and hence the term partial response.

Sketch a block diagram that shows the functionality of precoding, coding, tri-level conversion (offset), and decoding.

A binary sequence $d(k) = \{0\ 0\ 1\ 1\ 0\ 1\ 0\ 0\ 1\}$ is applied to the input of the duo-binary coder. Determine the data sequences $b(k)$, $c(k)$, and $c'(k)$ in the electrical domain that can be used to modulate an optical modulator.

Assuming that there is no dispersion in the transmission of the duo-binary data sequence, find the output pulse sequence at the output of the decoder. What is the physical realization of the decoder? Thence sketch the sequence at the output of a decision circuit.

Now the electrical signals are applied to a microwave amplifier that would condition the signals to appropriate signal levels so as to modulate the optical modulator. The measured spectra are recorded as shown in the diagrams below.

Determine where in the block diagram (as per attached diagram) that each of the spectrum belongs to the points of the diagram of the transmission system.

PROBLEM 6.11

DQPSK

DQPSK is a 2 bit per symbol modulation, that is, 2 bits/symbol; thus the scheme is spectrally efficient.

Give a brief account of the modulation schemes DPSK and DQPSK.

Give a structure of a precoder for DPSSK—that is, give a differential modulation with phase as the codes for "1" and "0."

Now extend this precoder and the phase quadrature modulation technique for the structure of a DQPSK optical transmitter.

PROBLEM 6.12

SSB and DSB Modulation

Referring to the diagrams shown in Figure P.6.12 for the generation of optical signals with SSB, state the functionality of the Hilbert transformer. Hence, could you deduce a general principle for the suppression of a sideband to generate SSB signals.

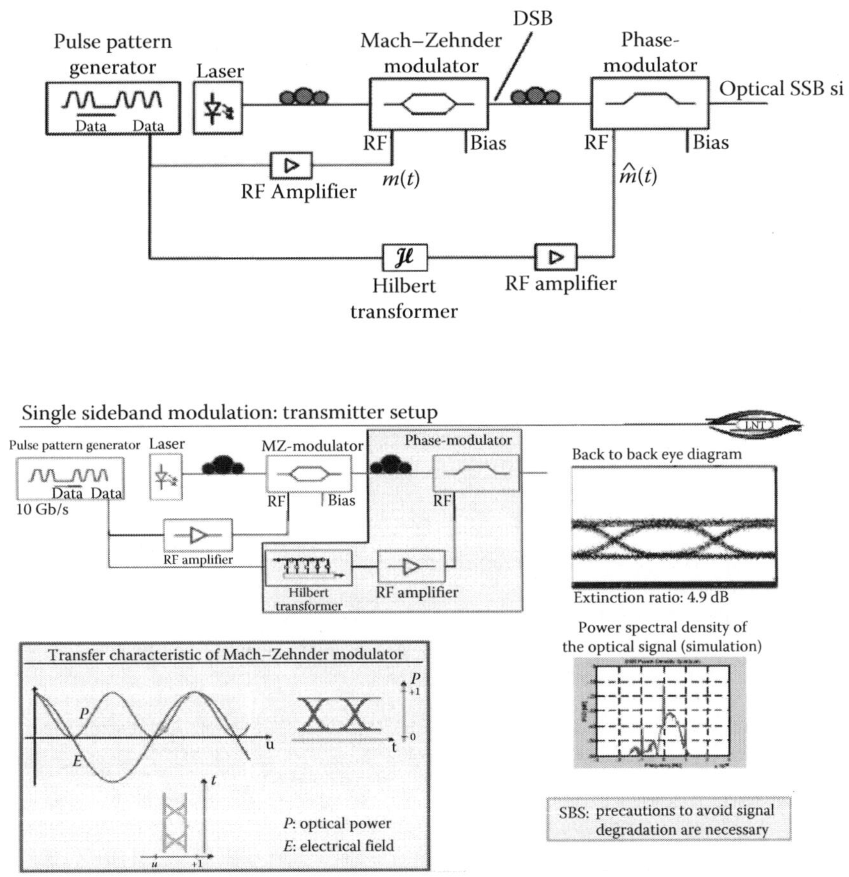

FIGURE P.6.12
SSB modulation and generation using (a) transform in optical domain, (b) transform in electrical domain, and (c) the realization of an SSB optical transmitter.

What is the role of the phase shifter π/2.

Explain the operation of the optical SSB transmitter, in both the time and frequency domain. Confirm that the spectrum is correct.

PROBLEM 6.13

Coherent Optical Communication Systems

Sketch a structure of an optical coherent receiver. Give a brief description of the roles of each component in your system.

What is the typical modern linewidth of the laser that acts as the local oscillator?

Give a distinction between the homodyne and heterodyne coherent systems.

A homodyne optical receiver has the following parameters:

A photodetector with a responsibility of 0.9 that is followed by an electronic preamplifier whose total equivalent noise spectral density is $5\,\mathrm{pA/(Hz)^{1/2}}$ and an electrical bandwidth of 15GHz. The transmission bit rate is 10Gb/s.

The local oscillator is a tunable laser source with a linewidth of 100MHz. The wavelength in vacuum of both the signal and the local oscillator is 1550.92. The average optical power of the local oscillator coupled to the photodetector is 0dBm. Sketch the structure of the receiver and then its equivalent small signal circuit that includes the generated electronic signal current at the output of the photodetector, the total noise currents looking from the input of the electronic preamplifier. What is the dominant noise source of the receiver?

For an optical signal with an average power of −20dBm, estimate the SNR at the output of the photodetector. Recalculate the SNR of the receiver if the frequency of the local oscillator is 20GHz away from that of the signal carrier frequency.

Appendix: Structures of Mach–Zehnder Modulator

It is clear that electric field lines are along the Z axis and the electrode of the driving signal is located at either side of the optical waveguide. This configuration is driven by only one electrical drive (data), and it is chirp free due to the symmetrical structure that leads to intrinsic balance, as shown in Figures 6.45 and 6.46.

There is a distinct difference from X-cut where the electrode of driving signal is located exactly on top of the optical waveguide. Z-cut has better coupling efficiency between electric field and optical waveguide. Dual-drive configuration or push-pull structure employs two electrical drives (data and inverted data) to perform the modulation. If the electrical fields are perfectly matched, that is, the driving signal amplitude has to be of the same amplitude and of the opposite phase, the dual-drive configuration will be chirp free. Nonetheless, Z-cut single-drive configuration has nonzero chirp coefficient because electric field lines that pass through two optical waveguides are often not similar. Table 6.6 gives typical configurations and applications of various types of optical modulators.

MATLAB and Simulink models of these optical modulators are presented in Chapter 12 in association with the modulation formats such as ASK for amplitude shift keying, DPSK, DQPSK as discrete differential phase modulation, and MSK and CPFSK as continuous-phase modulation.

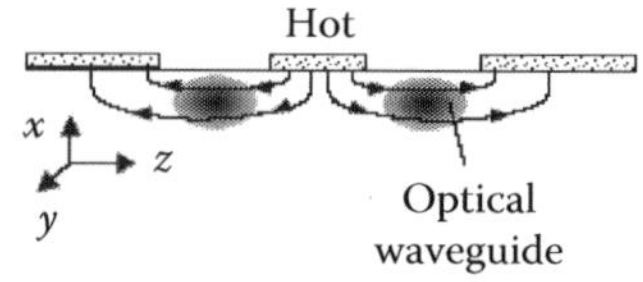

FIGURE 6.45
X-cut single-drive configuration. (From Nagata, H. et al., *IEEE Photonics Technol. Lett.*, 16(10), 2233, October 2004. With permission.)

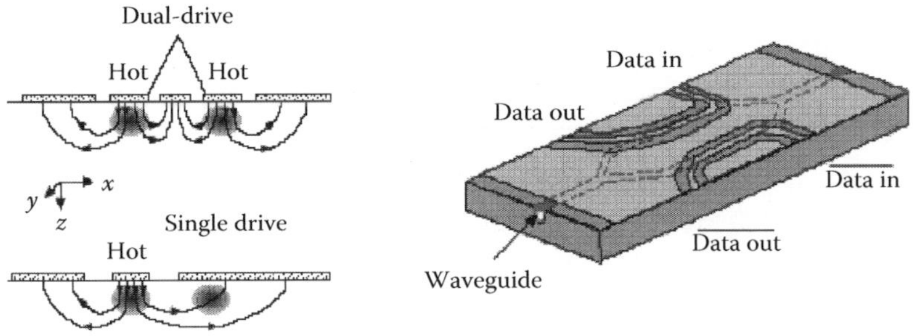

FIGURE 6.46
Z-cut single-drive and dual-drive configuration. (From Nagata, H. et al., *IEEE Photonics Technol. Lett.*, 16(10), 2233, October 2004; Lee, W.S., 80+Gb/s ETDM systems implementation: An overview of current technology, in *Proceedings of OFC 2006*, Anaheim, CA, Paper no. OTuB3, 2006. With permission.)

TABLE 6.6
Summary of Advantages, Disadvantages, and Application of X-Cut, Z-Cut, and EA Modulator

Configuration	Advantage	Disadvantage	Application
$LiNbO_3$ Z-cut	Lower and well-behaved frequency chirp than EA modulator Lower V π	Chirped (unchirped only by balanced driving) More DC drift than X-cut Buffer layer required Single drive not for PSK	Traditionally used for RZ and NRZ (balanced driven)
$LiNbO_3$ X-cut	Intrinsically balanced—chirp free No need to align the delay (phase) of two signals	Electro-optical efficiency lower than Z-cut Buffer layer required Higher driving voltage	Suitable for modulation required absolute chirp free
EA modulator	Large linear region	Chirp is high and dynamically changing with bias High insertion loss modulation <1.0	Cannot be used to PSK

References

1. (a) R.C. Alferness, Optical guided-wave devices, *Science*, 234(4778): 825–829, November 14, 1986; (b) M. Rizzi and B. Castagnolo, Electro-optic intensity modulator for broadband optical communications, *Fiber and Integrated Optics*, 21: 243–251, 2002; (c) H. Takara, High-speed optical time-division-multiplexed signal generation, *Optical and Quantum Electronics*, 33(7–10): 795–810, July 10, 2001; (d) E.L. Wooten, K.M. Kissa, A. Yi-Yan, E.J. Murphy, D.A. Lafaw, P.F. Hallemeier, D. Maack, D.V. Attanasio, D.J. Fritz, G.J. McBrien, and D.E. Bossi, A review of lithium niobate modulators for fiber-optic communications systems, *IEEE Journal of Selected Topics in Quantum Electronics*, 6(1): 69–80, January/February 2000; (e) K. Noguchi, O. Mitomi, H. Miyazawa, and S. Seki, A broadband Ti: $LiNbO_3$ optical modulator with a ridge structure, *Journal of Lightwave Technology*, 13(6): 1164–1168, June 1995.
2. Y. Shi, L. Yan, and A.E. Willner, High-speed electrooptic modulator characterization using optical spectrum analysis, *IEEE J. Lightwave Technol.*, 21: 2358–2364, 2003.

3. M. Suzuki, Y. Noda, H. Tanaka, S. Akiba, Y. Kuahiro, and H. Isshiki, Monolithic integration of InGaAsP/InP distributed feedback laser and electroabsorption modulator by vapor phase epitaxy, *IEEE Journal of Lightwave Technology*, LT-5(9): 127, September 1987.
4. P.K. Tien, Integrated optics and new wave phenomena in optical waveguides, *Reviews in Modern Physics*, 49: 361–420, 1977.
5. A. Yariv, C.A. Mead, and J.V. Parker, *IEEE Journal of Quantum Electronics*, QE-2: 243, 1966.
6. A. Mahapatra and E.J. Murphy, Electro-optic modulators, in *Optical Fiber Telecommunications*, I. Karminow and T.P. Li (Ed.), Academic Press, San Diego, Chapter 6, pp. 258–294, 2002.
7. H. Nagata, Y. Li, W.R. Bosenberg, and G.L. Reiff, DC drift of X-cut $LiNbO_3$ modulators, *IEEE Photonics Technology Letters*, 16(10): 2233–2335, October 2004.
8. H. Nagata, DC drift failure rate estimation on 10Gb/s X-cut lithium niobate modulators, *IEEE Photonics Technology Letters*, 12(11): 1477–1479, November 2000.
9. R.E. Epworth, K.S. Farley, and D. Watley, Polarization mode dispersion compensation, U.S. Patent 398/152, 398/202, 398/65.
10. C. Langton, Basic concept of modulation, www.complextoreal.com (access date: December 2007).
11. K.K. Pang, *Digital Transmission*. Melbourne, Australia: Mi-Tec Publishing, 2002, p. 58.
12. R. Redner and H. Walker, *SIAM Review*, 26(2): 195–239, 1984.
13. T. Kawanishi, S. Shinada, T. Sakamoto, S. Oikawa, K. Yoshiara, and M. Izutsu, Reciprocating optical modulator with resonant modulating electrode, *Electronics Letters*, 41(5): 271–272, 2005.
14. R. Krahenbuhl, J.H. Cole, R.P. Moeller, and M.M. Howerton, High-speed optical modulator in $LiNbO_3$ with cascaded resonant-type electrodes, *Journal of Lightwave Technology*, 24(5): 2184–2189, 2006.
15. P. Kaminow and T. Li, *Optical Fiber Communications*, Vol. IVA. New York: Elsevier Science, 2002, Chapter 16.
16. F. Amoroso, Pulse and spectrum manipulation in the minimum frequency shift keying (MSK) format, *IEEE Transactions on Communications*, 24: 381–384, March 1976.
17. E. Lach and K. Schuh, Recent advances in ultrahigh bit rate ETDM transmission systems, *IEEE Journal of Lightwave Technology*, 24(12): 4455–4467, 2006.
18. W.S. Lee, 80+Gb/s ETDM systems implementation: An overview of current technology, in *Proceedings of OFC 2006*, Anaheim, CA, 2006, Paper no. OTuB3.
19. I.B. Djordjevic and B. Vasic, 100-Gb/s transmission using orthogonal frequency-division multiplexing, *IEEE Photonics Technology Letters*, 18(15): 1576–1578, August 1, 2006.
20. A.J. Lowery, L. Du, and J. Armstrong, Orthogonal frequency division multiplexing for adaptive dispersion compensation in long haul WDM systems, OFC 2006 Postdeadline Sessions, Anaheim, CA, March 9, 2006, Paper PDP39.
21. F. Amoroso, *IEEE Transactions on Communications*, 24: 381–384, 1976.
22. H. Gnauck, X. Liu, X. Wei, D.M. Gill, and E.C. Burrows, Comparison of modulation formats for 42.7-gb/s single-channel transmission through 1980km of SSMF, *IEEE Photonics Technology Letters*, 16(3): 909–911, 2004.
23. G.P. Agrawal, *Fiber-Optic Communication Systems*, 3rd edn. New York: Wiley, 2002.
24. A. Hirano, Y. Miyamoto, and S. Kuwahara, Performances of CSRZ-DPSK and RZ-DPSK in 43-Gbit/s/ch D G.652 single-mode-fiber transmission, in *Proceedings of OFC'03*, vol. 2, Atlanta, GA, 2003, pp. 454–456.
25. A.H. Gnauck, G. Raybon, P.G. Bernasconi, J. Leuthold, C.R. Doerr, and L.W. Stulz, 1-Tb/s (6/spl times/170.6Gb/s) transmission over 2000-km NZDF using OTDM and RZ-DPSK format, *IEEE Photonics Technology Letters*, 15(11): 1618–1620, 2003.
26. Y. Yamada, H. Taga, and K. Goto, Comparison between VSB, CS-RZ and NRZ format in a conventional DSF based long haul DWDM system, in *Proceedings of ECOC'02*, vol. 4, Copenhagen, Denmark, 2002, pp. 1–2.
27. O. Painter, P.C. Sercel, K.J. Vahala, D.W. Vernooy, and G.H. Hunziker, Resonant optical modulators, U.S. Patent WO/2002/050575, 27.06.2002.

28. C. Schubert, R.H. Derksen, M. Möller, R. Ludwig, C.-J. Weiske, J. Lutz, S. Ferber, A. Kirstädter, G. Lehmann, and C. Schmidt-Langhorst, Integrated 100-Gb/s ETDM receiver, *IEEE Journal of Lightwave Technology*, 25(1): 122–130, January 2007.
29. L.N. Binh, T.L. Huynh, K.K. Pang, and T. Sivahumara, MLSE equalizers for frequency discrimination receiver of MSK optical transmission systems, *IEEE Journal of Lightwave Technology*, 26(12): 1586–1595, 2008.

7

Direct Detection Optical Receivers

This chapter describes the fundamental understanding of the detection of optical signals using direct techniques. Coherent detection will be described in Chapter 8. Optical preamplification at the input of the optical receivers is described in Chapter 10. MATLAB® and Simulink® models are described wherever appropriate.

7.1 Introduction

Optical receivers are important components in optical fiber communication systems and networks. Optical receivers are normally placed at the far end of the transmission links or at the front end of an optical repeater in terrestrial optical systems and at terminal front ends in optical networks. Their principal function is the conversion of optical signals into electronic forms for further electronic amplification and signal processing, such as clock recovery, retiming, digital level detection, and sampling.

The typical arrangement of an optical receiver is that the optical signals are detected by a photodiode (a pin diode or an avalanche photodiode (APD) or a photon-counting device). Electrons generated in the photodetector (PD) are then electronically amplified through a front-end electronic amplifier. The electronic signals are then decoded for the recovery of the original format.

Optical coherent communication systems have also been developing dramatically due to electronic processing and the availability of stable narrow-linewidth lasers. The coherent detection techniques are described in the next chapter to keep the readers updated with this emerging trend in the twenty-first century.

This chapter deals with the analysis and design of electronic amplifiers incorporated into the photodetection device as the fundamental elements of an optical receiver. In recent years, the advance in optical amplifiers, particularly the rare-earth-doped optical fiber amplifiers, has reduced them to a compact size, which enables these fiber amplifiers to be used as a standard photonic component in most terrestrial optical fiber communication systems. With this in mind, we need to consider the optical amplifiers as a standalone system or an in-line optical receiver acting as an optical repeater or as a preamplifier with its front end placed before the PD, followed by an electronic amplifier.

Noises generated in the electronic preamplifier lower the signal-to-noise ratio or reduce the eye opening of the received digital signals at its output. The equivalent noise power of the electronic preamplifier increases as a square function of the frequency. Thus, it is necessary to reduce this frequency-dependent noise. A noise-matching network can be placed at the front end, which would act as a noise filter. An example of a high-frequency electronic preamplifier is given in Section 7.5.

Optical amplifiers, the rare-earth-doped optical fiber amplifiers, have advanced much to provide a boost to optoelectronic detection. The integration of an OA would further improve the coherent detection; a treatment of this optically preamplified detection process is also given. The invention of optical amplifiers in the 1990s has thus revolutionized the design of systems used for optical transmission over a long haul, as they offer significant optical gain that can compensate for fiber losses and other losses of optical components. Optically preamplified receivers will be explained in Chapter 9.

This chapter is organized as follows. Section 7.2 gives an overview of the roles of optical receivers in various optical transmission systems and their applications in networking. Section 7.3 then gives an account of optical components of an optical receiver for direct detection. Noises in PDs and integration with electronic preamplifiers are described in Section 7.4. Methods for calculations of receiver noises are also explained in this section. The performance evaluation of a binary optical receiver given in terms of noise and signal power at the output of the receiver is described in Section 7.5. Two examples of thedesign and noise estimation when an FET or a BJT (bipolar junction transistor) is used as the front-end electronic amplification device are illustrated here. Section 7.6 gives an example of a noise-matching network integrated at the front end of an electronic preamplifier. Finally, Section 7.7 gives the concluding remarks.

7.2 Optical Receivers in Various Systems

In modern optical and photonic communication systems, optical amplifiers can be incorporated in front of a front-end optoelectronic receiver to obtain an optically amplified preamplifier. This type of optical receiver is discussed in Chapter 10, which deals with optical amplification.

The ultimate goal of the design of an optical receiver is to determine the minimum optical energy in terms of the number of photons per bit period required at the input of a PD so that it would satisfy a certain optical signal-to-noise ratio for an analog optical system or determine the minimum optical energy in terms of sensitivity so that it would satisfy a certain bit error rate (BER) for a digital optical communication system, in other words, the minimum optical power required at a certain bit rate in a digital communication system so that the decision circuitry can be detected with a specified BER, e.g., $\mathrm{BER}=10^{-9}$ or 10^{-12}.

Table 7.1 shows the important requirements for optical receivers in various communication systems from terrestrial to local and wide area networks to undersea submarine systems.

The electronic amplifier is considered as a block diagram rather than a detailed circuit configuration. However, to present a clear picture, a section is dedicated to the design of the electronic front-end amplifier. To provide a generic approach for the design of optical receivers is the ultimate objective of this chapter.

7.3 Receiver Components

7.3.1 Structure

The design of an optical receiver depends on the modulation format of the signals that are transmitted through the transmitter. In particular, it depends on the modulation in

TABLE 7.1

Optical Receiver Characteristics and Their Relative Importance in Typical Optical Fiber Communication Systems

	Features	Undersea Optical Communications	Terrestrial Communications and WAN	Point to Point	LAN
High receiver sensitivity	Maximum repeater spacing	Critical	Moderate critical	Moderate	Moderate
Wide dynamic range	Flexible and convenient system configurations	Moderate	Moderate critical	Moderate critical	Critical
Bit rate transparency	Variable bit rate operation	Not required	Not required	Desirable	Desirable
Bit rate dependency	Flexible	Accommodated by the use of appropriate line codes and scrambling/descrambling		Desirable	Desirable
Fast acquisition time	Short preamble bit sequence	Not required	Not required	Moderate critical	Moderate critical

analog or digital, Gaussian or exponential pulse shape, on–off keying (OOK) or multiple levels, etc.

Optical signals treated in this chapter are combined with intensity-modulated and direct detection (IM/DD) OOK systems. They are thus treated with an assumption that lightfaces that arrive at the receiver are polarization independent and their polarization plays no part in the degradation of the optical signals, except due to the polarization mode dispersion, which is independently treated in Chapter 3. In coherent optical systems, which have been treated in many other textbooks on optical communication,the polarization and the coherence of the optical source play a very significant role in the design and implementation of optical fiber communication systems. These coherent optical systemswere of great interest to the research and development community from early to mid-1980s, but are no longer of interest due to the invention of the erbium-doped optical amplifiers (EDFA) in the late 1980s.

Figure 7.1 shows the schematic diagram of a digital optical receiver. An optical receiver front end consists of a PD, for converting lightwave energy into electronic currents; an electronic preamplifier, for further amplification of the generated electronic current, usually in voltage form; a main amplifier, for further voltage amplification; a clock recovery circuitry, for regenerating the timing sequence; and a voltage-level decision circuit, for sampling the waveform for the final recovery of the transmitted and received digital sequence. Therefore, the optoelectronic preamplifier is followed by a main amplifier with an automatic control to regulate the electronic signal voltage to be filtered and then sampled by a decision circuit with synchronization by a clock recovery circuitry.

An in-line optical fiber amplifier can be incorporated in front of a PD to obtain an optical receiver with the optical amplifier at the front end, to improve its receiving sensitivity. This optical amplification at the front end of an optical receiver will be discussed in Chapter 9, which deals with optical amplification processes.

The structure of the receiver consists of three parts: the front-end section, the linear channel of the main amplifier and AGC, and the data recovery section. These sections are described in order in the following text.

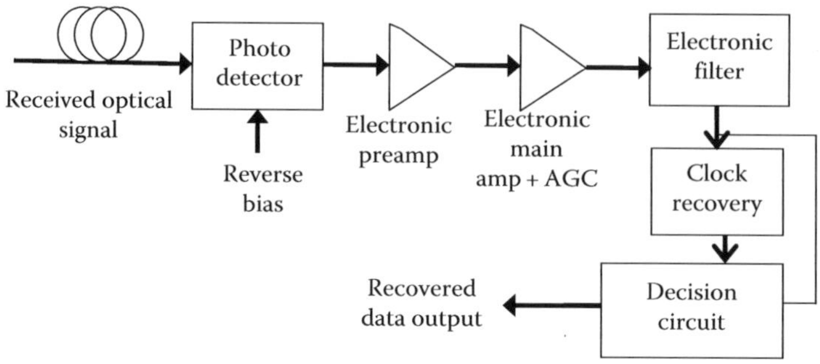

FIGURE 7.1
Schematic diagram of the digital optical receiver IM/DD.

7.3.2 Photodiodes

A PD detects and converts the input optical power into an electric current output. An ideal PD would be highly quantum efficient, would add no noises to the received signals, would respond uniformly to all signals with different wavelengths within the transparent windows of optical fibers around 1300–1550 nm, and, finally, would not be saturated and behave linearly as a function of signal amplitude. There are several different types of PDs that are commercially available. Among the semiconductor-based types, the photodiode is used almost exclusively for high-speed fiber-optic systems due to its compactness and, thus, extremely high bandwidth. The two most commonly used types of photodiodes are PIN and APD. Detailed reviews of these photodiodes have been presented in well-known published literature [1–3]. The fundamental characteristics of these two types of devices are briefly described for completeness.

7.3.2.1 p-i-n Photodiodes

A pin PD consists of three regions of semiconductors: the heavily doped, p; the intrinsic, i; and the n-doped regions. It is essential that the p–n junction is reverse biased to cause mobile electrons and holes to move away from the junction, thus increasing the width of the depleted layer and hence producing a high electric field by the immobile charges at both ends. If lightwaves are incident on the PIN surface and absorbed in the *high-field depleted* region, the electron–hole pairs generated will move at a saturation-limited velocity. No significant contribution of photons falls in other regions. Thus, since the width of the p region, W_p, is inversely proportional to its doping concentration, a decrease in the doping level leading to intrinsic behavior would widen W_p; hence, the depleted region now becomes wider and compatible with the absorption region to produce a large photon-generated current. A p^+ region must be added in order to make good ohmic contact. A wider intrinsic region gives higher quantum efficiency, η, but lower response rate.

7.3.2.2 Avalanche Photodiodes

A high-field multiplication region is placed adjacent to the lightly doped depletion region—intrinsic in contrast with the PIN structure as shown in Figure 7.2. Similarly the structure

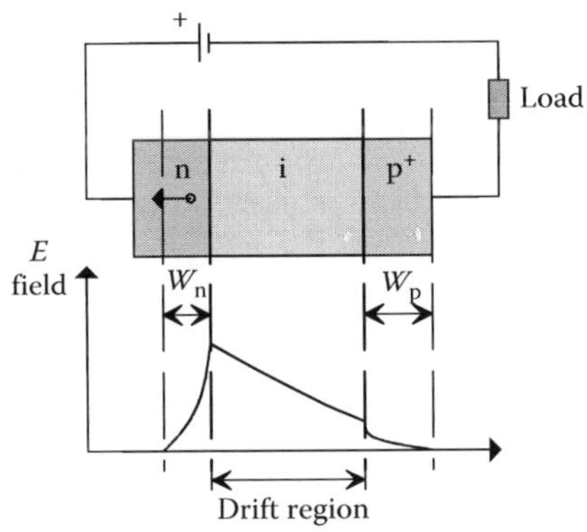

FIGURE 7.2
Schematic diagram of a *p-i-n* photodiode.

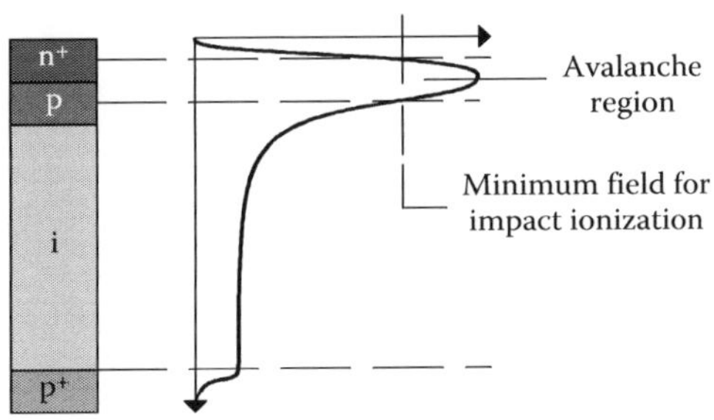

FIGURE 7.3
Schematic diagram of an APD.

and field distribution of an APD is shown in Figure 7.3. Photons absorbed in the intrinsic region produce electron–hole pairs. These pairs drift to the high-field region whereby they gain sufficient energy (E field must be above the threshold for impact ionization to occur) to ionize bound electrons in the valence band upon colliding with them. Thus, the avalanche effect occurs only when the electric field in the high-field multiplication region is above the threshold for impact ionization. A guard ring is usually added to prevent the leakage surface current, I_{surf}.

7.3.2.3 Quantum Efficiency and Responsivity

The two most important characteristics of a PD are its quantum efficiency and its response speed. These parameters depend on the material bandgap, the operating wavelength, and the doping and thicknesses of the p, i, and n regions of the device. The quantum efficiency, η, is the number of electron–hole carrier pairs generated per incident photon of energy, $h\nu$, and is given by

$$\eta = \frac{\text{number of electron pairs generated}}{\text{number of incident photons}} = \frac{I_p/q}{P_o/h\upsilon} = \frac{I_p h\upsilon}{P_o q} \tag{7.1}$$

where
I_p is the photocurrent
P_o is the incident optical power

The performance of a photodiode is often characterized by its responsivity, $\Re$, which is related to the quantum efficiency by

$$\Re = \left(\frac{I_P}{P_0}\right)G = \left(\frac{\eta q}{h\upsilon}\right)G = \left(\frac{\eta q\lambda}{hc}\right)G \tag{7.2}$$

where G is the APD *average multiplication factor* and $G=1$ for the non-APD case. The optical power is given by

$$P_o = \frac{i_s}{G\Re} \tag{7.3}$$

where it means that for the same incident optical power, APD could produce a higher photocurrent than PIN. Thus, based on Equations 7.1 and 7.2, we could model PIN and APD.

7.3.2.4 High-Speed Photodetectors

When the speed of a PD is increased, for example, to be used in a high-speed system above 10 Gb/s, a pin detector is normally used and the effective receiving area is smaller in order to reduce the junction capacitance. This decreased area then forces the detector to produce a lower breakdown voltage.

7.4 Detection and Noises

The front-end subsystem is the most important component because the noise contributed by this part severely affects the overall performance of the system. The design of an amplifier front end depends on the sensitivity required and the bandwidth of the system. For example, an optical front-end amplifier for an OTDR (optical time domain reflectometer) would require a very highly sensitive front end while its bandwidth is not critical. On the other hand, a front-end optical amplifier for a high-bit-rate optical communication requires a wide bandwidth amplifier as well as a reasonably high sensitivity and low noise contribution. In Table 7.1, optical front-end amplifiers and demands on their characteristics with their limits are given.

In general, optically amplified front ends are required for two types of optical systems: the short-link systems in local area or metropolitan area networks and the terrestrial-link systems. The former requires a high data accumulation rate and the latter requires a fast acquisition rate and reasonable sensitivity.

Thus, electronic preamplifiers can take two most typical structures: the high-impedance (HI) and the transimpedance (TI) front end. The difference is that in the TI configuration there is a shunt feedback from the output to the input. This TI configuration offers a wide bandwidth and the HI configuration provides a high sensitivity because its noise is

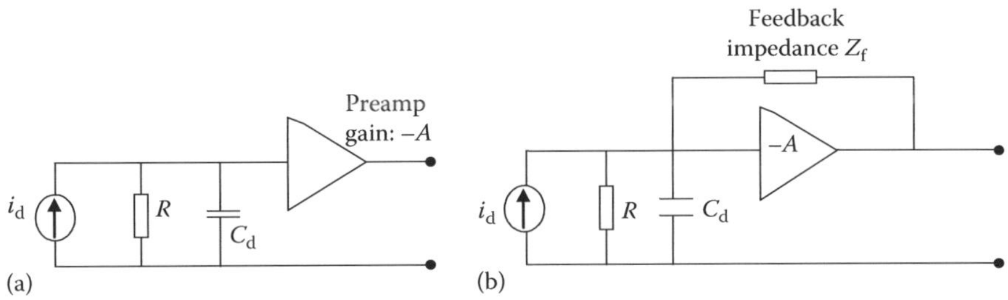

FIGURE 7.4
Schematic diagram of the electronic preamplifier in an optical receiver. (a) HI optical front-end receiver and (b) TI optical front-end receiver. The current source represents the electronic current generated in the PD. C_d is the photodiode capacitance.

almost negligible compared with the TI model. The TI and HI configurations are shown in Figure 7.4. A PD is represented by a current source and a capacitor in parallel. The noise current (quantum shot noise) is not shown. A diode capacitance due the capacitance dropped across the depletion region of the p–n junction is also included. In fact this diode capacitance, C_d, is a critical component in the design of the PD.

The HI front end would have a very high impedance, Z_i, as the input impedance of the electronic amplifier. This high impedance would limit the bandwidth of the overall amplifier. Usually Z_i would be a few hundreds ohms. While a TI front end offers a lower receiver sensitivity, the bandwidth is increased by a factor of A, the amplifier linear gain.

7.4.1 Linear Channel

The linear channel in optical receivers consists of a main electronic amplifier and a low-pass filter or an equalizer whose objectives are to reduce the noise or distortion in the channel without incurring further intersymbol interference (ISI). The preamplifier and the main amplifier have the bandwidth normally wider than that of the system. Thus, in calculating the total noise contribution, the noise spectra density must be integrated over the electronic amplifier bandwidth.

7.4.2 Data Recovery

Once the optical signals are detected and amplified, the data streams are sampled with respect to the time interval recovered by a clock recovery circuit, normally by using a surface acoustic wave filter if the bit rate is high, and then amplified and fed into a triggering circuit. In several high-speed optical receivers, the components are usually integrated in a chip or by using a hybrid circuit if the bit rate is higher than 2.5 Gbits/s.

7.4.3 Noises in Photodetectors

The most important noise source in a phtotodetector is the fundamental quantum or shot noise caused by the intrinsic fluctuations in the photoexcitation of electronic carriers. In any interval of T seconds, the probability of exactly N primary electrons generated is given by the time-varying Poisson distribution:

$$P[N,(t_0,t_0+T)] = \frac{\Lambda^N e^{-\Lambda}}{N!} \tag{7.4}$$

where $\Lambda = \int_{t_0}^{t_0+T} \lambda(t)dt \xrightarrow{\text{where}} \lambda(t) = \frac{\eta}{\hbar\nu} p(t) + \lambda_d$ and λ_d is the number of electrons generated under dark condition, or the dark current electrons of the PD; thus, Λ is the average number of primary electrons produced during the interval.

For an avalanche photodetector (APD), the internal avalanche multiplication factor is also a random process. The variation of the avalanche gain gives rise to excess noise. The probability of distribution of the avalanche gain depends on the type of APD. In particular, it is a function of the ratio of hole ionization probability to that of the electron, the factor k. The extra noise is represented by an excess noise factor, $F(g)$, given by

$$\langle g^2 \rangle = F(g)G^2, \quad \text{with } G = \langle g \rangle \tag{7.5}$$

The term $F(g)$ is usually determined by experimental works and can be approximated as

$$F(G) \simeq kG + \left(2 - \frac{1}{G}\right)(1-k) \tag{7.6}$$

and $F(G)$ can usually be estimated by

$$F(G) \simeq G^x \tag{7.7}$$

For a Ge APD, $x \sim 1$, while for a well-designed Si APD, $x \sim 0.4$ or 0.5.

7.4.4 Receiver Noises

Before proceeding to the system's calculations to determine the performance of optical receivers, it is necessary that the noises generated in the PD and the preamplifier front ends are considered.

Readers can in fact consider investigating the system calculations and returning back to the noise calculations provided that they take into account either the total equivalent noise spectral density at the input of the detector or its noise (see Figures 7.5–7.7). This section describes all noise mechanisms related to the photodetection process, including the electronic noise associated with the receiver.

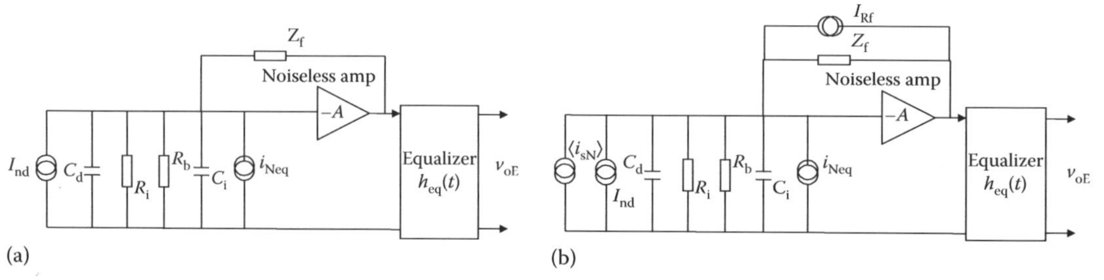

FIGURE 7.5
Optical receiver small-signal model and associated noise sources.

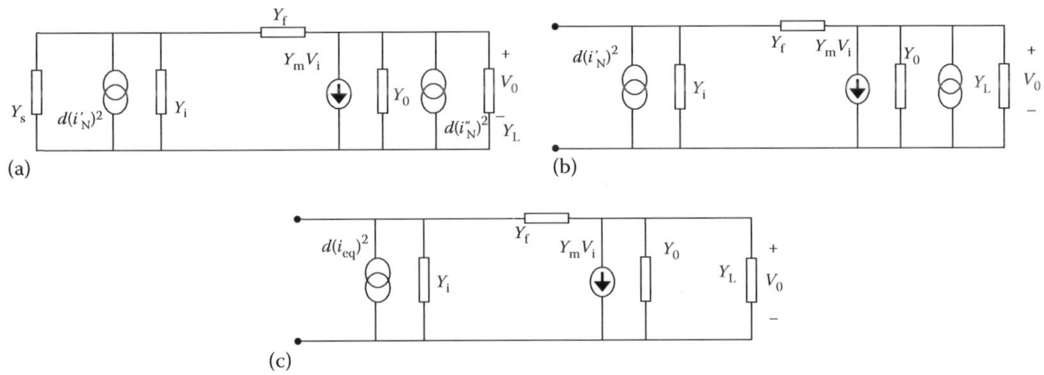

FIGURE 7.6
Small equivalent circuits including noise sources. Y-parameter model (a) representing the ideal current model and all current noise sources at the input and output ports, (b) with noise sources at input and output ports, and (c) with a total equivalent noise source at the input.

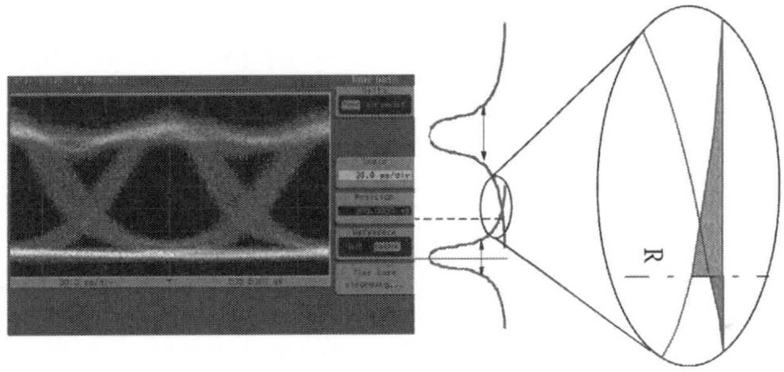

FIGURE 7.7
Time-dependent fluctuating signal in the receiver in response to the random digital bit stream and the probability density distribution as a function of the output voltage level and the decision level for recovery of the transmitted signals.

Shot noises and thermal noises are the two most significant noises in optical detection systems. Shot noises are either generated by a quantum process or by electronic biasing. The noises are calculated in terms of noise spectral density, which is the square of noise current per unit frequency (in hertz). Thus, the noise spectral density is to be integrated over the total amplifier bandwidth to obtain the equivalent noise currents.

7.4.4.1 Shot Noises

Electrical shot noises are generated by the random generation of streams of electrons (current). In optical detection, shot noises are generated by (1) biasing currents in electronic devices and (2) photo currents generated by the photodiode.

For a bias current, *I*, the spectral current noise density, S_I, is given by

$$S_I = \frac{d(i_I^2)}{df} = 2qI \quad \text{in A}^2/\text{Hz} \tag{7.8}$$

where q is the electronic charge. The current i_I represents the noise current due to the biasing current, *I*.

7.4.4.2 Quantum Shot Noise

The average current, $\langle i_s^2 \rangle$, generated by the PD by an optical signal with an average optical power, P_{in}, is given by

$$S_Q = \frac{d\langle i_s^2 \rangle}{df} = 2q\langle i_s^2 \rangle \quad \text{in A}^2/\text{Hz} \tag{7.9a}$$

In the case where the APD is used, the noise spectral density is given by

$$S_Q = \frac{d\langle i_s^2 \rangle}{df} = 2q\langle i_s^2 \rangle \langle G_n^2 \rangle \quad \text{in A}^2/\text{Hz} \tag{7.9b}$$

where $\langle G_n^2 \rangle$ is the average avalanche gain of the detector. It is noted here again that the dark currents generated by the PD must be added to the total equivalent noise current at the input after it is evaluated. These currents are generated even in the absence of the optical signal. The dark currents can be eliminated by cooling the PD to at least below the temperature of liquid nitrogen (77 K).

7.4.4.3 Thermal Noise

At a certain temperature, the conductivity of a conductor varies randomly. This random movement of electrons generates a fluctuating current even in the absence of an applied voltage. The thermal noise of a resistor R is given by

$$S_R = \frac{d(i_R^2)}{df} = \frac{2k_B T}{R} \quad \text{in A}^2/\text{Hz} \tag{7.10}$$

where
- k_B is Boltzmann's constant
- T is the absolute temperature (in K)
- R is the resistance (in ohms)
- i_R denotes the noise current due to the resistor R

7.4.5 Noise Calculations

In this section, the design of a receiver for use in a digital communication system and the methods for noise calculations are described. Binary and multilevel operations are also provided.

Schematics of an optical preamplifier of the receiver and an electronic analog equalizer for the detection of optical digital modulated signals are shown in Figure 7.5. An equalizer is considered in order to extend the bandwidth of the receiver to the range of several gigahertz for ultra-high-speed operations of these digital optical receivers. This is similar to earlier design considerations of the optical receiver when multimode fibers were used in the first generation of optical fiber communication systems [4–6]. The noise sources and the small-amplification circuit model can be simplified, as shown in Figure 7.6, in which all the noise sources of the electronic amplifier and the PD are presented and grouped into total noise current sources at the input and the output of the amplifier. These sources are then transferred to total equivalent current sources, as seen at the input of the amplifier. Thus, it is very straightforward to find out the signal-to-noise ratio and the contribution of noises from the electronic amplification process and the quantum shot noise process in the detection of optical signals.

Our goal is to obtain an analytical expression of the noise spectral density equivalent to a source looking into the electronic amplifier, including the quantum shot noises of the PD. A general method for deriving the equivalent noise current at the input is by representing the electronic device by a Y-equivalent linear network, as shown in Figure 7.6. The two current noise sources, $d(i_N'^2)$ and $d(i_N''^2)$, represent the summation of all noise currents at the input and at the output of the Y network. This can be transformed into a Y circuit with the noise current at the input as follows:

The output voltages, V_o, for Figure 7.6a can be written as

$$V_o = \frac{i_N'(Y_f - Y_m) + i_N''(Y_i + Y_f)}{Y_f(Y_m + Y_i + Y_o + Y_L) + Y_i(Y_o + Y_L)} \tag{7.11a}$$

and for Figure 7.6b, as

$$V_o = \frac{(i_N')_{eq}(Y_f - Y_m)}{Y_f(Y_m + Y_i + Y_o + Y_L) + Y_i(Y_o + Y_L)} \tag{7.11b}$$

By comparing these two equations, we can deduce the equivalent noise current at the input of the detector to be

$$i_{Neq} = i_N' + i_N'' \frac{Y_i + Y_f}{Y_f - Y_m} \tag{7.12a}$$

Then reverting to the mean-square generators for a noise source, we have

$$d(i_{Neq})^2 = d(i_N')^2 + d(i_N'')^2 \left| \frac{Y_i + Y_f}{Y_f - Y_m} \right|^2 \tag{7.12b}$$

It is therefore expected that if the Y matrix of the front-end low-noise amplifier is known, the equivalent noise at the input of the amplifier can be obtained using (7.12b).

7.5 Performance Calculations for Binary Digital Optical Systems

As described above, the noises of the optical receivers consist of thermal noises and quantum shot noises due to the bias currents and the photocurrent generated by the PD with and without the optical signals. Thus, quantum shot noises are strongly signal dependent.

These noises degrade the sensitivity of the receiver, and thus a penalty can be estimated. Another source of interference that may also result in signal penalty is the ISI. The goal of this section is to obtain an analytical expression for the sensitivity of the optical receiver in an IM/DD pulse-code-modulated (PCM) system.

7.5.1 Signals Received

Assume that the received signal power is

$$p(t) = \sum_{j} a_j h_p(t - jT_B) \tag{7.13}$$

where T_B is the bit period. The average output voltage is thus given by

$$\langle v_o \rangle = \Re \langle G_n \rangle \left[\sum_{j} a_j \frac{1}{T_B} \int_{-T_b/2}^{T_b/2} h_p(t - jT_B)\mathrm{d}t \right] R_I A \tag{7.14}$$

where

$h_p(t - jT_B)$ is the impulse response of the system evaluated at each time interval

R_I is the input resistance of the overall amplifier of the system, including both the front-end and linear channel amplifiers

It is assumed that the overall amplifier has a flat gain response, A, over the bandwidth of the system:

$$a_j \Big|_{t=0} = \begin{cases} b_o \approx 0 \\ b_1 \end{cases} \tag{7.15}$$

where

b_o is the energy when a transmitted "0" is received

b_1 is the energy when a transmitted "1" is received

The summation over a number of periods is necessary to take into account the contribution of adjacent optical pulses.

We now have to distinguish between two cases: when a "0" or a "1" is transmitted and received.

Case (a): OFF or a transmitted "0" is received

Using (7.14) and (7.15), we have

$$\langle v_o \rangle_0 = v_{oo} = \Re \frac{b_0}{T_B} G R_I A \cong 0 \tag{7.16}$$

with the total equivalent noise voltage at the output, v_{NTo}^2, given by

$$v_{NTo}^2 = v_{NA}^2 = i_{Neq}^2 R_I^2 A^2 \tag{7.17}$$

Case (b): ON or a transmitted "1" is received
In this case, the average signal voltage received at the output is

$$\langle v_o \rangle_1 = v_{01} = \Re \frac{b_1}{T_b} \langle G_n \rangle R_I A \tag{7.18}$$

with the total equivalent noise mean voltage at the output given by

$$v_{NT1}^2 = v_{oSN}^2 + v_{NA}^2 \tag{7.19}$$

where
- v_{oSN}^2 is the signal-dependent shot noise
- v_{NA}^2 is the amplifier noise at the output, given by

$$v_{NA}^2 = i_{Neq}^2 R_I^2 A^2 B \tag{7.20}$$

The signal-dependent noise is in fact the quantum shot noise, and is given by

$$v_{oSN}^2 = \int_0^B 2q \langle i_s \rangle_1 \langle G_n^2 \rangle R_I^2 A^2 df \tag{7.21}$$

where
- B is the 3 dB bandwidth of the overall amplifier
- $\langle i_s \rangle_1$ is the average photocurrent received when a "1" was transmitted

This current is estimated as follows:

$$\langle i_s \rangle_1 = \sum_{-\infty}^{\infty} \Re \frac{b_1}{T_B} \int_{-T_b/2}^{T_b/2} h_p(t - jT_B) dt \tag{7.22}$$

or

$$\langle i_s \rangle_1 = \Re \frac{b_1}{T_B} \int_{-\infty}^{\infty} h_p(t) dt \tag{7.23}$$

Using normalization with $\int_{-\infty}^{\infty} h_p(t)dt = 1$, (7.23) becomes

$$\langle i_s \rangle_1 = \Re \frac{b_1}{T_B} \tag{7.24}$$

7.5.2 Probability Distribution

The optical systems under consideration are typical IM/DD systems in which the optical energy of each pulse period is equivalent to that of at least a few hundred photons.

This value is large enough for a Gaussian distribution of the probability density to be warranted.

The probability density function for a "0" transmitted and received by the optical receiver is thus given by (illustrated in Figure 7.7)

$$p\left[v_o|\text{“}0\text{”}\right] = \frac{1}{(2\pi v_{NT0}^2)^{1/2}} \exp\left[\frac{-(v_0 - v_{00})^2}{2v_{NT0}^2}\right] \tag{7.25}$$

and, similarly, for a "1" transmitted is given by

$$p\left[v_o|\text{“}1\text{”}\right] = \frac{1}{(2\pi v_{NT1})^{1/2}} \exp\left[\frac{-(v_0 - v_{01})^2}{2v_{NT1}^2}\right] \tag{7.26}$$

The total probability of error or a BER is defined [7,8] as

$$\text{BER} = p(1)p(0/1) + p(0)p(1/0) \tag{7.27}$$

where

$p(1)$ and $p(0)$ are the probabilities of receiving "1" and "0," respectively

$p(1/0)$ and $p(0/1)$ are the probabilities of deciding "1" when "0" is transmitted and vice versa

In a binary bit stream, "1" and "0" are likely to occur equally, that is, $p(1) = p(0) = 0.5$, hence (7.27) becomes

$$\text{BER} = \frac{1}{2}[p(0/1) + p(1/0)] \tag{7.28}$$

Thus, as equal probabilities of transmitting "0" and "1" are assumed, for a decision voltage level of d, as indicated in Figure 7.5, the total probability of error, P_E, is the summation of the errors of deciding "0" or "1" by integrating the probability density function over the shaded regions, given by

$$\text{BER} = P_E = \frac{1}{2}\int_d^{\infty} p\left[v_0|\text{"}0\text{"}\right]dv_0 + \frac{1}{2}\int_{-\infty}^{d} p\left[v_0|\text{“}1\text{”}\right]dv_0 \tag{7.29}$$

Substituting for the probability distribution using (7.27a and b) leads to

$$\text{BER} = \frac{1}{2\sqrt{\pi}}\int_{\frac{d-v_{00}}{v_{NT0}}}^{\infty} e^{-\frac{x^2}{2}}dx + \frac{1}{2\sqrt{\pi}}\int_{\frac{v_{01}-d}{v_{NT1}}}^{d} e^{-\frac{x^2}{2}}dx \tag{7.30}$$

The functions in Equation 7.30 have the standard form of the complementary error function, $Q(\alpha)$, defined as

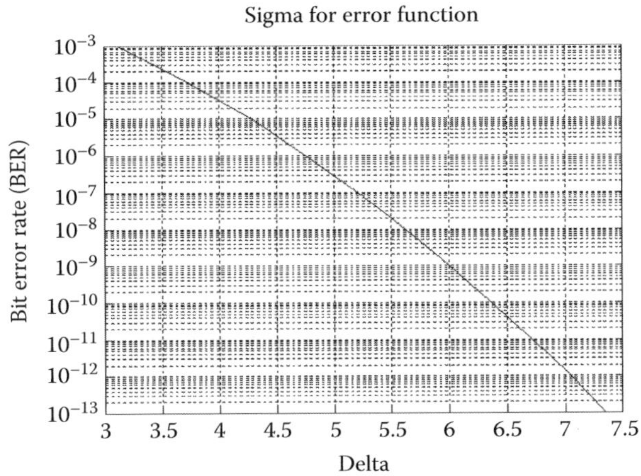

FIGURE 7.8
BER as a function of the δ parameter. Sometimes δ is assigned as the system Q Magnum function.

$$Q(\alpha) = \frac{1}{\sqrt{2\pi}} \int_{\alpha}^{\infty} e^{-x^2/2}\,dx \tag{7.31}$$

Then Equation 7.30 becomes

$$\text{BER} = \frac{1}{2}\left[Q\left(\frac{d}{v_{\text{NT0}}}\right) + Q\left(\frac{v_{01} - d}{v_{\text{NT1}}}\right)\right] \tag{7.32}$$

The $Q(\delta)$ function is a standard function, as shown by the curve in Figure 7.8. It is noted that for a BER = 1e-9, the value of δ is about 6, which is the normal standard for communications at a bit rate of 155 Mb/s to 40 Gb/s. However, in a laboratory demonstration, a BER of 10^{-12} is required that corresponds to a δ value of 7.

7.5.3 Minimum Average Optical Received Power

Using the condition $p[v_o/\text{"0"}] = p[v_o/\text{"1"}]$, from (7.32) we have

$$\frac{d}{v_{\text{NT0}}} = \frac{v_{01} - d}{v_{\text{NT1}}} \equiv \delta \tag{7.33}$$

Now assuming $v_{00} = 0$,

$$\text{BER} = Q(\delta) \tag{7.34}$$

Thus, by eliminating the variable d from (7.33), we obtain

$$\delta = \frac{v_{01} - \delta v_{\text{NT0}}}{v_{\text{NT1}}} \tag{7.35}$$

or, alternatively,

$$v_{01} = \delta(v_{NT0} + v_{NT1}) \tag{7.36}$$

Substituting v_{01}, v_{NT1}, and v_{NT0}, we have

$$\Re\frac{b_1}{T_b}\langle G_n\rangle R_I A = \partial\left\{\left[2\Re q\frac{b_1}{T_B}\langle G_n^2\rangle BR_I^2A^2 + v_{NA}^2\right]^{1/2} + v_{NA}\right\} \tag{7.37}$$

However, the amplifier noise voltage, v_{NA}, is given by $v_{NA} = i_{Neq}R_I AB^{1/22}$. Thus, by substituting this noise voltage and eliminating $R_I A$ and solving for the energy essential for the "1" transmitted and received at the PD, the energy b_1 can be written as

$$b_1 = \frac{q}{\Re}\left[\delta^2 G^x + \frac{2\delta i_{Neq} T_B}{qG}\right] \tag{7.38}$$

where we have used the approximations $\langle G_n\rangle = G$ and $\langle G_n^2\rangle = \langle G\rangle^{2+x}$ where x is the factor dependent on the ionization ratio in an APD. For a pin PD, $G = 1$.

The optical receiver sensitivity, denoted by RS, can thus be defined as

$$\text{Receiver sensitivity} = RS = 10\log_{10}\frac{P_{av}}{P_0}\ \text{in dBm} \tag{7.39}$$

where

$P_{av} = b_1/T_b$

$P_o = 1.0\,\text{mW}$ (the reference power level for evaluating power in dBm)

This is the optical receiver sensitivity, which is defined as the minimum optical power required for the receiver to operate reliably with a BER below a specified value.

In Equation 7.39, there are two terms clearly specifying the dependence of the signal, δ, and the amplifier noise contribution. It is reminded that the term b_1 represents the optical energy of the "1" required for the optical receiver to detect with a certain BER. The term q/R in the RHS of (7.38) is equivalent to the optical power required to generate one electron, or the number of photons required at the optical receiver to generate one electron. The typical measure for an optical receiver is the number of photons required for it to operate with a specified BER. Thus, the term $[\delta^2 G^x]$ indicates the number of photons required to overcome the opening of the eye at a particular BER, and the term $2\delta i_{Neq}T_b/qG$ indicates the number of photons required to overcome the noises of the electronic preamplifier at a specific BER. With an APD, the gain of the avalanche gain also contributes to an increase in the noise level, and hence higher power. However, the current avalanche gain compensates for this.

We note that the analysis here has not taken into account the effects of pulse shape and transfer function characteristics of the receiver. Thus, in this case, an equivalent noise current power, as seen from the input port of the electronic preamplifier, would simplify the analysis of the optical receiver sensitivity, or SNR. The effects of the pulse shape and transfer function of the receiver are considered in the following sections.

7.5.3.1 Fundamental Limit: Direct Detection

Referring to (7.38), we can see that in the case that there are no electronic noises, the first term is the number of photons in the quantum limit of a direct detection receiver. Thus, we have

$$n_p = [\delta^2 G^x] \tag{7.40}$$

For a PIN detector and a BER of 1*e*–9, the number of photons in the quantum limit is 36.

Examples:
For an ASK-modulated optical signal at a bit rate of 10 Gb/s, estimate the number of photons required for the incoming symbol "1" so that a BER of 1e–9 can be achieved.

7.5.3.2 Equalized Signal Output

We recall the optical signal $p(t)$ fallen on the input of the PD:

$$p(t) = \sum_{k=-\infty}^{\infty} a_k h_p(t' - kT_B) \tag{7.41}$$

where
T_B is the signal bit period
$h_p(t)$ is the input optical shape constrained by

$$\int_{-\infty}^{+\infty} h_p(t)dt = 1 \tag{7.42}$$

It is noted that this function represents the optical power variation as a function of time, and thus it is always positive.

Now neglecting the DC component due to the dark current, the signal at the output of the equalizer can be written as

$$\langle v_{0E}(t) \rangle = \Re G p(t) * h_{fe}(t) * h_{eq}(t) \tag{7.43}$$

where
$h_{fe}(t)$ is the impulse response of the optical receiving subsystem
$h_{eq}(t)$ is the impulse response of the electronic equalizer
$*$ indicates the convolution

The optical receiver is usually dominated by the pole contributed by the diode capacitance, C, and associated resistance of the network. It can be shown without difficulty that (refer to Figure 7.5) the frequency domain transfer function is given by

$$H_{\text{fe}}(\omega) = -\frac{R_{\text{f}}}{1 + j\omega\dfrac{R_{\text{f}}C}{A}} \tag{7.44}$$

where

A is the open-loop voltage gain of the electronic preamplifier
R_{f} is the feedback resistance
C is the total equivalent input capacitance at the input port of the amplifier

Thus knowing the input optical shape, the equalizer transfer function for a given pulse shape can be determined by multiplying the transfer function of the receiver and that of the equalizer. It is noted that the 3 dB bandwidth of a receiver with feedback is given by $A/R_{\text{f}}C$, which is about A times the bandwidth of a receiver without feedback. Thus, the demand on the equalizer is much less for this case.

7.5.3.3 Photodiode Shot Noise

The output noise power due to the random fluctuation of the multiplied Poisson nature of the current is (t) produced by the PD can be written [4] as

$$\left\langle i_{\text{SN}}^2 \right\rangle = \int_{-\infty}^{+\infty} qG^{2+x}\Re \left\{ \sum_{k=-\infty}^{\infty} a_k h_{\text{p}}(t' - kT_{\text{B}}) + \lambda_{\text{d}} \right\} h_{\text{I}}^2(t - t')dt' \tag{7.45}$$

where

$h_{\text{I}}(t) = h_{\text{fe}}(t) * h_{\text{eq}}(t)$ is the overall current impulse response of the optical preamplifier and the equalizer in cascade
λ_{d} is the electrons/s generated by the PD when no lightwave is incident on the PD

It is important to note that the quantum shot noise power depends on the signal levels, $\{a_{\text{k}}\}$, and the time t,that is, the noise is signal dependent and nonstationary.

Without a loss of generality, we can consider the output quantum shot noise at the instant $t = 0$; then, by substituting $t = 0$ and $t' = \tau$, (7.45) becomes

$$\left\langle i_{\text{SN}}^2(0) \right\rangle = \int_{-\infty}^{+\infty} qG^{2+x}\Re \left\{ \sum_{k=-\infty}^{\infty} a_k h_{\text{p}}(\tau - kT) h_{\text{I}}^2(-\tau) \right\} d\tau + \int_{-\infty}^{+\infty} qG^{2+x}\Re\lambda_{\text{d}} h_{\text{I}}^2(\tau) d\tau \tag{7.46}$$

The first term is contributed by the signal amplitude and the signal-dependent shot noise, and the second term indicates the contribution of the dark current of the PD.

The shot noise power due to the signal sequence, $\{a_k\}$, at the time t = 0 can thus be rewritten as

$$\left\langle i_{\text{SN}}^2(0) \right\rangle = qG^{2+x}\Re \left\{ \sum_{k=-\infty}^{\infty} a_k (h_{\text{p}} * h_{\text{I}}^2)(-kT_{\text{B}}) \right\} \tag{7.47}$$

$$\left\langle i_{SN}^2(0)\right\rangle = qG^{2+x}\Re\left\{\sum_{k=-\infty}^{\infty} a_k(h_p * h_I^2)(-kT_B)\right\} = qG^{2+x}\Re a_0 g(0) + qG^{2+x}\Re\left\{\sum_{k=-\infty}^{\infty} a_k g(-kT_B)\right\} \quad (7.48)$$

where $(h_p * h_I^2)(-kT_B) = g(-kT_B)$.

The first term is the quantum shot noise due the pulse under consideration, and the second term is due to the preceding and succeeding signal pulses. The effects of the succeeding pulses on the noise at $t=0$ is likely since there is some delay due to signal traveling through the system. This dependence of noise power on the preceding and succeeding pulses is similar to the ISI. However, it is important to note that this effect cannot be eliminated by equalizing the pulse to obtain zero ISI, since $h_0(kT)=0$ does not imply that $g(kT)=0$. The exact estimation of the noise power contributed to the considered pulse, $\left\langle i_{SN}^2(0)\right\rangle$, requires statistical knowledge of the distribution of the pulse sequence. This is commonly unknown, and thus some criteria must be used. The worst case that would contribute the largest noise effect is when the sequence $\{a_k\}$ has the adjacent symbols of the maximum values, b_M. In this case, we have

$$\left\langle i_{SN}^2(0)\right\rangle = qG^{2+x}\Re a_0 g(0) + qG^{2+x}\Re\left\{\sum_{\substack{k=-\infty \\ k\neq 0}}^{\infty} b_M g(-kT_B)\right\} \quad (7.49)$$

The frequency domain of the noise power can be estimated as

$$\left\langle [i_{SN}^F(0)]^2\right\rangle = qG^{2+x}\Re\left\{\frac{b_M}{T_B}\sum_{\substack{k=-\infty \\ k\neq 0}}^{\infty} b_M G\left(\frac{2\pi N}{T_B}\right) - \frac{b_M - a_0}{2\pi}\int_{-\infty}^{\infty} G(\omega)d\omega\right\} \quad (7.50)$$

where

$$G(\omega) = \Im[g(t)]$$

$$G(\omega) = \frac{1}{2\pi} H_p(\omega)[H_I(\omega) * H_I(\omega)] \quad (7.51)$$

$$\sum_{k=-\infty}^{\infty} g(-kT_B) = \frac{1}{T_B}\sum_{n=-\infty}^{\infty} b_M G\left(\frac{2\pi N}{T_B}\right)$$

The worst-case total noise power is therefore given by

$$\left\langle [i_{SN}^F(0)]^2\right\rangle = \left\langle i_{sd}^2\right\rangle + \left\langle i_{SN}^2(0)\right\rangle \quad (7.52)$$

where $\left\langle i_{sd}^2\right\rangle$ is the shot noise due to the dark current of the PD

7.5.4 Total Output Noises and Pulse Shape Parameters

It is now advantageous to normalize the input optical pulse and the output equalized pulse as follows:

$$\begin{aligned} h'_{\mathrm{p}}(t) &= T_{\mathrm{B}} h_{\mathrm{p}}(tT_{\mathrm{B}}) \\ h'_{\mathrm{o}}(t) &= T_{\mathrm{B}} h_{\mathrm{o}}(tT_{\mathrm{B}}) \end{aligned} \tag{7.53}$$

or, equivalently, in the frequency domain as

$$\begin{aligned} H'_{\mathrm{p}}(f) &= H_{\mathrm{p}}\left(\frac{2\pi f}{T_{\mathrm{B}}}\right) \\ H'_{\mathrm{o}}(f) &= \frac{1}{T_{\mathrm{B}}} H_{\mathrm{o}}\left(\frac{2\pi f}{T_{\mathrm{B}}}\right) \end{aligned} \tag{7.54}$$

These temporal and spectral functions depend only on the shapes of the input and output pulses, not on the epoch T. The normalization can be achieved by setting

$$H'_{\mathrm{p}}(0) = 1 \tag{7.55}$$

while the requirement on $H'_{\mathrm{o}}(f)$ for zero ISI is

$$\sum_{-\infty}^{\infty} H'_{0}(f+k) = 1 \quad \text{for } |f| < 0.5 \tag{7.56}$$

Thus, all the output noises can be expressed in terms of these normalized functions. We can thus obtain from (7.51),

$$H_{\mathrm{I}}\left(\frac{2\pi f}{T_{\mathrm{B}}}\right) = \frac{T_{\mathrm{B}}}{\Re G} H'(f), \quad \text{with } H'(f) = \frac{H'_{\mathrm{o}}(f)}{H'_{\mathrm{p}}(f)} \tag{7.57}$$

The normalized effects of pulse shaping are given as

$$\begin{aligned} I_1 &= \int_{-\infty}^{\infty} H'_{\mathrm{p}}[H' * H'(f)] df \\ I_2 &= \int_{-\infty}^{\infty} |H'(f)|^2 df \\ I_2 &= \int_{-\infty}^{\infty} |H'(f)|^2 f^2 df \end{aligned} \tag{7.58}$$

The term I_1 influences the quantum shot noise of the PD from the optical signal received at the input, the term I_2 influences the dark current shot noises and the thermal noises due to the input resistance of the preamplifier, and the term I_3 influences the frequency-dependent parts of the transfer function of the electronic preamplifier.

7.5.4.1 FET Front-End Optical Receiver

As an example of the estimation of noises at the output of the optical receiver, we illustrate an optical receiver with a PD followed by an FET front end. Referring to Figure 7.5 for an FET front end, we have the following parameters:

$$C = C_a + C_d \tag{7.59}$$

$$R_p = R_b R_i/(R_b + R_i) \tag{7.60}$$

where
- R_b is the bias resistance
- R_i is the input resistance of the amplifier at mid-band
- C_d is the diode capacitance under reverse bias
- C_a is the input capacitance of the amplifier

The spectral densities of thermal noises due to the feedback resistance, R_f, and the bias resistance are given by

$$\begin{aligned} S_R &= \frac{2k_BT}{R_f} \\ S_{R_b} &= \frac{2k_BT}{R_b} \end{aligned} \tag{7.61}$$

where
- T is the temperature
- k_B is Boltzmann's constant

Then the noise power at the equalizer output is given by

$$\left\langle i_{RN}^2 \right\rangle = 2k_BT\left(\frac{1}{R_f}+\frac{1}{R_b}\right)\cdot\int_{-\infty}^{\infty}\left|H_I(2\pi f)\right|^2 df = 2k_BT\left(\frac{1}{R_f}+\frac{1}{R_b}\right)\cdot\left(\frac{1}{\Re G}\right)^2 I_2T_b \tag{7.62}$$

where
- T_b is the bit period
- T is the absolute temperature

The noise due to the transconductance of the FET is given by

$$\begin{aligned} \left\langle i_{g_mN}^2 \right\rangle &= \frac{2k_BT\Gamma P}{g_m}\cdot\frac{1}{2\pi}\int_{-\infty}^{\infty}\left|H_I(2\pi f)\right|^2\left(\frac{1}{R_f^2}+\omega^2C^2\right)d\omega \\ &= \frac{2k_BT\Gamma P}{g_m}\cdot\left(\frac{1}{\Re G}\right)^2\left(\frac{T_bI_2}{R_f^2}+\frac{(2\pi C)^2}{T_b}I_3\right) \end{aligned} \tag{7.63}$$

where

g_m is the transconductance of the FET

$\Gamma \sim 0.7$ is the factor related materials of type GaAs

P is the imperfection factor of the FET device

The quantum shot noise of the PD is given by

$$\left\langle \left[i_{SN}^{F}(0) \right]^2 \right\rangle = qG^{2+x}\Re \left\{ \frac{b_M}{T} \sum_{\substack{k=-\infty \\ k\neq 0}}^{\infty} b_M G\left(\frac{2\pi N}{T} \right) - \frac{b_M - a_0}{2\pi} \int_{-\infty}^{\infty} G(\omega)d\omega \right\}$$

$$= \frac{q}{\Re} G^x \left[a_0 I_1 + b_M \left(\sum_1 - I_1 \right) \right] \quad (7.64)$$

where $\sum_1 = \sum_{-\infty}^{\infty} H_p'(k)\left[H'(k) * H'(k) \right]$

7.5.4.2 *BJT Front-End Optical Receiver*

A BJT can also be employed as the front-end amplifier of the optical receivers. A typical circuit design of such a receiver is shown in Figure 7.9, in which three BJTs have been used with a direct coupled pair and a peaking stage to extend the bandwidth of the overall gain and then a shunt feedback to form a TI configuration.

7.5.4.2.1 *Noise Generators*

Electrical shot noises are generated by the random generation of streams of electrons (current). In optical detection, shot noises are generated by (1) biasing currents in electronic devices and (2) photo currents generated by the photodiode.

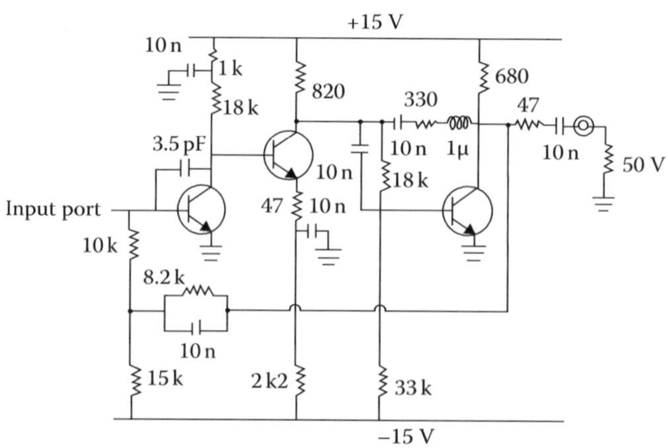

FIGURE 7.9
Design circuit of the electronic preamplifier for the balanced optical amplifier. All BJTs are either of type Phillips BFR90A or BFT24.

A biasing current, I, generates a spectral current density, S_I, given by

$$S_I = \frac{d(i_I^2)}{df} = 2qI \quad \text{in A}^2/\text{Hz} \tag{7.65}$$

where q is the electronic charge. The quantum shot noise, $\langle i_s^2 \rangle$, generated by the PD by an optical signal with an average optical power, P_{in}, is given by

$$S_Q = \frac{d\langle i_s^2 \rangle}{df} = 2q\langle i_s^2 \rangle \quad \text{in A}^2/\text{Hz} \tag{7.66}$$

In the case where the APD is used, the noise spectral density is given by

$$S_Q = \frac{d\langle i_s^2 \rangle}{df} = 2q\langle i_s^2 \rangle\langle G_n^2 \rangle \quad \text{in A}^2/\text{Hz} \tag{7.67}$$

It is noted here again that the dark currents generated by the PD must be added to the total equivalent noise current at the input after it is evaluated. These currents are generated even in the absence of the optical signal. These dark currents can be eliminated by cooling the PD to at least below the temperature of liquid nitrogen (77 K).

At a certain temperature, the conductivity of a conductor varies randomly. The random movement of electrons generates a fluctuating current even in the absence of an applied voltage. The thermal noise spectral density of a resistor R is given by

$$S_R = \frac{d(i_R^2)}{df} = \frac{4k_B T}{R} \quad \text{in A}^2/\text{Hz} \tag{7.68}$$

where

k_B is Boltzmann's constant
T is the absolute temperature (in K)
R is the resistance (in ohms)
i_R denotes the noise current due to the resistor R

7.5.4.2.2 Equivalent Input Noise Current

Our goal is to obtain an analytical expression of the noise spectral density equivalent to a source looking into the electronic amplifier including the quantum shot noises of the PD. A general method for deriving the equivalent noise current at the input is by representing the electronic device by a Y-equivalent linear network, as shown in Figures 7.6 and 7.7. The two current noise sources, $di_N'^2$ and $di_N''^2$, represent the summation of all noise currents at the input and at the output of the Y network. This can be transformed into a Y-network circuit with the noise current referred to the input as follows:

The output voltages, V_0, as referred to Figure 7.6, can be written as

$$V_0 = \frac{i_N'(Y_f - Y_m) + i_N''(Y_i + Y_f)}{Y_f(Y_m + Y_i + Y_o + Y_L) + Y_i(Y_o + Y_L)} \tag{7.69}$$

and then using the equivalent model, we have

$$V_0 = \frac{(i'_\mathrm{N})_\mathrm{eq}(Y_\mathrm{f} - Y_\mathrm{m})}{Y_\mathrm{f}(Y_\mathrm{m} + Y_\mathrm{i} + Y_\mathrm{o} + Y_\mathrm{L}) + Y_\mathrm{i}(Y_\mathrm{o} + Y_\mathrm{L})} \tag{7.70}$$

By comparing these two equations, we can deduce the equivalent noise current at the input of the detector to be

$$i_\mathrm{Neq} = i'_\mathrm{N} + i''_\mathrm{N}\frac{Y_\mathrm{i} + Y_\mathrm{f}}{Y_\mathrm{f} - Y_\mathrm{m}} \tag{7.71}$$

Then reverting to mean-square generators for a noise source, we have

$$d(i_\mathrm{Neq})^2 = d(i'_\mathrm{N})^2 + d(i''_\mathrm{N})^2\left|\frac{Y_\mathrm{i} + Y_\mathrm{f}}{Y_\mathrm{f} - Y_\mathrm{m}}\right|^2 \tag{7.72}$$

It is therefore expected that the Y matrix of the front-end low-noise amplifier is known as the equivalent noise at the input of the amplifier, which can be obtained by using (7.72).

For a given source, the input noise current power of a BJT front end can be found by using (see also Appendix)

$$i^2_\mathrm{Neq} = \int_0^B d(i_\mathrm{Neq})^2 = a + \frac{b}{r_\mathrm{E}} + cr_\mathrm{E} \tag{7.73}$$

where
- B is the bandwidth of the electronic preamplifier
- r_E is the emitter resistance of the front-end transistor of the preamplifier
- a, b, and c are the parameters depending on the circuit elements and the amplifier bandwidth

Thus, an optimum value of r_E can be found; hence, an optimum biasing current is to be set for the collector current of the BJT such that i^2_Neq is at a minimum:

$$r_\mathrm{Eopt} = \sqrt{\frac{b}{c}} \quad \text{hence} \quad \rightarrow i^2_\mathrm{Neq}\Big|_{r_\mathrm{E}=r_\mathrm{Eopt}} = a + 2\sqrt{bc} \tag{7.74}$$

If two types of BJTs are considered, Phillips BFR90A and BFT24, then a good approximation of the equivalent noise power can be found as

$$\begin{aligned} a &= \frac{8\pi B^3}{3}\left\{r_\mathrm{B}C_\mathrm{s}^2 + (C_\mathrm{s} + C_\mathrm{f} + C_\mathrm{tE})\tau_\mathrm{T}\right\} \\ b &= \frac{B}{\beta_\mathrm{N}} \\ c &= \frac{4\pi B^3}{3}(C_\mathrm{s} + C_\mathrm{f} + C_\mathrm{tE})^2 \end{aligned} \tag{7.75}$$

The theoretical estimation of the transistors can be derived from the measured scattering parameters given by the manufacturer as

$$
\begin{aligned}
&\text{BFR90A}\\
&r_{\text{Eopt}} = 59\,\Omega\\
&I_{\text{Eopt}} = 0.44\,\text{mA}\\
&I_{\text{eq}}^2 = 7.3\times 10^{-16}\,\text{A}^2 \longrightarrow \frac{I^2_{\text{eq}}}{B} = 4.9\times 10^{-24}\,\text{A}^2/\text{Hz}\\
&I_{\text{eq}} = 27\,\text{nA} \longrightarrow \frac{I_{\text{eq}}}{\sqrt{B}} = 2.21\,\text{pA}/\sqrt{\text{Hz}}
\end{aligned}
\tag{7.76}
$$

$$
\begin{aligned}
&\text{BFT24}\\
&r_{\text{Eopt}} = 104\,\Omega\\
&I_{\text{Eopt}} = 0.24\,\text{mA}\\
&I_{\text{eq}}^2 = 79.2\times 10^{-16}\,\text{A}^2 \longrightarrow \frac{I^2_{\text{eq}}}{B} = 6.1\times 10^{-24}\,\text{A}^2/\text{Hz}\\
&I_{\text{eq}} = 30.2\,\text{nA} \longrightarrow \frac{I_{\text{eq}}}{\sqrt{B}} = 2.47\,\text{pA}/\sqrt{\text{Hz}}
\end{aligned}
\tag{7.77}
$$

Note that the equivalent noise current depends largely on some not well-defined values, such as the capacitance, the transit times, the base-spreading resistance, and the short-circuit current gain, β_N. The term $I_{\text{eq}}/\sqrt{B}$ is usually specified as the noise spectral density equivalent referred to the input of the electronic preamplifier.

7.6 HEMT-Matched Noise Network Preamplifier

7.6.1 Structure

A third-order noise-matching network (shown in Figure 7.10) has been obtained by Park [9] to tailor for a 10 GHz low-noise optical receiver. This network is inserted into our preamplifier model and the electronic circuit is connected, as shown in Figure 7.11.

For high receiving bit rate, we need to design a preamplifier with high-frequency response. In electronics, the drain-to-source HEMT parasitic capacitance, C_{ds}, is negligibly small at high frequency. Furthermore, we can reduce the complexity of solving this circuit by considering the *Miller effect* capacitance. Thus, we are able to construct an equivalent circuit of the small-signal, high-frequency response for the electronic circuit, and the simplified equivalent circuit is shown in Figure 7.12.

The equivalent circuit (Figure 7.12) can be further simplified to that shown in Figure 7.13. Thus, the required preamplifier's transfer function can be obtained as follows:

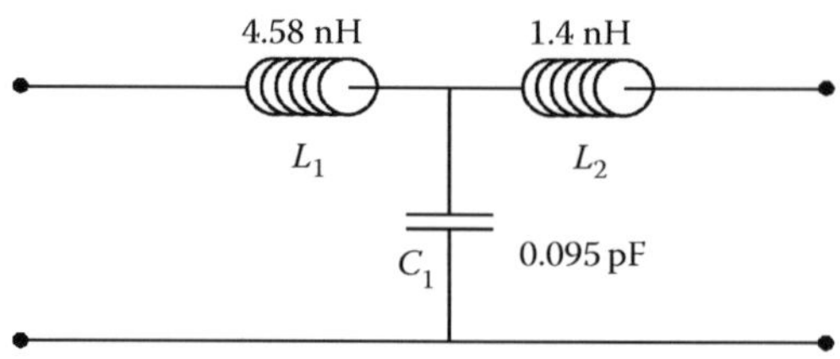

FIGURE 7.10
Third-order noise-matching network.

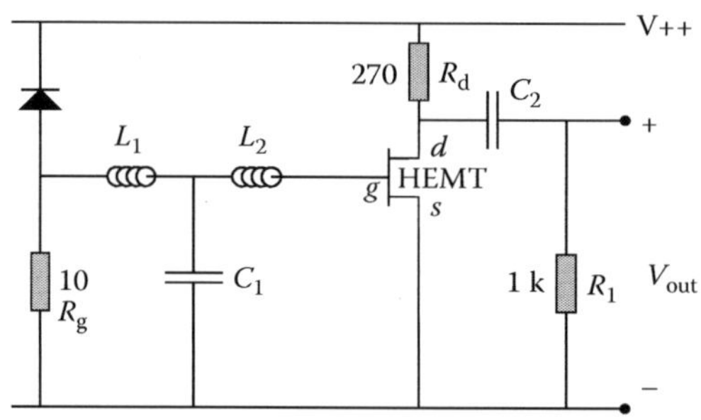

FIGURE 7.11
Electronic circuit of HEMT noise-matched network preamplifier.

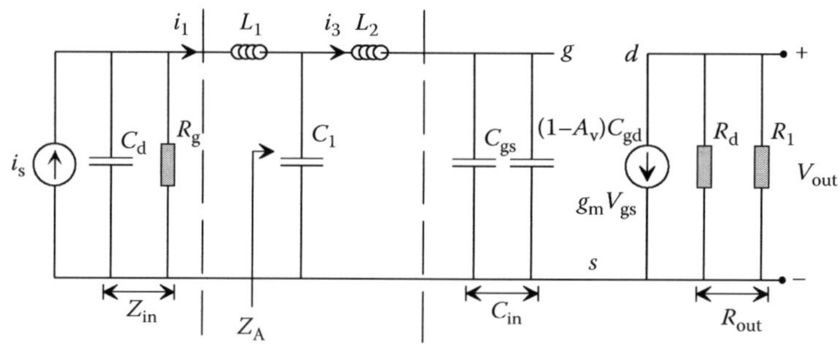

FIGURE 7.12
Equivalent circuit of HEMT noise-matched network preamplifier.

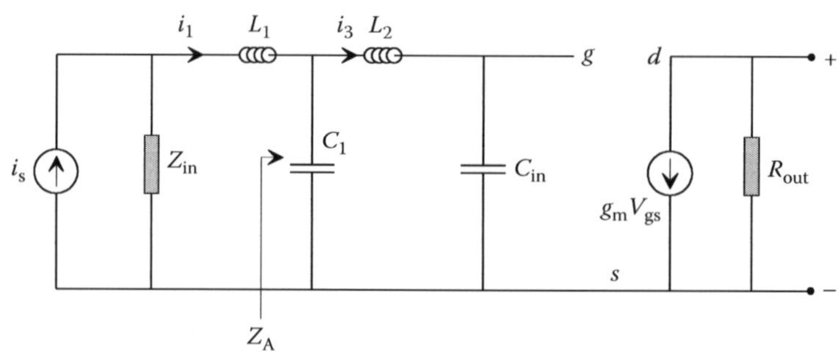

FIGURE 7.13
Simplified equivalent circuit of HEMT noise-matched network preamplifier.

$$i_3 = \left(\frac{sC_{\text{in}}}{s^3 L_2 C_1 C_{\text{in}} + sC_1 + sC_{\text{in}}} \right) \cdot i_1 \tag{7.78}$$

$$i_1 = \left(\frac{Z_{\text{in}}}{Z_{\text{in}} + Z_{\text{A}}} \right) \cdot i_s \tag{7.79}$$

By substituting (7.79) into (7.78), we obtain the transfer function of the HEMT-matched noise network preamplifier:

$$H_{\text{HEMT}} = \frac{V_{\text{out}}}{i_{\text{s}}} = -g_{\text{m}} R_{\text{out}} \left(\frac{1}{s^3 L_2 C_1 C_{\text{in}} + sC_1 + sC_{\text{in}}} \right) \cdot \left(\frac{Z_{\text{in}}}{Z_{\text{in}} + Z_{\text{A}}} \right) \tag{7.80}$$

where

$$R_{\text{out}} = \frac{R_{\text{d}} R_{\text{l}}}{R_{\text{d}} + R_{\text{l}}} \tag{7.81}$$

$$Z_{\text{in}} = \frac{R_{\text{g}}}{sR_{\text{g}} C_{\text{d}} + 1} \tag{7.82}$$

$$Z_{\text{A}} = sL_1 + \frac{\left(\frac{1}{sC_1} \right) \cdot \left(sL_2 + \frac{1}{sC_{\text{in}}} \right)}{\frac{1}{sC_1} + \frac{1}{sC_{\text{in}}} + sL_2} = \frac{s^4 L_1 L_2 C_1 C_{\text{in}} + s^2 L_1 C_1 + s^2 L_1 C_{\text{in}} + s^2 L_2 C_{\text{in}} + 1}{s^3 L_2 C_1 C_{\text{in}} + s(C_1 + C_{\text{in}})} \tag{7.83}$$

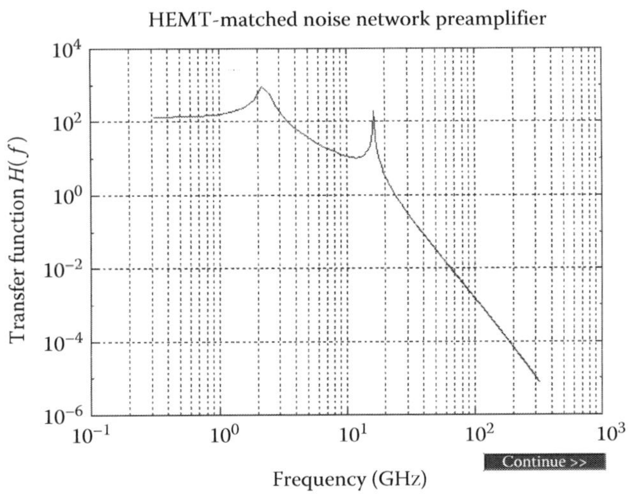

FIGURE 7.14
Frequency response of the HEMT noise-matched network preamplifier.

7.6.2 Noise Theory and Equivalent Input Noise Current

The circuit diagrams in Figure 7.20 can be represented with reflection coefficients shown in Figure 7.15. We denote the reflection coefficients of Z_{ph}, Z_s, and Z^*_{opt} by Γ_{ph}, Γ_s, and Γ^*_{opt}, which are normalized to Z_o and defined by

$$\Gamma_x = \frac{Z_x - Z_0}{Z_x + Z_0} \tag{7.84}$$

where x = ph, s, and opt. It is useful to obtain the S parameters of the third-order noise-matching network defined in the matrix form in Equation 7.85:

$$S = \begin{bmatrix} S_{11} & S_{12} \\ S_{21} & S_{22} \end{bmatrix} \tag{7.85}$$

From Figure 7.16, we find Z_1, which is given by

$$Z_1 = sL_1 + \left(sC_1 + \frac{1}{1+sL_2}\right)^{-1} = \frac{s^3C_1L_1L_2 + s^2C_1L_1 + s(L_1+L_2)+1}{s^2C_1L_2 + sC_1 + 1} \tag{7.86}$$

Applying the circuit theory of the T network and substituting (7.85), we obtain

$$S_{11} = \frac{Z_1 - 1}{Z_1 + 1} = \frac{s^3C_1L_1L_2 + s^2(C_1L_1 - C_1L_2) + s(L_1 + L_2 - C_1)}{s^3C_1L_1L_2 + s^2(C_1L_1 + C_1L_2) + s(L_1 + L_2 + C_1) + 2} \tag{7.87}$$

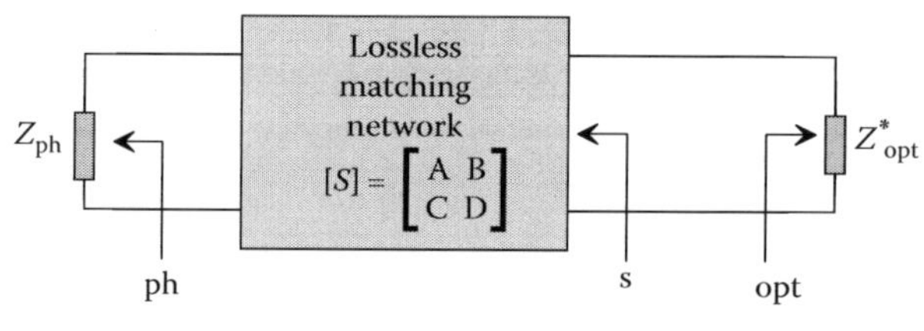

FIGURE 7.15
Lossless matching network with source and load impedances.

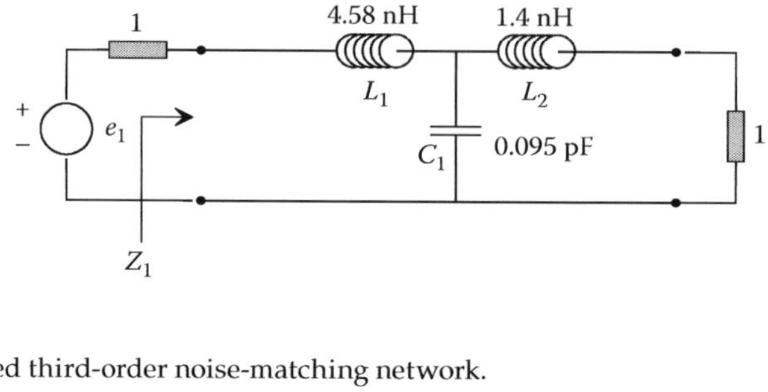

FIGURE 7.16
Schematic of the matched third-order noise-matching network.

From the circuit theory, S_{21} is given by

$$S_{21} = 2\sqrt{\frac{r_1}{r_2}}\left(\frac{V_2}{e_1}\right) \tag{7.88}$$

From Figure 7.16, we can obtain

$$\frac{V_2}{e_1} = \frac{1}{s^3C_1L_1L_2 + s^2(C_1L_1 + C_1L_2) + s(L_2 + L_1 + C_1) + 2} \tag{7.89}$$

Hence, by substituting (7.89) into (7.87), we have

$$S_{21} = \frac{2}{s^3C_1L_1L_2 + s^2(C_1L_1 + C_1L_2) + s(L_2 + L_1 + C_1) + 2} \tag{7.90}$$

Thus, using the symmetrical property, the other two S parameters can be obtained as follows:

$$S_{22} = S_{11} = \frac{s^3C_1L_1L_2 + s^2(C_1L_1 - C_1L_2) + s(L_1 + L_2 - C_1)}{s^3C_1L_1L_2 + s^2(C_1L_1 + C_1L_2) + s(L_1 + L_2 + C_1) + 2} \tag{7.91}$$

$$S_{12} = S_{21} = \frac{2}{s^3C_1L_1L_2 + s^2(C_1L_1 + C_1L_2) + s(L_2 + L_1 + C_1) + 2} \tag{7.92}$$

By using Equation 7.84, we can derive all the required reflection coefficients. First of all, consider the equivalent photodiode circuit shown in Figure 7.17.

Then, Z_{ph} can be found as

$$Z_{ph} = R_s + \frac{1}{sC_d} \tag{7.93}$$

From Equation 7.93, we have

$$\Gamma_{ph} = \frac{Z_{ph} - Z_0}{Z_{ph} + Z_0} \tag{7.94}$$

Then, we substitute Z_{ph} from Equation 7.93:

$$\Gamma_{ph} = \frac{sC_d(R_s - 1) + 1}{sC_d(R_s + 1) + 1} \tag{7.95}$$

Knowing the S parameters and Γ_{ph}, we can obtain

$$\Gamma_s = S_{22} + \frac{S_{12}S_{21}\Gamma_{ph}}{1 - \Gamma_{ph}S_{11}} \tag{7.96}$$

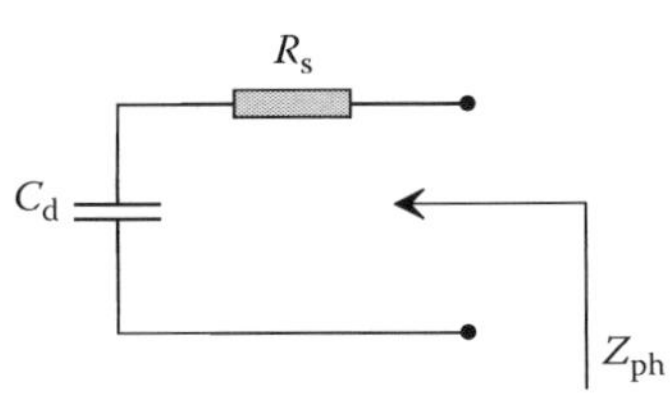

FIGURE 7.17
Equivalent circuit for the photodiode.

and

$$\Gamma_{opt} = \Gamma_s = S_{22} + \frac{S_{12}S_{21}\Gamma_{ph}}{1-\Gamma_{ph}S_{11}} \tag{7.97}$$

The equivalent input noise current density appearing across the photodiode junction capacitance can be expressed in terms of reflection coefficients [9] as

$$i_{eq}^2 = 4kTR_s(\omega C_d)^2 \frac{\left(1-\left|\Gamma_{opt}^*\right|^2\right)}{G_M\left|1-\Gamma_s\Gamma_{opt}^*\right|^2}\left\{F_{min}\left(1-\left|\Gamma_s\right|^2\right)+\frac{4R_n}{Z_0}\frac{\left|\Gamma_s-\Gamma_{opt}\right|^2}{\left|1+\Gamma_{opt}\right|^2}\right\}\Delta f \tag{7.98}$$

where
F_{min} and R_n can be obtained from the data given in Table 7.2
G_M is the transducer power gain of the lossless matching network

Our goal is to minimize the equivalent input noise current density. This can be done by minimizing the photodiode junction capacitance, C_d; the series resistance of the photodiode, R_s; the minimum noise figure, F_{min}; and the noise resistance, R_n; and by maximizing the transducer power gain. Among these parameters in Equation 7.98, only the transducer power gain, G_M, and the output reflection coefficient, Γ_s, are related to the noise-matching network, and these may be optimized by design. The transducer power gain can be expressed in terms of reflection coefficients [9] and is defined by

$$G_M = 1-\left|\frac{(1+\Gamma_s)(1-\Gamma_{opt})-(1-\Gamma_s)(1+\Gamma_{opt})}{(1+\Gamma_s)(1-\Gamma_{opt}^*)+(1-\Gamma_s)(1+\Gamma_{opt}^*)}\right|^2 \tag{7.99}$$

The transducer power gain has its maximum value of unity when $\Gamma_s=\Gamma_{opt}$. This condition is equivalent to the output admittance of the noise-matching network being conjugately matched to the complex conjugate of the optimum source admittance. Figure 7.18 shows a typical equivalent input noise current density of a HEMT noise-matched front-end receiver.

TABLE 7.2

Noise Parameters of the Packaged HEMT with 0.3 μm Gate Length

	Γ_{opt}			
Frequency (GHz)	**(Mag)**	**(Ang)**	**F_{min} (dB)**	**R_n (Ω)**
2	0.79	30	0.33	29
4	0.73	59	0.35	21
6	0.68	87	0.44	14.5
8	0.63	119	0.55	9.5
10	0.59	139	0.66	6
12	0.55	164	0.75	4

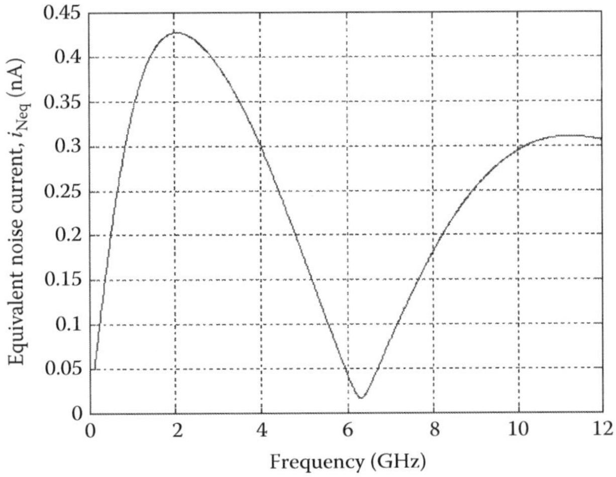

FIGURE 7.18
Equivalent input noise current density of the HEMT noise-matched front-end receiver.

7.7 Concluding Remarks

This chapter has addressed the considerations of optical receivers in which the electronic currents are generated when the optical energy/power from the modulated optical signals is absorbed by the photosensitive regions directly. The noise generation process and related small-signal models of electronic preamplifiers are described. An equivalent noise current as seen at the input port of the electronic preamplifier is presented and the method to derive it for the front-end amplifier is described. Both FET and BJT types are given. Further, the effects of the pulse shape on the output signals are also given, related to the transfer function of an electronic preamplifier. A noise-matching network at the front end of the optical preamplifier would reduce the noise effects at high frequency, and this is crucial in the design of an ultra-wide-band optical receiver.

These noise models can be integrated into coherent receivers, which will be described in Chapter 8. For a coherent optical receiver, a local oscillator, a powerful laser, mixes the optical signals with its output field and gives a beating signal in the PD due to its square-law property. The phase and amplitude of the optical signals are preserved in the electronic domain, and thus processing in the electronic domain can be performed to extract the digital and analog properties of the original signals.

7.8 Problems

PROBLEM 7.1

Suppose that we have an ideal PD that produces no dark current, i.e., no electron–hole pairs are generated in the absence of an optical pulse. Assume that an optical pulse of an energy b_i falls on the PD in an interval T. This can be interpreted by the receiver as a "0" pulse if no electron–hole pairs are generated with the pulse present. Show that if the probability of this pulse that is interpreted as a "0" is 10^{-9}, then $b_1 = 36h\nu/\eta$, where η is the quantum efficiency of the PD.

The average power, $p(t)$, detected with this PD is given by $b_1 B_r/2$, where B_r is the system bit rate, if we assume that equal numbers of "1" and "0" are received. Show that if the quantum efficiency is 1.0, the minimum optical power required at a BER of 10^{-9} for a bit rate of 155 Mb/s is about −63 dBm at an operating wavelength of 810 nm. What is the required optical power for a system operating at 1300 nm?

PROBLEM 7.2

A 1310 nm digital optical receiver is operating at 2.2 Gbit/s and has an effective noise bandwidth of 1.5 GHz. The pin photodiode has negligible dark current and 90% quantum efficiency. The equivalent input resistance is 100 Ω and the equivalent noise current is 2.5 pA/(Hz)$^{1/2}$. Calculate the receiver sensitivity corresponding to a BER of 10^{-9}. How much does it change if the receiver is designed to operate reliably up to a BER of 10^{-12}?

PROBLEM 7.3

Consider an optical receiver operating at 140 Mbit/s in a PCM system. It has an APD photodiode with an x factor of 0.8 and a responsivity of 0.7 A/W; its avalanche gain depends on the bias reverse voltage. An electronic amplifier that follows the PD has an equivalent noise current density of 10.0 nA/(Hz)$^{1/2}$ and a 3 dB bandwidth of 100 MHz. For a system BER of 10^{-9}, calculate and plot the receiver sensitivity as a function of the avalanche gain between 10 and 70. Comment on the receiver performance with this range of avalanche gain.

PROBLEM 7.4

An optical receiver consists of a PIN photodiode and an APD followed by an electronic amplifier with the following characteristics:

PIN-PD

- InGaAs type with an average quantum efficiency of 0.8 over the wavelength region of 1.2–1.7 μm
- Dark current noise of 0.05 nA
- Total rise time and fall time of 1.0 ns
- Diode capacitance at a typical reverse bias of 0.5 pF

APD

- Ge type with an x factor of 0.9
- An average avalanche gain of 30 over the wavelength range of 1200–1700 nm
- Total dark current of 7.0 nA
- Total rise and fall time of 1.0 ns

Electronic Amplifier

It has an HI-type front end followed by a main amplifier and an equalizer with a total equivalent noise spectral density of 0 pA/(Hz)$^{1/2}$ and an overall bandwidth of 200.0 MHz.

For a PCM system operating at a bit rate of 140 Mbps and a BER of 10^{-9}, find the energy of the "1" pulse required at the input of the PD. You may assume that the probability density distributions of receiving "1" and "0" are Gaussian and equal. Hence, deduce the receiver sensitivity in dBm.

The above receiver is to be considered for a new receiver configuration that is to be specified for two optical communication links with the following requirements:

(b1) Link 1: Optical communications between two supercomputers operating at a bit rate of 1.2 Gb/s.

(b2) Link 2: PCM optical communication transmission systems operating at 140 Mb/s, using a single-mode optical fiber for linking two towns A and B 200 km apart. Bidirectional transmission at 1300 and 1550 nm wavelength regions. The specification of the standard fiber is given in the ECS4344 lectures.

A number of optoelectronic components such as PDs and sources by Fujitsu are available. Select appropriate components and design the link.

PROBLEM 7.5

An optical fiber communication link has the following components:

- Optical transmitter: InGaAsP laser diode operating in a region about 1320 nm with a spectral width of 2.0 nm.
- The laser power coupled to a single-mode fiber of 10 mm diameter pigtail of 1.0 mW.
- The fiber is single mode at 1300 nm wavelength with a cutoff wavelength of 1190 nm.
- The total fiber loss is 0.35 dB/km.
- The receiver consists of an InGaAs pin photodiode, which has a sensitivity of P dBm, approximately given by $P = 18.5 \log 10\,(B) - 68.5$ dBm, where B is the bit rate in Mb/s.

For a bit rate of 1–1000 Mb/s, plot the attenuation-limited transmission distance, including a 0.1 dB connector loss at each end and a 3.0 dB system operating margin.

Appendix: Noise Equations

Refer to the small-signal and noise models given in Figures 7.19 and 7.20 The noises generated in a transistor can be expressed as

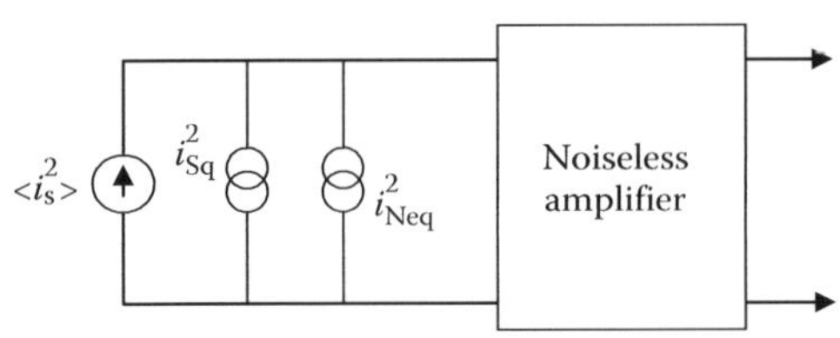

FIGURE 7.19
Equivalent noise current at the input and noiseless amplifier model. i_{Sq}^2 is the quantum shot noise that is signal dependent and i_{Neq}^2 is the total equivalent noise current referred to the input of the electronic amplifier.

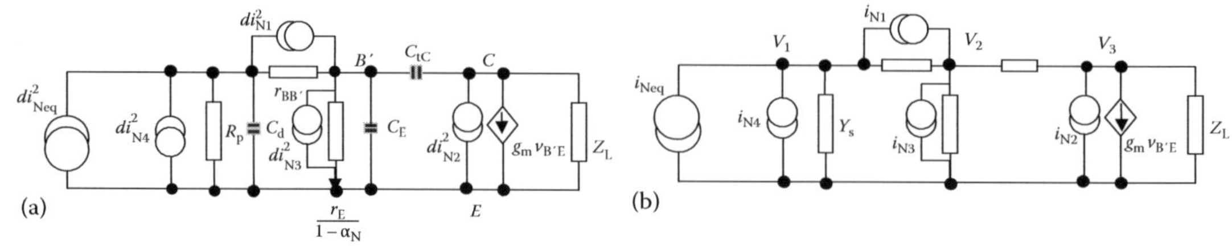

FIGURE 7.20
(a) Approximated noise equivalent and small-signal model of a BJT front end. (b) Generalized noise and small-signal model circuit. Note that $r_B = r_{sd} + r_{BB}$ and $C_d = C_p + C_i$, where r_{BB} is the base resistance, r_{sd} is the diode resistance, C_d is the photodiode capacitance, and C_i is the input capacitance.

$$
\begin{aligned}
di_1^2 &= 4k_B T g_B df \\
di_2^2 &= 2qI_c df \simeq 2k_B T g_m df \\
di_3^2 &= 2qI_B df \simeq 2k_B T(1-\alpha_N) g_m df \\
di_4^2 &= \text{shot noise of diodes and thermal noise of bias in } g_B \text{ resistors}
\end{aligned}
\tag{7.100}
$$

where
- g_B is the base conductance
- I_C is the collector bias current
- I_B is the base bias current

From the nodal analysis of the small-signal equivalent circuit given in Figure 7.20, we can obtain the relationship

$$
\begin{bmatrix} Y_s + g_B & -g_B & 0 \\ -g_B & g_B + y_1 + y_f & -y_f \\ 0 & g_m - y_f & y_f + y_2 \end{bmatrix}
\begin{bmatrix} V_1 \\ V_2 \\ V_3 \end{bmatrix} =
\begin{bmatrix} i_{eq} - i_{N1} \\ i_{N1} + i_{N3} \\ i_{N2} \end{bmatrix}
\tag{7.101}
$$

Hence, V_3, V_2, and V_1 can be found by using Euler's rule for the matrix relationship

$$
V_3 = \frac{\Delta_{13}}{\Delta}(i_{eq} - i_{N1}) + \frac{\Delta_{23}}{\Delta}(i_{N1} + i_{N3}) + \frac{\Delta_{33}}{\Delta}(i_{N2})
\tag{7.102}
$$

Hence, the noise currents as referred to the input are

$$
d(i_1'^2) = \left| \frac{Y_s}{g_B} \right|^2 dI_1^2 = \omega_2 C_s^2 r_B 4kTdf
\tag{7.103}
$$

$$d(i_2'^2) = \left|(Y + y_1 + y_f) + \frac{Y_s}{g_B}(y_1 + y_f)\right|^2 \left|\frac{1}{y_f - g_m}\right|^2 = \left|\frac{1}{y_f - g_m}\right|^2 dI_2^2$$

$$= \left(\frac{1}{\beta_N r_E^2} + \omega^2\left(C_0^2 - \frac{2C_B r_B}{\beta_N r_E}\right) + \omega^4 2C_s r_B C_B^2\right) 2kTr_E df \tag{7.104}$$

$$d(i_3'^2) = \left|\frac{Y_s}{g_B} + 1\right|^2 dI_3^2 = (\omega\, C_s^2 r_B^2 + 1)\frac{2kT}{\beta r_E} df \tag{7.105}$$

in which we have assumed that $\Omega C_f \ll g_m$.

References

1. R.C. Alferness, Waveguide electrooptic modulators, *IEEE Journal of Microwave Theory and Technology*, MTT-30: 1121–1137, 1982.
2. R.C. Booth, $LiNbO_3$, integrated optic devices for coherent optical fiber systems, *Thin Solid Films*, 126: 167–176, 1985.
3. S. Ritchie and A.G. Steventon, The potential of semiconductors for optical-integrated circuits, in *Proceedings of Conference on Digital Optical Circuit Technology*, vol. 362, NATO, France, pp. 1111–1120, 1985.
4. S.D. Personick, Receiver design for digital fiber optic communications systems, *Bell System Technical Journal*, 52: 843–886, 1973.
5. S.D. Personick, Optical detectors and receivers, *IEEE Journal of Lightwave Technology*, 26(9): 1005–1020, 2008.
6. J.L. Hullet and T.V. Muoi, A feedback receive amplifier for optical transmission systems, *IEEE Transactions on Communications*, 24(10): 1180–1185, 1976.
7. R. Dogliotti, G. Luvison, and G. Pirani, Error probability in optical fiber transmission systems, *IEEE Transactions on Information Theory*, IT-25(2): 170–178, 1979.
8. S.D. Personick, *Optical Fiber Transmission Systems*. New York: Plenum Publishers, 1980.
9. Y. Yamamoto, Receiver performance evaluation of various digital optical modulation-demodulation systems in the 0.5–1.0 μm wavelength region, *IEEE Journal of Quantum Electronics*, QE-16: 1251–1259, 1980.

8

Optical Coherent Detection

Chapter 7 discussed the detection of optical signals by the direct conversion of optical signal power to electronic current in the photodiode followed by electronic amplification. This chapter gives a fundamental understanding of coherent detection of optical signals that requires the mixing of optical fields of the optical signals and that of the local oscillator, a high-power laser, so that its beating product would result in modulated signals preserving both its phase and amplitude characteristics in the electronic domain. Optical preamplification in coherent detection will be discussed in Chapter 10.

8.1 Introduction

With the exponential increase in data traffic, especially due to the demand for ultra-broad bandwidth driven by multimedia applications, cost-effective ultra-high-speed optical networks have become highly desirable. It is expected that the Ethernet technology will not only dominate access networks but will also become the key transport technology of next-generation metro/core networks. One-hundred-gigabit Ethernet (100 GbE) is currently considered to be the next logical evolution step after 10 GbE. Based on the anticipated 100 GbE requirements, 100 Gb/s data rate of serial data transmission per wavelength is required. To achieve this data rate while complying with current system design specifications, such as channel spacing, chromatic dispersion, and polarization mode dispersion (PMD) tolerance, coherent optical communication systems with multilevel modulation formats will be desired since they can provide high spectral efficiency, high receiver sensitivity, and potentially high tolerance to fiber dispersion effects [1]. Compared to conventional direct detection in intensity-modulation/direct-detection (IMDD) systems, which only detect the intensity of the light of the signal, coherent detection systems can retrieve the phase information of light and, therefore, can tremendously improve the receiver sensitivity.

Coherent optical receivers are important components in long-haul optical fiber communication systems and networks in order to improve the receiver sensitivity and thus the extra transmission distance. Coherent techniques were considered for optical transmission systems in the 1980s when the extension of repeater distance between spans was pushed to 60 km instead of 40 km for a single-mode optical fiber at a bit rate of 140 Gb/s. However, in the late 1980s, the invention of optical fiber amplifiers has overcome this attempt. Recently, interests in coherent optical communications have attracted significant research activities for ultra-bit-rate DWDM optical systems and networks. The motivation has been possible due to the following reasons: (1) the uses of optical amplifiers in cascade fiber spans have added significant noises and thus limit the transmission distance, (2) the advances of digital signal processors whose sampling rate can reach few tens of gigasamples per second allowing the processing of beating signals to recover the phase or phase estimation, (3) the availability of advanced signal processing algorithms such as Viterbi and Turbo, and (4) the differential coding and the modulation and detection of such signals may not require optical phase lock loop (OPLL), hence self-coherent and digital signal processing, to recover transmitted signals.

As shown in Chapter 7, the typical arrangement of an optical receiver is in such a way that the optical signals are detected by a photodiode (a pin diode or APD or a photon-counting device). Electrons generated in the photodetector are then electronically amplified through a front-end electronic amplifier. The electronic signals are then decoded for the recovery of original format. However, when the fields of incoming optical signals are mixed with those of a local oscillator whose frequency can be identical to or different from that of the carrier, the phase and frequency properties of the resultant signals reflect those of the original signals. Coherent optical communication systems have also been developing dramatically due to electronic processing and the availability of stable narrow linewidth lasers.

This chapter deals with the analysis and the design of coherent receivers with OPLL and the mixing of optical signals and that of the local oscillator in the optical domain and thence detection by the optoelectronic receivers following this mixing. Thus, both optical mixing and photodetection devices act as the fundamental elements of a coherent optical receiver. Depending on the frequency difference between the lightwave carrier of the optical signals and that of the local oscillator, the coherent detection can be termed as heterodyne or homodyne detection. For heterodyne detection, there is a difference in the frequency, and thus the beating signal falls in a passband region in the electronic domain. Thus, all the electronic processing at the front end must be in this passband region. For homodyne detection, there is no frequency difference, and thus the detection is in the baseband of the electronic signal. Both cases would require a locking of the local oscillator and carrier of the signals. An OPLL is thus discussed in this chapter.

This chapter is organized as follows: Section 8.2 gives an account of the components of coherent receivers, Section 8.3 outlines principles of optical coherent detection under heterodyne or homodyne techniques, and Section 8.4 gives details of an OPLL, which is a very important development for modern optical coherent detection. Finally, Section 8.8 provides some concluding remarks.

8.2 Coherent Receiver Components

The design of an optical receiver depends on the modulation format of the signals and thus transmission through the transmitter. The modulation of the optical carrier can be in the form of amplitude, phase, and frequency. Furthermore, the phase shaping also plays a critical role in the detection and the bit error rate (BER) of the receiver and hence the transmission systems. In particular, it is dependent on the modulation in analog or digital, Gaussian or exponential pulse shape, on–off keying or multiple levels, etc.

Figure 8.1 shows the schematic diagram of a digital coherent optical receiver that is similar to the direct detection receiver but with an optical mixer at the front end. However, the phase of the signals at the baseband or passband of the detected signals in the electrical domain would remain in the generated electronic currents and voltages at the output of the electronic preamplifier. An optical front end is an optical mixer combining the fields of the optical waves of the local laser and the optical signals so that the envelope of the optical signals can be beating with each other to a product with the summation of the frequencies and the difference of the frequencies of the lightwaves. Only the lower frequency term that falls within the absorption range of the photodetector is converted into the electronic current preserving both the phase and the amplitude of the modulated signals.

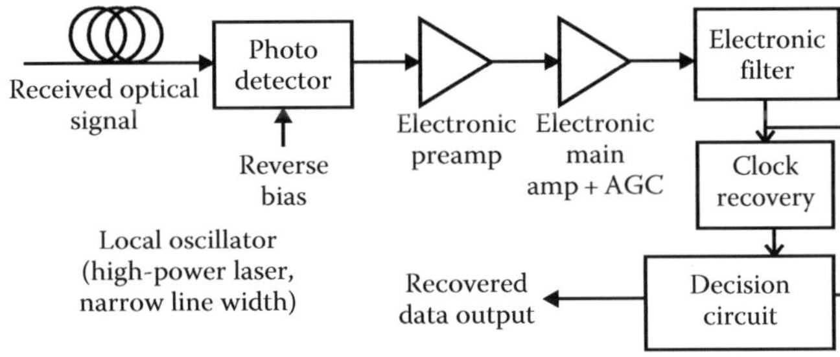

FIGURE 8.1
Schematic diagram of a digital optical coherent receiver with an additional local oscillator mixing with the received optical signals before being detected by an optical receiver.

Thence an optical receiver front end, very much the same as that of the direct detection, is connected following the optical processing front end consisting of a photodetector for converting lightwave energy into electronic currents; an electronic preamplifier for further amplification of the generated electronic current followed by an electronic equalizer for bandwidth extension, usually in voltage form; a main amplifier for further voltage amplification; a clock recovery circuitry for regenerating the timing sequence, and a voltage-level decision circuit for sampling the waveform for the final recovery of the transmitted and received digital sequences. Therefore, the optoelectronic preamplifier is followed by a main amplifier with an automatic control to regulate the electronic signal voltage to be filtered and then sampled by a decision circuit with synchronization by a clock recovery circuitry.

In an in-line fiber, an optical amplifier can be incorporated in the front of the photodetector to form an optical receiver with an optical amplifier front end to improve its receiving sensitivity. This optical amplification at the front end of an optical receiver will be treated in Chapters 9 and 10 dealing with optical amplification processes.

The structure of the receiver thus consists of four parts: the optical mixing front and front-end section, the electronic pre-amplification, the linear channel of the main amplifier and AGC if necessary, and the data recovery section. The optical mixing front end performs the summation of the optical fields of the local oscillator and that of the optical signals. The polarization orientation between these lightwaves is very critical in order to maximize the beating of the additive field in the photodiode. Depending on the finite or null frequency difference between these fields, the resulting electronic signals are derived from the detector; the electronic signals can be in the baseband or the passband and the detection technique is termed as heterodyne or homodyne, respectively (Figure 8.2).

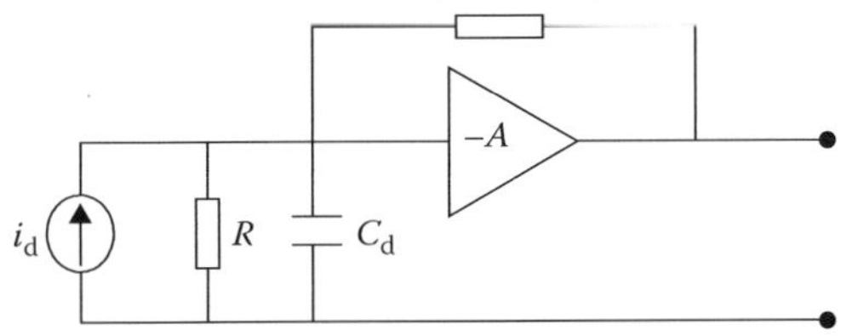

FIGURE 8.2
Schematic diagram of an electronic preamplifier in an optical receiver of transimpedance electronic amplifier at the front end. The current source represents the electronic current generated in the photodetector due to the beating of the local oscillator and the optical signals. C_d is the photodiode capacitance.

8.3 Coherent Detection

Optical coherent detection can be distinguished by the "demodulation" scheme in communications techniques in association with the following definitions: (1) coherent detection is the mixing between two lightwaves or optical carriers, one is information-bearing lightwaves and the other a local oscillator with an average energy much larger than that of the signals and (2) demodulation refers to the recovery of baseband signals from the electrical signals.

A typical schematic diagram of a coherent optical communication employing guided wave medium and its components is shown in Figure 8.3, in which a narrowband laser incorporating an optical isolator cascaded with an external modulator is usually the optical transmitter. Information is fed via a microwave power amplifier to an integrated optic modulator, of the commonly used $LiNbO_3$ or EA type. The coherent detection is the principal feature of coherent optical communications, which can be further distinguished with heterodyne and homodyne techniques depending on whether there is a difference or not between the frequencies of the local oscillator and that of the carrier of the signals. A local oscillator is a laser source whose frequency can be tuned to be approximately equivalent to a monochromatic source; a polarization controller would also be used to match its polarization with that of the information-bearing carrier. The local oscillator and the transmitted signal are mixed via a polarization-maintaining coupler and then detected by coherent optical receiver. Most of the previous coherent detection schemes are implemented in a mixture of photonic domain and electronic/microwave domain.

Coherent optical transmission returns into the focus of research. One significant advantage is the preservation of all the information of the optical field during detection, leading to enhanced possibilities for optical multilevel modulation. This chapter investigates the generation of optical multilevel modulation signals. Several possible structures of optical M-PSK and M-QAM transmitters are shown and theoretically analyzed. Differences in the optical transmitter configuration and the electrical driving lead to different properties of the optical multilevel modulation signals. This is shown by deriving general expressions applicable to every M-ary-PSK and M-ary-QAM modulation format and exemplarily clarified for square-16-QAM modulation.

Coherent receivers are distinguished as synchronous and asynchronous. Synchronous detection requires an OPLL that recovers the phase and frequency of the received signals to lock the local oscillator (LO) to that of the signal to measure the absolute phase and

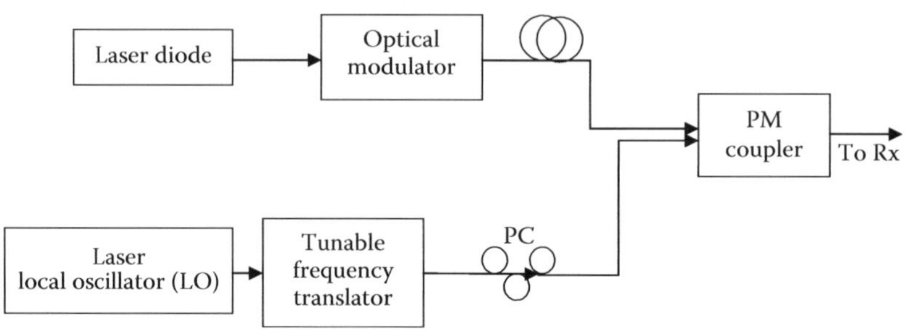

FIGURE 8.3
The typical arrangement of coherent optical communications systems [2,3]. LD/LC is a very narrow linewidth laser diode as a local oscillator without any phase locking to the signal optical carrier. PM coupler is polarization maintaining fiber-coupled device, PC is the polarization controller.

frequency of the signals relative to that of the LO. Thus, synchronous receivers allow direct mixing of the band-pass signals and the baseband; this technique is termed as homodyne reception. For asynchronous receivers, the frequency of the LO is approximately the same as that of the receiving signals and no OPLL is required. In general, the optical signals are first mixed with an intermediate frequency (IF) oscillator that is about two to three times of the 3 dB passband. The electronic signals can then be recovered using electrical PLL at lower carrier frequency in the electrical domain. The mixing of the signals and a local oscillator of an IF frequency is referred to as heterodyne detection.

If no LO is used for demodulating the digital optical signals, then differential or self-homodyne reception may be utilized, or classically termed as autocorrelation reception process or self-heterodyne detection.

Coherent communications have been an important technique in the 1980s and the early 1990s, but then research was interrupted with the advent of optical amplifiers that offered up to 20 dB gain without difficulty in the late 1990s. Nowadays, however, coherent systems emerge again into the focus of interest due to the availability of digital signal processing and low-priced components, the partly relaxed receiver requirements at high data rates, and several advantages that coherent detection provides. The preservation of the temporal phase of the coherent detection enables new methods for adaptive electronic compensation of chromatic dispersion. In WDM systems, coherent receivers offer tunability and allow channel separation via steep electrical filtering. Furthermore, only the use of coherent detection permits convergence to the ultimate limits of spectral efficiency. To reach higher spectral efficiencies, the use of multilevel modulation is required. Considering this, coherent systems are also beneficial, because all the information of the optical field is available in the electrical domain. That way, complex optical demodulation with interferometric detection—which has to be used in direct detection systems—can be avoided, and the complexity is transferred from the optical to the electrical domain. Several different modulation formats based on the modulation of all four quadratures of the optical field were proposed in the early 1990s, describing the possible transmitter and receiver structures and calculating the theoretical BER performance. However, a more detailed and practical investigation of multilevel modulation coherent optical systems for today's networks and data rates is missing so far.

Currently, coherent reception has attracted significant interests due to the following reasons: (1) the received signals of the coherent optical receivers are in electrical domain, which is proportional to that in the optical domain. This, in contrast to the direct detection receivers, allows exact electrical equalization or exact phase estimation of the optical signals and (2) using heterodyne receivers, DWDM channels can be separated in the electrical domain by using electrical filters with sharp roll of the passband to the cutoff band. Presently, the availability of ultrahigh-sampling rate digital signal processors (DSP) allows users to conduct filtering in the DSP in which the filtering can be changed with ease.

However, there are disadvantages that coherent receivers would suffer from: (1) coherent receivers are polarization sensitive that requires polarization tracking at the front end of the receiver; (2) homodyne receivers require OPLL and heterodyne receivers require electrical PLL that would need control and feedback circuitry, optical or electrical, which may be complicated; and (3) for differential detection, the compensation may be complicated due to the differentiation receiving nature.

In Chapter 12, when some advanced modulation formats are presented for optically amplified transmission systems, the use of photonic components are extensively exploited to take advantage of the advanced technology of integrated optics, planar lightwave circuits. Modulation formats of signals depend on whether the amplitude, the phase, or the frequency of the carrier is manipulated, as mentioned in Chapter 2. In this chapter, the detection is coherently converted to the intermediate frequency range in the electrical domain and the signal envelop. The down-converted carrier signals are detected and then recovered. Both binary level and multilevel modulation schemes employing amplitude, phase, and frequency shift keying modulation are described in this chapter.

8.3.1 Optical Heterodyne Detection

The basic configuration of optical heterodyne detection is shown in Figure 8.4. The local oscillator whose frequency can be higher or lower than that of the carrier is mixed with the information-bearing carrier thus allowing down- or up-conversion of the information signals to the intermediate frequency range. The down-converted electrical carrier and signal envelope is received by the photodetector. This combined lightwave is converted by the PD into electronic current signals, which are filtered by an electrical band-pass filter (BPF) and then demodulated by a demodulator. A low-pass filter (LPF) is also used to remove higher-order harmonics of the nonlinear photodetection process, the square-law detection. Under the use of an envelop detector, the process is asynchronous; hence the name asynchronous detection. If the down-converted carrier is recovered and then mixed with IF signals, then this is synchronous detection. It is noted that the detection is conducted at the intermediate frequency range in electrical domain, hence there is a need for controlling the stability of the frequency spacing between the signal carrier and that of the local oscillator. That means the mixing of these carriers would result in an IF carrier in the electrical domain prior to the mixing process or envelop detection to recover the signals.

The coherent detection thus relies on the electric field component of the signal and the local oscillator. The polarization alignment of these fields is critical for optimum detection. The electric field of the optical signals and the local oscillator can be expressed as follows:

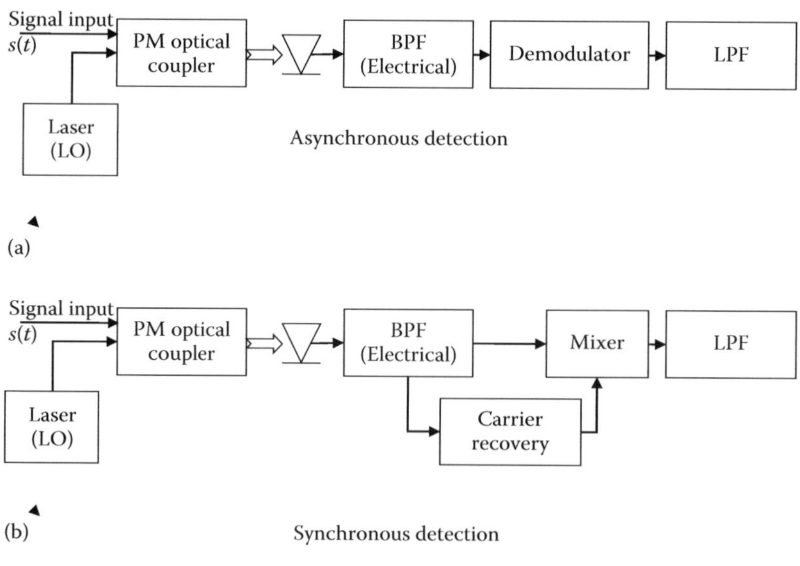

FIGURE 8.4
Schematic diagram of optical heterodyne detection: (a) asynchronous and (b) synchronous, receiver structures. LPF is a low-pass filter. BPF is a band-pass filter. PD is the photodiode.

$$E_s(t) = \sqrt{2P_s(t)}\cos\{\omega_s t + \phi_s + \varphi(t)\} \tag{8.1}$$

$$E_{LO} = \sqrt{2P_{LO}}\cos\{\omega_{LO} t + \phi_{LO}\} \tag{8.2}$$

where

$P_s(t)$ and P_{LO} are the instantaneous signal power and average power of the signals and local oscillator, respectively
$\omega_s(t)$ and ω_{LO} are the signal and local oscillator angular frequencies
ϕ_s and ϕ_{LO} are the phase including any phase noise of the signal and the local oscillator
$\varphi(t)$ is the modulation phase

The modulation can be of amplitude with the switching on and off (amplitude shift keying [ASK]) of the optical power or phase or frequency with the discrete or continuous variation of the time-dependent phase term. For discrete phase, it can be phase shift keying (PSK), differential PSK (DPSK), or differential quadrature PSK (DQPSK), and when the variation of the phase is continuous, we have frequency shift keying if the rate of variation is different for the bit "1" and bit "0," as given in Chapter 4.

Under an ideal alignment of the two fields, the photodetection current can be expressed by

$$i(t) = \frac{\eta q}{h\upsilon}\left[P_s + P_{LO} + 2\sqrt{P_s P_{LO}}\cos\{(\omega_s - \omega_{LO})t + \phi_s - \phi_{LO} + \varphi(t)\}\right] \tag{8.3}$$

where the higher frequency term (the sum) is eliminated by the photodetector frequency response, η is the quantum efficiency, q is the electronic charge, h is Planck's constant, and υ is the optical frequency.

Thus, the power of the local oscillator dominates the shot noise process and at the same time boosts the signal level, hence enhancing the signal-to-noise ratio (SNR). The oscillating term is the beating between the local oscillator and the signal, which is proportional to the amplitude, and is the square root of the product of the power of the local oscillator and the signal.

The electronic signal power S and shot noise N_s can be expressed as follows:

$$\begin{aligned} S &= 2\Re^2 P_s P_{LO} \\ N_s &= 2q\Re(P_s + P_{LO})B \\ \Re &= \frac{\eta q}{h\upsilon} = \text{responsivity} \end{aligned} \tag{8.4}$$

with B is the 3 dB bandwidth of the electronic receiver. Thus, the optical SNR (OSNR) can be written as follows:

$$\text{OSNR} = \frac{2\Re^2 P_s P_{LO}}{2q\Re(P_s + P_{LO})B + N_{eq}} \tag{8.5}$$

where N_{eq} is the total electronic noise equivalent power at the input to the electronic preamplifier of the receiver. From this equation, we can observe that if the power of the local oscillator is significantly increased so that the shot noise dominates over the equivalent noise, at the same time increasing the SNR, the sensitivity of the coherent receiver can only be limited by the quantum noise inherent in the photodetection process. Under this quantum limit, the $\mathrm{OSNR_{QL}}$ is given by

$$\mathrm{OSNR_{QL}} = \frac{\Re P_s}{qB} \tag{8.6}$$

8.3.1.1 ASK Coherent System

Under the ASK modulation scheme, the demodulator of Figure 8.4 is an envelope detector (in lieu of the demodulator) followed by a decision circuitry. That is, the eye diagram is obtained and a sampling instant is established with a clock recovery circuit. The synchronous detection would require that a locking between the carrier frequencies is obtained and then tuned to the local oscillator frequency in order to track the frequency component of the signal. The amplitude demodulated envelope can be expressed as follows:

$$\begin{gathered} r(t) = 2\Re\sqrt{P_s P_{LO}}\cos(\omega_{IF})t + n_x\cos(\omega_{IF})t + n_y\sin(\omega_{IF})t \\ \omega_{IF} = \omega_s - \omega_{LO} \end{gathered} \tag{8.7}$$

The IF frequency ω_{IF} is the difference between those of the LO and the signal carrier, n_x and n_y, are the expected values of the orthogonal noise power components, which are random variables.

$$r(t) = \sqrt{[2\Re P_s P_{LO} + n_x]^2 + n_y^2}\cos(\omega_{IF}t + \Phi)t \tag{8.8}$$

with

$$\Phi = \tan^{-1}\frac{n_y}{2\Re P_s P_{LO} + n_x}$$

8.3.1.1.1 *Envelope Detection*

The noise power terms can be assumed to follow a Gaussian probability distribution and are independent of each other with a zero mean and a variance σ, the probability density function (PDF) can thus be given as follows:

$$p(n_x, n_y) = \frac{1}{2\pi\sigma^2}e^{\frac{-(n_x^2+n_y^2)}{2\sigma^2}} \tag{8.9}$$

With respect to the phase and amplitude this equation can be written as follows [3]:

$$p(\rho, \phi) = \frac{\rho}{2\pi\sigma^2}e^{\frac{-(\rho^2+A^2-2A\rho\cos\phi)}{2\sigma^2}} \tag{8.10}$$

where

$$\rho = \sqrt{[2\Re\sqrt{P_s(t)P_{LO}} + n_s(t)]^2 + n_y^2(t)}$$
$$A = 2\Re\sqrt{P_s(t)P_{LO}} \tag{8.11}$$

The PDF of the amplitude can be obtained by integrating the phase amplitude PDF over the range of 0–2π and is given as follows:

$$p(\rho) = \frac{\rho}{\sigma^2} e^{\frac{-(\rho^2+A^2)}{2\sigma^2}} I_0\left\{\frac{A\rho}{\sigma^2}\right\} \tag{8.12}$$

where I_0 is the modified Bessel's function. If a decision level is set to determine the "1" and "0" level, then the probability of error and the BER can be obtained, assuming an equal probability of error between the "1s" and "0s," as follows:

$$\text{BER} = \frac{1}{2}P_e^1 + \frac{1}{2}P_e^0 = \frac{1}{2}\left[1 - Q\left(\sqrt{2\delta}, d\right) + e^{-\frac{d^2}{2}}\right] \tag{8.13}$$

where Q is the Magnum function (given in Chapter 7) and δ is given by

$$\delta = \frac{A^2}{2\sigma^2} = \frac{2\Re^2 P_s P_{LO}}{2q\Re(P_s + P_{LO})B + i_{Neq}^2} \tag{8.14}$$

When the power of the local oscillator is much larger than that of the signal and the equivalent noise current power, then this SNR becomes

$$\delta = \frac{\Re P_s}{qB} \tag{8.15}$$

The physical representation of the detected current and the noises current due to the quantum shot noise and noise equivalent of the electronic preamplification can be shown in Figure 8.5 in which the signal current can be general and derived from the output of the detection scheme, that from a photodetector or a back-to-back pair of photodetectors of a balanced receiver for detecting the phase difference of DPSK, DQPSK, or CPFSK signals and converting to amplitudes.

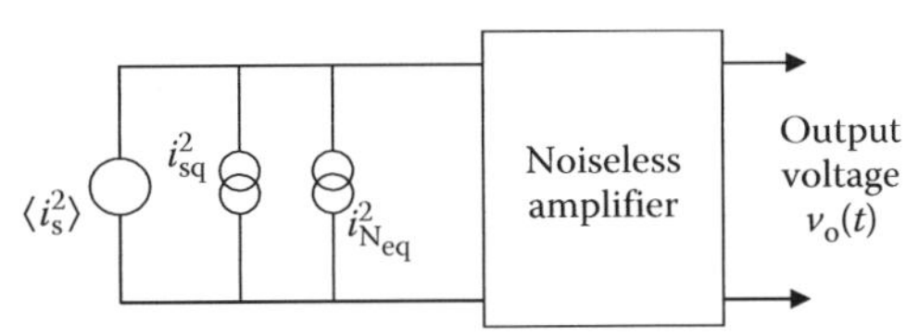

FIGURE 8.5
Equivalent current model at the input of the optical receiver, average signal current, and equivalent noise current of the electronic preamplifier, as seen from its input port.

The BER is optimum when setting its differentiation with respect to the decision level δ; an approximate value of the decision level can be obtained as follows:

$$d_{\text{opt}} \cong \sqrt{2+\frac{\delta}{2}} \Rightarrow \text{BER}_{\text{ASK-e}} \cong \frac{1}{2}e^{-\frac{\delta}{4}} \tag{8.16}$$

8.3.1.1.2 Synchronous Detection

ASK can be detected using synchronous detection [4], and the BER is given by

$$\text{BER}_{\text{ASK-S}} \cong \frac{1}{2}\text{erfc}^{\frac{\sqrt{\delta}}{2}} \tag{8.17}$$

8.3.1.2 PSK Coherent System

Under the phase shift keying modulation format, the detection is similar to that in Figure 8.6 for heterodyne detection, but after the BPF an electrical mixer is used to track the phase of the detected signal. The received signal is given by

$$r(t) = 2\Re\sqrt{P_s P_{\text{LO}}}\cos[(\omega_{\text{IF}})t + \varphi(t)] + n_x \cos(\omega_{\text{IF}})t + n_y \sin(\omega_{\text{IF}})t \tag{8.18}$$

The information is contained in the time-dependent phase term $\varphi(t)$.

When the phase and frequency of the voltage control oscillator (VCO) are matched with those of the signal carrier, then the received electrical signal can be simplified to

$$\begin{aligned} r(t) &= 2\Re\sqrt{P_s P_{\text{LO}}}\, a_n(t) + n_x \\ a_n(t) &= \pm 1 \end{aligned} \tag{8.19}$$

Under Gaussian statistical assumption, the probability of the received signal of a "1" is given by

$$p(r) = \frac{1}{\sqrt{2\pi\sigma^2}} e^{-\frac{(r-u)^2}{2\sigma^2}} \tag{8.20}$$

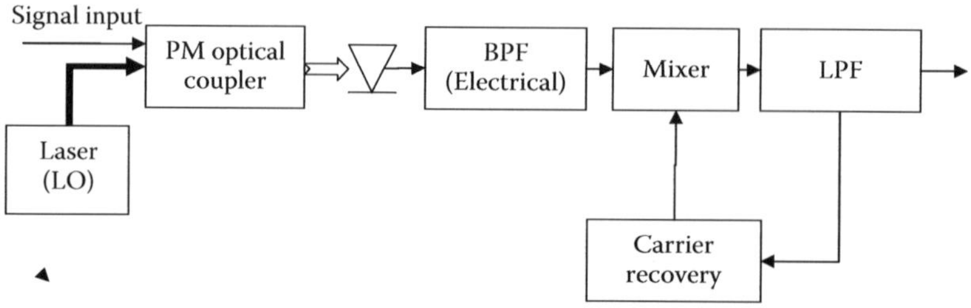

FIGURE 8.6
Schematic diagram of optical heterodyne detection for PSK format.

Furthermore, the probability of the "0" and "1" is assumed to be equal. We can obtain the BER as the total probability of the received "1" and "0" as follows:

$$\mathrm{BER}_{\mathrm{PSK}} = \frac{1}{2}\mathrm{erfc}(\delta) \tag{8.21}$$

As observed in the synchronous detection, there is a need for carrier recovery circuitry, usually implemented using a phase lock loop (PLL) that complicates the overall receiver structure. It is possible to detect the signal by a self-homodyne process by beating the carrier of 1 bit period to that of the next consecutive bit; this is called the differential detection. The detection process can be modified as shown in Figure 8.7 in which the phase of the IF carrier of 1 bit is compared with that of the next bit, and a difference is recovered to represent the bit "1" or "0." This requires a differential coding at the transmitter and an additional phase comparator for the recovery process. In Chapter 12 on differential PSK, the differential decoding is implemented in the photonic domain via a photonic phase comparator in the form of a MZ interferometer incorporating a thermal tuning optical delay line. The BER can be expressed as follows:

$$\mathrm{BER}_{\mathrm{DPSK\text{-}e}} \cong \frac{1}{2}e^{-\delta} \tag{8.22}$$

$$r(t) = 2\Re\sqrt{P_s P_{\mathrm{LO}}}\cos[\pi A_k s(t)] \tag{8.23}$$

where

$s(t)$ is the modulating waveform

A_k represents bit "1" or "0"

This is equivalent to the baseband signal and the ultimate limit is the BER of the baseband signal.

The noise is dominated by the quantum shot noise of the local oscillator with its square noise current given by

$$i^2_{\mathrm{N\text{-}sh}} = 2q\Re(P_s + P_{\mathrm{LO}})\int_0^{\infty}\left|H(j\omega)\right|^2 d\omega \tag{8.24}$$

where $H(j\omega)$ is the transfer function of the receiver system, normally a trans-impedance of the electronic preamp and that of a matched filter. As the power of the local oscillator is

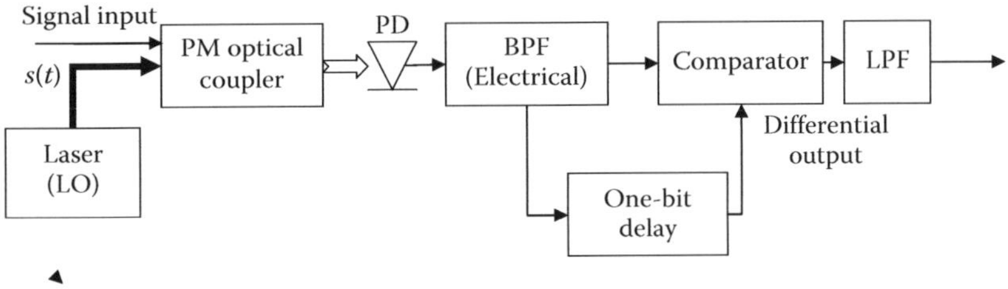

FIGURE 8.7
Schematic diagram of optical heterodyne and differential detection for PSK format.

much larger than the signal and integrating over the dB bandwidth of the transfer function, this current can be approximated by

$$i^2_{\text{N-sh}} \simeq 2q\Re P_{\text{LO}}B \tag{8.25}$$

Hence, the SNR (power) is given by

$$\text{SNR} \equiv \delta \simeq \frac{2\Re P_s}{qB} \tag{8.26}$$

The BER is the same as that of a synchronous detection and is given by

$$\text{BER}_{\text{homodyne}} \cong \frac{1}{2}\text{erfc}\sqrt{\delta} \tag{8.27}$$

The sensitivity of the homodyne process is at least 3 dB better than that of the heterodyne, and the bandwidth of the detection is half of its counterpart due to the double sideband nature of the heterodyne detection.

8.3.1.3 FSK Coherent System

The nature of FSK is based on the two frequency components that determine the bits "1" and "0." There are a number of formats related to FSK depending on whether the change of the frequencies representing the bits is continuous or noncontinuous, the FSK or CPFSK modulation formats. For noncontinuous FSK, the detection is usually performed by a structure of dual frequency discrimination, as shown in Figure 8.8, in which two narrowband filters are used to extract the signals. For CPFSK, both the frequency discriminator and the balanced receiver for PSK detection can be used. The frequency discrimination is indeed preferred as compared to the balanced receiving structures because it would eliminate the phase contribution by the local oscillator or optical amplifiers, which may be used as an optical preamp. We highlight the performance of these detection schemes for modulation format using noncoherent techniques in the following chapters.

8.3.2 Optical Homodyne Detection

Optical homodyne detection matches the transmitted signal phases to that of the local oscillator phase signal. A schematic of the optical receiver is shown in Figure 8.9. The field

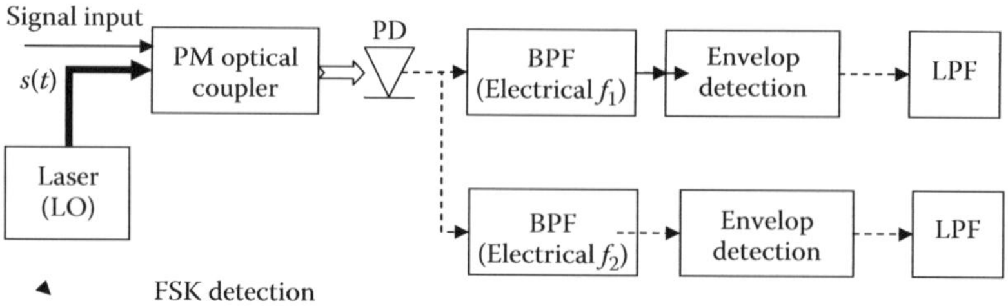

FIGURE 8.8
Schematic diagram of optical homodyne detection of FSK format.

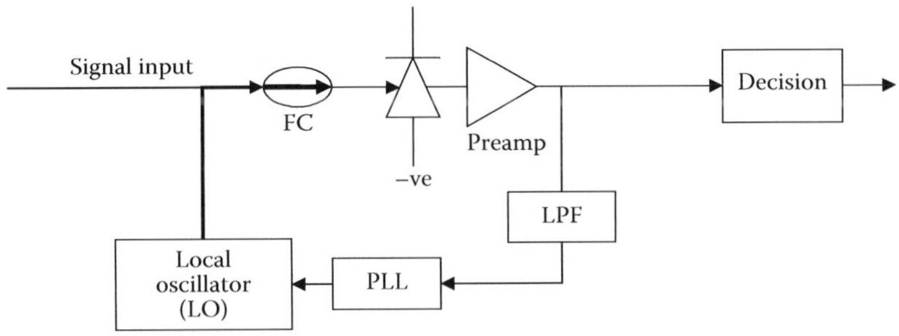

FIGURE 8.9
General structure of an optical homodyne detection system. FC is the fiber coupler, LPF, the low-pass filter, and PLL, the phase lock loop.

of the incoming optical signals is mixed with the local oscillator whose frequency and phase are locked with that of the signal carrier waves via a phase lock loop. The resultant electrical signal is then filtered and thence a decision circuitry.

8.3.2.1 Detection and Optical PLL

Optical homodyne detection requires the phase matching of the frequency of the signal carrier and that of the local oscillator. This type of detection would give a very sensitive result, in principle, of 9 photons/bit. Implementation of such system would normally require an OPLL whose structure of a recent development [5] is shown in Figure 8.10. The local oscillator frequency is locked into the carrier frequency of the signals by shifting to the modulated sideband component via the use of the optical modulator. A single sideband optical modulator is preferred. However, a double sideband may also be used. This modulator is excited by the output signal of a voltage-controlled oscillator whose frequency is determined by the voltage level of the output and electronic BPF condition to meet the required voltage level for driving the electrode of the modulator. The frequency of the local oscillator is normally tuned to the region such that the frequency difference

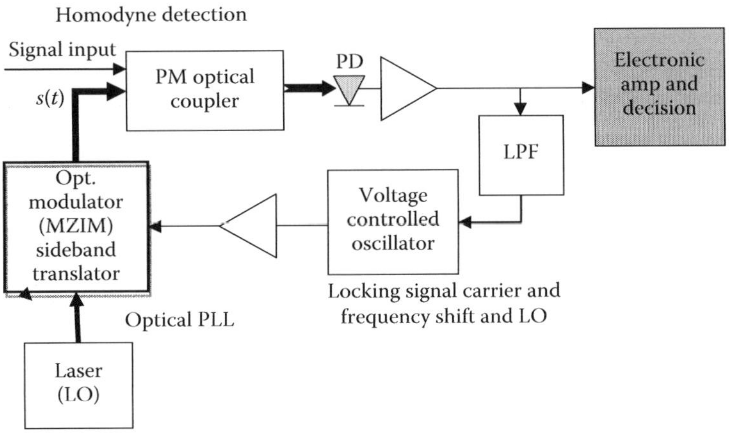

FIGURE 8.10
Schematic diagram of optical homodyne detection: electrical line (dashed), optical line (continuous and solid) using an optical phase lock loop.

with respect to the signal carrier falls within the passband of the electronic filter. When the frequency difference is zero, then there is no voltage level at the output of the filter, and thus the OPLL has reached the final stage of locking. The bandwidth of the optical modulator is important so that it can extend the locking range between the two optical carriers.

Any difference in frequency between the LO and the carrier is detected and noise filtered by the LPF. This voltage level is then fed to a voltage control oscillator to generate a sinusoidal wave that is then used to modulate an intensity modulator modulating the lightwaves of the LO. The output spectrum of the modulator would exhibit two sidebands and the LO lightwave. One of these components would then be locked to the carrier. A close loop would ensure a stable locking. If the intensity modulator is biased at the minimum transmission point and the voltage level at the output of the VCO is adjusted to 2 Vπ with driven signals of π/2 phase shift with each other, then we would have carrier suppression and sideband suppression. This eases the confusion of the close loop locking.

Under a perfect phase matching, the received signal is given by

$$i_s(t) = 2\Re\sqrt{p_s P_{LO}}\cos\left\{\pi\, a_k s(t)\right\} \tag{8.28}$$

where a_k takes the value ±1 and $s(t)$ is the modulating waveform. This is a baseband signal; thus the error rate is the same as that of the baseband system.

The shot noise power induced by the local oscillator and the signal power can be expressed as follows:

$$i_{NS}^2 = 2q\Re(p_s + P_{LO})\int_0^{\infty}\left|H(j\omega)\right|d\omega \tag{8.29}$$

where $|H(j\omega)|$ is the transfer function of the receiver whose expression, if under a matched filtering, can be

$$\left|H(j\omega)\right|^2 = \left[\frac{\sin(\omega T/2)}{\omega T/2}\right]^2 \tag{8.30}$$

where T is the bit period. Then the noise power becomes

$$i_{NS}^2 = q\Re(p_s + P_{LO})\frac{1}{T} \simeq \frac{q\Re P_{LO}}{T}, \quad \text{when } p_s \ll P_{LO} \tag{8.31}$$

Thus the SNR is

$$\text{SNR} = \frac{2\Re p_s P_{LO}}{q\Re P_{LO}/T} = \frac{2p_s T}{q} \tag{8.32}$$

thence the BER is

$$P_E = \frac{1}{2}\operatorname{erfc}(\sqrt{SNR}) \rightarrow BER = \operatorname{erfc}(\sqrt{SNR}) \quad (8.33)$$

8.3.2.2 Quantum Limit Detection

For homodyne detection, a super quantum limit can be achieved. In this case, the local oscillator is used in a very special way that matches the incoming signal field in polarization, amplitude, and frequency and is assumed to be phase locked to the signal. Assuming that the phase signal is perfectly modulated such that it acts in-phase or counter-phase with the local oscillator, the homodyne detection would give a normalized signal current of

$$i_{sC} = \frac{1}{2T}\left[\mp\sqrt{2n_p} + \sqrt{2n_{LO}}\right]^2 \quad \text{for } 0 \le t \le T \quad (8.34)$$

Assuming further that $n_p = n_{LO}$ the number of photons for the LO for the generation of detected signals, then the current can be replaced with 4 n_p for the detection of a "1" and nothing for a "0" symbol.

8.3.2.3 Linewidth Influences

8.3.2.3.1 Heterodyne Phase Detection

When the linewidth of the light sources is significant, the intermediate frequency deviates due to a phase fluctuation and the probability density function is related to this linewidth conditioned on the deviation δω of the intermediate frequency. For a signal power of p_s, the total probability of error is given as

$$P_E = \int_{-\infty}^{\infty} P_C(p_s, \partial\omega) p_{IF}(\partial\omega)\partial\omega \quad (8.35)$$

The probability density function of the intermediate frequency under a frequency deviation can be written as [6] follows:

$$p_{IF}(\partial\omega) = \frac{1}{\sqrt{\Delta\upsilon BT}} e^{-\frac{\partial\omega^2}{4\pi\Delta\upsilon B}} \quad (8.36)$$

where
Δυ is the full IF linewidth at FWHM of the power spectral density
T is the bit period

8.3.2.3.2 Differential Phase Detection with Local Oscillator

The DPSK detection requires a Mach Zehnder delay interferometer (MZDI) and a balanced receiver either in the optical domain or in the electrical domain. If the detection is in the electrical domain then the beating signals in the PD between the incoming signals and the LO would give the beating electronic current, which is then split and one branch is delayed by one bit period and then summed up. The heterodyne signal current can be expressed as [7]

$$i_s(t) = 2\Re\sqrt{P_{LO}p_s}\cos(\omega_{IF}t + \phi_s(t)) + n_x(t)\cos\omega_{IF}t - n_y(t)\sin\omega_{IF}t \tag{8.37}$$

The phase $\phi_s(t)$ is expressed by

$$\phi_s(t) = \varphi_s(t) + \{\varphi_N(t) - \varphi_N(t+T)\} - \{\varphi_{pS}(t) - \varphi_{pS}(t+T)\} - \{\varphi_{pL}(t) - \varphi_{pL}(t+T)\} \tag{8.38}$$

The first term is the phase of the data and takes the value 0 or π. The second term represents the phase noise due to shot noise of the generated current, and the third and fourth terms are the quantum shot noise due to the local oscillator and the signals. The probability of error is given by

$$P_E = \int_{-\pi/2}^{\pi/2}\int_{-\infty}^{\infty} p_n(\phi_1 - \phi_2)p_q(\phi_1)\partial\phi_1\partial\phi_1 \tag{8.39}$$

where

$p_n(.)$ is the probability density function of the phase noise due to the shot noise

$p_q(.)$ is for the quantum phase noise generated from the transmitter and the local oscillator [8]

The probability or error can be written as follows:

$$p_N(\phi_1 - \phi_2) = \frac{1}{2\pi} + \frac{\rho e^{-\rho}}{\pi}\sum_{m=1}^{\infty} a_m \cos\left(m(\phi_1 - \phi_2)\right)$$

$$a_m \sim \left\{\frac{2^{m-1}\Gamma\left[\frac{m+1}{2}\right]\Gamma\left[\frac{m}{2}+1\right]}{\Gamma[m+1]}\left[I_{m-1/2}\frac{\rho}{2} + I_{(m+1)/2}\frac{\rho}{2}\right]\right\}^2 \tag{8.40}$$

where

$\Gamma(.)$ is the gamma function

It is the modified Bessel function of the first kind

The PDF of the quantum phase noise can be given as [9] follows:

$$p_q(\phi_1) = \frac{1}{\sqrt{2\pi D\tau}}e^{\frac{\phi_1^2}{2D\tau}} \tag{8.41}$$

where D is the phase diffusion constant and the standard deviation from the central frequency is given as follows:

$$\Delta\upsilon = \Delta\upsilon_R + \Delta\upsilon_L = \frac{D}{2\pi} \tag{8.42}$$

which is the sum of the transmitter and the local oscillator FWHM linewidth. Substituting (8.41) and (8.40) into (8.39), we obtain

$$P_E = \frac{1}{2} + \frac{\rho e^{-\rho}}{2} \sum_{n=0}^{\infty} \frac{(-1)^n}{2n+1} e^{-(2n+1)^2 \pi \Delta \upsilon T} \left\{ I_{n-1/2} \frac{\rho}{2} + I_{(n+1)/2} \frac{\rho}{2} \right\}^2 \tag{8.43}$$

This equation gives the probability of error as a function of the received power. The probability of error is plotted against the receiver sensitivity and the product of the linewidth with the bit rate (or the relative bandwidth of the laser linewidth and the bit rate) are shown in Figure 8.11 for DPSK modulation format at 140 Mb/s bit rate and the variation of the laser linewidth from 0 to 2 MHz.

8.3.2.3.3 *Differential Phase Detection under Self-Coherent*

Recently, the laser linewidth requirement for DQPSK modulation and differential detection for DQPSK has also been studied; no local oscillator is used for self-coherent detection. It has been shown that for the linewidth of up to 3 MHz, the transmitter laser would not significantly influence the probability of error, as shown in Figure 8.13 [9].

8.3.2.3.4 *Differential Phase Coherent Detection of Continuous Phase FSK Modulation Format*

The probability of error of CPFSK can be derived by taking into consideration the delay line of the differential detector, the frequency deviation, and phase noise [11]. Similar to Figure 8.8, the differential detector configuration is shown in Figure 8.14a, and the conversion of frequency-to-voltage relationship is shown in Figure 8.14b. If heterodyne detection is employed, then a BPF is used to bring the signals back to the electrical domain.

The detected signal phase at the shot noise limit at the output of the LPF can be expressed as follows:

$$\phi(t) = \omega_c t + a_n \frac{\Delta\omega}{2} \tau + \varphi(t) + \varphi_n(t) \tag{8.44}$$

with $\omega_c = 2\pi f_c = (2n+1)(\pi/2\tau)$

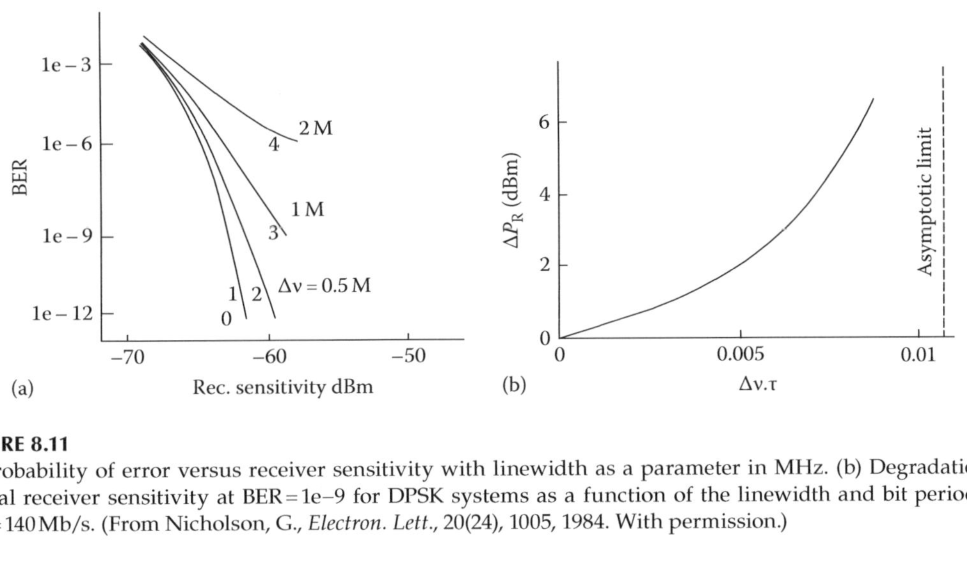

FIGURE 8.11
(a) Probability of error versus receiver sensitivity with linewidth as a parameter in MHz. (b) Degradation of optical receiver sensitivity at BER = 1e–9 for DPSK systems as a function of the linewidth and bit period; bit rate = 140 Mb/s. (From Nicholson, G., *Electron. Lett.*, 20(24), 1005, 1984. With permission.)

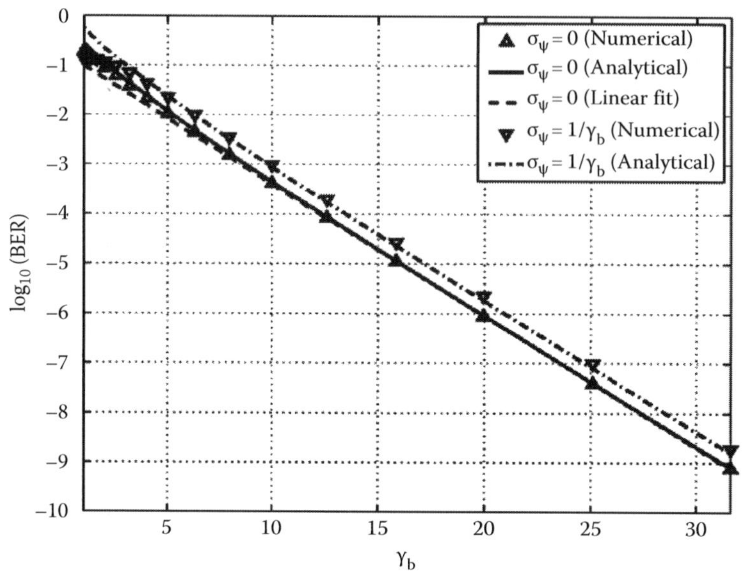

FIGURE 8.12
Analytical approximation (solid line) and numerical evaluation (triangles) of BER for cases of zero linewidth and that required to double the BER. The dashed line is the linear fit for zero linewidth. Bit rate 10Gb/s per channel. (Extracted from Savory, S. and Hadjifotiou, A., *IEEE Photonics Technol. Lett.*, 16(3), 930, March 2004.)

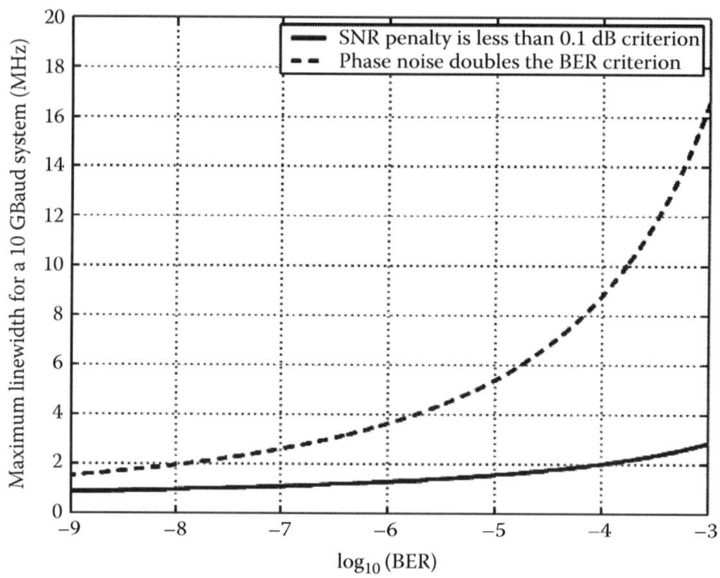

FIGURE 8.13
Criteria for neglecting linewidth in a 10-GSymbols/s system. The loose bound is to neglect linewidth if the impact is to double the BER with the tighter bound being to neglect linewidth if the impact is a 0.1 dB SNR penalty. Bit rate 10 GSymbols/s. (Extracted from Yamamoto, Y. and Kimura, T., *IEEE J. Quantum Electron.*, QE-17, 919, 1981.)

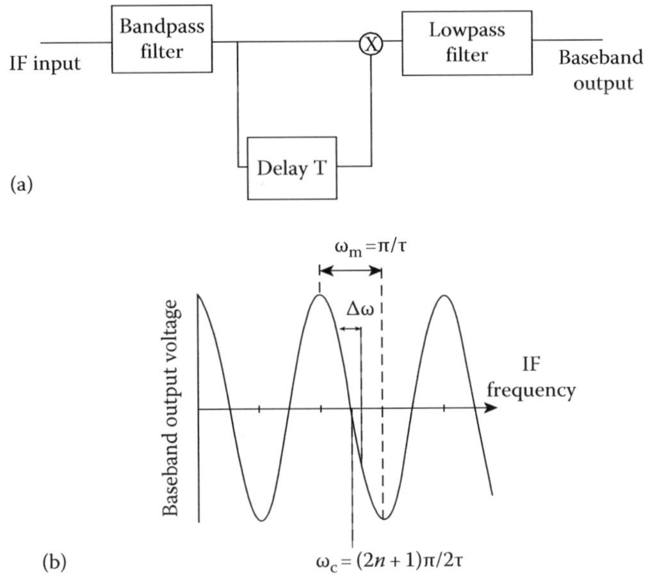

FIGURE 8.14
(a) Configuration of a CPFSK differential detection and (b) frequency-to-voltage conversion relationship of FSK differential detection. (Extracted from Iwashita, K. and Masumoto, T., *IEEE J. Lightwave Technol.*, LT-5/4, 452, 1987, Fig. 1.)

where
- τ is the differential detection delay time
- $\Delta\omega$ is the deviation of the angular frequency of the carrier for the "1" or "0" symbol
- $\varphi(t)$ is the phase noise due to the shot noise
- $n(t)$ is the phase noise due to the transmitter and the local oscillator quantum shot noise
- a_n takes the values of ±1, the binary data symbol

Thus by integrating the detected phase from $-(\Delta\omega/2)\tau \to \pi - (\Delta\omega/2)\tau$ we obtain the probability of error as follows:

$$P_E = \int_{-\frac{\Delta\omega}{2}\tau}^{\pi-\frac{\Delta\omega}{2}\tau/2} \int_{-\infty}^{\infty} p_n(\phi_1 - \phi_2) p_q(\phi_1) \partial\phi_1 \partial\phi_1 \tag{8.45}$$

Similar to the case of DPSK system, substituting (8.40) and (8.41) into (8.45) we obtain

$$P_E = \frac{1}{2}\frac{\rho e^{-\rho}}{2} \sum_{n=0}^{\infty} \frac{(-1)^n}{2n=1} e^{-(2n+1)^2\pi\Delta\upsilon\tau} \left\{ I_{n-1/2}\frac{\rho}{2} + I_{(n+1)/2}\frac{\rho}{2} \right\}^2 e^{-(2n+1)^2\pi\Delta\upsilon\tau} \cos\{(2n+1)\alpha\}$$

$$\alpha = \frac{\pi(1-\beta)}{2} \quad \text{and} \quad \beta = \frac{\Delta\omega}{\omega_m} = 2\pi\tau/T_0 \tag{8.46}$$

ω_m is the deviation of the angular frequency with "m" as the modulation index
T_0 is the pulse period or bit period

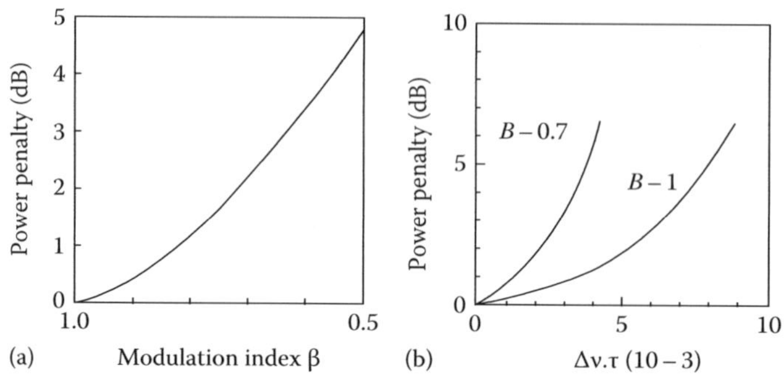

FIGURE 8.15
(a) Dependence of receiver power penalty at BER of 1e-9 on modulation index β (ratio between frequency deviation and maximum frequency spacing between f_1 and f_2) (b) Receiver power penalty at BER 1e-9 as a function of product of beat bandwidth and bit delay time—effects excluding LD phase noise (Extracted from Yamamoto, Y. and Kimura, T., *IEEE J. Quantum Electron.*, QE-17, 919, 1981; Iwashita, K. and Masumoto, T., *IEEE J. Lightwave Technol.*, LT-5/4, 452, 1987, Figs. 2 and 3.)

The modulation index parameter *b* is defined as the ration of the actual frequency deviation to the maximum frequency deviation (Figure 8.15).

8.4 Self-Coherent Detection and Electronic DSP

Coherent techniques described earlier would offer significant improvement but face a setback due to the availability of stable local oscillator and an OPLL for locking the frequency of the local oscillator and that of the signal carrier.

Digital signal processing (DSP), which has been widely used in wireless communications, has a key role to play in the implementation of coherent optical communication systems. Applying DSP techniques to coherent optical communication systems overcomes the difficulties of OPLL, and also improves the performance of the transmission systems in the presence of fiber-degrading effects including chromatic dispersion, PMD, and fiber nonlinearities.

Coherent optical receivers have the following advantages: (1) the shot-noise limited receiver sensitivity can be achieved with a sufficient local oscillator (LO) power; (2) closely spaced WDM channels can be separated with electrical filters having sharp roll-off characteristics; and (3) the ability of phase detection can improve the receiver sensitivity compared with the IMDD system [12]. In addition, any kind of multilevel phase modulation formats can be introduced by using the coherent receiver. While the spectral efficiency of binary modulation formats is limited to 1 bit/s/Hz/polarization (which is called the Nyquist limit), multilevel modulation formats with N bits of information per symbol can achieve the spectral efficiency of *N* bit/s/Hz/polarization. Recent research has focused on M-ary phase shift keying (M-ary PSK) and even quadrature amplitude modulation (QAM) with coherent detection, which can increase the spectral efficiency by a factor of $\log_2 M$ [13–15]. Moreover, for the same bit rate, since the symbol rate is reduced, the system can have higher tolerance to chromatic dispersion and PMD.

However, one of the major challenges in coherent detection is to overcome the carrier phase noise when using a local oscillator (LO) to beat with the received signals to retrieve the modulated phase information. Phase noise can result from lasers, which will cause a power penalty to the receiver sensitivity. A self-coherent multi-symbol detection of optical differential M-ary PSK is introduced to improve the system performance; however, higher analog-to-digital conversion resolution and more digital signal-processing power are required compared to a digital coherent receiver [16]. Further, differential encoding is also necessary in this scheme. As for the coherent receiver, initially, an OPLL is an option to track the carrier phase with respect to the LO carrier in homodyne detection. However, an OPLL operating at optical wavelengths in combination with distributed feedback (DFB) lasers may be quite difficult to be implemented because the product of laser linewidth and loop delay is too large [17]. Another option is to use electrical PLL to track the carrier phase after down-converting the optical signal to an intermediate frequency (IF) electrical signal in a heterodyne detection receiver, as mentioned above. Compared to heterodyne detection, homodyne detection offers better sensitivity and requires a smaller receiver bandwidth [18]. On the other hand, coherent receivers employing high-speed analog-to-digital converters (ADCs) and high-speed baseband digital signal processing (DSP) units are becoming increasingly attractive rather than using an OPLL for demodulation. Conventional block *M*th power phase estimation (PE) scheme is proposed in [12,17] to raise the received M-ary PSK signals to the *M*th power to estimate the phase reference in conjunction with a coherent optical receiver. However, this scheme requires nonlinear operations, such as taking the *M*th power and the $\tan^{-1}(\cdot)$, and resolving the $\pm 2\pi/M$ phase ambiguity, which incurs a large latency to the system. Such nonlinear operations would limit further potential for real-time processing of the scheme. In addition, nonlinear phase noises always exist in long-haul systems due to the Gordon–Mollenauer effect [19], which severely affect the performance of a phase-modulated optical system [20]. The results in [21] show that such *M*th power PE techniques may not effectively deal with nonlinear phase noise.

The maximum likelihood (ML) carrier phase estimator derived in Ref. [22] can be used to approximate the ideal synchronous coherent detection in optical PSK systems. The ML phase estimator requires only linear computations, thus it is more feasible for online processing for real systems. Intuitively, one can show that the ML estimation receiver outperforms the *M*th power block phase estimator and conventional differential detection, especially when the nonlinear phase noise is dominant, thus significantly improving the receiver sensitivity and tolerance to the nonlinear phase noise. The algorithm of ML phase estimator is expected to improve the performance of coherent optical communication systems using different M-ary PSK and quadrature amplitude modulation (QAM) formats. The improvement by DSP at the receiver end can be significant for the transmission systems in the presence of fiber-degrading effects, including chromatic dispersion, PMD, and nonlinearities for both single-channel and DWDM systems.

8.5 Digital Signal Processing Coherent Optical Receiver

8.5.1 Theory of DSP-Assisted Coherent Detection

The schematic of synchronous coherent receiver using maximum likelihood (ML) phase estimation for coherent optical communications is shown in Figure 8.16 [24]. The optical differential QPSK (DQPSK) system can be used as an example. Phase-diversity receivers are usually employed to retrieve in-phase (I) and quadrature-phase (Q) signals in coherent lightwave systems with phase modulation. Such a receiver composes of a 90° optical

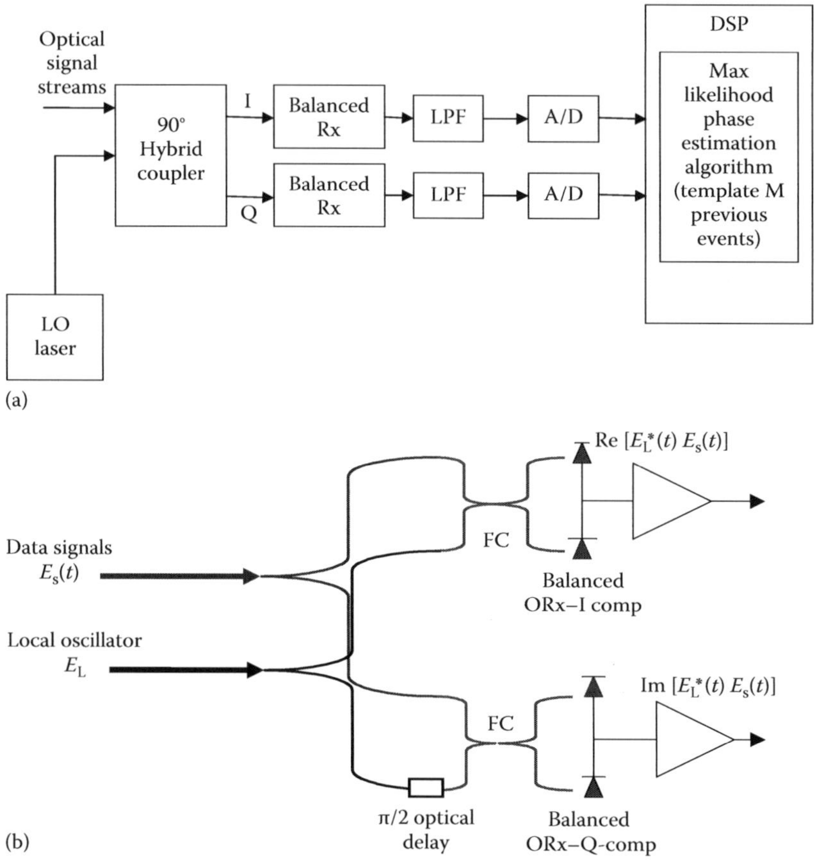

FIGURE 8.16
Scheme of a synchronous coherent receiver using digital signal processing for phase estimation for coherent optical communications. (Adapted from Zhang, S. et al., A comparison of phase estimation in coherent optical PSK system, Photonics Global'08, Singapore, December 2008, Paper C3-4A-03; Zhang, S. et al., Adaptive decision-aided maximum likelihood phase estimation in coherent optical DQPSK system, *OptoElectronics and Communications Conference (OECC)'08*, Sydney, Australia, July 2008, pp. 1–2, Paper TuA-4.) (a) Generic scheme and (b) detailed optical receiver.

hybrid coupler to mix the incoming signal with the four quadruple states associated with the LO inputs in the complex field space. A p/2 phase shifter is used to extract the quadrature component of the optical signal. The optical hybrid then provides four lightwave signals to two pairs of balanced photodetectors to reconstruct I and Q information of the transmitted signal; this is equivalent to self-homodyne detection, as in the balanced receiver there would be a delay interferometer to compare the phases of two consecutive bits. We can assume a complete matching of the polarization between the signal and the local oscillator fields so that only the impact of phase noise should be considered. Indeed, the maximum likelihood algorithm is the modern maximum likelihood sequence estimation (MLSE) technique [25].

The output signal reconstructed from the photocurrents can be represented by

$$r(k) = E_0 \exp[j(\theta_s(k) + \theta_n(k))] + +\tilde{n}(k) \tag{8.47}$$

where k denotes kth sample over the time interval $[kT, (k+1)T]$ (T is the symbol duration), $E_0 = \Re\sqrt{P_{LO}p_s}$, $\Re$ is the photodiode responsivity, P_{LO} and p_s are the powers of LO and received signal, respectively; $\theta_s(k) \in \{0, \pi/2, \pi, -\pi/2\}$ is the modulated phase, the phase difference carrying the data information; $\theta_n(k)$ is the phase noise during the transmission. $\{\tilde{n}(k)\}$ is a complex white Gaussian random variable with $E[\tilde{n}(k)] = 0$ and $E[|\tilde{n}(k)|^2] = N_0$.

In order to retrieve information from the phase modulation $\theta_s(k)$ at time $t = kT$, the carrier phase noise $\theta_n(k)$ is estimated based on the received signal over the immediate past L symbol intervals, that is, $\{r(l), k-L \le l \le k-1\}$. In the decision-feedback strategy, a complex phase reference $v(k)$ is computed by the correlation of the L received signal samples $r(l)$ and the L symbol decisions $\{\hat{m}(l), k-L \le l \le k-1\}$, where $\hat{m}(l)$ is the receiver's decision on $\exp(j\theta_s(l))$:

$$v(k) = \sum_{l=k-L}^{k-1} r(l)\hat{m}^*(l) \tag{8.48}$$

Here, * denotes the complex conjugation. An initial L-symbol long-known data sequence is sent to initiate the processor/receiver. Alternatively, the maximum likelihood decision processor can be trained prior to the transmission of data sequence. It is important that not only the $\pi/4$-radian phase ambiguity can be resolved, but also the decision-aided method is now totally linear and efficient to implement based on (8.48). To some extent, the reference $v(k)$ assists the receiver to acquire the channel characteristics. With the assumption that $\theta_n(k)$ varies slowly compared to the symbol rate, the computed complex reference $v(k)$ from the past L symbols forms a good approximation to the phase noise phasor $\exp(j\theta_n(k))$ at time $t = kT$.

Using the phase reference $v(k)$ from (8.48), the decision statistic of the ML receiving processor is given as follows:

$$q_i(k) = r(k)C_i^* \cdot v(k), \quad i = 0,1,2,3 \tag{8.49}$$

where

$C_i \in \{\pm 1, \pm j\}$ is the DQPSK signal constellation

$(\cdot)$ denotes the inner product of two vectors

The detector declares the decision $\hat{m}(k) = C_k$ if $q_k = \max q_i$. This receiver/processor has been shown to achieve coherent detection performance if the carrier phase is a constant and the reference length L is sufficiently long.

The performance of ML, the receiving processor, can be evaluated by using Monte Carlo simulations in two cases: a linear phase noise system and a nonlinear phase noise system. For comparison, the Mth power phase estimator and differential demodulation can also be also employed.

Nonlinearity in an optical fiber can be ignored when the launch optical power is below the threshold, so the fiber optic can be modeled as a linear channel. The phase noise difference in two adjacent symbol intervals, that is, $\theta_n(k) - \theta_n(k-1)$, obeys a Gaussian distribution with the variance σ^2 determined by the linewidth of the transmitter laser and the LO, given by the Lorentzian linewidth formula $\sigma^2 = 2\pi\Delta v T$, where Δv is the total linewidth of transmitter laser and the LO [7]. In simulations, σ can be set to 0.05, corresponding to the overall linewidth $\Delta v = 8$ MHz when the bit rate is 40 Gsymbol/s.

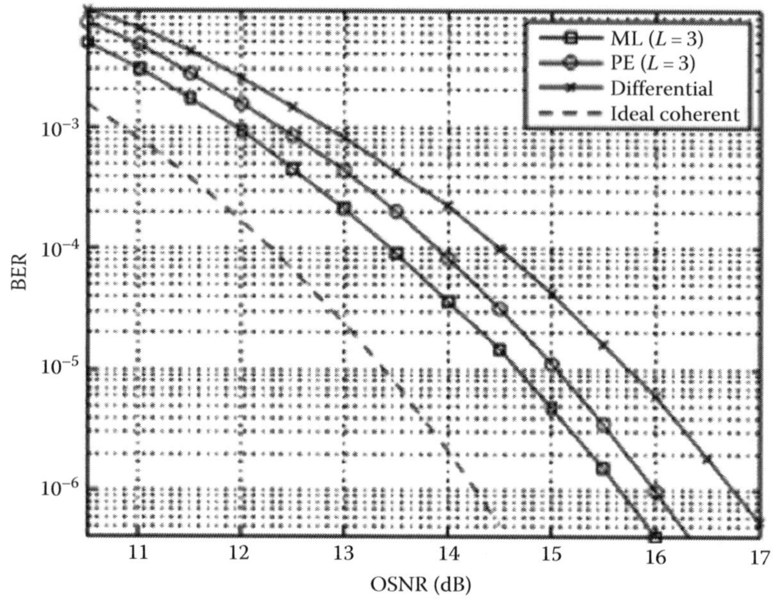

FIGURE 8.17
Simulated BER performances of a 10-Gsymbols/s DQPSK signal in linear optical channel with laser linewidth 2 MHz ($\sigma = 0.05$) and decision-aided length $L = 3$. (Extracted from Zhang, S. et al., A comparison of phase estimation in coherent optical PSK system, Photonics Global'08, Singapore, December 2008, Paper C3-4A-03.)

As observed in Figure 8.17, the ML processor outperforms the phase estimator and differential demodulation by about 0.25 and 1 dB, respectively. Although the performance gap between ML processor and Mth power phase estimation is not very large in the case of the MPSK system, the Mth power phase estimator requires more nonlinear computations, such as an arctan(·), which incurs a large latency in the system and leads to phase ambiguity in estimating the block phase noise. On the contrary, ML processor is a linear and efficient algorithm, and there is no need to deal with the $\pi/4$ radian phase ambiguity; thus it is more feasible for online processing for the real systems. We also extend our ML phase estimation technique to QAM system. The conventional Mth-power phase estimation scheme suffers from the performance degradation in QAM systems since only a subgroup of symbols with phase modulation $\pi/4 + n \cdot \pi/2$ ($n = 0, 1, 2, 3$) can be used to estimate the phase reference. The maximum tolerance of linewidth per laser in Square 16-QAM can be improved 10 times compared to the Mth-power phase estimator scheme.

Let us now consider the case with nonlinear phase noise. In a multispan optical communication system with erbium-doped fiber amplifiers (EDFAs), the performance of optical DQPSK is severely limited by the nonlinear phase noise converted by amplified ASE noise through the fiber nonlinearity [19]. The laser linewidth is excluded to only consider the impact of the nonlinear phase noise on the receiver. The transmission system can comprise a single-channel DQPSK-modulated optical signal transmitted at 10–40 Gsymbol/s over N 100 km equally spaced amplified spans that are fully compensated for the chromatic dispersion and equalized gain of optical power along the fiber, as shown in Figure 8.18. We set the following parameters: $N = 20$, fiber nonlinear coefficient $\gamma = 2\,W^{-1}\,km^{-1}$, fiber loss $\alpha = 0.2$ dB/km, the amplifier gain is $G = 20$ dB, the NF figure is 6 dB, the optical wavelength $\lambda = 1553$ nm, and the bandwidth of the optical filter and the electrical filter are set to $\Delta\lambda = 0.1$ nm and 7 GHz, respectively. The OSNR is defined as the ratio between the signal

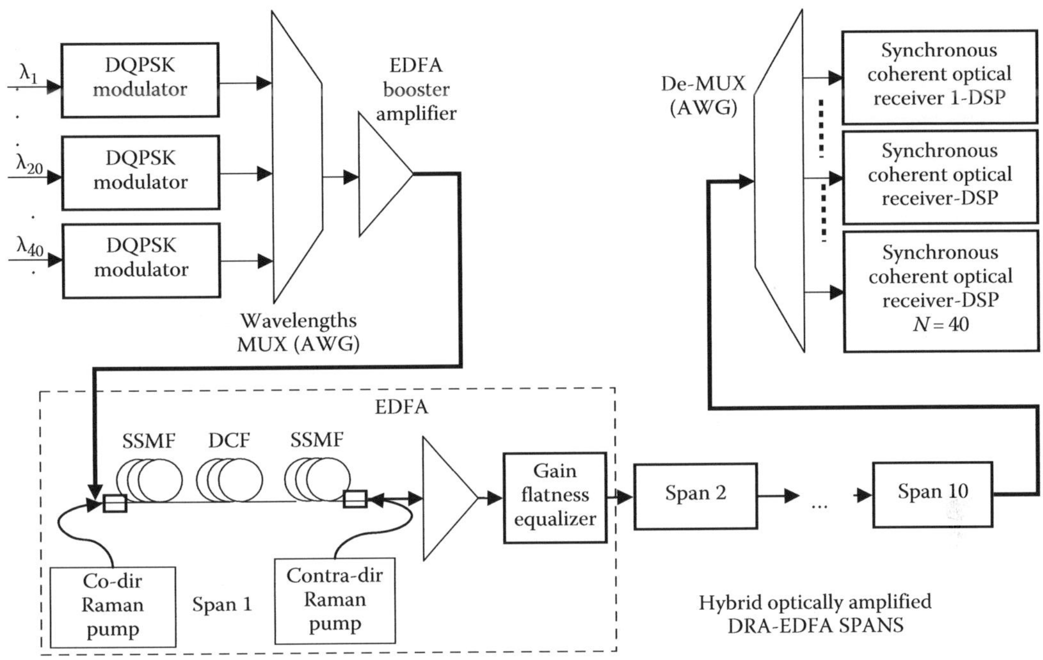

FIGURE 8.18
DWDM optical DQPSK system with N-span optic fibers using synchronous coherent receiver to compensate nonlinear phase noise under both Raman amplification and EDF amplification.

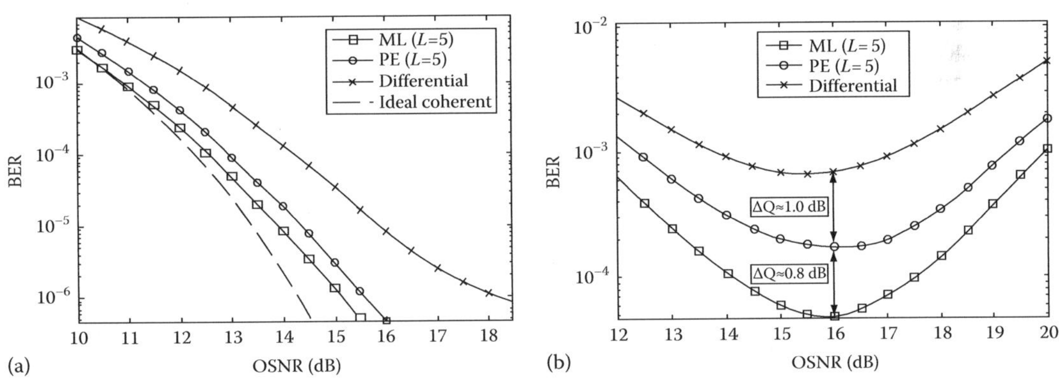

FIGURE 8.19
Monte Carlo-simulated BER of a 10-Gsymbol/s DQPSK signal in an N-span optical fibers system with template length $L = 5$: (a) $N = 20$; (b) $N = 30$ under different detection schemes of ML estimator, phase estimator, differential demodulation, and ideal coherent detection. ML is the maximum likelihood, PE is the phase estimation. (Extracted from Zhang, S. et al., A comparison of phase estimation in coherent optical PSK system, Photonics Global'08, Singapore, December 2008, Paper C3-4A-03.)

power and the noise power in two polarization states within a 0.1 nm bandwidth. As shown in Figure 8.19a, the receiver sensitivity of the ML processor has improved by about 0.5 and 3 dB over phase estimation and differential demodulation, respectively, at the BER level of 10^{-4}. Besides, it is noteworthy that differential demodulation has exhibited an error floor due to severe nonlinear phase noise with high-launch optical power at the transmitter.

To obtain the impact of the nonlinear phase on the demodulation methods and the noise loading effects, the number of amplifiers can be increased to $N = 30$. The Q-factor is used to show a numerical improvement of ML processor receiver. As shown in Figure 8.19b, with the increase of launch power, the ML receiver/processor approximates the optimum performance because of the reducing variance of the phase noise. At the optimum point, the ML receiver/processor outperforms the Mth power phase estimator by about 0.8 dB while having about 1.8 dB receiver sensitivity improvement compared to differential detection. It is also found that the optimum performance occurs at an OSNR of 16 dB when the nonlinear phase shift is almost 1 rad. With the launch power exceeding the optimum level, nonlinear phase noise becomes severe again. Only the ML phase estimator can offer BER beyond 10^{-4}, while both Mth power phase estimator and differential detection exhibit error floor before BER reaches 10^{-4}. From the numerical results, the ML processor carrier phase estimation shows a significant better receiver sensitivity improvement than Mth power phase estimator and conventional differential demodulation for optical DQPSK signal in nonlinear phase noise dominant systems. Again, we want to emphasize that the ML estimator is a linear and efficient algorithm, and there is no need to deal with the $\pi/4$-radian phase ambiguity, thus, it is feasible for online processing for the real systems.

In summary, the performance of synchronous coherent detection with ML processor carrier phase estimation may offer a linear phase noise system and a nonlinear optical noise system separately. The ML DSP phase estimation can replace the OPLL, and the receiver sensitivity is improved compared to conventional differential detection and Mth power phase estimator. The receiver sensitivity is improved by approximately 1 and 3 dB in these two channels (20 spans for the nonlinear channel) respectively, compared to differential detection. For the 30-span nonlinear channel, only the ML processor can offer a BER beyond 10^{-4} (using Monte Carlo simulation) while both Mth power phase estimator and differential detection exhibit error floor before BER reaching 10^{-4}. At the optimum point of the power, the Q-factor of ML receiver/processor outperforms the Mth power phase estimation by about 0.8 dB. In addition, an important feature is the linear and efficient computation of the ML phase estimation algorithm, which enables the possibility real-time online DSP processing.

8.5.2 DSP-Based Phase Estimation and Correction of Phase Noise and Nonlinear Effects

Kikuchi's group [12] at the University of Tokyo uses electronic signal processing based on Mth power phase estimation to estimate the carrier phase. However, DSP circuits for Mth power phase estimator need nonlinear computations, thus impeding the potential possibility of real-time processing in the future. Furthermore, the Mth power phase estimation method requires dealing with the $\pi/4$ radian phase ambiguity when estimating the phase noise in adjacent symbol blocks. The electronic DSP is based on the ML processing for carrier phase estimation to approximate the ideal synchronous coherent detection in optical phase modulation system, which requires only linear computations, thus eliminating the OPLL and is more feasible for online processing for the real systems. Our initial simulation results show that the ML receiver/processor outperforms the Mth power phase estimator especially when the nonlinear phase noise is dominant, thus significantly improving the receiver sensitivity and tolerance to the nonlinear phase noise.

Liu's group [26,27] at Alcatel-Lucent Bell Labs uses optical delay differential detection with DSP to detect differential BPSK (DPSK) and DQPSK signals. Since direct detection only detects the intensity of the light, the improvement of DSP is limited after the direct detection. After the synchronous coherent detection of our technique, DSPs such as electronic equalization of chromatic dispersion and PMD offer better performance since the phase information is retrieved. And as the level of phase modulation increases, it becomes more and more difficult to apply the optical delay differential detection because of the rather complex implementation of the receiver and the degraded SNR of the demodulated signal, while the ML estimator/processor can still demodulate other advanced modulation formats such as 8-PSK, 16-PSK, and 16-QAM.

8.5.3 DSP-Based Forward Phase Estimation Optical Coherent Receivers of QPSK Modulation Format

Recent progresses in digital signal processing (DSP) [17,28] with the availability of ultrahigh sampling rate allow the possibility of DSP-based phase estimation and polarization management techniques that make coherent detection robust and practical. This section is dedicated to the new emerging technology, which will significantly influence the optical transmission and detection of optical signals at ultrahigh speed.

Recalling the schematic of a coherent receiver of Figure 8.16, Figure 8.20 shows a coherent detection scheme for QPSK optical signals in which they are mixed with the local oscillator field. In the case of modulation format QPSK or DQPSK, a $\pi/2$ optical phase shift is needed to extract the quadrature component, thus the real and imaginary parts of a phase shift keyed signals can be deduced at the output of a balanced receiver from a pair of identical photodiodes connected back to back. Note that for a balanced receiver the quantum shot noise contributed by the photodetectors is double, as noise is always represented by the noise power and therefore no current must be applied (Figure 8.21).

When the received signal is raised to the quadratic power, the phase of the signal disappears because $e^{j4\theta_s(t)} = 1$ for DQPSK whose states are 0, $\pi/2$, π, and 2π. The data phase is excluded and the carrier phase can be recovered with its fluctuation with time. This estimation is computed using the DSP technique. The estimated carrier phase is then subtracted with the detected phase of the signal and the carrier to give only the phase states of the signals, as indicated in the diagram. This is a feedforward phase estimation and suitable for DSP implementation.

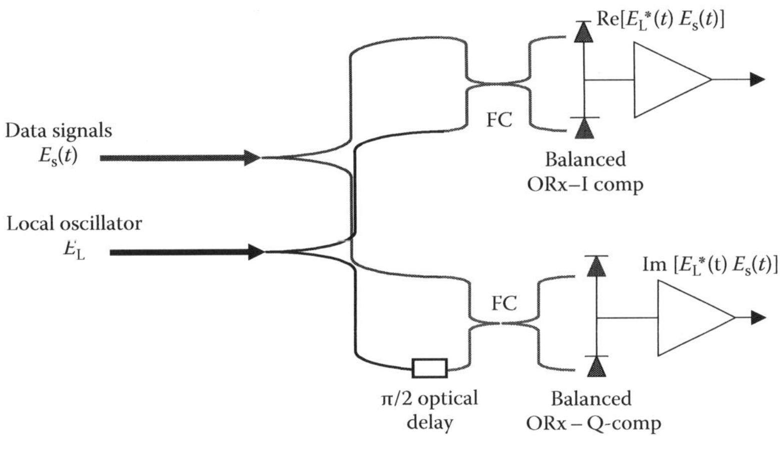

FIGURE 8.20
Schematic of a coherent receiver using balanced detection techniques for I–Q component phase and amplitude recovery.

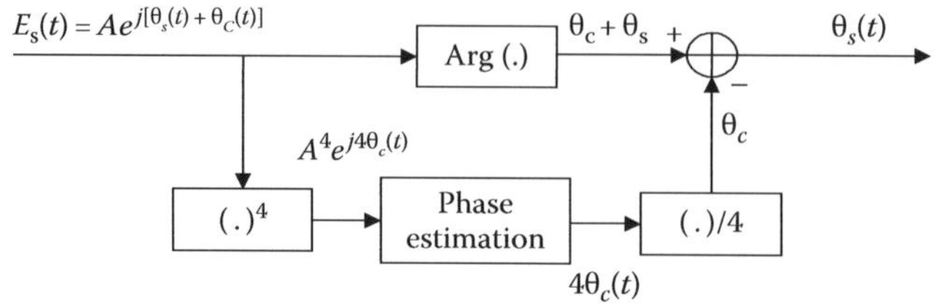

FIGURE 8.21
Schematic of a QPSK coherent receiver.

8.6 Coherent Receiver Analysis

At ultrahigh bit rate the laser must be externally modulated; thus the phase of the lightwave is conserved along the fiber transmission line. The detection can be direct detection, self-coherent, or homodyne and heterodyne. The sensitivity of the coherent receiver is also important for the transmission system, especially the PSK scheme under both homodyne and heterodyne transmission techniques. This section gives the analysis of receiver for synchronous coherent optical fiber transmission systems. Consider that the optical fields of the signals and local oscillator are coupled via a fiber coupler with two output ports 1 and 2. The output fields are then launched into two photodetectors connected back to back and then the electronic current is amplified using a transimpedance type and further equalized to extend the bandwidth of the receiver. Our objective is to obtain the receiver penalty and its degradation due to imperfect polarization mixing and unbalancing effects in the balanced receiver.

The following parameters are used in the analysis:

E_s	Amplitude of signal optical field at the receiver
E_L	Amplitude of local oscillator optical field
P_s, P_L	Optical power of signal and local oscillator at the input of the photodetector
$s(t)$	The modulated pulse
$\langle i_{NS}^2(t)\rangle$	Mean square noise current (power) produced by the total optical intensity on the photodetector
$\langle i_s^2(t)\rangle$	Mean square current produced by the photodetector by $s(t)$
$S_{NS}(t)$	Shot noise spectral density of $\langle i_s^2(t)\rangle$ and local oscillator power
$i_{N_{eq}}^2(t)$	Equivalent noise current of the electronic preamplifier at its input
$Z_T(\omega)$	Transfer impedance of the electronic preamplifier
$H_E(\omega)$	Voltage transfer characteristic of the electronic equalizer followed the electronic preamplifier (Figure 8.22)

The combined field of the signal and local oscillator via a directional coupler can be written with their separate polarized field components as follows:

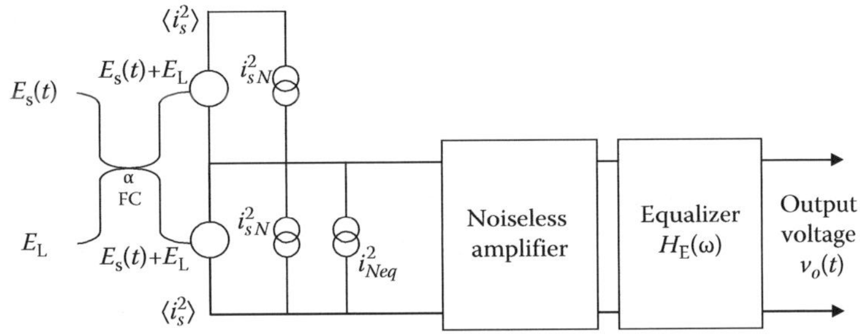

FIGURE 8.22
Equivalent current model at the input of the optical balanced receiver under coherent detection, average signal current, and equivalent noise current of the electronic preamplifier, as seen from its input port and equalizer. FC is the fiber coupler.

$$\begin{aligned}
E_{sX} &= \sqrt{K_{sX}}E_s\cos(\omega_s t-\phi_{m(t)})\\
E_{sY} &= \sqrt{K_{sY}}E_s\cos(\omega_s t-\phi_{m(t)}+\delta_s)\\
E_{LX} &= \sqrt{K_{LX}}E_L\cos(\omega_L t)\\
E_{LY} &= \sqrt{K_{LY}}E_L\cos(\omega_L t+\delta_L)\\
\phi_{m(t)} &= \frac{\pi}{2}K_m s(t)
\end{aligned} \tag{8.50}$$

where
- $\phi_{m(t)}$ represents the phase modulation
- K_m is the modulation depth
- $K_{sX}\ K_{sY}\ K_{LX}\ K_{LY}$ are the intensity fraction coefficients in the X and Y directions of the signal and local oscillator fields, respectively

Thus, the output fields at ports 1 and 2 of the coupler in the X-plane can be obtained using the matrix transfer as follows:

$$\begin{bmatrix} E_{R1X}\\ E_{R2X}\end{bmatrix}=\begin{bmatrix}\sqrt{K_{sX}(1-\alpha)}\cos(\omega_s t-\phi_{m(t)}) & \sqrt{K_{LX}\alpha}\sin(\omega_L t)\\ \sqrt{K_{sX}\alpha}\sin(\omega_s t-\phi_{m(t)}) & \sqrt{K_{LX}(1-\alpha)}\cos(\omega_L t)\end{bmatrix}\begin{bmatrix}E_s\\ E_L\end{bmatrix} \tag{8.51}$$

$$\begin{bmatrix} E_{R1Y}\\ E_{R2Y}\end{bmatrix}=\begin{bmatrix}\sqrt{K_{sY}(1-\alpha)}\cos(\omega_s t-\phi_{m(t)}) & \sqrt{K_{LY}\alpha}\sin(\omega_L t+\delta_L)\\ \sqrt{K_{sY}\alpha}\sin(\omega_s t-\phi_{m(t)}) & \sqrt{K_{LY}(1-\alpha)}\cos(\omega_L t)+\delta_L\end{bmatrix}\begin{bmatrix}E_s\\ E_L\end{bmatrix} \tag{8.52}$$

where α is the intensity coupling ratio of the coupler. Thus the field components at ports 1 and 2 can be derived by combining the X and Y components from (8.51) and (8.52), thence the total power at ports 1 and 2 are given as follows:

$$
\begin{aligned}
P_{R1} &= P_s(1-\alpha)+P_L\alpha)+2\sqrt{P_sP_L\alpha(1-\alpha)K_p}\,\sin(\omega_{IF}t+\phi_{m(t)}+\phi_p-\phi_c)\\
P_{R2} &= P_s\alpha+P_L(1-\alpha)+2\sqrt{P_sP_L\alpha(1-\alpha)K_p}\,\sin(\omega_{IF}t+\phi_{m(t)}+\phi_p-\phi_e+\pi)\\
\text{with}\quad K_p &= K_{sX}K_{LX}+K_{sY}K_{LY}+2\sqrt{K_{sX}K_{LX}K_{sY}K_{LY}}\cos(\delta_L-\delta_s)\\
\phi_p &= \tan^{-1}\left[\frac{\sqrt{K_{sX}K_{LY}}\sin(\delta_L-\delta_s)}{\sqrt{K_{sX}K_{LX}}+\sqrt{K_{sY}K_{LY}}\cos(\delta_L-\delta_s)}\right]
\end{aligned}
\tag{8.53}
$$

where ω_{IF} is the intermediate angular frequency and is equal to the difference between the frequencies of the local oscillator and the carrier of the signals. ϕ_e is the phase offset, and $\phi_p-\phi_e$ is the demodulation reference phase error. In (8.53), the total field of the signal and the local oscillator are added; the product of the field vector and its conjugate gives the power. Only the term with frequency falling within the range in which the photodetector is sensitive would produce the electronic current. Thus, the term with the sum of the frequency of the wavelength of the signal and the local oscillator would not be detected and only the product of the two terms would be detected as given.

Now assuming a binary PSK (BPSK) scheme, the pulse has a square shape with amplitude +1 or −1, the PD is a p-i-n type, and the PD bandwidth is larger than the 3 dB signal bandwidth, an equalized electronic preamplifier. The signal at the output of the electronic equalizer or the input of the decision circuit is given by

$$
\hat{v}_D(t)=2K_HK_p\sqrt{P_sP_L\alpha(1-\alpha)K_p}\int_{-\infty}^{\infty}H_E(f)df\int_{-\infty}^{\infty}(t)dt\,\sin\left(\frac{\pi}{2}K_m\right)\cos(\phi_p-\phi_e)\xrightarrow{\text{substituting the integral}}
$$

$$
\hat{v}_D(t)=2K_HK_p\sqrt{P_sP_L\alpha(1-\alpha)K_p}\,\sin\left(\frac{\pi}{2}K_m\right)\cos(\phi_p-\phi_e) \tag{8.54}
$$

$K_H=1$ for homodyne; $K_H=1/\sqrt{2}$ for heterodyne

For a completely balanced receiver, $K_B=2$ and $\alpha=0.5$ otherwise $K_B=1$. The integrals of the first line in (8.54) are given by

$$
\begin{aligned}
&\int_{-\infty}^{\infty}H_E(f)df=\frac{1}{T_B}\qquad \text{since}\,H_E(f)=\sin c(\pi T_B f)\\
&\int_{-\infty}^{\infty}s(t)dt=2T_B
\end{aligned}
\tag{8.55}
$$

$H_E(f)$ is the transfer function of the matched filter for equalization with T_B is the bit period.

The total noise voltage as a sum of the quantum shot noise generated by the signal and the local oscillator and the total equivalent noise of the electronic preamplifier at the input of the preamplifier at the output of the equalizer is given by

$$\langle v_N^2(t)\rangle = \frac{[K_B\alpha S'_{IS} + (2-K_B)S_{IX} + S_{IE}]\int_{-\infty}^{\infty}|H_4(f)|^2\,df}{K_{IS}^2}$$

or

$$\langle v_N^2(t)\rangle = \frac{[K_B\alpha S'_{IS} + (2-K_B)S_{IX} + S_{IE}]}{K_{IS}^2 T_B} \tag{8.56}$$

For homodyne and heterodyne detections, we have

$$\langle v_N^2(t)\rangle = \Re q\frac{P_L}{\lambda T_B}[K_B\alpha S'_{IS} + (2-K_B)S_{IX} + S_{IE}] \tag{8.57}$$

where the spectral densities S'_{IX}, S'_{IE} are given by

$$S'_{IX} = \frac{S_{IX}}{S'_{IS}}$$
$$S'_{IE} = \frac{S_{IE}}{S'_{IS}} \tag{8.58}$$

Thus, the receiver sensitivity for binary PSK and equiprobable detection and Gaussian density distribution is as follows:

$$P_e = \frac{1}{2}\operatorname{erfc}\left(\frac{\delta}{\sqrt{2}}\right) \tag{8.59}$$

where δ is given by

$$P_e = \frac{1}{2}\operatorname{erfc}\left(\frac{\delta}{\sqrt{2}}\right)$$

with

$$\delta = \frac{\hat{v}_D}{2\sqrt{\langle v_N^2\rangle}} \tag{8.60}$$

Thus, the receiver sensitivity is given as follows:

$$P_s = \langle P_s(t)\rangle = \frac{\Re q\delta^2}{4\lambda T_B K_H^2}\,\frac{[K_B\alpha S'_{IS} + (2-K_B)S_{Ix} + S_{IE}]}{\eta K_p(1-\alpha)\alpha K_B^2\sin^2\left(\frac{\pi}{2}K_m\right)\cos^2(\phi_p-\phi_e)} \tag{8.61}$$

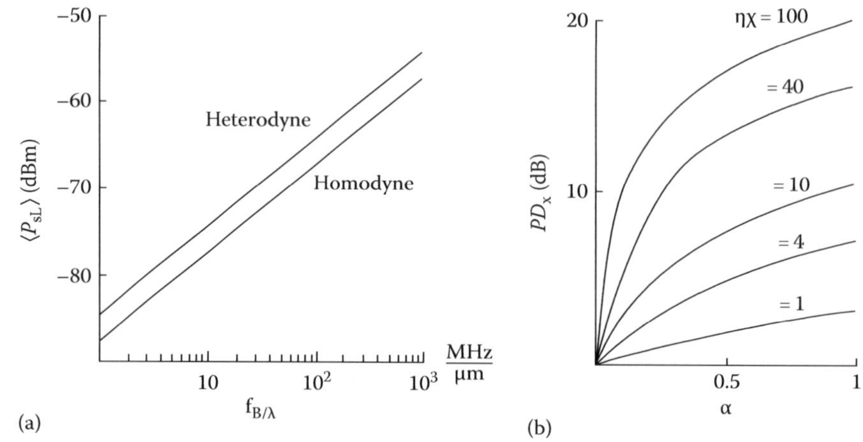

FIGURE 8.23
(a) Receiver sensitivity of coherent homodyne and heterodyne detection, signal power versus bandwidth over the wavelength. (b) Power penalty of the receiver sensitivity from the shot noise limited level as a function of the excess noise of the local oscillator. (Extracted from Hodgkinson, I., *IEEE J. Lightwave Technol.*, 5(4), 573, 1987, Figs. 1 and 2).

8.6.1 Shot Noise–Limited Receiver Sensitivity

In the case that the power of the local oscillator dominates the noise of the electronic preamplifier and the equalizer then the receiver sensitivity is given as follows:

$$P_s = \langle P_{sL} \rangle = \frac{\Re q \delta^2}{4\lambda T_B K_H^2} \tag{8.62}$$

This shot noise–limited receiver sensitivity can be plotted as shown in Figure 8.23.

8.6.2 Receiver Sensitivity under Nonideal Conditions

Under nonideal conditions, the receiver sensitivity departs from the shot noise–limited sensitivity and characterized by the receiver sensitivity penalty PD_T as follows:

$$PD_T = 10\,\mathrm{Log}\frac{\langle P_s \rangle}{\langle P_{sL} \rangle} \text{ in dB} \tag{8.63}$$

$$\begin{aligned} PD_T = {} & 10\,\mathrm{Log}_{10}\left[\frac{K_B \alpha S'_{IS} + (2-K_B)S_{Ix} + S_{IE}}{K_B \alpha}\right] - 10\,\mathrm{Log}_{10}\left[K_B(1-\alpha)\right] \\ & -10\,\mathrm{Log}_{10}\left([\eta][K_p]\sin^2\left(\frac{\pi}{2}K_m\right)\cos^2(\phi_p - \phi_e)\right) \end{aligned} \tag{8.64}$$

with η is the local oscillator excess noise factor.

The receiver sensitivity is plotted against the ratio f_B/λ for the case of homodyne and heterodyne detections in Figure 8.23a, and the power penalty of the receiver sensitivity is plotted against the excess noise factor of the local oscillator in Figure 8.23b. Receiver power penalty can be deduced as a function of the total electronic equivalent noise spectral density, and as a function of the rotation of the polarization of the local oscillator; it can be found in Ref. [29]. Furthermore in Ref. [28] the receiver power penalty and the normalized heterodyne center frequency can vary as a function of the modulation parameter and as a function of the optical power ratio at the same polarization angle can also be found.

8.7 MATLAB® and Simulink® Models

8.7.1 Phase Detection

One possibility to recover the phase modulation is by using the MZDI and binary or M-ary PSK demodulator. In this method, the differential decoding of the received signal is done at MZDI before sampling, and the received phase information is demodulated by a M-ary PSK demodulator.

Figure 8.24 shows the MATLAB® and Simulink® structures of the phase modulation receiver, *MZDI_pi/4_Rx*. In this block, the input signal is divided into two by using splitter. The first branch of the signal goes to the MZDI1, while the second branch goes to MZDI2. Additionally, the output signal from MZDI is fed to *Elec_Rx* block. The role of this block is explained as amplitude detection. Then the two signals from the MZDI are compared to recover the in-phase and quadrature part of the signal. Afterward, the signal is sampled by using two unit delay blocks. The sampled phase signal is collected at phase_est block above. This is required for calculating the phase BER of the Star 16-QAM in this model.

The phase estimation can be performed after sampling. This is very probable in the current progress of electronic processing system. The phase signal at the output of phase estimator is collected. The optical demodulation process is executed by an M-ary PSK demodulator baseband block. After the demodulation, the demodulated phase signals are combined with the amplitude signals.

8.7.2 Mach Zehnder Delay Interferometer

The main purpose of MZDI is to perform differential decoding of the phase modulation. The structure of the MZDI is modified for this system.

Figure 8.25 above shows the structure of the MZDI1 block. The differential decoding in MZDI is performed by delaying the input signal by 1 bit, then comparing the phase difference between input signal and delayed input signal. Additionally, the delayed signal is phase shifted by some appropriate value, so that the in-phase (real) and quadrature (imaginary) part of the signal can be recovered back. In this model, the first MZDI carries out phase shifting of $\pi/4$, while the second MZDI does phase shifting of $-\pi/4$. By comparing (adding and subtracting) those signals after the MZDI, the in-phase and quadrature part of the signal can be recovered.

After the phase shifting, the signal is then detected by photodiode. The constructive and destructive port of the photodiode acts as the balanced receiver. Figure 8.26 shows the structure of the constructive and destructive photodiode.

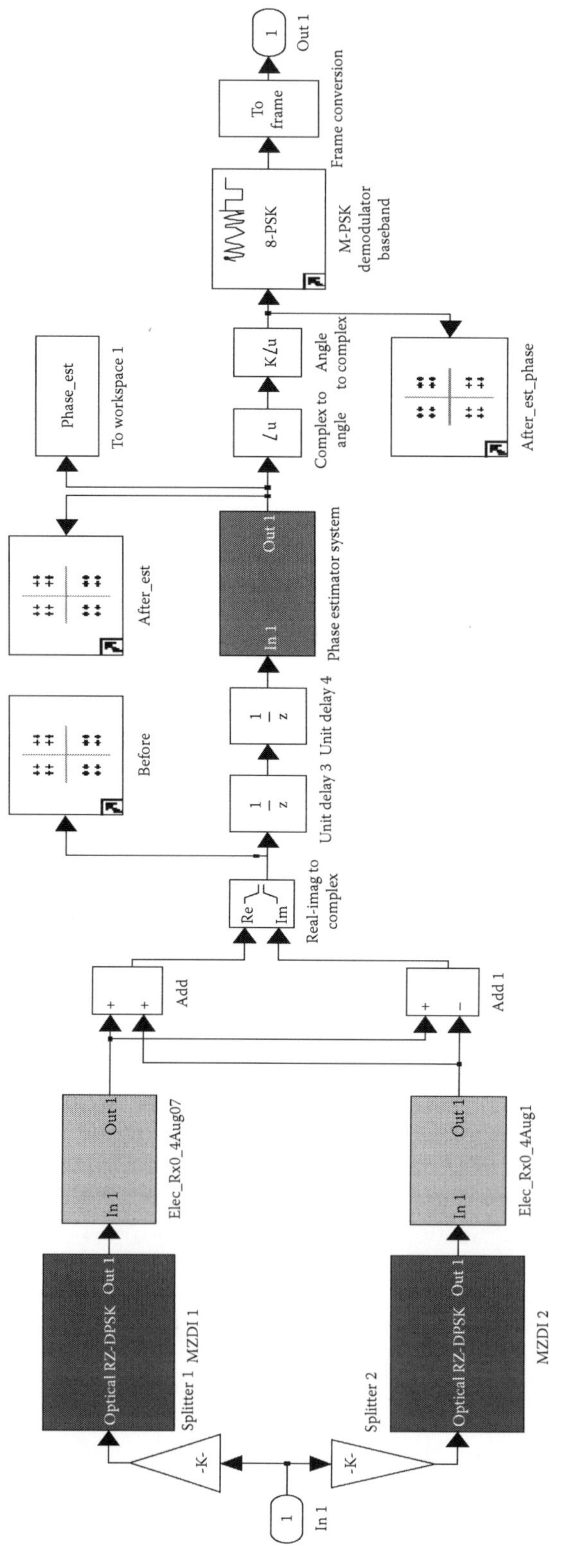

FIGURE 8.24
Structure of the MZDI_pi/4_Rx block.

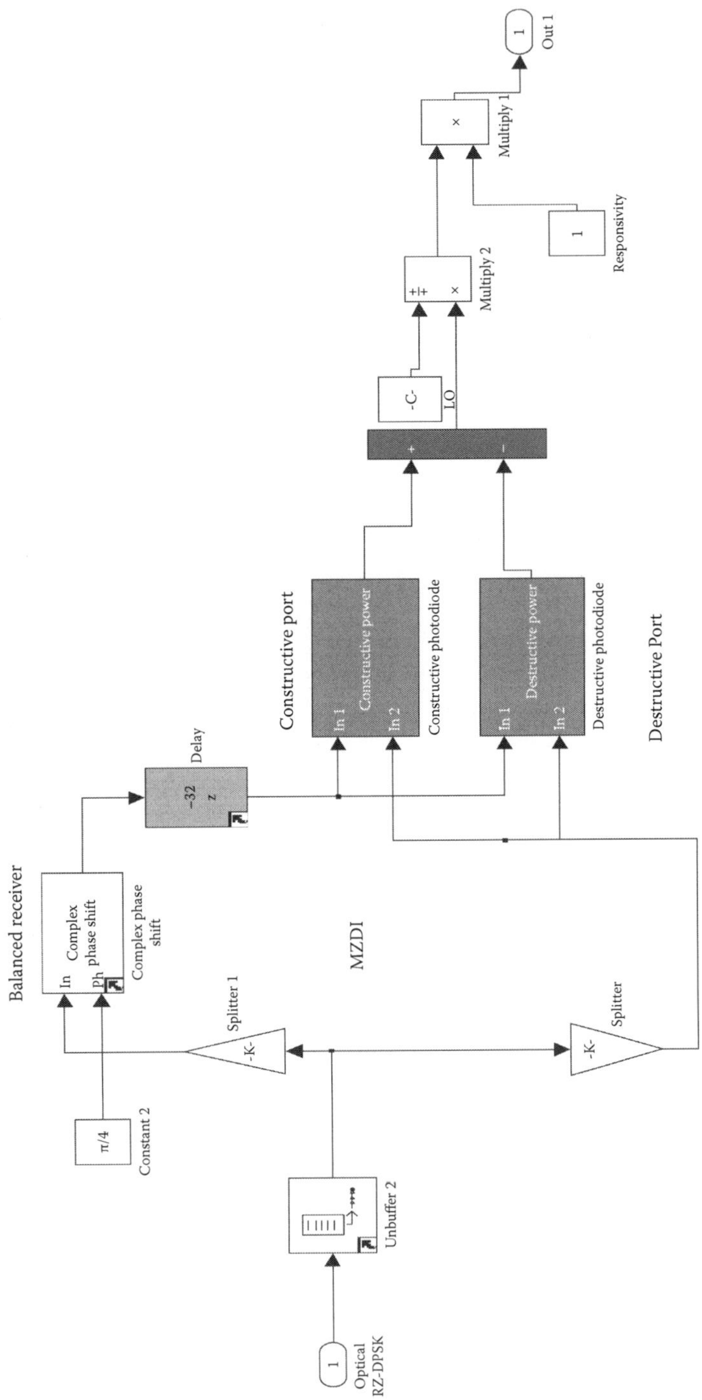

FIGURE 8.25
Structure of the MZDI1 block.

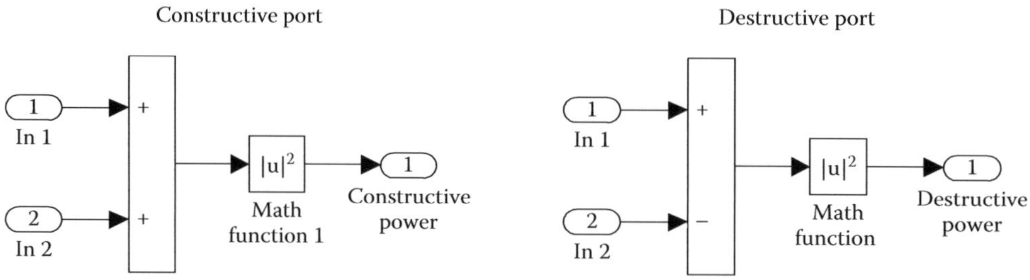

FIGURE 8.26
Structure of constructive port(left); structure of destructive port (right).

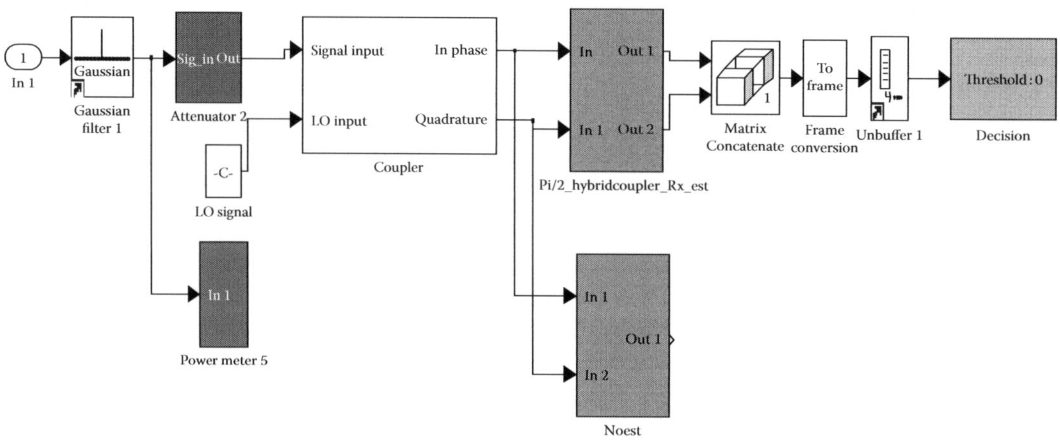

FIGURE 8.27
Structure of the receiver with π/2 hybrid coupler.

8.7.3 Receiver with Coupler

Another possibility to recover the phase information is by applying a $2 \times 4\pi/2$ hybrid coupler. This kind of receiver is employed in a coherent receiver as described earlier. In this receiver, the differential decoding and demodulating are executed after sampling in the M-ary DPSK demodulator. Figure 8.27 demonstrates the structure of the receiver applying this method.

Referring to Figure 8.27, the signal received is attenuated by the attenuator, so that it can be detected by the receiver. The signal input is mixed with the local oscillator (LO) input of 15 mW at the *coupler* block. The output of coupler block is then fed to *pi/2_hybridcpupler_Rx_est* to be detected. The outputs of this block are the amplitude-detected and phase-detected signals. After that, the matrix concatenate would construct those signals in order, and the threshold block would compare the received signals with the transmitted signals

8.7.4 π/2 Optical Hybrid Coupler

Figure 8.28 illustrates the coupler block. This coupler block consists of a π/2 hybrid coupler and balanced photodetector. In the π/2 hybrid coupler, the input signal and the LO signal are mixed together. Accordingly, the output field of the 90° hybrid coupler can be represented as follows:

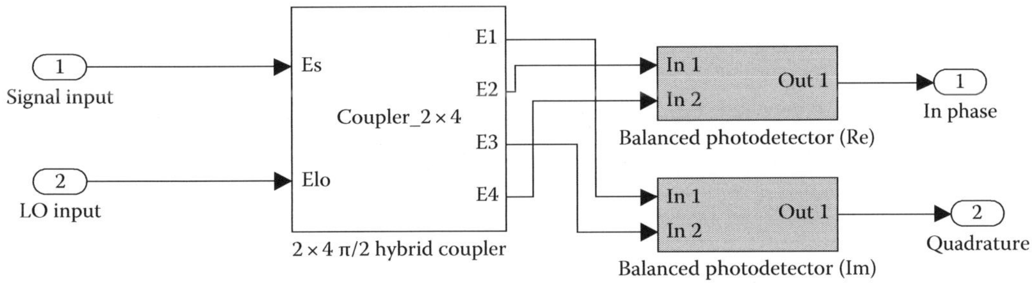

FIGURE 8.28
Structure of the 2 × 4 π/2 hybrid coupler block.

$$E_1(t) = \frac{1}{2}[jE_s(t) + E_{LO}(t)] \tag{8.65}$$

$$E_2(t) = \frac{1}{2}[-E_s(t) + E_{LO}(t)] \tag{8.66}$$

$$E_3(t) = \frac{1}{2}[-jE_s(t) + E_{LO}(t)] \tag{8.67}$$

$$E_4(t) = \frac{1}{2}[E_s(t) + E_{LO}(t)] \tag{8.68}$$

By using this equation, this coupler is created by using the embedded MATLAB code in Simulink. The embedded MATLAB code for the 2×4 90° hybrid coupler is available in Appendix. From the coupler output, the signal is then detected at the balanced photodetector. E_2 and E_4 are fed to the first photodetector to produce in-phase signal, while E_1 and E_3 are fed to the second photodetector to produce quadrature signal. Note that a square of the magnitude represents the optical power to electronic current detection of a photodetector. In simulation, the structure of photodetector is shown in Figure 8.29:

Figure 8.30 shows the structure of *pi/2_hybridcpupler_Rx_est*. The input of this block is the in-phase and quadrature signal obtained after the photodetection. The in-phase and

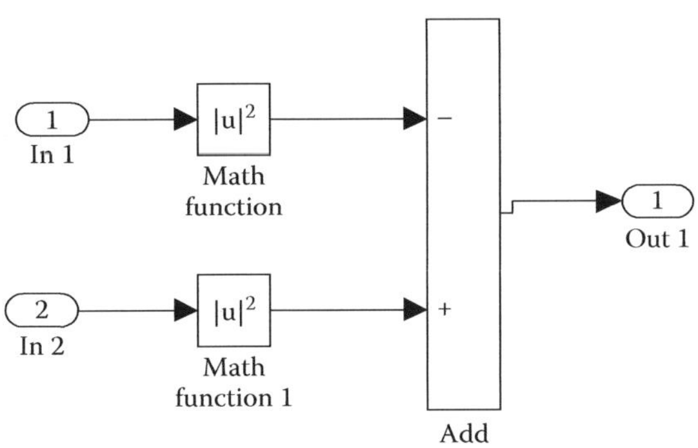

FIGURE 8.29
Structure of the balanced photodetector.

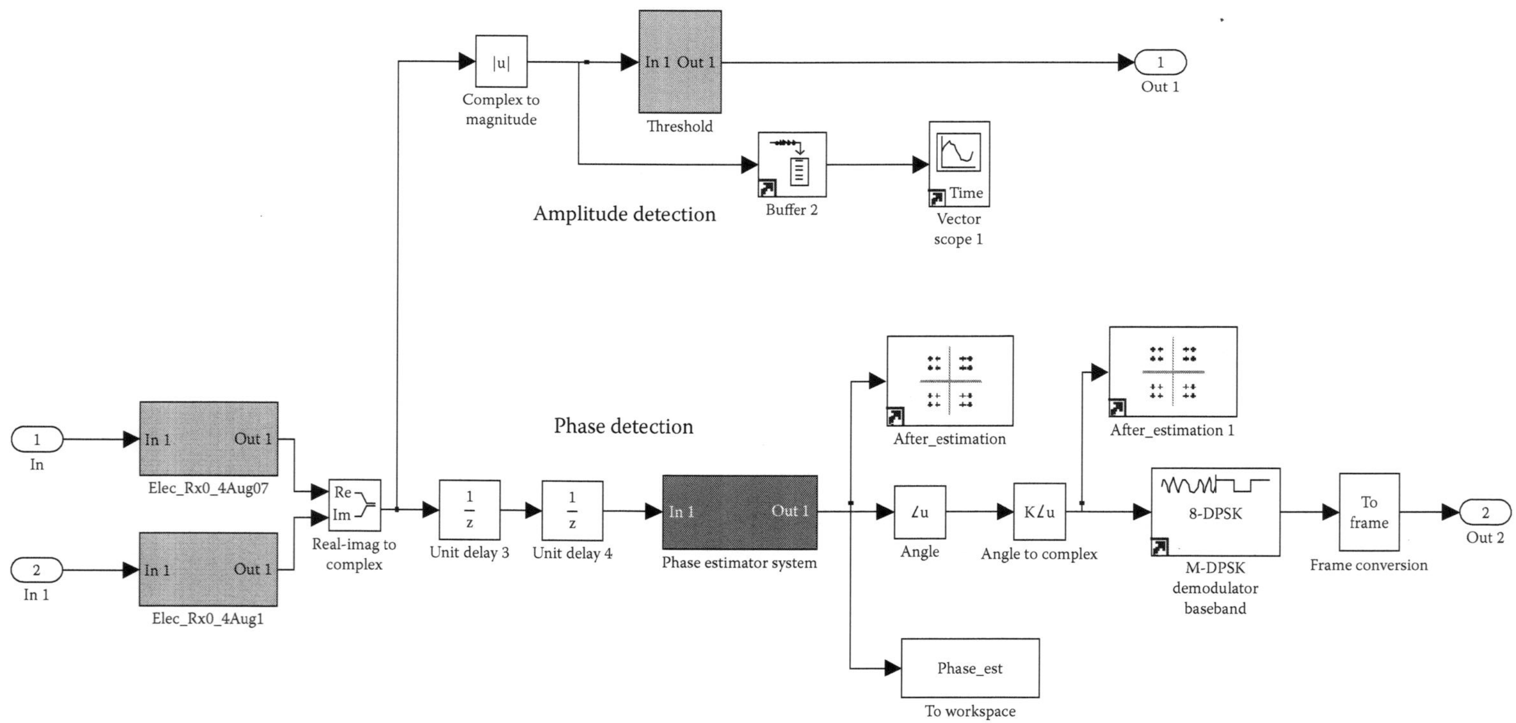

FIGURE 8.30
Structure of the 2 × 4 π/2 hybrid coupler_Rx.

quadrature of the input signal is fed to *Elec_Rx* block. The role of this block is explained in Section 8.3.5.1. In order to recover the signal, the in-phase and quadrature parts are combined. Then, based on the transmitted signal constellation, the decision must be determined. This decision consists of two parts. The first part is amplitude detection, while the second part is phase detection.

8.7.5 Amplitude Detection

The amplitude detection in this part is similar to amplitude detection in Chapter 7. First, the magnitude of the symbol is taken, and then based on the received signal, the decision threshold is made. If the sampled magnitude is lower than the threshold, it is decided as bit 0, otherwise it is decided as bit 1.

8.7.6 Phase Detection

In phase detection, first, the signal is sampled by using two unit delay blocks. The phase estimation is performed after sampling. However, the phase estimator for this kind of receiver is different from the phase estimator used in the previous part. After the phase estimation, the phase signal at the output of phase estimator is then collected and the differential decoding and demodulation process is performed by the 8-DPSK demodulator baseband block. Afterward, the demodulated phase signals are ready to be combined with the amplitude signal (Figure 8.27).

8.7.7 Electronic Noise Model of Electronic Preamplifier

As described in the noise processes of Chapter 7, the electronic noise of the optical receiver is contributed mainly by the preamplifier followed the photodetector as it is the most significant component. Contributions by the main amplifier or other electronic subsystems remain negligible as compared to that of the preamp stage. MATLAB and Simulink models are shown in Figure 8.31. They consist of the quantum shot noise block, which is a signal-dependent shot noise, an electronic noise source representing all equivalent noise sources, as explained in Chapter 7 and Figure 8.5; a transimpedance block that transfers the electronic current to voltage level (transimpedance transfer).

8.8 Concluding Remarks

We have described the principles of coherent detection, homodyne, heterodyne, and self-homodyne techniques. The receiver sensitivity and probability of error of these optical receivers have been discussed. OPLL and methods of its implementation in modern optical transmission systems are described. Coherent optical receivers without using a local oscillator, the self-coherent detection technique suitable for differential phase modulation, has emerged as a potential detection system if associated with digital signal processing and advanced equalization and phase estimation such as the maximum likelihood sequence estimation can offer significant improvement in the receiver sensitivity reducing the BER. Coherent detection will thus offer significant development in modern optical communications. Theoretically, coherent PSK detection

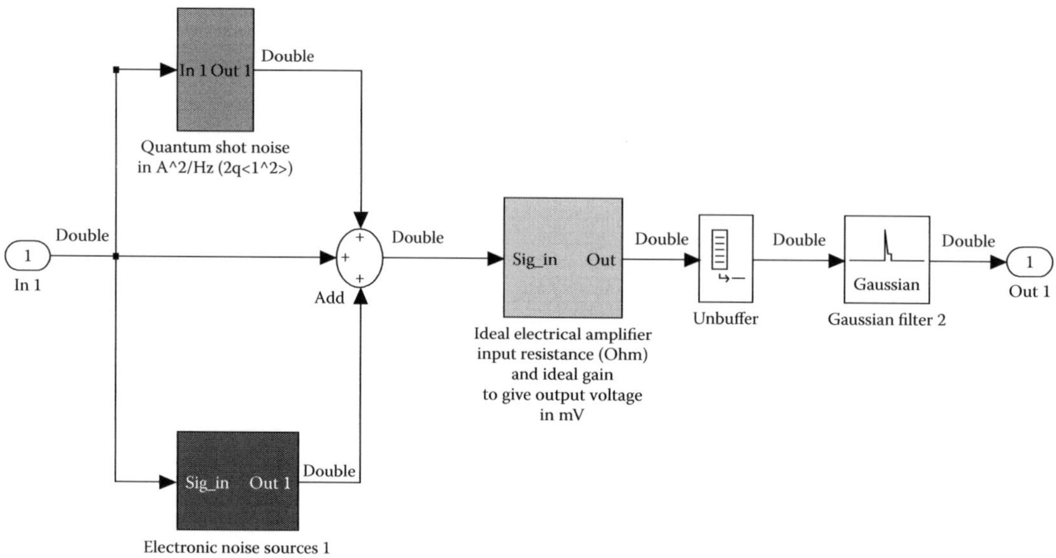

FIGURE 8.31
Structure of the noises in the electronic preamplification.

can reach the quantum limit of about 16 photons/bit; this remains to be proven in practice.

8.9 Problems

PROBLEM 8.1

Coherent Optical Communication Systems

(a) Sketch a structure of an optical coherent receiver employing either a homodyne or a heterodyne technique. Give a brief description of the roles of each component in your system.
(b) What is the typical modern linewidth of the laser that acts as the local oscillator?
(c) Give a distinction between the homodyne and heterodyne coherent system.
(d) A homodyne optical receiver has the following parameters:

A photodetector with a responsibility of 0.9 that is followed by an electronic preamplifier whose total equivalent noise spectral density is 5 $pA/(Hz)^{1/2}$ and electrical bandwidth is 15 GHz. The transmission bit rate is 10 Gb/s.

The local oscillator is a tunable laser source with a linewidth of 100 MHz. The wavelength in vacuum of both the signals and the local oscillator is 1550.92. The average optical power of the local oscillator coupled to the photodetector is 0 dBm.
(e) Sketch the structure of the receiver and then its equivalent small signal circuit that includes the generated electronic signal current at the output of the photodetector, the total noise currents looking from the input of the electronic preamplifier. What is the dominant noise source in this receiver?
(f) For an optical signal with an average power of −20 dBm, estimate the SNR at the output of the photodetector.
(g) Recalculate the SNR of the receiver if the frequency of the local oscillator is 20 GHz away from that of the signal carrier frequency.

PROBLEM 8.2

Concept on Homodyne and Heterodyne Coherent Detection

A binary PSK modulation transmission uses an optical carrier at $\lambda = 1550\,\text{nm}$. The average optical power is 0 dBm. The bit rate is 10 Gb/s assuming that the signals are dispersion compensated perfectly and gain equalized by optical amplifiers with negligible ASE noises. The receiver responsivity is 0.9 A/W. At the receiver, a local oscillator is employed to mix with the optical signals. The frequency of the LO can be assumed to be identical to that of the signal carrier.

(a) Sketch a schematic diagram of the receiver and indicate the signal magnitude level in either optical or electrical domain.
(b) Repeat (a) if the frequency difference between the LO and the carrier is 40 GHz.
(c) Estimate the current magnitude of the optical signals at the output of the photodetector for (a) and (b).

PROBLEM 8.3

Homodyne ASK

Calculate the sensitivity of a homodyne receiver for binary ASK system with BER = 1e-9. The data bit rate is 10 Gb/s and the wavelength is 1550 nm. It uses a PIN photodetector of a responsivity of 0.8 A/W and a dark current of 2 nA. The PIN load resistance is 100 Ω and the receiver noise bandwidth is 15 GHz. The amplifier total noise equivalent at the input is about five times that of the thermal resistance at the input. You may assume equal probability of the "1" and "0." The relative phase between the local oscillator and the incoming lightwave is not larger than 10°.

Appendix: π/2 Hybrid Coupler Embedded MATLAB® Code

```
function [E1, E2, E3, E4] = Coupler_2x4 (Es,Elo)
% This block supports the Embedded MATLAB subset.
% See the help menu for details.
%This equation is obtained from :
% Digital Optical Communications. pp. 131-168 by L.N. Binh.
% in chapter "Coherent Photonic Transmission"
E1 = 0.5*(1i*Es+Elo);
E2 = 0.5*(-Es+Elo);
E3 = 0.5*(-1i*Es+Elo);
E4 = 0.5*(Es+Elo);
```

References

1. A.H. Gnauck and P.J. Winzer, Optical phase-shift-keyed transmission, *Journal of Lightwave Technology*, 23: 115–130, 2005.
2. R.C. Alferness, Guided wave devices for optical communication, *IEEE Journal of Quantum Electronics*, QE-17: 946–959, 1981.
3. W.A. Stallard, A.R. Beaumont, and R.C. Booth, Integrated optic devices for coherent transmission, *IEEE Journal of Lightwave Technology*, LT-4(7): 852–857, July 1986.

4. T. Okoshi and K. Kikuchi, *Coherent Optical Communications*, Kluwer, Tokyo, Chapter 5, 1988.
5. V. Ferrero and S. Camatel, Optical phase locking techniques: An overview and a novel method based on single sideband sub-carrier modulation, *Optics Express*, 16(2): 818–828, January 21, 2008.
6. J. Garrett and G. Jacobsen, Theoretical analysis of heterodyne optical receivers for transmission systems using (semiconductor) lasers with nonneglible linewidth, *IEEE Journal of Lightwave Technology*, LT-3/4: 323–334, 1986.
7. G. Nicholson, Probability of error for optical heterodyne DPSK systems with quantum phase noise, *Electronics Letter*, 20(24): 1005–1007, 1984.
8. S. Shimada, *Coherent Lightwave Communications Technology*. London, U.K.: Chapman and Hall, 1995, p. 27.
9. Y. Yamamoto and T. Kimura, Coherent optical fiber transmission system, *IEEE Journal of Quantum Electronics*, QE-17: 919–934, 1981.
10. S. Savory and A. Hadjifotiou, Laser linewidth requirements for optical DQPSK optical systems, *IEEE Photonics Technology Letters*, 16(3): 930–932, March 2004.
11. K. Iwashita and T. Masumoto, Modulation and detection characteristics of optical continuous phase FSK transmission system, *IEEE Journal of Lightwave Technology*, LT-5/4: 452–462, 1987 (Fig. 1).
12. D.-S. Ly-Gagnon, S. Tsukamoto, K. Katoh, and K. Kikuchi, Coherent detection of optical quadrature phase-shift keying signals with carrier phase estimation, *Journal of Lightwave Technology*, 24: 12–21, 2006.
13. E. Ip and J.M. Kahn, Feedforward carrier recovery for coherent optical communications, *Journal of Lightwave Technology*, 25: 2675–2692, 2007.
14. L.N. Binh, Dual-ring 16-star QAM direct and coherent detection in 100 Gb/s optically amplified fiber transmission: Simulation, *Optical and Quantum Electronics*, 40(10): 707–732, August 5, 2008. Accepted December 2008. Published on line December 2008.
15. L.N. Binh, Generation of multi-level amplitude-differential phase shift keying modulation formats using only one dual-drive Mach–Zehnder interferometric optical modulator, *Optical Engineering*, 48(4): 045005-1–045005-10, 2009.
16. M. Nazarathy, X. Liu, L. Christen, Y.K. Lize, and A. Willner, Self-coherent multisymbol detection of optical differential phase-shift-keying, *Journal of Lightwave Technology*, 26: 1921–1934, 2008.
17. R. Noé, PLL-free synchronous QPSK polarization multiplex/diversity receiver concept with digital I&Q baseband processing, *IEEE Photonics Technology Letters*, 17: 887–889, 2005.
18. L.G. Kazovsky, G. Kalogerakis, and W.-T. Shaw, Homodyne phase-shift-keying systems: Past challenges and future opportunities, *Journal of Lightwave Technology*, 24: 4876–4884, 2006.
19. J.P. Gordon and L.F. Mollenauer, Phase noise in photonic communications systems using linear amplifiers, *Optical Letters*, 15: 1351–1353, 1990.
20. H. Kim and A.H. Gnauck, Experimental investigation of the performance limitation of DPSK systems due to nonlinear phase noise, *IEEE Photonics Technology Letters*, 15: 320–322, 2003.
21. S. Zhang, P.Y. Kam, J. Chen, and C. Yu, Receiver sensitivity improvement using decision-aided maximum likelihood phase estimation in coherent optical DQPSK system, in *Conference on Lasers and Electro-Optics/Quantum Electronics and Laser Science and Photonic Applications Systems Technologies*, San Jose, CA, Technical Digest (CD), Optical Society of America, 2008, Paper CThJJ2.
22. P.Y. Kam, Maximum-likelihood carrier phase recovery for linear suppressed-carrier digital data modulations, *IEEE Transactions on Communications*, COM-34: 522–527, June 1986.
23. (a) S. Zhang, P.Y. Kam, J. Chen, and C. Yu, A comparison of phase estimation in coherent optical PSK system, Photonics Global'08, Singapore, December 2008, Paper C3-4A-03. (b) S. Zhang, P.Y. Kam, J. Chen, and C. Yu, Adaptive decision-aided maximum likelihood phase estimation in coherent optical DQPSK system, in *OptoElectronics and Communications Conference (OECC)'08*, Sydney, Australia, July 2008, pp. 1–2, Paper TuA-4.
24. S. Zhang, P.Y. Kam, J. Chen, and C. Yu, A comparison of phase estimation in coherent optical PSK system, Photonics Global'08, Singapore, December 2008, Paper C3-4A-03.

25. A.P. Clark, *Equalizers for Digital Modem*. London, U.K.: Pentec Press, 1985.
26. X. Liu, X. Wei, R.E. Slusher, and C.J. McKinstrie, Improving transmission performance in differential phase-shift-keyed systems by use of lumped nonlinear phase-shift compensation, *Optical Letters*, 27(18): 1351–1353, September 2002.
27. X. Wei, X. Liu, and C. Xu, Numerical simulation of the SPM penalty in a 10-Gb/s RZ-DPSK system, *IEEE Photonics Technology Letter*, 15(11): 1636–1638, November 2003.
28. E. Ip and J.M. Kahn, Digital equalization of chromatic dispersion and polarization mode dispersion, *IEEE Journal of Lightwave Technology*, 25: 2033–2043, 2007.
29. I. Hodgkinson, Receiver analysis for optical fiber communications systems, *IEEE Journal of Lightwave Technology*, 5(4): 573–587, 1987.

9

Erbium-Doped Fiber Optical Amplifiers and Simulink® Models

Optical fiber amplifiers are dominating the loss compensation in fiber-optic communication systems, especially the erbium-doped types, for communication spectra based on silica fibers. Several models have been developed for such photonic amplification devices. However, there are no comprehensive models developed on the MATLAB® and Simulink® platforms. This chapter describes in detail the development of a Simulink model for erbium-doped fiber amplifiers (EDFAs) operating under steady state and dynamic conditions.

The equations that govern the pump and signal propagation in the doped fiber are derived and tabulated using corresponding amplifier parameters. The model blocks are then described under the Simulink environment. Simulation results are compared with experimental values, and their close agreements indicate the validity of the Simulink EDFA model. This model is demonstrated so that it can be proven to be user friendly in research and teaching laboratories.

9.1 Introduction

Since 1989, the rare-earth metal-doped fibers have been exploited as the optical amplification medium and laser sources, both in experimental and theoretical domains. There is no doubt that EDFAs are the amplifying devices employed in commercial optical transmission systems. Several models have also been reported and developed for the design and system simulation [1–5].

The mathematical laboratory, MATLAB, has also been progressively developed and improved over the last two decades. It is now considered as the standard mathematical toolbox in teaching and R&D laboratories around the world. Recently, the Simulink toolbox has been added to it. This gives an additional and powerful tool for the simulation of physical systems. The Laboratory for Optical Communications and Applied Photonics of Monash University has developed several packages under MATLAB and Simulink for the simulation of photonic devices and optical communication systems for both research and training purposes.

Why should Simulink be developed while there are a number of commercial simulation packages available, such as VPI systems. Simulink is a modular toolbox attached to MATLAB that is now used as a universal computing tool in universities worldwide. Its communication blockset contains many toolboxes for digital communications in the baseband as well as in the passband. Thus, these toolboxes can be used as a modular set for the generation and detection of lightwave signals. As the frequency of the lightwaves is much higher than the baseband signals, it is not likely that the carrier is contained under the envelope of the signals, because it requires an extremely high sampling rate, which is currently not possible in computing systems. Thus, complex envelope signals are used to represent the phase of the carrier. Therefore, the simulation models for digital optical communications can be modeled very accurately and with ease due to the modularity of Simulink. Simulink techniques and samples of models may be made available as a toolbox in MATLAB in the future so that global laboratories can equip and facilitate teaching and research activities. Thus, the Simulink models described in this book would be easily integrated into a standard MATLAB software package in university, research, and engineering laboratories, as well as in the field, especially the toolset. The digital optical communications MATLAB and Simulink models can thus be the most universally affordable for researchers and students of advanced level of applied physics and telecommunications engineering.

This chapter thus describes the comprehensive and successful implementation of the EDFA to the Simulink and MATLAB 7 R14. It has been organized as follows. The fundamental and theoretical expressions for representing the optical amplification and the rate of transitions of the erbium ions and lightwave propagation over the amplifying medium are described in Section 9.2. Both forward- and backward-pumping configurations are given. Section 9.3 gives an account of the Simulink models for the amplifiers in the steady state and dynamic environments. It then illustrates the samples of the Simulink simulator; it also presents details of the simulation results to illustrate the validity of the developed models and confirm them with the experimental data obtained from the EDFA constructed in the laboratory as well as those specified by commercial manufacturers. Finally, Section 9.4 concludes the chapter.

The described MATLAB and Simulink models are proven to be user friendly. It would be a good tool for researchers and teaching graduates of advanced levels in the fields of photonics and optical communications.

9.2 Fundamental and Theoretical Issues of EDFAs

In this section, basic theoretical analyses of the EDFA are described for the development of EDFA simulation modeling. Several technical terms such as pumping scheme, population inversion factor, and amplified spontaneous emission (ASE) noise are defined.

9.2.1 EDFA Configuration

The EDFA is mainly constructed with a short length of optical fiber whose core has been doped with less than around 0.1% erbium, an optically active rare earth element, which has many unique intrinsic properties for optical amplification [6,7]. The optical fiber used is named the erbium-doped fiber (EDF). The optical amplifier engineer chooses the appropriate EDF during the EDFA development. Different doped fibers have different characteristics of the absorption and emission factors, which will be identified and discussed later in this chapter.

An EDFA consists of an erbium-doped silica fiber and other photonic components such as the pump source, the optical isolators (OIs), and the wavelength division multiplexing (WDM) coupler, as shown in Figure 9.1 [9–12]. An enhanced EDFA module consists of an EDF, a wavelength-dependent coupler (WSC), and a polarization-insensitive OI. An

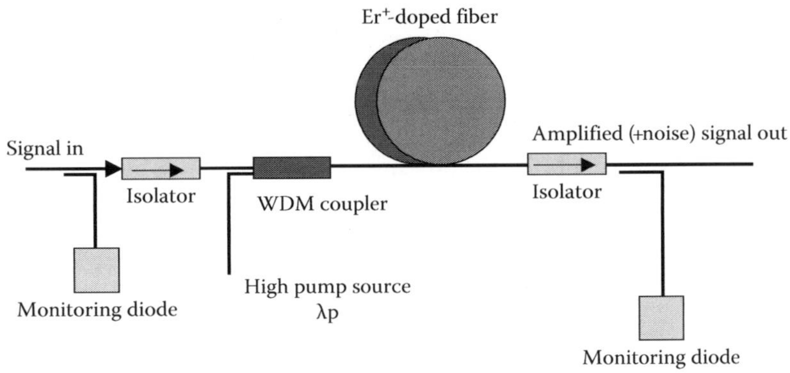

FIGURE 9.1
Standard forward-pumping EDFA.

optical filter may be required to improve EDFA performance [13] depending on specific applications. The optical filter is used to reduce the ASE noise and protect the amplifier from saturation caused by the ASE accumulated in an in-line amplification system. The OI is used for achieving a stable amplified operation to prevent spurious oscillations and reflection. The pump source is used for pumping the erbium ions to an upper energy level; several wavelength-pumping laser diodes exist at present. Both 1480 and 980 nm pumping laser diodes can provide a high pump gain; the investigation of these two pumping wavelengths will be described later in this chapter. The WSC/WDM is used to combine the pump sources and the incoming signal. The pump lightwaves are guided and propagate along the EDF length, and the depleted erbium ions then rise to an excited state by absorption the energy. As the signal propagates through the EDF, it stimulates the emission of light from the excited ions; thus, the signal is amplified. The tap coupler is used as a separator taping the pumping power in order to protect the receiver, normally used in preamplifiers.

Two pumping configurations can be used in EDFAs: backward pumping, and dual backward and forward pumping, as shown in Figure 9.2.

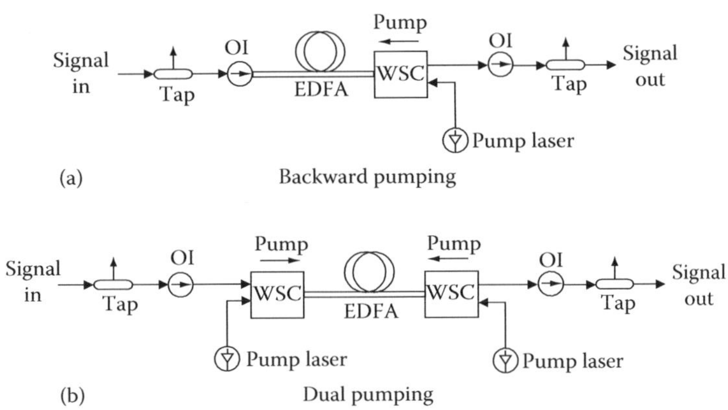

FIGURE 9.2
(a) Backward pumping and (b) dual pumping.

The only difference between the three configurations is the direction of pumping propagation. As per the names of these configurations, in forward pumping, the signal light co-propagates with the pump light, whereas in backward pumping it counter-propagates relative to the pump light, and in dual pumping, it pumps bidirectionally. Regarding noise performance, forward pumping can produce optical amplification with low noise, but may offer a lower amplification gain due to the fluctuations of the pump power and the signal power. On the other hand, backward pumping may be effective for high power output but it suffers a higher ASE noise. Thus, an enhanced model can be obtained in which bidirectional pumping or dual pumping can be combined.

9.2.2 EDFA Operational Principles

The basic operation principle of the EDFA is founded on the fact that the erbium ions can exist in several energy states (Figure 9.3). When they are at a high energy state, photons could be stimulated to give up some of their energy (as light) and return to a lower-energy (more stable) state; thence, photonic amplification would take place [14].

9.2.2.1 *Pump Wavelength and Absorption Spectrum*

The pump wavelength used is either 980 or 1480 nm due to the availability of these laser sources. Why do most EDFAs use only these two as their pumping wavelengths and not any other wavelengths? Before we go looking for the answer, let us see the energy levels of the triply ionized erbium ion, Er^+, the absorption band of Er^+, and the pump efficiency (Figure 9.4) [14].

There are several states to which the erbium ions can be pumped using sources of different wavelengths, such as those operating at 1490, 980, and 800 nm. However, in real lightwave

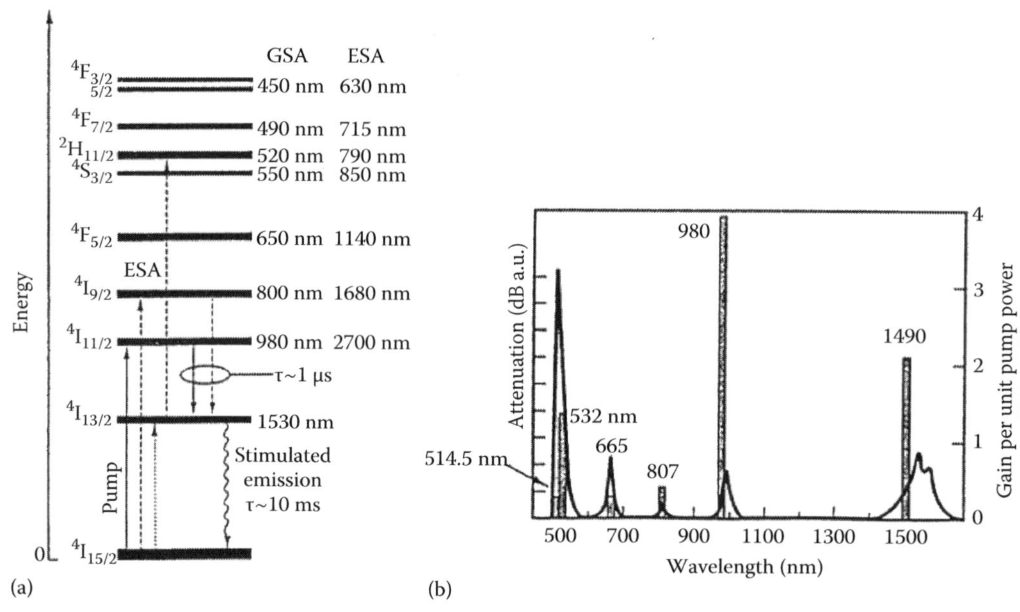

FIGURE 9.3
(a) Energy levels of erbium ions and (b) gain and attenuation spectra.

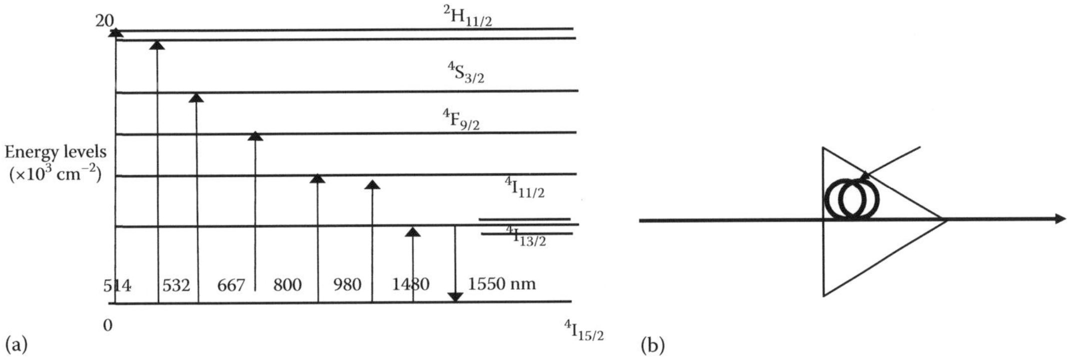

FIGURE 9.4
Erbium ion energy levels and pump spectra: (a) details and (b) simplified model.

systems, the pump wavelength must provide a high power to achieve a high gain per pump. The commonly available laser diodes can operate at 800, 980, and 1480 nm, but the pump efficiency can go more than 1 dB/mW with low attenuation depending on the pump frequency. The only pump wavelength laser sources that can give a high pumping efficiency with lower attenuation are those operating at 980 and 1490 nm. In practice, the 980 nm pumping source is commonly used due to its high gain coefficient (4 dB/mW). The model demonstrated in this chapter employs both 980 and 1480 nm pumping sources. The difference in the effects of these two wavelength sources is mainly caused by the absorption and emission factors.

9.2.2.2 *Pump Mechanism*

The amplification mechanism is the key for developing the EDFA simulation model. The optical amplification gain is supplied by the excited erbium ions (Er^{+}) when they are pumped to achieve population inversion. Depending on the energy states of the dopant, pumping schemes can be classified as a two-level scheme or a three-level scheme. The main difference between the two-level and three-level pumping schemes is related to the energy state occupied by the stimulated-emission event. In the case of a three-level scheme, the lower level is the ground level, whereas in the two-level scheme, the state is an excited state with a fast relaxation time.

As shown in Figure 9.5, erbium ions are pumped to an upper energy level by the absorption of lightwave energy from the pump source at either 980 or 1480 nm to regain the equilibrium distribution [13].

Stage 1

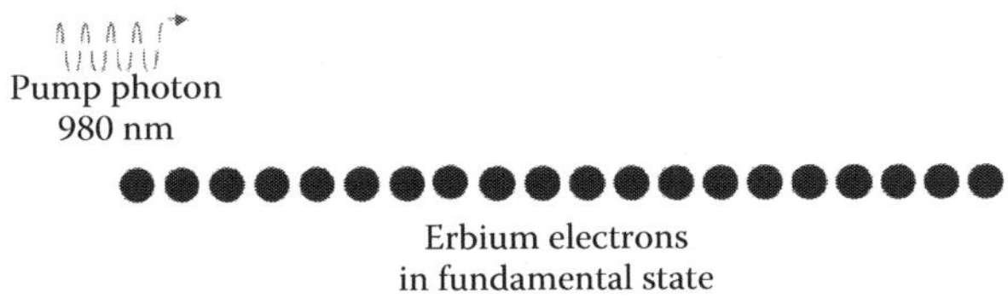

FIGURE 9.5
The pump photon starts propagating through the erbium fiber, all the erbium electrons are now in the fundamental state (ground state).

Stage 2

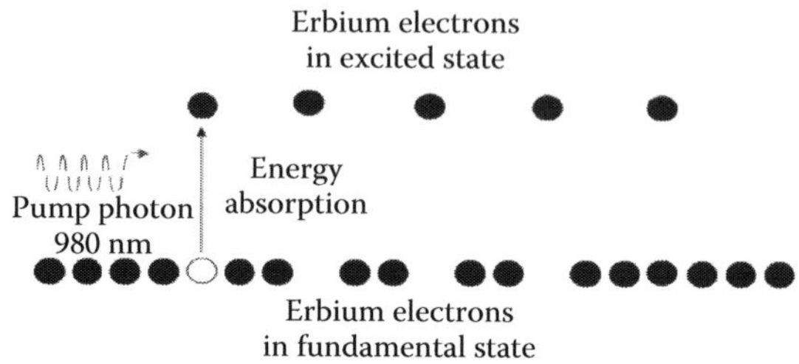

FIGURE 9.6
The erbium electrons pumped to higher-level energy states.

Stage 3

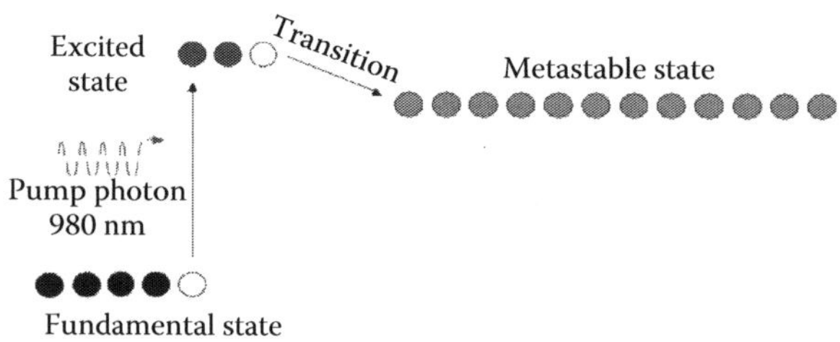

FIGURE 9.7
The erbium electrons now revert to lower states (metastable state) spontaneously after some mean lifetime t – 1 ms (spontaneous emission).

Stage 4

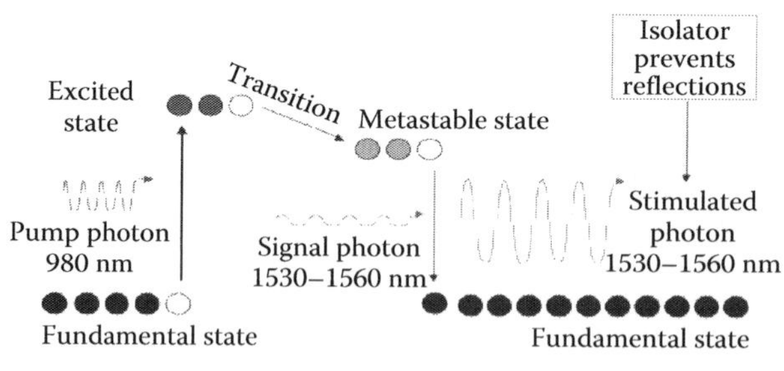

FIGURE 9.8
The erbium electrons revert to the fundamental state spontaneously (spontaneous emission) and are stimulated to do so by some incident photons (signal) of exactly the energy corresponding to the drop to lower state (stimulated emission).

The coherence of pump sources and signals means the incident photon and the stimulated photon have the same frequency, phase, propagation direction, and polarization [15]. The optical gain is achieved when the population inversion reaches a sufficiently high level at which the absorption from the ground state is considerably reduced. This can be readily achieved with the available pump lasers because the metastable state exists in Er^{3+}, thus rendering the population inversion possible. In addition, the photon lifetime from the metastable state to the fundamental state must be longer than that from the excited state to the metastable state, normally around 10 ~ 10.5 ms; the lifetime from the excited state to the metastable state is normally about 1 ms. Therefore, the ions are accumulated in the metastable state; hence, the ion population in the excited stated is negligibly small. This fact is critical in the design of the simulator.

9.2.3 General Amplifier Noises and Gain in Transmission

Noises are generated during the signal amplification process. The dominant amplification noise source is the ASE (amplifier stimulated emission) noise. While ions are undergoing transition from the excited state to the fundamental state, some excited erbium ions decay to the fundamental state with spontaneous emission before amplifying the injected signal photons (Figure 9.9). Thus, photons are emitted with random phase and direction [13].

The ASE spectral range is very wide and covers almost the C band and L band including the gain spectrum of the amplifier, as shown in Figure 9.10.

It reduces the available gain for the signal field. Figure 9.11 shows that the saturated gain has been reduced by the ASE. This is due to the incoherence of the excited erbium ions with the signal photons. The ASE noises cause degradation to the signal and noise ratio (SNR).

EDFAs are employed in long-haul transmission systems; the ASE level grows after a cascade of amplifiers and begins to saturate the optical amplifiers, hence the reduction in

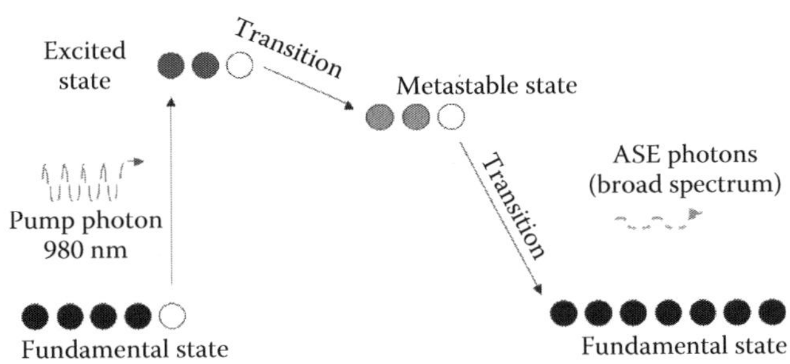

FIGURE 9.9
ASE photons demonstration.

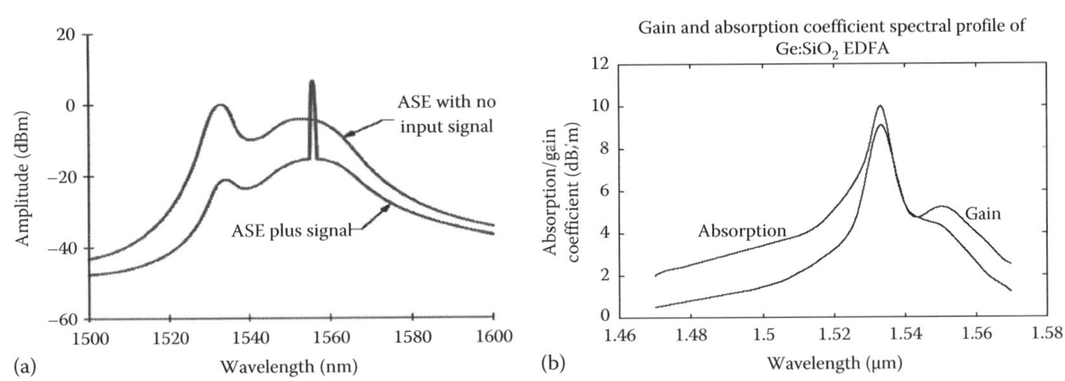

FIGURE 9.10
(a) ASE noise spectrum and (b) absorption and gain spectra.

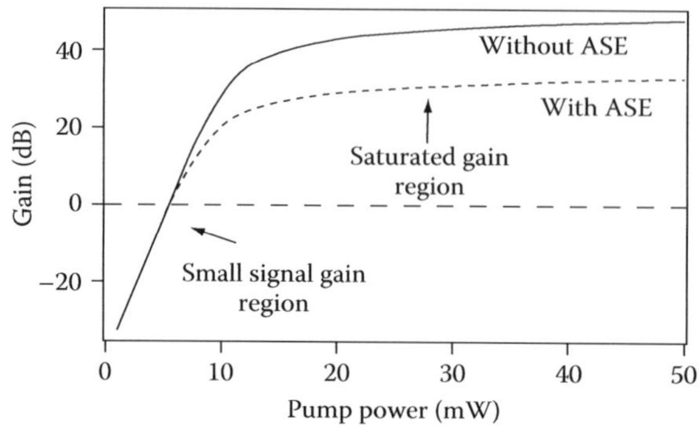

FIGURE 9.11
Optical gain reduction by the ASE.

the signal gain. This is because the total power is almost equal to the pump power plus the ASE power.

9.2.3.1 Amplifier Gain Modulation

The gain modulation of EDFAs can provide an attractive approach for implementing status and control communication in modern dense wavelength division multiplexing (DWDM) communication networks. The desired control signal is imposed as amplitude over modulation on the payload communication signals propagating down the system via modulation of the EDFA pump power [2]. We can describe herewith in this chapter the impacts of pump modulation and signal modulation effects on the EDFA gain.

9.2.3.2 EDFAs in Long-Haul Transmission Systems

Optical amplifiers can be employed as three different types for optical transmission systems, as a booster amplifier, an in-line amplifier regenerator, and as an optical preamplifier at the receiver, as shown in Figure 9.12.

- Booster amplifier: The booster amplifier increases and provides a high optical power signal prior to transmission. In long-haul optical communication systems, the use of a booster amplifier can increase the link power budget and reduce the number of regenerators required.
- In-line optical amplifier: The main function of an in-line amplifier is to compensate for the fiber loss of the fiber span or the dispersion-compensating fibers, thereby overcoming the need for O/E regeneration.
- Optical preamplifier: The main function of the optical preamplifier is to increase the power level of a signal before detection, and hence the receiver sensitivity.

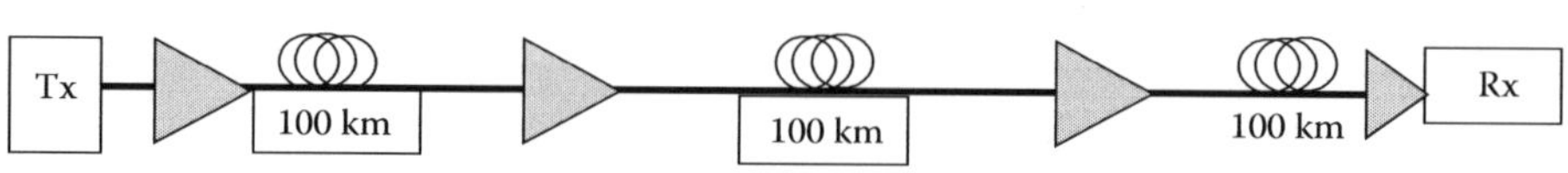

FIGURE 9.12
The employment of three types of amplifiers and related monitoring equipment.

9.3 EDFA Simulation Model

In this section, the mathematical expressions for developing the EDFA simulation model are described. In general, the simulation models can be classified as steady state and dynamic models and/or perturbation models [6,7]. In this chapter, only the first two models are used for EDFA simulation. The models of a single EDFA are investigated; a cascade of EDFAs can be implemented without difficulty and left for users to construct.

9.3.1 Amplifier Parameters

A number of parameters can clarify the principles of the simulation model, as shown in Tables 9.1 and 9.2.

Bononi and Rusch have developed a dynamic model based on the one-dimensional ordinary nonlinear differential equation for the time-dependent population of excited erbium ions in EDFAs [3]. With the three-level states, the amplification in the EDF is based on the stimulated erbium ions; thus, the state population would have a discrepancy. Simplifying the scheme to be a three-level pumping, the three populations of erbium atoms are the ground state 1 (the fundamental state with population density n_1), level state 2 (the excited state population density, also called the metastable state with signal frequency n_2), and level state 3 (the excited state population with pump frequency n_3). The rate equation for these three states can be defined and given as follows [6,7,14]:
Ground state 1 population equation:

$$\frac{\partial n_1}{\partial t} = -W_{1-3} * n_1 - W_{1-2} * n_1 + W_{2-1} * n_2 + \frac{n_2}{\tau_{2-1}} + W_{3-1} * n_3 + \frac{n_3}{\tau_{3-2}} \tag{9.1}$$

TABLE 9.1
Erbium-Doped Fiber Parameters

Physical Meaning	Expression	Symbol and Unit
Erbium-doped fiber parameters (Part I)		
Core radius of EDF		r (μm)
Core area of EDF		A (m^2)
Overlap factor of EDFA at wavelength (λ)		Γ_λ
Ion density of EDF		ρ (ions/m^3)
Florescence lift time of EDFA		τ (ms)
Absorption cross section at wavelength (λ)		$\sigma_\lambda^\alpha(\mathrm{m}^2)$
Emission cross section at wavelength (λ)		$\sigma_\lambda^e(\mathrm{m}^2)$
Absorption coefficient at wavelength (λ)	$\rho\Gamma_\lambda\sigma_\lambda^\alpha$	α_λ (m^{-1})
Absorption over length L at wavelength (λ)	$\alpha_\lambda \times \mathrm{L}$	
Plank constant	6.626e−34	h (J·s)
Speed of light		c (m s^{-1})

TABLE 9.2

Glossary of EDFA Parameters

Physical Meaning	Expression	Symbol and Unit
Erbium-doped fiber parameters (Part II)		
Length of the erbium fiber		L (m)
Population density in fundamental state		n_1 (ions/m^3)
Population density in excited state (assume 980 nm pump)		n_3 (ions/m^3)
Population density in metastable state		n_2 (ions/m^3)
Signal power		P_{in} (mW)
Pump power		P_{out} (mW)
Signal power in simulation	$N_s = \frac{P_{signal}^{in} c}{h\lambda_s}$	N_s (photons/s)
Pump power in simulation	$N_p = \frac{P_{pump}^{in} c}{h\lambda_p}$	N_p (photons/s)

State 2 population equation:

$$\frac{\partial n_2}{\partial t} = W_{1-2} * n_1 - W_{2-1} * n_2 + W_{3-2} * n_3 - \frac{n_2}{\tau_{2-1}} \tag{9.2}$$

And, finally, state 3 population equation:

$$\frac{\partial n_3}{\partial t} = W_{1-3} * n_1 - W_{3-2} * n_3 + W_{3-1} * n_3 - \frac{n_3}{\tau_{3-2}} \tag{9.3}$$

The transition from state 3 to state 2 (W_{3-2}) is much more influential on the EDFA gain than the transition down to the fundamental state (W_{3-1}) (Table 9.3). The transition rate W_{3-1} can be neglected. The transition W_{1-3} is associated with the pump source. Furthermore, we can refer to Figure 9.9; the florescence lifetime for the transition rate W_{3-2} is short enough (1 μs) compared with that of the transition rate W_{2-1} (10 ms); thus, the erbium ions can be accumulated in state 2. Hence, the transition rate W_{1-3} should be very close to the transition rate W_{3-2} due to shorter lifetime. We, thus, have $W_{3-2} \cong W_{1-3}$. Substituting this into Equation 9.2 we have

$$\frac{\partial n_2}{\partial t} = W_{1-2} * n_1 - W_{2-1} * n_2 + W_{1-3} * n_3 - \frac{n_2}{\tau_{2-1}} \tag{9.4}$$

In fact, n_3, n_2, and n_1 are independent. However, as mentioned above, the erbium ions are accumulated in the upper state 2; the population density in n_3 can thus be negligible. The total population density, ρ, of the fiber is a constant value. Therefore, if $n_3 \approx 0$, then n_1 and n_2 are no longer independent of each other under the constraint that the sum of the two equals the total population density of the erbium ions of the fiber, $n_1 + n_2 = \rho$. In particular, the population is a function of both time and linear distance along the EDF [3,6,7,14].

TABLE 9.3

Transition Rates between States of a Stimulated Er^{3+} Ion

Physical Meaning	Symbol	Expression	Comment
Transition rate from state 1 to state 2	W_{1-2}	$W_{1-2} = \frac{\Gamma_s \sigma_{1-2} N_s}{A}$	Transition associated to the signal
Transition rate from state 1 to state 3	W_{1-3}	$W_{1-3} = \frac{\Gamma_p \sigma_{1-3} N_p}{A}$	Transition associated to the pump (980 nm)
Transition rate from state 2 to state 1	W_{2-1}	$W_{2-1} = \frac{\Gamma_s \sigma_{2-1} N_s}{A}$	Transition associated to the signal (C band) 1525–1565 nm (L band) 1566–1605 nm
Transition rate from state 3 to state 2	W_{3-2}	$W_{3-2} = \frac{\Gamma_p \sigma_{3-2} N_p}{A}$	Transition associated to the pump (980 nm)
Transition rate from state 3 to state 1	W_{3-1}	$W_{3-1} = \frac{\Gamma_p \sigma_{3-1} N_p}{A}$	Transition associated to the pump (980 nm)

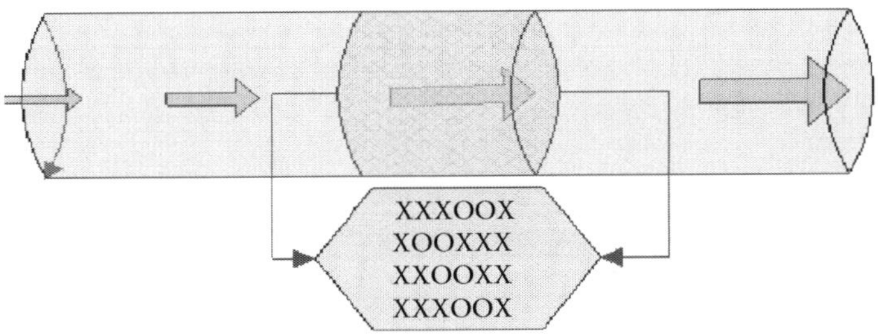

FIGURE 9.13
Power evolution of the signal under the amplification process along the EDF.

Furthermore, the relationship between the input power and the population has to be determined next. Figure 9.13 illustrates how the input power evolutes along the EDF.

Referring to Figure 9.13, we see that the input light is amplified along the distance during the time. Hence, the optical power through the EDF is determined by [2]

$$\frac{\partial P_p}{\partial z} = -\Gamma_p * \sigma_{1-3} n_1 * P_p - \gamma_p * p_p \tag{9.5}$$

$$\frac{\partial P_s}{\partial z} = -\Gamma_s * \sigma_{1-2} n_1 * P_s + \Gamma_s * \sigma_{2-1} n_2 * P - \gamma_s * p_s \tag{9.6}$$

where γ denotes the scattering and other losses of the EDF. As shown in Figure 9.13, X refers to the codirectional photon. O is the photon simulation in a different direction, hence generating noises, reflection, power, and scattering.

9.3.2 EDFA's Dynamic Model

In our simulator, the focus is on EDFA's dynamic modeling rather than steady state modeling as "EDFAs with high output powers for multiple-channel optical networks are strongly saturated and the resultant effective time constant is reduced to the order of tens of microseconds, which is 1000 times lesser than their lift time" [6,7]. The time-dependent problem is especially important in optical networks, where the optical channels are constantly reconfigured. Dynamic modeling can predict the EDFA gain more accurately during multichannel amplification. This section describes the essential Bononi–Rusch approach and develops it to accommodate the time-varying pump and communication signals. The multichannel amplification modeling principle will also be considered in this section.

First, we note that Equations 9.1 through 9.6 can be typically solved by obtaining the population in different states as a function of the transition rates $W_{1\text{-}2} = \Gamma_s \sigma_{1\text{-}2} N_s / A$, $W_{1\text{-}3} = \Gamma_p \sigma_{1\text{-}3} N_p / A$, and $W_{2\text{-}1} = \Gamma_s \sigma_{2\text{-}1} N_s / A$.

Substituting these into Equations 9.5 and 9.6, we have

$$\frac{\partial P_p}{\partial z} = -W_{1\text{-}3} * n_1 * A - \gamma_p * p_p \tag{9.7}$$

$$\frac{\partial P_s}{\partial z} = -W_{1\text{-}2} * n_1 * A + W_{2\text{-}1} * n_2 * A - \gamma_p * p_p \tag{9.8}$$

Multiplying Equation 9.4 by the effective core area of the fiber and substituting it in (9.7) and (9.8) [12], we have

$$\frac{\partial N_2}{\partial t} = P_s(0,t) * (1 - e^{G_s}) + P_s(0,t) * (1 - e^{G_p}) - \frac{N_2}{\tau} - \gamma_p * p_p - \gamma_s * p_s \tag{9.9}$$

where G_s and G_p are the amplification gains of the power evolution, respectively. The increment of the power along the EDFA is nonlinear and assumed to be varying exponentially; thus, we have [1]

$$P_p(L,t) = P_p(0,t) e^{\frac{\Gamma_p \sigma_{1\text{-}3}}{A} N_2 - \Gamma_p \sigma_{1\text{-}3} \rho L - \gamma_p} \tag{9.10}$$

$$P_s(L,t) = P_s(0,t) e^{\frac{\Gamma_s (\sigma_{1\text{-}2} + \sigma_{2\text{-}1})}{A} N_2 - \Gamma_s \sigma_{1\text{-}2} \rho L - \gamma_s} \tag{9.11}$$

Assuming $\gamma = 0$ leads to

$$G_p = \frac{\Gamma_p (\sigma_{1\text{-}3} + \sigma_{3\text{-}1})}{A} N_2 - \Gamma_p \sigma_{1\text{-}3} \rho L \tag{9.12}$$

$$G_s = \frac{\Gamma_s (\sigma_{1\text{-}2} + \sigma_{2\text{-}1})}{A} N_2 - \Gamma_s \sigma_{1\text{-}2} \rho L \tag{9.13}$$

Input channels with different wavelengths have different EDFA gains due to the spectral dependence of the absorption and emission cross sections and the overlap factor. For the three-level transition schemes (980 nm pump), $\sigma_{1\text{-}2}$ and $\sigma_{2\text{-}1}$ are the absorption and emission factors, respectively, at the signal wavelength and σ_{1-3} and σ_{3-1} correspond to the absorption and emission due to the pump wavelength; σ_{3-1} is assumed to be negligible over the short transition time. For the two-level scheme, the pump source normally operates in the spectral region 1480–1490 nm; the rate equations are similar to those obtained for the three-level scheme, but the absorption and emission factors $[\sigma_{1-3} \quad \sigma_{3-1}]$ are to be matched with the correspondent wavelength. Rewrite these matrices to $[\sigma_a \quad \sigma_e]$. The $[\sigma_a \quad \sigma_e]$ matrix for the pump wavelength is $[\sigma_{a_p} \quad \sigma_{e_p}]$, where σ_{e_p} is nonzero and refers to the spectrum of the absorption and emission factors. For multichannel amplification, the modeling equations are integrated and summarized as

$$\frac{\partial N_2}{\partial t} = \sum_{\lambda_0}^{\lambda_n} P_\lambda(0,t) * (1 - e^{G_\lambda}) - \sum_{\lambda_0}^{\lambda_n} \gamma_\lambda * p_\lambda - \frac{N_2}{\tau} \tag{9.14}$$

9.3.3 EDFA's Steady-State Modeling Principles

It is well known that one of the advantages of the EDFA is its slow dynamics as a result of the long, spontaneous lift time of around 10 ms. This is also the reason why steady state models play an important role in the dynamic analysis of EDFA properties for system applications [5]. EDFA steady state modeling has been widely used. An EDFA operates in a steady state condition when $\partial N_2/\partial t = 0$, that is, when there is no population increment and the EDFA gain reaches the saturation mode. Thus, in order to model the EDFA in steady state, we have

$$0 = \sum_{\lambda_0}^{\lambda_n} P_\lambda^{\text{saturation}}(0,t) * \left(1 - e^{G_\lambda^{\text{saturation}}}\right) - \sum_{\lambda_0}^{\lambda_n} \gamma_\lambda * p_\lambda^{\text{saturation}}(0,t) - \frac{N_2^{\text{steady-state}}}{\tau} \tag{9.15}$$

which can be simplified to

$$P_p(L,t) = \sum_{\lambda_0}^{\lambda_n} P_{\lambda_p}^{\text{steady}}(0,t) e^{\frac{\Gamma_p \sigma_{a\lambda p}}{A} N_2^{\text{steady}} - \Gamma_p \sigma_{a\lambda_p} \rho L - \gamma_p} \tag{9.16}$$

$$P_s(L,t) = \sum_{\lambda_0}^{\lambda_n} P_{\lambda_s}^{\text{steady}}(0,t) e^{\frac{\Gamma_s(\sigma_{a\lambda s} + \sigma_{e\lambda s})}{A} N_2^{\text{steady}} - \Gamma_s \sigma_{a\lambda s} \rho L - \gamma_s} \tag{9.17}$$

This model, thus, can predict the onset of the steady state of the amplification gains and pump absorption in an optical amplifier containing an arbitrary number of pumps and signals for co- and contra-propagating scenarios.

9.3.4 Population Inversion Factor

Population inversion indicates the number of ions inverted from state 1 to state 2 excited by the pumping energy. In order to obtain the amplification gain, the number of photons gained through simulated emission from the metastable state to the fundamental state must be greater than that gained through absorption from the fundamental state to state 2. This condition implies that the population in the metastable state must be maintained at a higher level than that at the fundamental state. The degree of population inversion can thus be expressed by the population inversion factor, n_{sp} [14], as

$$n_{sp} = \frac{N_2}{N_2 - \dfrac{\sigma_e}{\sigma_a} N_1} \tag{9.18}$$

where n_{sp} is related to the pump and signal powers and is the key factor of the ASE (amplifying spontaneous emission) noise.

9.3.5 Amplifier Noises

In the previous section, the production of the ASE photons is described. The noise statistics of the amplified lightwaves follows a Gaussian distribution. By considering noises as random variables generated by various independent sources, it can be assumed that in general the probability density function of their sum also approaches that of the Gaussian. The effect of the amplifier noise can therefore be modeled as a Gaussian random process with a noise variance.

9.3.5.1 ASE Noise Model

The ASE noise, considered as a random noise, can be modeled by generating complex spectral components whose real and imaginative random variables follow a Gaussian probability density function with the variance P_{ase} corresponding to the average noise. The ASE noise spectral power can be estimated over a bandwidth Δf as

$$P_{ase} = n_{sp}(G-1)hf_s\Delta f = N_0\Delta f \tag{9.19}$$

where it is assumed that a sufficiently large bandwidth for ASE detection is used that is sensitive within the spectral region 1480–1605 nm; G is the small signal amplification gain. To model the ASE noise, a Gaussian wave with the ASE mean power and variance are used, which are described in the next section.

9.3.5.2 Other Noise Sources

In practice, an optical band-pass filter (BPF) is usually incorporated in the EDFA to reduce the ASE noise contribution. ASE-induced noises are contributed by two noise sources: (1) spontaneous–spontaneous beat noise: sp–sp beat noise is caused by the self-interference of the generated spontaneous photons and is given as $P_{sp-sp} = 2n_{sp}^2(G-1)^2e^2\Delta f$; and (2) signal–spontaneous beat noise: s–sp beat noise is produced by the beating between the signal photons and those generated by spontaneous emission and is given as $P_{s-sp} = 4e^2\dfrac{P_{in}}{hf}n_{sp}(G-1)G$.

As the s–sp noise dominates the ASE noise, we have

$$N_{s-sp} = 4e^2N_{in}n_{sp}(G-1)G \tag{9.20}$$

9.3.6 Simulation Model

This section gives the evaluation of the numerical data resulting from our EDFA simulator and EDFA module, experimental results are compared with those generated by our models. The principal advantage of the Simulink EDFA simulator is its versatility and ease of use. These results are by no means exhaustive.

According to the simulation modeling principle from the previous section, here we focus on how to implement EDFA simulation to MATLAB and Simulink. This section documents the Simulink simulation design and developments, explaining how each of the blocks within the simulator are brought together. Figures 9.14 and 9.15 illustrate the EDFA simulator and EDFA modules and the associated monitoring equipment. The explorer views of the subsystem within the Simulink model are available to users on request [18].

9.3.6.1 Simulator Design Outline

Simulink can be easier for a user unfamiliar with the programmer's methods or general programming techniques, i.e., those of operators or system technicians. This reduces the likelihood of human-induced errors when modeling the system of choice, which is important in the rapidly changing area of information transmission.

The Simulink simulation model described in this chapter is basically divided into two parts: (1) EDFA simulator and (2) EDFA module. The EDFA simulator simulates the details of amplification along the EDF for investigating and scoping how the signal can be pumped along the fiber, also observing the effect on gain modulation during the amplification. The EDFA simulator development is based on the EDFA dynamic modeling. And the EDFA

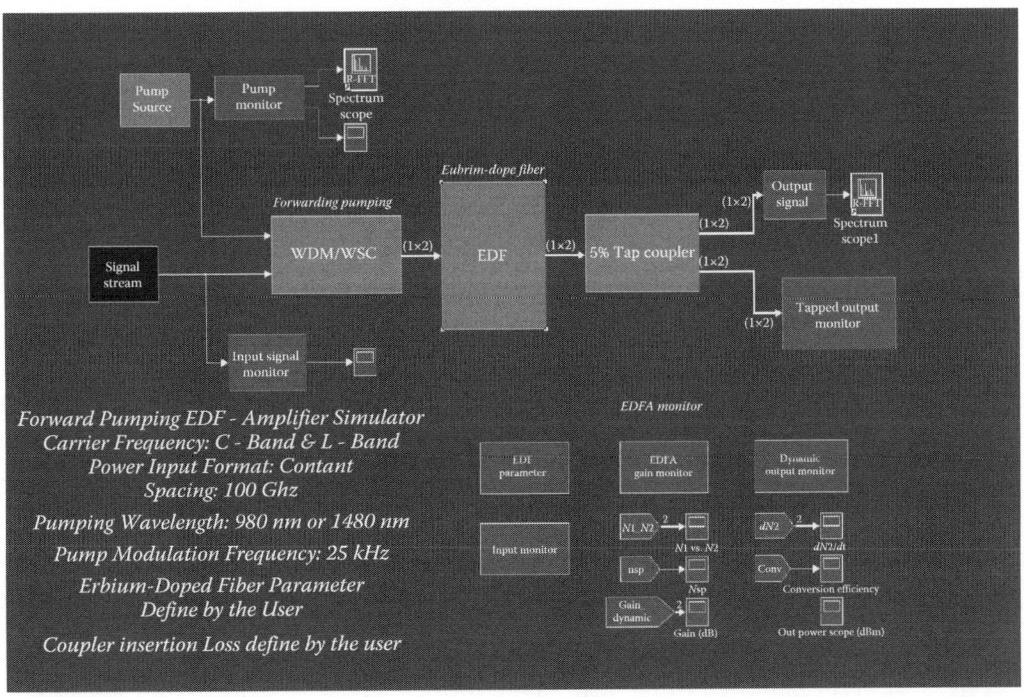

FIGURE 9.14
EDFA simulator model amplifier: general layout.

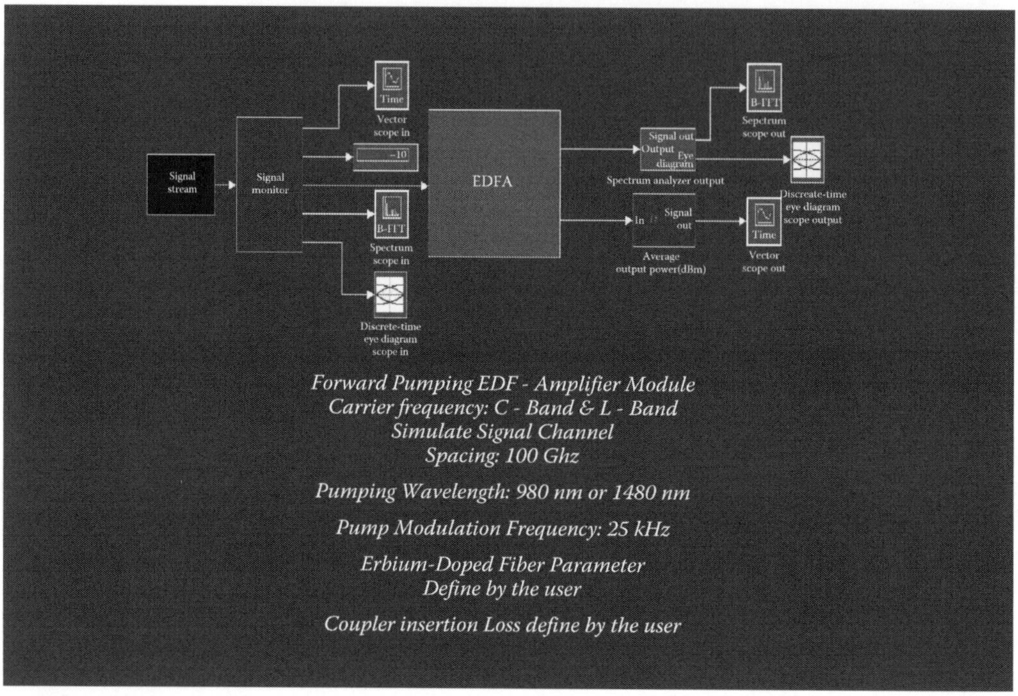

FIGURE 9.15
EDFA module amplifier and monitoring systems.

module is used with the whole optical system simulation for predicting the amplification output and its noise scoping. The EDFA module development is based on the EDFA steady state modeling. Thus, the functions of these two simulations is different. The EDFA model can be easily combined with the transmitter simulator and the fiber simulator to form an optical communication link.

9.3.6.2 Simulator Design Process

Intrinsically, MATLAB and Simulink are very powerful design tools [19–21]. The initial task of the model setup focuses on how to simulate the pumping schemes of the EDF. This is the primary goal, and also the principal part of our simulator. The simulator is divided into a number of stages. The flowchart shown in Figure 9.16 outlines the design process.

9.3.6.3 Simulator Requirement

In this section, the requirements related to the basic issues, such as software compatibility and simulation efficiency, of the MATLAB and Simulink softwares are described. First, the EDFA Simulink simulator is designed based on Simulink and MATLAB 7. Running the simulator in MATLAB and Simulink are the initial requirement. As there are many versions of MATLAB and Simulink available, running an incorrect simulator version may cause bugs/hangs or jams.

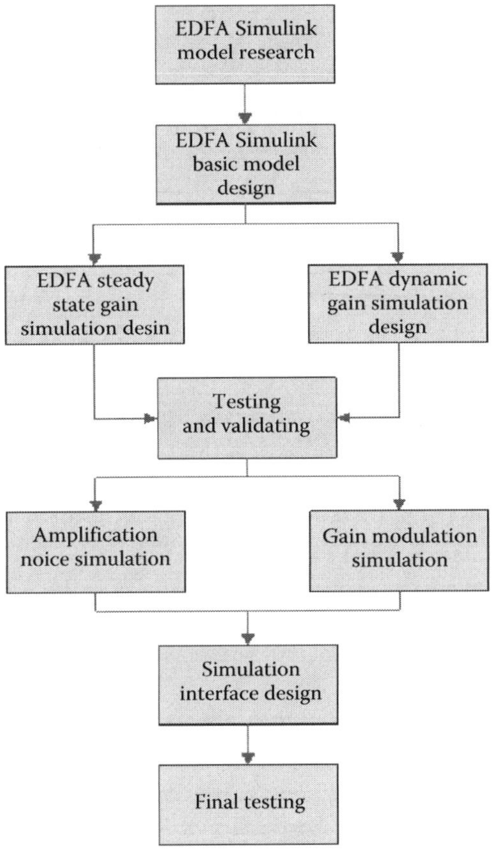

FIGURE 9.16
Simulation design flowchart.

Second, the blocksets of the Simulink library must be included and activated, the Simulink simulation can then be opened. The specific blocksets needed to assure the operation are Simulink DSP blocksets, Simulink Sink, Simulink Communications blocksets, and Simulink Extras blocksets.

9.3.6.4 Simulator Design Assumptions

9.3.6.4.1 Sampling Time Assumption

For designing the simulator, it is decided that a discrete rather than continuous sampling method is employed, primarily for practical reasons. In particular, a discrete sampling system would decrease the processing power required to execute simulations and reduce the execution time.

According to the well-known sampling theorem, all data are sampled at or above the Nyquist frequency, that is, all sampling is performed at least at twice the highest frequency within the simulator (the sampling frequency applied for this chapter was up to seven times the highest frequency). The use of discrete sampling periods ensures the synchronization of various system components as the data signal is propagated, an important consideration in optical systems. The considerations of optical transmission system parameters show the highest frequency to be that of the carrier waveform. For the purpose of this model, the carrier frequency is located in the C band and the L band, which are at around 0.2 PHz (Peta-Hz). Thus, sampling within the system needs to be performed at seven times this rate, i.e., at approximately 1.4 PHz, in order to ensure signal aliasing does not affect sampling and also ensure more accurate results; all blocksets require a discrete sampling time at which to generate data. Unfortunately, this sampling rate is very high and may cause slow execution, thus increasing execution time.

9.3.6.4.2 Signal Streams

Signal streams in the EDFA simulator are not defined during the simulation; the EDFA simulator assumes the average signal power. Thus, the signal stream simulating in the EDFA simulator would be a constant value with a superimposed noise on Gaussian pulses. Hence, the output power also is the average signal power.

The signal stream in the EDFA module is developed as the PSK (phase shift key) modulation signal with Gaussian noise. The carrier is defined to be a 192.5 THz sinusoidal waveform. The data stream is assumed to be at a 10 Gb/s (or even higher to 40 Gb/s) binary pulse.

9.3.6.5 Further Assumptions

In the development of the EDFA dynamic gain simulator, a number of assumptions have been made. (1) As the pumping wavelengths of 980 and 1480 nm can provide a high gain amplification and lower noise, we assume that the EDFA can be pumped with either a 980 or a 1480 nm source. (2) Since the transient lifetime from the excited state to the metastable state is sufficiently short, we can assume that the erbium atoms operate at only two levels relevant for the amplification process. (3) The effective area of the erbium-doped active region is small compared to that of the optical mode at each wavelength of interest. This assumption for the erbium-doped core allows us to treat the overlapping factor to be a constant value, independent of energy level and optical power. (4) As the amplified spontaneous emission is randomly generated, thus, in the simulator, we assume that the stimulated ions propagate in one direction, to estimate the average mean ASE noise power. (5) Finally, no reflection occurs during the amplification process.

Assumptions for the modeling of the steady state EDFA gain are also made. (1) As the pumping wavelengths of 980 and 1480 nm can provide a high amplification gain and lower noise, it is assumed that the EDFA is pumped with either a 980 or a 1480 nm source. (2) Since the transient lifetime from the excited state to the metastable state is short, we again assume two levels of operation of the doped ions. (3) As the excited-state absorption only affects the pump absorption along the fiber, it can be neglected. (4) The response of each individual ion in the collection is equally broadened, at a picosecond time scale. (5) The ASE noise is assumed to be small so that it does not saturate the EDFA amplifier. (6) The effective area of the erbium-doped active region is small compared to that of the optical mode at each wavelength of interest. This assumption for the erbium-doped core allows the assumption of the constant overlapping factor, which is independent of the energy level. (7) Finally, as in the steady state model, the population increment, ∂N_2, is assumed to be zero.

9.3.6.5.1 System Initialization

To correctly operate the simulator, an initialization process is required. Written as an m-file, this loads parameters and variables required for the simulator operation. Titled "initialization," the user needs to run this file before starting the Simulink simulation. Failing to do so will result in an error warning. It is within the initialization file that the user can modify system parameters. There are several data to be defined in the initialization file. Table 9.4 gives the input data that can be defined before the simulation. Table 9.5 gives the glossary for input signals.

The use of the initialization file simplifies the operation for users. All variable parameters are contained here and automatically updated in the Simulink platform upon execution (saved in the MATLAB memory called "workspace"). Not incorporating the initialization file would require the user to input the same variables in multiple blocks, thus increasing the possibility of errors and complicating the operation of the simulator.

9.3.7 EDFA Simulator

The EDFA simulator is combined with various segments, as illustrated by the prototype in Figure 9.14. Each block presents a different segment of the EDFA simulator. For example, the input signal stream is represented by a block, which is named [Signal Stream],

TABLE 9.4

Parameter Set for the Erbium-Doped Fiber

Physical Meaning	Symbol Defined in Initialization File
Erbium-doped fiber parameters	
Core radius of EDF	r (μm)
Overlap factor of EDFA at wavelength (k)	Gamma
Ion density of EDF	p
Peak absorption at 1530 nm	Peak_absorsion
Peak absorption at 980 nm	Peak_absorsion_pump
Length of the fiber	L (m)

TABLE 9.5

Input Parameter Set

EDFA Input Initialize	Symbol Defined in Initialization File
Input signal power	Signal_Power (dBm)
Input pump power	Pump_Power (mW)
Input wavelength	Index
Pump wavelength	Pump_wavelength
Number of channel inputs	N
Pump modulation index	mod
Pump modulation frequency	Mod_freq

and each block consists more than one subsystem inside to check the design just by double-clicking the block. The EDFA in the Simulink simulation is implemented as these subsystems and they are called from a main program. The following text demonstrates the input and output ports in the EDFA simulator and the EDFA module. The structure of the Simulink model of the EDFA simulator is shown in Figure 9.17, in which the central part is the EDFA simulator (Figure 9.18).

9.3.7.1 Using the EDFA Simulator

Input:
(1) Signal stream—power of the signal (includes C-band and L-band signals) and (2) pump source—power of the pump (includes pump modulation, 980 and 1480 nm pump sources)

Output:
All the output displayers have been put into the block named ### Monitor, and readers can check the output values by opening the monitor block. (1) EDF parameters (EDF parameters monitor)—the parameters of the EDF, such as radius and ions' density; (2) EDFA gain monitor—dynamic gain and steady state gain scope (all the gain data have been put in order [Ascent]); (3) EDFA dynamic gain monitor—output signal power, output pump power, average ASE power, noise figure, SNR, other noise powers; and (4) also, scopes display the evolution of the signal. Segment by segment of the simulator can be illustrated in the rest of this section.

9.3.7.2 Signal Data Stream Modeling

As illustrated in Figure 9.19, the signal stream basically models the C-band and L-band lightwave signals. As we have assumed the system using a 100 GHz optical frequency spacing (0.8 nm) between lightwave carriers, the C band is also assumed for the wavelength window of 1525–1565 nm. Thus, there are around 51 channels that can be used as optical carriers over this band. We can see from Figure 9.20 that there are five signal power blocks with bold lines except the last one. The bold lines are multidimensional vector time series associated with 10 channels of each signal block. The wavelength section is used for the conversion of the signal power in mW into the number of photons (photon energy per bit period). The input signal power of each channel has been defined as a constant mean power.

The value of the mean signal power is defined in the file [Simulation_initailise.m]. The signal carrier frequency is ultrahigh, around 192.5 THz. If we inject a signal generator with a real 192.5 THz signal (e.g., sinuous wave at 192.5 THz), the Simulink generates the output based on each sampling point of this signal wave, and it may take a long time (maybe more than 2 days!) to generate the output result, which is, naturally, very inconvenient. On the other hand, it predicts the average output power after the amplifier is good enough for the design. Therefore, an average mean input signal power can be used under our considerations.

9.3.7.3 Pump Source

Pump power also has been set as a constant mean power for simulation. We have already identified that there are three different EDFA configurations: forward pumping, backward pumping, and dual (or bidirectional) pumping. Only forward pumping is described in this section as the others can follow easily. The Simulink model for the pump source is shown in Figure 9.21.

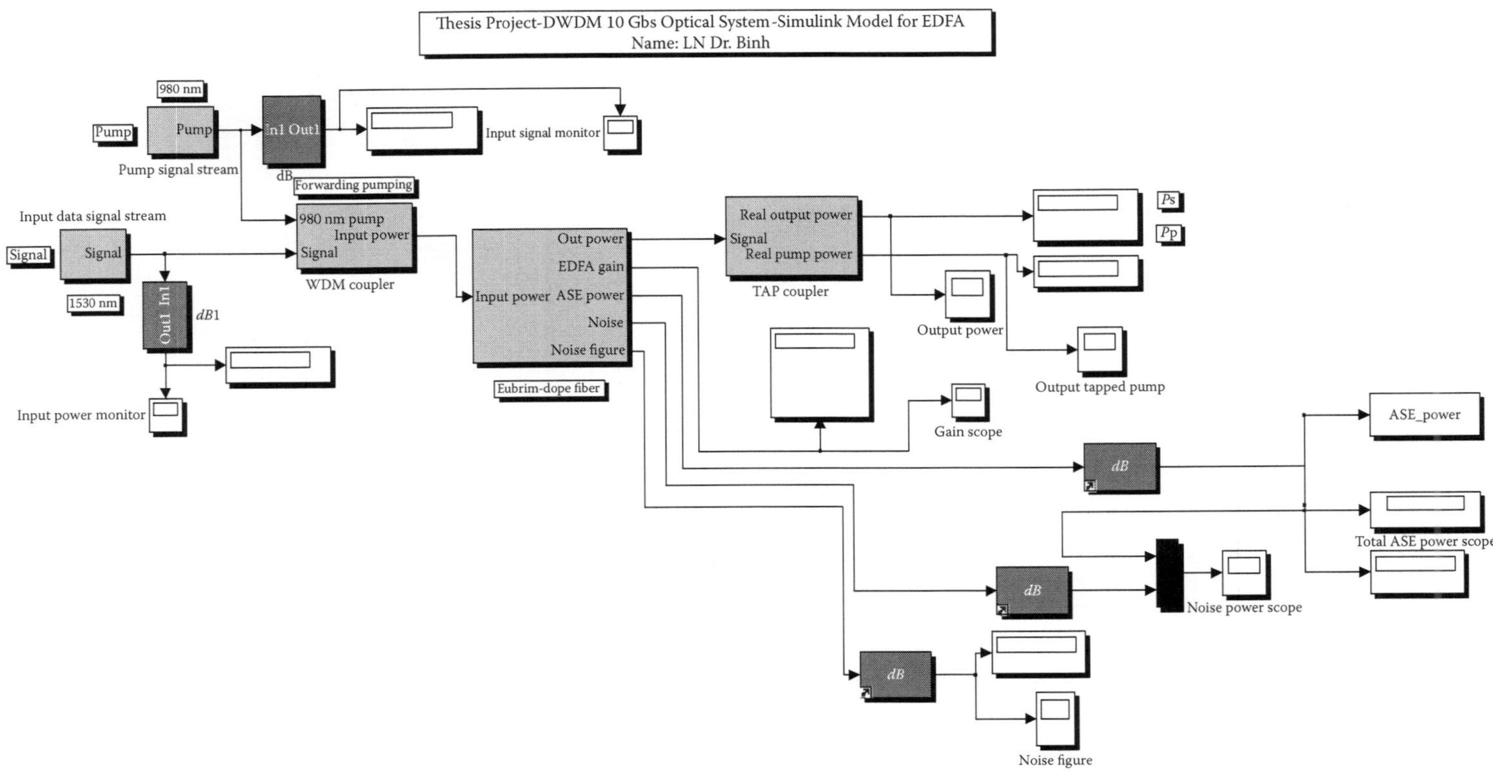

FIGURE 9.17
Structures of the EDFA simulator using Simulink.

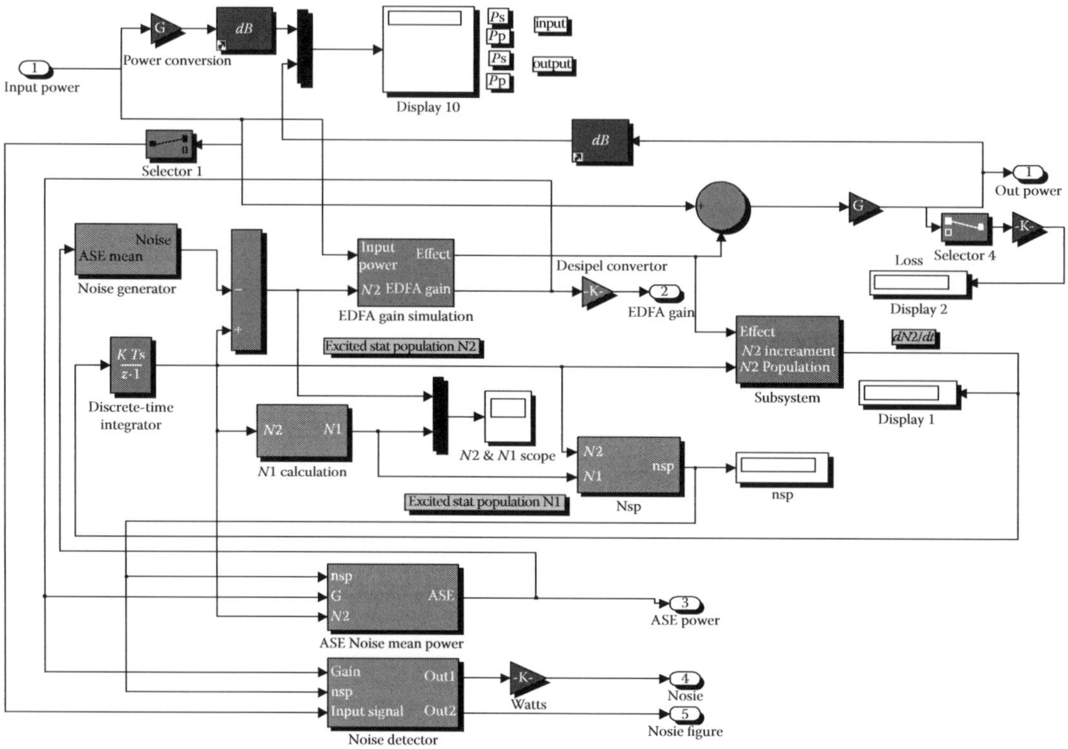

FIGURE 9.18
Structures of the EDFA simulation central block of Figure 9.17.

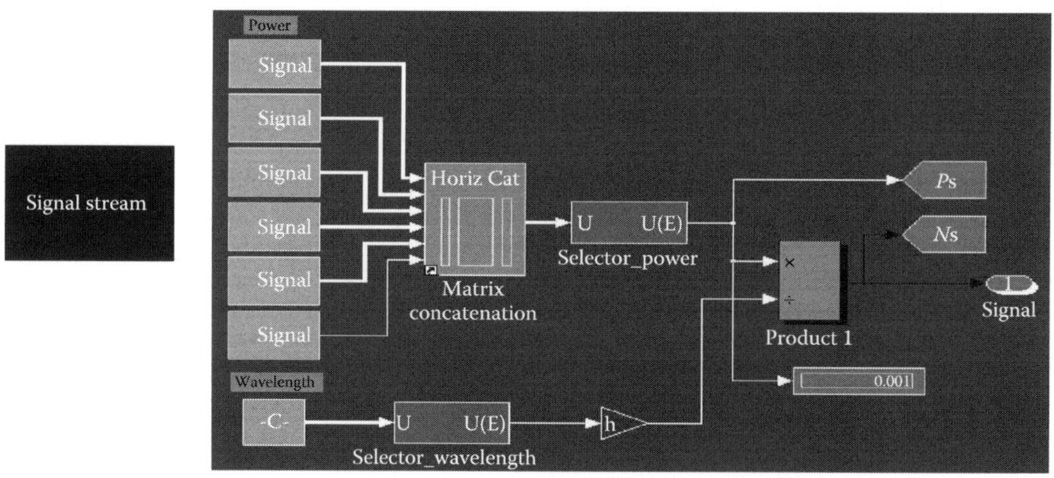

FIGURE 9.19
Signal stream.

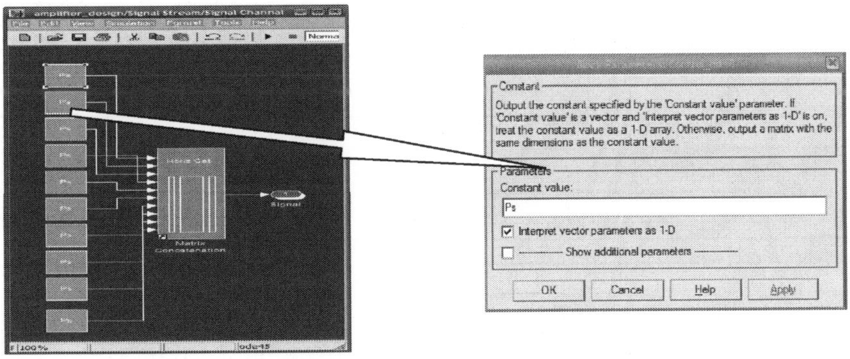

FIGURE 9.20
Signal block.

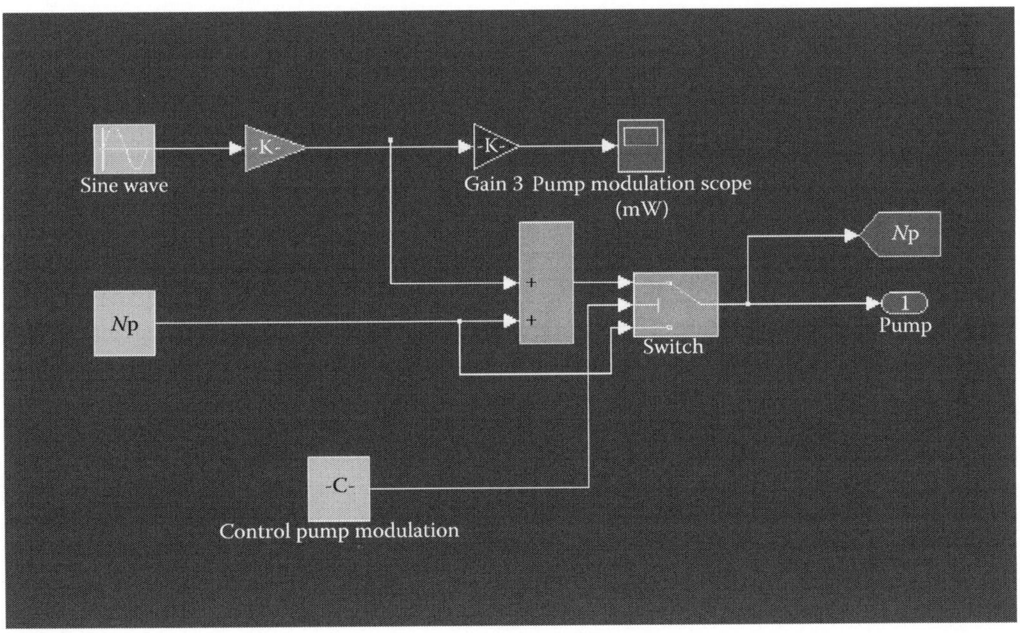

FIGURE 9.21
Pump source.

9.3.7.4 Pumping Wavelength

As described above, there are two pump wavelengths that can offer a high amplifier gain. In our simulation, wavelength modeling is implemented within the initialization file. The instruction code for setting the event when different wavelength channels propagate through the EDFA is as follows:

Algorithm:

If Pump wavelength = 980e−9
Absorption factor 980 nm
Emission factor 980 nm

Else if Pump wavelength = 1480e–9
Absorption factor 1480 nm
Emission factor 1480 nm
Else
Reset

According to the absorption and emission characteristics, the absorption and emission properties of the fiber change with different wavelength propagation. The absorption factor is normally given by the fiber manufacturer or is measured.

9.3.7.5 Pump Modulation

Pump modulation can be added and its effects on gain modulation can be investigated. Freeman and Conradi [2] use a dimensionless number called "modulation index" to measure the power in the modulated signal. Using this definition, the input pump time series has the form $P_p(0,t) = P_p(0) + P_p^0(0) m_p \cos \omega t$, where $P_p^0(0)$ is the initial mean pump power (in watts) and m_p is the modulation index. As the mean pump power given here has a constant value, it is added to the term $m_p P_p^0(0) \cos \omega t$ in the summation box, as illustrated in Figure 9.22.

$P_p(0) m_p \cos \omega t$ is the output of the sine wave generator in Figure 9.22. The gain labeled [M] is the modulation index, m_p. N_p is the constant mean pump power; here, the unit of the pump power is photons/second. The method for converting watts to number of photons/second has been explained in the previous section.

9.3.7.6 EDF Modeling

We have described theoretical EDF modeling. This section gives the Simulink simulation model design. Simulink simulation models are based on the rate equations given in Section 9.2. The characteristic of the EDF depends on two factors: the EDFA gain and the ASE noise characteristics. The gain characteristic depends on the absorption and emission factors at different wavelengths. Due to the fact that different EDFs provide different absorption and emission characteristics, simulating the absorption and emission spectra would be the first steps.

The existing absorption spectra given in [2,6–8,17] can be used. The absorption spectrum is illustrated in Figure 9.10. Although all the data from these references are different, yet the slopes of these spectrums are quite similar. According to this assumption, the normalization of these curves would be the first choice for the simulation of the spectrum. The simulation of the normalized spectra is shown in Figure 9.23.

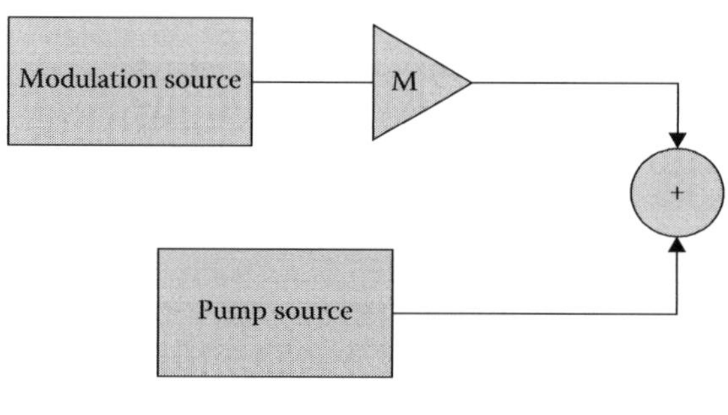

FIGURE 9.22
Pump modulation module.

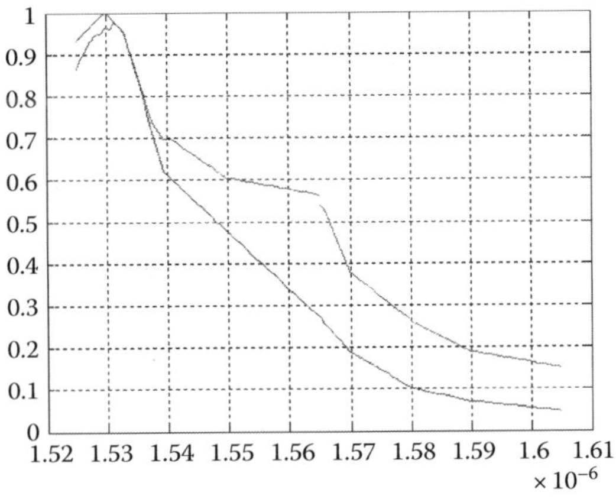

FIGURE 9.23
Absorption and emission factors of the normalized spectra (C band and L band).

The EDF simulation can be started after the simulation of these spectra. Figure 9.24 illustrates the prototype of the EDF simulation.

9.3.7.7 EDFA's Dynamic Gain Model

The equations for solving the dynamic gain model have been described in the above section. This section describes how all the blocks are integrated. As we have mentioned in Section 9.4, the gain of the EDFA is based on the equations

$$G_p = \frac{\Gamma_p(\sigma_{1-3}+\sigma_{3-1})}{A} N_2 - \Gamma_p\sigma_{1-3}\rho L - \gamma_p G_p \text{ and } G_s = \frac{\Gamma_s(\sigma_{1-2}+\sigma_{2-1})}{A} N_2 - \Gamma_s\sigma_{1-2}\rho L - \gamma_s G_s.$$

In our simulation, the Simulink model representing these expressions is given in Figure 9.25.

In Figure 9.25, B_s and B_p are expressed by the matrix of $\Gamma(\sigma+\sigma)/A$ and, similarly, C_s and C_p are expressed by the matrix of $\Gamma\sigma\rho L - \gamma G$. Due to the absorption and emission characteristics, different wavelength signals propagating through the EDF would cause a change in the absorption and emission factors. So, C_s and B_s *act as* the memory locations that store all the absorption and emission values. The following selector is an artificial intelligent block, which can select the corresponding values for matching the wavelengths. The selector block works by outputting with selected reference indices that form the B_s matrix, like $[Bs_1 \quad Bs_2 \quad Bs_3 \quad \cdots \quad Bs_{51}]$ Selected [index 1] => Output – Bs_1. Accordingly, the absorption and emission factors change with respect to different wavelengths (Figure 9.26).

As shown in the block diagram of Figure 9.25, the loop continues the execution till the difference of the input power and the output power is the same as that of N_2/τ, which means N_2 is saturated. Thus, the gain remains in the steady state. The data inside the discrete time integrator block sets the step time for each increment; here, we set 3.25e–7 as the sampling step. The output of the model is given in Figure 9.27.

9.3.7.8 EDFA's Steady-State Gain Model

At steady state, $dN_2/dt = 0$. The Simulink model can be illustrated as in Figure 9.28.

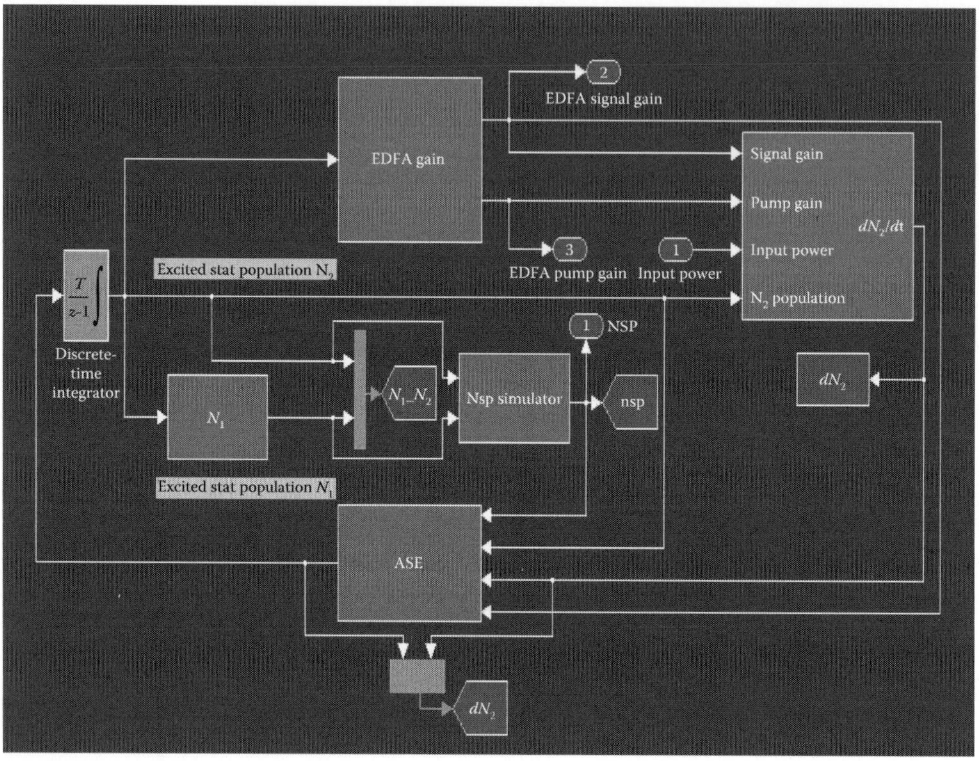

FIGURE 9.24
EDFA simulation block.

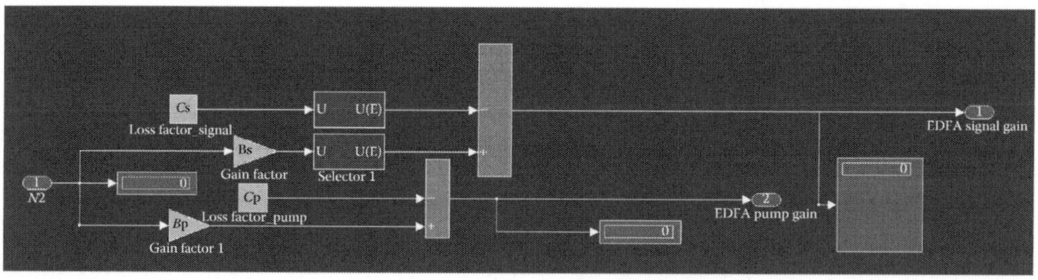

FIGURE 9.25
EDFA's dynamic gain modeling corresponding to Figure 9.24.

Here, the signal gain is equal to $B_s * N_2$ (*saturate*) − C_s and the pump gain is equal to $B_p * N_2$ (*saturate*) − C_p. The Simulink development is similar to EDFA dynamic model. The output would be the same as given in Figure 9.27.

9.3.7.9 Population Inversion Factor Modeling

- N_1 simulation

Recall the three state populations and the assumption of N_3 to be zero. The relation between N_1 and N_2 is then $N_1 = N_{total} - N_2$ (Figure 9.29).

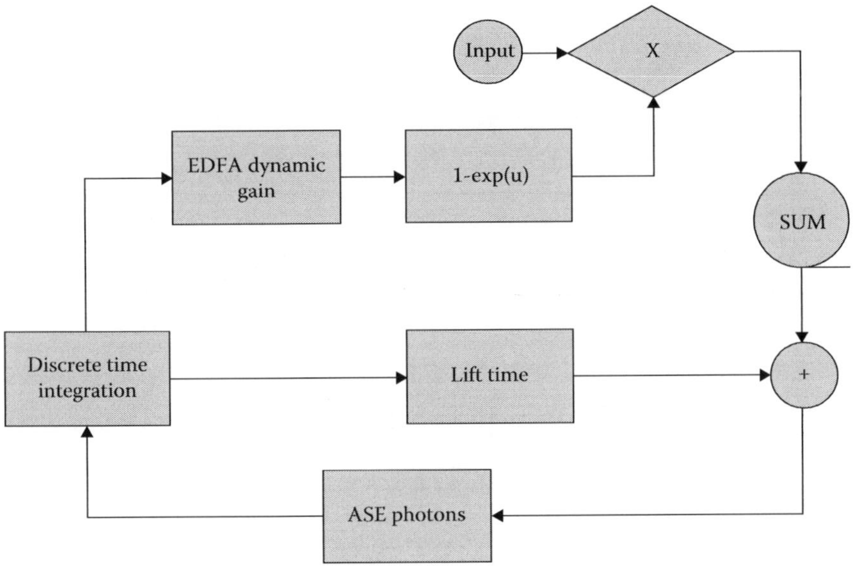

FIGURE 9.26
dN_2/dt simulation block diagram.

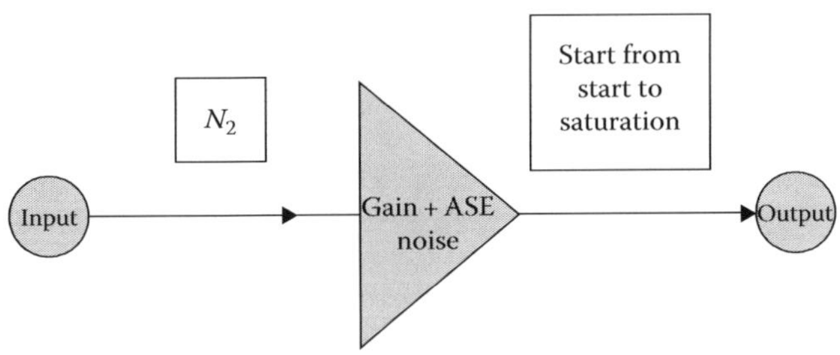

FIGURE 9.27
Output algorithm.

- N_{sp} simulation (Figure 9.30)

According to the population inversion factor equation, N_1 would be equal to N_2. This would lead to a result approaching infinity. Therefore, the [if statement] could make the result limited under 10. Also, the population inversion factor would be greater than 10 in practice.

9.3.7.10 Amplifier Noise Modeling

In the previous section, we have mentioned the method for simulating the amplification noise. Therefore, to simulate the ASE noise, determining the average mean ASE power will be the first action. For dynamic gain modeling in Simulink, the population amplification process is derived in three serial sections (Figure 9.31).

Thus, the calculation for determining the ASE is conducted in three different statements. In the simulation, the algorithm for simulating the ASE mean power is given in Figure 9.32.

After the determination of the ASE mean power, the Gaussian assumption of the ASE mean power is used and the variance is set to model the ASE power and the number of randomly generated photons, as shown in Figures 9.33 and 9.34.

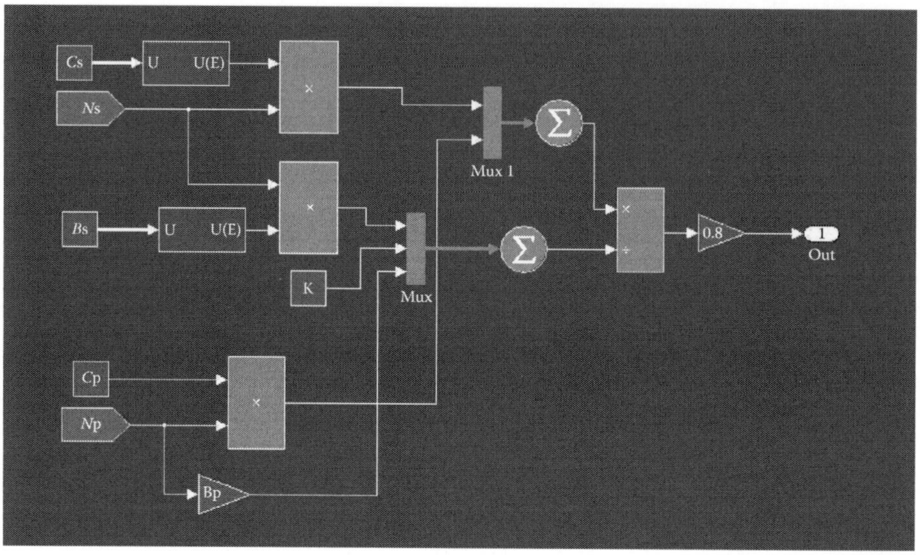

FIGURE 9.28
EDFA steady state N_2 model.

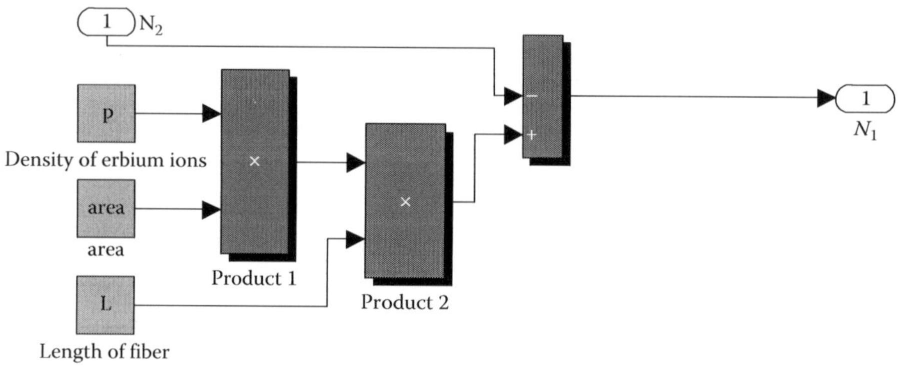

FIGURE 9.29
Representation of N_1 in the simulator.

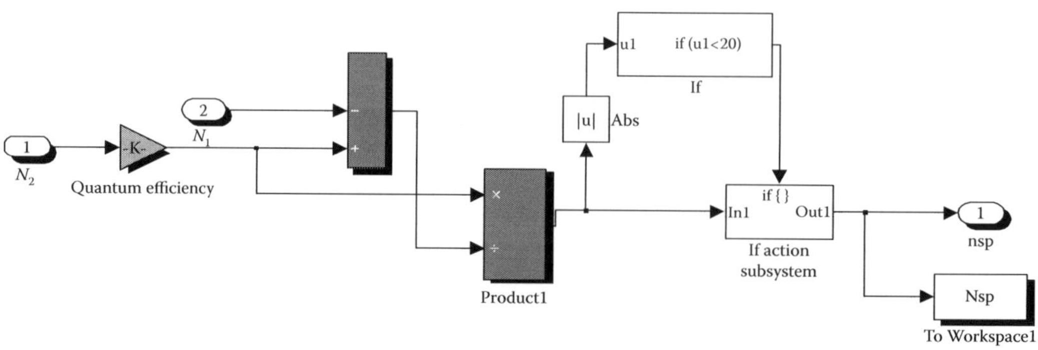

FIGURE 9.30
Representation of n_{sp} in the simulator.

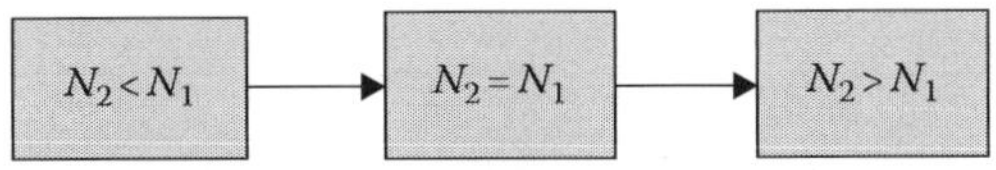

FIGURE 9.31
Amplified population in sequential process.

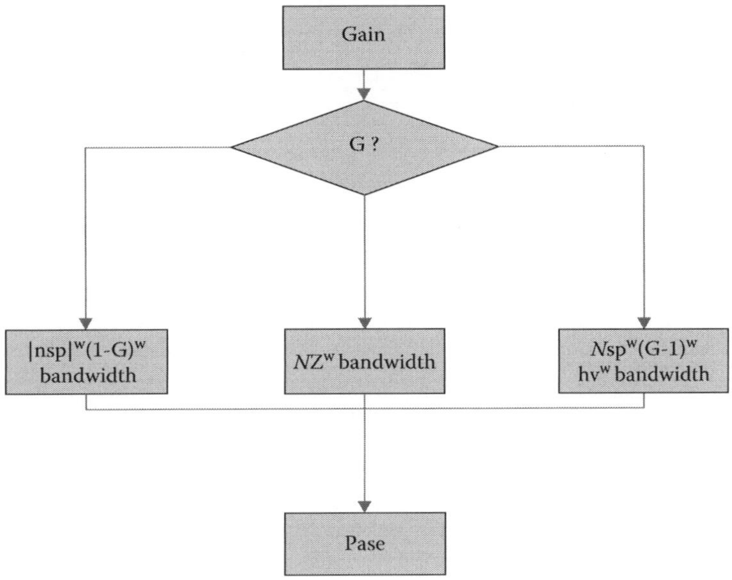

FIGURE 9.32
Modeling flowchart for the ASE noise power; $G<1$.

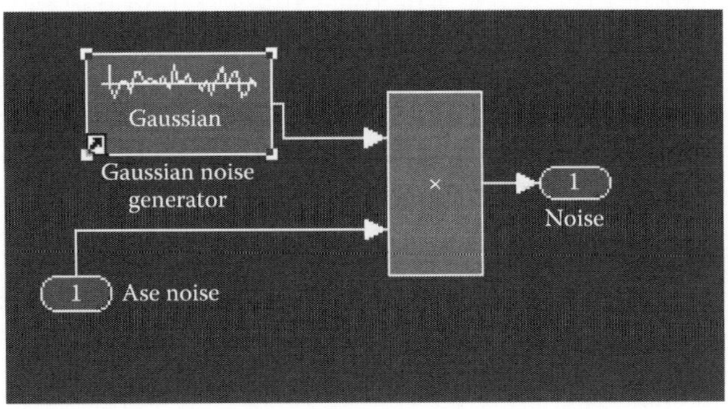

FIGURE 9.33
ASE model.

9.3.7.11 Simulink® EDFA Simulator: Execution Procedures

The simulator consists of three major parts: the initialization file [EDFA_initialisation.m], the EDFA simulator [EDFA_simulator.mdl], and the EDFA module [EDFA_module.mdl].

Typical steps for executing a simulation:

1. Access MATLAB 7.0 R14 or any latest version. Open the initialization file from its stored disk, use the file, and then open it as normal.

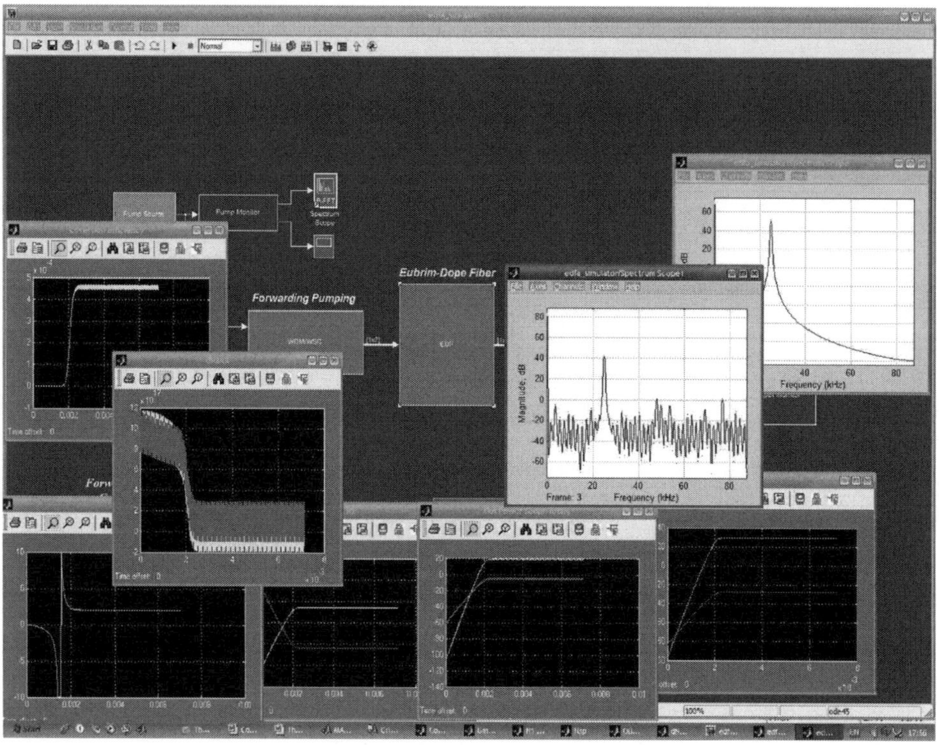

FIGURE 9.34
Samples of the screen shots of the EDFA simulator.

2. The step upon brings the MATLAB-coded initialization file on the screen. Refer to the appendices for a hard copy of this file. This file allows the user to define all variables pertaining to the optical simulator. The user is free to modify any value within this file. The file has been commented so as to instruct the user and future designers of the simulator's operation.
3. To run a simulation, enter the required parameter values; then, from the debug menu, select save and run. This will run the file and load the relevant values into both the MATLAB workspace and the Simulink simulator. The simulator should also open in the Simulink workspace. To analyze the initialized system, simply open the Input Monitor block and EDF Parameter; the initialized input will be present there.
4. To inspect the result of the EDFA simulator, double-click the Output Monitor block or either open the scope to see the dynamic evolution. Indeed, all results are saved in the workspace for further investigation.
5. To inspect the results of the EDFA module, scopes are popped up automatically; if not, just double-click the scope.

Notes: As the EDFA module can combine with other simulation blocks, thence, the input data stream can be replaced by any other device simulator block. To increase the execution time, the run type in [normal mode], [accelerating mode], and [from outside] in the option menu can be changed.

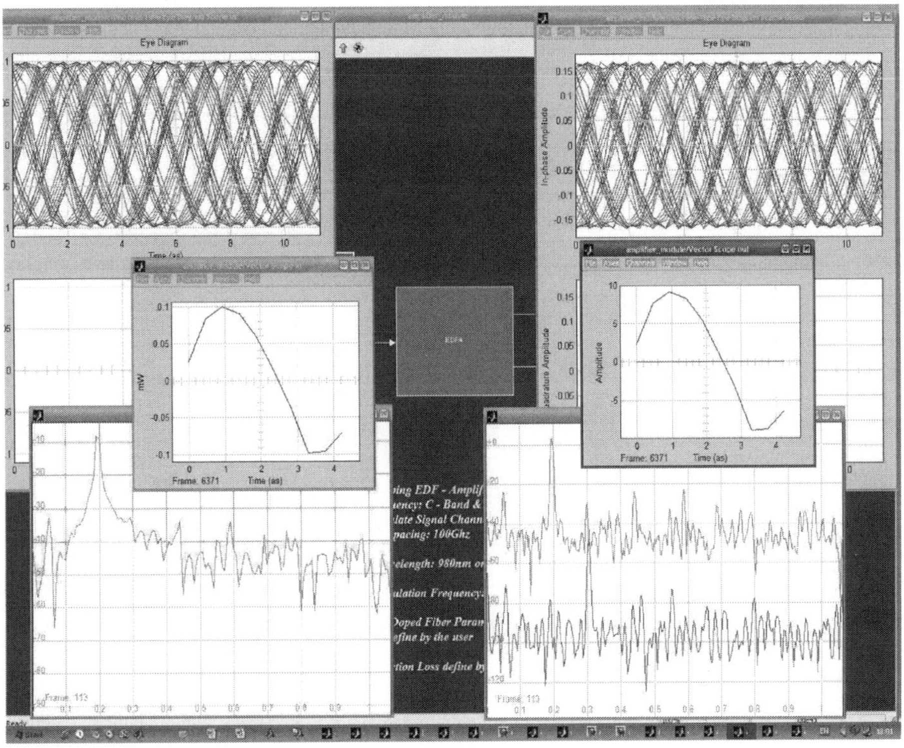

FIGURE 9.35
Samples of the screen shots of the EDFA simulator.

Typical simulator screen shots of the Simulink EDFA simulator are given in Figures 9.34 and 9.35.

The simulated results are based on practical EDFA parameters constructed in our laboratory and some commercial amplifiers operating in the C band. Table 9.6 gives a set of the parameters employed in the simulator, and the power generated at the output of the EDFA is plotted versus the input signal power with the wavelength of the input channel as a parameter, as shown in Figure 9.36. The saturation is set at around −14 dBm, as expected (Figure 9.37).

Table 9.6 gives the testing parameters to obtain the results illustrated in Figures 9.36 to 9.48.

The EDFA gain obtained from the simulation agrees with that of the experiment; the only discrepancy is during low signal input power—this might be due the internal setting of the commercial device, which is unknown. All the other factors, such as the absorption and emission spectra, in the commercial EDFA may be slightly different from the simulator (Figure 9.38).

The spectral gain is dependent on the stimulated-emission factor. As the population inversion rate changes with the signal wavelength, the gain is peaked at 1528 nm (Figure 9.39).

The gain does not increase infinitely with the increasing pump power, as the ions' density in EDFs is constant. Thus, the pump power just produces the energy for pumping ions to the higher-energy state. Thus if all the ions have been pumped to the high state, then even more pump power, the number of simulated photons still remain the same, hence the gain saturation as illustrated (Figure 9.40).

TABLE 9.6

EDFA Simulator Parameters

Single Channel	C Band	EDF Parameter		
Pump wavelength (nm)	980	MFD = 5.9 μm at 1550 nm		
EDF length (m)	16	Cutoff wavelength	940 nm	
Pump power (mW)	195	Peak absorption	6.09 dB/m	at 1530 nm
Without gain modulation			5.24 dB/m	at 979 nm

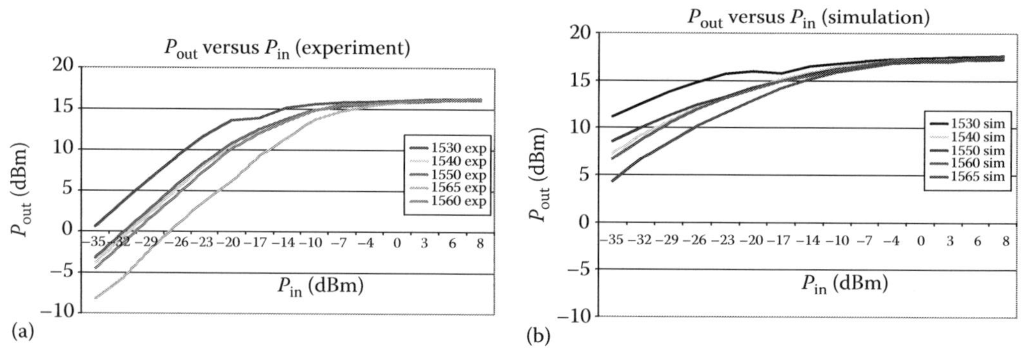

FIGURE 9.36
(a) Experimental results: EDFA output power versus input power level with channel wavelength as a parameter. (b) Simulation results: EDFA output power is plotted against the input signal level with the input channel wavelength as a parameter.

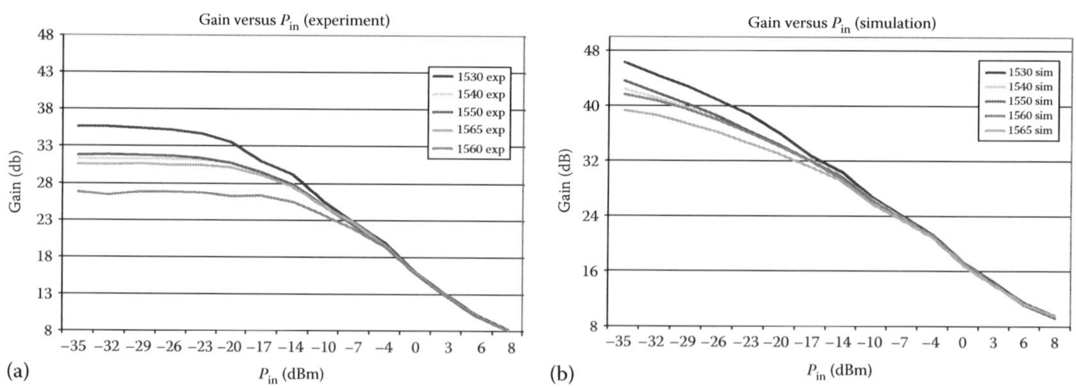

FIGURE 9.37
(a) Experimental results of the amplification gain. (b) Simulation results of the amplification gain.

9.3.7.12 Amplification in the L Band

We can compare the L-band signal with the C-band signal and tabulate, as shown in Table 9.7. The gain is plotted over the C band and the L band, as shown in Figure 9.41.

The pump power can be varied and the EDFA gain in the L band is plotted in Figure 9.42 with the parameters of the amplifier given in Table 9.6 (Figure 9.43) and with the variation of the input power.

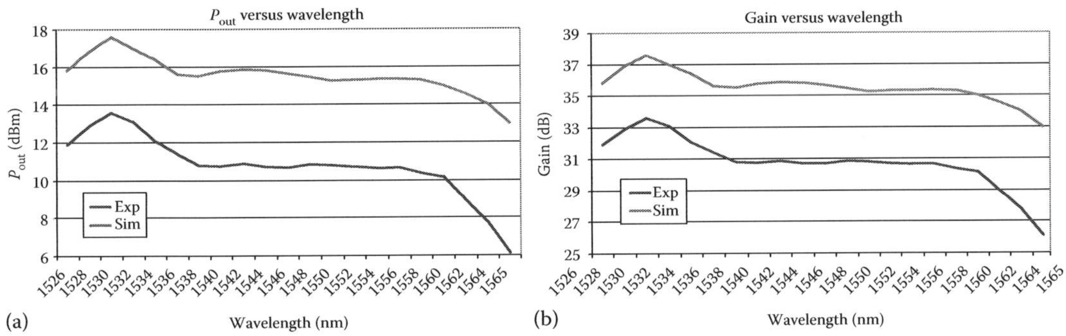

FIGURE 9.38
Spectral properties of (a) output signal power and (b) EDFA gain.

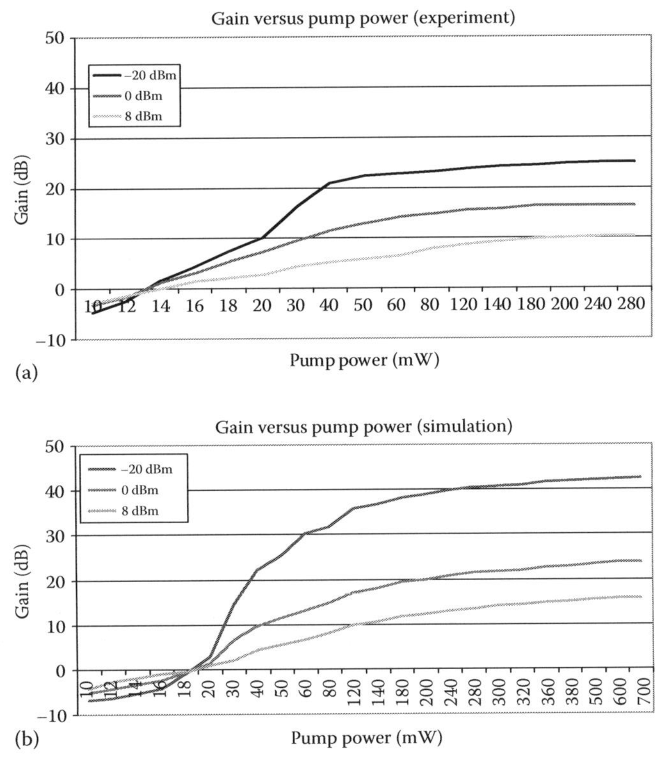

FIGURE 9.39
(a) and (b) Pump power plotted against the amplification gain with the signal power as a parameter.

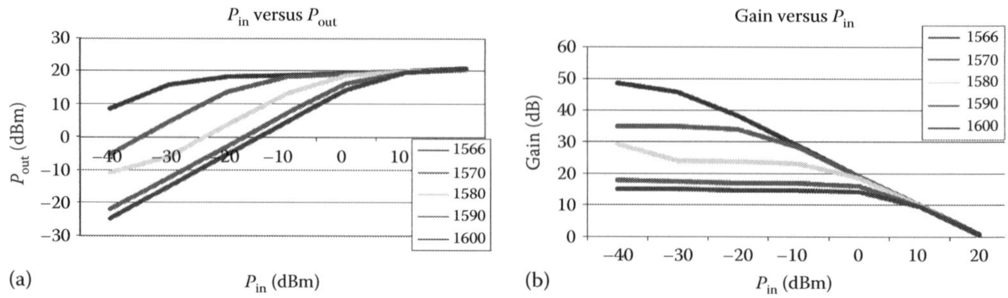

FIGURE 9.40
EDFA gain versus input signal power level for L band. (a) Output power versus input level. (b) Gain against input power level.

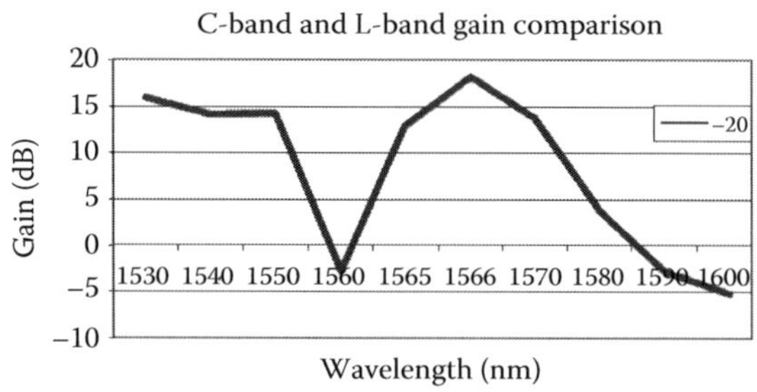

FIGURE 9.41
EDFA gain spectrum.

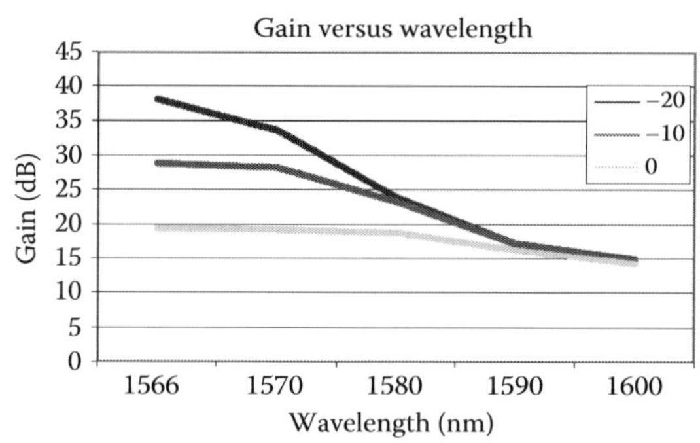

FIGURE 9.42
L-band gain spectrum.

TABLE 9.7
Signal Input and Output Power Levels at Different Wavelength Regions

	Wavelength				
Signal Level	**1530**	**1540**	**1550**	**1560**	**1565**
–20	16.01	14.22	14.3	–2.709	12.97
	1566	1570	1580	1590	1600
–20	18.24	13.8	3.838	–2.709	–5.115

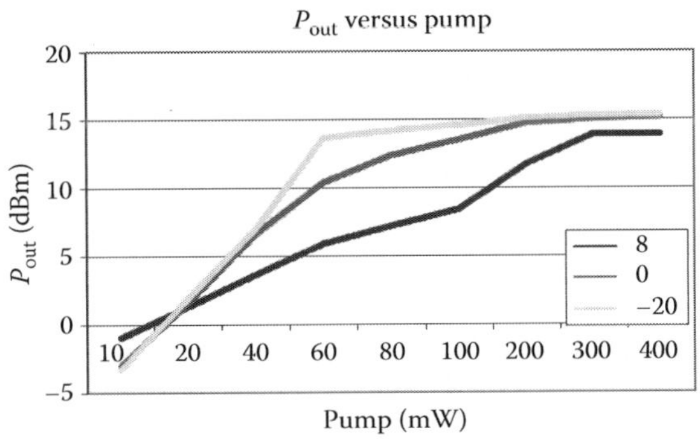

FIGURE 9.43
Output power of the EDFA as a function of the pump power with input power level as a parameter.

9.3.8 Multichannel Operation of the EDFA

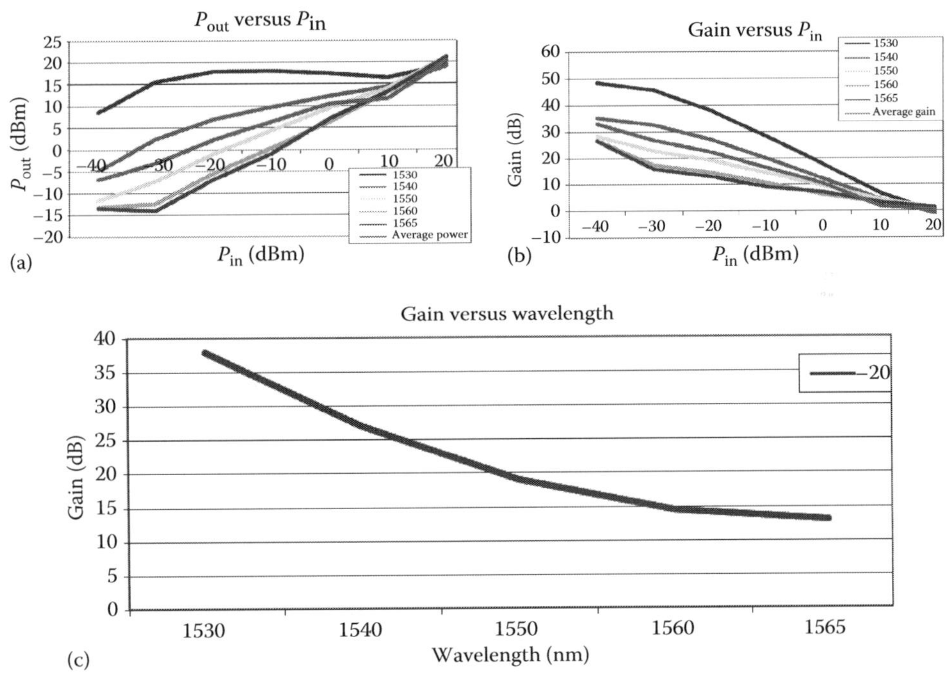

FIGURE 9.44
EDFA characteristics: (a) output power as a function of input signal power level for different channel wavelengths, (b) gain variation as a function of input level, and (c) gain spectra.

9.3.8.1 ASE Measurement

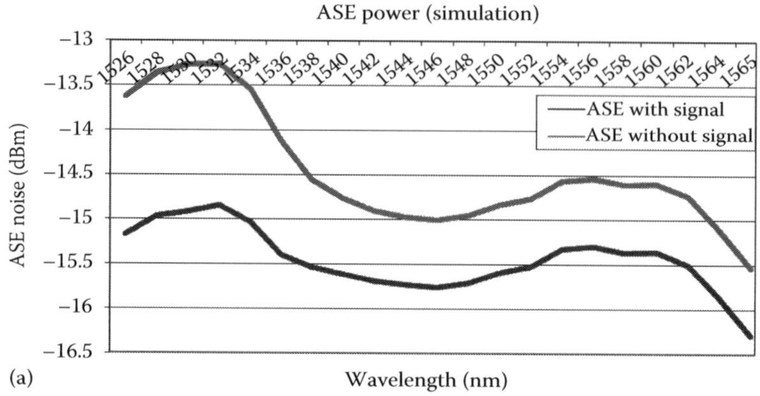

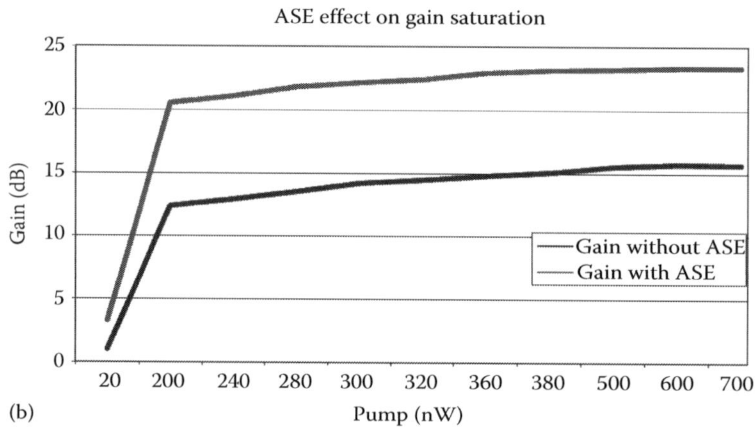

FIGURE 9.45
ASE noise: (a) spectral property and (b) as a function of pump power.

9.3.8.2 Pump Wavelength Testing

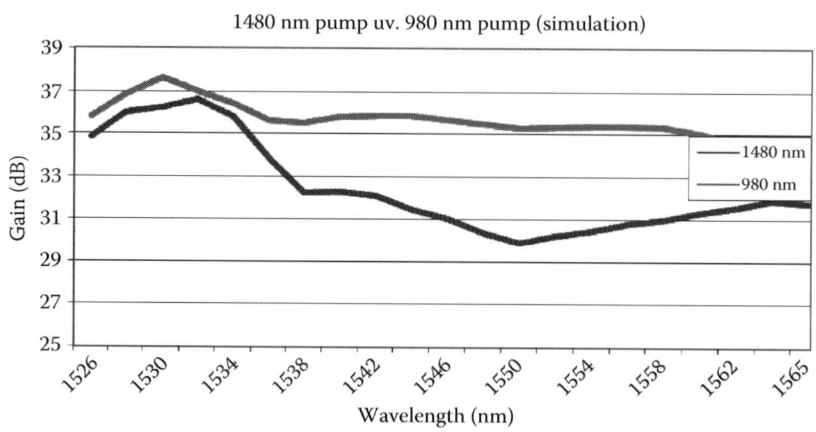

FIGURE 9.46
EDFA gain variation with pumps at 980 and 1480 nm.

9.3.8.3 Gain Pump Modulation Effect

Figures 9.47 and 9.48 depicts the variation of the output power of the amplifier under the cases with and without modulating the pump beam. Clearly the ripple of the output power happens when the pump beam is modulated. However the gain is the we use less average pump power.

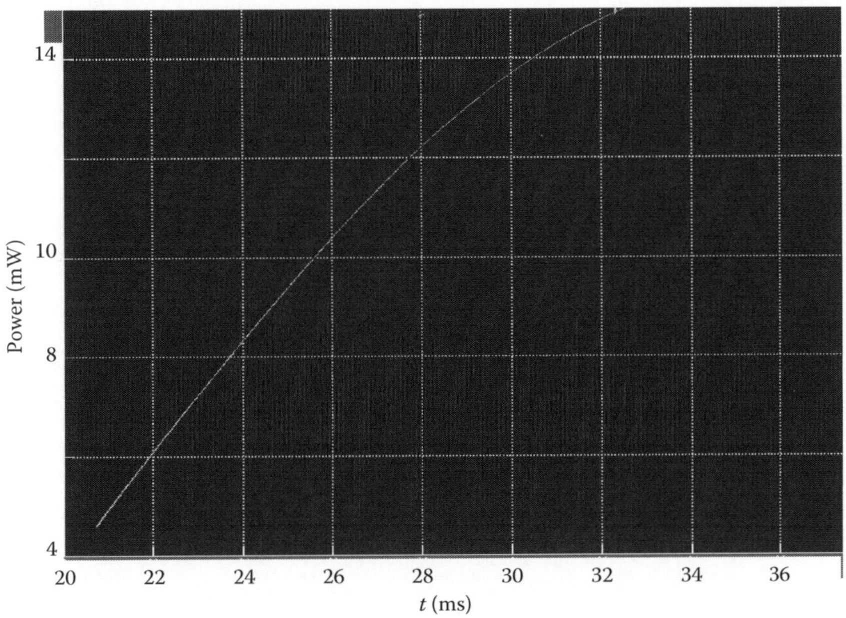

FIGURE 9.47
Output signal without pump source modulation (time dependent).

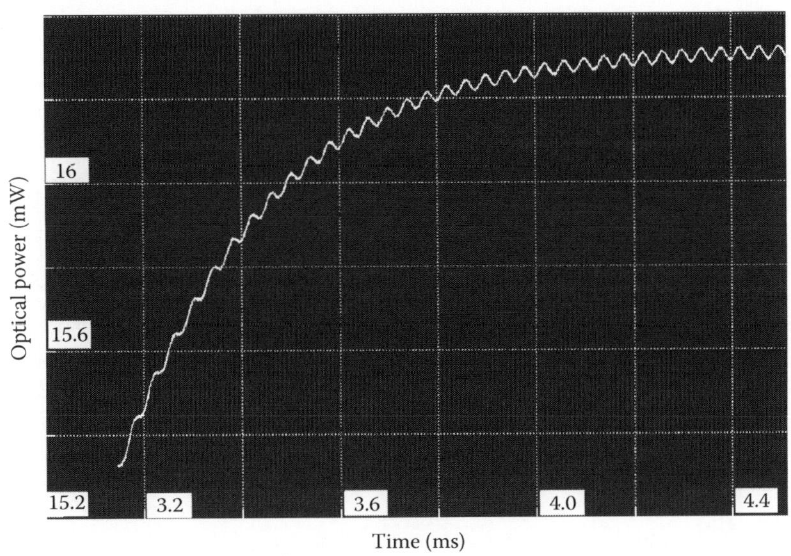

FIGURE 9.48
Output signal with pump source modulation (time dependent).

9.3.8.4 Samples of the Simulink® Simulator

This section shows the output scopes observed from the Simulink simulator. These can be divided into two subsections: EDFA simulation scoping and EDFA module scoping. Here, we demonstrate the operation of the Simulink EDFA simulator as an example incorporating simulated results. This provides the user with an example to follow when using the simulator. The example assumes that the system parameters are defined in the initialization file. These parameters are loaded using the run command in the debug menu. This allows loading the initialized simulation values in the MATLAB platform.

9.3.8.4.1 EDFA Simulator (Figure 9.49)

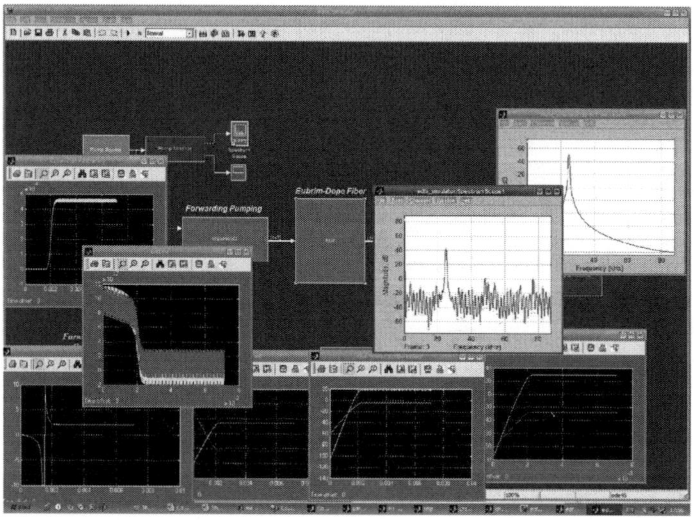

FIGURE 9.49
Samples of the screen shots of the EDFA simulator.

9.3.8.4.2 EDFA Simulator Inspection Scopes (Figure 9.50)

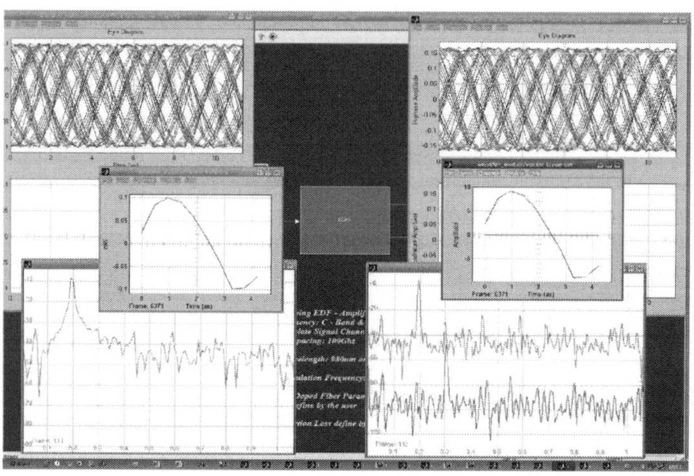

FIGURE 9.50
Samples of the screen shots of the EDFA simulator.

9.4 Concluding Remarks

This chapter demonstrates the successful implementation of the modeling of the EDFA operating under dynamic and steady states using MATLAB and Simulink. This model has been used to investigate the characteristics of the EDFAs over the C-band and L-band spectral regions. Predicted results by varying the parameter settings are shown to be in good agreement with those obtained from experiments. The MATLAB and Simulink toolboxes for simulating the dynamic system offer significant advantages as these are well known and available in worldwide teaching and research laboratories. Multiple wavelength signal propagations and ASE noise simulations have been taken into account in our EDFA simulator. Although we demonstrate only the Simulink [EDFA module] using wavelength, one for the pump and one for the signal, it is straightforward to modify the simulation to achieve the multiple-channel propagation as [EDFA Simulator]. The only limitations are the computer memory and the CPU runtime. Cascaded EDFA systems can also be connected, and the effects of propagation loss and noise in such systems can be modeled with ease. The gain modulation of the EDFA has not been discussed in this chapter but will be reported in a future article.

Primarily, a powerful simulator of modern optical amplifiers' module and its various components has been established. Future works will aim to incorporate a number of optimization components into the EDFA simulator, thence, further enhancing the uses of the simulator. Multiple-channel propagation simulation in the EDFA module will also be incorporated. The developed EDFA simulator and the EDFA modules will provide the framework for the near-future development of the Raman amplifier simulation using MATLAB and Simulink.

References

1. I. Nusinsky and A.A. Hardy, Analysis of the effect of up conversion on signal amplification in erbium-dope fiber amplifier, *IEEE Journal of Quantum Electronics*, 39(4): 548–554, April 2003. (b) A. Hardy and R. Oron, Signal amplification in strongly pumped fiber amplifiers, *IEEE Journal of Quantum Electronics*, 33(3): 307–313, March 1997.
2. S. Novak and R. Gieske, Simulink model for EDFA dynamics applied to gain modulation, *IEEE Journal of Light Wave Technology*, 20(6): 986–992, June 2002.
3. S. Novak and R. Gieske, Analytic model for gain modulation in EDFAs, *IEEE Journal Light Wave Technology*, 20(6): 975–985, June 2002.
4. S. Yamashita and M. Nishihara, L-band erbium-doped fiber amplifier incorporating an inline fiber grating laser, *IEEE Journal of Quantum Electronics*, 7(1): 44–53, January/February 2001.
5. Y. Sun, J.L. Zyskind, and A.K. Srivastava, Average inversion level, modelling, and physics of erbium-doped fiber amplifiers, *IEEE Journal of Selected Topics in Quantum Electronics*, 3(4): 991–1007, August 1997.
6. A.A. Saleh, R.M. Jopson, J.D. Evankow, and J. Aspell, Modelling of gain erbium-doped fiber amplifier, *IEEE Photonic Technology Letters*, 2(10): 714, 1990.
7. K. Izuka, Experiment on erbium-doped fiber amplifiers, Advanced Labs for Special Topic in Photonics (ECE 1640H), University of Toronto, Toronto, Ontario, Canada, April 28, 1998.
8. K. Izuka, Erbium-doped fiber amplifier, ECE 1640H, Advanced Labs for Special Topic in Photonics, University of Toronto, Toronto, Ontario, Canada, April 28, 2003.

9. L.N. Binh, Modelling of dispersion managed WDM optical communication systems, Technical Report, Electrical and Computer System Engineering Faculty, Monash University, Clayton, Victoria, Australia, 2003.
10. R. Scholten, Quantum and advanced optics laser, Lecture notes, School of Physics, University of Melbourne, Melbourne, Victoria, Australia, 2003.
11. V. Mehta, Amplified spontaneous emission tutorial, ECE 1654, Optical Networks, University of Toronto, Toronto, Ontario, Canada, 2003.
12. Overview of optical fiber transmission using DWDM technology, White paper, Celion Network, Inc., Richardson, TX, 2003.
13. W.J. Miniscalco, Erbium-doped glass for fiber amplifiers at 1500 nm, *IEEE Journal of Light Wave Technology*, 9(22): 234–250, 1991.
14. G.P. Agrawal, *Fiber-Optic Communication Systems*, 2nd edn. John Wiley & Sons, Sydney, New South Wales, Australia, 1997.
15. E. Kreyszig, *Advanced Engineering Mathematics*, 7th edn. John Wiley & Sons, New York, 1993.
16. ADVA EDFA specification data sheet, Optical Service & Solution (ADVA), Ramsey, NJ.
17. L.N. Binh, Optical transmitter and receiver for Gb/s broadband communication system and network, ECE 4405, Lecture notes, Monash University, Clayton, Victoria, Australia, 2004.
18. L.N. Binh, Optical repeater, regenerator and amplifier, ECE 4405, Lecture notes, Monash University, Clayton, Victoria, Australia, 2004.
19. MATLAB Simulink Help Files, refer Simulink program.
20. MATLAB Simulink Reference Books.
21. *MATLAB Simulink Dynamics Model Design Handbook.*

10

MATLAB® and Simulink® Modeling of Raman Amplification and Integration in Fiber Transmission Systems

The dynamic modeling of optical transmission systems is very important in ultra-high-speed optically amplified communication systems. Systems operating at bit rates above 40 Gb/s require detailed and accurate modeling. Therefore, the modeling of the Raman gain amplification must be integrated into the propagation of modulated optical signals over the models of multi-span transmission link.

This chapter thus describes the Raman amplification in various types of transmission and dispersion-compensating fibers (DCFs) including (1) the physical properties and operational characteristics and (2) its implementation in the lightwave-modulated envelop propagation equation for the transmission of 40 Gb/s signals in dense wavelength division multiplexing (DWDM) optical systems. We demonstrate the integration of the Raman amplification gain and noises into transmission system with the following: (1) Numerical techniques to solve these design parameters for a single-channel bidirectionally pumped system operating in the C-band (1525–1565 nm) range at 1550 nm. (2) Investigations of the Raman amplification of different commercially available fibers for obtaining the Raman design parameters of gain, noise figure (NF), and dispersion in distributed and discrete Raman configurations. Comparisons are contrasted with published documents and existing optical amplification technology. (3) The numerical solution adapted from average power to a Gaussian pulse operating at 40 Gb/s and propagated along the length of the fiber using the split-step Fourier method (SSFM) to allow for integration into the amplitude shift keying (ASK) and multiple shift keying MATLAB® and Simulink® models. (4) The description of the improvement of transmission link distance and noise performance.

10.1 Introduction

Stimulated Raman scattering (SRS), or the Raman effect, is an optical phenomenon that until only recently has had commercial interest in its potential technologies in engineering. It has been sparked by the increasing demands of the Internet, as a communication tool for both business and personal needs. However, the increased demands have elucidated the limitations of existing modern systems in terms of higher-capacity data transport [1–6].

The Raman scattering phenomenon is associated with the molecular diffraction of light, where photons interact with molecules to induce transitions in energy states. Typically, the photons scattered from these interactions are of the same energy and display the same characteristics as the incident light, in particular, wavelength. This type of occurrence is known as Rayleigh, or elastic, scattering. However, a small fraction of light, approximately 1 in 10^7, is scattered at different optical frequencies and are of a much lower energy level than that of the incident photons. This inelastic effect is referred to as Raman scattering [7]. The optical effects associated with Raman are of particular importance in the design of modern lightwave systems. The requirements for upgrading existing systems to accommodate for increased signal transmission and transmission quality over longer distances are becoming increasingly important [8].

In modern long-haul fiber-optic communication systems, the transmission distance is limited by fiber loss and dispersion. Traditional methods that use the electrical conversion of the optical signal to overcome this limitation, such as repeaters to retransmit signals at progressive stages, are becoming increasingly complex and expensive. In the 1990s, optical amplifiers, which directly amplified the transmission signal, became widespread, minimizing system intricacies and cost, while upgradation in transmission fiber design, in particular DCFs, minimized linear phase distortions in the signal. In modern systems, existing erbium-doped fiber amplification (EDFA)-lumped optical amplifiers are employed to ensure the quality of the transmitted signals [8–11].

Raman scattering has become important in the application of optical amplification because of several important reasons in comparison to other similar methods [11,12]. Raman optical amplifier (ROA) can be described simply as a pump laser that emits lightwaves down an optical fiber; this signifies that it can be compatible with most available transmission systems. The operation of the pump laser is dependent on the gain that is achieved, in particular, the pump wavelength. This means that the medium of transmission is completely independent, in contrast to the EDFA. The fact that the gain is pump wavelength dependent theoretically means that amplification is achievable for any frequency. It is important to note that utilizing a number of lasers at variable frequencies in a system will provide a broad gain bandwidth. There are also advantages to ROA from EDFA in low noise characteristics, which can improve the overall signal quality.

The simulation of the amplification processes and the generation of gain and noises of ROA are so important for the integration of these properties into optical transmission systems, especially in multi-span optically amplified transmission system employing dense WDM technique. Furthermore, advanced modulation formats have attracted significant interests in the combat of impairments such as dispersion and nonlinear effects. The performance of these modulation format schemes under ROA of long-haul transmission systems has not been demonstrated when both forward and backward amplification sections are used in the same span. Thus it is necessary to integrate the ROA into the transmission model for the simulation of the transmission performance of advanced modulation formats in multi-span ROA and EDFA optically amplified long-haul transmission link.

This chapter has three separate components: (1) the physical model of Raman amplification, (2) the system transmission model integrating Raman amplification, and (3) the integration of ROA in the transmission link model operating under advanced modulation formats.

Therefore, the objectives of this chapter are the following: (1) to investigate the physical properties of Raman amplification, apply it to a system model, such as ASK or multiple shift keying, and compare results against different system configurations to determine if ROA is beneficial in expanding span transmission reach and capacity; (2) in particular, to focus on the physical modeling of ROA in terms of operational principles of ROA, especially the optically distributed gain and its NF; (3) to use MATLAB to numerically solve and analyze these parameters, and to compare results of modeling with published documents and other commercially available forms of optical amplification, such as EDFA; (4) to transport these results into a dynamic Simulink transmission system model employing amplitude, phase, or frequency shift keying techniques (such as ASK), discrete and continuous phase shift keying techniques (such as differential phase shift keying (DPSK)), and minimum shift keying (MSK) to prove the effectiveness of our modeling techniques and the integration of ROA in long-haul transmission systems; and (5) to illustrate with the transmission performance a case study of ROA long-haul transmission system along the coastline of Vietnam using hybrid EDFA-ROA optically amplified spans of an average distance of 240 km undersea.

10.2 ROA versus Erbium-Doped Fiber Amplifiers

The comparisons between existing lumped EDFA and distributed ROA systems can be imagined by signal power evolution with the loss and gain compensated with each other along the fiber length. Both forms of optical amplification provide gain to a signal power and are iterated over the span of the fiber. However, there are important differences. In lumped EDFA (1) gain characteristics depend on a large number of device parameters and (2) gain is amplified to the signal at the start of the segment and, because of the effective length of EDFA (a few 10s of meters), is amplified immediately. In distributed ROA (1) the transmission medium, silica fiber, is responsible for the gain characteristics and (2) gain is delivered backward from the end of the first lumped circuit and, because of the effective length of ROA (in kilometers), is gradually distributed.

The advantages of distributed ROA are that gain is obtainable in almost all types of transmission fibers, ease of gain clamping, low noise characteristics, improved optical signal-to-noise ratio (OSNR), gain nonresonance, very wide gain bandwidth Δv_g of more than 6 THz (or about 50 nm in the 1550 nm spectral region) achievable by using multiple pumps at different wavelengths, and the possibility of extending power budget margins. However, the disadvantages of the ROA are the requirement of high pump powers, poor pumping efficiency at lower powers; fast response time, hence new noises; and nonlinear effects that are not negligible.

The advantages of lumped EDFA are the following: (1) they are integrated well into existing systems due to intrinsic properties of silica, (2) they have good pumping efficiency at lower powers, and (3) they are widely studied and deployed in WDM systems. The disadvantages of lumped EDFA are (1) gain resonance and bandwidth limited to C and L bands (1525–1605 nm), (2) gain equalization is necessary for ultra-wideband DWDM, (3) fiber nonlinearities and noises, and (4) difficulty in gain clamping.

It is important to realize that [13–15] while there are some disadvantages to ROA, advancements in fiber technology that minimize these issues have been experimented. A design issue to be considered in this chapter is the incorporation of a DCF into the system model. The advantages in DCF for ROA include (1) high negative dispersion characteristics that compensate for the dispersion induced in the fiber from group velocity dispersion (GVD) and self-phase modulation (SPM), (2) small effective area—hence lower pump power necessary, and (3) high germanium (Ge) concentration in the silica fiber.

10.3 Raman Amplification

10.3.1 Principles

SRS is a nonlinear effect due to interactions between lightwaves with molecular vibrations in silica fiber. It exists when the transmitted power of a channel exceeds the threshold power. For a single channel

$$P_{th} = \frac{16 \cdot A_{eff}}{K_p \cdot L_{eff} \cdot g_r} \tag{10.1}$$

where K_p is the polarization constant and is assumed to be equal to 2 for a completely unscrambled signal. Scattering occurs when the polarization of the molecule changes with the vibrational motion. This is caused when an electric field distorts the electrons of the molecule to create an induced dipole moment, $\mu = \alpha \cdot E$, and for a diatomic molecule $\alpha = \alpha_0 + (\partial\alpha/\partial q)_0 q + L$. If the molecule is vibrating with frequency v_0, then $q = A\cos(2\pi v_0 t)$, and there is an oscillating electric field at v_i frequency, $E = E_0 \cos(2\pi v_i t)$, then the induced dipole moment becomes

$$\mu = E_0 \cdot \alpha_0 \cos(2\pi v_i t) + E_0 \left(\frac{\partial\alpha}{\partial q}\right)_0 A\cos(2\pi v_i t)\cos(2\pi v_0 t)$$

$$= E_0 \cdot \alpha_0 \cos(2\pi v_i t) + \frac{1}{2} E_0 A \left(\frac{\partial\alpha}{\partial q}\right)_0 [\cos 2\pi(v_i - v_0)t + \cos 2\pi(v_i + v_0)] \tag{10.2}$$

This shows that the moment can oscillate at three frequencies and can therefore emit light at these frequencies that correspond to Rayleigh scattering (v_i), Stokes scattering ($v_i - v_0$), and anti-Stokes scattering (AS) ($v_i + v_0$). The emission of light is in the form of energy transitions, which will depend on the interaction or collision between the incident light and the quantum of vibrational energy. In SRS, the scattered photons will have energy gained or lost by the incident photon. If the incident photon excites a transition of vibration energy in the molecule, the scattered photons experience a gain and downshift to the Stokes frequency. In contrast, if vibration energy already exists in the molecule, the incident photon can absorb it, which causes a loss in the scattered photons and upshifts to the AS frequency. The excited vibration level is populated according to a Boltzmann distribution. As the anti-Stoke scattering is initiated in the excited vibration level, its transition rate is also proportional to this distribution. The relative intensity of the Stokes shift is an essential element in the design of Raman amplifiers. The physical model of ROA including the Raman gain, the associated noises in the system, and dispersion incurred along the fiber are briefly presented.

The frequency difference, $\Omega_R = v_i - v_0$, known as the Stokes shift is an essential element in Raman gain. The intrinsic properties of glass in the transmission medium, usually fused silica, also provide a range within which the frequency of the pump, $v_p = v_i$, and frequency of the signal, $v_s = v_0$, can differ.

The Raman gain in silica-based fiber can reach up to 20 THz in which the Stokes shift can happen. The maximum Raman gain, g_R is observed at 13.2 THz; this corresponds to a wavelength difference of 100 nm between the pump and the signal. As $f = c/\lambda$, wavelength is inversely proportional to frequency; thus, larger frequencies result in smaller wavelengths. The Raman gain bandwidth $\Delta v_g = 6\,\text{THz}$ is defined as the full-width half maximum (FWHM) of the dominant peak in the gain spectrum. The overall optical gain can be expressed in terms of the pump intensity I_P:

$$g(v) = g_R(v) \cdot I_P = g_R(v) \cdot \left(\frac{P_P}{A_{eff}} \right) \tag{10.3}$$

This gain is thus dependent on the frequency and the pump power. The parameter Raman gain efficiency, $g_{R\_eff} = g_R / A_{eff}$, is a critical design parameter. Different fibers have different gain efficiencies due to smaller effective areas and higher germanium (Ge) concentrations, which increase the overall Raman gain efficiency and hence the optical gain, and lower the threshold power required to initiate lasing and the SRS process.

10.3.2 Raman Amplification Coupled Equations

The power evolution of the pump and the signal power along a length of a fiber, z, can be modeled by the following two coupled equations, respectively:

$$\begin{aligned} \frac{dP_S}{dz} &= -\alpha_S P_S + \left(\frac{g_R}{A_{eff}} \right) P_P P_S \\ \xi \frac{dP_P}{dz} &= -\alpha_P P_P - \left(\frac{v_P}{v_S} \right) \left(\frac{g_R}{A_{eff}} \right) P_S P_P \end{aligned} \tag{10.4}$$

It illustrates that the pump power provides the energy for amplification and depletes as signal power increases. As pump power approaches the signal power, the optical gain is reduced and gain saturation occurs. In ROA, the laser can be pumped from either end of the fiber by changing the polarization, $\xi = -1$. The schematic drawing of the pumping configurations is demonstrated in Figure 10.1. Pumping in different directions changes the signal power evolution along the length of the fiber.

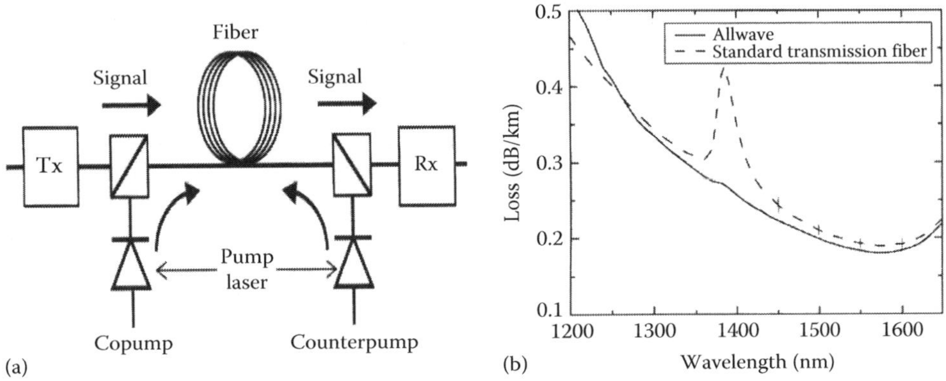

FIGURE 10.1
(a) Schematic of the ROA using the bidirectional configuration and (b) attenuation spectrum of two types of transmission fibers.

The power along the length of the fiber is reduced by the fiber losses, which occur in the medium due to intrinsic properties. These result in microscopic fluctuations in the density of the core, which affects the refractive index. Doping the fiber with dopants such as germanium further increases these losses. It is also a limiting factor in transmission distance lengths as optical receivers require minimum amounts of power to recover the signal [15,16].

This attenuation coefficient, $\alpha_{P,S}$, includes all sources of power attenuation and is expressed in units of decibels per kilometer by the relation $\alpha(\mathrm{dB/km}) = 10\log(e)\alpha \approx 4.343\alpha$. If pump depletion can be neglected, that is, in small-signal amplification, the pump power for forward pumping becomes

$$P_P(z) = P_0 \exp(-\alpha_P z) \tag{10.5}$$

with P_0 as the input pump power. Hence, the signal intensity at the output of an amplifier of length L is determined as follows:

$$P_S(L) = P_S(0)\exp\left(\frac{g_R P_0 L_{\mathrm{eff}}}{A_{\mathrm{eff}}} - \alpha_S L\right) \tag{10.6}$$

The effective length, L_{eff}, is the length where the nonlinearities exist or SRS occurs in the fiber and is defined as follows:

$$L_{\mathrm{eff}} = \frac{1 - \exp(-\alpha_P L)}{\alpha_P} \tag{10.7}$$

As fiber losses exist at the pump wavelength, it is less than the actual length and can be approximated by $L_{\mathrm{eff}} \approx 1/\alpha_P$ if $\alpha_P L \gg 1$. Hence the gain amplification, the ratio of the power of the signal with and without Raman amplification, is given by

$$G_A = \frac{P_S(L)}{P_S(0)\exp(-\alpha_S L)} = \exp(g_o L) \tag{10.8}$$

This is referred to as the on–off Raman gain and can be determined from Figure 10.4 from the difference in dBm of the Raman pumped fiber and the passive fiber without amplification. The difference from the figure is approximately 20 dB. Hence the small signal gain, g_o, can be obtained. The signal power evolution along the total length of fiber can be imagined to be gradually distributed and exponentially distributed to equalize that of the fiber attenuation.

In Raman amplification of the signal, the noises are contributed by the following processes. In the SRS process, noise is added to the amplified signal because of random phases associated with spontaneous generated photons. The spontaneous scattering factor is dependent on the temperature of the amplifier given as $n_{sp} = 1/(1 - \exp(-h\Omega_R/k_B T))$. For a Raman amplifier, $n_{sp} \sim 1.13$ as it is always a fully inverted amplifier. The noise incurred by the spontaneous Raman scattering across the length of the fiber is accumulated and is known as amplified spontaneous emissions (ASE).

Rayleigh backscattering occurs in all fibers and is the fundamental loss in signal power. Typically, it is seen as ASE propagating in the backward direction and is small in comparison to the power of the signal and, therefore, of no concern. In ROA, however, as a result of large transmission spans, it can cause the ASE to be reflected back in the forward direction and be re-amplified enhancing the overall noise; this is known as double Rayleigh backscattering (DBS), and because it can occur anywhere along the fiber it is referred to as multi-path interference (MPI).

In ROA, the short upper state lifetime results in instantaneous gain, which leads to coupling of the pump fluctuations with the signal. To avoid this, backward-pumping configurations are adopted, which changes the upper state lifetime to the transit time. If forward pumping is to be used, pumps with low noise are required. NF is the determination of the signal denigration over the length of the transmission span. It is the SNR of input over output, and in ROA it is dependent on the pumping power and the net gain of the system and is given by

$$\mathrm{NF} = 2n_{\mathrm{sp}}\frac{g_{\mathrm{R}}}{A_{\mathrm{eff}}}\int_0^L \frac{P_{\mathrm{p}}dz}{G(z)}dz + \frac{1}{G_{\mathrm{L}}} \tag{10.9}$$

where

$G(z)$ is the net gain at distance z along the fiber

G_{L} is the net gain at the end of the fiber

10.3.3 Raman and Fiber Propagation under Linear and Nonlinear Fiber Dispersion

10.3.3.1 Propagation Equation

In optical amplifiers, lightwaves are beamed down a length of fiber to transmit a signal. The properties of the fiber refractive index profile, the waveguide geometrical structures cause the dispersion of the transmitted pulse sequence. Nonlinear dispersion is caused by the nonlinear effects that occur in the fiber because of increased signal power. In geometrical-optics description, dispersion or pulse broadening is caused by different rays following different paths and not arriving at the output end of the fiber at the same time. In modal description, dispersion is related to the group velocities associated with different modes. The GVD is caused by chromatic dispersion, which exists in fibers that support only the fundamental mode. It causes pulse broadening whereby different spectral components of the pulse are propagating at different velocities and arrive at the output of fiber length at different times. It is expressed by $v_{\mathrm{g}} = (d\beta/dw)^{-1}$, where β is the propagation constant. The derivative of the propagation constant in terms of frequency determines the GVD parameter: $\beta_2 = d^2\beta/bw^2$. This parameter determines the pulse-broadening effects especially when the pulse is short in ultra-high-speed transmission over the propagating long distance. This can be related to the dispersion parameter D. As dispersion can be both negative and positive, it can also be designed at a wavelength of zero dispersion λ_{ZD}, which is approximately 1.3 µm. It is important that while dispersion is minimized in design, it cannot be completely removed as it also minimizes the other nonlinear effects in the fiber, such as SPM and cross-phase modulation (XPM).

The SPM causes a nonlinear phase shift (NLPS) in the fiber; the accumulated NLPS can be solved as follows:

$$\varphi_{NL} = \gamma P_{S0} \int_0^L G(z)dz \tag{10.10}$$

where γ, the nonlinear parameter responsible for SPM, is defined as $\gamma = 2\pi n_2/\lambda_S A_{eff}$, where $n_2 = 2.6 \times 10^{-20}$ 1/m^2 is the cladding refractive index in fused silica. The NLPS due to ROA, that is, the ratio with and without ROA, can be determined by

$$R_{NL} = \frac{\varphi_{NL}(\text{on})}{\varphi_{NL}(\text{off})} = \frac{1}{L_{eff}} \int_0^L G(z)dz \tag{10.11}$$

Pulse broadening can be described by using the wave equation, expressed as follows:

$$\frac{\partial A}{\partial z'} + \frac{i}{2}\beta_2 \frac{\partial^2 A}{\partial t'^2} - \frac{1}{6}\beta_3 \frac{\partial^3 A}{\partial t'^3} = 0 \tag{10.12}$$

where
β_3 is the dispersion slope
$A(t, z)$ is a slowly varying amplitude

To incorporate nonlinear effects into the pulse broadening equation and compensate for dispersion solitons are integrated in the form of the Schrödinger equation.

$$\frac{\partial A}{\partial z} = \frac{i}{2}\beta_2 \frac{\partial^2 A}{\partial t^2} + \frac{1}{6}\beta_3 \frac{\partial^3 A}{\partial t^3} + i\left|A\sqrt{\gamma}\right|^2 + \frac{1}{2}\left[g(\xi) - \alpha\right] + iTR\frac{\partial \left|A\sqrt{\gamma}\right|^2}{\partial t} \tag{10.13}$$

This is the un-normalized nonlinear Schrödinger equation (NLSE), which includes the effects of distributed amplification and dispersion management.

10.3.3.2 Standard Single-Mode Fiber and DCF as Raman Fibers

The Raman gain coefficients of various types of fibers such as the standard SMF (SSMF) have been measured and tabulated, as shown in Table 10.1.

In long-haul transmission, it becomes necessary to compensate for dispersion. Dispersion can be seen as a linear process whereby pulse spreading can be negated by introducing a fiber with the same degree of GVD but opposite polarization. DCF exists to manage the impact of dispersion with an opposite sign of dispersion and dispersion slope in comparison to SMF. It also has a higher attenuation in comparison to SMF and therefore will add greater loss, which will result in increased noise power. The power evolutions of the pump and signal in Raman amplification can be described as a set of ordinary differential equations (ODEs):

TABLE 10.1

Some Different Commercial Fibers and the Parameters Used to Numerically Solve Raman Amplification

Type of Fiber	Raman Gain Efficiency (1/W km)	Effective Area (μm^2)
Allwave	0.35	84.95
Corning NZ-DSF	0.72	80
LEAF (NZ-DSF)	0.45	72
SMF-28 (NDSF)	0.38	84.95
Truewave RS (NDSF)	0.58	60
Truewave reach fiber	0.6	55
OFS Raman fiber	2.5	18.7

$$
\begin{aligned}
\frac{\partial P_S}{\partial z} &= -(\alpha_{RS}+\alpha_S)P_S + \frac{g_R}{A_{eff}}P_P P_S \\
\frac{\partial N_S}{\partial z} &= \left(-\alpha_{RS}-\alpha_S+\frac{g_R}{A_{eff}}P_P\right)N_S + \frac{g_R}{A_{eff}}(2P_P N_P)^{1/2}P_S + \left(\alpha_{RS}+\alpha_S+\frac{g_R}{A_{eff}}P_P\right)\frac{h\omega_S}{2}B_0 \\
\frac{\partial P_P}{\partial z} &= -(\alpha_{RP}+\alpha_P)P_P - \frac{\omega_S}{\omega_P}\frac{g_R}{A_{eff}}\left(P_S+N_S+\frac{h\omega_S}{2}B_0\right)P_P \\
\frac{\partial N_P}{\partial z} &= -(\alpha_{RP}+\alpha_P)N_P - \left(\frac{\omega_S}{\omega_P}\frac{g_R}{A_{eff}}P_S\right)(2P_P N_P)^{1/2} + (\alpha_{RP}+\alpha_P)\frac{h\omega_S}{2}B_0 \\
&\quad + \left(\frac{\omega_S}{\omega_P}\frac{g_R}{A_{eff}}P_S\right)\left(2P_P\frac{h\omega_S}{2}B_0\right)^{1/2}
\end{aligned}
\tag{10.14}
$$

To model these equations, classical numerical techniques need to be utilized as these equations represent a complex system that cannot be solved analytically. The particular choice of numerical approximation (Runge–Kutta, etc.) depends on the method needed to solve (i.e., initial value ODEs), accuracy required, and the consideration of simulation time in processing. This last consideration is especially important when implementing from MATLAB to systems in Simulink. The function ode45.m of MATLAB can be used to obtain the initial value by using ODEs employing the Dormand–Prince formula that meets the requirements of accuracy in approximation and in processing speed.

Assumptions made in our modeling platform include the following: (1) forward and backward noise powers are typically less than input signal powers by 30 dB; (2) backscattering powers of pumps and signals are usually about 20–30 dB lower than their original powers; (3) noise effects such as spontaneous Raman scattering, Rayleigh backscattering, and thermal factor can be reasonably ignored when simulating the Raman gain profile; and (4) $\alpha_{RS}=\alpha_{RP}$ for simplicity.

The coupled noise Equations 10.2 and 10.4 include the fundamental noise sources including zero-point fluctuations, amplification of noise, partition noise, and Rayleigh scattering. The Raman gain coefficient is dependent on the difference between the pump and the signal frequencies. In the case where the pump is co-pumped with the signal at 1450 nm in nonzero dispersion-shifted fiber (NZDSF), the effects of different shifting frequencies for the signal on the Raman gain are highlighted as shown in Table 10.2.

Figure 10.2 confirms that Raman amplification is possible over a wide range of signal frequencies for a single pump, with a peak coefficient existing at 13.2 THz from the pump or 100 nm wavelength. Thence, due to the fact that optical amplification is directly proportional to the Raman coefficient, gain is also distributed across a large bandwidth of signal frequencies. In the single pumped configuration, to determine the performance the gain ripple was determined over the Raman bandwidth. Figure 10.2b shows that the gain ripple is approximately 6 dB.

As Raman amplification is theoretically possible at any wavelength, careful consideration of pumping wavelengths and Raman pumping power when using multiple pump lasers can provide broadband gain. This is possible because individual gain spectra accumulate at the different wavelengths to provide the total system gain.

To model ROA in all configurations, forward, backward, and bidirectional, two ode45 functions in MATLAB are implemented for backward and forward pumping, respectively. This was achieved by changing the pump direction. The signal power always propagates in the same direction. To achieve bidirectional pumping, the effects of the pump power co- and counter-propagating are accumulated to give the total pump power. Figure 10.3 shows the evolution of the distributed gain under both backward and forward pumping.

TABLE 10.2

Effects of Signal Frequency Shifting on the Effective Raman Coefficient and Total Raman Gain

Shifting Frequency THz	Shifting Frequency 1/cm	Signal Frequency THz	Signal Wavelength nm	Effective Raman 1/W km	Raman Gain dB
6	200	200.9	1493.3	0.24	6.61
10.5	350	196.4	1527.5	0.5	13.86
13.2	450	193.7	1548.8	0.72	20.13
16.5	550	190.4	1575.6	0.22	6.06

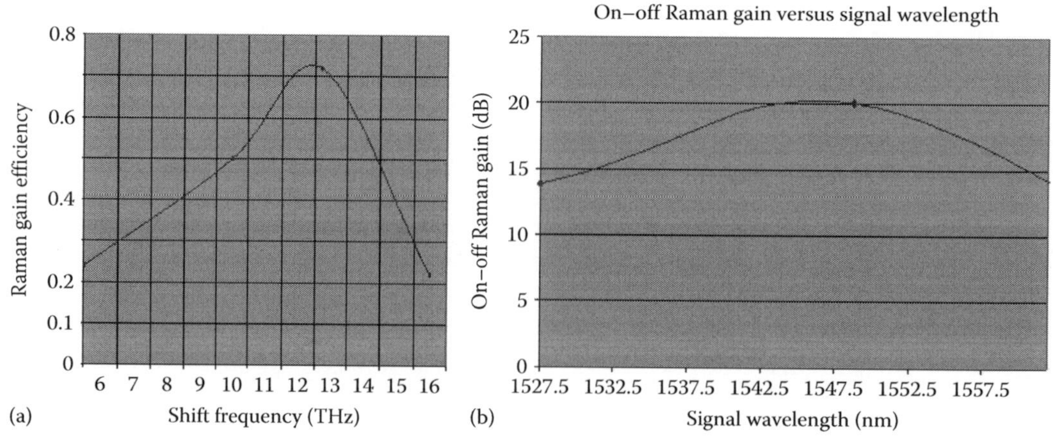

FIGURE 10.2
(a) Effective Raman gain coefficient versus shifting frequency for NZDSF and (b) gain over the Raman bandwidth.

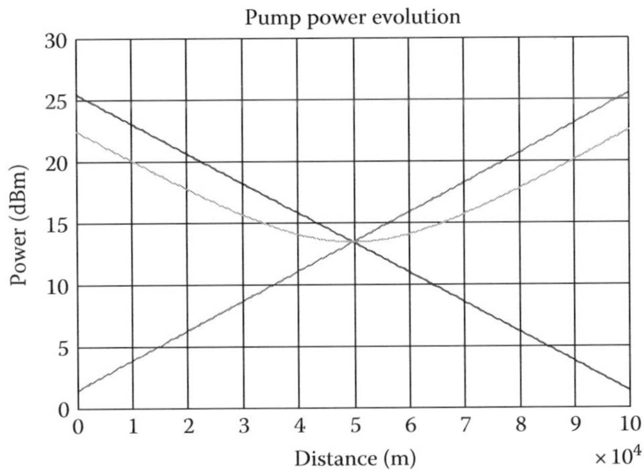

FIGURE 10.3
The power evolution of the pump along the total length of the Corning NZ-DSF for forward, backward, and bidirectional configurations.

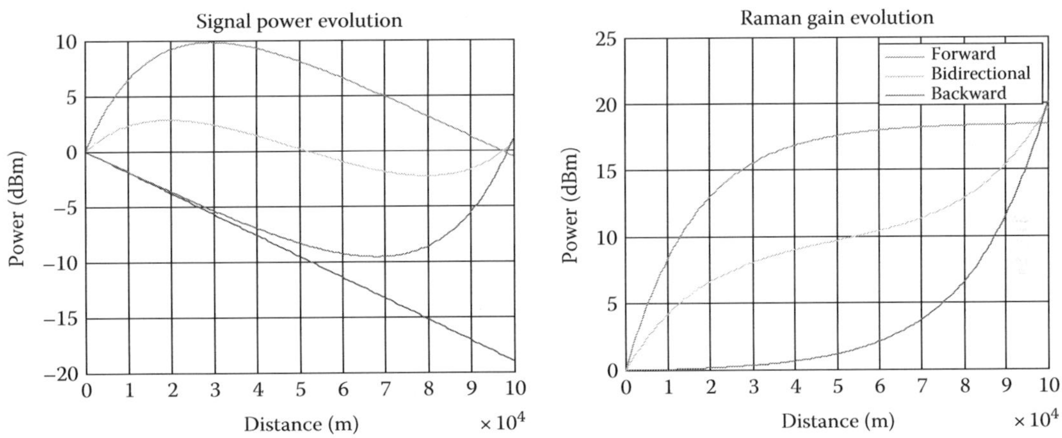

FIGURE 10.4
(a) The power evolution of the signal along the total length of the fiber for forward, backward, and bidirectional configurations, upper, middle, and lower respectively. The red line represents a passive fiber with no gain, $\alpha = 0.19$ dB/km. (b) Gain evolution over the total length of the fiber due to Raman amplification for forward, backward, and bidirectional configurations.

The pump power is shown to reduce to provide gain for the signal power; this is due to pump depletion, as shown in Figure 10.4.

The signal power evolution shows how it changes the attenuation map in comparison to the passive fiber, without Raman amplification. In the forward configuration, there is a large gain in power at the beginning of the fiber length. This results in increasing effects of nonlinearities in the fiber, because of the power dependence of the refractive indices of the core and the cladding. $n'^j = n^j + \bar{n}_2(P/A_{eff})$ with $j = 1, 2$. In the backward configuration, the gain occurs toward the end of the fiber after a substantial power loss. This power loss will increase the possibility of noises altering the quality of the signal. The bidirectional configuration shows a balanced result in terms of noises and nonlinearities. Another important

TABLE 10.3

Final Net Gain for Different Commercial Fibers and the Required Pump Power to Pump the Fiber over the Span Just Compensating Fiber Losses

	Average Power (dBm)			Required Pump Power
Type of Fiber	**Forward**	**Backward**	**Bidirectional**	**(Net Gain = 0) ~mW**
Allwave	−9.502	−9.327	−9.397	700
Corning NZ-DSF	−0.551	1.128	0.707	350
LEAF (NZ-DSF)	−6.879	−6.529	−6.654	550
SMF-28 (NDSF)	−8.707	−8.491	−8.574	650
Truewave RS (NDSF)	−3.638	−2.864	−3.099	425
Truewave reach fiber	−3.165	−2.297	−2.554	425

aspect to consider is the on–off Raman gain provided by amplification, in other words, the gain experienced with Raman in comparison to the gain experienced without Raman. The evolution of gain along the total length of the fiber can be solved through the decibel difference.

The on–off Raman gain evolution is the important parameter that needs to be implemented using the SSFM for Gaussian pulse and ASK testing in Simulink. It also needs to compensate for the passive fiber losses (net gain = 0). This is important because if the signal has a net gain or loss it will be effected detrimentally when it propagates through numerous spans. Table 10.3 shows the net gain and pumping conditions in different types of fibers over a 100 km span; typical Raman lasers operate at <500 mW. This limits the transmission spans of LEAF, SMF-28, and Allwave.

To illustrate the numerical solutions, an analytical solution was derived from the coupled equations ignoring the pump depletion. It was determined that pump depletion reduces the transfer of power from the pump to the signal as the pump power depletes sooner. To further investigate the initial input, signal power was increased to approach the initial pump power; it was observed that there was a greater reduction in signal gain due to gain saturation. Figure 10.5a and b shows the signal power evolution along the fiber under the Raman pumped gain and loss.

Numerically solving the coupled equations using the ode45 function in MATLAB, the signal and power evolution for average power only can be obtained. To implement into the Simulink ASK model, this solution needed to be applied to a Gaussian pulse at the correct bit rate of 40 Gb/s, which could be propagated using the SSFM.

To implement the Raman in the beam propagation, the NLSE file is modified to include Raman amplification. This is achieved by changing the SSFM half-step parameter, which is the gradient of the attenuation iterated over a number of segments along the fiber. In the passive fiber case, the half-step is constant; to include ROA the Raman gain contribution at each point is added to this half-step.

10.3.3.3 Noise Figure

Modeling the noise profile, the noise power evolution of the signal along the total length of the fiber was determined. The resulting signal noise agrees with published results [13] (Figure 10.6); discrepancies are a result of the fact the numerical equations modeled included Rayleigh scattering and pertinent noises other than ASE. Investigating the signal noise evolution along the fiber is essential because it provides a NF of the system. It is a value of the signal denigration over the total span and is proportional to the attenuation

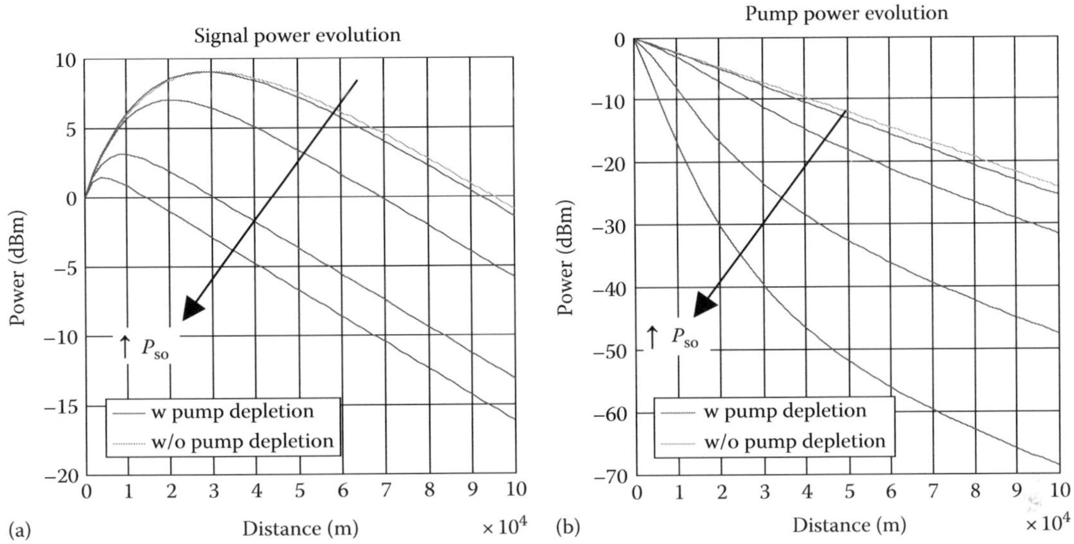

FIGURE 10.5
(a) Power evolution of the signal along the total length of the fiber with and without the effects of pump depletion. (b) Power evolution of the pump along the total length of the fiber with and without the effects of pump depletion.

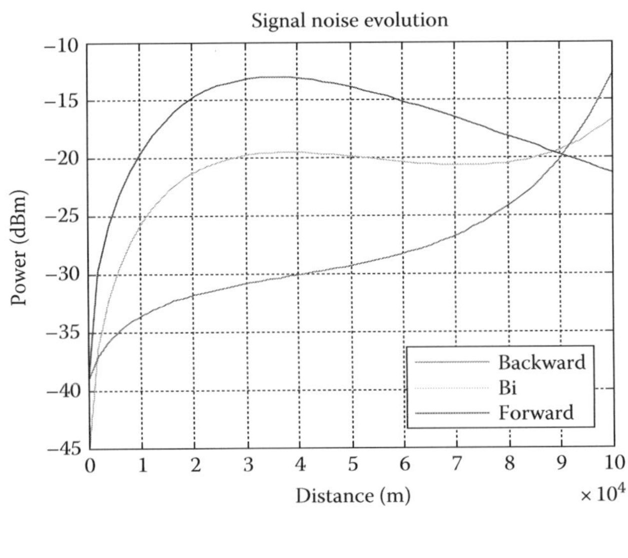

FIGURE 10.6
The noise power associated with the signal along the total length of the fiber for forward, backward, and bidirectional configurations.

and length. Typically for distributed Raman amplification (DRA), the transmission link is greater than 50 km, and the net gain of the signal is below 1. Therefore, from Figure 10.7, it is shown that the NF is typically large for amplification over these lengths.

Distributed amplification is analogous to numerous discrete amplifiers providing gain along the length of the fiber; ideally, if the loss at each point in the span could be corrected, the ideal NF could be determined. Theoretically, it is described by $NF|_{dB} \approx 10 \log_{10}(2\alpha + 1)$.

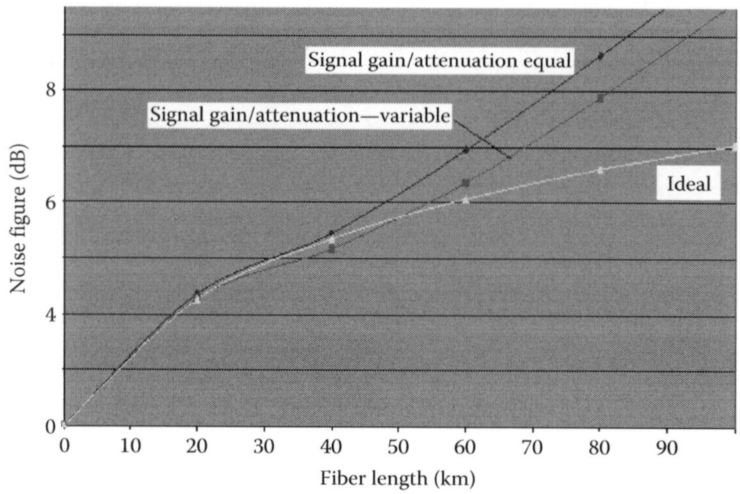

FIGURE 10.7
Noise figure versus on–off Raman gain for backward-pumping case and investigating practical and ideal configurations.

For a passive fiber with no DRA (on–off gain = 0) over 100 km, the NF degrades by 20 dB. By counter-pumping with an on–off gain of 20 dB, the NF is reduced to 17 dB. Although there is a high NF, it does not mean that distributed amplifiers are noisier than lumped.

To evaluate the performance of the DRA in relation to a passive fiber or discrete Raman/EDFA amplifier, it is useful to define the *effective NF*: $NF_{eff} = NF_R - \alpha_S L$. It accounts for the attenuation of the fiber and numerically equates the result to an equivalent discrete NF. Therefore, from Figure 10.8 it can be observed that the NF_{eff} for backward DRA is equal to −3 dB for Corning NZ-DSF. Compared to a lumped amplifier with an NF limit of at least 3 dB, the overall improvement of distributed amplifiers is at least 6 dB for backward pumping.

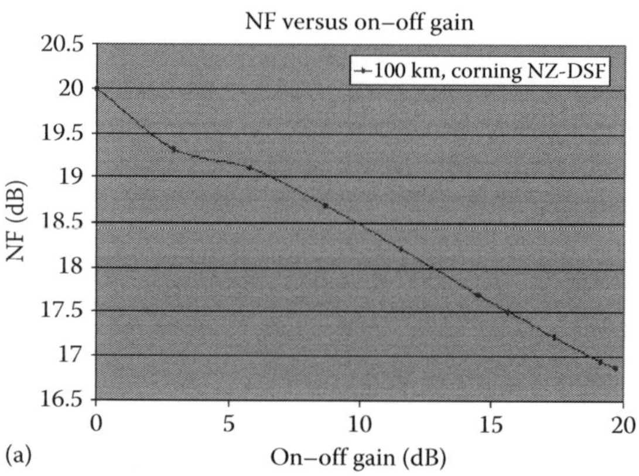

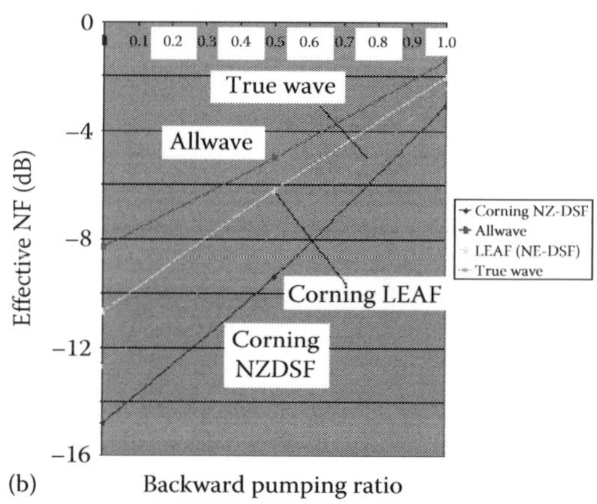

FIGURE 10.8
(a) Noise figure versus on–off Raman gain for a 100 km fiber varying input pump power using Corning NZ-DSF. (b) Noise figure versus backward-pumping ratio for different fibers over a 100 km transmission link.

The NF is also dependent on the pumping configuration, as it is dependent on the accumulation of pumping power and net gain along the fiber. In backward-pumping methods, the power reduction in signal and therefore net gain is much more dominant, and therefore the NF is expected to be much higher. Figure 10.8 illustrates the impact these configurations have on total NF. It shows an improvement of over 10 dB for forward-pumping configurations. When varying the backward-pumping ratio for different fibers over the span, the different values of NF were determined, and it was shown that for reduced Raman gain coefficient, g_R, values, the impact of the pumping scheme, forward to backward, on the NF was reduced. As the NF is dependent on the pumping power and the net gain, if it is used in a hybrid system configuration with an EDFA pumping gain at the start of the fiber, the NF for ROA can be lowered (Table 10.4). These results assume an ideal EDFA; hence, NF = 3 dB is included in the total NF.

TABLE 10.4

Noise Figure of 100 km Backward-Pumped ROA in Hybrid Configuration with Ideal EFDA Gain Using Corning NZ-DSF

Ideal EDFA Gain (dB)	NF (dB)
0	17
1	19.13
2	18.27
3	17.46
4	16.6
5	15.8
6	15
7	14.09
8	13.25
9	12.45

10.3.3.4 Dispersion

Another important aspect to consider is the effect of dispersion in the fiber. This was tested for all types of system configurations for two fibers of very different fiber properties (Table 10.5).

It demonstrates that in the forward-pumping scheme Raman amplification causes the most nonlinear effects due to the increased signal power gained throughout the fiber.

10.4 Nonlinear Raman Gain/Scattering Schrödinger Equation

As can be seen, the model incorporates an SSMF and a DCF. The system revolves around the NLSE, as it governs the propagation of optical pulses inside fibers. Thus if any distributed amplification is to take place, this equation must be transformed to incorporate Raman amplification. The NLSE as solved for the ASK model before manipulation is

TABLE 10.5

Dispersion Characteristics for Two Commercial Fibers with Different Properties in a 100 km Span

	Corning NZ-DSF		Allwave	
	NLPS (rad)	Raman NLPS (dB)	NLPS (rad)	Raman NLPS (dB)
Forward	7.00e−04	13.79	8.00e−04	14.28
Bidirectional	1.52e−04	7.09	1.53e−04	7.35
Backward	4.28e−05	1.58	4.14e−05	1.7

Source: Headley, C. and Agrawal, G.P., *Raman Amplification in Fibre Optical Communication Systems*, Elsevier, London, U.K., 2005. With permission.

$$\frac{\partial A}{\partial z} = -\frac{i}{2}\beta_2 \frac{\partial^2 A}{\partial t^2} + \frac{1}{6}\beta_3 \frac{\partial^3 A}{\partial t^3} - \frac{\alpha}{2} A + i\gamma\left(|A|^2 A + \frac{i}{\omega_0}\frac{\partial}{\partial T}\left(|A|^2 A\right) - T_R A \frac{\partial |A|^2}{\partial T}\right) \tag{10.15}$$

The NLSE incorporates fiber losses through the use of the attenuation factor α and second- and third-order dispersion (TOD) effects through the use of β_2 and β_3, respectively. T_R becomes negligible as it is almost insignificant. The NLSE that incorporates distributed amplification and dispersion management is

$$\frac{\partial u}{\partial \xi} = \frac{j}{2}\frac{\partial^2 u}{\partial \tau^2} + \delta \frac{\partial^3 u}{\partial t^3} + j|u|^2 u + \frac{L_D}{2}\left(g(\xi) - \alpha\right) - j\tau_R \frac{\partial |u|^2}{\partial \tau} \tag{10.16}$$

The first Schrödinger equation is un-normalized, meaning that it can be subjected to modeling techniques as it is represented in its raw entirety. The second equation is normalized, meaning that it has to go under the microscope and transformed into its un-normalized state. This alone was a hard task and took hours upon hours of complex formulation. This formulation can be seen in the appendices, but the overall result is shown in the following equation:

$$\frac{\partial A}{\partial z} = \frac{i}{2}\beta_2 \frac{\partial^2 A}{\partial t^2} + \frac{1}{6}\beta_3 \frac{\partial^3 A}{\partial t^3} + i\left|A\sqrt{\gamma}\right|^2 + \frac{1}{2}\left[g(\xi) - \alpha\right] + iTR\frac{\partial \left|A\sqrt{\gamma}\right|^2}{\partial t} \tag{10.17}$$

where $\gamma = 2\pi n_2/\lambda A_{\text{eff}}$, $\xi = z/L_D$, and $L_D = T_0^2/|\beta_2|$. Due to operating parameters, the T_R term can be neglected, resulting in the equations shown earlier. To make calculations simpler, the $\beta_1 z$ term cancels as β_1 equals zero. It must also be noted that the $g(\xi)$ term can be solved for the three cases of forward, bidirectional, and backward pumping.

10.4.1 Fiber Nonlinearities

Fiber nonlinearities are very important when considering Raman amplification within transmission fibers. The refractive index of germano-silicate fibers changes with the intensity of the light propagating through the fiber. The nonlinear propagation effects relevant to the evolution of the electric field amplitude involve a cubic term in the electric field, which can be seen in the NLSE. In a communication system with multiple signal channels, the nonlinear effects are typically classified by the number of field amplitudes involved in the process, each associated with its own frequency. Three fiber nonlinearities that are caused due to the power dependency on refractive index are as follows: (1) SPM—this term is used if all the three fields belong to the same frequency as the channel of interest. If SPM occurs, the phase of a given channel is modulated by the signal itself. SPM causes broadening of the pulse in the frequency domain if there is an absence of GVD. In certain conditions, SPM can be used to counteract the effect of GVD on a soliton. In other conditions, it causes unwanted pulse distortion. (2) XPM—this is used when a phase change appearing on a signal channel may also be caused by signals, a tone, or two other channels. The phase modulation caused by other signal channels causes a signal distortion and hence degradation of the system performance. The XPM is strongly dependent on the walk-off between the signal channel and the channels causing the XPM. Thus, it depends on the channel's spacing, the GVD, and the polarization of the channels. (3) Four wave mixing (FWM)—this occurs when all three amplitudes in the cubic nonlinear term belong to channels other than the one considered. In this process, power is transferred to new frequencies from the signal channels. The appearance of additional waves and the depletion of the signal channels will degrade the system performance through both cross talk and depletion. The efficiency of the FWM depends on channel dispersion and channel spacing.

Other types of nonlinearities are the two major scattering effects that take place due to the intrinsic properties of optical fibers: stimulated Brillion scattering (SBS) and SRS. Both scattering effects shift the frequency of the signal downward. These effects are considerably more dominant at high pump powers as the intensity of scattered light grows exponentially once a power threshold is exceeded. Throughout this chapter, SRS has been the theoretical foundation behind the amplification process and requires a large pump power to be produced. This is the reason why such high pump powers are required when implementing Raman amplification techniques. SRS does not contribute to fiber losses like SBS does. Input powers less than 10 mW are low enough for SRS to not take place, whereas SBS does not need high pump powers to provide loss to the signal.

10.4.2 Dispersion

From their early invention, single mode optical fibers have been exploited to their transmission limits. The limits of an optical fiber were not realized until its full implementation and greater transmission rates were passed through. Dispersion is a term that has been rarely mentioned in this document, but is of great importance in transmission systems. Dispersion can be classified as two separate optical effects: chromatic dispersion and polarization mode dispersion (PMD).

Chromatic dispersion represents the fact that different colors or wavelengths travel at different speeds, even within the same mode. Chromatic dispersion is the result of material dispersion, waveguide dispersion, or profile dispersion. PMD is another complex optical effect that can occur in single-mode optical fibers. SMFs support two perpendicular polarizations of the original transmitted signal. If both polarization modes were perfectly round and free from all stresses, they would propagate at exactly the same speed, resulting in zero PMD. However, practical fibers are seldom perfect, thus the two perpendicular polarizations may travel at different speeds and, consequently, arrive at the end of the fiber at different times [10]. Generally, the effects of PMD are smaller than any given data rate; thus they can be ignored.

10.4.3 Split-Step Fourier Method

The numerical method used to solve the NLSE is known as the SSFM. This method accurately models the fiber nonlinearities within the system. It only applies to the un-normalized expression of the Schrödinger equation and requires that the NLSE be transformed into

$$\frac{dA}{dz} = A(\hat{D} + \hat{N}) \tag{10.18}$$

where

$\widehat{D}$ represents the differential operator that accounts for dispersion and absorption in a linear regime

$\widehat{N}$ is a nonlinear operator that governs the effect of fiber nonlinearities

The resulting un-normalized NLSE in split-step form can be represented as follows:

$$\widehat{D} = \frac{i}{2}\frac{\partial^2 |\beta_2|}{\partial t} + \frac{1}{6}\frac{\partial^3 \beta_3}{\partial t} + \frac{1}{2}\left[g(\xi) - \alpha\right]$$
$$\widehat{N} = i\left|A\sqrt{\gamma}\right|^2 + iTR\frac{\partial\left|A\sqrt{\gamma}\right|^2}{\partial t} \tag{10.19}$$

In theory, the nonlinear effects and dispersion work in conjunction within transmission fibers. As SSFM is only an approximation method, these two factors can be broken up and solved individually. The fiber transmission distance is partitioned into a large number of segments of width h. Within a segment, the effect of nonlinearity is included at the mid-plane (shown by a dashed line). Both dispersion and nonlinearity can be solved through an analytical approach that involves converting from time domain to Fourier domain then inverse Fourier domain and then back into time domain. This will be shown in the MATLAB code in the appendix.

10.4.4 Gaussian Pulses, Eye Diagrams, and Bit Error Rate

Testing the SSFM was quite a challenge, as the runtime in Simulink is quite slow. Another way of testing the propagation program is to send a theoretical Gaussian pulse through the fiber and check whether the Gaussian pulse that is theoretically transmitted is coming back out with either a gain or a loss, depending on initialization parameters. The full code for this process is shown in the appendices. A depiction of the typical Gaussian curve that illustrates the FWHM and the amplitude is given in the following text; the eye diagram can also be used to examine signal integrity [11]. The resultant Gaussian pulses from the propagation program are compared to the eye diagrams from the ASK model, which are both compared, in turn, to that of the numerical solution. This numerical solution had to be added to give gain to both the system and propagation models.

10.5 Raman Amplification and Gaussian Pulse Propagation

10.5.1 Fiber Profiles

Before any system design can take place, the right fiber parameters have to be implemented into the program. Table 10.6 depicts the typical optical fibers found in industry and the required pump powers to make the signal reach back to net gain zero for a length of 100 km of fiber.

Table 10.6 offers no surprises, as the required pump power decreases when the Raman gain coefficient is increased. Thus the ideal fiber for the modeling application is the

TABLE 10.6

Fiber Properties

Type of Fiber	Raman Gain Efficiency (g_{eff})	Effective Area (μm^2)	Required Pump Power for Net Gain = 0 (mW)
Allwave	0.35	84.95	700
Corning NZ-DSF	0.72	80	350
LEAF (NZ-DSF)	0.45	72	550
SMF-28 (NDSF)	0.38	84.95	650
Truewave RS (NDSF)	0.58	60	425
Truewave reach fiber	0.6	55	425

Corning NZDSF. Testing for these parameters took place in my colleague's program, which numerically solved the coupled equations that govern Raman amplification. This program was later incorporated into the Gaussian propagating program and then the ASK model.

For Corning fiber, the dispersion factor can be estimated using the following characteristics: $D_\lambda = \frac{S_0}{4}\left(\lambda - \left(\lambda_0^4/\lambda^3\right)\right)$ (see Chapter 1), where $S_0 = 0.093\,\text{ps/(nm}^2\cdot\text{km)}$, $\lambda_0 = 1302$–$1322\,\text{nm}$, and $\lambda = 1550\,\text{nm}$ (operating wavelength). Thus the calculated dispersion is approximately equal to $17\,\text{ps/(nm}\cdot\text{km)}$. The dispersion effects on other wavelength channels can be estimated.

The gain profile obtained from MATLAB is transferred into the propagation program in the Simulink model to help add gain to the NLSE, which in turn underwent the SSFM. Some of the results achieved by the numerical solution of the coupled equations are given in the following text. Note: For all testing $\alpha_S = 0.19\,\text{dB/km}$, $\alpha_P = 0.24\,\text{dB/km}$, effective area $(A_{eff}) = 80\,\mu\text{m}^2$, Raman gain coefficient $= 0.72\,\text{W}^{-1}\,\text{km}^{-1}$. The plot shown earlier depicts the power over the length of the fiber (Figure 10.9). Forward (upper), backward (lower), and

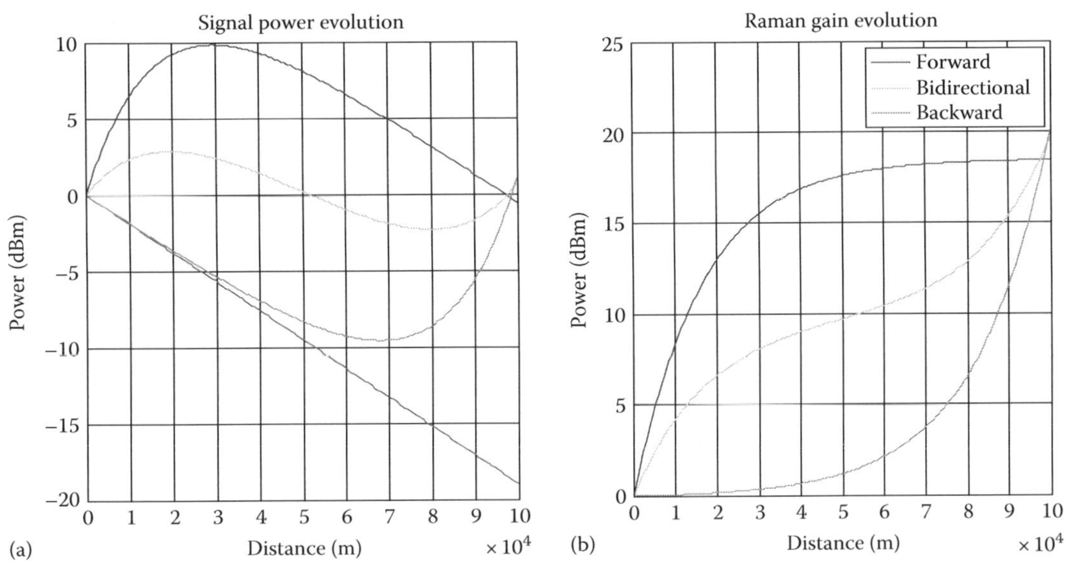

FIGURE 10.9

(a) The power evolution of the signal along the total length of the fiber. (b) The gain evolution over the total length of the fiber due to Raman amplification.

bidirectional (middle) pumping techniques are shown, along with the passive fiber (red). The typical net gain at the end of the fiber is approximately zero.

Figure 10.9b shows the gain profile over the length of the fiber, which is of great importance as this is the result that must be implemented into the propagation program that uses SSFM to simulate the transmission of the signal over the fiber. As the attenuation due to the signal is constant over the fiber, it can be considered as a constant gradient. To incorporate the gain into the NLSE, it too had to be converted into gradients, so it could be simply added to the attenuation map. This can be seen in any of the Raman ssprop programs in the appendices.

10.5.2 Gaussian Pulse Propagation

Once the NLSE incorporated the Raman gain and was solved using the SSFM propagation program, it needed to be tested in a system setting, before being implemented into the ASK model. The best way to test the SSFM is with a single Gaussian pulse. Before testing the single pulse on the Raman-induced fiber, it was tested on a passive fiber, where the end result is known. Thus the following plots are obtained (Figure 10.10):

In all the Gaussian testing cases presented throughout the results section, the input pulse is depicted in blue and the output in green. In this first case, the input pulse was 1 mW and the output pulse was 0.01258 mW. The power lost over 100 km of SMF should be 19 dB. Actual power lost in Gaussian system transmission is $10*\log_{10}(0.01258/1) = -19.0032$ dB. As the result was within an acceptable range, the Gaussian program was ready for further testing. The next type of testing with the Gaussian pulse inputs was the three configurations of Raman amplification. The lengths were left the same, and the pump power maximum used was 350 mW in either forward or backward cases, or 175 mW at both ends when bidirectional.

10.5.2.1 Bidirectional Pumping Case

This case shows the signal power evolution over the length of the fiber. As can be seen, the fiber experiences backward and forward power gains, due to the power from pumps

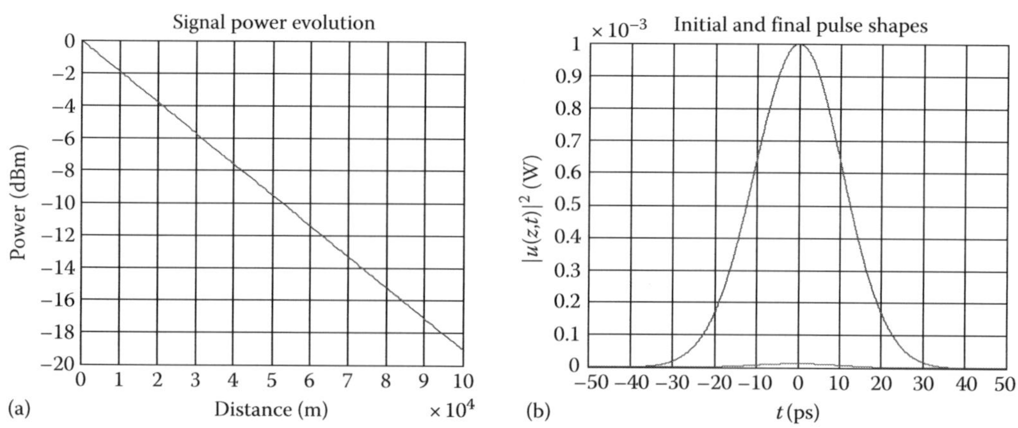

FIGURE 10.10
(a) Attenuation map of a passive fiber over the total length without amplification, $\alpha = 0.19$ dB/km and (b) Gaussian pulse before and after transmission.

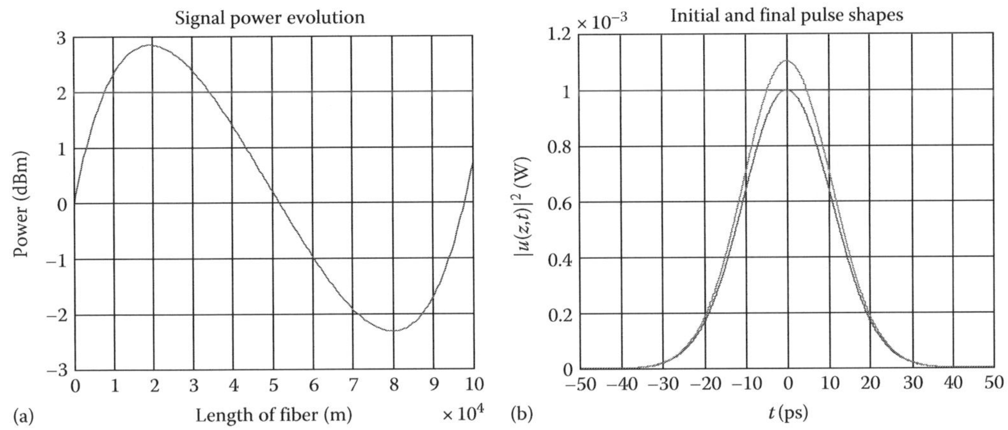

FIGURE 10.11
(a) Signal power evolution: bidirectional pumping and (b) Gaussian: bidirectional pumping.

being implemented in the forward and backward direction, hence the term bidirectional. At the end of the fiber, the expected output is 0.7075 dB. To check to see whether the output obtained from the split-step method is correct, one must look at the output Gaussian pulse. The peak at the output is 1.104 mW, which in turn correlates to $10\ \log_{10}(1.104/1) = 0.4297$ dB (Figure 10.11). The expected output is 0.7075 dB; thus there is a little bit of a discrepancy between the two functions.

10.5.2.2 Forward-Pumping Case

The output of the signal obtained from the end of the fiber is −0.5503 dB after undergoing forward gain. The output as seen by the graph given earlier is 0.8803 mW. This in turn correlates to a gain of $10\ \log_{10}(0.8803/1) = -0.5537$ dB. This means the forward case is almost identical to the numerical model result of −0.5503 dB (Figure 10.12).

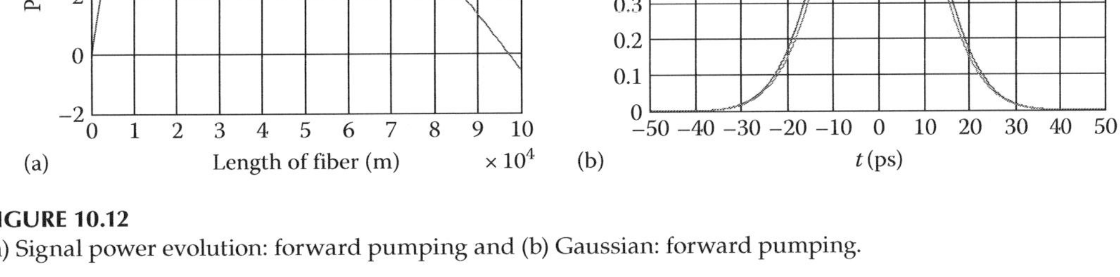

FIGURE 10.12
(a) Signal power evolution: forward pumping and (b) Gaussian: forward pumping.

10.5.2.3 Backward-Pumping Case

The output of the signal obtained from the end of the fiber is 1.128 dB after undergoing backward gain whereas a final pulse of 1.141 mW on the Gaussian correlates to 10 Log_{10} (0.8803/1) = 0.5729 dB (Figure 10.13). As observed, there are inaccuracies in the three pumping techniques presented. This is due to a number of factors including the numerical approximation of the gain curves, the gradients used from the gain curves, and the implementation of the fast Fourier transform (FFT) used to solve the NLSE. All three estimation factors contribute to significant dissimilarities between propagating techniques.

10.5.2.4 Back-to-Back Performance

The initial characteristics of the ASK model need to be tested before simulation can take place. The first back-to-back eye diagram is an illustration of the input from the system (Figure 10.14).

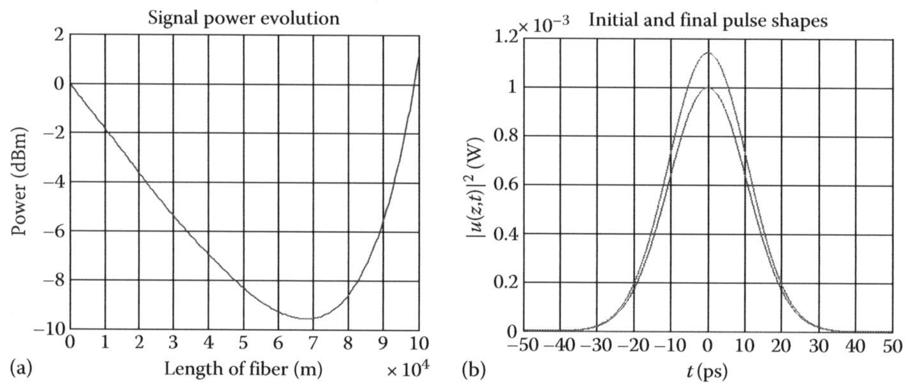

FIGURE 10.13
(a) Signal power evolution: backward pumping and (b) Gaussian: forward pumping.

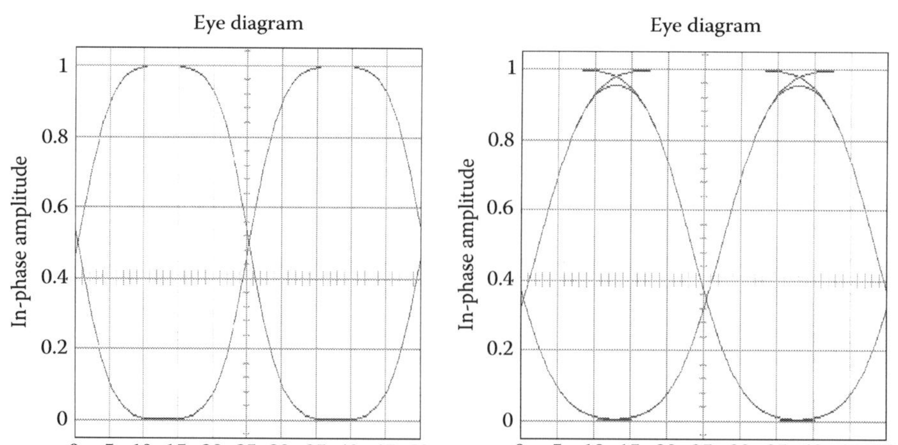

FIGURE 10.14
(a) Input signal from the ASK transmitter and (b) output signal at the ASK receiver.

The output eye diagram is as expected even though the output is dilated, the eye is clearly distinguishable, and thus what is transmitted is being received. Note that the pulse width is as initialized, 25 ps.

10.5.2.5 Propagation under No Amplification

The following plots represent the ASK model working over a fiber, but no amplification is taking place. The first plot represents the Gaussian output over 3 km of fiber with $P_0 = 1$ mW and a total length of fiber = 3 km (Figure 10.15).

As can be observed, both the eye diagram at the output stage and the Gaussian pulse correlates well. At the output stages, both diagrams show amplitudes of approximately 0.88 mW correlating to a loss of 0.5552 dB. Theoretical foundation depicts the loss at 0.57 dB, leaving only a small room of error.

10.5.2.6 Propagation under Fiber Raman Amplification

Forward pump power = 0.4 W and backward pump power = 0 W.

As can be seen from the signal power graph (Figure 10.16), the expected output is 2.883 dB. On the Gaussian plot, the output obtained is 10 $\log_{10}(1.928) = 2.8511$ dB, which also matches that of the output eye diagram.

The output shows that the amplitude of the eye diagram is correlating with that of the Gaussian pulse. This is of no surprise as the signal runs through the same process as the Gaussian pulse. As indicated earlier, the noise experienced in the ASK model is due to the numerical approximation of the gain curves, the gradient functions used from the gain curves, and the implementation of the FFT used to solve the NLSE, and therefore cannot be solved and can only be minimized when using this estimation technique.

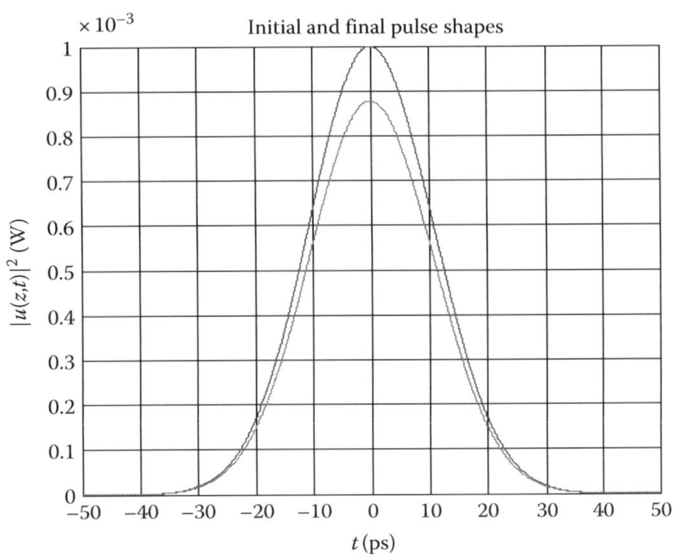

FIGURE 10.15
Gaussian: 3 km passive fiber.

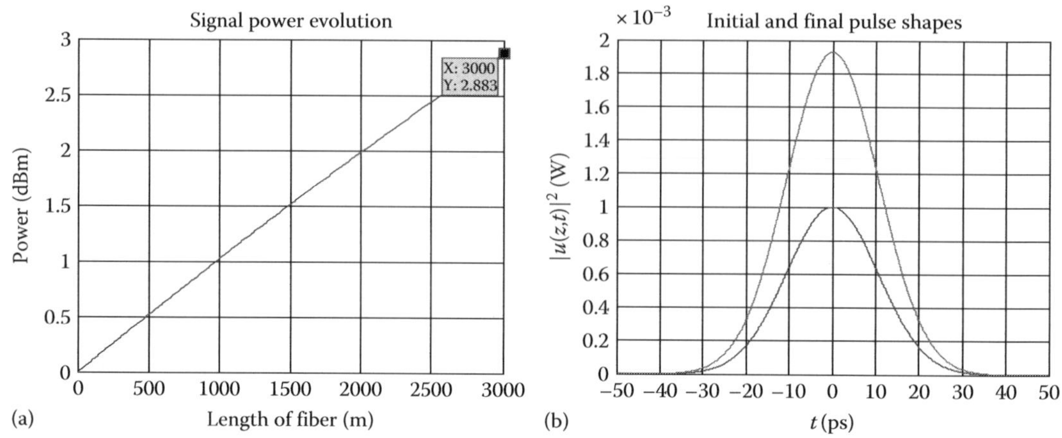

FIGURE 10.16
(a) Signal power evolution: forward pumping and (b) Gaussian: 3 km forward pumped fiber.

10.6 Optically Amplified Transmission

10.6.1 EDFA Amplification and Distributed Raman Amplification over 99 km Fiber (1 km Mismatch)

10.6.1.1 EDFA

As an example, the following results are obtained using the ASK Simulink model with EDFAs. As it was, it contained 50 km SMF, an EDFA, 49 km DCF, and then another EDFA. The eye diagram for the input is given in the following text. The reason that there is a mismatch between the fibers is due to the system requiring dispersion to reduce nonlinear effects within the fiber. The pulse width is 25 ps, which is the expected result after the initialization file is run. This corresponds to a 40 Gb/s transmission rate. The plot shown on the next page depicts what is received after transmission is complete.

The power input into this system is 1 mW; thus there is quite a large gain as the output is approximately 0.13 W, which equates to an overall system gain of 10 log(0.13/0.001) = 21.1394 dB. This is quite unrealistic, and exceeds the power needed at the receiving end; thus the overall system gain should be dropped back down to approximately zero. There are two ways of achieving this: increasing the length of the fiber or decreasing the amount of gain that the amplifiers provide. As there needs to be a comparison between the ASK model with EDFAs and DRA, the length was kept consistent, and the EDFA gains were changed from 20 to 9.5 dB and noise kept at the same ratio, thus decreasing it to 2.375 dB. The eye diagram is also obtained, showing a signal peak of approximately 1 mW. This is more realistic, as the input signal is the same as the received signal and falls within a reasonable operating range. It must also be noted that the signal becomes a lot less noisy due to the fact that the EDFA is not providing too much gain, which in turn provides greater noise in the system.

10.6.1.2 Distributed Raman Amplification

Figure 10.17c depicts the output of the ASK model when using DRA (input is the same as that of the previous model). The output peak power of this particular model correlates

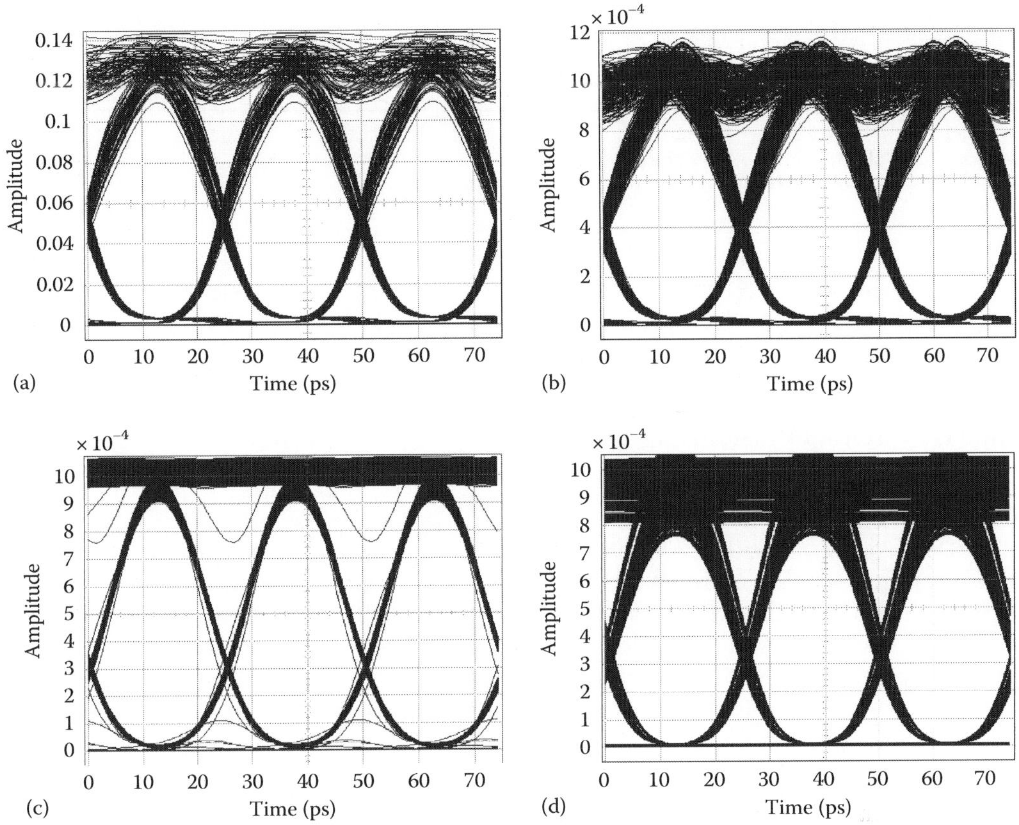

FIGURE 10.17
(a) Output signal of ASK model (20 dB gain EDFA only), (b) output signal of ASK model (9.5 dB gain EDFA only), (c) output signal of ASK model (DRA model), and (d) output signal of ASK model (hybrid model).

to that of the input. The accuracy of the result could be made a lot clearer than what is presented. But, as mentioned before, the use of the FFT degrades the accuracy of the algorithm taking place, the SSFM. It must be noted that the Raman amplification model implemented in this section did not account for any noises that could have been added to the transmission signal.

When taking bit error rates (BERs) into account, the noise of Raman amplification must be added. To check the differences in BER between Raman models with and without noise, both of them have been included with the EDFA model. This process gives an estimated description as to what the effects of noise are on the transmitted signal. The BERs for the three systems have been calculated using the equations as given by Agrawal [3]. To make the process of calculation easier, these equations were made into a MATLAB program, which can be seen in the appendices. Table 10.7 shows the BERs for the two systems and the effect that mismatching has on the system. It illustrates the differences between the error rates of three models after undergoing changes in mismatching. All three amplification techniques have low BERs when the mismatch is zero, then starts to increase as the mismatch increases. This correlates in accordance with theoretical foundations—as the Q parameter increases, the BER improves; thus Q is inversely proportional to BER.

TABLE 10.7
Bit Error Rate and *Q* Factor for Different Mismatching

Kilometer Mismatch	BER			*Q* Parameter		
	EDFA	Raman without Noise	Raman with Noise	EDFA	Raman without Noise	Raman with Noise
0	1.33e−77	8.03e−268	4.34e−117	23.8123	47.4096	31.9322
1	4.76e−62	7.01e−164	3.04e−94	22.9173	33.6381	26.9262
2	2.05e−54	1.21e−144	1.22e−71	19.1478	28.6273	20.9189
3	1.04e−41	5.77e−83	2.79e−49	14.174	23.3518	16.129
4	9.41e−17	3.90e−22	3.49e−19	11.3113	11.5046	11.7835
5	9.85e−15	4.90e−13	4.93e−13	8.939	7.5701	7.4943

10.6.2 Hybrid Amplification

Hybrid testing of the system involved the addition of Raman amplification and EDFAs to coexist in a single system model. The EDFA was implemented after both the SMFs and DCFs. The length of the fiber was increased to 100 km (span = 200 km), and the pump power was decreased to 100 mW in the backward direction; therefore the gain required by the EDFAs is 13.26 dB. This was found out by first testing the Raman amplification in the Gaussian program and then deducing the amount of gain required to boost the signal back to a net gain of zero. The eye diagram obtained at the end of a single span for hybrid amplification is shown in the following text. There was no mismatch between the two fibers (Figure 10.18).

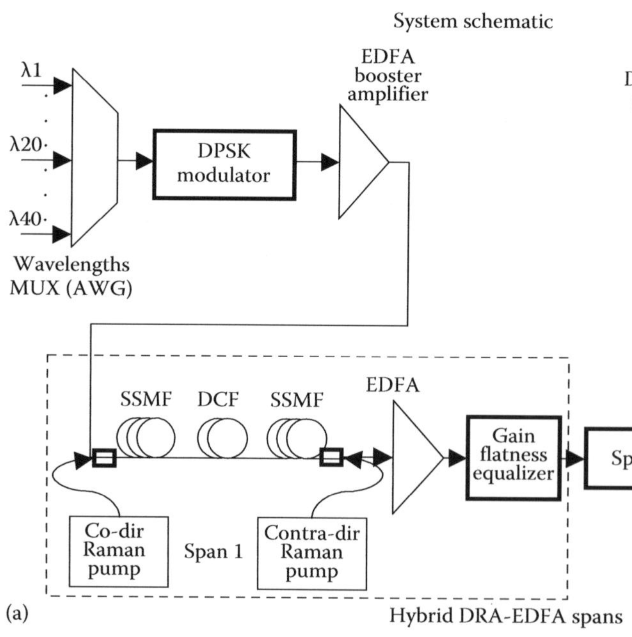

FIGURE 10.18
Case study: Generic schematic of the hybrid optically amplified hut-skipped fiber transmission systems along the coast of Vietnam. (a) System arrangement and (b) corresponding Vietnam North–South submarine hut-skipped backbone (——) undersea transmission links and spans.

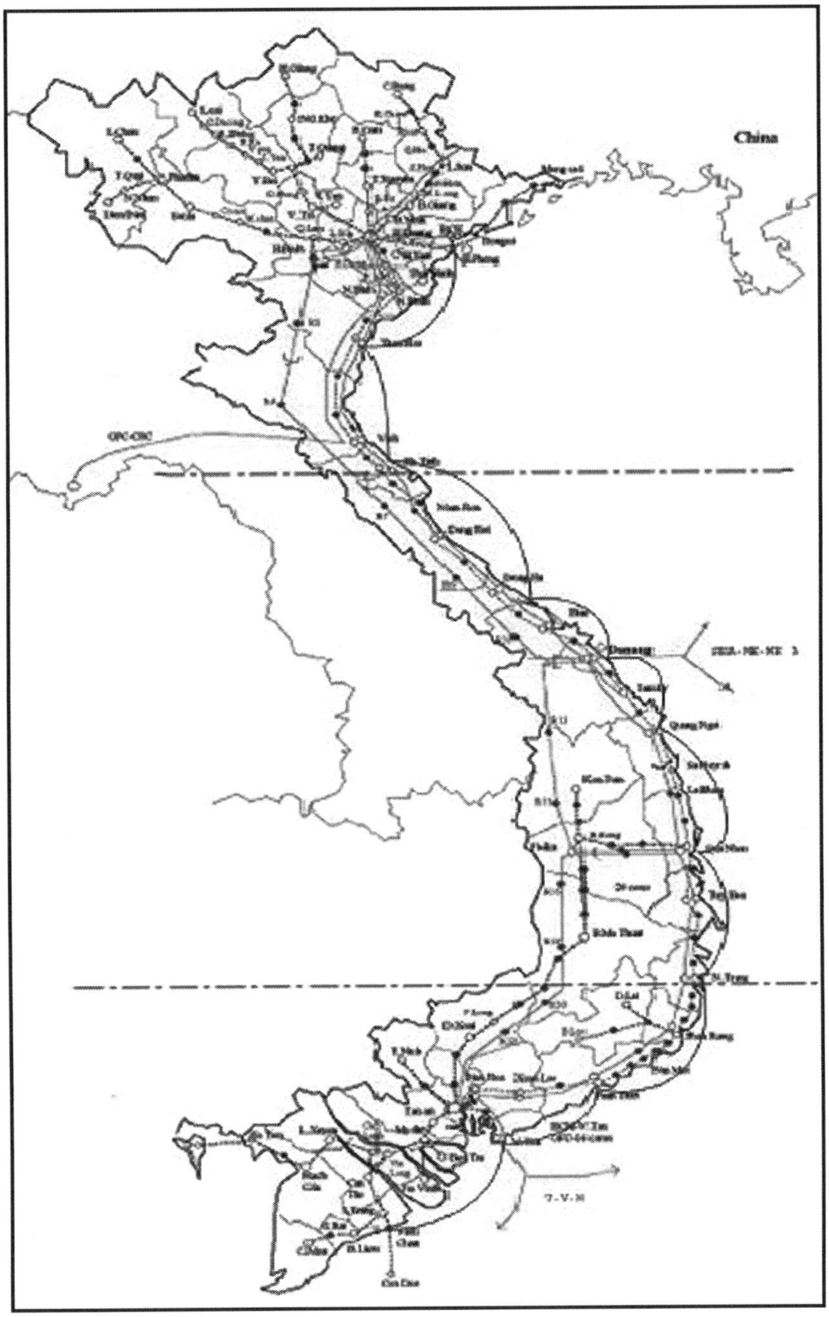

(b)

FIGURE 10.18 (continued)

The resultant Q parameter and BER are 10.8830 and 4.8712e−013, respectively. This is noticeably higher than that of the Raman and EDFA models when they were in the same position. This could be due to a number of reasons like the amount of noise added to the system being considerably higher due to both amplification techniques working in tandem. Thus, from the results obtained it would be safe to say that only Raman amplification is a good model for use in transmission.

It must be noted that this model had lengthy spans of 100 km, which should not be compared to that of the previous section. When the length of the fibers were decreased to 50 km each (span = 100 km), the total BER and Q parameter were 4.8712e−013 and 9.8450, respectively. This is considerably higher, but gives a much more accurate depiction as to what values should be obtained.

10.6.3 Long-Haul Optically Amplified Transmission

The final set of demonstration of the effectiveness of our design of the ROA and EDFAs as hybrid optical amplification distributed over transmission fiber spans over a long-haul undersea hut-skipped jump transmission backbone. Raman amplification can be achieved by pumping from the two ends of the spans. It is found to be approximately 2000 km with the modulation scheme of DPSK and an average hut-skipped distance of 240 km composed of Corning Vascade fibers of positive and negative dispersion types. Thus this section presents the methodology for obtaining the results derived from the simulation models described in previous sections. Initially the method for the evaluation of the BER is described by the statistical evaluation of the eye diagram. The quality factor (Q factor) is then obtained under the assumption that the probability distribution function follows a Gaussian type. The transmission performance can then be evaluated by setting a mismatching distance of the DCF and that of the transmission fiber. The SPM nonlinear effects are evaluated by setting the launched power at the input of the first span and the distortion is measured.

10.6.3.1 Eye Diagram and BER Evaluation under Noise Loading Effects

In this study, the dispersion of individual span and cascaded span transmission is completely compensated; thus only noises of cascaded optical amplifiers contributed to the degradation of the detected eye diagram at the output of the optical receiver. BER and its corresponding quality factor (Q factor) are used to determine system performance by evaluating its eye diagram. In practice, BER has been calculated by the probability of the incorrect received bit in a working transmission system. However, in our Simulink model, the BER is evaluated, assuming a Gaussian probability density function as given in (10.5) and (10.6) [3]:

$$\text{BER} = \frac{1}{2}\text{erfc}\left(\frac{Q}{\sqrt{2}}\right) \tag{10.20}$$

$$Q = \frac{I_1 - I_2}{\sigma_1 + \sigma_2} \tag{10.21}$$

where

I_1 and I_2 are average voltages of received high and low voltage levels, respectively

σ_1 and σ_2 are the standard deviations of received high and low levels, respectively

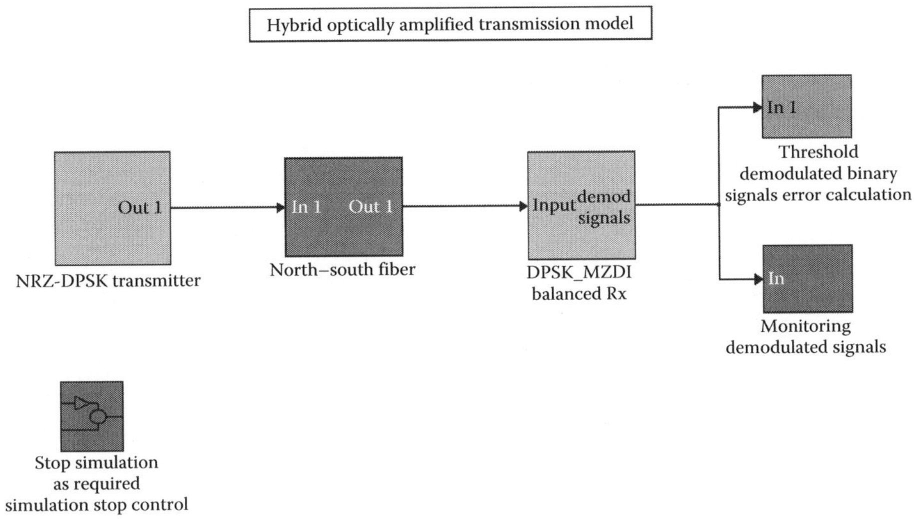

FIGURE 10.19
Transmission system setup with general blocks in MATLAB and Simulink.

Our model automatically records the received eye diagram by the time scope set at persistence mode as provided in MATLAB and Simulink (Figures 10.19 through 10.25) and depicted in Figure 10.26.

Figure 10.26 illustrates the eye diagram of the system working with 12 dBm total launched power, perfect dispersion compensation, and evaluating at center wavelength (1550 nm).

10.6.3.2 Launched Power versus BER

Increasing launched power to the transmission system improves the sensitivity of the photodiode. But when over the SPM threshold, the nonlinear effects become serious. Therefore, there is a trade-off between high launched power and low nonlinear effects in order to improve the performance of the optical transmission systems. It is thus useful to study launched power versus BER because the best launched power can be determined.

Launched power is the peak power of whole 40 wavelengths channels. Launched power of each channel is thus calculated as *total power* (dBm) + log 10 (1/4). In our MATLAB and Simulink models, the launched power is a global parameter and is declared in the model initial file. BER is evaluated against the launched power, which is a parameter that varies from 5 to 23 dBm. Also note that nonlinear threshold power for peak power is set at 15 dBm (32 mW) for return-to-zero (RZ) format. Although typical nonlinear threshold power is about 3 dBm, using RZ pulse shape can improve the threshold power to about 3 dB. Applying the methodology described earlier, center channel at wavelength 1550 nm is analyzed. The nonlinearity performance result is plotted in Figure 10.27.

As observed in Figure 10.27, launched power from 5 to 12 dBm is proportional to the Q factor in the log scale. From 12 to 15 dBm, it is saturated; Q factor reaches its peak value. Launched power keeps increasing over 15 dBm and leads to rapid decrease in the Q factor. Hence, 12 dBm is relatively the best launched power of our long-haul transmission system under hybrid Raman distributed amplification and lumped EDFA across all spans of the 1994 km distance.

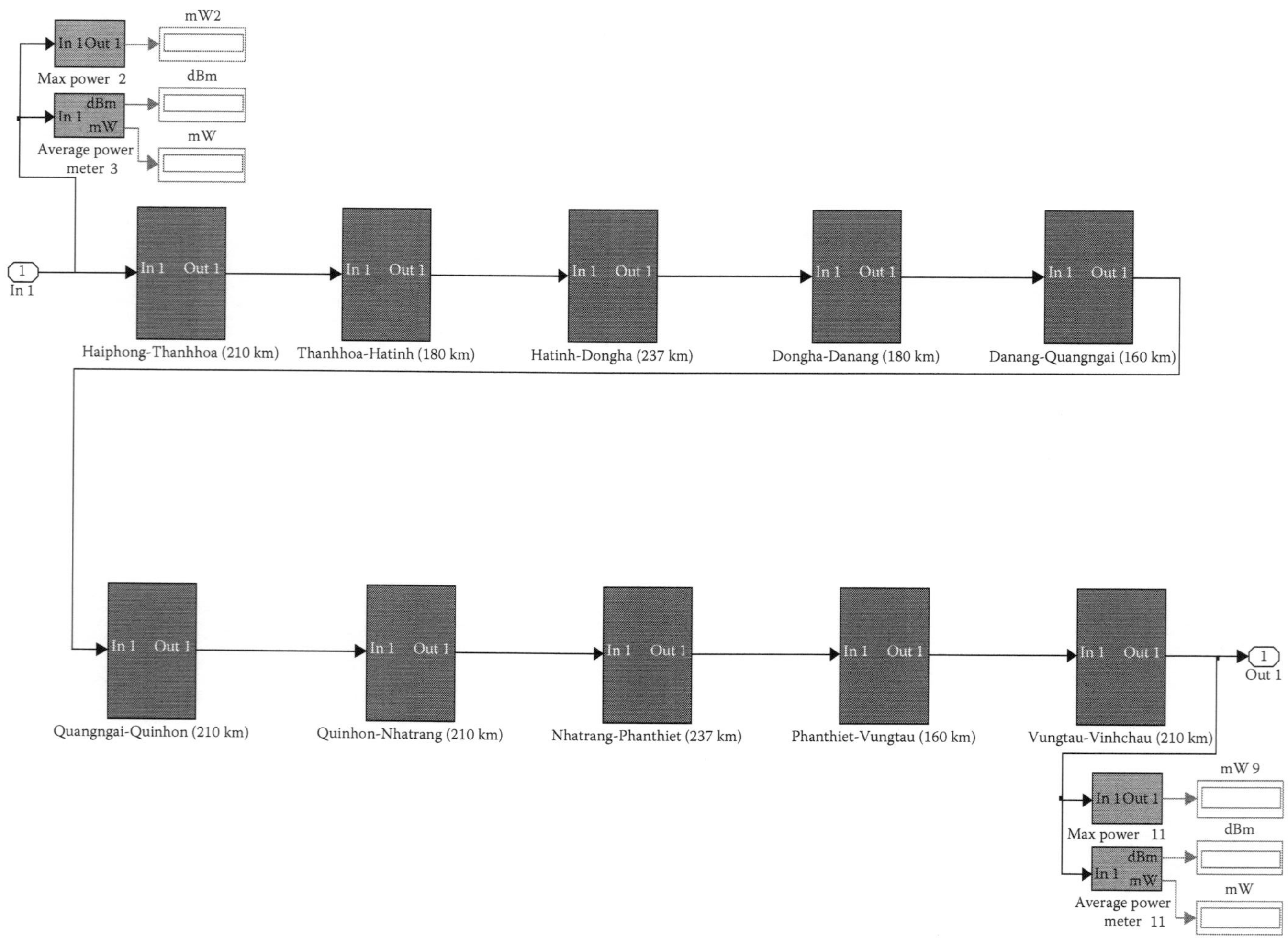

FIGURE 10.20
MATLAB and Simulink models of the multi-span transmission block. The names indicated below each block are the start and stop locations of undersea spans. Optical powers are monitored in each spans for optimization and at the transmitter and receiver for measuring of launched power and OSNR.

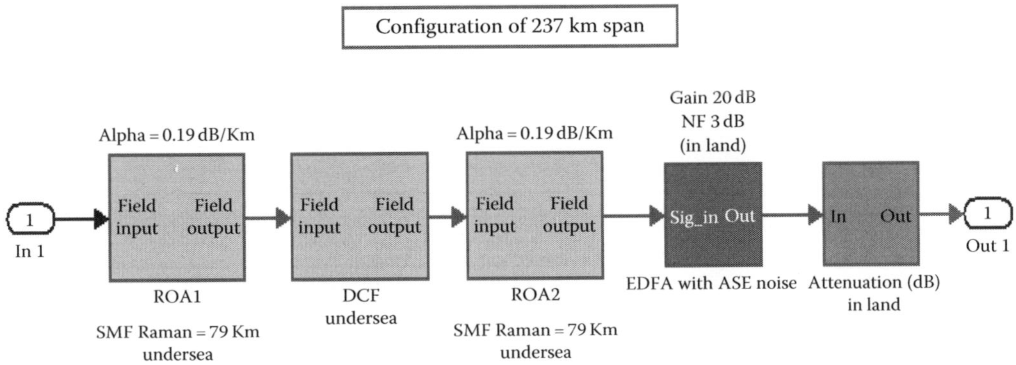

FIGURE 10.21
MATLAB and Simulink models of hybrid 237 km spans using bidirectional DRA and EDFA located at the end of the span.

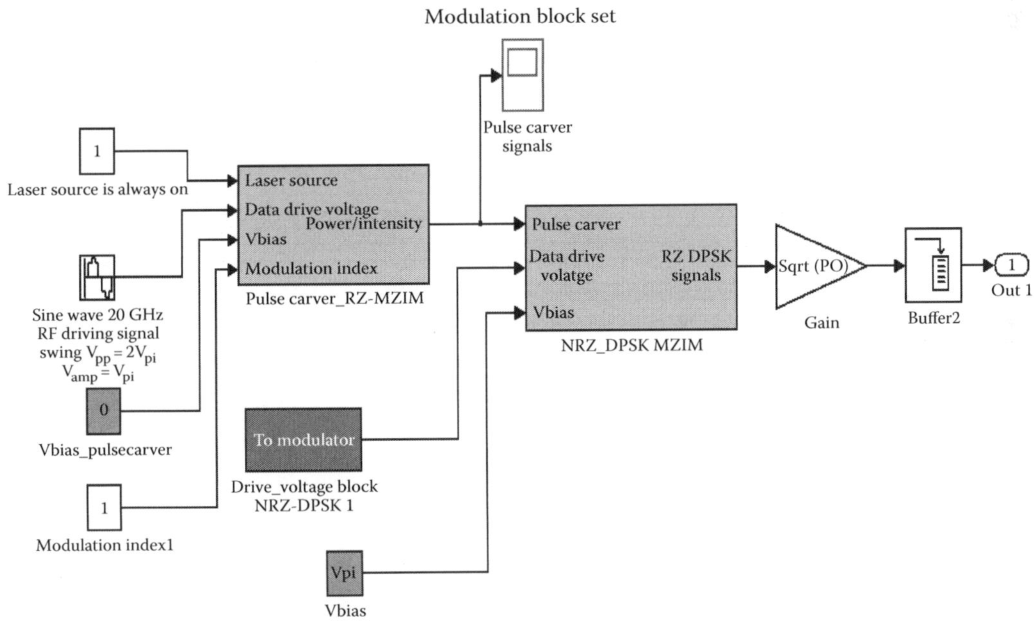

FIGURE 10.22
MATLAB and Simulink modulation blocksets.

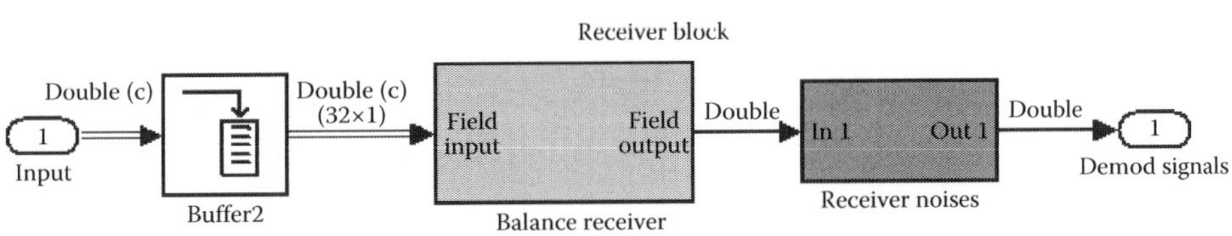

FIGURE 10.23
MATLAB and Simulink receiver blocks.

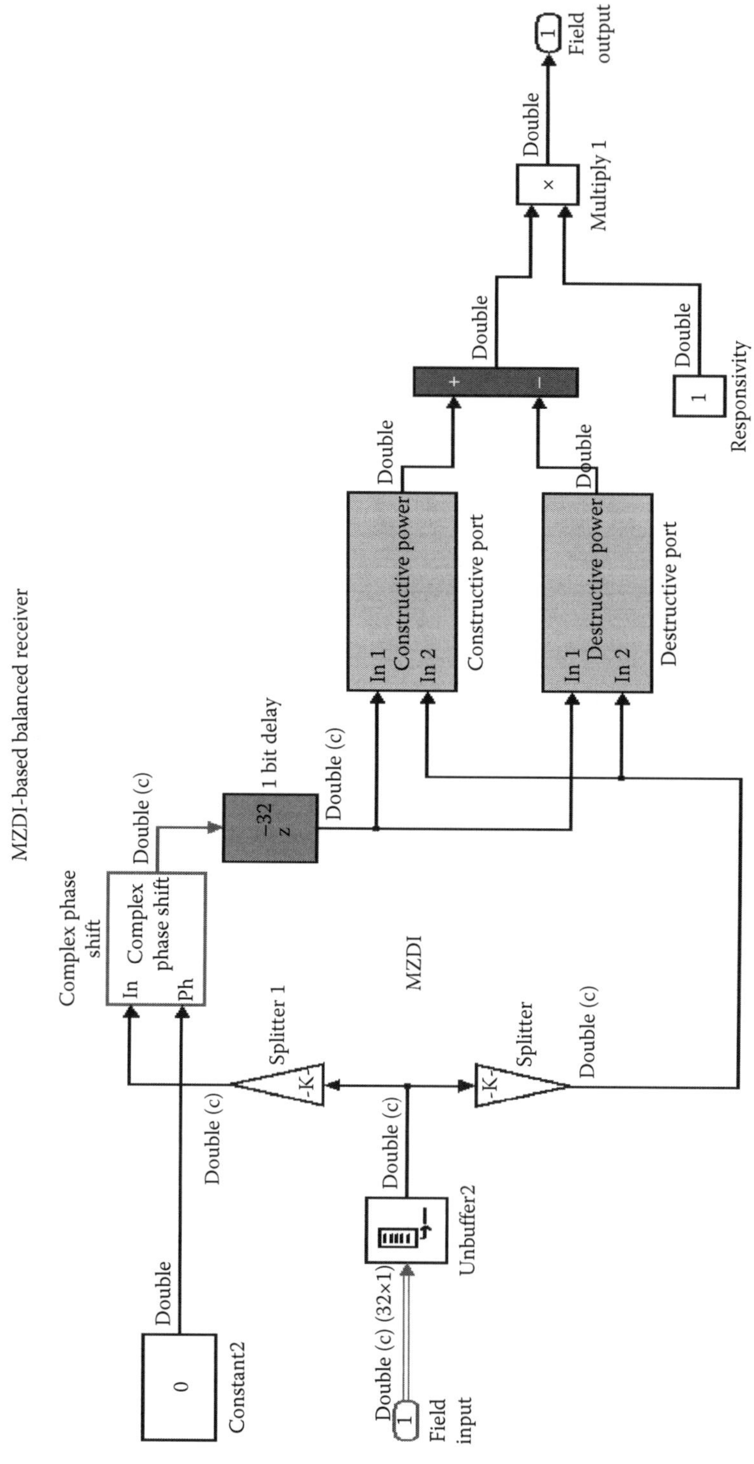

FIGURE 10.24
MATLAB and Simulink models of balanced receiver.

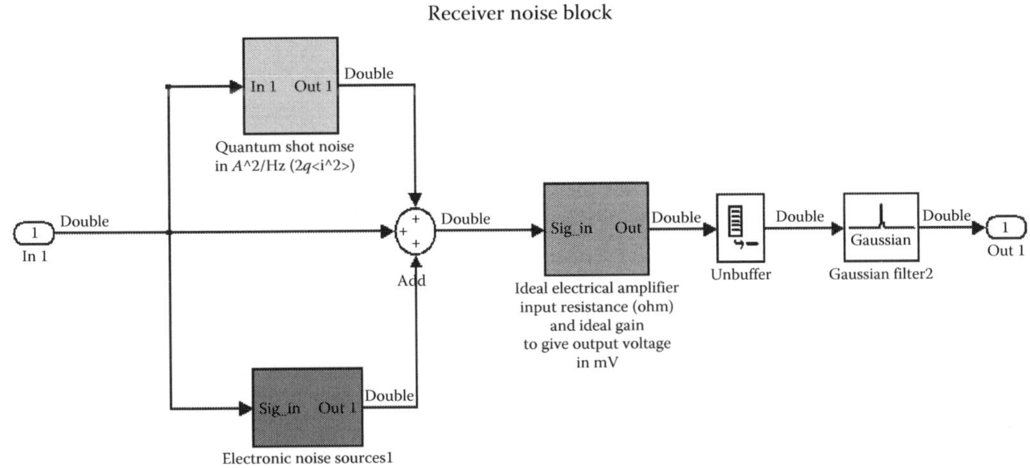

FIGURE 10.25
MATLAB and Simulink receiver noise blocks.

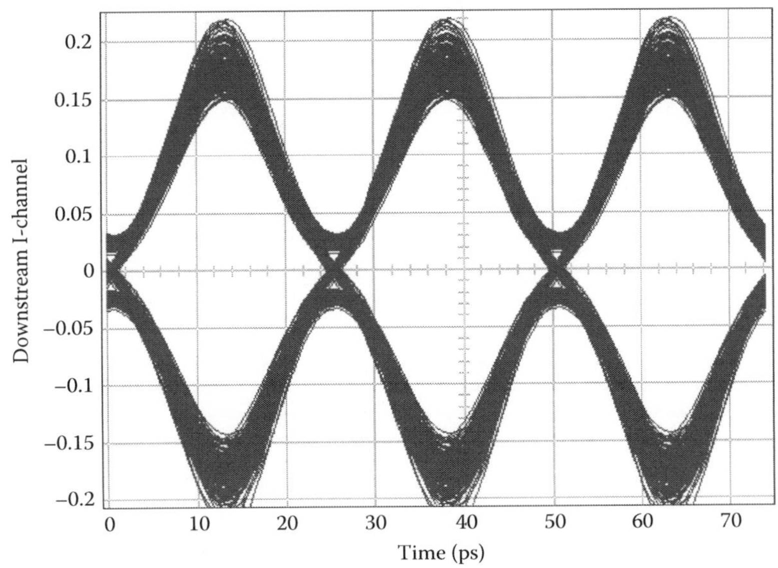

FIGURE 10.26
Typical eye diagram received after the balanced receiver under perfect dispersion compensation over 1994 km hybrid DRA-EDFA optically amplified transmission system.

In the Simulink transmission system, 40 wavelength channels are transmitted simultaneously. Each wavelength channel suffers different dispersion factor due to the fiber dispersion slope and the gain slope of the DRA and EDFA. It is assumed that EDFA gain is flat within transmitting spectral region by using optical gain equalizer associated at each amplifier. The Q factor is plotted against the launched power as shown in Figure 10.27 for three optical channels located in the C-band, in the central region, and the two extreme boundaries.

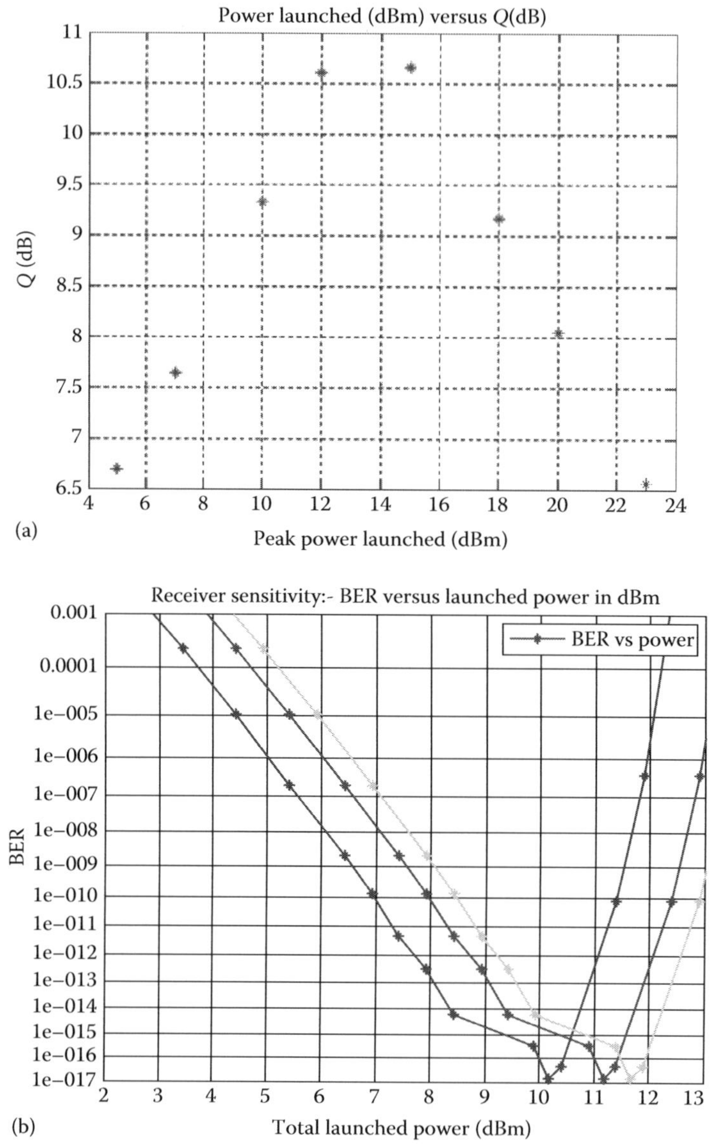

FIGURE 10.27
Selected channels—Q (dB) versus peak launched power (dBm): (a) single channel and (b) channel 1 (far left), channel central, and channel 40 (far right).

The transmission system benefits from using the dispersion-matched pair fiber. Corning Vascade L1000 and Vascade S1000 can be seen with this long-haul transmission. The performance of three channels located at the center and two border channels is almost similar and is plotted in Figure 10.27.

10.6.3.3 Dispersion Tolerance

Practically, it is impossible to install exactly the calculated fiber length; perfect dispersion compensation is thus not expected. As shown in $\beta_{L1000}L_{L1000}+\beta_{S1000}L_2=0$, if a mismatching is

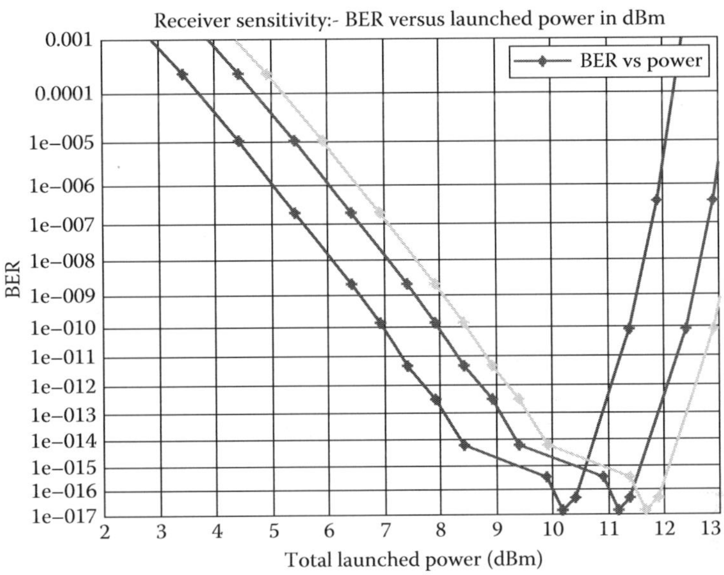

FIGURE 10.28
Setting up of dispersion mismatch by inserting an additional fiber section in the last span as indicated by the block "Mismatching." ROA1: first Raman gain fiber propagation section; ROA2: Raman gain fiber section under contra-directional pump. In1: optical signals input port. Out1: output signals to be launched into EDFA in-land section or to balanced optical receiver. Attenuator is for adjusting the signal level to below threshold level.

added, the residual dispersion is increased by the dispersion factor of the fibers S1000 and L1000. It is useful to investigate the mismatching dispersion tolerance in order to determine the maximum allowance of the difference in length of the two types of fibers over the total transmission distance. The maximum tolerance is determined as the distance over which the system performance is still error free with BER < 1e–9.

In order to measure the dispersion tolerance, an extra mismatching fiber distance is inserted at the end of the last span of the transmission model, as shown in Figure 10.28.

The mismatch fiber distance is varied and the corresponding *Q* factors are obtained until reaching a 3 dB penalty level. The NF of all EDFAs is set at 3 and 5 dB for the worst case. Although latest EDF amplifier technology has made available low NF of 3 dB, the ageing of the EDF amplifier in practical optical transmission could degrade the NF of EDFA to 5 dB [17]. It is thus useful to obtain the performance in the case of 5 dB NF ASE (amplification stimulated emission). The variation of the *Q* factor as a function of transmission mismatch distance is plotted in Figure 10.29. Typical eye diagrams of the decoded DPSK signals at the output of the balance receiver are shown in Figure 10.30 under perfect match and 2.5 km dispersion mismatch.

In Section 10.6.3.1, BER is evaluated at threshold level of zero and at the center sample where the "eye" is opened at maximum. Practically, a voltage comparator can be used for determining the high and low levels. Obviously, the voltage reference of the comparator is not perfectly zero, and the sampling may not be at the center of the pulses due to propagation delay or the jittering effects due to Brillouin scattering. Hence, it is worth studying how BER changes versus delay offset and threshold offset. The bit rate of the system model is 40 Gb/s, and in a one bit period there are 32 samples. BER is evaluated at different received samples and different voltage threshold levels. A BER contour plot is shown in Figure 10.31 for DPSK under noise loading.

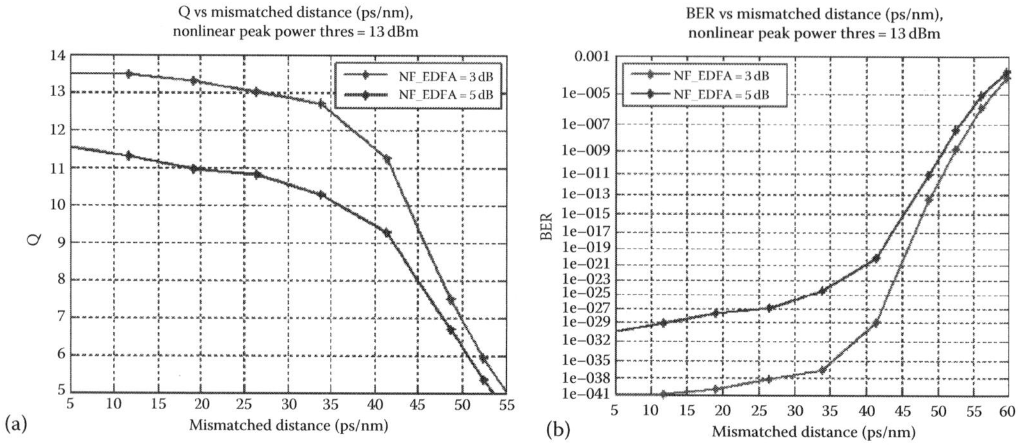

FIGURE 10.29
Dispersion tolerance of the hybrid EDFA-ROA optically amplified transmission system: (a) quality factor variation and (b) BER variation. Modulation format DPSK 40 Gb/s over a total transmission distance of 1994 km over 10 spans of 240 km on average hut-skipped distance. *Q* factor versus dispersion mismatch distance (km) as evaluated at the output of the balanced receiver. * denotes 3 dB NF performance and + denotes 5B NF performance.

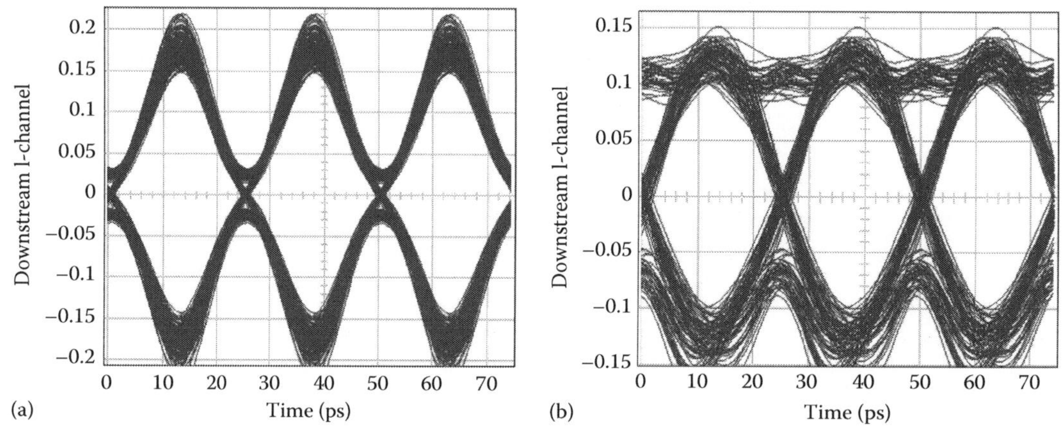

FIGURE 10.30
Eye diagrams obtained at the output of the balanced receiver under (a) perfect dispersion compensation—noise loading; (b) 2.5 km dispersion mismatch with a peak power launch of 12 dBm at the input of the first span and 3 dB noise figure of all EDFA sections.

10.6.3.4 Remarks

This section has presented the transmission performance of a multi-span hybrid EDFA-DRA amplification structure over 10 spans of Vascade transmission and DCFs with the hut-skipped distance averaged at 240 km. This is very much suitable for Vietnam North South fiber backbone with hut-skipped 240 km distance along its coastline for the transport 40 Gb/s over 40 wavelength channels in the C-band.

The simulation results obtained by the MATLAB and Simulink models for the optically amplified submarine hut-skipped system have proven that (1) the EDF amplifier can enhance the distributed gain of DRA so as to increase span hut-skipped distance to 240 km;

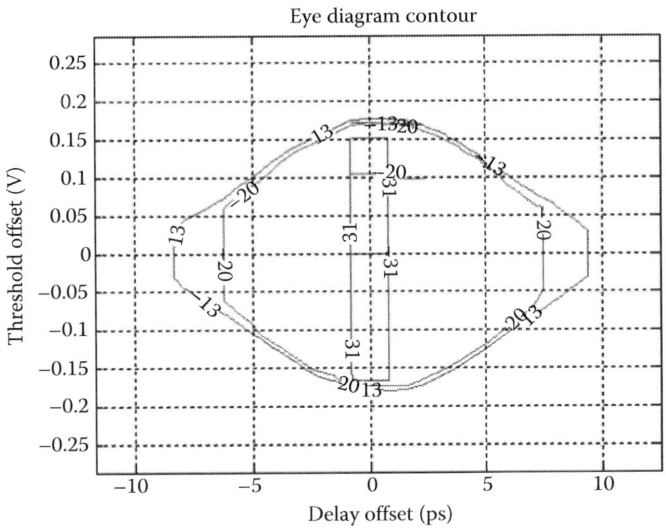

FIGURE 10.31
Eye contour diagram of noise-loaded transmission. The numbers are BER (1e−*N*).

(2) the NF of the EDF amplifier is important as it would limit the noise floor to acceptable level for long-haul transmission; and (3) the effects of SPM limit the launched power level. The maximum peak launched power for carrier-suppressed RZ (CSRZ) DPSK is 12 dBm.

A dispersion tolerance of 3 km SSMF at 1550 nm or 51 ps/nm dispersion factor is achieved for the DPSK CSRZ modulation format. This tolerance is reasonable under simulation. But in practice this puts a very stringent demand on the installation of the cables undersea. This tolerance does not include the PMD effects, which can be compensated at the receiver. Coherent detection with digital signal processing may boost the receiver sensitivity [18] and extend the dispersion tolerance and additional spans.

Further works can be employed to improve the transmission performance including electronic equalization at the output of the balanced receiver to extend the dispersion tolerance and coherent phase estimation of the detected signals.

10.7 Concluding Remarks

By numerically solving the set of ODEs for pump and signal power and noises in both distributed and discrete ROA, the design concerns of gain, NF, and dispersion can be examined. Distributed amplification was shown to pump for 100 km spans of different commercial fibers. The required power for each fiber for net gain to compensate for fiber losses was examined, with pump lasers typically operating at 500 mW. It showed that the transmission span length for Corning NS-DSF and Truewave could be extended well beyond 100 km in a purely distributed ROA.

Typically, SMF transmission spans of 50 km are used because the NF is close to that of an ideal distributed NF, that is, if the loss is corrected at every point along the fiber. In 100 km long spans of the fiber, the NF for distributed ROA in comparison to lumped amplifiers is improved by 6–16 dB depending on the configuration. It was determined that fibers with higher values of g_R provide lower system NFs. Forward pumping achieves an NF improvement of nearly 10 dB in comparison to backward pumping. It can be further improved by using a hybrid configuration, such as an EDFA, to provide a fraction of the gain to the signal. It increases the overall net gain along the fiber and lowers the required pump power for ROA.

Dispersion was also determined, and it was shown that for the different types of fibers there is no real benefit; however, in using different pumping configurations, the total NLPS along the fiber could be reduced. There is an approximate benefit of about 12.5 dB in using a backward-pumping configuration. For discrete Raman amplifiers, typically less than 25 km high gains of up to 40 dB could be achieved using specially designed Raman fiber or DCF with high Raman gain efficiencies and small affective areas. The comparison of NF in discrete Raman amplifiers for different pumping schemes showed that larger span transmissions resulted in a greater discrepancy of up to 7 dB. It was also determined that larger gain efficiencies cause less net gain and more gain saturation to occur. The parameters of gain and NF were integrated into a Gaussian pulse and propagated down the fiber using the SSFM to convert the average powers of the numerical solutions to pulses. This was achieved by changing the half-step SSFM parameter to include ROA by adding the on–off Raman gain gradient at each step size along the fiber. The associated NFs were added to the system as a discrete value. They were then analyzable using MATLAB and Simulink models.

Furthermore, the most reliable form of pumping in a system model setting is the bidirectional configuration. This pumping scheme finds the trade-off between adding too much noise (found when the signal loses power) and adding too much nonlinearity (found when the signal power goes too high). The overall distance in which the signal can propagate without regeneration was found to be 2000 km. The accuracy of this figure depends on many factors. These factors include the SSFM, the sampling of gradients, the accuracy of results, dependency on other programs, the runtime of programs, etc. Such factors affect not only the distance found, but many of the other values and plots found.

We have shown that through estimation and simulation there can be an accurate depiction of DRA in the optical fiber, as it utilizes the fiber in which the signal propagates to perform amplification. All the error rates found concur with theoretical models within texts and were well below that of the minimum required to acquire an accurate eye diagram.

Finally, a multi-span ROA and EDFA optically amplified transmission system is demonstrated to prove the effectiveness of ROA over a multi-span transmission link. This is essential for a system that requires undersea hut-skipped spans such as in the Vietnam coastal backbone to avoid problems on land acquisition for cable installation.

10.8 Problems

PROBLEM 10.1

Give an account of the optical amplification of lightwaves in an Er-doped optical fiber amplifier including the structure of the fiber amplification medium, the energy diagram of the erbium ion, and the amplification and noise generation process.

PROBLEM 10.2

Similarly, repeat Question 1 for a Raman-distributed optical fiber amplifier.

PROBLEM 10.3

(a) Search and download from Avenex.com or jdsu.com technical specifications of EDFAs that would be used as in-line optical amplifiers, booster amplifiers, and optical preamplifiers for optical receiver.
(b) Contrast their optical properties, for example, optical gain, NF, saturation region, minimum input power range, and total maximum output power, spectral gain region, and gain ripple.

PROBLEM 10.4

Similar to Question 3, for Raman optical amplifiers, especially the pump laser source.

PROBLEM 10.5

From the gain and the optical input and output power, sketch the curve representing the input and output power of the optical amplifiers of problems 10.3 and 10.4.

PROBLEM 10.6

A span in a long-haul optical communication system consists of a transmission fiber of 100 km SSMF and a DCF with dispersion factor at 1550 nm of −85 ps/nm/km.

(a) What is the length of the DCF for full compensation at 1550 nm?
(b) Assuming that the DCF has a core diameter of 2.5 μm. Referring to the technical specification of SSMF (SMF-28) as given in Chapter 2, state the core diameter of the SSMF and hence the spot size. Calculate the nonlinear SPM threshold of SSMF and then DCF.
(c) Show that two optical amplifiers would be needed to equalize the losses in SSMF and DCF. Obtain the gain and the minimum and maximum optical power of the two amplifiers.

Appendix: Raman Amplification and Split-Step Fourier Method—MATLAB® Program

```
nt=input(1);
u0=input(2:nt+1);
dt=input(nt+2);
dz=input(nt+3);
nz=input(nt+4);
alpha_indB=input(nt+5);
betap=input(nt+6:nt+9);
gamma=input(nt+10);
P_non_thres=input(nt+11);
maxiter=input(nt+12);
tol=input(nt+13);
%Ld=input(nt+14);
%Aeff=input(nt+15);
%Leff=input(nt+16);
```

```
tic;
%tmp=cputime;
%--------------------
%
%
% The following effects are included in the model: group velocity
% dispersion (GVD), higher order dispersion, loss, and self-phase
% modulation (gamma).
%
% USAGE
%
% u1=ssprop(u0,dt,dz,nz,alpha,betap,gamma);
% u1=ssprop(u0,dt,dz,nz,alpha,betap,gamma,maxiter);
% u1=ssprop(u0,dt,dz,nz,alpha,betap,gamma,maxiter,tol);
%
% INPUT
%
% u0 - starting field amplitude (vector)
% dt - time step - [in ps]
% dz - propagation stepsize - [in km]
% nz - number of steps to take, ie, ztotal=dz*nz
% alpha - power loss coefficient [in dB/km], need to convert to linear
to have P=P0*exp(-alpha*z)
% betap - dispersion polynomial coefs, [beta_0… beta_m] [in ps^(m-1)/km]
% gamma - nonlinearity coefficient [in (km^-1.W^-1)]
% maxiter - max number of iterations (default=4)
% tol - convergence tolerance (default=1e-5)
% OUTPUT
% u1 - field at the output
%-----
% Convert alpha_indB to alpha in linear domain
%-----
alpha=1e-3*log(10)*alpha_indB/10; % alpha (1/km) - see Agrawal p57
%-----
%P_non_thres=0.0000005;
ntt=length(u0);
w=2*pi*[(0:ntt/2-1),(-ntt/2:-1)]'/(dt*nt);
%t=((1:nt)'-(nt+1)/2)*dt;
gain=numerical_gain_hybrid(dz,nz);
for array_counter=2:nz+1
  grad_gain(1)=gain(1)/dz;
  grad_gain(array_counter)=(gain(array_counter)-gain(array_counter-1))/
dz;
end
gain_lin=log(10)*grad_gain/(10*2);
clear halfstep
  halfstep=-alpha/2;
  for ii=0:length(betap)-1;
  halfstep=halfstep - j*betap(ii+1)*(w.^ii)/factorial(ii);
  end
  square_mat=repmat(halfstep, 1, nz+1);
  square_mat2=repmat(gain_lin, ntt, 1);
  size(square_mat);
  size(square_mat2);
```

```
  total=square_mat+square_mat2;
clear LinearOperator
  % Linear Operator in Split Step method
  LinearOperator=halfstep;
  halfstep=exp(total*dz/2);
u1=u0;
ufft=fft(u0);
% Nonlinear operator will be added if the peak power is greater than the
% Nonlinear threshold
  iz=0;
while (iz<nz) && (max((gamma*abs(u1).^2+gamma*abs(u0).^2))>P_non_thres)
  iz=iz+1;
  uhalf=ifft(halfstep(:,iz).*ufft);
  for ii=1:maxiter,
  uv=uhalf.*exp((-j*(gamma)*abs(u1).^2+(gamma)*abs(u0).^2)*dz/2);
  ufft=halfstep(:,iz).*fft(uv);
  uv=ifft(ufft);
  if (max(uv-u1)/max(u1)<tol)
  u1=uv;
  break;
  else
  u1=uv;
  end

  end
  % fprintf('You are using SSFMn');
  if (ii== maxiter)

  fprintf('Failed to converge to %f in %d iterations',tol,maxiter);
  end
  u0=u1;
  end

if (iz<nz) && (max((gamma*abs(u1).^2+gamma*abs(u0).^2))<P_non_thres)
% u1=u1.*rectwin(ntt);
  ufft=fft(u1);
  ufft=ufft.*exp(LinearOperator*(nz-iz)*dz);
  u1=ifft(ufft);
  %fprintf('Implementing Linear Transfer Function of the Fibre
Propagation');
end
%toc;
output=u1;
```

References

1. C. Headley and G.P. Agrawal, *Raman Amplification in Fibre Optical Communication Systems*. London, U.K.: Elsevier, 2005.
2. M.N. Islam, *Raman Amplifiers for Telecommunications 1-Physical Principles*. New York: Springer-Verlag, 2004.

3. G.P. Agrawal, *Fiber-Optic Communications Systems*, 3rd edn. New York: John Wiley & Sons, 2002.
4. P.L. Kelley, *Nonlinear Fiber Optics*, 3rd edn. New York: Academic Press, 1995.
5. D.J. Blumenthal, Distributed Raman amplification for ultra-high speed fiber cross-phase-modulation wavelength converters, Project Report 2001–02.
6. Y. Aoki, Properties of fiber Raman amplifiers and their applicability to digital optical communication systems, *IEEE Journal of Lightwave Technology*, 6: 1225–1239, 1988.
7. M.L. Dakss and P. Melman, Amplified spontaneous Raman scattering and gain fiber Raman amplifiers, *IEEE Journal of Lightwave Technology*, 6: 806–813, 1985.
8. M. Menif, M. Karasek, and L.A. Rusch, Cross-gain modulation in Raman fiber amplifier: Experimentation and modeling, *IEEE Photonics Technology Letters*, 14(9): 1261–1263, September 2002.
9. J. Bromage, Raman amplification for fiber communications systems, *IEEE Journal of Lightwave Technology*, 6: 79–93, 2004.
10. S. Namiki and Y. Emori, Ultrabroad-band Raman amplifiers pumped and gain-equalized by wavelength-division-multiplexed high-power laser diodes, *IEEE Journal of Selected Topics in Quantum Electronics*, 7(1): 3–16, January/February 2001.
11. R.H. Stolen, Relation between the effective area of a single-mode fiber and the capture fraction of spontaneous Raman scattering, *Journal of the Optical Society of America B*, 19(3): 498–501, 2002.
12. (a) H. Suzuki, N. Takachio, H. Masuda, and K. Iwatsuki, Super-dense WDM transmission technology in the zero-dispersion region employing distributed Raman amplification, *IEEE Journal of Lightwave Technology*, 4(4): 973–981, April 2003. (b) H. Suzuki, J.-I. Kani, H. Masuda, N. Takachio, K. Iwatsuki, Y. Tada, M. Sumida, "1-Tb/s (100×10Gb/s) super-dense WDM transmission with 25-GHzchannel spacing in the zero-dispersion region employing distributed Raman amplification technology, *IEEE Photonics Technology Letters*, 12(7): 903–905, July 2000.
13. D. Dahan and G. Eisenstein, The properties of amplified spontaneous emission noise in saturated fiber Raman amplifiers operating with CW signals, *Optics Communications*, 236: 279–288, 2004.
14. P. Gallion, V. Dallot, and E. Bridoux, Study of noise properties in optical distributed Raman amplifiers using a semiclassical model, in *Proceedings SPIE, Photonics West, Optoelectronics*, vol. 4646, San Jose, CA, Jan 2002, pp. 618–629.
15. A.V. Wesson and R.I. Killey, Calculation of the effect of pump depletion in WDM systems with distributed Raman gain, in *Proceedings of London Communication Symposium* (*LCS 2001*), London, U.K., September 2001, pp. 59–62.
16. L. Xueming, Optimization for various schemes of distributed fibre Raman amplifiers, *Journal of Optics A: Pure and Applied Optics*, 6: 1017–1026, 2004.
17. H.A. Haus, Optimum noise performance of optical amplifiers, *IEEE Journal of Quantum Electronics*, 37(6): 813–823, June 2001.
18. Guifang Li, Recent advances in coherent optical communications, *Advances in Optics and Photonics*, 1(2): 279–307, 2009.

11

Design of Optical Communication Systems and Simulink® Models

11.1 Introduction

Chapters 2 through 6 have been dedicated to the study of optical fibers and their operation parameters, optical passive and active components, optoelectronic receivers, as well as optical transmitters. An optical transmission system combines all these components with appropriate optical powers at the transmitter and at the receiver so as to satisfy the bit-error-rate (BER) requirement or the optical signal-to-noise ratio (OSNR). Shown in Figure 11.1a is the basic structure of the transmission system of multi-span and multiwavelength channels with mid-span optical amplifiers in a bidirectional arrangement. Figure 11.1b and c also shows the working-fiber and protection-fiber structures including add/drop mux insertion.

This chapter thus provides a general approach to the design of optical transmission systems, especially when optical fibers are used as the transmission medium. Both nondispersion-compensating fibers and dispersion-compensating fibers (DCFs) are employed in the transmission systems described.

Readers may consider summarizing the transmission properties of various photonic and optical components that make up the transmission system, mainly their insertion losses and dispersion factors as well as spectral properties. For example, for optical amplifiers, one needs to know the minimum input power level, the optical gain, the saturated output power level, and the noise figure. Naturally, for single-mode optical fibers, the dispersion factor, the attenuation factor, as well as the nonlinear threshold factor need to be known. Thus, readers can bypass the details of the optical components, but must take into account the parameters essential for system design. Simulink® models are based on the operational principles of each component that are formed in blocks. These blocks are interconnected together to form a transmission system. Thus, the principles of operation of each physical component are very important and must be represented faithfully in Simulink in order to obtain accurate simulation results.

The following sections give a generic outline of the design strategy for the transmission systems, followed by a detailed study of the attenuation and dispersion budgets for different optical transmission structures. First, a single-span transmission system is developed; thence, multi-span optical amplified fiber transmission systems are developed. The techniques for the generation of the dispersion and power budgets are given for unidirectional and bidirectional transmissions. Some typical design examples are given for the bidirectional transmission of dense wavelength division multiplexed (DWDM) channels in the same fiber (see Figure 11.1a and b). Bidirectional transmission means the wavelength channels are transmitted and received from both ends of the fiber link. Thus, some optical components, such as optical circulators, are required for routing the lightwave channels. Figure 11.1c and d shows the arrangement of a bidirectional amplifier and transmitter, respectively.

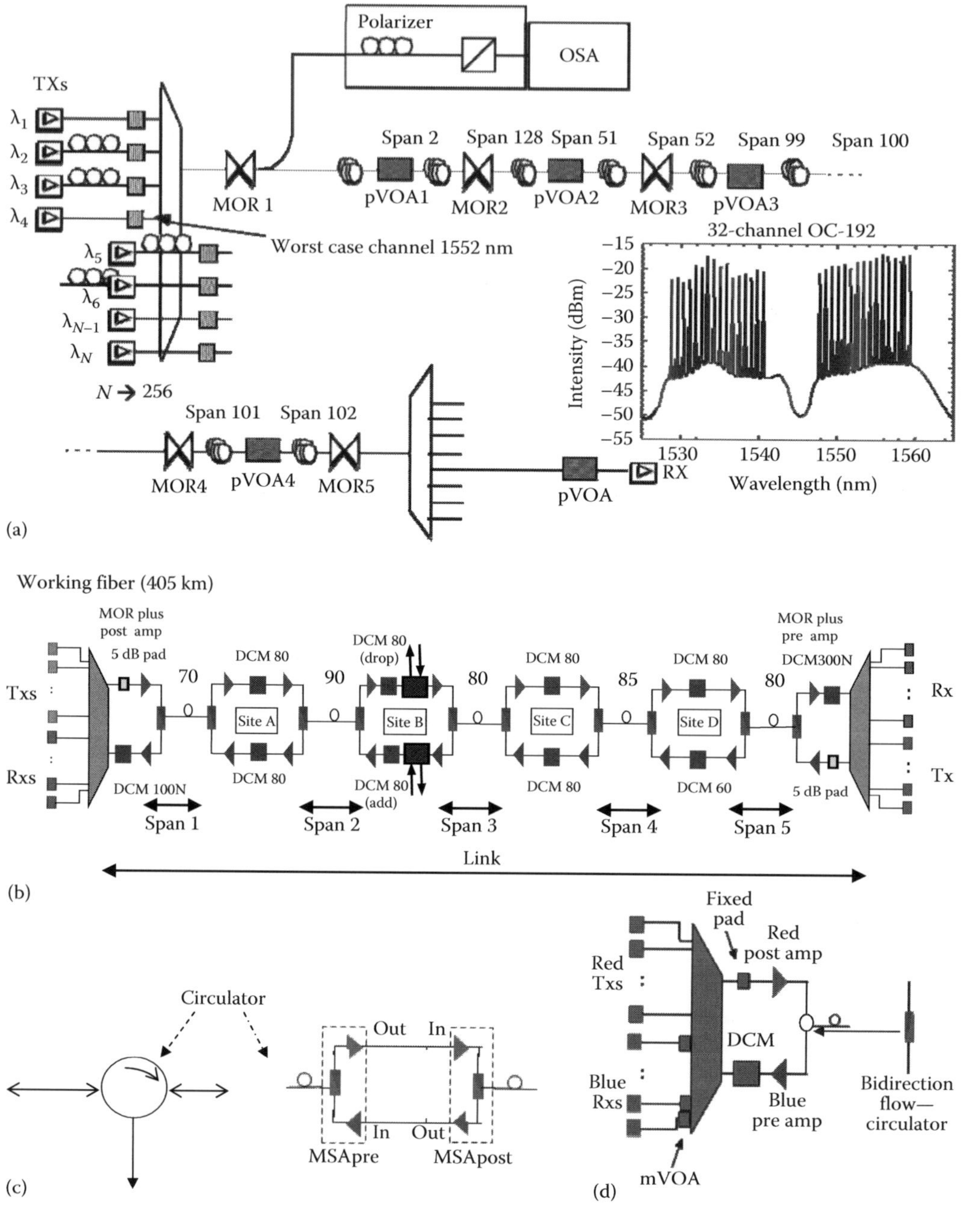

FIGURE 11.1
Optical transmission link with mid-spans and optical amplifiers—bidirectional transmission in fibers and multiwavelength channels. (a) Basic structure for DWDM channels—inserted is the spectra of DWDM channels. (b) 405 km fiber structures. MOR = mid-span optical repeater, DCM = dispersion-compensating module, DCMxxx = dispersion-compensating module for xxx km of SSMF, and pad = adjustable optical attenuator. (c) Arrangement of optical amplifiers—mid-span amplifier (MSA) for bidirectional amplification. (d) Details of bidirectional transmitter. MSAPre and MASPost = pre- and post-optical mid-span amplifications.

It should be noted that only digital systems are described in this chapter using both intensity modulation and direct detection (IM/DD). Advanced optically amplified transmission systems employing coherent detection or DD under external modulation formats like DPSK, ASK, and FSK are also given in combination with Simulink models.

11.2 Long-Haul Optical Transmission Systems

A long-haul transmission system by definition is one in which the distance between the transmitter and the receiver is an important factor. This means that the maximum transmission between these subcomponents of the systems must be extended without the use of optical–electrical repeaters, but optical amplifiers in mid-spans can be employed. The total distance is normally divided into a number of spans. Each span is connected to the next one by mid-span optical repeaters (MORs). System transponders would only be considered briefly, as they are now very much standard subsystems.

11.2.1 Intensity-Modulation Direct-Detection Systems

A DD optical receiver requires a minimum received signal power to sustain an adequate signal-to-noise ratio (SNR) to satisfy the BER requirement. Since the total fiber attenuation is dependent on the fiber length, there is a maximum path length beyond which the received signal strength will be inadequate to satisfy the SNR requirement of the communication systems.

If in system design we start with specifications including the transmitted power and the receiver sensitivity (RS), we can estimate the maximum attenuation allowable for the transmission medium. This total attenuation includes distributed losses in the fiber, coupling losses into and out of the fiber, other coupling losses at the signal-monitoring site, optical add/drop multiplexer (OADM), etc. In addition to these average losses, a system operating margin is also added to protect the system against aging components, occasional misalignment of fiber components, or additional repairs of broken fibers by splicing.

Example 11.1

An optical transmission system consists of (1) an optical transmitter: direct IM output power = −0.5 dBm at the output of the fiber pigtail; (2) an optical receiver: RS = −41.5 dBm; (3) optical fibers: average fiber attenuation and splicing losses = 0.4 dB/km; (4) coupling losses at the receiver = 0.5 dB; and (5) an operating margin = 3 dB. What is the maximum distance between the transmitter and the receiver?

Solution

Optical transmitter	−0.5 dBm
Optical receiver	−41.5 dBm
Operating margin	3 dBm
Coupling loss at receiver	0.5 dB
Hence, total loss allowable for fibers	41.5 − 3 − 0.5 = 38 dB
Fiber attenuation (average)	0.4 dB/km
Thus, total maximum distance	38/0.4 = 38 × 2.5 km = 95 km

The RS is a critical parameter of the link and can be determined by the acceptable probability of error in the detection and the decision level, as described in Chapter 7. We recall the energy b_1 required for bit "1" received under IM/DD as follows:

$$b_1 = \frac{q}{\Re}\left[\partial^2 G^x + \frac{2\partial i_{\mathrm{Neq}} T_{\mathrm{b}}}{qG}\right] \tag{11.1}$$

Thence, the RS is given by

$$\mathrm{RS} = 10\,\mathrm{Log}_{10}\frac{P_{\mathrm{av}}}{P_0}\cdot \quad \text{in dBm when } P_0 = 1\,\mathrm{mW} \tag{11.2}$$

where $P_{av} = b_1/Tb$ and $P_0 = 1.0$ mW (the reference power level for evaluating power in dBm)

The term δ is specified by the BER curve and the statistical properties of the distribution of the received "1" and "0." Hence, with a specific BER, bit period, and photodetector type, we can estimate the equivalent noise current at the input of the electronic preamplifier so that a minimum detection power is required. This would allow us to decide whether an optical amplifier preamplifier is needed at the front end of the optical receiver. The decision threshold also affects the BER, and this is normally used to set the level. Typical eye diagrams at different bit rates and a modulation format NRZ or RZ are shown in Figure 11.2.

We note that the two terms within the bracket of Equation 11.1 represent the number of photons required to overcome the requirement due to the delta Q-factor (i.e., the BER) and the electronic noise current. This is due to the fact that the term $q/\Re$ is indeed the photon energy.

Example 11.2

For a 10 Gb/s IM/DD transmission system, determine the following:

(a) What is the RS of the optical receiver if the following conditions must be met: BER = 1e−9 and the photodetector is PIN type with a quantum efficiency of 0.8?

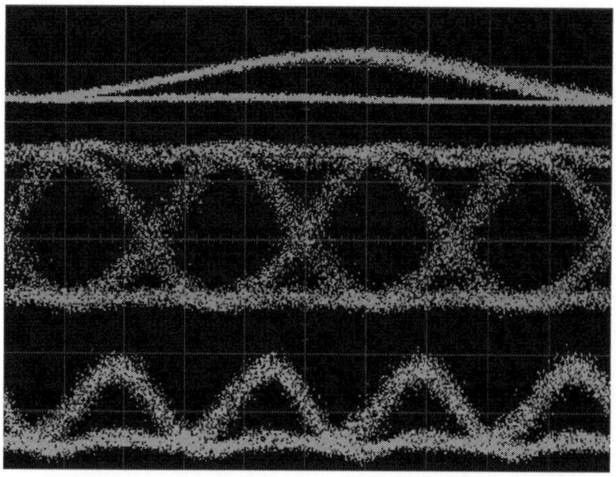

FIGURE 11.2
Typical eye diagrams after the optical receiver for different formats (NRZ or RZ) at different bit rates.

(b) What is the equivalent number of photons per bit that must be available at the receiver?
(c) What is the number of photons required to overcome the BER requirement?
(d) What is the number of photons per bit to overcome the electronic noise current requirement?
(e) Repeat (a)–(d) for an avalanche photodiode (APD) with an avalanche gain of 20.

Thus, we can see that the RS varies with respect to the bit rate. Naturally, the shorter the pulse width, the higher the peak power of the pulse to satisfy the average power required, as estimated from the RS.

11.2.2 Loss-Limited Optical Communication Systems

The fiber loss plays an important role in the design and implementation of optical communication systems. Considering an optical transmitter generating an optical power of P_{tr} and that the optical signals are received with an optical receiver having an ability of detecting a minimum optical power, P_r, at a bit rate B_R, then the maximum transmission distance would be given by

$$L = \frac{10}{\alpha_F} \mathrm{Log}_{10}\left(\frac{P_{tr}}{P_r}\right) \tag{11.3}$$

where α is the total fiber loss per unit length (in dB/km) including splice and connector losses. The bit-rate dependence of L arises from the linear dependence of P_r on the bit rate B_R by $P_r = N_p h\upsilon B_R$ with N_p as the average number of photons/bit required for the receiver to detect with a certain SNR.

EXERCISE 11.1

Given that the fiber losses for the wavelength regions 810, 1300, and 1550 nm are 2.5, 0.4, and 0.15, respectively, plot the maximum transmission distance versus the bit rate when the transmitter average power is 0 dBm and the average number of photons that the receiver can detect is 300.

11.2.3 Dispersion-Limited Optical Communication Systems

We have described the dispersion effect in optical fibers, which is very important for optical transmission systems when the loss is no longer a problem if optical amplifiers are employed. When the loss-limited distance is longer than the dispersion-limited transmission distance, then these systems are called dispersion-limited transmission systems.

The limit for a system with a bit rate of B_R and a total dispersion factor, D_T, of the fiber is given by

$$B_R L \leq \frac{1}{4|D_T|\sigma_\lambda} \tag{11.4}$$

where σ_λ is the FWHM of the source spectrum. Thus, for a standard optical fiber operating at 1550 nm, the total dispersion factor is about −15 ps/(nm · km), the source FWHM is 1.0 nm, and the bit rate–distance product is about 166.6 (Gb/s) · km, that is, if a system having an operating bit rate of 10 Gb/s would be able to transmit signals for 16.6 km. If a dispersion-shifted fiber is used with a typical dispersion factor of about 1–2 ps/(nm · km), then the transmission distance can be increased by about 15–7.5 times that of the SSMF or more than 100 km up to 240 km. When these fibers are employed for extremely long distances, the polarization dispersion becomes significant and the PMD and DGD parameters must be taken into account to estimate the total allowable transmission distance, especially for an operating bit rate of 10 Gb/s.

Several schemes currently under investigation include the use of fiber gratings, dispersion compensators, optical signal processors/filters, and optical solitons, where the nonlinear effects of fibers through self-phase modulation are used to counter the linear effects of chromatic dispersion.

11.2.4 System Preliminary Design

In the design and commissioning of optically amplified fiber communication systems, there are normally three steps to be followed:

1. Preliminary design in which the power budget and dispersion budget (rise-time and fall-time budgets) must be established.
2. Justification and tuning of the preliminary design by simulation and/or in-house hardware demonstration. A detailed transmission performance can be achieved here, e.g., BER versus dispersion tolerance and BER variation with the sampling position or the timing quality factor.
3. Commissioning and installation of the transmission system and trial operation. A statistical performance of the system can be obtained, and thus the operating margin can be set.

In this section, we describe the preliminary design of the transmission systems under both linear and nonlinear dispersive effects of the fiber. More detailed designs would be given in the remaining sections of this chapter. We also show how to estimate the maximum transmission distance of multi-span systems in which cascade optical amplifiers are employed.

11.2.4.1 Single-Span Optical Transmission Systems

In practice, the design of an optical communication system should meet two principal design requirements: the power budget and the rise-time budget, which account for the loss-limited and dispersion-limited systems, respectively.

Specifications for the system would be the system bit rate, B_R, and the design transmission distance. Thus, there are commonly two main cases: the point-to-point links and the ultra-long or terrestrial/undersea links. The performance of the optical fiber systems is determined by the BER and the operating wavelength. The cost of operating in the 810 nm window would incur the lowest loss, which increases with the wavelength windows of 1300 and 1550 nm. Normally, for systems operating at 100 Mb/s at a distance less than 20 km, single-mode fibers can be used, and this is the case for most LAN and Intranet applications. These systems and networks can be wired at the 810 nm window. For longer transmission systems, the two wavelength windows of 1310 and 1550 nm must be used.

11.2.4.2 Power Budget

The power budget of a transmission link can be represented as shown in Figure 11.3. Depending on the availability of the power at the output of the transmitter or on the RS

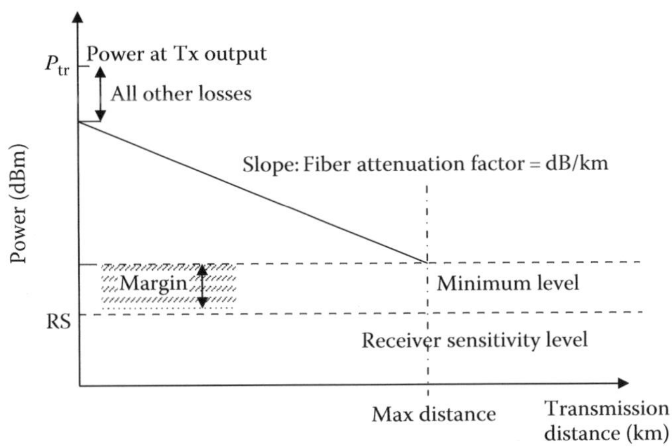

FIGURE 11.3
Power budget by graphical representation for a single-span transmission system.

of the optical receiver, one can determine the operating conditions of other subsystems or optical components, for example, the fiber distance, the number of spans, the noise of the electronic preamplifiers, the system operating margin, etc. Either a table of the power budget of the transmission link or a graph, as shown in Figure 11.3, can be established.

11.2.4.3 Rise-Time/Dispersion Budget

We have previously established an approximate relationship between the dispersion, the bit rate, and the bandwidth. For on–off keying (OOK), the bandwidth of the detection system, Δf, can be about half the bit rate (pulse repetition rate). The relationship $B_R L \le \frac{1}{4|D_T|\sigma_\lambda}$ can be used or the bit rate, B_R, can be approximately written as

$$B_R \frac{1}{4\Delta\tau} = 2\Delta f \tag{11.5}$$

However, when writing the relationship between all the components and subsystems of the transmission system, we must remember that the summation of all the RMSs (root mean squares) of all the rise times* and broadening factors must be less than 70% of the bit period (NRZ or RZ bit time). This can be written as

$$\sigma = \sqrt{\sum_{i=1}^{N} (t_{ri}^2 + t_{fi}^2 + \Delta\tau_j^2)} \tag{11.6}$$

* Rise time is measured from 10% to 90% of the rising edge of a rectangular pulse; this is usually the step function of the subsystem. Thus, the rule of thumb is that the broadening factor is approximately 50% of the rise time. If the bandwidth of the subsystem is measured, then the pulse broadening can be roughly one-fourth of the bandwidth.

where

t_{ri} and t_{fi} are the rise time and the fall time, respectively, of the ith component in the transmission link

$\Delta\tau$ is the pulse broadening of the ith fiber in the link taken for both sides of the pulse after transmission through an optical fiber of length L

Thus, both the rise time and the fall time must be accounted for. Thence, we must have

$$\sigma = \sqrt{\sum_{i=1}^{N} (t_{ri}^2 + t_{fi}^2 + \Delta\tau_j^2)} \leq 70\% \frac{1}{B_R} \tag{11.7}$$

that is, the total broadening of the pulse must be less than 70% of the bit period. The relationship between the RMS pulse width and the bandwidth of the system specified by the pulse shape can be found as shown in Table 11.1.

The RMS pulse width, σ, of a pulse shape $h(t)$ can be found by writing the normalized pulse shape and the RMS definitions as

$$\int_{-\infty}^{+\infty} h(t)dt = 1 \tag{11.8}$$

$$\sigma = \langle t^2 \rangle - \langle t \rangle^2 \tag{11.9}$$

where

$$\langle t \rangle = \int_{-\infty}^{+\infty} t \cdot h(t)dt$$

$$\langle t \rangle^2 = \int_{-\infty}^{+\infty} t^2 \cdot h(t)dt$$

TABLE 11.1

RMS Pulse Width, σ, and Spectral Width, Δf, of Three Typical Pulse Shapes

	$h(t)$	σ	$H(f)$	Bandwidth
Exponential pulse	$\frac{1}{\tau}e^{-\frac{t}{\tau}}$	τ	$\frac{1}{1+j\omega\tau}$	$\frac{0.159}{\sigma}$
Rectangular pulse	$\frac{1}{\tau}$ $-\frac{\tau}{2} \leq t \leq \frac{\tau}{2}$	$\frac{\tau}{2\sqrt{3}}$	$\frac{\sin(f\tau)}{f\tau}$	$\frac{0.402}{\sigma}$
Gaussian pulse	$\frac{1}{\sqrt{2\pi\tau}}e^{-\frac{t^2}{2\tau^2}}$	τ	$e^{-2(\pi f\tau)^2}$	$\frac{0.133}{\sigma}$

When there are a number of linear systems (e.g., optical fiber spans) in cascade, the overall impulse response can be found by convolving the impulse response of each individual fiber span. The resultant RMS pulse broadening can be found by adding the RMSs of all pulse-broadening factors. That is

$$\sigma^2 = \sigma_1^2 + \sigma_2^2 + \sigma_3^2 + \cdots + \sigma_N^2 \tag{11.10}$$

Example 11.3

The rise times and fall times of the transmitter and the optical detector are 1 and 0.5 ns, respectively. The bandwidth of the optical receiver is 2 GHz. The total pulse broadening of the fiber is 1.5 ns. What is the maximum bit rate of the OOK system?

Solution

Broadening due to optical transmitter	0.5 ns
Broadening due to optical receiver	(1/2)0.5 ns
Fiber broadening	1.5 ns
Broadening due to electronic amplifiers of the optical receiver	1/(4 × 2 × 1e9) = 1/8 ns
Total RMS pulse broadening (pulse width)—impulse response	$\sigma = (0.5)^2 + (0.25)^2 + (1.5)^2 + (1/8)^2 = (0.763)$^2 ns
Thus, the maximum bit rate	$B_R = 1/4(\sigma) = 328.8$ MHz

11.2.4.4 Multi-Span Optical Transmission Systems

For multi-span transmission systems, the power budget should be designed for the case of upgrading an existing installed optical fiber system or a new system that is to be installed. For the upgrading case, normally for non-amplified single-mode transmission systems, dispersion-compensating modules (DCMs) are to be inserted at the end of a fiber transmission span, usually in multiple lengths of 40 km section. Thus, the span length must be of an order of 80 or 100 km depending on the quality of the single-mode fibers. Two optical amplifiers are to be inserted, one at the end of the transmission fiber (normally, standard single-mode optical fibers) and the other at the output of the DCM. The optical power at the output of the first optical amplifier must be less than the nonlinear threshold of the DCM, which should be very low, about 0 dBm, due to the small effective area of the DCF. The optical power at the output of the optical amplifier after the DCM must be high enough, about the launched power level into the first span. Thus, this optical amplifier must be a booster optical amplifier type. The power distribution of the optical channels along the multi-span transmission distance is shown in Figure 11.4. Ideally, the length of each span should be the same and any excess power can be attenuated using attenuation pads. This is preferred for manufacturing almost identical optical amplifiers in all spans, and hence minimizing the transmission system management. Also shown in this figure is the minimum level, so as to ensure that this meets the minimum power level for optical amplification.

Figure 11.5 shows the dispersion budget for multi-span transmission. The minimum dispersion tolerance is to be determined depending on the modulation formats. Initially, there is no dispersion at the output of the transmitter. The dispersion gradually increases as a function of the transmission distance in the first span. If the DCM is used at the end

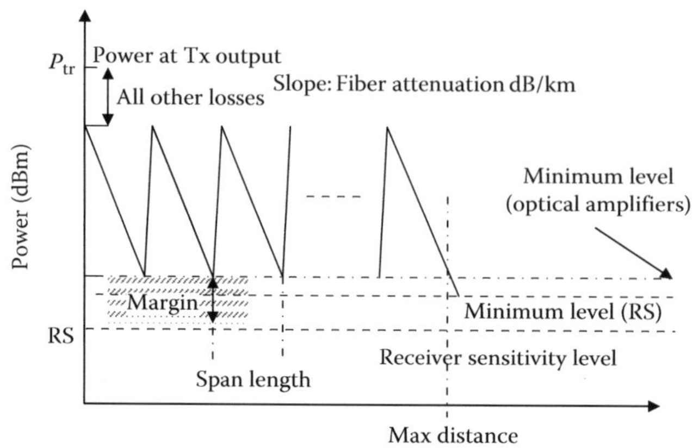

FIGURE 11.4
Power budget by graphical representation for a multi-span transmission system.

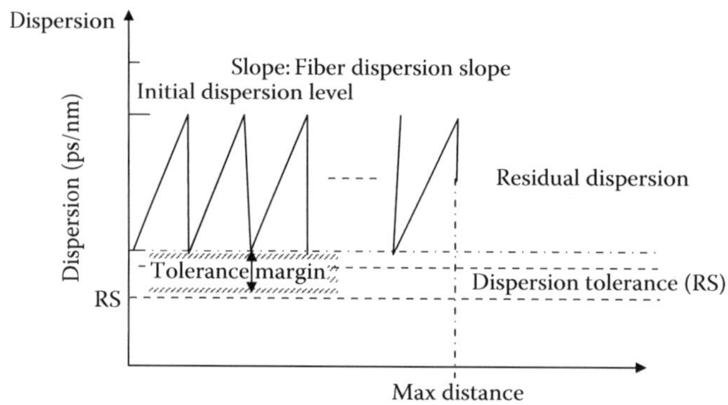

FIGURE 11.5
Dispersion budget by graphical representation for a multi-span transmission system.

of each span, then the dispersion can be fully compensated before the transmission to the next span and so on. This dispersion map varies with respect to the change of the wavelength channel. This change can be estimated using the spectral dependence of the dispersion factor. Thus, at the end of the multi-span transmission, there is always a residual dispersion at each wavelength channel. This residual dispersion can be compensated for each channel at the output of a wavelength demultiplexer before the optical signals are received. Tunable fiber Bragg gratings (FBG) can be used for tuning in order to match the residual dispersion. Alternatively, this dispersion can be compensated in the electronic domain by digital signal processors.

11.2.4.5 Maximum Transmission Distance of Cascaded Optically Amplified Multi-Span Systems

In long-haul transmission system cascading, using several optical amplifiers is very common in order to extend the total transmission distance. The question is what is the maximum transmission distance for which one can continue to insert optical amplifiers in the mid-span so that the SNR would be satisfied at the front of the optical receiver, which would satisfy the BER and the RS for a certain modulation format?

To answer this question we must acquire the following knowledge: What is the required OSNR for a particular modulation format? What is the noise figure or the number of spontaneous photons generated in the optical amplifiers prior to the stimulated emission, and is this the average effective ASE noise? What is the number of optical amplifiers to be cascaded?

The SNR at the end of a cascaded optically amplified link can be estimated as

$$\mathrm{OSNR} = \frac{P_i 10^{-(\alpha-\alpha_C)L/10}}{\frac{L}{L_T} n_{sp} h\nu B} \tag{11.11}$$

where

α is the fiber loss
α_C is the average connector losses per km in a span
L is the span length
L_T is the total length of the transmission link
n_{sp} is the number of spontaneous photons, normally equal to 2 for 3 dB NF
P_i is the optical power launched into the input of the transmission link
OSNR is the optical SNR required for a certain modulation format

For ASK and for a BER of 1e–9, the quality factor is 6, which is the SNR at the output of the optical receiver in voltage–amplitude ratio. Then the $\mathrm{OSNR} = 10\ \mathrm{Log}_{10}(6 \times 6) = 15.5\ \mathrm{dB}$. In Equation 11.11, we have assumed that the noise spectral density generated in an optical amplifier is given by

$$S_{ASE}(f) = n_{sp} h\nu df \tag{11.12}$$

This noise power is to be integrated over the bandwidth of the system, B.

Thus, the maximum number of cascaded amplifiers is given by $N = L_T/L$; thence, the maximum distance can be determined.

Example 11.4

A long-haul optical transmission system operating at 1550 nm uses EDFAs as lumped optical amplifiers spaced at an interval of 100 km. Both +D and −D dispersion fibers are employed in alternating spans. The mode spot sizes of these fibers are the same. The modulation format is NRZ-ASK. A BER of 1e−12 is required at an operating bit rate of 10 Gb/s.

Assume that the power launched into the first span is 0 dBm. The fiber loss is 0.2 dB/km. In addition, the average splicing loss per km of the fiber is 0.03 dB/km.

(a) What is the required OSNR for this modulation format at the front of the optical receiver?
(b) If the number of spontaneous photons is 2 for the ASE noise of the EDFAs and all optical amplifiers are identical, estimate the maximum transmission distance of the system after which no further amplification is allowed.

Solution

(a) For a BER = 1e−12, $\delta = 7$; hence, the OSNR is

$\mathrm{OSNR} = 10\ \mathrm{Log}_{10}(49) = 17\ \mathrm{dB}$

The maximum number of spans is given by Equation 11.11 with the following values: $n_{sp} = 2$, $\alpha = 0.2$ dB/km, $\alpha_C = 0.03$ dB/km, $B = 10$ Gb/s, $P_i = 1$ mW, and $N = 39$ or the maximum distance is 3900 km.

11.2.5 Gaussian Approximation

The impulse responses of the single-mode optical fiber can be approximated as Gaussian pulse shapes in the far-field region, while they are strongly chirped in the near-field region, as shown Figure 11.6. And when they are in cascade, the overall impulse response of the transmission system can be represented as Gaussian. Thus, the impulse response and its Fourier transform, which gives the frequency response of the system, are given by* (see Figures 11.6 and 11.7 and Appendix)

$$h(t) = e^{-\frac{t^2}{2\tau^2}} \rightarrow H(\omega) = e^{-\frac{(\omega\tau)^2}{2}} \tag{11.13}$$

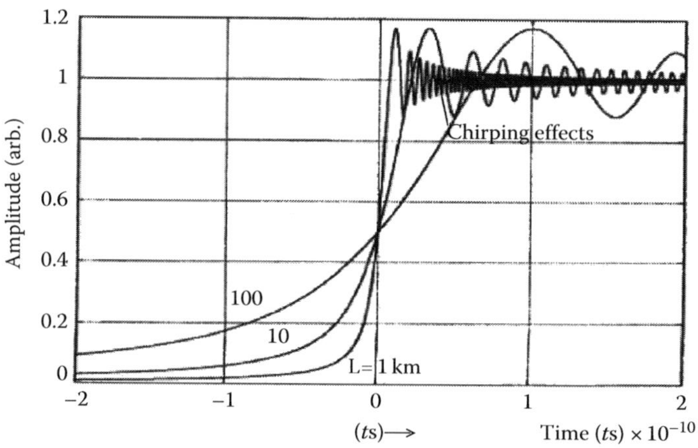

FIGURE 11.6
Fiber step responses for a single-mode optical fiber of distances 1, 10, and 100 km and chirp effects near the edge.

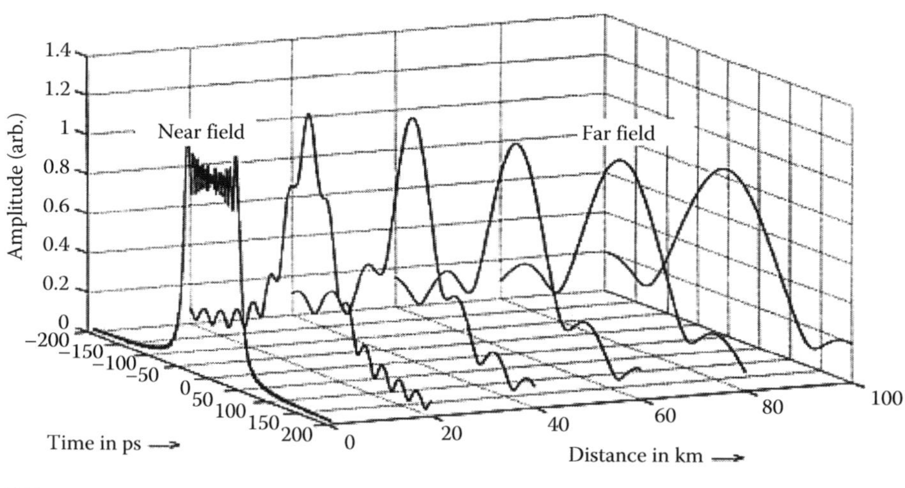

FIGURE 11.7
Gaussian impulse response of single-mode optical fibers—near field and far field.

* See Section 3.4 on transfer function and impulse responses of single-mode optical fibers.

The bandwidth of this system can then be found by setting the absolute value of $H(\omega)$ to 1/(sqrt2) and solving for ω. Thence, the 3 dB bandwidth and the rise time of the system can be found with the relationship

$$\omega_{3\text{dB}} = \frac{0.83}{\tau} \tag{11.14}$$

If the link is made up by N fiber spans of the same dispersion factor, then each fiber length in tandem has a transfer function of

$$H_n(\omega) = e^{-\frac{N(\omega\tau)^2}{2}} \tag{11.15}$$

The dependence of the transfer function as a square of the frequency indicates that whenever the bit rate is double, the transmission distance is reduced by a factor of 4. This is the "rule of thumb" for fiber transmission, for setting the dispersion tolerance as a function of the bit rate.

Note that the design described in this section is given as a preliminary design so as to perform the selection of optical systems based on the electronic properties of the receiver and the electrical driving unit of the transmitter.

The power budget and the dispersion budget are used commonly for the preliminary design. Accurate design details must be obtained by simulation and/or by laboratory demonstration prior to field installation.

11.2.5.1 Link Budget Measurement

Link budget measurement is done based on measuring and estimating margins on a close-to worst-case system. It is required that a guarantee is obtained for end-of-life (EOL) BER > 10^{-12} (Figure 11.8). Figure 11.9 shows the layout of the power budget in which an OSNR of 3.88 dB is accounted for a Q factor of 7 (BER = 1e–12) and allowances for aging, corrections of the transmitter and the receiver, and distortion as well as the error in power control are accounted for.

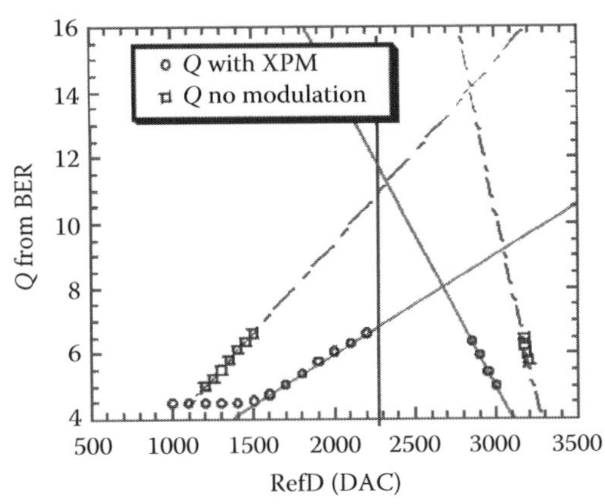

FIGURE 11.8
Details of Q factor as a function of sampling and decision point (voltage) of the eye diagram.

Stressed span loss (dB)	24.95
OSNR margin to $Q=7$	3.88
Aging allowance	–4.00
Worst-case Rx/Tx correction	–1.90
Worst-case distortion dBQ × 2	–0.60
Power control error	–1.50
Sum degradations (dB)	–8.00
System margin (dB)	–4.12
Guaranteed maximum span loss (dB)	20.83

FIGURE 11.9
Table arrangement of loss per span in optical transmission (in dB).

11.2.5.2 System Margin Measurement

The magnitude of any impairment that degrades the system to $Q=7$ can be used as a metric for system margin. Thus, any distortion by linear and nonlinear dispersion effects and/or noise contribution to the transmission of lightwave channels must be estimated and an allowance of the degradation of the eye diagram must be allocated in the power budget. Figure 11.10 shows various effects that can degrade the eye diagram and the aspects of noise measurement for link budget estimation.

Figure 11.10 shows the effects of the noise margin on the quality factor, Q, especially the slope of the BER. The noises are loaded when cascaded optical amplifiers (MOR and MSA) are active. These noises are accumulated along the multi-span transmission line and, hence, superimposed on the received voltage levels. The bandwidth of the ASE noises is very wide, and these are mixed with the optical signals, which results in high noise levels in the detected eye diagram. This decreases the degree of eye opening, and some day the degradation may be measured by using the term eye-opening penalty (EOP).

11.2.5.2.1 Noise Margin

A noise margin measurement can be conducted by using an ASE noise source superimposed on the optical signals, as shown in Figure 11.11. Thence, the BER and the Q factor can be determined under unloaded and loaded environments (Figure 11.12).

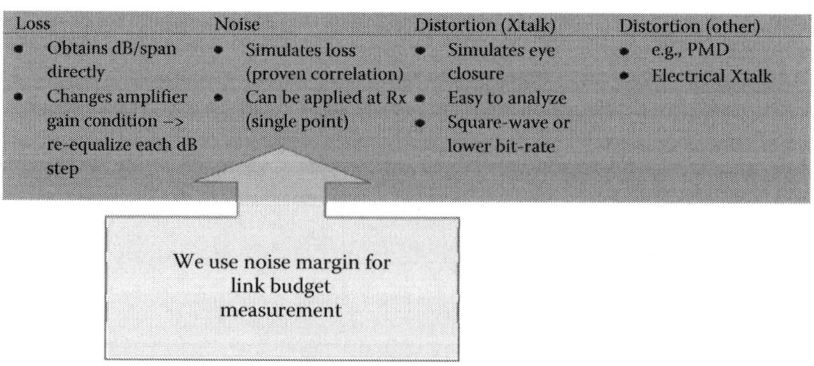

FIGURE 11.10
Budget measurement.

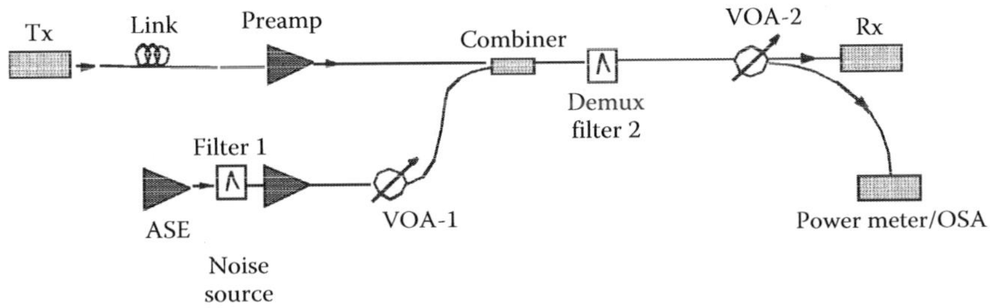

FIGURE 11.11
Arrangement for noise margin measurement—ASE = noise sources, VOA = variable optical attenuator, and OSA = optical spectrum analyzer.

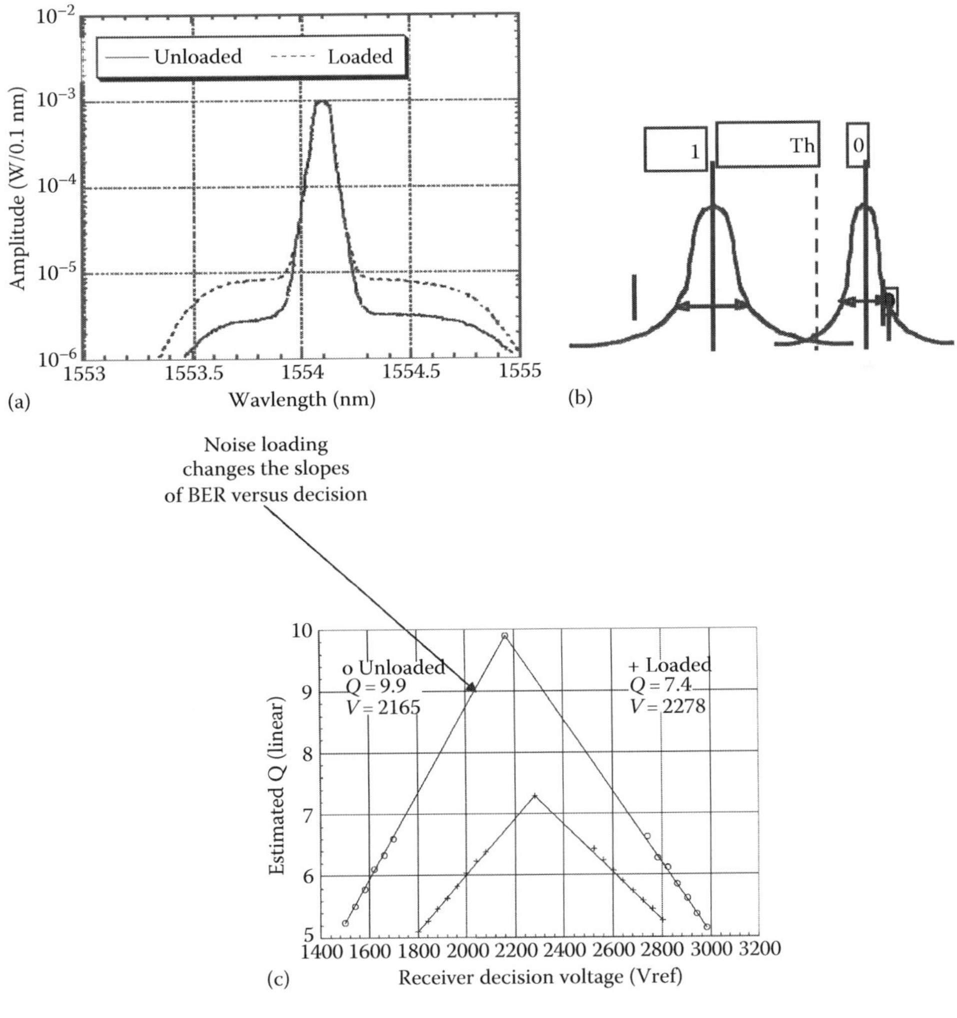

FIGURE 11.12
Noise margin measurement. (a) Noise spectra under unloading and loading conditions, (b) probability distribution, and (c) BER versus decision voltage level.

11.2.5.2.2 *Dispersion Map*

The dispersion map of the transmission system under a noise margin is also important, and an arrangement for such measurement can be done, as shown in Figure 11.13a. A dispersion compensator can be inserted after the transmitter whose dispersion factor can be varied so as to determine the dispersion tolerance and the noise-margin level at a specific BER. Figure 11.13b shows a typical noise margin versus dispersion for a wavelength channel. These measurements can be done by simulation. That is why a good and accurate simulation package is required for the evaluation of modern optical fiber communication systems.

11.2.5.2.3 *Worst-Case Distortion*

The worst case includes "inaccessible distortion" for a worst-case model, e.g., FWM and power fluctuation. This worst case is very critical for DWDM, in which the wavelength channels are equally spaced, when the interaction of three waves to create the fourth wave located exactly on another wavelength channel takes place. This generates unwanted cross talk. Two rows of the worst-case penalty are added in the table of Figure 11.14 to account for these circumstances.

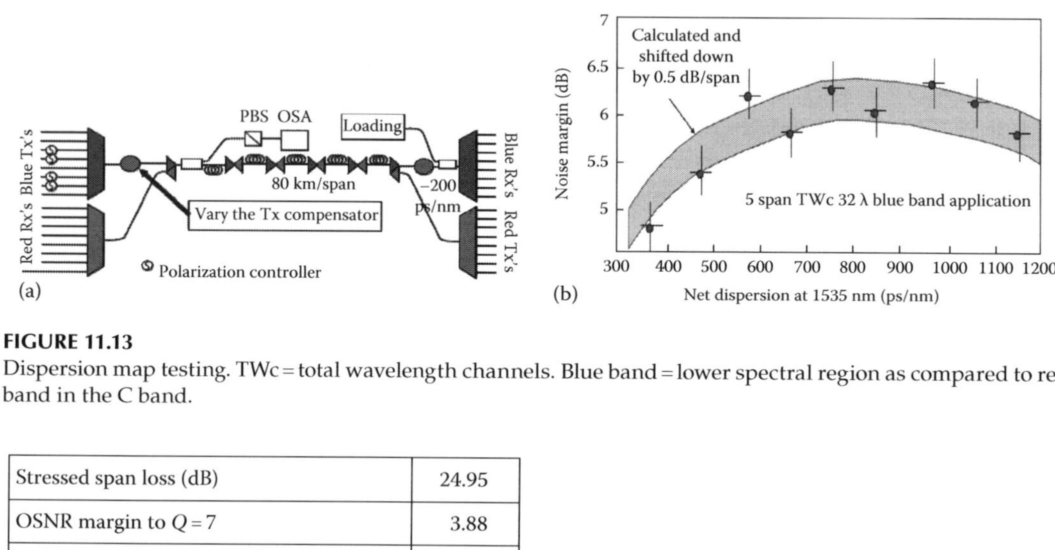

FIGURE 11.13
Dispersion map testing. TWc = total wavelength channels. Blue band = lower spectral region as compared to red band in the C band.

Stressed span loss (dB)	24.95
OSNR margin to $Q = 7$	3.88
Aging allowance	−4.00
w.c Rx/Tx correction	−1.90
w.c. distortion dBQ × 2	−0.60
Power control error	−1.50
Sum degradations (dB)	−8.00
System margin (dB)	−4.12
Guaranteed max. span loss (dB)	20.83

(a)

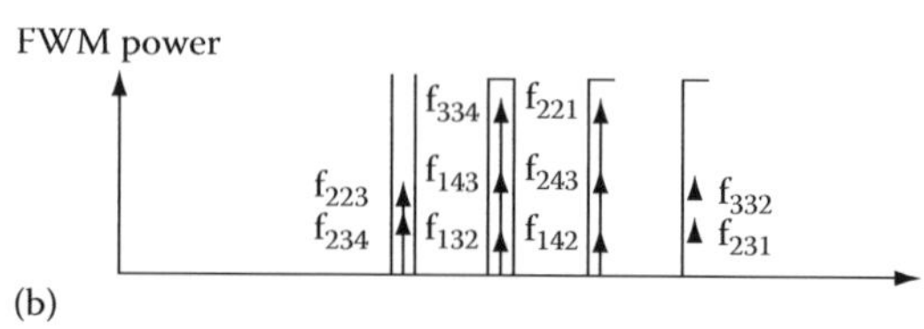

(b)

FIGURE 11.14
Worst-case distortion illustrated by (a) table of power budget and (b) FWM and variation of cross talk due to the fluctuation of power level.

11.2.5.2.4 Guarantee of Maximum Span Loss

Finally, the maximum span loss should be determined so as to ensure that the loss of the span during installation must be met in order to achieve quality transmission (Figure 11.15). This can also be proven if the verification agrees with the analysis so that the loss/span is within ±0.5 dB/span (1-sigma), the dispersion window is within ±30 ps/nm (1-sigma), and the provisioned peak power is within ±0.25 dBm (1-sigma). The verifications are reproducible within ±0.3 dB/span and, on an average, the verification is 0.5 dB/span more than the analysis.

11.2.5.2.5 From Modeled Budget to Installation Budget

Verified modeled budgets are further adjusted to account for equipment behavior and installation and operating procedures, for example, super-decoder penalty and equalization penalty. The verifications of the modeled budget and the installed budget can be confirmed from the experience of installed transmission systems. The system operating margin can also be reduced from statistical data obtained from these installed transmission systems.

11.2.6 Some Notes on the Design of Optical Transmission Systems

In modern optical transmission systems, the design procedures depending on specific applications need to (1) employ basic DWDM (Tb/s) for transmission principles and clarify the DWDM (Tb/s) modulation for Tx; (2) deal with dispersion and losses (Tb/s) in transmission system design: advanced fibers; (3) set engineering rules for impacts due to nonlinearity effects; and (4) examine optical amplifiers—distributed type and lumped type—for system design rules.

11.2.6.1 Allocations of Wavelength Channels

The DWDM channel spacing and locations of the frequencies of the channel follow the ITU standard. Such locations can be observed from Table 11.2. The frequency and frequency spacing between channels can be estimated by

Stressed span loss (dB)	24.95
OSNR margin to $Q = 7$	3.88
Aging allowance	−4.00
w.c Rx/Tx correction	−1.90
w.c distortion dBQ × 2	−0.60
Power control error	−1.50
Sum degradations (dB)	−8.00
System margin (dB)	−4.12
Guaranteed max. span loss (dB)	20.83

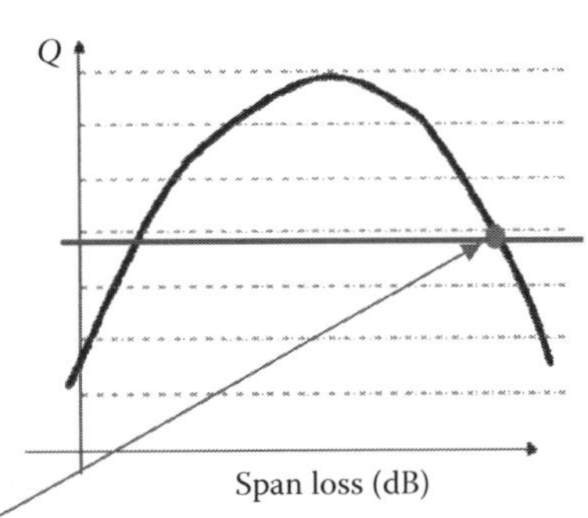

FIGURE 11.15
Guarantee of span loss for quality transmission.

TABLE 11.2

ITU Wavelength Grid for Dense Wavelength Division Multiplexing

	L Band				C Band				S Band			
	100 GHz Grid		50 GHz Offset		100 GHz Grid		50 GHz Offset		100 GHz Grid		50 GHz Offset	
	THz	nm	THz	nm	THz	nm	THz	nm	THz	nm	THz	nm
1	186.00	1611.78	186.05	1611.35	191.00	1569.59	191.05	1569.18	196.00	1529.55	196.05	1529.16
2	186.10	1610.92	186.15	1610.48	191.10	1568.77	191.15	1568.36	196.10	1528.77	196.15	1528.38
3	186.20	1610.05	186.25	1609.62	191.20	1567.95	191.25	1567.54	196.20	1527.99	196.25	1527.60
4	186.30	1609.19	186.35	1608.76	191.30	1567.13	191.35	1566.72	196.30	1527.21	196.35	1526.82
5	186.40	1608.32	186.45	1607.89	191.40	1566.31	191.45	1565.90	196.40	1526.43	196.45	1526.04
6	186.50	1607.46	186.55	1607.03	191.50	1565.49	191.55	1565.08	196.50	1525.66	196.55	1525.27
7	186.60	1606.60	186.65	1606.17	191.60	1564.67	191.65	1564.27	196.60	1524.88	196.65	1524.49
8	186.70	1605.74	186.75	1605.31	191.70	1563.86	191.75	1563.45	196.70	1524.11	196.75	1523.72
9	186.80	1604.88	186.85	1604.45	191.80	1563.04	191.85	1562.63	196.80	1523.33	196.85	1522.94
10	186.90	1604.02	186.95	1603.59	191.90	1562.23	191.95	1561.82	196.90	1522.56	196.95	1522.17
11	187.00	1603.16	187.05	1602.73	192.00	1561.41	192.05	1561.01	197.00	1521.78	197.05	1521.40
12	187.10	1602.31	187.15	1601.88	192.10	1560.60	192.15	1560.20	197.10	1521.01	197.15	1520.63
13	187.20	1601.45	187.25	1601.02	192.20	1559.79	192.25	1559.38	197.20	1520.24	197.25	1519.86
14	187.30	1600.60	187.35	1600.17	192.30	1558.98	192.35	1558.57	197.30	1519.47	197.35	1519.09
15	187.40	1599.74	187.45	1599.31	192.40	1558.17	192.45	1557.76	197.40	1518.70	197.45	1518.32
16	187.50	1598.89	187.55	1598.46	192.50	1557.36	192.55	1556.95	197.50	1517.93	197.55	1517.55
17	187.60	1598.04	187.65	1597.61	192.60	1556.55	192.65	1556.15	197.60	1517.16	197.65	1516.78
18	187.70	1597.18	187.75	1596.76	192.70	1555.74	192.75	1555.34	197.70	1516.40	197.75	1516.01
19	187.80	1596.33	187.85	1595.91	192.80	1554.94	192.85	1554.53	197.80	1515.63	197.85	1515.25
20	187.90	1595.48	187.95	1595.06	192.90	1554.13	192.95	1553.73	197.90	1514.86	197.95	1514.48
21	188.00	1594.64	188.05	1594.21	193.00	1553.32	193.05	1552.92	198.00	1514.10	198.05	1513.72
22	188.10	1593.79	188.15	1593.36	193.10	1552.52	193.15	1552.12	198.10	1513.33	198.15	1512.95
23	188.20	1592.94	188.25	1592.52	193.20	1551.72	193.25	1551.31	198.20	1512.57	198.25	1512.19
24	188.30	1592.10	188.35	1591.67	193.30	1550.91	193.35	1550.51	198.30	1511.81	198.35	1511.43
25	188.40	1591.25	188.45	1590.83	193.40	1550.11	193.45	1549.71	198.40	1511.05	198.45	1510.67

26	188.50	1590.41	188.55	1589.98	193.50	1549.31	193.55	1548.91	198.50	1510.28	198.55	1509.90
27	188.60	1589.56	188.65	1589.14	193.60	1548.51	193.65	1548.11	198.60	1509.52	198.65	1509.14
28	188.70	1588.72	188.75	1588.30	193.70	1547.71	193.75	1547.31	198.70	1508.76	198.75	1508.38
29	188.80	1587.88	188.85	1587.46	193.80	1546.91	193.85	1546.51	198.80	1508.01	198.85	1507.63
30	188.90	1587.04	188.95	1586.62	193.90	1546.11	193.95	1545.72	198.90	1507.25	198.95	1506.87
31	189.00	1586.20	189.05	1585.78	194.00	1545.32	194.05	1544.92	199.00	1506.49	199.05	1506.11
32	189.10	1585.36	189.15	1584.94	194.10	1544.52	194.15	1544.12	199.10	1505.73	199.15	1505.36
33	189.20	1584.52	189.25	1584.10	194.20	1543.73	194.25	1543.33	199.20	1504.98	199.25	1504.60
34	189.30	1583.68	189.35	1583.27	194.30	1542.93	194.35	1542.53	199.30	1504.22	199.35	1503.84
35	189.40	1582.85	189.45	1582.43	194.40	1542.14	194.45	1541.74	199.40	1503.47	199.45	1503.09
36	189.50	1582.01	189.55	1581.60	194.50	1541.34	194.55	1540.95	199.50	1502.71	199.55	1502.34
37	189.60	1581.18	189.65	1580.76	194.60	1540.55	194.65	1540.16	199.60	1501.96	199.65	1501.59
38	189.70	1580.35	189.75	1579.93	194.70	1539.76	194.75	1539.37	199.70	1501.21	199.75	1500.83
39	189.80	1579.51	189.85	1579.10	194.80	1538.97	194.85	1538.58	199.80	1500.46	199.85	1500.08
40	189.90	1578.68	189.95	1578.27	194.90	1538.18	194.95	1537.79	199.90	1499.71	199.95	1499.33
41	190.00	1577.85	190.05	1577.43	195.00	1537.39	195.05	1537.00	200.00	1498.96	200.05	1498.58
42	190.10	1577.02	190.15	1576.61	195.10	1536.60	195.15	1536.21	200.10	1498.21	200.15	1497.83
43	190.20	1576.19	190.25	1575.78	195.20	1535.82	195.25	1535.42	200.20	1497.46	200.25	1497.09
44	190.30	1575.36	190.35	1574.95	195.30	1535.03	195.35	1534.64	200.30	1496.71	200.35	1496.34
45	190.40	1574.54	190.45	1574.12	195.40	1534.25	195.45	1533.85	200.40	1495.97	200.45	1495.59
46	190.50	1573.71	190.55	1573.30	195.50	1533.46	195.55	1533.07	200.50	1495.22	200.55	1494.85
47	190.60	1572.88	190.65	1572.47	195.60	1532.68	195.65	1532.28	200.60	1494.47	200.65	1494.10
48	190.70	1572.06	190.75	1571.65	195.70	1531.89	195.75	1531.50	200.70	1493.73	200.75	1493.36
49	190.80	1571.23	190.85	1570.82	195.80	1531.11	195.85	1530.72	200.80	1492.99	200.85	1492.61
50	190.90	1570.41	190.95	1570.00	195.90	1530.33	195.95	1529.94	200.90	1492.24	200.95	1491.87

Note: The International Telecommunication Union established the grid for DWDM systems. ITU recommends a 100 GHz (0.8 nm) spacing for DWDM operation. A 50 GHz offset (0.4 nm spacing) offers twice the bandwidth. The grid is centered at 193.10 THz (1552.52 nm).

$$f = \frac{c}{\lambda}$$

$$\Delta f = -\frac{c}{\lambda^2}\Delta\lambda = -\frac{f}{\lambda}\Delta\lambda \tag{11.16}$$

Thus, by selecting a reference wavelength, e.g., the absorption wavelength, of a chemical substance and, thence, the frequency of such a channel, the frequencies of other wavelength channels can be estimated. A channel spacing of 50 GHz is now the normal standard for bit rates of 10 and 40 Gb/s, in which 44 channels can be accommodated in the C band.

The number of possible DWDM channels depends on (1) channel spacing, (2) amplifier spectral range, (3) source spectral width, (4) bit rate per optical channel and modulation formats, and (5) fiber dispersion and multiplexer/demultiplexer range.

The number of channels can be estimated by $N = 1 + \frac{\lambda_1 - \lambda_2}{\Delta\lambda}$, where $\Delta\lambda$ = channel spacing, and λ_1 and λ_2 are the two extreme wavelengths of the end channels.

Exercise: If $\lambda_1 = 1540$ nm and $\lambda_2 = 1560$ nm for a channel spacing of 1.6 nm (200 GHz), then $N = 12.5$ channels. Similarly, prove that if $\lambda_1 = 1535$ nm and $\lambda_2 = 1565$ nm for a channel spacing of 50 GHz, then $N \sim 44$ channels.

11.2.6.2 Multiwavelength MATLAB® and Simulink® Models

In modeling DWDM optical fiber transmission systems, normally the central wavelength channel is fully dispersion compensated; the other channels, in particular, those at the far ends of the spectrum are evaluated with the effects of the residual dispersion. As far as the modeling is concerned, the optical carriers are not included in the model, because if they are included, then the sampling rate must be in hundreds of tera-samples/second. This would drain the computing resources. Thus, this kind of problem is resolved by modeling the amplitude of the modulated lightwave as a complex amplitude, and thus the phase of the carrier can be found from this complex quantity.

The multiplexing of lightwaves of different wavelengths can be justified by including the carrier frequency and the observation of the spectrum of multiplexed channels, as shown in Figure 11.16. The spectrum of Figure 11.16c can be observed by setting the spectrum analyzer in the usual mode. Note that the multiplexer is to be set as shown in Figure 11.17.

11.2.6.3 Link Design Process

Choosing an application means

- Knowing your line rate and fiber type, whether it is a currently installed fiber system requiring upgrading or a newly installed transmission system.
- Determining what amplifier configuration meets the requirements.
- Checking to see if the application is supported, e.g., if link budgets and OADM rules are available.
- Remembering to consider the minimum span loss in span loss calculations: If you have to pad to reach the minimum span loss, the value after padding will be used in all calculations.

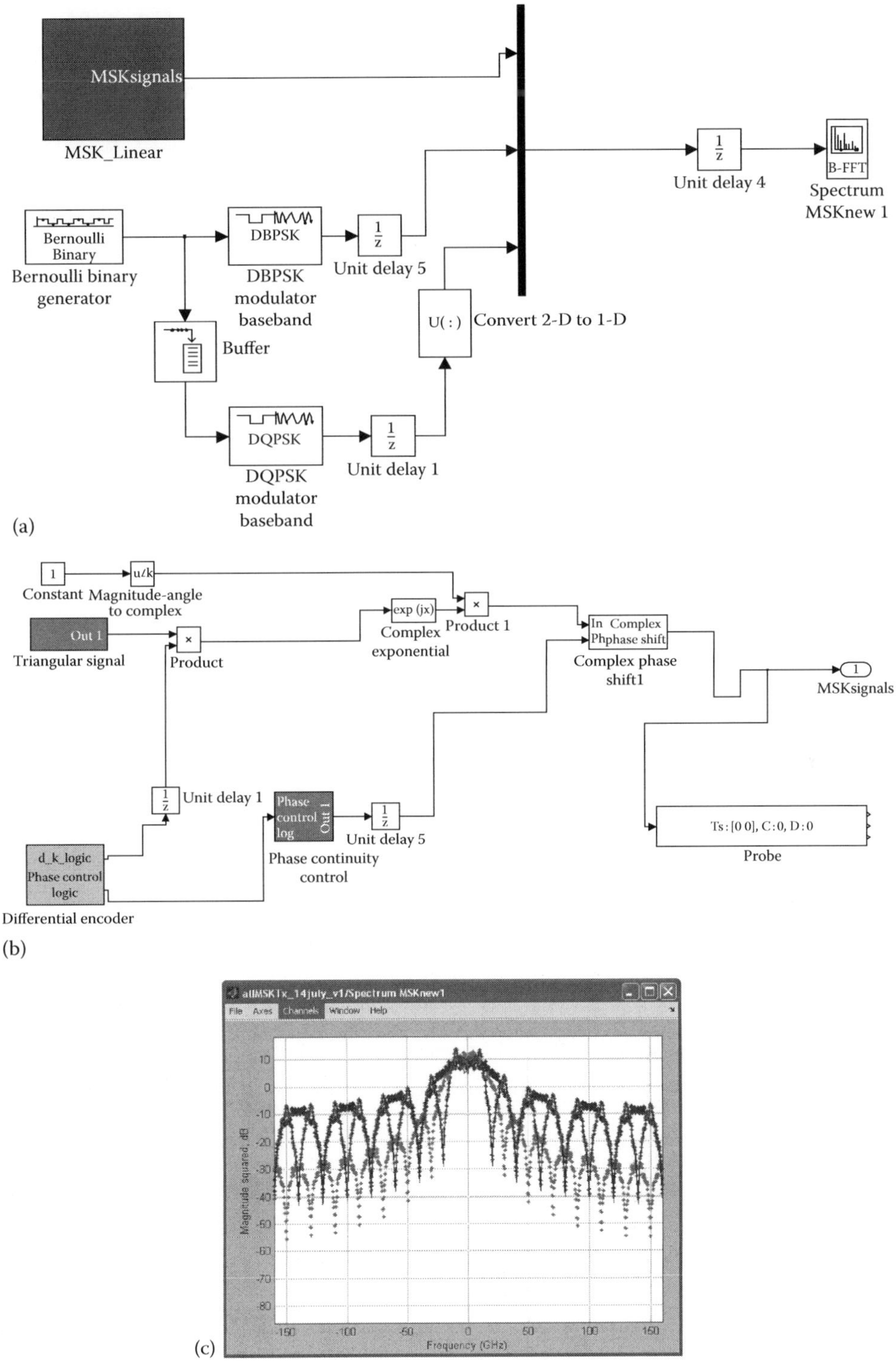

FIGURE 11.16
Simulink models of wavelength multiplexed channels. (a) Overall model. (b) Under mask of sub-model linear MSK. (c) Screen shot of the multiplexed channels in the baseband as observed by the spectrum analyzer.

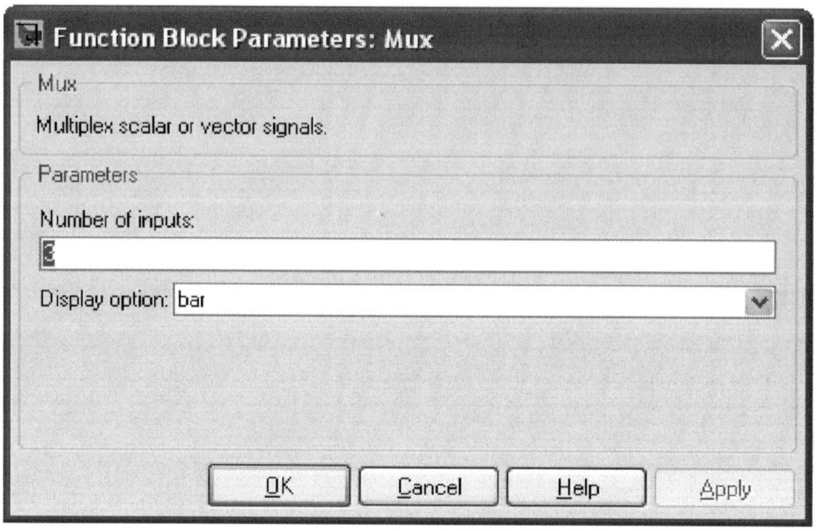

FIGURE 11.17
Setting of the mux block of Figure 11.16.

11.2.6.4 Link Budget Considerations

11.2.6.4.1 Forward Error Coding

Forward error coding (FEC) is normally used at the transmitter to increase the chances of detection without error. In this process, a number of redundant bits are used and coded in the data sequence:

- FEC adds dB/span and reduces BER.
- FEC benefits are system dependent and may vary from application to application.
- If FEC is to be used, check standards for specific information.
- FEC Reed–Solomon codes—42.7 Gb/s for 40 Gb/s base rate; 10.7 Gb/s for 9.953 Gb/s base rate.

11.2.6.4.2 Excess Loss: Operating Margin

- Calculating the excess loss (EL) (in dB),

$$\mathrm{EL} = \sum_{L_{\max} < L < L_{\max+2}} L - L_{\max} \tag{11.17}$$

L = actual span loss, L_{total} = sum of actual losses (including fixed pads, if any), n = number of spans, and $L_{\max}$ = link budget max dB/span (with/without FEC).

Then the EL rules can be stated as

- $L < L_{\max} + 2$.
- $L_{\text{total}} < n \times L_{\max}$.

- EL > 4 dB is not supported. FEC may be used to improve the SNR, and thus increase the operating margin.

$$\text{If } 2 < \text{EL} < 4PL_{\text{total}} < n'(L_{\max} - 1).$$

11.2.6.4.3 Basic Steps in the Design of Transmission Systems

The basic steps in the design of transmission systems for fiber-type modules and DCMs are as follows:

Define the system's span losses and lengths: (1) What budgets and windows are available for this application? (2) What extra margin (if any) must be added to the span losses? (3) What BER is required? (4) Is FEC an option or a requirement? (5) Where should regenerators be placed? For example, to eliminate mixed fiber links where possible: (i) Check the length windows to ensure there are DCM solutions available. (ii) Check the losses using the EL rules: attenuation pad span losses to minimum.

If there is an OADM site: (1) Check all sub-links for DCM windows and check mid-span-access losses. This may place a limit on sites and on the number of add/drop wavelengths supported. (2) Determine the DCMs required and their locations. (3) Take into account special OADM placement rules.

DCMs may be required at the Add Tx or Drop Rx: (1) Add Txs may be required to have a specific chirp. (2) Use a MidSpan-Opt-Regen Plus if DCMs are required at a line-amplifier site.

Calculate the MOR losses and add fixed pads where required.

Find the transmitter fixed-pad information and note the fixed-pad attenuator upgrade strategy.

Corning LEAF and enhanced LEAF perform differently. (1) Using LEAF budgets provides a conservative but safe estimate for ELEAF (enhanced LEAF) links; however, DCMs and pads will be different. (2) TrueWave Classic, TrueWave-RS, TrueWave+, and TrueWave– are not equivalent. (3) Treat them as completely separate fiber types, as both link budgets and DCMs will vary between these different fiber types.

Note: Link designs for 100 GHz or 50 GHz do not apply to 200 GHz as the DWDM filter specifications are different. Use array waveguide gratings (AWG)—bidirectional filters, multiplexing, and demultiplexing.

11.3 Link Budget Calculations under Linear and Nonlinear Effects

Under the nonlinear effects on the transmission performance, we need to consider (1) distortion versus noise, (2) high optical power: distortion of the signal, and (3) low power: low SNR.

Figure 11.18 shows the variation of the Q factor as a function of the launched power. When the launched power reaches the nonlinear threshold, distortion happens, leading to the degradation of system performance. Figure 11.19 shows the effects of the decision level on the Q factor. Gaussian distribution of the amplitude and noise levels are assumed. Optimum performance is achieved when the probability of error for the detection of the received "1" and "0" is the same.

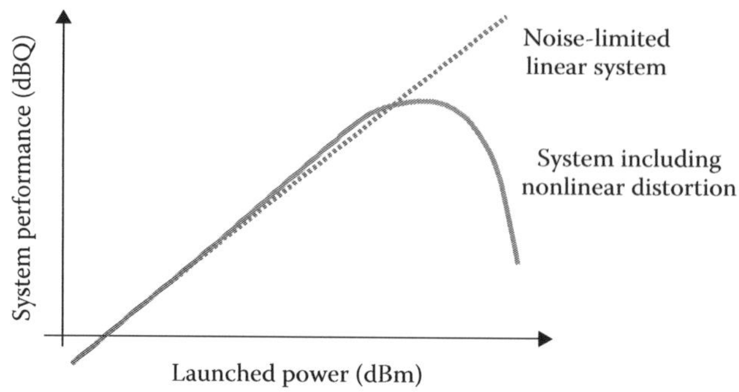

FIGURE 11.18
Variation of system performance as a function of the launched power—note linear and nonlinear region.

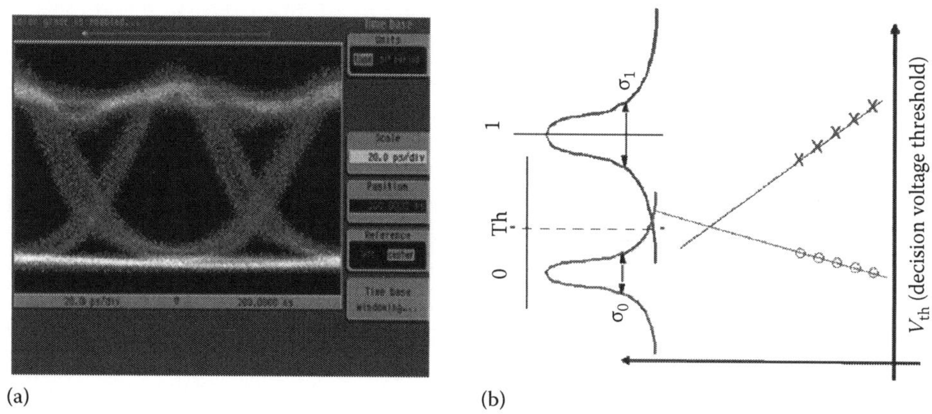

FIGURE 11.19
Eye diagram and extrapolation of the Q factor. (a) Received eye diagram. (b) Extrapolation of Q factor.

11.3.1 Budget

When designing the transmission system, we must keep in mind that

1. Link budgets are evaluated and verified for the system at (i) worst case, (ii) EOL, and (iii) reference equalization (RefEQ).
2. Budget guarantees include worst-case equalization (EOL).
3. Based on worst-case implementation and power control errors over life.
4. Worst case incorporates the following: (i) *Transmitter*: waveform, chirp, power, wavelength, noise, and jitter. (ii) *Receiver*: complex transfer function, power, noise, jitter, and sampling. (iii) *Fiber*: dispersion, DGD, and loss. (iv) *Amplifiers*: noise figure, gain ripple and tilt, and output power. (v) *System*: net dispersion, disp. map, channel separation, and polarization.

11.3.2 System Impairments

The system impairments fall into two groups: distortion impairment and noise impairment.

1. *Distortion impairment*:
   Any bounded reduction of the inner part of the eye diagram within the sample window includes (i) Rx distortions: transfer function related; (ii) Tx distortions: drive related, extinction ratio; (iii) optical path distortions: dispersion, SPM, FWM, and XPM < 5 channels; and (iv) PMD, cross talk, and MPI.
2. *Noise impairment*:
   Any unbounded stochastic reduction of the inner eye within the sample window includes (i) receiver noise, (ii) Tx noise, (iii) optical amplifier noise, (iv) modulation instability noise, and (v) FWM and XPM > 4 channels.

Signal-dependent noise contribution to Q is a function of distortion. However, noise and distortion can be separated for budgeting purposes.

11.3.2.1 Power and Time Eyes

The BER must include both the amplitude or the power eye and the time eye, which may be due to jittering of the waveform on the sampling instant for the evaluation of system performance. This is illustrated in Figure 11.20. Figure 11.21 shows the eye diagram and its degradation at both the transmitter and at the end of the transmission link.

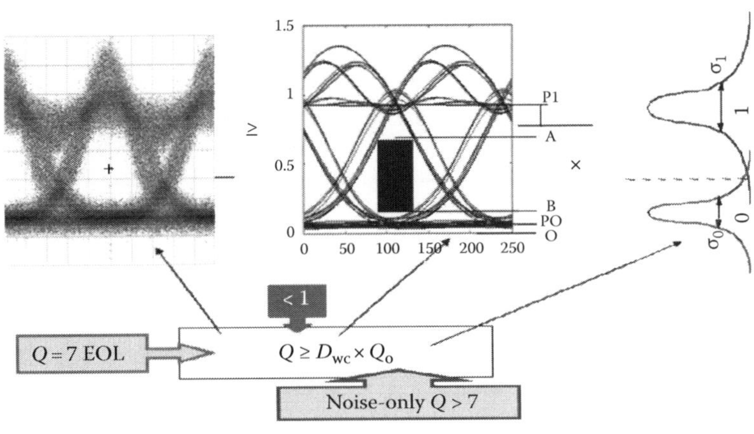

FIGURE 11.20
Illustrations of the eye diagram expected at the EOF with the original Q factor and effects of jittering for time Q.

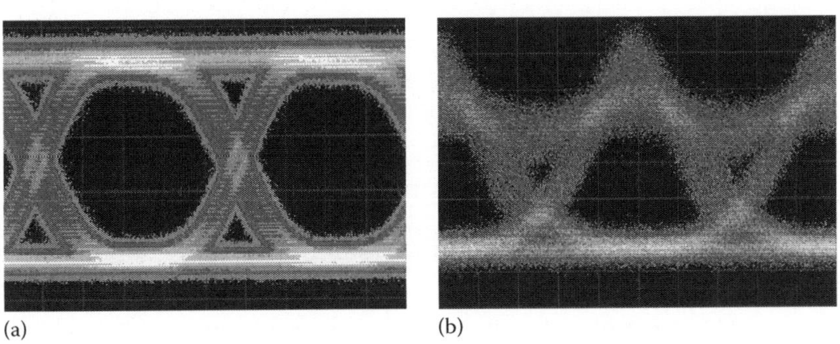

FIGURE 11.21
Eye at (a) the transmitter output and (b) at the end of the transmission link.

11.3.2.2 Dispersion Tolerance due to Wavelength Channels and Nonlinear Effects

We must note that (a) Distortion depends on net dispersion and, hence, on wavelength and bundled wavelength groups. (b) Self-phase modulation makes distortion curves power dependent. (c) Budgets are guaranteed over a net dispersion window meaning that the dispersion windows are reduced by any errors in link dispersion values.

11.3.2.3 Dependence on Wavelength Channels and Launched Power

Figure 11.22 shows examples of the dispersion tolerance in terms of the variation of EOP as a function of wavelength and launched power. It is expected that the worst performance must be targeted for the design that is the worst-case scenario.

The variation of the quality factor, Q, may also vary with respect to the total span length, that is, the optical amplification would contribute to the ASE noises, and, hence, the degradation of the quality of the system would occur. This can be illustrated in Figure 11.23. Thus, the design of optical amplifiers is very important in order to achieve optimum performance of the transmission systems.

11.3.2.4 Budget

Budget is based on measuring and estimating margins on a close-to-worst-case system, and guarantee can be obtained for EOL BER $< 10^{-12}$. A power budget table can be set as shown in Table 11.3:

The OSNR is based on ASK modulation and NRZ pulse shaping. This number can be updated for DPSK or any other modulation format as the dispersion would be less due to a narrower spectral width and, hence, the quality factor.

Example 11.5: Six-Span Optically Amplified Link: Upgrading of the Existing Transmission System with Standard Single-Mode Optical Fibers

A transmission system requires the following:

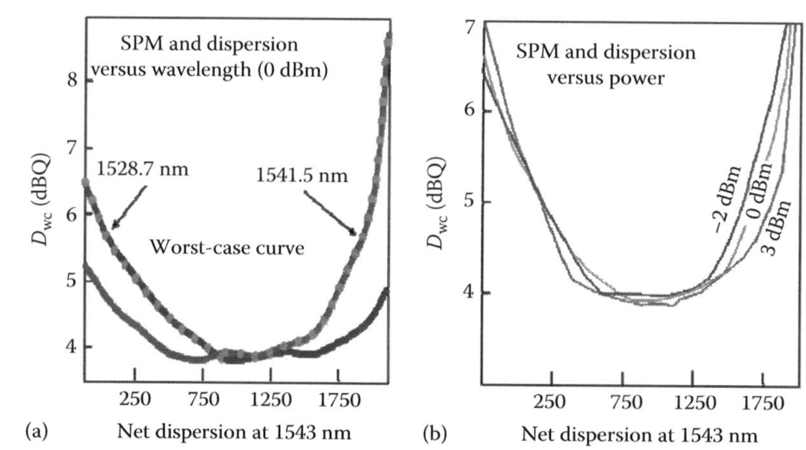

FIGURE 11.22
EOP versus dispersion as a function of (a) wavelength and (b) launched power.

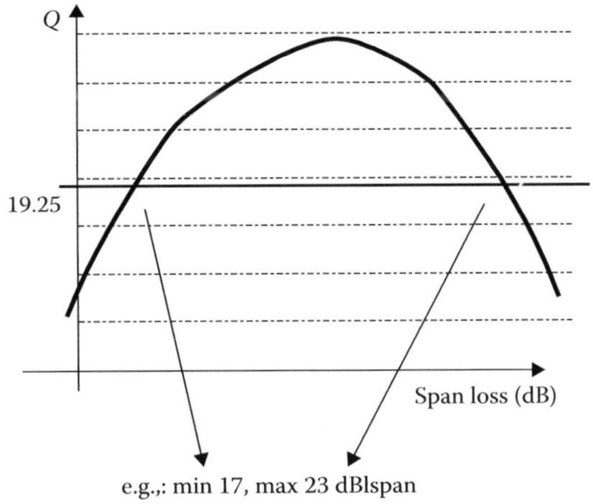

FIGURE 11.23
Variation of the quality factor, Q, with respect to the total span loss.

TABLE 11.3
Power Budget Including System Margin and Degradation Based on Statistical Measurement

Stressed span loss (dB)	24.95
OSNR margin to $Q=7$	3.88
Aging allowance	−4.00
Worst-case Rx/Tx correction	−1.90
Worst-case distortion dBQ × 2	−0.60
Power control error	−1.50
Sum degradations (dB)	−8.00
System margin (dB)	−4.12
Guaranteed maximum span loss (dB)	20.83

- Six-span NZDSF, 16 wavelength channels at 200 GHz spacing or 32 wavelength channels at 100 GHz spacing, and OC192 × 476 km total length
- Bidirectional transmission using one fiber only
- Addition of 1.4 dB margin to each calculated span loss
- Fiber attenuation of 0.22 dB/km—single optical fiber
- Not using 1625 nm, as it is reserved for online continuity monitoring

Two options, A and B, of the transmission systems can be given, as shown in Figure 11.24.

- Compare option A and option B.
- Choose an application type.
- Find DCMs, fixed pad (attenuator), and provisioning information for the final link design; thence, select the cheapest-possible solution (minimizing the number of mid-span optical regenerators as much as possible).
- Bidirectional transmission and compensation requires optical circulators, which would then allow the dispersion compensation for wavelength channels of different spectral regions (red and blue regions).

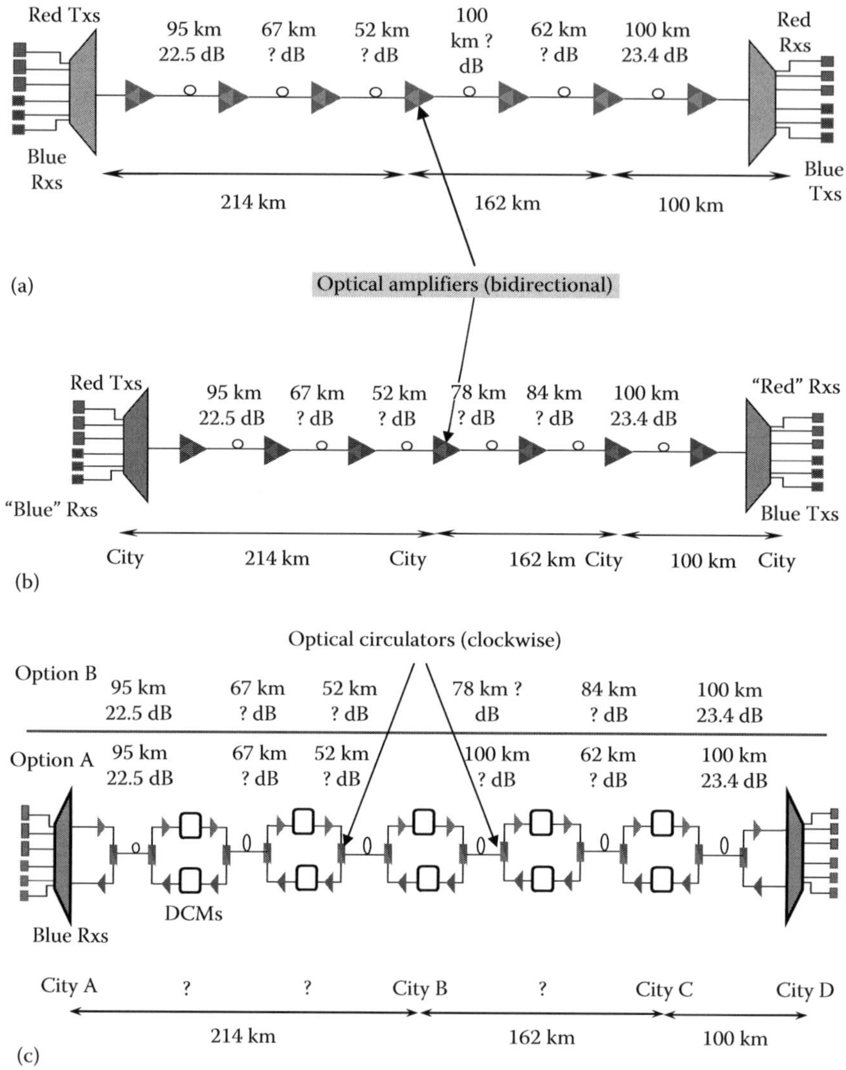

FIGURE 11.24
Design structures of multi-span link using bidirectional techniques. (a) Option A. (b) Option B. (c) Details of bidirectional transmission arrangement.

Solution: Six-span link

- Mid-span amplification: For DCM60, the loss is 7.3 dB, 11–7.3 = 3.7; therefore, use a 4 dB attenuation pad. For DCM80, the loss is 9.3 dB; therefore, use a 2 dB attenuation pad.
- Attenuation pads are required for minimum span loss requirements as well. Example: A loss of 16.4 requires 2 dB attenuation pads, etc. (Figure 11.25).

Solution with provisioning (Figure 11.26).

Option B	95 km 22.5 dB	67 km 18 dB	52 km 18 dB	78 km 18.8 dB	84 km 20.1 dB	100 km 23.4 dB
Option A	95 km 22.5 dB	67 km 18 dB	52 km 18 dB	100 km 23.7 dB	62 km 18 dB	100 km 23.4 dB

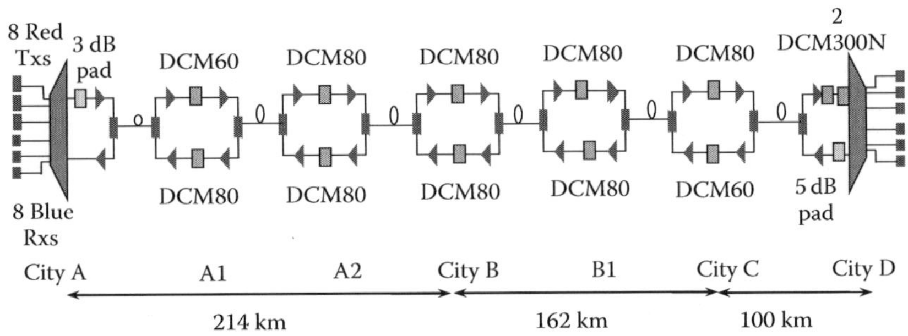

FIGURE 11.25
Solution of a six-span link for a 405 km distance via three cities.

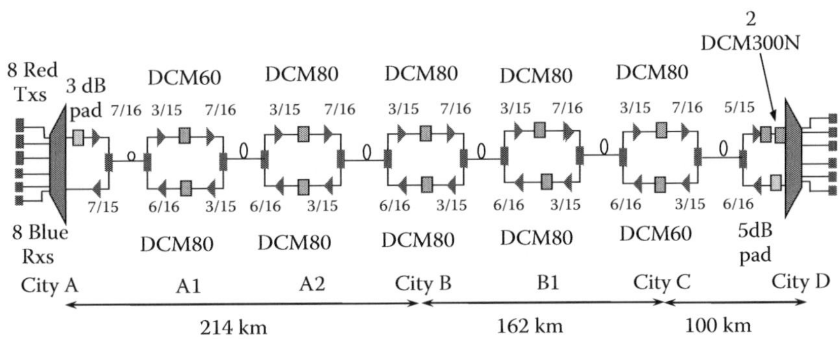

FIGURE 11.26
Transmission system solution with provisioning.

Example 11.6: 16 Wavelength Channel, Five-Span NDSF, MOR Bidirectional Optical Transmission

The same transmission distance, as in the example given above, is used, but with only five spans. Readers should inspect the design below and check to see whether they can confirm the design (Figure 11.27).

- Site A is the line-amplifier site closest to the RED transmitter (Tx)
- In each case, the chirp on the add/drop wavelengths must be positive (to compensate for the large negative DCMs within the add/drop sub-link).
- Two-span windows must be 154–166 km in length (155 km).
- Three-span windows must be 231–249 km in length (245 km).
- Rules apply for FEC with a BER of 10^{-12}.

11.3.3 Engineering an OADM Transmission Link

The structure of an OADM is shown in Figure 11.28. Selected wavelength channels of the incoming muxed channels can be dropped or wavelength channels can be added to the

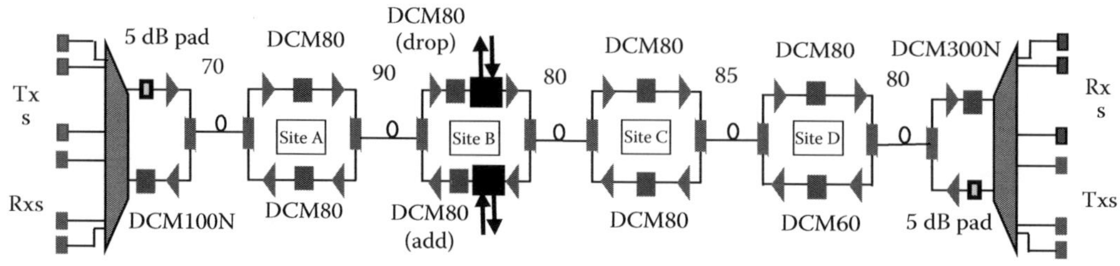

FIGURE 11.27
Transmission with provisioning for OADM insertion.

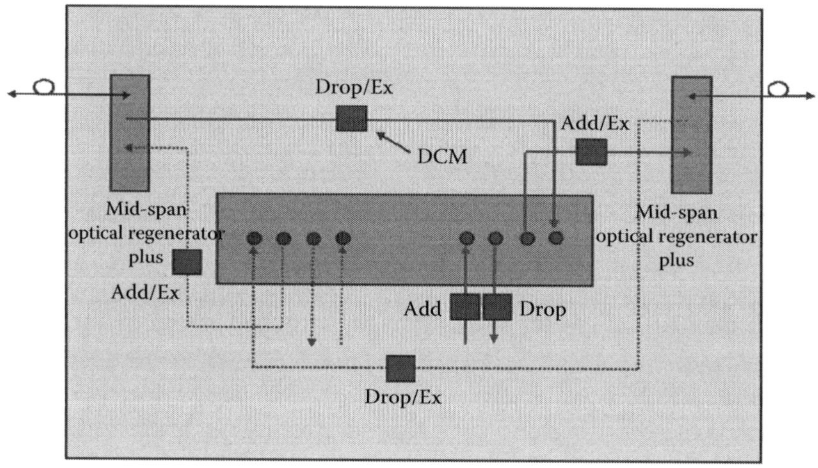

FIGURE 11.28
Structure of an OADM.

transmission system, which can be observed from this diagram. The details of the optical components, such as optical filters and circulators, are not shown in this section but observed from the system point of view.

Knowing the fiber and amplifier type, and the number of wavelength channels, we can determine the compensation rules for the link. The MSA loss now includes DCMs and OADM couplers (attenuation pad up, if required). Furthermore, provisioning of powers of sites, which changed the amplifier type as a result of the OADM, should also be made. Thence, determine the OADM compensation rules (in both power budget and dispersion budget) for the link. One should also apply these rules for both the working and protection fiber transmission paths.

11.4 Problems

PROBLEM 11.1

The distance between the transmitter and the receiver in an optical fiber transmission link using IM/DD is 100 km. The fiber link is made by splicing similar fibers of 22 km length each. The splicing loss is 0.15 dB per splice. The fiber attenuation is 0.4 dB/km.

(a) How many splices are required for 100 km link? Hence, estimate the total splice loss.
(b) The coupling loss is 1.5 dB in total at the transmitter and the receiver.

If the optical receiver has an RS of −30 dBm, what is the required power at the output of the optical transmitter if a power operating margin of 5 dB is used?

PROBLEM 11.2

An optical fiber transmission system over 1500 km distance is to be designed with a span length of 100 km. The following components are to be selected from the Web sites: jdsuniphase.com, corning.com, and avenex.com. The following parameters of the system are specified as

- Bit rate = 10 Gb/s with BER= 1e−12.
- Modulation format: OOK or DPSK, DQPSK or MSK or Star QAM.
- The number of wavelength channels in the C band must be at least 40 with a 50 GHz spacing between the channels, with a DWDM transmission.
- Operating margin = 3 dB.

Fiber: single-mode type, either standard or nonzero dispersion-shifted fiber.

(a) Select appropriate optical subsystems: optical transmitter including a laser and an external optical modulator. Give reasons why an external modulation must be used. Use the 3 dB bandwidth of the modulator to estimate the pulse-broadening factor of the transmitter.
(b) Select an integrated optical receiver and state the rise time and the fall time of the receiver, and, thence, estimate the broadening factor of the receiver. State the total equivalent noise current at the input of the receiver.
(c) Select either SMF-28 or LEAF or any other fibers that you could find. Corning Vascade fibers should be used for undersea routes. Hence, select DCF modules to match the dispersion of the transmission fiber per 100 km span.
(d) Specify the wavelength grid of the channels and the dispersion factor at the wavelength position of the channels. Hence, calculate the residual dispersion of the channels at the extreme wavelength locations.
(e) Select optical amplifiers, EDFAs, etc. State their minimum and maximum input optical powers, the maximum optical power at the output, the optical gain, and the noise figure. Hence, what is the power level that would drive the amplifier into saturation?
(f) Establish the structure of the transmission systems, including spans, the optical transmitter, and the receiver. Make sure that the details of the optical transmission spans must be specified.
(g) Establish the rise-time budget table. Sketch the dispersion/broadening of the pulse over each span. Make sure that the broadening satisfies the 70% bit period criterion for all channels.
(h) Establish an attenuation budget table. Make sure that the RS can be employed to recover the signals.

PROBLEM 11.3

Mini-Project Design: Optical Communication System Design

Network distribution:	Australian ultra-wide band optical fiber backbone networks
Transmission link distance:	Melbourne–Sydney, Melbourne–Hobart, Melbourne–Adelaide
Transmission techniques:	DWDM techniques with synchronous digital hierarchy technology STM-64, bit rate = 9953 Mb/s or 40 Gb/s return-to-zero format or Ethernet 100 Gb/s
Modulation formats:	ASK, DPSK, DQPSK, or MSK with pulse shaping NRZ or RZ to be considered
BER	$<10^{-12}$—to be confirmed with numerical simulation
Network configuration:	SDH ring, —e.g., double ring with optical cross connect near Canberra
Total capacity:	40 Gb/s × 32 wavelength channels or 80 Gb/s × 16 wavelength channels
Wavelength regions:	Suggested C band, L band, or S band of 1485–1625 nm
Optical channel frequency spacing:	50 or 100 GHz spacing between channels
Transmission medium:	SMF-28 (G.652 ITU standard) with dispersion-compensating techniques (fiber or chirped transmitters) or NZDSFs, e.g., LEAF or TrueWave
Fault-monitoring systems:	OTDR and wavelength channel fault identification—specify OTDR management system
Optical amplifiers:	Erbium-doped fiber amplifiers, and/or Raman fiber amplifiers
Photonic components:	DWDM mulitplexers/demultiplexers, couplers, etc., available from Web sites or Laboratory of Optical Communications (Monash)
Operating margin:	3.5 dB
Transmitters and receivers:	Refer to commercial products at Web sites of manufacturers, e.g., www.nel.co.jp, www.jds-uniphase.com, www.nortelnetworks.com, www.lucent.com, and www.fujitsu.com

Suggestions:

- Build a technical case for the optical fiber transmission systems and networks.
- Build a business case for a rollout of a new network linking the capital cities in Australia. You should consider the following: (a) Base the case on carrying bulk data (IP over SDH [synchronous digital hierarchy] to remove the cost of provision of an ATM layer) and (b) assume the incumbent is currently operating on a 2.5 Gb/s equipment over fibers up to 15 year old.
- Based on the technical design of the network, present a business case requesting funding from a venture capitalist to build the network.

PROBLEM 11.4

A long-haul optically amplified fiber transmission system operating at a wavelength of 1550 nm is envisaged to use EDFAs spaced at intervals of 100 km. Alternating +D and –D single-mode optical fibers are used. The mode spot size of these fibers is the same. The power launched into the link at the transmitter is 0 dBm. The fiber attenuation is 0.22 dB/km. In addition, there are splice losses, which average out at 0.03/km.

An OSNR of 17 dB is required at the receiver to provide a BER of 1e–9 at an operating bit rate of 10 Gb/s. Assume that the system bandwidth is equal to that of the bit rate and the noise figure of the amplifier, NF = 3 dB, for an optical gain of 20 dB. Estimate the maximum transmission length of the system such that a satisfactory performance can be maintained.

PROBLEM 11.5

1. Sketch the structure of an optical transmission system consisting of
   a. An optical transmitter that can multiplex all optical channels allocated in the C band with a 50 GHz spacing between the channels
   b. Optical demultiplexers and receivers for each wavelength channel
   c. A transmission distance of 200 km with a span length of 100 km on an average; dispersion compensators as optical fibers can be installed along the transmission path
   d. Optical amplifiers in each span
2. Design the transmitters, the optical receivers, and the fiber spans with optical amplifiers having the following parameters:
   a. The fiber nonlinear threshold for SMF-28 of 5–10 dBm for the total average power of all channels.
   b. The diameter of the spot size of the DCF of about 30% of SMF-28 and its dispersion factor of –68 ps/(nm km); the attenuation of this DCF of about 0.35 dB/km.
   c. Optical amplifiers can have two types—type 1: gain = 25 dB, saturation output power = 15 dBm, minimum input power = –25 dBm, and maximum input power = 0 dBm; type 2: gain = 20 dB, saturation output power = 10 dBm, minimum input power = –25 dBm, and maximum input power = 0 dBm.
   d. An optical RS of –28 dBm and an OSNR of 15 dB for a bit rate of 10 Gb/s.
   e. An EL of multiplexer and demultiplexer of about 1 dB.

It is suggested that you should conduct the following:

- Following the example in Problem 11.1, estimate the maximum transmission distance
- Obtain a power budget and distribution along all spans of the transmission system
- Obtain the dispersion map of the system along all spans

PROBLEM 11.6

- For a system bit rate of 10 Gb/s; an ASK-NRZ transmission format with a BER of 1e–12; the photonic and optoelectronic components specified and given, respectively, in Problem 11.2; and any other necessary optical components; design an optical transmission system for a section of the network that has a total transmission distance of 1000 km. In case you cannot obtain the RS of part (b), it is suggested to assume that the electronic preamplifier is adjusted so that the RS is –20 dBm; the rise time and the fall time remain the same.

- An optical transmitter using distributed feedback lasers have a typical laser linewidth of 0.001 nm and an external modulator. The average optical power of –10 to 2 dBm at the output of the fiber pigtail of the modulator is normally available for such a transmission system. The transmitter would have a total rise time and fall time of not higher than 2 ps.
- Optical fibers can be selected from a Corning fiber product, SMF-28 or LEAF, that is, the dispersion factor at 1550 nm is 17ps/(nm · km) and 4 ps/(nm · km), respectively. The nonlinear SPM coefficient for silica is about 2.3 × 10–20 m2/W and the attenuation of the fibers is 0.2 dB/km. The spot sizes are about 4 mm for SMF-28 and 5 mm for LEAF.
- Now, if a number of optical channels with a wavelength spacing of 50 GHz over the entire C band, 1525–1560 nm, carrying NRZ data at 40 Gb/s are multiplexed into the same fiber, taking the fiber nonlinearity into consideration, describe the details of considerations for any change in the design of part (a).

PROBLEM 11.7

An undersea optical transmission is to be designed. The undersea distance is 220 km. The bit rate is 40 Gb/s and an optical receiver with a sensitivity of –20 dBm is available. Optical amplifiers EDFA and ROA are considered with the following parameters—fiber types: SSMF and DCF (with a dispersion factor of –85 ps/(nm km) and an effective area of 17 μm^2); EDFA: gain = 20 dB and NF = 5 dB; output power: saturated output power = 15 dBm for booster amplifier type and +10 dBm for optical preamplification type; ROA: pump power = 500 mW, gain in SSMF = 12 dB over a 20 km distributed distance of the fiber, and NF = 3 dB, for both co- and contra-directional pumps.

(a) Sketch a proposed structure of the transmission link
(b) Obtain a diagram of the signal power or the power budget over the link
(c) Obtain the dispersion budget over the link

12

Simulink® Models of Optically Amplified Digital Transmission Systems

This chapter gives examples and simulation models of a number of optically amplified transmission systems in MATLAB® and Simulink®, employing advanced modulation formats such as amplitude shift keying (ASK), differential phase shift keying (DPSK), differential quadrature phase shift key (DQPSK), minimum shift keying (MSK), and Star-quadrature amplitude modulation (QAM) with optically amplified spans using either erbium-doped fiber amplifiers (EDFAs) or distributed Raman optical amplifiers (ROAs). In the appendix, we provide instructions on the use of these models—the MATLAB and Simulink codes of these models can also be downloaded from CRC Press sites.

12.1 ASK Modulation Formats Transmission Models

Advanced optical fiber communications have attracted tremendous interest in recent years due to urgent demands on the upgradation of the fiber backbone transmission systems and networks. In these modern optical transmission systems, modulation techniques employing amplitude, phase, and/or frequency shift keying (FSK) modulation schemes are considered to be the keys for reducing the signal and carrier spectral properties. Together with these schemes, the formats of the pulse sequence such as return-to-zero (RZ), non-RZ (NRZ), carrier-suppressed RZ (CSRZ), or duo-binary (DB) or bi-phase shaping of the pulse sequence are considered to optimize the transmission energy contents of the transmitted sequence, and hence the enhancement of the total transmission distance without violating the nonlinear/linear limit of the transmission or dispersion-compensating fibers (DCFs).

This section outlines the modeling of the ASK in association with the pulse-shaping modes of RZ, NRZ, and CSRZ and DB formats so that the simulated results can be corroborated. The modeling platform is based on the MATLAB and Simulink as this mathematical and graphical modeling package offers sufficient all-signal representation and processing blocksets that allow the generation, modulation, and processing of lightwaves and signal schemes at optical frequency in the passband and baseband domains. Furthermore, it offers a range of equipment that could be employed to inspect signals in the temporal and spectral domain to study the impacts of the modulation spectra on the transmitted data sequence.

Models have been developed for ASK for all formats and their transmission, especially the dispersion tolerance, over standard single-mode fibers (SMFs) are demonstrated. Furthermore, a DB modulation scheme with alternating phase property is also demonstrated. The developed models have been proven effective in the transmission and receipt by interferometric delay schemes for differential detection and recovery of the precoded sequence. The optical modulation employing external modulators, such as the electro-absorption (EA) and electro-optic (EO) modulators, is very important for modulation schemes using both amplitude and phase. These are exploited in our study with the EA- and $LiNbO_3$-type modulators. We note that the models representing photonic components, such as the modulators and interferometric delay-balanced receivers, are operating in the photonic domain; that is, the interaction between guided optical waves and traveling wave electrical signals represents exactly the real EO operation. No mathematical representation is used. This contrasts the modeling of several packages developed for optical communication systems in which the mathematical languages have been used. The Simulink models are thus described wherever appropriate to prove these findings.

12.1.1 Introductory Remarks

Owing to the growing demand on high-capacity transmission over the Internet, high data rate of 40 Gb/s per channel has appeared to be an attractive feature in the next generation of lightwave communication systems. Under the current 10 Gb/s dense wavelength division multiplexing (DWDM) optical system, overlaying 40 Gb/s on the existing network can be considered to be the most cost-effective method for upgrading purposes. However, there are a number of technical difficulties confronted by communications engineers involving interoperability that requires 40 Gb/s line system to have signal optical bandwidth, tolerance to chromatic dispersion (CD), resistance to nonlinear cross talk, and susceptibility to accumulated noise over a multi-span of optical amplifiers to be similar to 10 Gb/s system.

In view of this, advanced modulation formats have been demonstrated as an effective scheme to overcome the impairments of 40 Gb/s systems. The DPSK modulation format has attracted extensive studies due to its advantages over conventional ASK or on–off keying (OOK)* signaling formats, including having 3 dB lower optical signal-to-noise ratio (OSNR) [1] at a given bit error rate (BER), being more robust to narrowband optical filtering, and being more resilient to some nonlinear effects such as cross-phase modulation (XPM) and self-phase modulation (SPM). Moreover, spectral efficiencies can be improved by using multilevel signaling. On top of that, coherent detection is not critical as differential detection requires the comparison of two consecutive pulses; hence the source coherence is required only over one bit period.

Nevertheless, the DPSK format involves rapid phase change causing intensity ripples due to CD that induce pattern-dependent SPM-GVD (group velocity dispersion) effect [2]. Therefore, RZ pulse can be employed in conjunction with DPSK to generate more tolerance to the data pattern-dependent SPM-GVD effect. In addition, RZ improves dispersion tolerance and nonlinear effects particularly in long-haul networks at high data rate. Specific RZ formats like CSRZ help to reduce the inherent larger spectral bandwidth.

At 40 Gb/s, the generation of RZ pulse is not as feasible as it is at 10 Gb/s due to the large bandwidth requirement. Thus, 40 Gb/s RZ signals are produced optically in "pulse carver" by driving the modulator with a 20 GHz RF signal. With the remarkable advancement in external modulators, especially in Mach–Zehnder interferometric modulators (MZIM), this is easily achieved by utilizing the microwave optical transfer characteristics of MZIM.

* The term amplitude shift keying (ASK) is preferred rather than OOK as it reflects the quasi-coherence of the detection scheme in which the amplitude and phase of the carriers do contribute to the sensitivity of the receiving system.

In this chapter, we present the viability of RZ-ASK modulation formats for the next generation of lightwave communication systems, especially for ultra-high-capacities and ultra-long-haul links. Simulations models have been described based on the MATLAB and Simulink platforms. A user-friendly environment in Simulink provides fast design and modeling tools to simulate components and essential devices such as MZIM. System performances, optical signal bandwidth, eye diagram, and error probabilities can be evaluated instantaneously by changing design parameters. In high-cost and expensive long-haul optical transmission systems, these simulation results are important for communication engineers to design network configurations prior to experiment and installation.

The principles of generation by using MZ interferometric structures based on biasing voltage and voltage swing of the modulating signals and characteristics such as signal spectra of different modulation formats are depicted. The receiver structures and working principles based on single-ended and balanced detection are explained.

This section is organized as follows: In Section 12.1.2, a general optical communication system comprising essential photonic components in single-channel and multichannel structures is described. In Section 12.1.3, the basic principles of components in photonic transmitter, such as laser sources and optical modulators, are discussed with reference to Simulink modeling. Section 12.1.4 briefly covers the principal propagation phenomena and constant and its second- and third-order parameters that would severely distort the carrier-modulated data sequence. This section also gives details of the propagation model. Note that the propagation model under ROA was covered in Chapter 10.

Section 12.2 gives models of different pulse-shaping formats. Thence, Sections 12.3 and 12.4 describe the generic structures of the Simulink model and especially DPSK schemes.

Sections 12.5 and 12.7 outline the Simulink models for DQPSK and DB modulation schemes. Section 12.6 illustrates the optical spectra of modulated signals using DPSK and MSK. Thence, Section 12.8 gives details of a typical continuous-phase SK scheme, the MSK, or alternatively an FSK scheme.

Section 12.9 deals with the modeling design process of multilevel (M-ary) in both phase and amplitude of the carrier. Detailed Simulink models of two typical M-ary levels, 16 amplitude DPSK (MADPSK) and 16 Star-QAM, are also described.

Finally, concluding remarks about novelties of the developed models and future works on improving the simulator are given. Each section is written so that it can be independent of the others so that readers can skip to different sections. Due to this independent arrangement of sections, this chapter is longer than any other chapter of this book.

12.1.2 Components for Advanced Optical Communication System

The transmitter consists of a modulator that is responsible for converting data in the electrical domain to that in the optical domain either directly or externally using EO or EA modulators. The modulated optical signal is then transmitted through a single-mode optical fiber using compensating fibers. In the long-haul link, propagation loss is considered to be significant and requires an optical amplifier to boost optical signal power at the span of 80–120 km at the regeneration point. At the receiver, the optical signal is detected using typical P-type-Intrinsic-N-type (PIN) photodiode or avalanche photodiode with gain equalization and chromatic compensation. The signal is decoded and regenerated in electrical domain for user output.

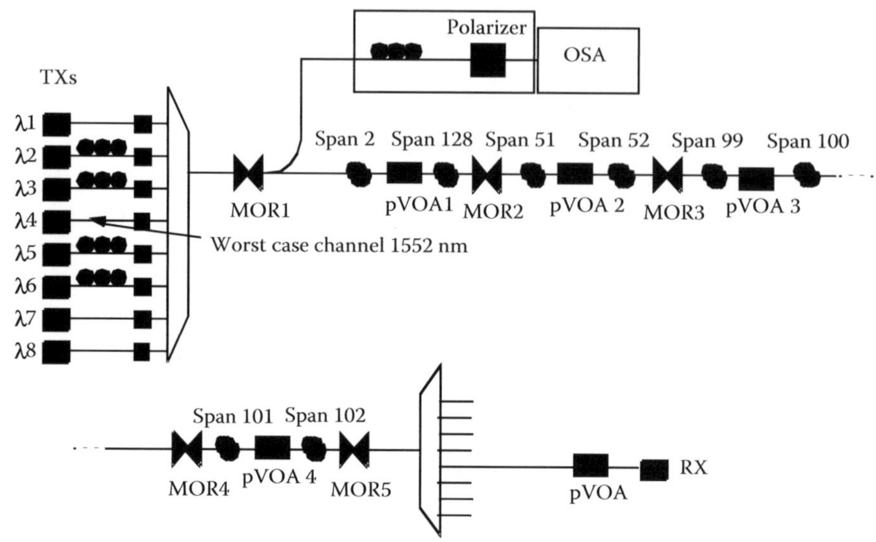

FIGURE 12.1
Typical long-haul DWDM multi-span optical fiber transmission systems. MOR, mid-span optical regenerator; VOA, variable optical attenuator; OSA, optical spectrum analyzer; Tx, transmitter; Rx, receiver; and VOA, variable optical attenuator.

The next level of complexity in optical transmission systems involved the addition of multiple wavelength of light on a single fiber to use costly bandwidth efficiently. Each channel consists of a wavelength of light, which is a distinct "color" that can be combined with the multiplexer for transmission and separated at the receiver with demultiplexer. This technique is called wavelength division multiplexing (WDM), and a typical WDM system is illustrated in Figure 12.1.

As the exponentially growing Internet traffic puts higher data rate on high demand, DWDM system is realized by multiplexing more channels within a band window with approximately 100 GHz spacing. With the advent of EDFAs, DWDM is made practical and economical by allowing a multitude of wavelengths in the fiber to be boosted simultaneously.

12.1.2.1 Optical Source

In this book, laser sources such as distributed feedback laser (DFB) are used throughout due to their superior performance and characteristics in long-haul communication link. DFB lasers have narrow linewidths typically about 100 MHz or even narrower. Linewidth is the range of wavelengths to describe how coherent real sources produce radiation. A narrower linewidth is attributed to less sensitivity to temperature variation than other laser sources. The wavelength shifts are under 0.1 nm/°C owing to stabilization from grating. For this reason, in this section, DFB lasers can be characterized using a single frequency sine wave function provided in MATLAB and Simulink that we assume as nearly monochromatic. The frequency of these optical sources is very high in the order of a few hundred terahertz. If sampling is used then the sampling rate is extremely high and no computing resources can meet this demand. Thus complex amplitudes are used. This means that the phase of the optical carrier (OC) is represented as the resultant angle between the real and complex parts of the sampled amplitude.

12.1.2.2 Optical Modulators

As presented in Chapter 5, it is important to note that a laser diode presents problems to optical modulation because it has a threshold current. Since threshold currents are dependent on age and temperature, degradation will result in an *L–I* curve. (*L–I* curve is a representation of relationship between output light signal and direct current of laser diode [3,4].) Generally, external modulators consist of EA modulators and EO modulators. The common operating principle is based on the EO effect (change of refractive index in material is proportional to the applied electric field). This is extremely useful in high-speed operation because refractive index changes quickly with respect to the applied electric field.

EA modulators are monolithic semiconductor planar waveguides. They operate on either the Fraz–Keldysh effect (FKE) or the quantum stark effect by forming multiple quantum wells that consist of multiple p-type and n-type layers. The advantages of using these modulators are chirp relaxation (even though there is some inherent chirp), high input impedance, and monolithical integration on the same semiconductor chip with the laser diode [4]. This offers greater advantage over EO modulators since insertion loss can be reduced. The modulators are diodes that operate in reverse bias. When there is no bias voltage, EA modulators are transparent to incoming optical signals. On the other hand, the applied voltage can generate a photocurrent that increases waveguide attenuation and signal losses. In other words, output power from EA modulators are the highest with no applied bias voltage and decreases as the bias voltage increases, as shown in Figure 12.2b. It is to be noted that the optical modulation index of EA modulators does not reach unity so that there are some residual optical power level, and this may cause an increase in the zero level and hence lower the quality *Q*-factor or the BER of the transmission system. For this reason, the MZIM, which has been described in Chapter 5 and briefly revisited in Section 12.1.2.3, is preferred.

12.1.2.3 Mach–Zehnder Intensity Modulator

For 10 or 40 Gb/s DWDM applications, EO modulators are the technology of choice due to better performance in terms of chirp, extinction ratio (higher), and modulation speed compared to EA devices. Over the years, lithium niobate ($LiNbO_3$) has been the most popular material for EO modulators because it enables low loss waveguide and high EO effect [5]. The most popular $LiNbO_3$ EO modulators that are currently employed particularly in high-performance long-haul optical transmission system are based on MZ interferometric

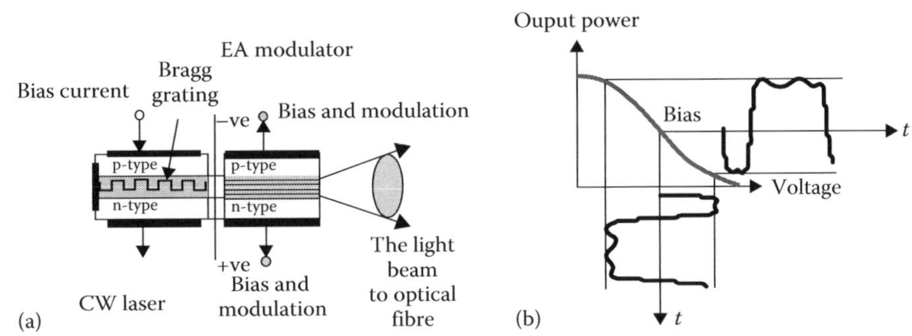

FIGURE 12.2
(a) Structure of EA modulator coupled with a DFB laser and (b) optical–electrical transfer characteristics.

optical waveguide structures. The MZ modulator (MZM) has two pairs of electrodes deposited along the arms of the interferometer, one for DC bias voltage and the other for modulating signals. Meanwhile, numbers of DC bias voltage categorized the operation of MZMs as single drive and dual drive.

12.1.2.3.1 Single-Drive MZIM

In an MZM, the incoming light is split equally into two paths at Y branch (3 dB splitter) and travel along the upper and lower traveling wave electrodes. An electric field is applied across the optical waveguide by an electrode that generates a change of refractive index of EO waveguide material via the EO effect. The change in the refractive index profile causes the speed of light wave in the waveguide to change according to c/n_r [5]; thus phase shift occurs. When the guided lightwaves from upper and lower waveguides recombine at the output Y branch, either a constructive or a destructive interference is generated depending on the relative phase shift of two branches. If the light waves from two arms are in phase, "ON" state of logic 1 is inferred due to constructive interference. The output is interpreted as "OFF" state or logic 0 if the light waves are π (or 180°) out of phase derived from destructive interference. If a coupler is placed at the Y-junction, then naturally the constructive and destructive interference outputs can be obtained. These two outputs are then detected by a back-to-back connected photodetectors acting as a push–pull receiver (see Figure 12.3, also Chapters 5 and 6).

The combined optical wave at the output of Y branch is given as [5] follows:

$$E_o = \frac{E_i}{2}\left[1 + e^{j\pi\frac{V(t)}{V_\pi}}\right] = E_i \cos\left(\frac{\pi}{2}\frac{V(t)}{V_\pi}\right)e^{j\pi\frac{V(t)}{V_\pi}} \tag{12.1}$$

To express in terms of optical intensity to derive input–output transfer characteristics of MZIM [5]

$$E_o^2 = E_i^2 \cos^2\left(\frac{\pi}{2}\frac{V(t)}{V_\pi}\right) \tag{12.2}$$

where

E_o and E_i are the output and input electric field of the lightwaves
$V(t)$ is the driving voltage or applied voltage at traveling wave electrode
V_π is the voltage required to cause π phase shift of optical wave

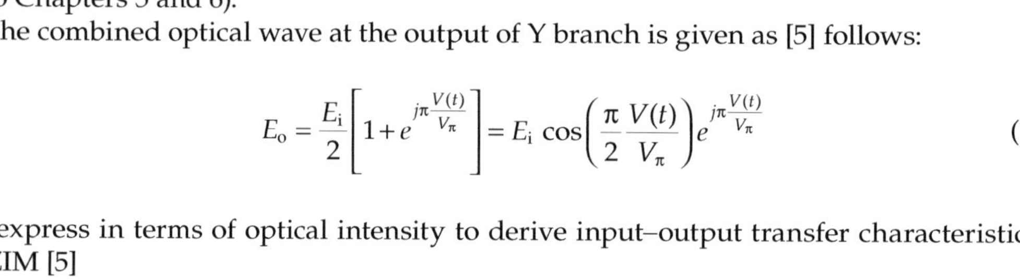

FIGURE 12.3
Illustration of (a) constructive and (b) destructive interference at single-drive MZ modulator output.

As the transfer characteristic is cos²-shaped, the modulator can be biased in different regions, and the modulating signal is superimposed onto the bias voltage. The nonlinear transfer function has been used to generate NRZ, RZ pulse with various duty cycle, and DPSK signal (see Figure 12.4).

12.1.2.3.2 Dual-Drive MZIM

MZIM can also be driven with dual electrode structure or commonly referred to as a push–pull modulator. In this section, we consider chirp-free driving by using two driving voltages with π phase difference (i.e., $u_1(t) = -u_2(t)$, $\delta = \pi$). The operational principle is similar to a single-drive MZIM, but with both arms applied with voltage to phase modulate the OC via the EO effect. The interference at the output of Y branch will produce phase modulation. If the phase modulation is exactly equal in each path but opposite in sign, by combining two optical signals, the phase modulation is converted into intensity modulation. As a consequence, a dual-drive MZIM has an attractive feature; it can be driven as a phase modulator (PM) and an intensity modulator by changing its driving voltage [6]. The phasor representation of dual-driven optical modulators has also been described in Section12.2 (Figure 12.5).

12.1.3 Transmission Loss and Dispersion Revisited

In an optical transmission system, optical fibers serve as a medium to transport optical signals from the source to the destination. Due to low attenuation and ultra-wide transmission bandwidth, optical fibers can transmit signals over long distances at high speed without amplification. Fiber losses impose a severe limitation on optical transmission because the signal power is greatly attenuated when reaching the receiver. With the advent

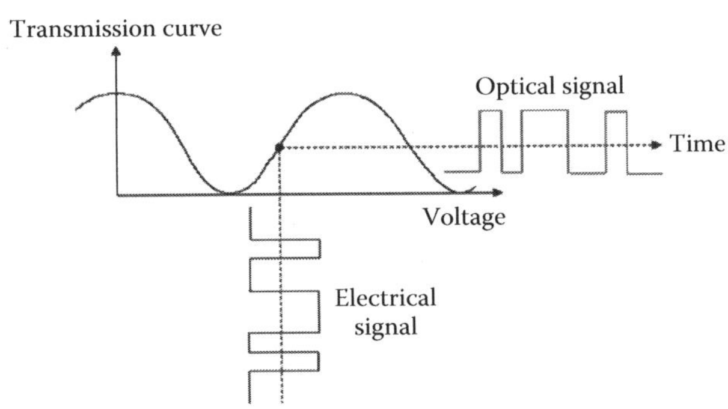

FIGURE 12.4
The transmission curve of MZIM forms the periodic waveform as a function of applied biasing voltage.

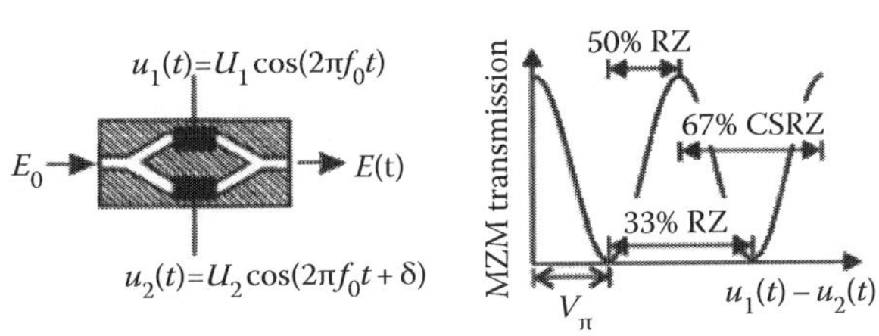

FIGURE 12.5
Schematic and intensity transmission curve of a dual-drive MZIM.

of optical amplifiers, long-haul transmission exceeding several thousand kilometers was achieved by compensating accumulated losses. However, the loss profile is imperative to determine spacing between amplifiers [7]. Fiber loss is expressed in units of decibels per kilometer and dispersion characteristics have been described in Chapters 2 and 3.

12.1.4 Nonlinear Effects

Nonlinear effects contribute to pulse distortion when optical power exceed several milliwatts in a single-channel optical transmission system due to their proportional relationship with the intensity of the lightwave field of propagating optical signal [4]. In spite of negative impacts, such as signal cross talk, optical signal power attenuation, and signal distortion, nonlinear effects can be exploited in curbing CD. Nonlinear effects can be categorized according to nonlinear refractive index or nonlinear optical signal scattering. The effects related to nonlinear refractive index comprise SPM, XPM, and four wave mixing (FWM). On the other hand, stimulated Raman scattering (SRS) and stimulated Brillouin scattering (SBS) are affected belonged to nonlinear optical signal scattering caused by an interaction between light and material. In this chapter, the nonlinear effects and their penalty on system performance have been given. These effects are also included in the NLSE, which can be revisited in Section 12.1.5 with given reference to the Simulink model and MATLAB.m files for the representation of the optical wave propagation.

12.1.5 Signal Propagation Model

12.1.5.1 Nonlinear Schrödinger Propagation Equation

Signal propagation in optical fibers is the most important issue to be addressed in lightwave transmission systems. Optical pulses are subjected to linearities and nonlinearities in the fibers. Linear degrading effects comprise of optical loss and GVD, while nonlinearities include SPM, FWM, SRS, and SBS effects. Interaction between these effects, including linear and nonlinear effects, can cause temporal and spectral properties of the optical pulse to change significantly. A mathematical model commonly used to describe the slowly varying envelope of the optical field is nonlinear Schrödinger equation (NLSE). The detailed derivations will not be discussed in this section but can be found in [8–10]. The general form of an NLSE can be expressed by

$$\frac{\partial A}{\partial z}=-\frac{j}{2}\beta_2\frac{\partial^2 A}{\partial T^2}+\frac{1}{6}\beta_3\frac{\partial^2 A}{\partial T^3}+\frac{j}{24}\beta_4\frac{\partial^4 A}{\partial T^4}-\frac{1}{2}\alpha A+j\gamma|A|^2A-a_1\frac{\partial}{\partial T}\left(|A|^2A\right)-j\alpha_2\frac{\partial}{\partial T}\left(|A|^2\right) \tag{12.3}$$

where

$A=A\ (z,t)$ is slowly varying *complex* envelope of the optical signal
α is the fiber loss
β_2 and β_3 are second- and third-order dispersion coefficients
γ is the nonlinearity coefficient of the fiber

When the pulses have a duration over 1 ps, the NLSE can be simplified to

$$\frac{\partial A}{\partial z}=-\frac{j}{2}\beta_2\frac{\partial^2 A}{\partial T^2}-\frac{1}{2}\alpha A+j\gamma|A|^2A \tag{12.4}$$

For a system to transmit at 10 Gb/s and above, the effect of third-order dispersion has to be taken into consideration. By contrast, for systems operating below 10 Gb/s, if the channel is not at λ_D, third-order dispersion can be neglected. The MATLAB.m file `[ssprop _ matlabfunction _ modified(input)]` was given and described in Chapter 3 and included in all models related to optical systems employing ASK, DPSK, Star-QAM, etc.

12.1.6 Low-Pass Equivalent Model: Linear Operating Region

In developing the low-pass equivalent model of signal propagation in optical fibers, first, we have to assume that linear and nonlinear terms do not interact with each other. This is achieved by setting the power under certain threshold to prevent nonlinearity effect taking place. Hence, we can set $\gamma = 0$. Then, simplified model can be solved directly in the frequency domain to obtain

$$A(z,\omega) = \exp\left(-\frac{\alpha z}{2} - j\frac{\beta_2 \omega^3 z}{2}\right) A(0,\omega) \tag{12.5}$$

Since $A(z,\omega)$ is the Fourier transform of $A(z, t)$, the low-pass equivalent model for SMF transfer function is [11]

$$H(f) = \exp(-j\phi(f)) = \exp(-j\alpha f^2) = \exp\left(-j\left[\alpha B^2\left(\frac{f}{B}\right)\right]^2\right) \tag{12.6}$$

where

$\alpha = \pi D(\lambda)\dfrac{\lambda^2}{c}L$

$f = \upsilon - \upsilon_c$

B is the data rate

f is the frequency

α is the attenuation constant

$D(\lambda)$ is the CD

λ is the operating wavelength

c is the speed of light

Hence, the fiber is effectively a linear filter with attenuation and phase shift dependent on the transmission length. This transfer function can be used whenever the nonlinear effects are not large in order to minimize the computation time.

12.1.7 Modulation Formats

Optimal modulation formats play an essential role to combat linear and nonlinear impairments, particularly in ultra-long-haul lightwave transmission systems. An optimal modulation format is responsible to carry the maximum bit rate with narrow optical spectrum to enable closer channel spacing in a DWDM system. Furthermore, high tolerance to the CD effect and robustness to nonlinearities such as SPM, XPM, and FWM are major properties to pursue. The format should also be compatible with the detection scheme in order to minimize receiver renovation and thus increase cost effectiveness.

In this section, the modulation formats covered were NRZ or NRZ-ASK, RZ or RZ-ASK, NRZ-DPSK, RZ-DPSK (including CSRZ-DPSK), and DB. Although NRZ and RZ are not advanced modulation formats, they have been used explicitly in conjunction with DPSK formats to develop a generic photonic transmitter based on the fundamental optical principles. In this section, the basic generation principles and characteristics of these five modulation formats will be discussed, and their applications in developing simulator will be described in Section 12.1.8.

12.1.8 NRZ or NRZ-ASK

For simplicity, we would refer to NRZ-ASK as NRZ, since they are fundamentally similar. NRZ has been the dominant modulation format in the last two decades because it is simple to generate and detect, and has compact spectra. On top of that, it can attain high sensitivity with EDFAs. Although higher bit rates demand is taxing the limit of NRZ signaling where new schemes are being deployed to replace NRZ, it served as a good reference for the purpose of comparison.

NRZ data is encoded by the following logic. A "zero" is a low logic level and "one" is a high logic. Figure 12.6 shows an NRZ signal corresponding to the logic level.

A continuous-wave (CW) laser is injected from a DFB laser source into an MZM. The modulator is biased at linear region and driven with V_π. The modulator acts as an intensity modulator to convert NRZ electrical signal with data rate B into ASK optical signal at the same rate. Due to nonlinear characteristics of the modulator, the NRZ optical signal usually has a better Q factor than the electrical signal. The optical spectrum of B data rate NRZ signal will have signal bandwidth of B Hz (see Figures 12.7 and 12.8).

12.1.9 RZ or RZ-ASK

RZ-ASK, also known as RZ pulses, has several advantages over NRZ signals. The advantages are attributed to how RZ handles the optical power. As illustrated in Figure 12.9, the power level of RZ pulses always returns to zero even when successive logic-one bits occur.

First, to generate optical RZ pulses, an MZM commonly known as "pulse carver" is fed with sinusoidal electrical signal to "carve" pulses from optical CW carrier wave. The optical RZ pulse train is then fed into another MZM known as intensity modulator to modulate the electrical data on the

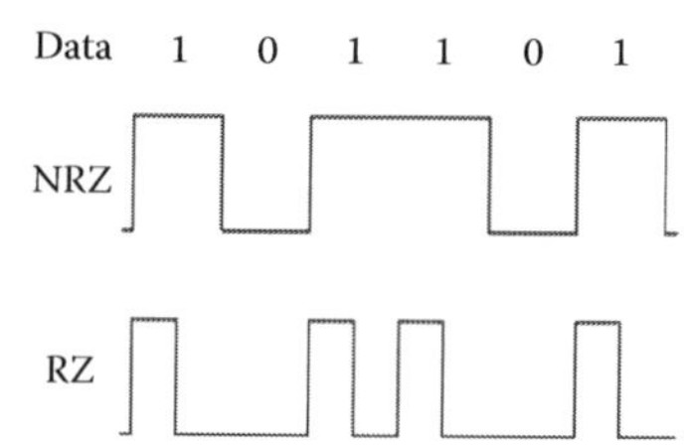

FIGURE 12.6
NRZ and RZ line coding for 101101 data sequence.

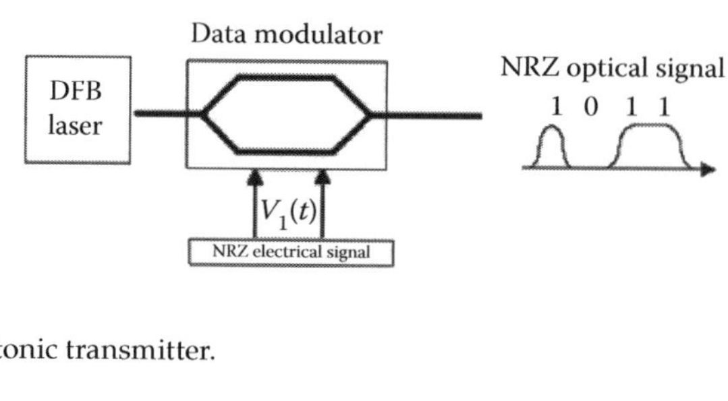

FIGURE 12.7
Block diagram of NRZ photonic transmitter.

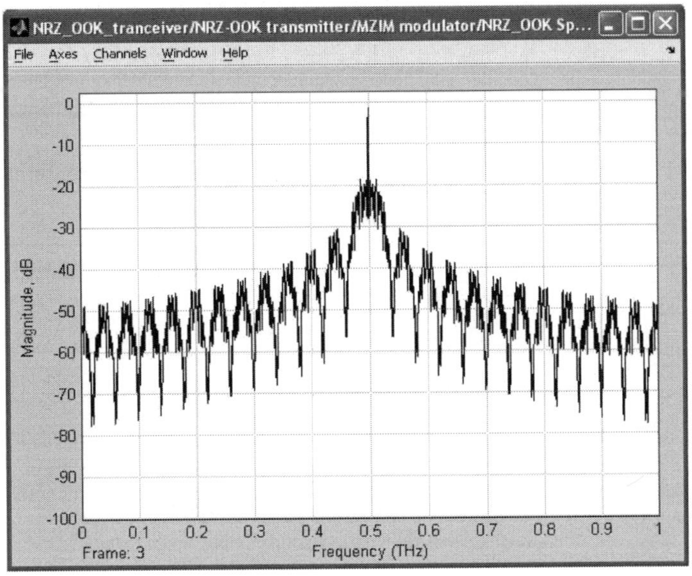

FIGURE 12.8
Optical spectrum of NRZ at carrier of 500 GHz.

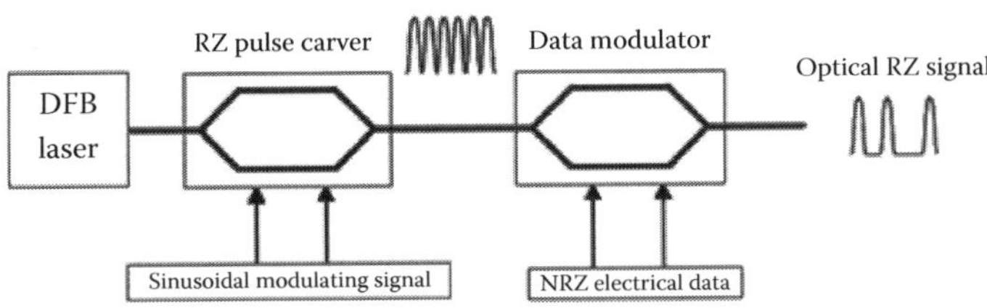

FIGURE 12.9
Block diagram of RZ photonics transmitter.

RZ pulses. There are a number of variations in RZ format based on the biasing point in the transmission curve shown in Table 12.1.

RZ pulse shape has a wider spectrum than NRZ due to its narrower pulse width, but CSRZ has eliminated this inherent disadvantage. CSRZ has an optical phase difference of π in adjacent bits, removes the OC component in optical spectrum, and reduces the spectral width (Figure 12.10).

The Simulink model for the pulse carver is shown in Figure 12.11. Figure 12.12 shows another Simulink model of the pulse carver using the principles of the operation of an intensity modulator, the MZIM. We can observe that two arms of the modulator have been modeled as phase shifters that are also biased in opposite sign. This is essential for operation as a dual-drive interferometric modulation device. The driving signals are also complementary with respect to each other so that the phase shift of the two optical arms act in a push–pull manner. The phasor diagram representing the pulse carver operation will be described in Section 12.2. Readers can modify the model to accommodate for the generation of these RZ pulses.

TABLE 12.1

Summary of RZ Format Generation and Characteristics of Single-Drive MZIM Based on Biasing Point, Drive Signal Amplitude, and Frequency

Biasing Point in Transmission Curve	RZ Generation Characteristics
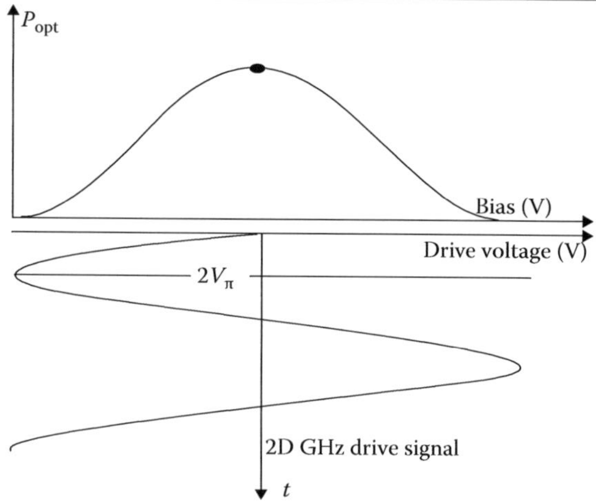	Biasing point: maximum, $V_{2\pi}$ Drive signal amplitude: $2V_\pi$ Drive signal frequency: 20 GHz Pulse width: 9–10 ps RZ pulse frequency: 40 GHz frequency doubling effect.
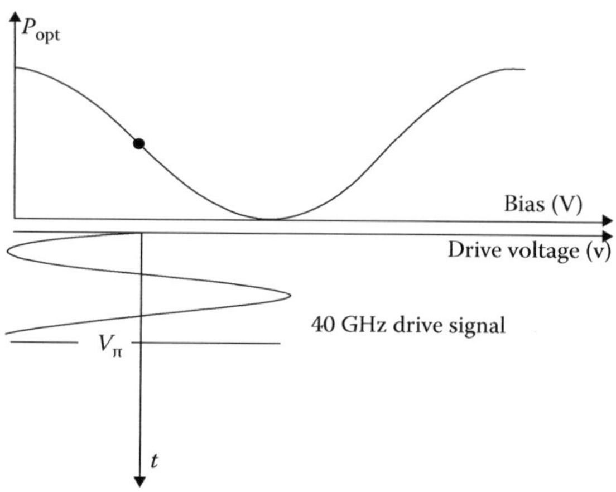	Biasing point: linear region, $V_{\pi/2}$ Drive signal amplitude: V_π Drive signal frequency: 20 GHz Pulse width: 9–10 ps RZ pulse frequency: 20 GHz
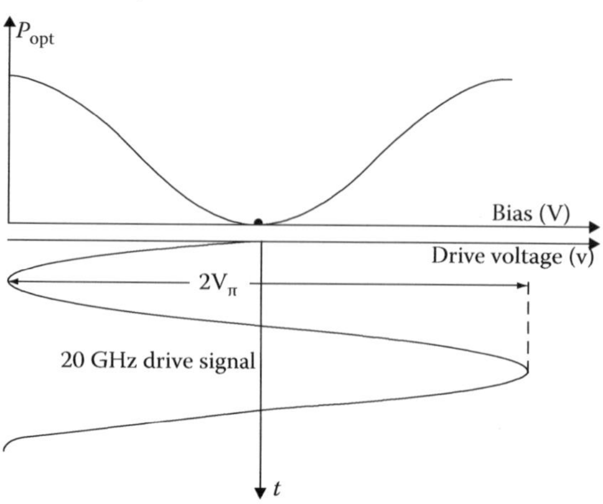	Biasing point: minimum, V_π Drive signal amplitude: $2V_\pi$ Drive signal frequency: 20 GHz Pulse width: 9–10 ps RZ pulse frequency: 40 GHz frequency doubling effect. (CSRZ is generated using this scheme.)

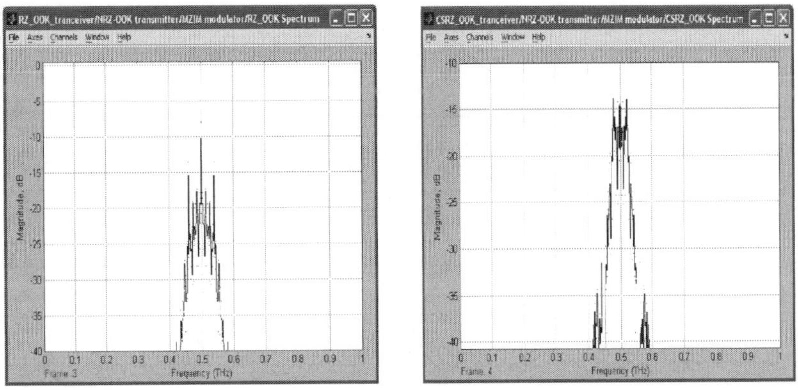

(a) (b)

FIGURE 12.10
Optical spectra of (a) RZ and (b) CSRZ at an artificial carrier of 500 GHz.

12.2 Return-to-Zero Optical Pulses

12.2.1 Generation

Figure 12.13 shows the conventional structure of an RZ-ASK transmitter in which two external $LiNbO_3$ MZIMs can be used. The MZIM shown in this transmitter can be either single- or dual-drive (push–pull) type. Operational principles of the MZIM were presented in Section 5.3 of Chapter 5 and Chapter 6. The optical OOK transmitter would normally consist of a narrow-linewidth laser source to generate lightwaves whose wavelength satisfies the ITU grid standard.

The first MZIM, commonly known as the pulse carver, is used to generate the periodic pulse trains with a required RZ format. The suppression of the lightwave carrier can also be carried out at this stage if necessary, commonly known as the CSRZ. Compared to other RZ types, CSRZ pulse shape is found to have attractive attributes for long-haul WDM transmissions including the π phase difference of adjacent modulated bits, suppression of the OC component in optical spectrum, and narrower spectral width.

Different types of RZ pulses can be generated depending on the driving amplitude of the RF voltage and the biasing schemes of the MZIM. The equations governing the RZ pulses electric field waveforms are

$$E(t)=\begin{cases}\sqrt{\dfrac{E_b}{T}}\sin\left[\dfrac{\pi}{2}\cos\left(\dfrac{\pi t}{T}\right)\right] & \text{67\% duty-ratio RZ pulses or CSRZ}\\[2ex] \sqrt{\dfrac{E_b}{T}}\sin\left[\dfrac{\pi}{2}\left(1+\sin\left(\dfrac{\pi t}{T}\right)\right)\right] & \text{33\% duty-ratio RZ pulses or RZ33}\end{cases} \tag{12.7}$$

where

E_b is the pulse energy per transmitted bit

T is one bit period

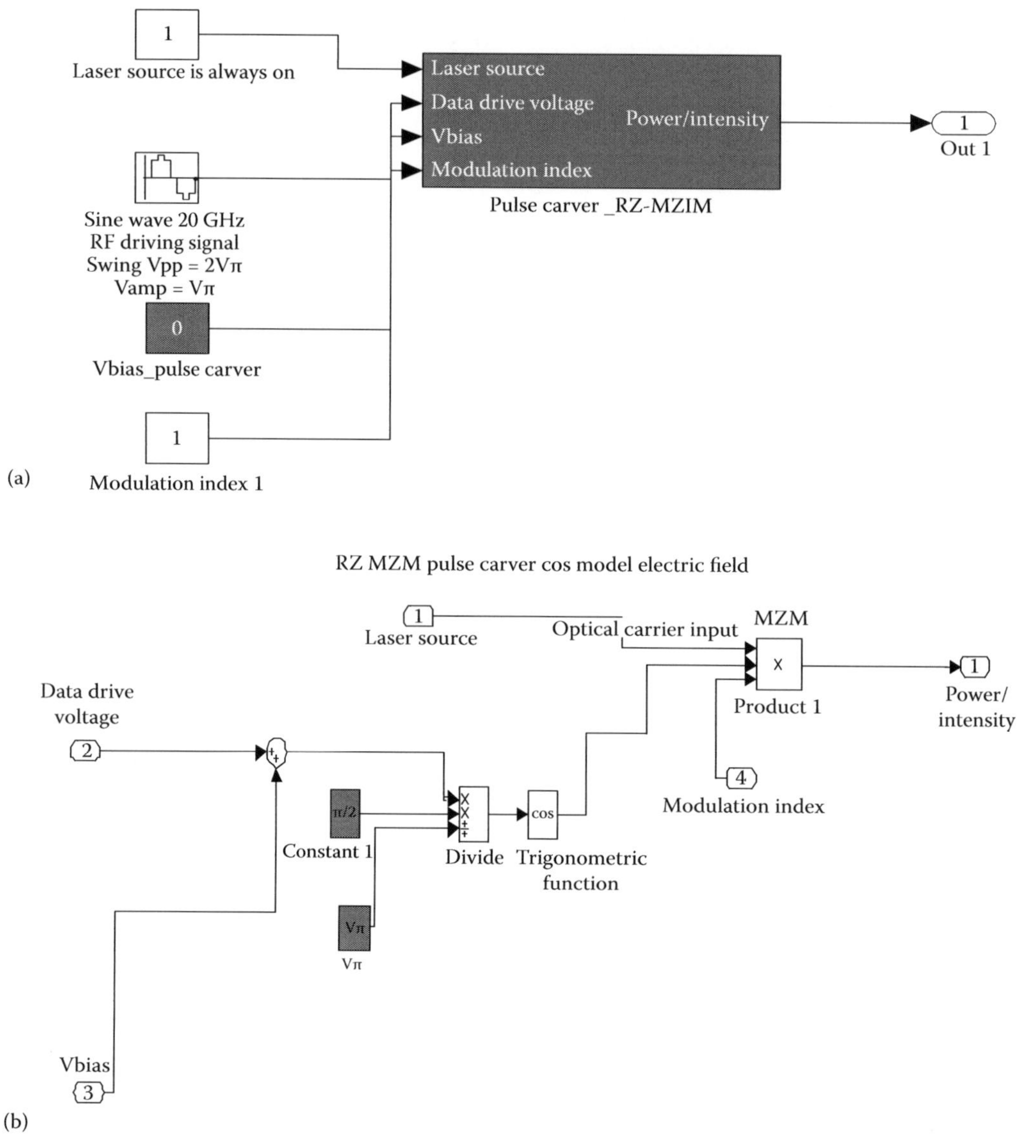

FIGURE 12.11
Simulink model of an optical pulse carver using mathematical model: (a) general block model and (b) inside the pulse carver.

The 33% duty-ratio RZ pulse is denoted as RZ33 pulse whereas the 67% duty cycle RZ pulse is known as the CSRZ type. The art in generation of these two RZ pulse types stays at the difference of biasing point on the transfer curve of an MZIM.

The bias voltage conditions and the pulse shape of these two RZ types, the carrier suppression and non-suppression of maximum carrier, can be implemented with the biasing points at the minimum and maximum transmission points of the transmittance characteristics of the MZIM, respectively. The peak-to-peak amplitude of the RF driving voltage is $2V_\pi$, where V_π is the required driving voltage to obtain a π phase shift of the lightwave carrier. Another important point is that the RF signal is operating at only a half of the transmission bit rate.

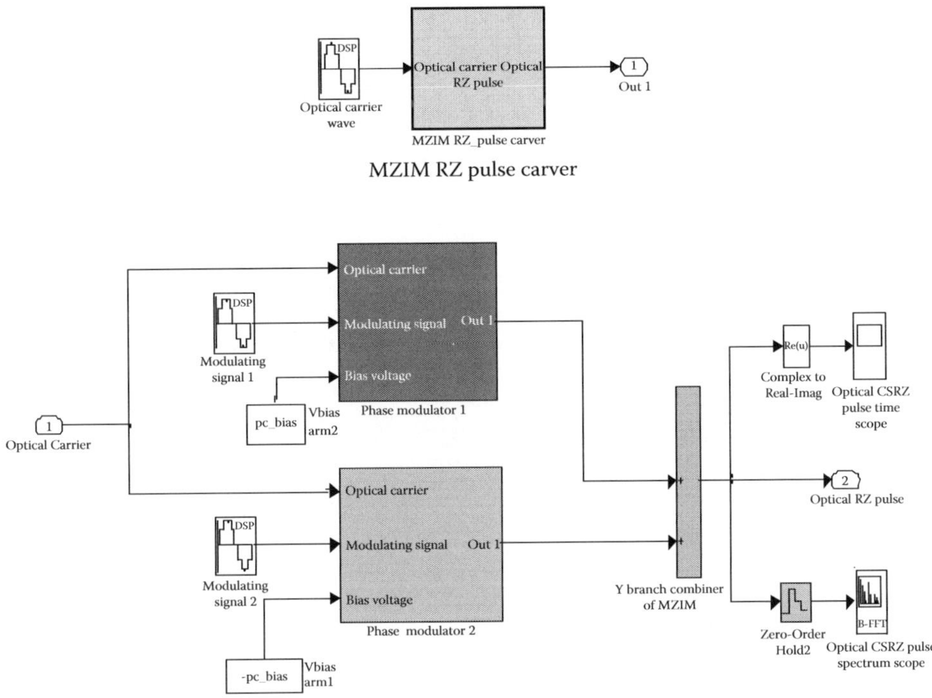

FIGURE 12.12
Simulink model of an optical pulse carver modeled using the principles of the operation of MZIM.

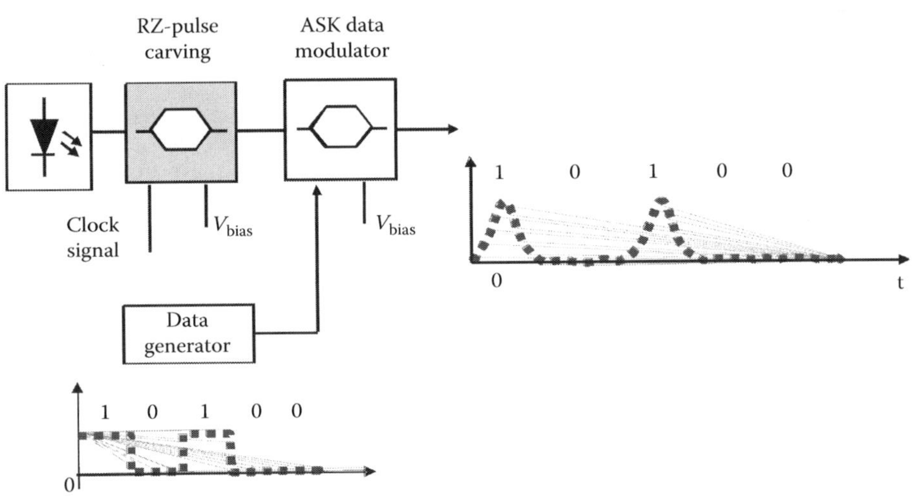

FIGURE 12.13
Conventional structure of an OOK optical transmitter utilizing two MZIMs.

Hence, pulse carving is actually implementing the frequency doubling. The generations of RZ33 and CSRZ pulse train are demonstrated in Figure 12.14a and b.

The pulse carver can also utilize a dual-drive MZIM that is driven by two complementary sinusoidal RF signals. This pulse carver is biased at $-V_{\pi/2}$ and $+V_{\pi/2}$ with the peak-to-peak amplitude of $V_{\pi/2}$. Thus a π phase shift is created between the states "1" and "0" of the

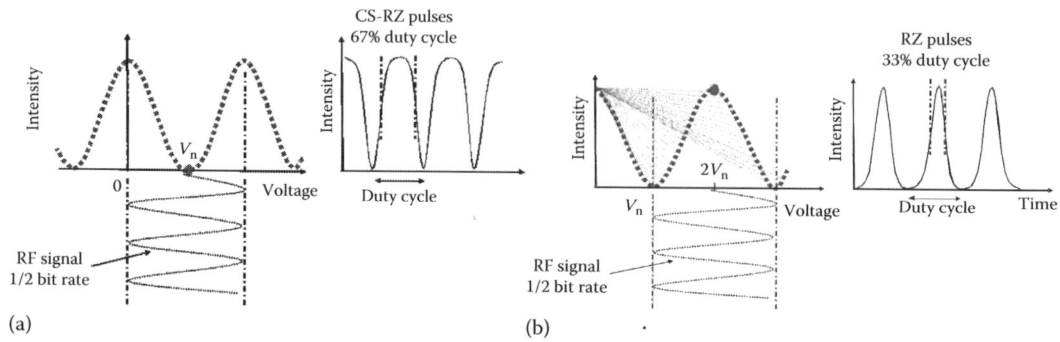

FIGURE 12.14
Bias point and RF driving signals for the generation of (a) CSRZ and (b) RZ33 pulses.

pulse sequence and hence the RZ with alternating phase 0 and π. If carrier suppression is required, then the two electrodes are applied with voltages V_π and swing voltage amplitude of V_π.

Although RZ modulation offers improved performance, RZ optical systems usually require more complex transmitters than those in the NRZ ones. Compared to only one stage for modulating data on the NRZ optical signals, two modulation stages are required for the generation of RZ optical pulses.

12.2.2 Phasor Representation

Recalling (12.1) we have

$$E_o = \frac{E_i}{2}\left[e^{j\varphi_1(t)} + e^{j\varphi_2(t)}\right] = \frac{E_i}{2}\left[e^{j\pi v_1(t)/V_\pi} + e^{j\pi v_2(t)/V_\pi}\right] \tag{12.8}$$

It can be seen that the modulating process for the generation of RZ pulses can be represented by a phasor diagram as shown in Figure 12.15. This technique gives a clear understanding of the superposition of the fields at the coupling output of two arms of the MZIM. Here, a dual-drive MZIM is used, that is, the data-driving signals $[V_1(t)]$ and inverse data ($\overline{\text{data}}$: $V_2(t) = -V_1(t)$) are applied into each arm of the MZIM, respectively, and the RF voltages swing in inverse directions. Applying the phasor representation, vector addition, and simple trigonometric calculus, the process of the generation of RZ33 and CSRZ is explained in detail and verified.

The width of these pulses are commonly measured at the position of full-width half maximum (FWHM). It is noted that the measured pulses are intensity pulses whereas we are considering the addition of the fields in the MZIM. Thus, the normalized E_o field vector has the value of $\pm 1/\sqrt{2}$ at the FWHM intensity pulse positions, and the time interval between these points gives the FWHM values.

12.2.3 Phasor Representation of CSRZ Pulses

Key parameters including the V_{bias}, the amplitude of the RF driving signal, are shown in Figure 12.16a. Accordingly, its initialized phasor representation is demonstrated in Figure 12.16b.

The values of the key parameters are outlined as follows: (1) V_{bias} is $\pm V_\pi/2$; (2) swing voltage of driving RF signal on each arm has the amplitude of $V_\pi/2$ (i.e., $V_{p-p} = V_\pi$); (3) RF signal

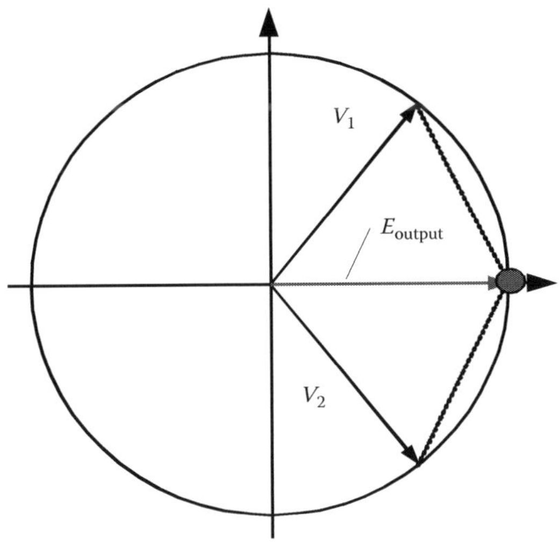

FIGURE 12.15
Phasor representation for the generation of output field in dual-drive MZIM.

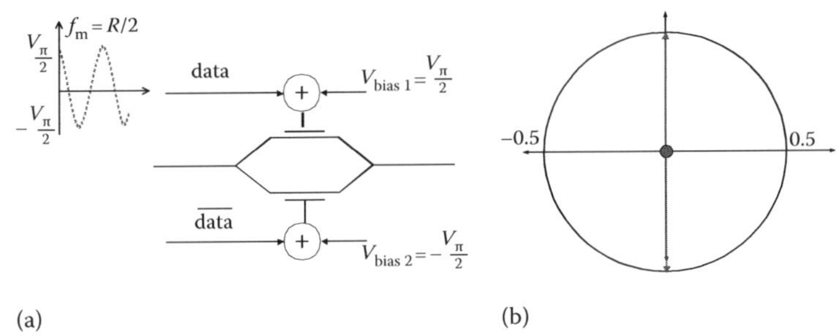

FIGURE 12.16
Initialized stage for the generation of CSRZ pulse: (a) RF driving signal and the bias voltages and (b) initial phasor representation.

operates at half of bit rate ($B_R/2$); (4) at the FWHM position of the optical pulse, the $E_{out} = \pm 1/\sqrt{2}$ and the component vectors V_1 and V_2 form a phase of $\pi/4$ with the vertical axis, as shown in Figure 12.17.

Considering the case for the generation of 40 Gb/s CSRZ optical signal, the modulating frequency is f_m ($f_m = 20\,\text{GHz} = B_R/2$). At the FWHM positions of the optical pulse, the phase is given by the following expressions:

$$\frac{\pi}{2}\sin(2\pi f_m) = \frac{\pi}{4} \Rightarrow \sin 2\pi f_m = \frac{1}{2} \Rightarrow 2\pi f_m = \left(\frac{\pi}{6}, \frac{5\pi}{6}\right) + 2n\pi \tag{12.9}$$

Thus, the calculation of TFWHM can be carried out and hence, the duty cycle of the RZ optical pulse can be obtained as given in the following expressions:

$$T_{FWHM} = \left(\frac{5\pi}{6} - \frac{\pi}{6}\right)\frac{1}{R2\pi} = \frac{1}{3}\pi \times \frac{1}{R} \Rightarrow \frac{T_{FWHM}}{T_{BIT}} = \frac{1.66\times 10^{-4}}{2.5\times 10^{-11}} = 66.67\% \tag{12.10}$$

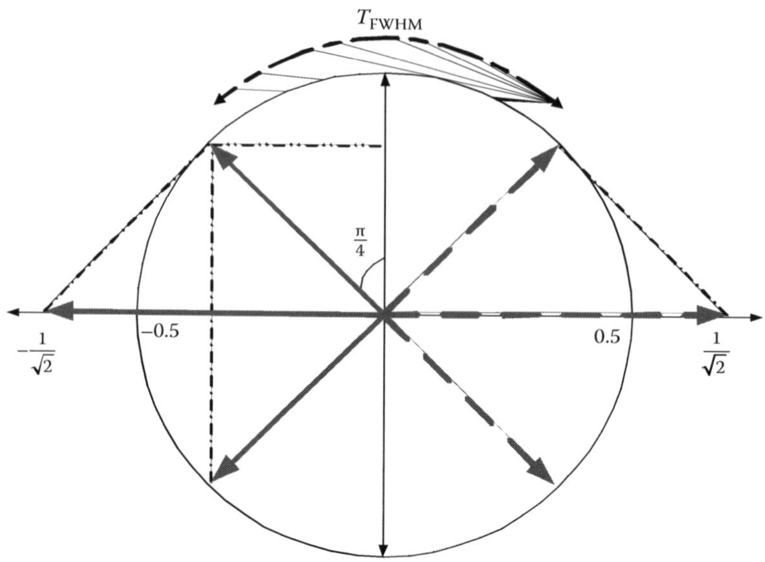

FIGURE 12.17
Phasor representation of CSRZ pulse generation using dual-drive MZIM.

The result obtained in (12.10) clearly verifies the generation of CSRZ optical pulses from the phasor representation.

12.2.4 Phasor Representation of RZ33 Pulses

Key parameters including the V_{bias}, the amplitude of driving voltage, and its correspondent initialized phasor representation are shown in Figure 12.18a and b, respectively.

The values of the key parameters are (1) V_{bias} is V_π for both arms; (2) swing voltage of driving RF signal on each arm has the amplitude of $V_\pi/2$ (i.e., $V_{p-p} = V_\pi$); (3) RF signal operates at half of bit rate ($B_R/2$).

At the FWHM position of the optical pulse, the $E_{output} = \pm 1/\sqrt{2}$ and the component vectors V_1 and V_2 form a phase of $\pi/4$ with the horizontal axis, as shown in Figure 12.19.

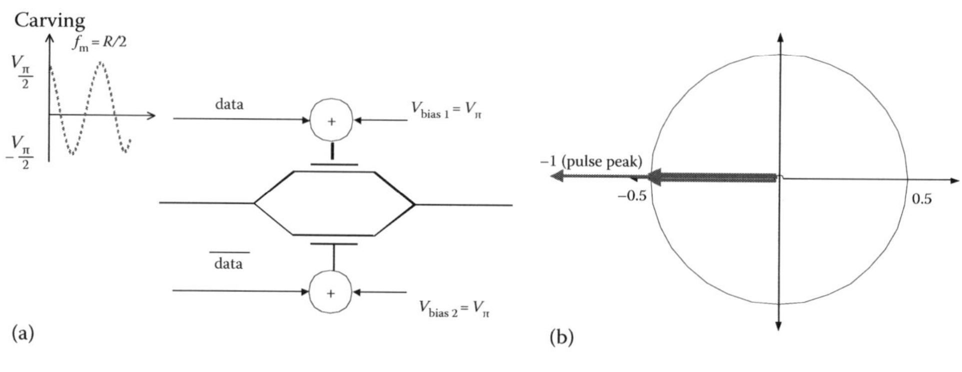

FIGURE 12.18
Initialized stage for the generation of RZ33 pulse: (a) RF driving signal and the bias voltage and (b) initial phasor representation.

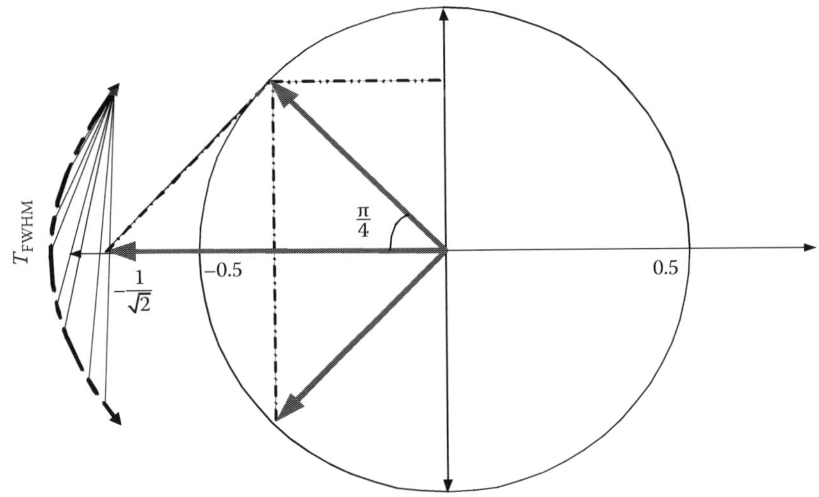

FIGURE 12.19
Phasor representation of RZ33 pulse generation using dual-drive MZIM.

Considering the scenario for the generation of 40 Gb/s CSRZ optical signal, the modulating frequency is $f_m (f_m = 20\,\text{GHz} = B_R/2)$. At the FWHM positions of the optical pulse, the phase is given by the following expressions:

$$\frac{\pi}{2}\cos(2\pi f_m t) = \frac{\pi}{4} \Rightarrow t_1 = \frac{1}{6f_m} \tag{12.11}$$

$$\frac{\pi}{2}\cos(2\pi f_m t) = -\frac{\pi}{4} \Rightarrow t_2 = \frac{1}{3f_m} \tag{12.12}$$

Thus, the calculation of TFWHM can be carried out, and hence, the duty cycle of the RZ optical pulse can be obtained as given in the following expressions:

$$T_{FWHM} = \frac{1}{3f_m} - \frac{1}{6f_m} = \frac{1}{6f_m} \therefore \frac{T_{FWHM}}{T_b} = \frac{1/6f_m}{1/2f_m} = 33\% \tag{12.13}$$

The result obtained in (12.8) clearly verifies the generation of RZ33 optical pulses from the phasor representation.

12.3 Differential Phase Shift Keying

Information encoded in the phase of an OC is commonly referred to as optical phase SK. In early days, PSK requires precise alignment of the transmitter and demodulator center frequencies [11]. Hence, PSK system is not widely deployed. With DPSK scheme introduced, coherent detection is not critical since DPSK detection only requires source coherence over a one-bit period by the comparison of two consecutive pulses.

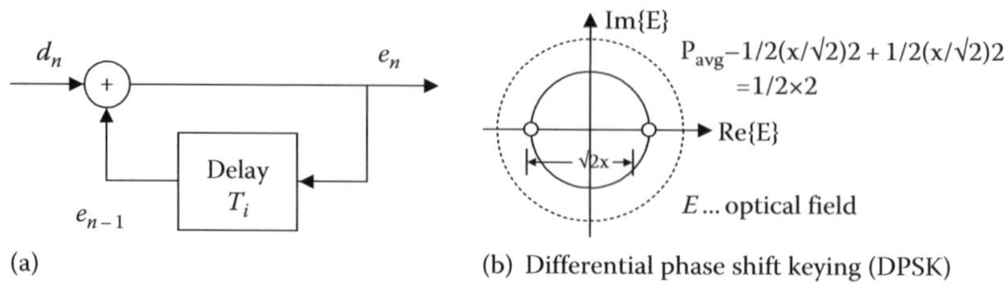

FIGURE 12.20
(a) The encoded differential data are generated by $e_n - d_n \oplus e_{n-1}$ and (b) signal constellation diagram of DPSK.

A binary 1 is encoded if the present input bit and the past encoded bit are of opposite logic and a binary 0 is encoded if the logic is similar. This operation is equivalent to exclusive OR (XOR) logic operation. Hence, an XOR gate is usually employed in differential encoder. NOR can also be used to replace XOR operation in differential encoding.

In optical applications, electrical data "1" is represented by a π phase change between the consecutive data bits in the OC, while state "0" is encoded with no phase change between the consecutive data bits. Hence, this encoding scheme gives rise to two points located exactly at π phase difference with respect to each other in signal constellation diagram (see Figure 12.20).

12.3.1 NRZ-DPSK

Figure 12.21 shows the block diagram of a typical NRZ-DPSK transmitter. Differential precoder of electrical data is implemented using the logic explained in Section 12.3. In phase modulation of the OC, MZM known as data PM is biased at minimum point and driven by data swing of $2V_\pi$. The modulator showed an excellent behavior so that the phase of the OC will be altered by π exactly when the signal transits the minimum point of the transfer characteristics.

12.3.2 RZ-DPSK

The setup of the RZ-DPSK transmitter is essentially similar to RZ-ASK with data intensity modulator replaced with data PM (Figure 12.22). The difference between them is the biasing point and the electrical voltage swing. Different RZ formats can also be generated according to Table 12.1.

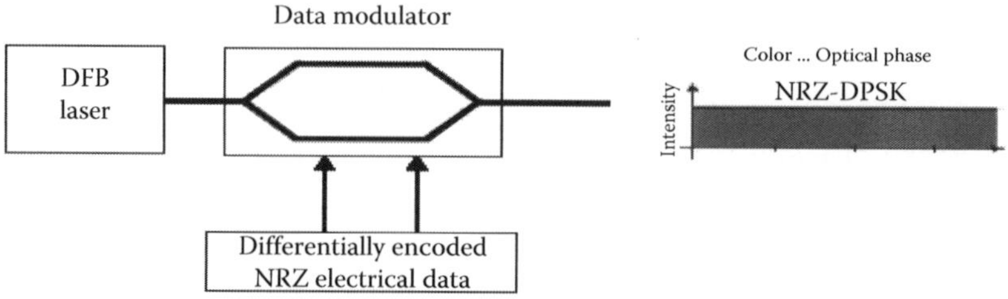

FIGURE 12.21
Block diagram of NRZ-DPSK photonic transmitter.

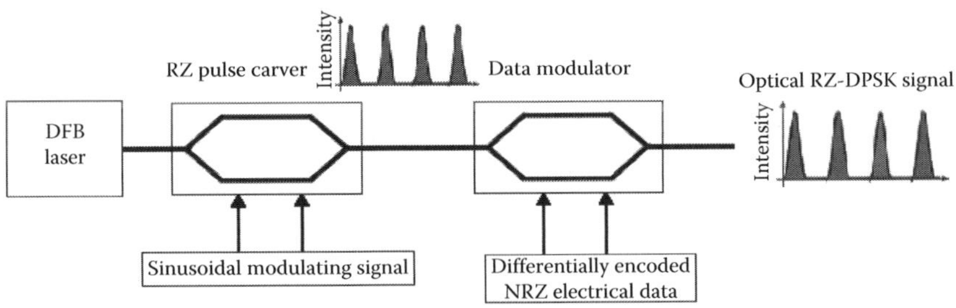

FIGURE 12.22
Block diagram of RZ-DPSK photonics transmitter.

12.3.3 Receiver

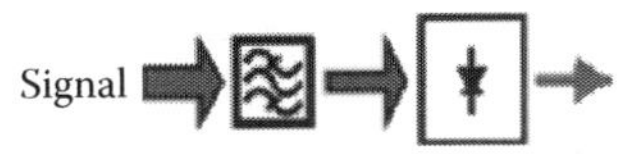

FIGURE 12.23
Single photodiode direct detection receiver after optical filtering for limiting noise effects. (From Hoshida, T. et al., *IEEE J. Lightwave Technol.*, 20, 1989, 2002. With permission.)

The receiver structure is dependent on the modulation format imposed on the OC. If ASK and DB formats are used, single photodiode direct detection receiver is usually employed to convert the optical signal power into electrical signal by photodiode. The single photodiode direct detection scheme is depicted in Figure 12.23.

DPSK receiver requires an optical preprocessing stage to accomplish demodulation. The optical signal is first passed through an MZ delay interferometer (MZDI) that has a differential delay equal to a bit period. The delay is achieved by extending the second path of the MZ arm approximately λ_{bit} as shown in Figure 12.24. Since in practice, exact delay is too difficult to construct, a heating element is applied on the interferometer arm to fine-tune the differential phase delay.

When there is no phase change between adjacent bits, maximum power appears at the constructive port. When the phase in adjacent bit differs by π, maximum power exists at the destructive port. Hence, the preceding bit served as a phase reference for demodulating the current bit [1]. As a result, the bit coherency is only required for two bit periods.

Under DPSK modulation and demodulation scheme, both constructive port and destructive port carry similar but logically inverted data stream. Therefore, they can be either detected by themselves ("single-ended detection") or balanced. Single-ended detection simply uses a photodiode to convert the optical signal power at either port into electrical signal. Balanced detection is completed by connecting two photodiodes and processed digitally.

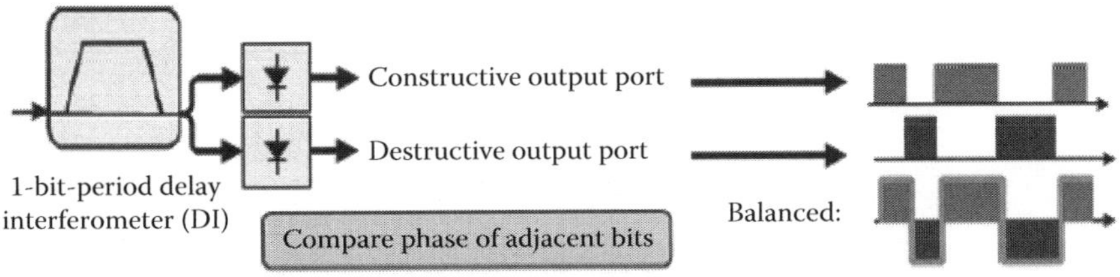

FIGURE 12.24
Balanced detection with delay demodulation using two photodiode direct detection. (From Hoshida, T. et al., *IEEE J. Lightwave Technol.*, 20, 1989, 2002. With permission.)

12.4 Simulink® Models

Analytical engineering approach to identify the optimization of parameters, especially in long-haul communication systems may impose limitations on the number of optical channels as well as distances. Since the interaction of these parameters may give hundreds or even thousands of combinations, computer-aided engineering approach provides assistance in terms of accuracy and speed from precise engineering point of view. There are a number of computer-based programs and simulation tools available to achieve optical system design modeling. These packages offer various benefits depending on the applications and situations investigated. However, some of them provide click-and-drop interface, and some of them are closed to user's intervention.

MATLAB and Simulink are two of the most user-friendly software platforms that can write the simulation software using different building blocks and tools. Simulink provides a set of libraries that contain digital signal processing blockset, communication blockset, mathematical fundamentals, and electronic modules. Hence, it is open for optical module modeling, which is critical in the modeling of optical components from physical to system level. The flexibility of the simulation by appropriate sampling techniques enhances the integration of optical module modeled at physical operating principle such as optical modulator with mathematical model that is required to solve fiber propagation equations.

The design processes depicted in Figure 12.25 illustrate how a 40 Gb/s photonic transmitter can be designed and tested. First, pseudorandom bit generator can be easily modeled using Bernoulli binary generator provided in Simulink communication blocksets. An optical CW carrier, which was assumed to be a DFB laser as explained earlier, is modeled readily by a sine wave in the signal processing blockset. Three circled processes in Figure 12.25

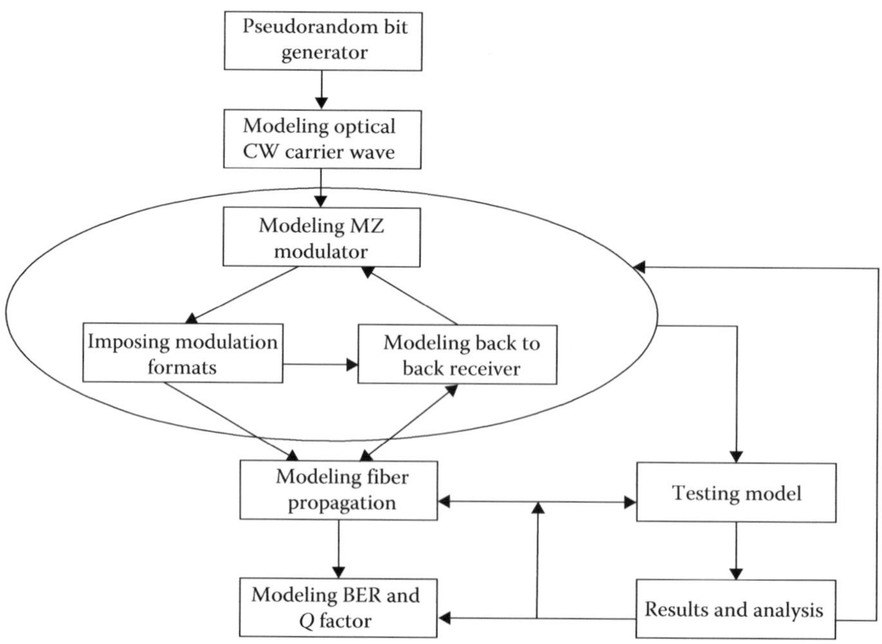

FIGURE 12.25
Flow diagram of simulator design processes.

form a circular procedure that requires constant testing and feedback from outcome results and analysis. Often, the design on the MZM had to be adapted, modified, and implemented both with mathematical model and physical operating principles. The modulator model would have to be confirmed with the modulation format as well as receiver to demodulate the signal. When modulation formats were compatible with the back-to-back receiver, propagation model was developed to test the system performance with BER model. This process would consume a lot of time as systems grew in complexity; a slight change in model would alter the result significantly. During the design process, results were collected to gain information and feedback on necessary model alterations and confirmations. The rest of this section will discuss modeling process and present the results from time to time to demonstrate the working principles of the model. Although the developed model can generate different modulation formats, only RZ-DPSK results including eye diagrams, time scope, and spectrum scope are used for the purpose of demonstration.

12.4.1 Bernoulli Binary Generator (Figures 12.26 and 12.27)

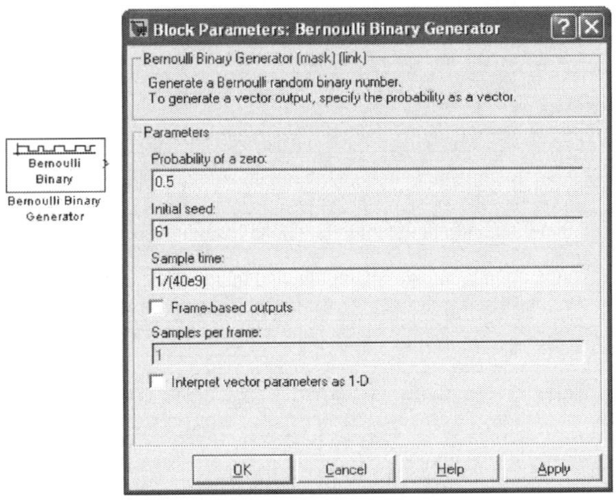

FIGURE 12.26
Block parameters setup and block diagram of Bernoulli binary generator in Simulink. Probability of a zero is set to 0.5 and sample time is set to 25 ps to output 40 Gb/s data sequence.

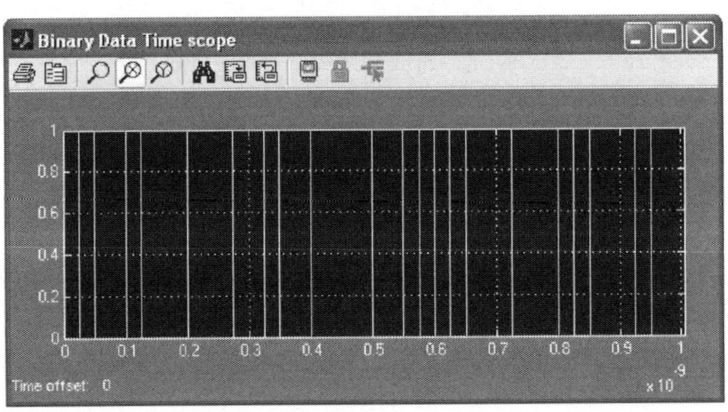

FIGURE 12.27
Binary datas in 1 ns generated by Bernoulli binary generator.

12.4.2 DFB Laser

OC frequency set in this case is 2.048 THz. It is different from 193 THz (1550 nm) because $2^n \times$ data rate of the OC will simplify the calculation in setting up eye diagram parameters since we can compute the exact number of cycles of sinusoidal sine wave in a bit period (Figure 12.28). Sample mode was set to discrete due to the similar reason stated previously. Sample time was relatively subject to how accurate we require the OC to be, typically 200 samples in a frame is substantial to produce a clean and nice wave (Figure 12.29).

12.4.3 Mach–Zehnder Interferometric Modulator

MZ interferometric modulator model was developed based on the principle explained in Chapter 6. Implementation can be divided into intensity modulation and phase modulation

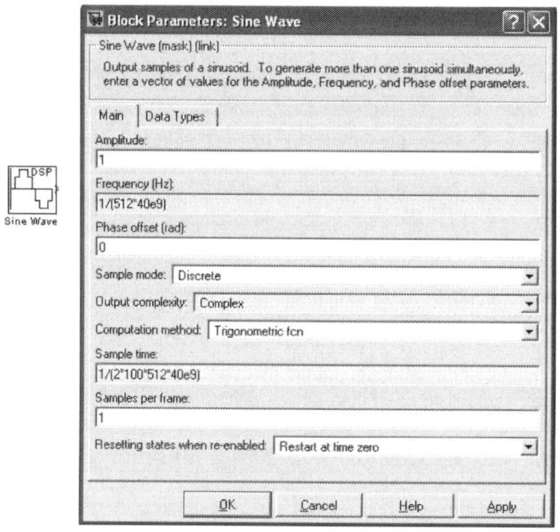

FIGURE 12.28
Block parameters setup and block diagram of DFB laser in Simulink.

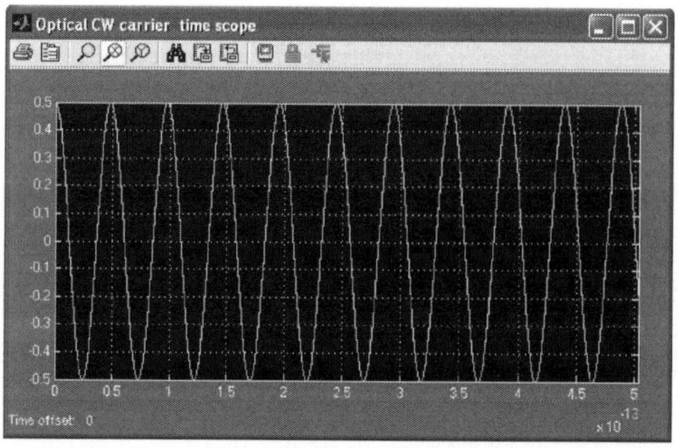

FIGURE 12.29
Optical CW carrier wave at 2.048 THz shown in 0.5 ps time span.

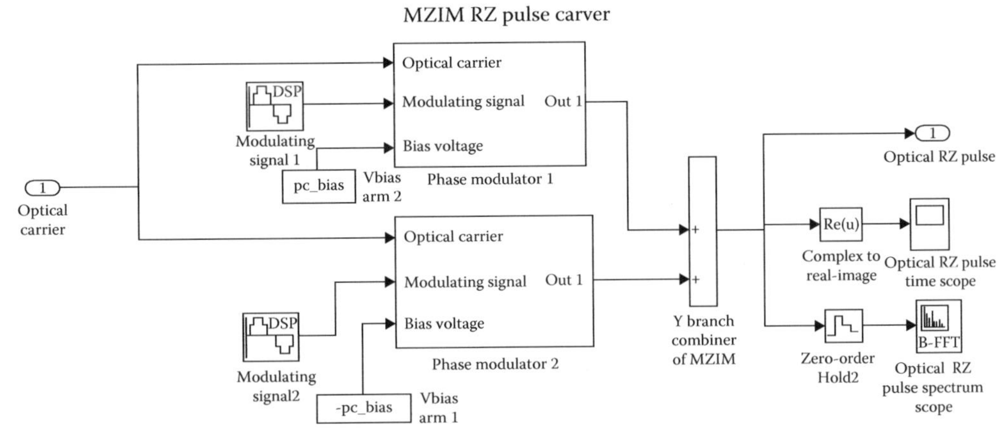

FIGURE 12.30
Dual-drive MZIM pulse carver.

depending on the biasing voltage and the voltage swing. Hence, the model was developed based on the pulse carver applications.

12.4.3.1 Pulse Carver

Since optical phase distortions (such as chirp) will affect DPSK receiver performance, dual-drive MZIM pulse carver (Figure 12.30) was required to operate in perfect push–pull operation. In other words, sinusoidal electrical signals amplitude has to be similar and of exact opposite phase. This requirement is difficult to achieve in practice but fairly straightforward in a simulation environment. Assuming extinction ratio as large, optical field exiting a dual-drive MZIM is described mathematically [1] as follows:

$$E(t) = \frac{E_0}{2}\left\{\exp\left[j\pi\frac{u_1(t)}{V_\pi}\right] + \exp\left[j\pi\frac{u_2(t)}{V_\pi}\right]\right\}$$

$$\approx E_0 \exp\left[j\pi\frac{u_1(t)+u_2(t)}{2V_\pi}\right] \times \cos\left[\pi\frac{u_1(t)-u_2(t)}{2V_\pi} + \varphi\right] \tag{12.14}$$

where
- E_0 is magnitude of input optical field
- V_π is the electrical voltage to produce π phase shift in either of the modulator's two arms
- φ is the relative optical phase between two arms without electrical drive signals
- $u_1(t) = U_1\cos(2\pi f_m t)$
- $u_2(t) = U_1\cos(2\pi f_m t + \theta)$
- f_m is the electrical signal frequency

Since chirp-free driving condition required $u_1(t) = -u_2(t)$, Equation 12.3 can be reduced to

$$E(t) \approx E_0 \cos\left[\pi\frac{u_1(t)-u_2(t)}{2V_\pi} + \varphi\right] \tag{12.15}$$

Let $\varphi = \frac{\varphi_1 - \varphi_2}{2}$ where $\varphi_1 = \pi\frac{V_{DC1}}{V_\pi}$; $\varphi_2 = \pi\frac{V_{DC2}}{V_\pi}$. By incorporating OC frequency into (12.16), optical field exiting dual-drive MZIM can be re-expressed as follows [12]:

$$E(t) = E_0\left\{\cos\left[2\pi f_c t + \alpha\pi\cos(2\pi f_m t) + \varphi_1\right] + \cos\left[2\pi f_c t + \alpha\pi\cos(2\pi f_m t + \theta) + \varphi_2\right]\right\} \quad (12.16)$$

where

f_c is the OC frequency

$\alpha = \pi\frac{U_{1,2}}{V_\pi}$ (assigned as the modulation index)

The mathematical analysis shows that the biasing voltage, V_π, is reduced by half with respect to single-drive MZIM, thus with driving voltage as well, an advantage using balanced modulator structure (Figure 12.31).

The MZIM had two arms biased at the voltage set by "pc_bias," and modulated by two modulating signal with exact π difference between them to create the push–pull configuration. The output of each arm was combined using a sum that acts like a coupler so that they can interfere constructively and destructively. The output was monitored using a time scope and spectrum scope.

Figure 12.32 shows the equally split OC signal enter one of MZIM path. Biasing voltage was fed to create a relative phase reference for electrical modulating signal. Electrical signal was used to modulate OC signal with phase relative to biasing voltage phase. This model essentially represented one of the two cosine terms in Equation 12.16 (Figure 12.33). Hence, if MZ interferometric structure was discarded, this model alone acted like a PM.

Table 12.2 tabulated the parameters required for setting the optical modulator to generate different RZ pulse shapes, especially the biasing point for superimposing the driving electrical signals.

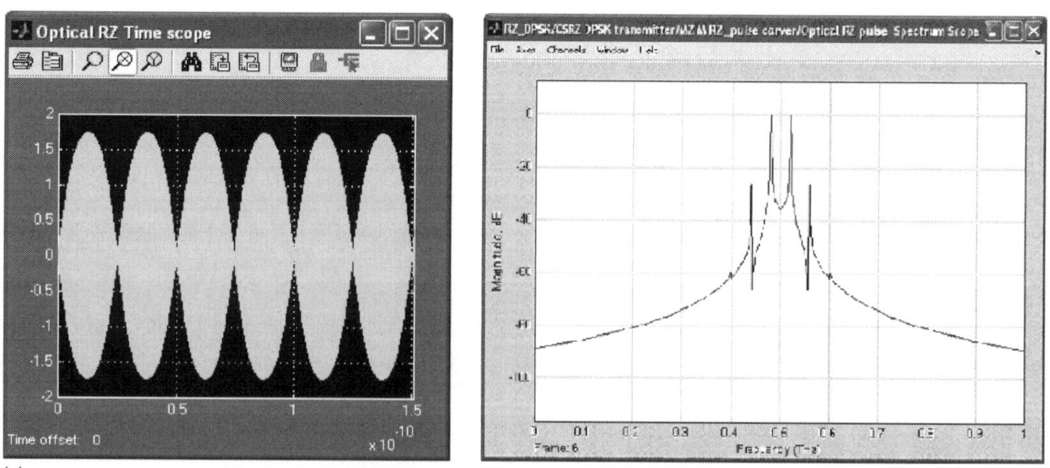

(a) (b)

FIGURE 12.31
(a) CSRZ pulse in time domain. The pulse is 25 ps wide with carrier reside in the pulse at 2.048 THz and (b) spectral characteristics of CSRZ pulse. The carrier was clearly shown as being eliminated at 500 GHz.

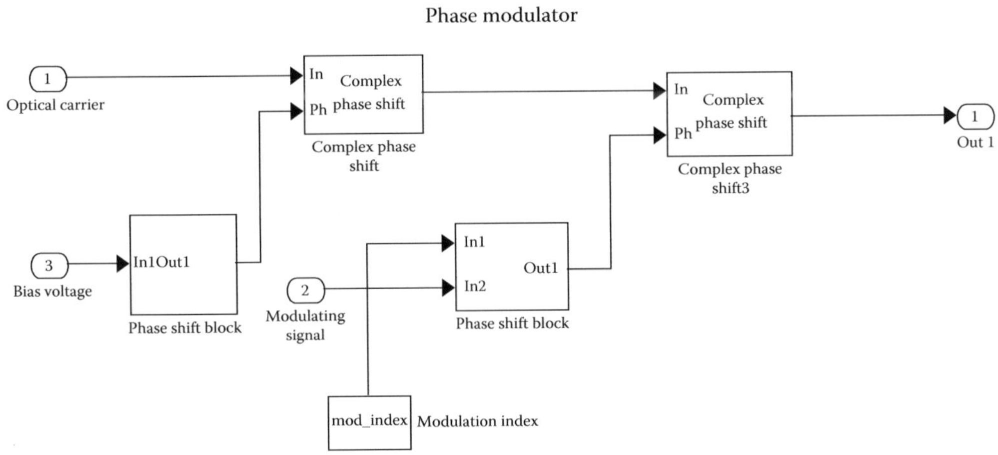

FIGURE 12.32
Phase modulation implemented in Simulink. One of the phase modulator arm in Figure 12.30.

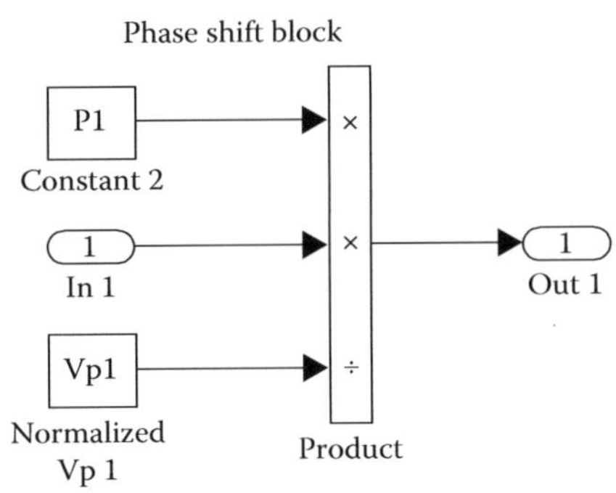

FIGURE 12.33
One of the phase shift block in Figure 12.32. A direct mathematical model derived for phase implementation.

TABLE 12.2
Parameters Setting for Three Formats of RZ Pulses

	Parameters Setting	
Biasing Point	**Arm 1**	**Arm 2**
Linear	Modulation index: $\pi/4$	Modulation index: $\pi/4$
	Signal frequency: 40 GHz	Signal frequency: 40 GHz
	Biasing voltage: $V_{\pi/4}$	Biasing voltage: $-V_{\pi/4}$
	Phase offset: $\pi/2$	Phase offset: $\pi+\pi/2$
Minimum (CSRZ)	Modulation index: $\pi/3$	Modulation index: $\pi/3$
	Signal frequency: 20 GHz	Signal frequency: 20 GHz
	Biasing voltage: $V_{\pi/2}$	Biasing voltage: $-V_{\pi/2}$
	Phase offset: 0	Phase offset: π
Maximum	Modulation index: $\pi/2$	Modulation index: $\pi/2$
	Signal frequency: 20 GHz	Signal frequency: 20 GHz
	Biasing voltage: V_{π}	Biasing voltage: $-V_{\pi}$
	Phase offset: $\pi/2$	Phase offset: $\pi/2$

12.4.3.2 Data Modulator

Data modulator based on MZIM structure employed similar models as in pulse carver (Figure 12.34). However, when the MZIM model was driven as PSK data modulator, the biasing voltage was always $V_{\pi/2}$ and $V_{\pi/2}$ data voltage swing. Phase modulation would be limited to 0 and π, equivalent to signal constellation diagram with two points at 0 and π and doubling in the frequency of driving data. If formats other than PSK modulation formats are used, then the biasing voltage and data voltage swing will be different and based on what were explained earlier (Figure 12.35).

Before feeding into PMs, there was a data-type conversion to convert data in Boolean form generated by Bernoulli generator into double type to be processed by PM model. PM

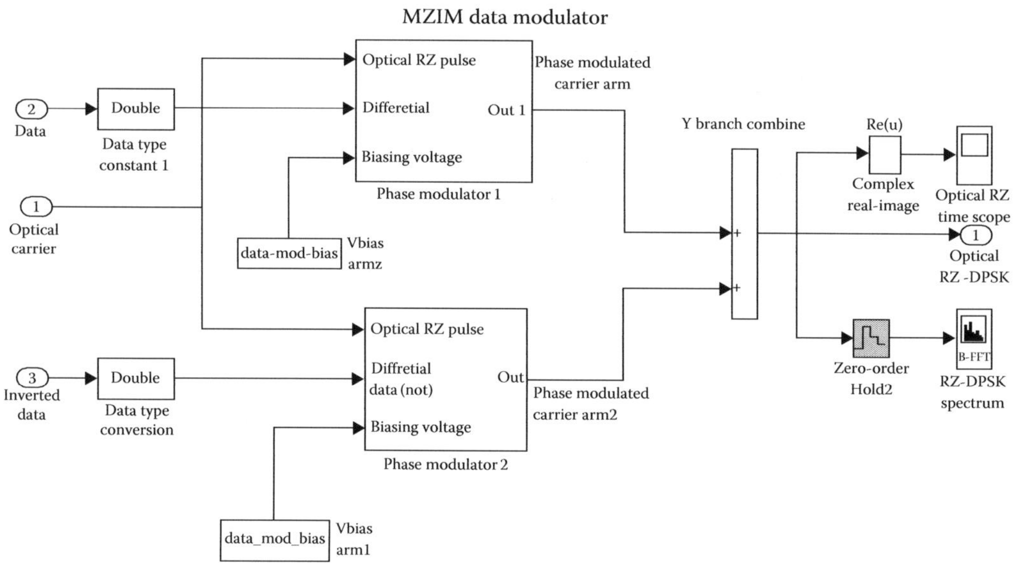

FIGURE 12.34
Data modulator based on dual-drive MZ interferometric structure driven by data and inverted data.

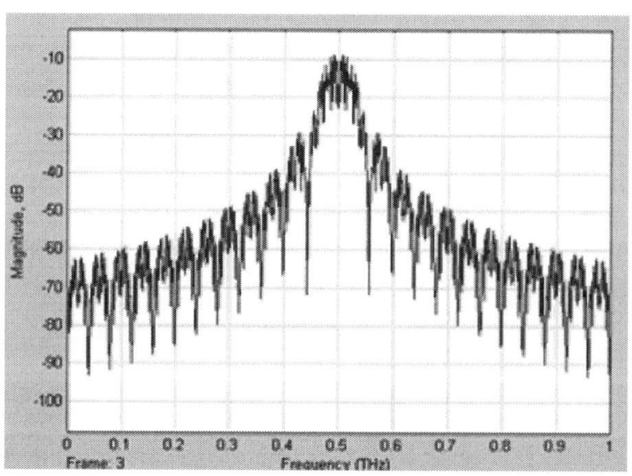

FIGURE 12.35
Spectrum scope monitored after data modulator using CSRZ-DPSK modulation format with a carrier of 500 GHz.

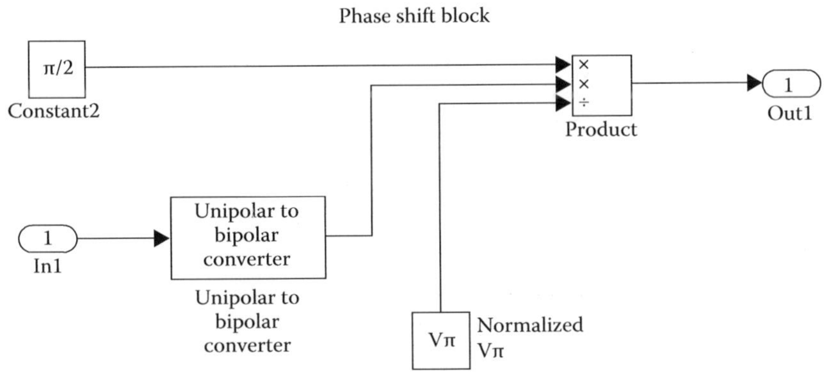

FIGURE 12.36
Second phase shift block in one of the phase modulator arm in Figure 12.34.

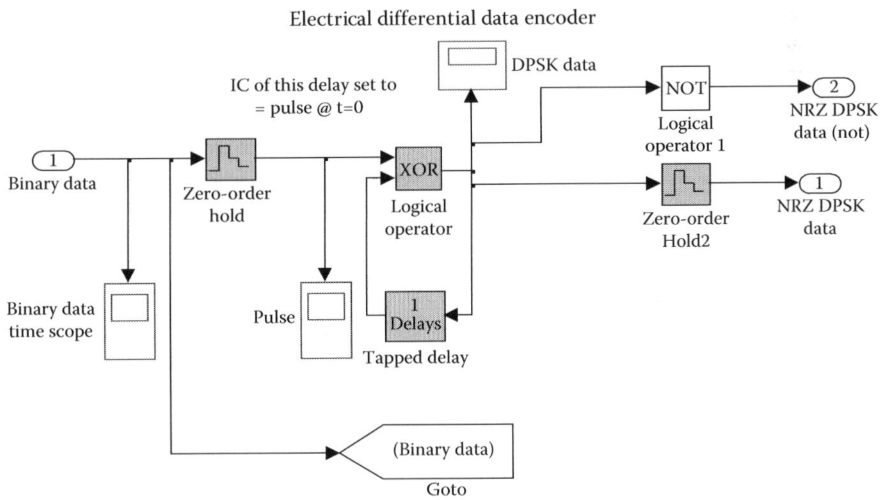

FIGURE 12.37
Electrical differential data encoder.

is similar to the one in Figure 12.34. Logically, they were similar, but phase shift block was slightly altered to assemble data modulation principle in this model (Figure 12.36).

12.4.3.3 Differential Data Encoder

Differential data encoder (Figure 12.37) was implemented using the differential encoding logic described in Section 12.3. Bernoulli binary generator was driven at 20 Gb/s and sampled at the same frame rate to XOR with the past encoded bit (Figure 12.38). This model consists of two NRZ DPSK data outputs complementing with each other that are necessary to operate dual-drive MZIM modulators (Figure 12.39).

12.4.4 Back-to-Back Receiver

Balanced receiver was constituted by an MZDI model, two photodiode models, and a comparator with time scopes and eye diagrams to evaluate performance (Figure 12.40). MZDI

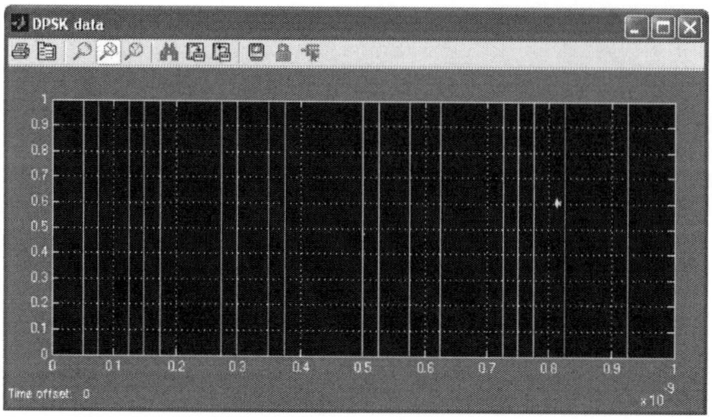

FIGURE 12.38
Differential encoded binary data in Figure 12.37.

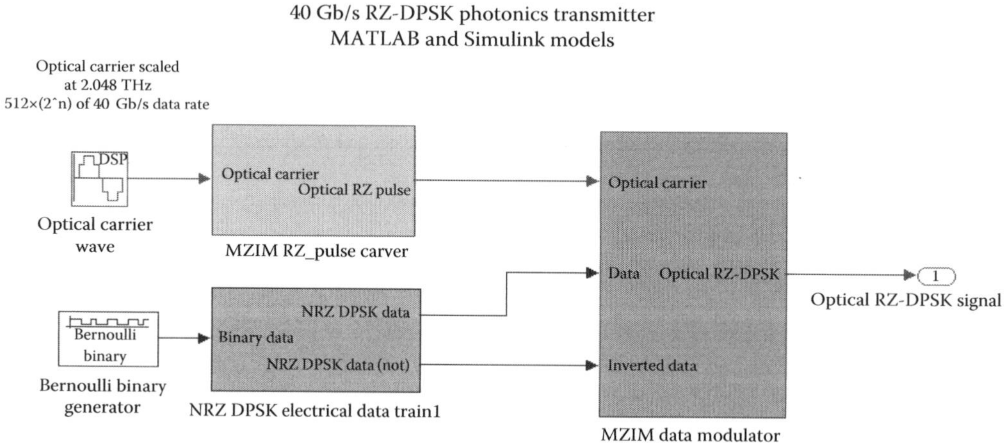

FIGURE 12.39
A complete Simulink model of photonics transmitter. (Created by Tiong, H.S. and under supervision of Binh, L.N. Copyrighted.)

was implemented by a delay block provided in the Simulink library. Since the OC wave was calculated in a discrete size of $2^n \times 40e^9$, it was simple to calculate exactly the number of samples required to delay a bit period. For instance, in this case, $2^n = 512$, therefore, the number of delay samples were 512 in order to delay a bit period.

Constructive and destructive interference of the received signal and a-bit-delay signal were implemented by addition and subtraction, respectively (Figure 12.41). When there is no phase change between adjacent bits, addition will constructively superpose two signals. Conversely, subtraction will superpose two signals with π phase shift. Since a photodiode converted optical power to electrical signal, it could be modeled as a square of the magnitude.

Single-ended detection would require either of the ports to be detected using eye diagram block. On the other hand, balanced detection would require subtraction (logically) between two outputs before detection by eye diagram (Figure 12.42).

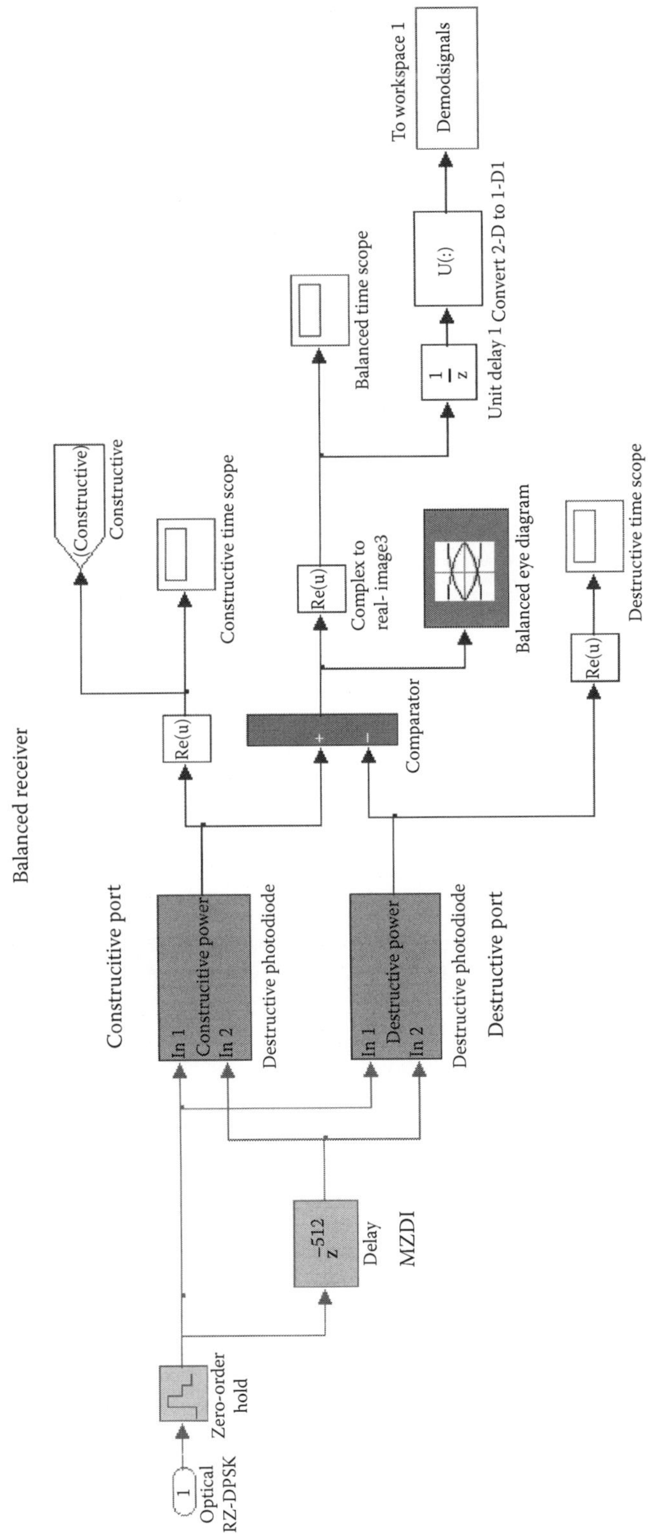

FIGURE 12.40
Simulink model of a balanced receiver.

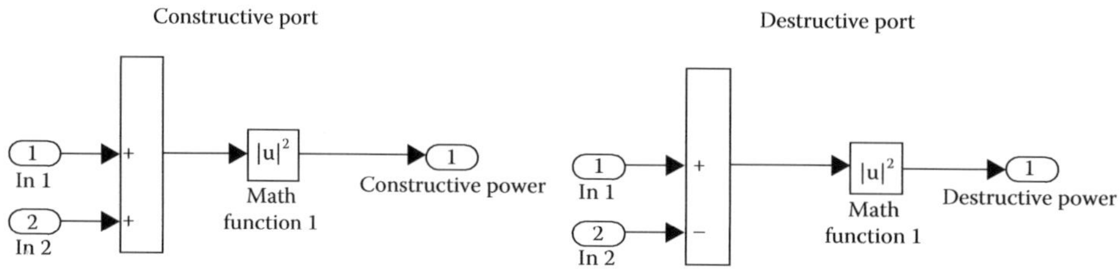

FIGURE 12.41
Constructive and destructive port in balanced receiver. Math function magnitude squared represents the photodiode.

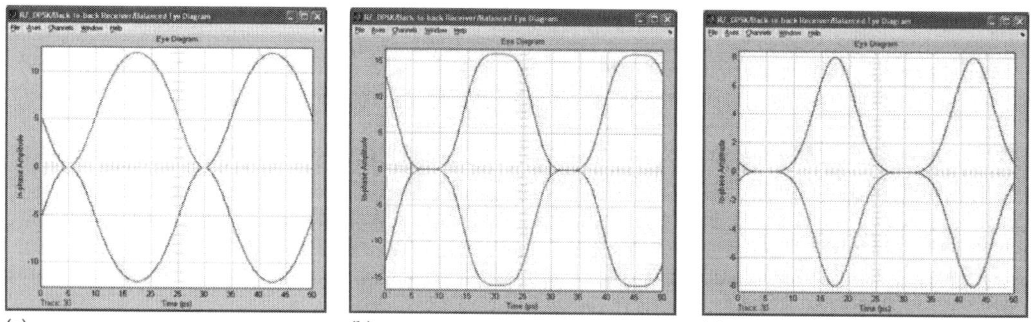

(a) (b) (c)

FIGURE 12.42
Eye diagrams of three different RZ modulation formats detected by back-to-back balanced receiver. (a) CSRZ, (b) 50% RZ, and (c) 33% RZ.

12.4.4.1 Eye Diagram

Samples per symbol can be set by carrier frequency × bit period. As shown in Figure 12.43, 2*eye_diagram implies that the eye diagram displays two bit periods. Hence, it can be adjusted accordingly to view over a number of bit periods.

12.4.5 Signal Propagation

The zero-order hold (sampler in green color) is used to obtain the pulse envelope of the optically modulated signal. The signal propagation model (Figure 12.44) could only deal with the pulse envelope since it was developed using split-step Fourier method to solve the NLSE equation in which the signal was assumed to have a slowly varying time envelope as explained in Section 3.7. Gain acted like an attenuator to control the power input to the fiber. Buffer was not in actual practice but it was necessary to specify the number of bit samples in a particular FFT window. The buffered signal was input to SMF block (Figure 12.45) to be processed by a MATLAB function ssprop_matlabfunction_modified.m.

In SMF blocks, the signal was concatenated by the parameters such as the number of FFT, time step, step size, the number of steps, fiber loss constant, propagation constants, and nonlinear coefficients (Figure 12.46). The signal was processed by the MATLAB function ssprop_matlabfunction_modified.m, which implemented the split-step Fourier method (SSFM) algorithm.

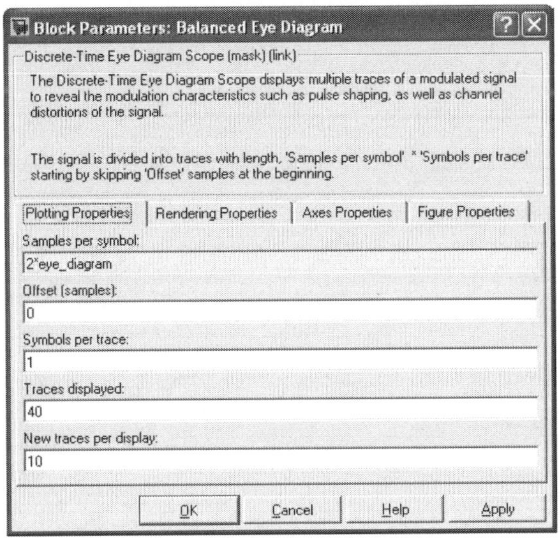

FIGURE 12.43
Parameter settings for eye diagram.

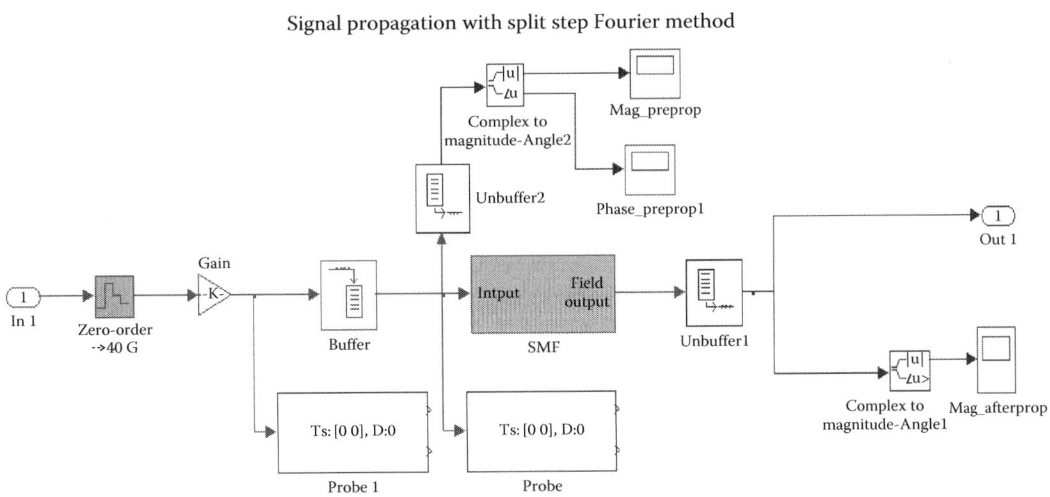

FIGURE 12.44
Simulink model of signal propagation in fiber.

12.4.6 Bit Error Rate

BER was calculated by using MATLAB programs listed in the appendix. The histogram was first computed using histogram.m, and a sample result of histogram after a 2 km fiber span was illustrated in the following diagram. Q factor and BER were calculated using BER_GaussianQ.m by assuming that the signal distortion due to fiber propagation was Gaussian distributed (Figure 12.47). This m-file is also listed in the appendix.

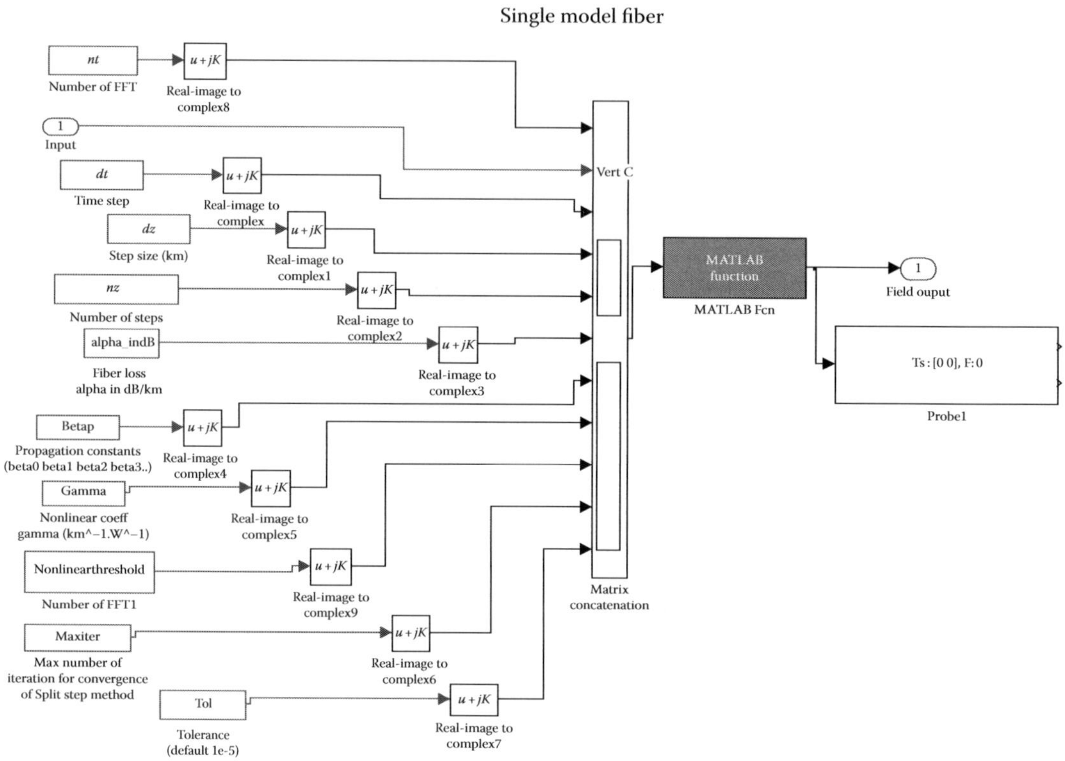

FIGURE 12.45
Parameters setting in SMF block.

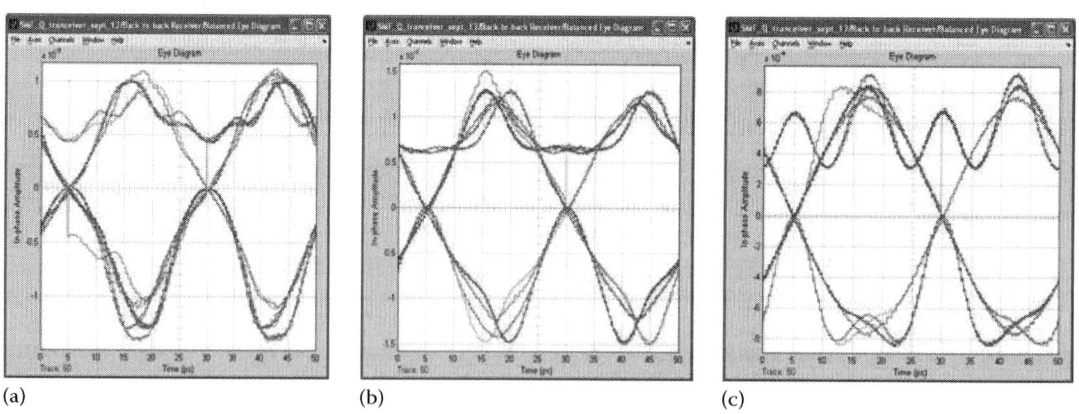

FIGURE 12.46
Eye diagrams of three different RZ modulation formats after 2 km fiber propagation using balanced receiver. (a) CSRZ-DPSK, (b) 50% RZ_DPSK, and (c) 33% RZ-DPSK.

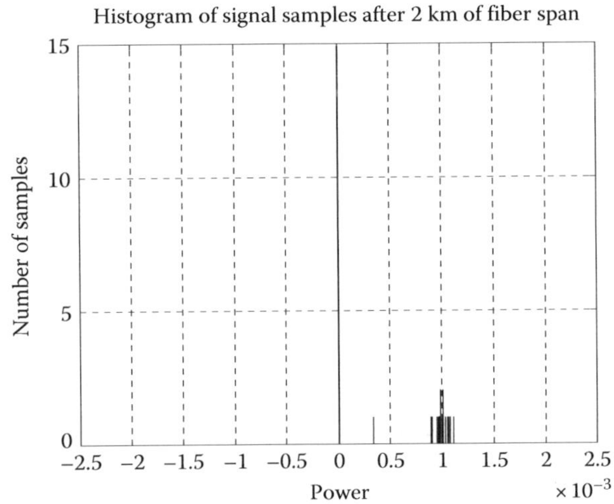

FIGURE 12.47
Histogram of signal samples after 2 km of fiber span detected at constructive port using CSRZ-DPSK modulation format.

12.5 DQPSK Modulation Formats Transmission Models

The required bandwidth availability to support modern-day optical fiber communication systems is now extending into the terabit per second region. Numerous system designs have been proposed with many focusing on multi-wavelength, DWDM implementation, allowing increase in channel capacity using available hardware. Certain modulation formats, having only been implemented in the electrical domain in the past, are now being transposed to the optical domain, allowing for increased transmission capacity. An example of such a modulation format is the DQPSK format. This format is even important for bit rates reaching 100 Gb/s. At this rate, the electronic detection is not possible and lowering the rate to the symbol rate plus polarization multiplexing may bring the detection speed down to 25 or 28 GSymbols/s.

A high-capacity 20 Gb/s (2 × 10 Gb/s) single-channel optical DQPSK simulator that is capable of doubling the bit rate compared to conventional OOK signaling techniques can be developed. The simulator has been developed in the MATLAB and Simulink. The versatile nature of the simulator is capable of showing the effects of incorrect dispersion management in fiber links and generates data that can be easily analyzed with the support of MATLAB. Optical components included in the model can be user-defined, enhanced, or removed as desired.

This chapter outlines the necessary theoretical background of the design of the optical DQPSK system. It also explains the detailed operation and purpose of each component model and its representation in Simulink. Each model developed in the simulator focuses on theoretical and practical design principles. As a result, the fundamental operation of each component in the system has been understood together with their impairments. This document also outlines the future extension of this simulator to support DWDM operation to allow terabit per second (Tb/s) or $N \times 40$ Gb/s systems to be simulated accurately before any physical implementation is performed.

12.5.1 DQPSK Optical System Components

Given that we now know what the general configuration of a DWDM system channel is composed of, we focus on the key optical components utilized in the transmitter (Tx), fiber, and receiver (Rx) of a single-channel RZ-DQPSK fiber link. We begin by considering the transmitter components first, we then present some optical fiber characteristics, and finally we consider the receiver components.

12.5.1.1 DQPSK Transmitter

Given that the system we are attempting to design can be tested in the Monash University Fiber Optics Laboratory, we need to familiarize ourselves with some of the equipment and their operating fundamentals. We present a brief description of key optical components that comprise the transmission system and are also later simulated.

12.5.1.1.1 Mach–Zehnder Modulator

MZMs are the most widely used optical devices for external modulation. Although several types of MZMs exist, in this chapter we consider the well-known lithium niobate ($LiNbO_3$) single-electrode modulator based on its common use in 10 Gb/s PSK systems. Its role in most DWDM systems, and in this chapter, is to externally phase modulate the OC and is thus usually placed in cascade with the DFB laser. The MZM consists of a Y splitter (3 dB), two waveguide arms, and a Y combiner as described in Chapter 5.

The physics behind the operation of this device is the EO effect. We only mention the core details here to gain an understanding of its operation. The EO effect allows the refractive index of the material to change as a response to an applied voltage $u_2(t)$ ($u_1(t) = 0$ in a single electrode MZM) also known as the modulating data stream. The phase of the signal as it passes through this region of the applied voltage (binary or sinusoidally varying) is changed. At the input, the power of the incident carrier is ideally split equally, one carrier wave experiences a phase change (assumed 0 or π rad in this chapter), while the other passes through unchanged; two waves emerge at the output. It should be noted that the OC that has experienced the 0 or π phase shift (in the lower arm of MZM) is of great importance since the differential phase of this waveform will represent the first encoded bit of the two bits (di-bit) to be transmitted. The driving voltage-to-power (P–V) characteristics of the MZM. To best simulate the MZM in this system design, we will assume that one of the MZMs are biased at the minimum transmission point with biasing voltage, V_π (~3.5 V in practice, with $2V_{\pi p-p}$ NRZ driving voltage swing) and $V_{\pi/2}$ for the second MZM (with $V_{\pi p-p}$ sinusoidal [approx RZ] driving voltage swing).

The MZM is also a useful device in its response to certain high bit rate pulse shapes (e.g., NRZ, RZ), which will be outlined later. They are normally used in the replacement of a normal PM biased to π phase shift. Other PMs simulated in this section operate under the same physical principles as the MZM; however, they are assumed to be biased to produce a π/2 phase shift in the OC.

12.5.1.1.2 Chirp

The role of chirp α, and in particular its influence, on the performance of modern fiber optic systems is to be considered. Chirp, in particular transmitter chirp, is known to be a main contributor to spectral broadening [13], also limiting transmission to short distances (for $\alpha > 0$).* Thus to minimize the effects of chirp, we have considered the sources of this effect. In this chapter, as outlined later, we simulate the MZ external modulator behavior. This EO device is integral to the design of the DQPSK transmitter and is used in many DWDM systems. The MZM is chosen as it provides superior signal integrity and normally has smaller chirping effects [5]. Frequency chirp is common in most optical transmitters; however, in the context of this chapter, we have also considered phase chirping effects.

12.5.1.2 DQPSK Receiver

In this section, wve consider the operation and theory of the optical components that are integral for the demodulation and detection of the DQPSK-transmitted signal.

12.5.1.2.1 Mach–Zehnder Delay Interferometer

The principle operation of this device is to convert phase-coded information, such as that generated by the DQPSK transmitter, into an intensity signal that can be detected by photodiode circuitry at the output of the DI arms. The intensity of the signal at the output of the DI arms is dependant on the phase difference between adjacent bit [14] $\Delta\phi_{mod}$ of the OC in both the upper and lower arms. The relationship between phase and intensity is expressed as follows:

$$I(\Delta\phi) = 0.5\cos\left(\Delta\phi_{mod} \pm \frac{\pi}{4} + \delta\phi_{DI}\right) + 0.5 \tag{12.17}$$

where $\delta\phi_{DI}$ is a phase offset. Note that this is an MZ device and thus operates on the principles of interference between the two arms, which is dependent on the phase difference. For the applications of DQPSK demodulation, the delay T is one bit in length (100 ps for 10 Gb/s operation) to allow for proper interference between two adjacent bits. Effectively, the delayed signal acts as a phase reference for the incoming symbol. The extra phase delay (phase $= \pm\pi/4$) is normally implemented in practice for DQPSK systems only using integrated thermal heaters.

The phase to intensity characteristic of the MZDI is shown in Figure 12.48 along with the expected eye diagram that will help in making performance measurements.

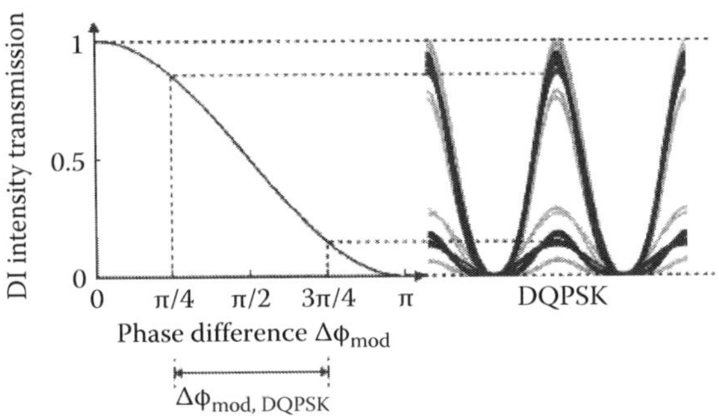

FIGURE 12.48
The phase to intensity characteristics of the MZDI and the expected eye diagram generated at the output of the MZDI based on the phase difference between the two arms, the lighter eye pattern is due to phase offset, $d\phi_{DI}$, from heater.

---

* Since $\alpha < 0$ (negative chirp) can actually increase the transmission distance by compressing the pulse, this technique is known as pre-chirp compensation.

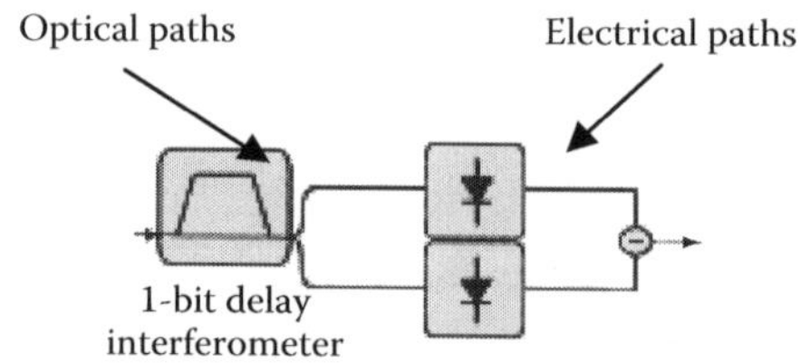

FIGURE 12.49
Balanced photodiode detection, diode placed at the output of each MZDI arm.

This optical eye diagram is then detected by a photodetector. These optical devices have been shown to operate in a stable manner using either fiber-based or planar-lightwave-circuit (PLC) technologies. The current waveform shape at the output of the photodiode due to the response of the incident power on the photodiode has a *cos*-like form. This pulse shape has been approximated by the *cos*-like I_{pulse} variable in our simulations.

12.5.1.2.2 The Photodiode

In order to detect the optical intensity signal at the output of the MZDI, shown in Figure 12.48, and to display this signal for later BER analysis, a photodiode diode or a pair of balanced diodes connected back to back is required [6]. The types most commonly used are the PIN and the avalanching photodiode (APD) [15]. The primary objective of the photodiode is to produce a conversion from optical photonic energy to electrical (electron flow) current. In this chapter, we are considering the PIN photodetector type. In practice, these PIN diodes (reversed-biased) are placed at the output arms of the MZDI (Figure 12.49). It is common practice to use either single- or balanced-diode detection at the receiver; however, in practice the balanced receiver configuration has shown higher sensitivity for DQPSK and other PSK modulation formats [16]. In this chapter, we consider the single-diode detection model for the sake of simplicity; however, we suggest that balanced detection be investigated at some stage in the future.

12.5.1.2.3 Noise Sources

In order to best simulate a true optical transmission system, we have considered various noise factors within the link, particularly at the receiver end. The noise sources implemented into the simulations include (1) MZDI phase offset $\delta\phi_{DI}$; (2) photodiode noise effects (i.e., quantum shot noise, photodiode dark current, and thermal noise); and (3) receiver electronic amplifier noise.

12.5.1.2.4 Digital Data Sampling

Most of the data supported by the blocksets can only support discrete data sources/values. Thus, a means of converting continuous waveforms to discrete values is required. Signaling theory states that a continuous signal $s(t)$ can be represented by a set of samples taken at a sampling frequency (f_s) [17]. This sampling frequency must be at least twice that of the highest frequency contained in the signal waveform, B. This is represented mathematically by $f_s \geq 2B$ samples/s; the value $f_s = 2B$ is referred to as the Nyquist rate. All samples made during the simulation are made at the Nyquist rate to ensure synchronization between data samples and to minimize sampling errors.

12.5.1.2.5 Pulse Shapes

We mentioned earlier the concept of pulse shapes as applicable to the MZIM. We consider now the RZ digital signal, the pulse shape used in the final design of this chapter. This pulse shape has a duration of half of the bit period ($T_b/2$) and thus has a 50% duty cycle. In most modern optical systems, this RZ pulse shape is often approximated by a sinusoid having frequency $1/T_b$ Hz [18]. The brief nature of the pulse makes the RZ pulse shape more tolerable to intersymbol interference (ISI) compared to the NRZ pulse shape; however, its power spectra occupies a larger bandwidth. In this chapter, we consider the RZ pulse shape as it is more compatible with the simulation development.

The NRZ pulse shape also has a data rate of $1/T_b$ b/s; however, its duration is now T_b seconds. The power spectrum is also reduced to half that of the RZ pulse spectrum. When digitizing data for transmission across many hundreds of kilometers, many digital modulation formats have been proposed and investigated. This chapter will investigate the DQPSK modulation format. This modulation scheme, although having been in existence for quite some time, had only been implemented in the electrical domain. Its application to optical systems proved difficult in the past as constant phase shifts in OCs were required to be maintained. However, with the improvement of optical technology and alternative transmitter design setup, these difficulties are eliminated.

One of the main attractive features of the DQPSK modulation format is that it offers both twice as much bandwidth and increased spectral efficiency compared to OOK. As an example, comparisons between 8×80Gb/s DQPSK systems and 8×40Gb/s OOK systems show that DQPSK modulation offers a more superior performance (i.e., spectral efficiency) [19]. The very nature of the signaling process also allows noncoherent detection at a receiver to be possible, thus reducing the overall cost of the system design.

As the name suggests, the idea behind DQPSK modulation format is to apply a differential form of phase shift modulation to the OC that encodes the data. DQPSK is an extension of the simpler DPSK format. Rather than having two possible symbol phase states (0 or π phase shift) between adjacent symbols, DQPSK is a four-symbol equivalent {0, π/2, π, or 3π/2}. Depending on the desired di-bit combination to be encoded, the difference in phase, $\Delta\phi_{mod}$, between the two adjacent symbols (OC pulses) is varied systematically. Table 12.3 outlines this behavior.

TABLE 12.3
DQPSK Modulation Phase Shifts

Di-bit	Phase Difference $\Delta\phi = \phi_2 - \phi_1$ (°)
00	0
01	90
10	180
11	270

In this table, ϕ_1 and ϕ_2 are the phases of consecutive symbols. Table 12.3 can also be represented in the form commonly known as a "constellation diagram" (see Figure 12.50); this graphically explains the signal states in both amplitude and phase.

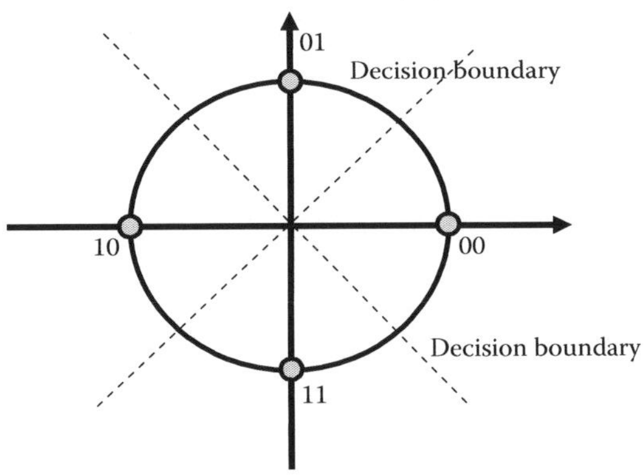

FIGURE 12.50
DQPSK signal constellation and decision boundaries for different states at π/4.

Ideally, we would maintain a constant amplitude and alter the phase of the signal during transitions; however, due to the characteristics of the optical equipment (i.e., MZM), some form of optical intensity modulation should also be observed in practice. This effect was investigated, but not implemented in the final simulator due to data pulse representation inaccuracies.

12.5.2 MATLAB® and Simulink® Simulators

With adequate knowledge and theory of the optical components required in the design of a single-channel 10 Gb/s DQPSK transmission link, we can now consider the development process of the optical simulator. This simulator considers only one channel implementing the DQPSK signaling technique (total 20 Gbauds/s baud rate); multiple channel configurations (DWDM) have been considered as future work. The primary design of a single-channel operating efficiently was considered critical. We have developed this simulator package in three main steps, each relating to the three main components in the transmission link. These include the transmitter, fiber, and receiver models. As this section focuses on an optical signaling technique (DQPSK) not attempted before, a significant amount of research into how this signaling technique is implemented in the optical domain using optical hardware was essential. The work done used a previously developed simulator [20], developed in Simulink, implementing the OOK format as a guide for initial development.

The transmitter performing the DQPSK transmission format was developed first and proved to be the most time-consuming as this digital signaling technique can be implemented in several ways using different hardware. The concept of modeling in the Simulink environment being a new experience resulted in additional time being allocated to familiarize ourselves with the Simulink package. The final technique chosen proved to be the simplest to design and simulate using conventional data-processing blocks in Simulink. The configuration of this transmission system is based on the research and development work performed in the field of DQPSK signaling transmission [19] and other available literature.

Figure 12.51 shows the top level of the single-channel optical DQPSK Simulink model developed. This model can be later altered and new blocks can be added to optimize the system to individual requirements.

The final results displayed during simulations include the di-bit NRZ electrical bit stream (2×10 Gb/s) that modulates the OC via the MZM and PM, the spectrum of the transmitted OC, the RZ time-domain pulse generated, and finally the eye diagrams both prior to and after fiber propagation to allow for BER testing and other system performance measurements (Figure 12.52). We now consider each block shown in Figure 12.51 and present the development of the simulator model.

12.6 Spectral Characteristics of Advanced Modulation Formats

Optical power spectra of 40 Gb/s NRZ/RZ33/CS-RZ DPSK and OOK optical signals are shown in Figures 12.53 and 12.54, respectively.

Several main points observed in these spectra are as follows:

- The optical power spectra of OOK formats have high power components at the carrier and signal frequencies, thus causing fiber nonlinear effects. On the contrary, DPSK spectra do not contain high power frequency components. Therefore, optical DPSK formats are more resilient to fiber nonlinear effects.

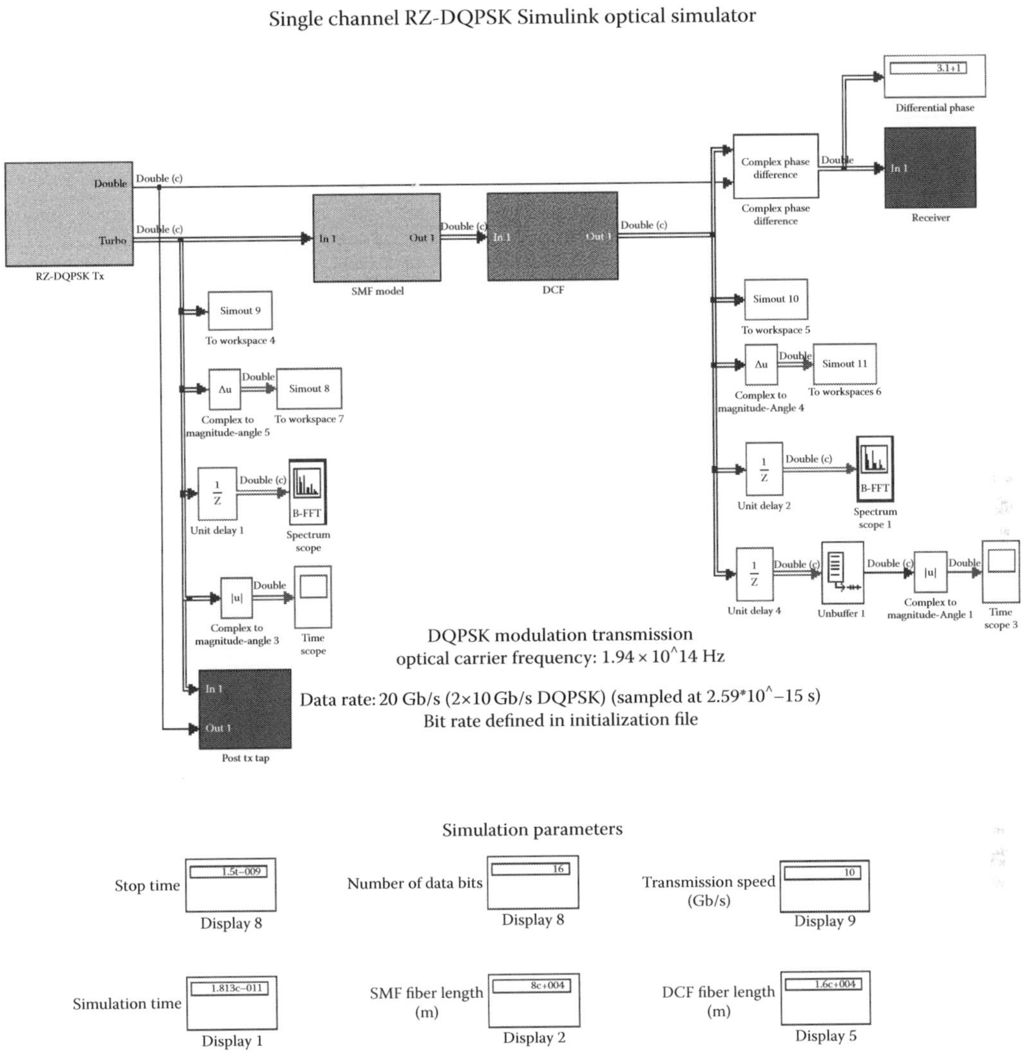

FIGURE 12.51
Single-channel RZ-DQPSK transmission Simulink optical simulator. (Designed by Laville, B. and supervisor, Binh, L.N. Version 1.1 September 2004.)

- RZ pulses are more susceptible to fiber CD due to their broader spectra compared to NRZ pulses. In particular, the RZ33-type pulse has the broadest spectrum (at 20 dB position down from the peak) and thus, RZ33 is least tolerant to fiber CD.
- However, the susceptibility of RZ33 and CS-RZ pulses to fiber CD provides a trade-off in the improvement of the resilience to fiber nonlinearities. This is because time-domain peaks of these fast-spreading optical pulses decrease rapidly when propagating along the fiber, hence lowering effects of fiber nonlinearities.

Optical power spectra of 40 Gb/s linear, weakly nonlinear, and strongly nonlinear MSK formats are compared in Figure 12.55. It can be observed that the power spectra of these formats are similar, with the exception point that side lobes in the power spectrum of strongly nonlinear MSKs are not highly suppressed as in the case of linear and

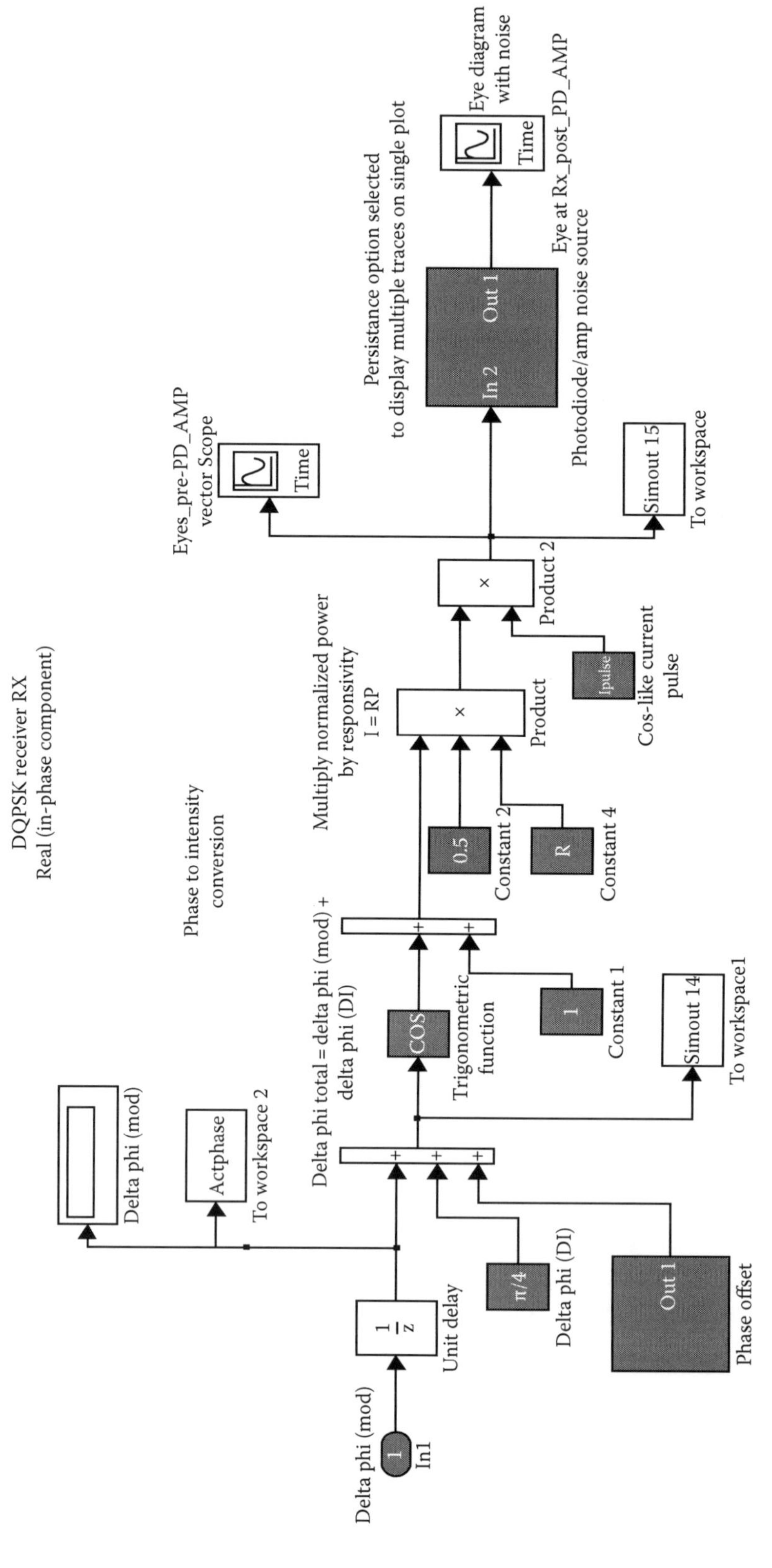

FIGURE 12.52
Structure of a DQPSK receiver.

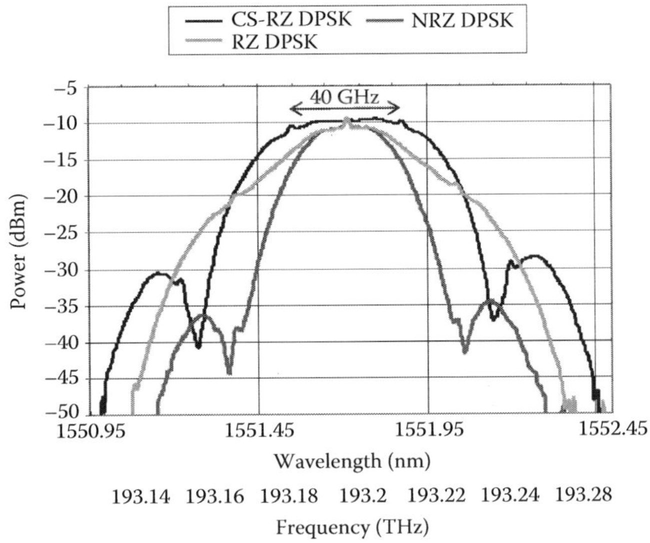

FIGURE 12.53
Spectra of 40 Gb/s CS-RZ/RZ33/NRZ DPSK optical signals.

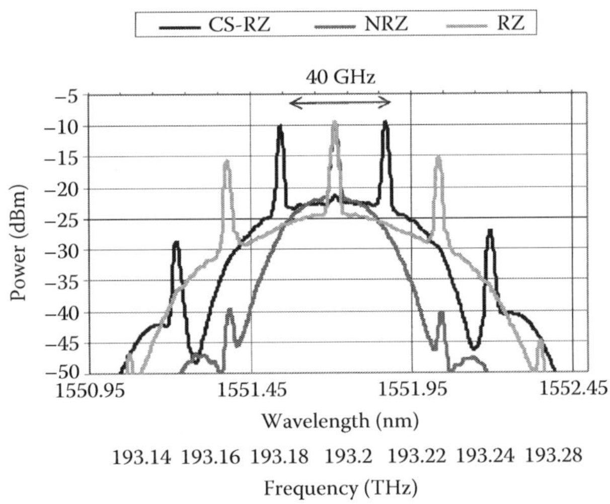

FIGURE 12.54
Spectra of 40 Gb/s CS-RZ/RZ33/NRZ OOK optical signals.

weakly nonlinear MSKs. All three formats offer better spectral efficiency compared to the DPSK counterpart.

Figure 12.56 compares the power spectra of three modulation formats: 80 Gb/s dual-level MSK, 40 Gb/s linear MSK, and 40 Gb/s NRZ DPSK. The intensity-splitting ratio for the optical dual-level MSK format is set at "0.8/0.2."

A number of key points are observed:

- The signal power spectrum of 80 Gb/s optical dual-level MSK has identical characteristics to that of the 40 Gb/s MSK format. The main lobe spectral widths of these

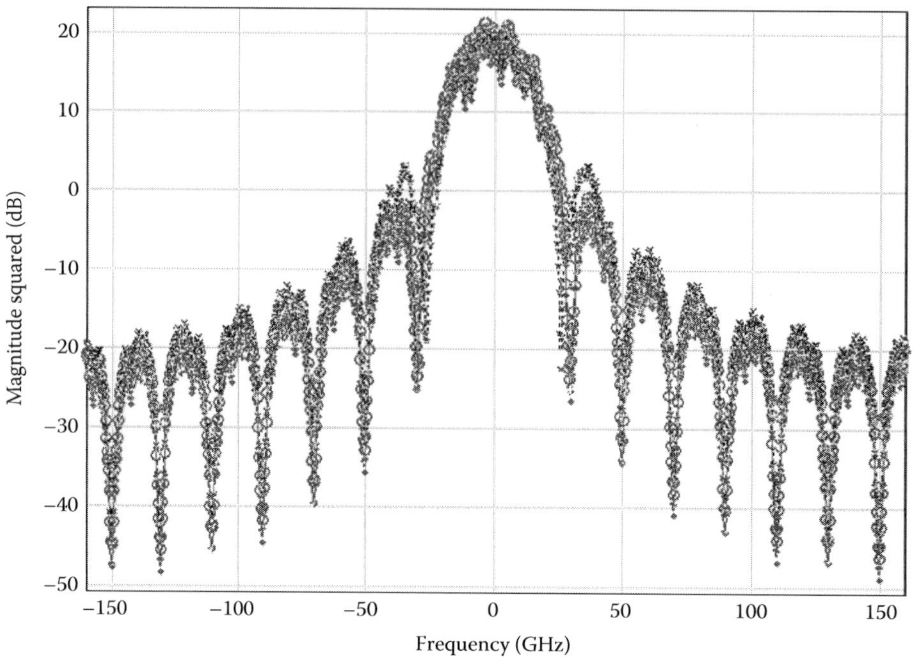

FIGURE 12.55
Optical power spectra of three 40 Gb/s optical MSK formats: linear, weakly nonlinear, and strongly nonlinear which are very close to each other.

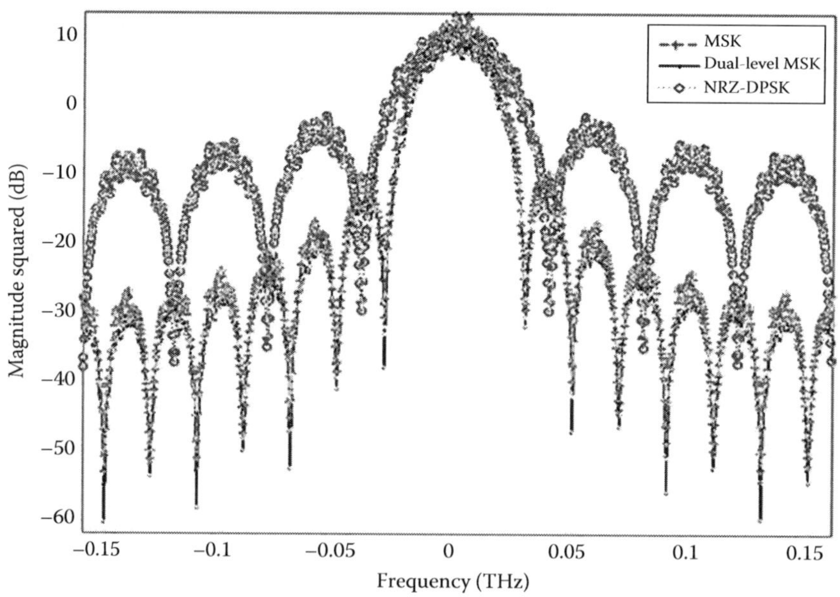

FIGURE 12.56
Spectral properties of three modulation formats: 40 Gb/s MSK, 80 Gb/s dual-level MS, and 40 Gb/s NRZ DPSK.

two formats are narrower than that of NRZ DPSK. More specifically, the base-width takes a value of approximately ±32 GHz on either side compared to ±40 GHz in the case of DPSK. Hence, it is expected that these MSK-based formats are more tolerant to fiber CD and more robust to tight optical filtering than NRZ DPSK.

- High suppression of side lobes and the confinement of signal energy in the main spectral lobe of 80 Gb/s dual-level MSK and 40 Gb/s MSK optical power spectra significantly mitigate interchannel cross talks between DWDM channels.

12.7 Partial Responses Duo-Binary Transmission Model

12.7.1 Remarks

Modulation format is important in determining the performance of 40 Gb/s optical fiber communication systems. DB and continuous-phase modulation (CPM) are shown to offer high spectral efficiency [35]. DB modulation formats minimize ISI impairments in a controlled way instead of trying to eliminate it. It is possible to achieve a signaling rate equal to the Nyquist rate of 2 Wsymbols/s in a channel of bandwidth W Hz. Optical DB technique has received much attention due to its high dispersion tolerance and high frequency-utilization efficiency by means of spectral narrowing. The DBM format is similar to the NRZ format, with the inclusion of phase coding. The phase characteristics of DBM signals compensate for the group velocity dispersion by reducing the spectral component in conventional NRZ modulation. ISI is reduced since bit patterns such as 101 are transmitted with the ones having the opposite phase. Therefore, if pulses spread out into the zero time slot, due to dispersion, they tend to cancel each other out. The recovery of signals at the receiver is relatively simple. Further more conventional direct detection receiver is applicable, hence the simple receiver structure. There are two types of DBM schemes, which are constant phase and alternating phase in blocks of logics "1s."

This section presents the models for photonic transmission with optical channels operating under DB modulation format. This includes the development and implementation of the photonic transmitter, the optical fiber propagation, and the optoelectronic receiver. DBM encodes two-level electrical signal to three-level electrical signal before modulating the lightwave carrier. The transmitter of the Simulink model will consist of a DB encoder and a dual-drive MZIM. A baseband modulation is first implemented in the DB encoder, which encodes the binary signal into three levels signals of "1," "0," and "–1." MZIM is an EO modulator that converts the electrical signal to optical signal.

The DB or phase-shape binary modulation formats can be generated by modulating a dual-drive MZIM. Recent works have shown that the driving voltages for the modulator can be reduced to generate variable pulse width DB optical signals. However, the pulse width of the DB DPSK has not been thoroughly investigated under the alternating phase of the "1" coded bits; that means the "0" "π" "0" "π" phases of consecutive "1" in contrary to conventional DB formats. We also present modeling performance of alternating phase DB modulation with an FWHM ratio with respect to the bit period of 100%, 50%, and 33% and compared with experimental transmission of carrier-suppressed DPSK over 50 km of SSMF and dispersion compensation. For the DB case, the transmission without dispersion compensation over the same SSMF length offers better performance for 50% FWHM DB modulation and slightly worst for 33%. The transmission performance, the

BER versus receiver sensitivity, of these DB modulation formats are compared with the carrier-suppressed DPSK experimental transmission.

Section 12.7.2 describes the fundamental aspects of the DBM format. It is then followed by the description of each component in the 40 Gb/s DBM photonic transmission systems. The implementation of Simulink model of the communication system is described.

12.7.2 The DBM Formatter

Modulation format aims to modulate one or more field properties to suit system needs. There are four types of field properties: intensity, phase, polarization, and frequency. Symbols are constellated in one or more dimensions, in order to carry more information and to travel further distances. Data modulation format is the information-carrying property of the optical field.

The DBM format has become an attractive modulation format over these recent years, compared to other formats, such as NRZ and RZ. This is due to the fact that it can overcome the fiber CD in high-capacity transmissions. It is the characteristics of the DBM format that makes it a preferred format.

DBM schemes can be described as correlative-level coding or a partial-response signaling schemes. Correlative-level coding schemes is adding ISI to transmitted signal in a controlled manner; a signaling rate equal to the Nyquist rate of $2W$ symbols/s in a channel of bandwidth W Hz can be achieved. "Duo" in the word "duo-binary" indicates the doubling of the transmission capacity of a conventional binary system. The DBM format is, in fact, NRZ modulation with an inclusion of phase coding. The one bits in the data input are phase modulated. For instance, for a bit pattern of 101, this data will be transmitted with the ones carrying opposite phase, 0 and π. If the pulse of the one bits spread out to the zero time-slot in between, they will cancel each other. This effect increases the dispersion tolerance, and allows the signal to be transmitted over a longer distance.

DB coding converts a two-level binary signal of 0s and 1s into a three-level signal of "−1," "0," and "+1." This is done by, first, applying the binary sequence to a pulse–amplitude modulator to produce two-level short pulses of amplitude of −1 and +1, with −1 corresponding to 0 and +1 corresponding to 1. This sequence is, then, applied to a DB encoder to produce a three-level output of "−2," "0," and "2."

As shown in the input sequence of Figure 12.57, $\{a_k\}$ of uncorrelated two-level pulses is transformed into $\{c_k\}$, which is a sequence of correlated three-level pulses. The correlation between adjacent pulses is equivalent to introducing ISI into the transmitted signal in an artificial manner. The DB encoder is simply a filter involving a single delay element and a summer, as shown in Figure 12.58. However, once errors are made, they tend to propagate through the output. This is because a decision made on the current input a_k depends on the decision made on the previous input a_{k-1}. Therefore, precoding is needed to avoid this error propagation phenomenon. The binary sequence $\{b_k\}$ is converted into another binary sequence $\{d_k\}$ by modulo-two addition, XOR of b_k and d_{k-1}, as shown in Figure 12.59.

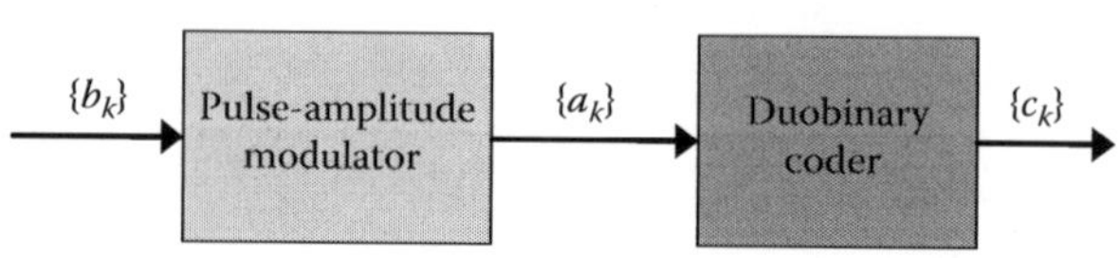

FIGURE 12.57
Brief overview of DB signaling.

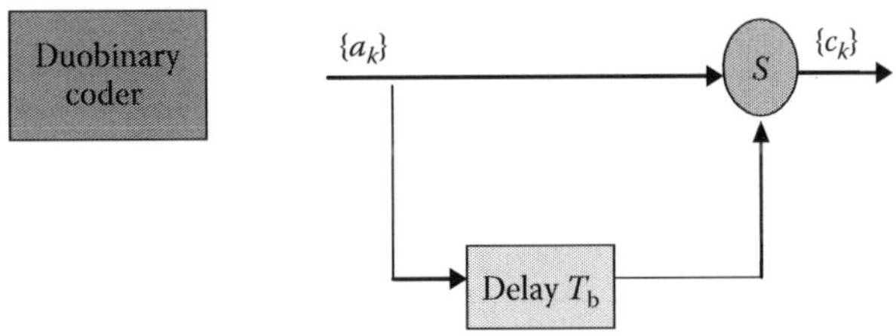

FIGURE 12.58
DB encoder—the block at the left is represented by the signal flow diagram shown in the right.

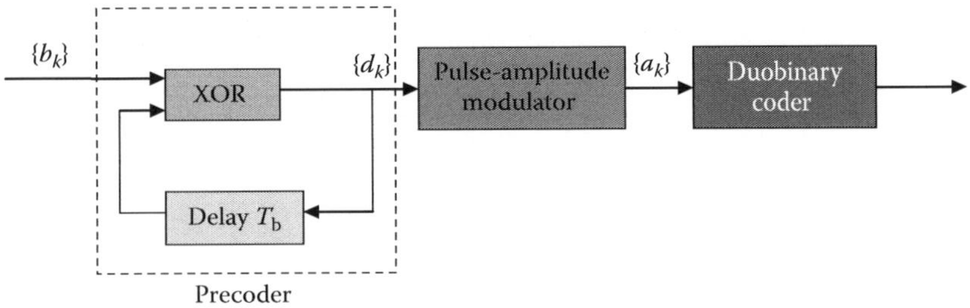

FIGURE 12.59
DBM scheme with precoder.

$$c_k = a_k + a_{k-1} \tag{12.18}$$

$$d_k = b_k \oplus d_{k-1} \tag{12.19}$$

The three-level DB output $\{c_k\}$ is then modulated into a two-level optical signal by an external modulator. The most commonly used external modulator is the MZIM. The optical DB signal has two intensity levels, "on" and "off." The "on" state signal can have one of the two optical phases, 0 and π. The two "on" states correspond to the logic states "1" and "−1" of the DB-encoded signal, $\{a_k\}$ and the "off" state correspond to the logic state "0." Figure 12.60a shows an example of the original binary signal, the DB-encoded signal, and optical DB signal, and then a summary of coding rule is shown in Figure 12.60b.

12.7.3 40 Gb/s DB Optical Fiber Transmission Systems

Ultra-long terrestrial networks transmitting signal at a bit rate of 40 Gb/s have matured over recent years. Various advanced modulation schemes, such as RZ, NRZ, NRZ-DPSK, and RZ-DPSK, have been proposed to achieve an extended reach and an improved capacity of the communication system.

Figure 12.61 shows the typical DWDM optical fiber communication system. Signals are modulated at the transmitters and are multiplexed together at the wavelength multiplexer before transmitting them into the fiber. The fiber link is divided into a number of spans. Each span consists of an SSMF and a DCF. The EDFA is used to compensate for the optical power loss of the transmission span. At the end of the fiber, the signals are demultiplexed and detected at the receivers.

The DBM format has become popular compared to other modulation formats because it extends the transmission distance as limited by fiber loss, without regenerative repeaters.

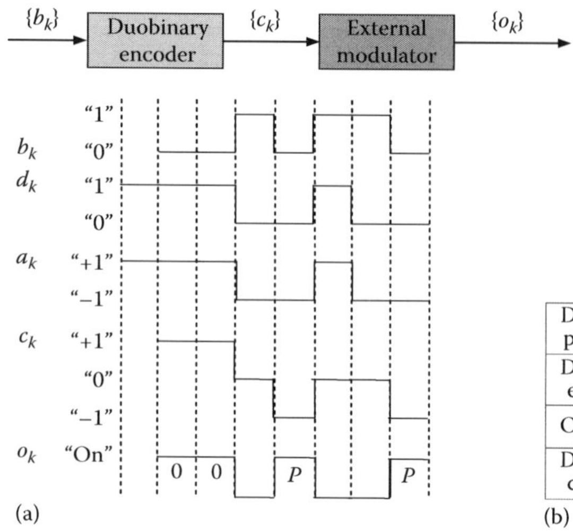

Duobinary precoding	$b(k) = \overline{d(k)} \oplus b(k-1)$
Duobinary encoding	$c(k) = b(k) + b(k-1)$
Offset	$c_{AC}(k) = c(k) - 1$
Duobinary decoding	$\hat{d}(k) = \lvert c_{AC}(k) \rvert$

(b)

FIGURE 12.60
(a) Example of original binary signal (b_k), precoded signal (d_k), DB-encoded signal (c_k), and optical DB signal (o_k); and (b) summary of coding rule.

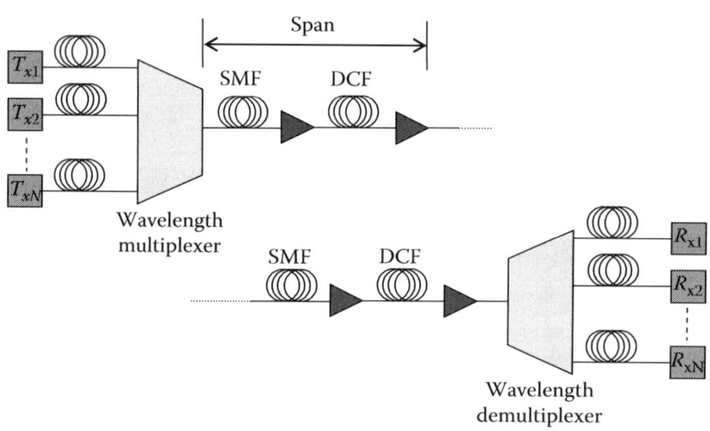

FIGURE 12.61
Ultra-long-haul fiber transmission.

It extends the dispersion limit without additional optical components, such as the DCF. CD has become a main effect that limits the transmission distance. The optical three-level transmission can overcome this limitation since the narrowband signal has higher tolerance to CD compared to the broadband signal. Furthermore, DB optical fiber communication systems can suppress SBS.

The main modules of the communication system are the transmitter, the optical fiber, and the optoelectronic receiver, as shown in Figure 12.62. The transmitter consists of the DB encoder and the MZIM. A series of 0s and 1s is modulated under the DBM scheme. These three-level electrical data is then used to modulate the laser source, producing a two-level optical signal. This modulated signal is transmitted along an optical fiber transmission link toward the EO receiver. The signal will be detected using a photodetector that

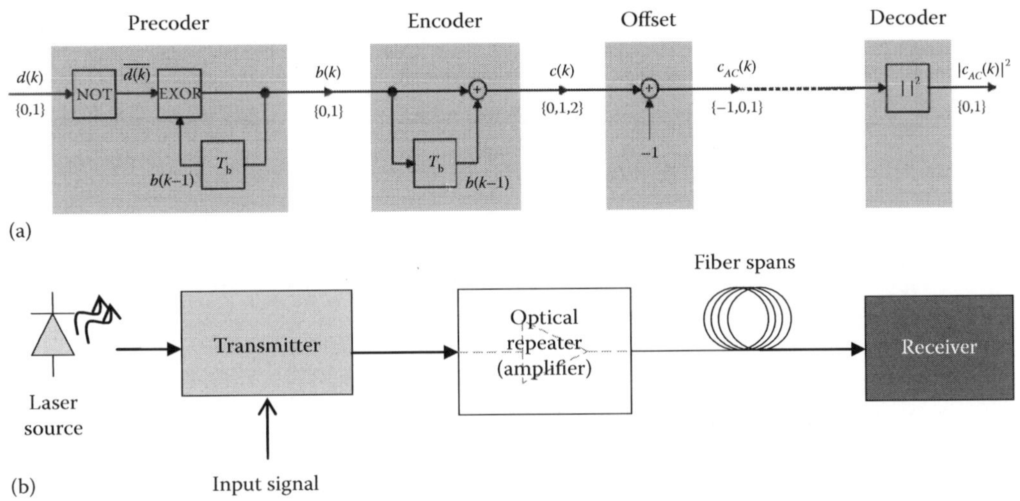

FIGURE 12.62
Main modules of a DBM optical communication system: (a) coder and decoder and (b) generic transmission system.

converts the two-level optical signal back to the electrical signal. Optical amplification can be done at some points along this transmission link to minimize the effect of fiber loss.

12.7.4 Electro-Optic Duo-Binary Transmitter

Transmitter modulates and converts the incoming electrical signal into optical domain. Depending on the nature of the signal, the resulting modulated light may be turned on and off, or may vary linearly in intensity between two levels. The output of the DB transmitter is the modulated lightwaves switched on and off at transitional instances of the input electrical signal. In general, a DBM transmitter, shown in Figure 12.63, consists of a monochromatic laser source, a coder, and a photonic modulator. Binary data is encoded by a DB encoder. This resulting three-level electrical signal is converted into two-level optical signal using the folding characteristics of an optical MZIM. It is then transmitted into the optical fiber.

There are two types of DB transmitters. The conventional DB transmitter, as previously mentioned, includes a dual-drive MZIM driven by three-level electrical signals in a push–pull configuration. Given the fact that an MZIM is normally driven by a two-level signal, the effect of driving it with a three-level signal has its uncertainties. It is proposed that three-level signals may experience significant distortion in electrical amplifiers operating in saturation leading to penalties for long word lengths. It may also cause the degradation of receiver sensitivity. For these reasons, the second type of DB transmitter has been proposed. This type of transmitter has the MZM driven by only two-level electrical signals. The optical DB signal generated is the same as the one of the DB transmitter type, that is, constant phase in blocks of 1s.

12.7.4.1 The Duo-Binary Encoder

The DB encoder encodes the binary signals, which is a sequence of 0s, to a three-level electrical signal. DB signal is a fundamental correlative coding in partial response signaling.

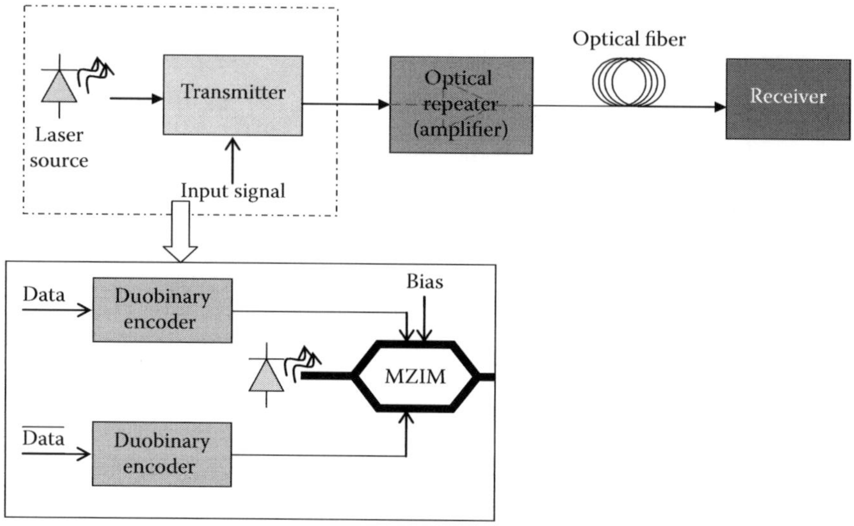

FIGURE 12.63
The transmitter of the 40 Gb/s DBM photonic transmission system.

A DB encoder consists of a precoder and a DB coder. A precoder is used before DB coding to allow for easier recovery of binary data at the receiver, and to avoid error propagation. The precoder is a simple binary digital circuit that consists of an XOR and a 1-bit delay feedback. The DB coder is a filter consisting of a single delay element and a summer (Figure 12.64).

Binary data input is precoded with the initialization of the 1-bit delay to 0. The output of the DB precoder is, then, modulated by a pulse-amplitude modulator, to produce a two-level electrical signal with amplitudes of –1 and 1. The DB signal is produced by adding data delayed by a 1 bit period to the present data. This DB signal is a three-level electrical signal with amplitudes of –2, 0, and 2. Finally, it is converted to a level of –1, 0, and 1. The three-level is mapped into optical domain by modulating both amplitude and phase. The "+1" and "–1" levels have the same optical intensity but opposite optical phase.

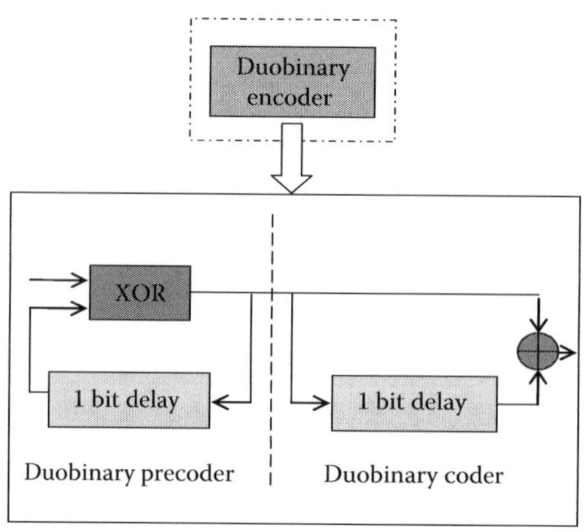

FIGURE 12.64
The DBM encoder of the 40 Gb/s DBM photonic transmission system.

12.7.4.2 The External Modulator

The EO modulator has already been described in Chapter 2. However, it is essential here to revisit the operation of the MZIM for DB operation. In an MZIM, the input OC is split into two paths via a Y junction. This Y junction splits the input signal into $E_i/\sqrt{2}$ each. The resultant signal is

$$E_o = \frac{E_i}{2}\left[1 + \exp\left(j\pi\frac{V(t)}{V_\pi}\right)\right] \tag{12.20}$$

where
V_π is the voltage to provide a π phase shift of each PM
$V(t)$ is the driving voltage

The input and output relationship of this MZIM is shown in Figure 12.65. It is accompanied by a phase modulation of $\exp(j\varphi(t))$ with $\varphi(t) = \pi V(t)$. For $V(t)$ from 0 to V_π, E_o and E_i have the same phase, and for $V(t)$ from V_π to $2V_\pi$, E_o and E_i have different phase.

MZIM can be single drive or dual drive. Single-drive X-cut $LiNbO_3$ MZM has no phase modulation along with the amplitude modulation. It follows the transfer characteristics of Figure 12.65. Dual-drive X-cut $LiNbO_3$ MZIM, on the other hand, has two paths phase modulated with opposite phase shifts in a push–pull operation. The V_π in Figure 12.65 is reduced by half in this case. For dual-drive y-cut $LiNbO_3$ MZIM, two paths are driven by complementary signals with V_1 equal to $-V_2$. The output electric field of a dual-drive MZIM is

$$E_o = \frac{E_i}{2}\left[\exp\left(j\pi\frac{V_1}{V_\pi}\right) + \exp\left(j\pi\frac{V_2}{V_\pi}\right)\right] \tag{12.21}$$

DB optical signal is generated by driving a dual-drive MZIM with push–pull operation, as shown in Figure 12.66. One arm is driven by the DB signal and the other arm is driven by the inverted DB signal. Figure 12.67 shows the operation of the MZIM. The output electric field $E_o(t)$ can be expressed as follows:

$$E_o(t) = E_i \cos\frac{\Delta\phi(t)}{2}\exp\left(-j\frac{\phi_0}{2}\right) \tag{12.22}$$

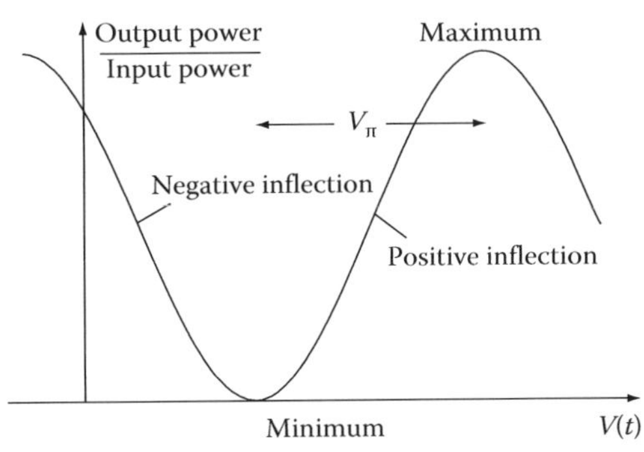

FIGURE 12.65
Input–output transfer characteristics of the MZIM.

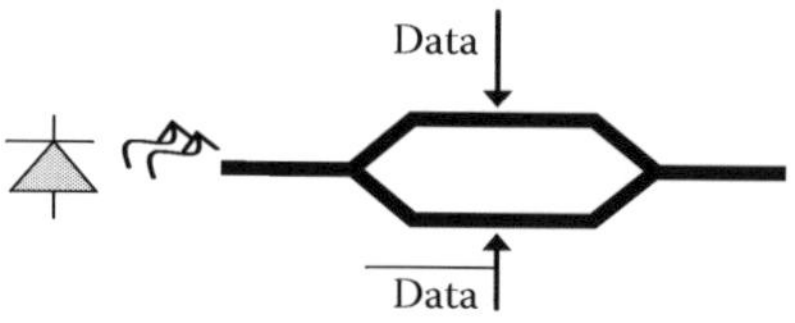

FIGURE 12.66
Dual-drive MZIM.

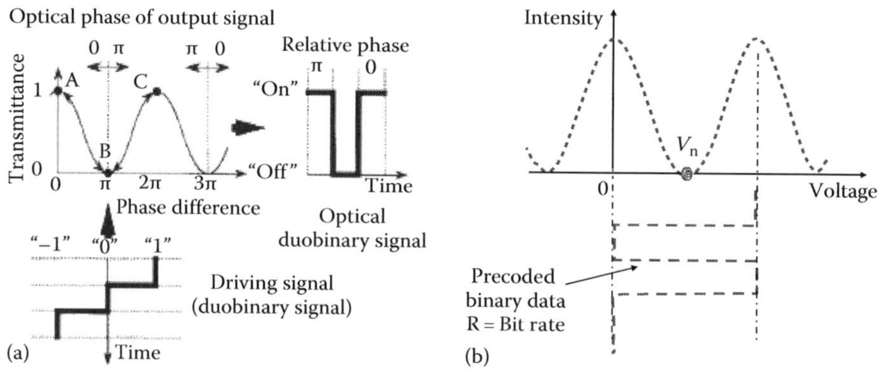

FIGURE 12.67
Driving operation of dual-drive MZM: (a) DB signaling and (b) DPSK.

where

E_i is the input electric field

$\Delta\varphi(t)$ is the phase difference between the lightwaves propagating in two optical waveguides

φ_0 is a constant when the MZIM is driven in a push–pull operation

At point B of Figure 12.67, the phase of the output optical signal is inverted. The optical DB signal is dependent on the biasing point of the driving signal, which is the electrical DB signal. By biasing at point B in Figure 12.67, "−1" and "+1" level of the electrical DB signal will correspond to the "on" state of the optical signal, while the "0" level will correspond to the "off" state. To achieve the effect of carrier suppressed, there must be a π phase different between the two arms.

12.7.4.3 Duo-Binary Transmitters and Precoder

The transmitter model, generally, consists of the DB coder and the MZIM (Figure 12.68). The DB coder encodes the incoming binary sequence of 0s and 1s to DB electrical signal. This signal is, then, used to drive the arms of the dual-drive MZIM. One arm is driven by the DB signal and the other arm is driven by the inverted DB signal. The Bernoulli binary generator generates a random sequence of binary electrical signal. It is set to generate data at a rate of 40 Gb/s. This signal is encoded by the DB encoder, which consists of a DB precoder and a DB coder. The first output of the encoder is shifted up by 1 to produced levels of "0," "1," and "2." This electrical DB signal is sent to the phase shift block, as shown in Figure 12.69, to represent these levels with a certain phase. This, in fact, represents the biasing point on the transmittance curve. For dual-drive MZIMs, the driving signal is biased

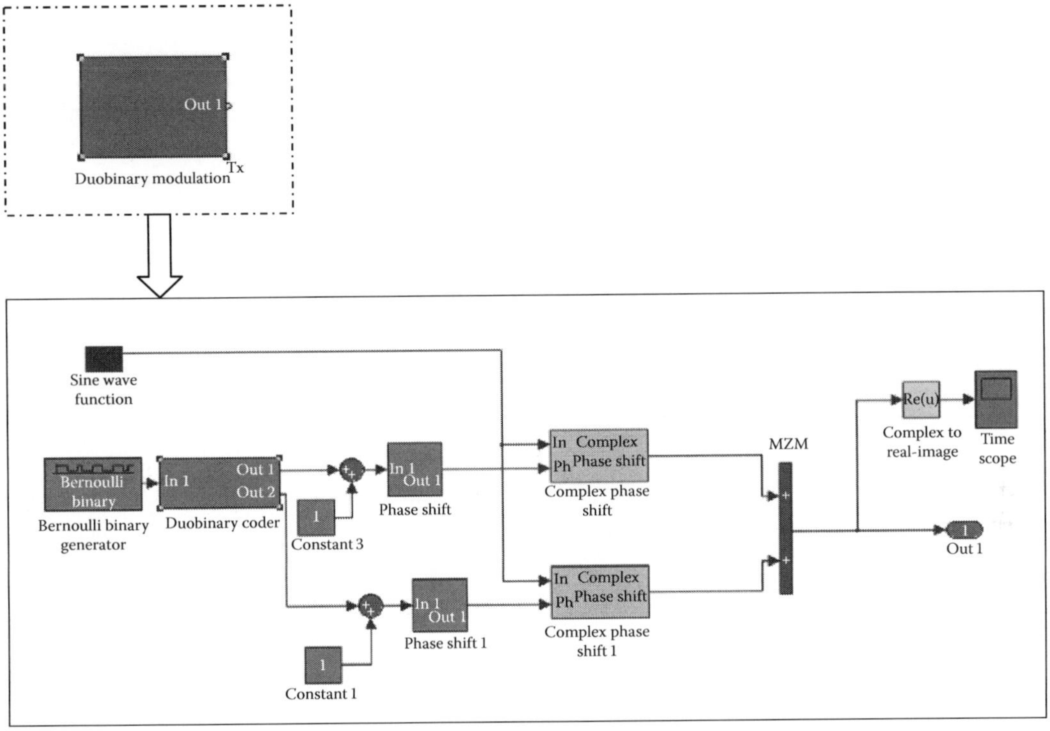

FIGURE 12.68
The conventional transmitter model of the 40 Gb/s DB optical fiber communication system.

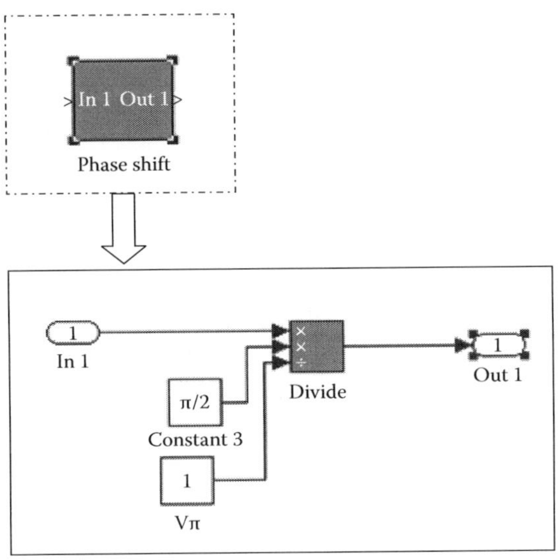

FIGURE 12.69
Phase shift block of the transmitter model.

at $V_{\pi/2}$. The second output of the DB encoder is the inversion of output 1. This output is used to modulate the second arm of the MZIM. The output 2 signal is shifted down by −1, to bias at the point $-V_{\pi/2}$ of the transmittance curve. MZIM is an amplitude modulator, accompanied by a shift of phase. This modulation is also called AM-PSK. This lightwave, which is the sine wave produced by the sine wave function, is modulated by the DB signal through the complex phase shift block. The input sine wave is shifted by the amount specified at the Ph input.

The DB coder consists of a precoder and a coder, as shown in Figure 12.70. The precoder is a differential coder, with an XOR gate and a 1-bit delay feedback path. The addition of −0.5 and division by 2 function as the amplitude modulator shifts levels of the signal from "0" and "1" to "−1" and "+1." The signal is, then, added to its 1-bit delay to produce a three-level DB signal of level "−2," "0," and "+2," followed by a conversion to a level of "−1," "0," and "+1." The summation of the signal with its 1-bit delay is the DB coder. For the second output, the output of the differential coder is inverted, before going through the same operation as Out1. Zero-order hold is placed before the output of the DB encoder functions to discretize the signals to have a fast-to-slow transition of signals. It holds and samples the signal before transmitting it out. If the signals are transmitted out without the zero-order hold, the transition to "0" level will be overseen. The signal will only have two levels, which are "−1" and "+1."

12.7.4.4 Alternative Phase DB Transmitter

Two types of DB transmitter models are proposed. The conventional DB transmitter, as mentioned previously, uses a dual-drive MZM driven by three-level DB electrical signals.

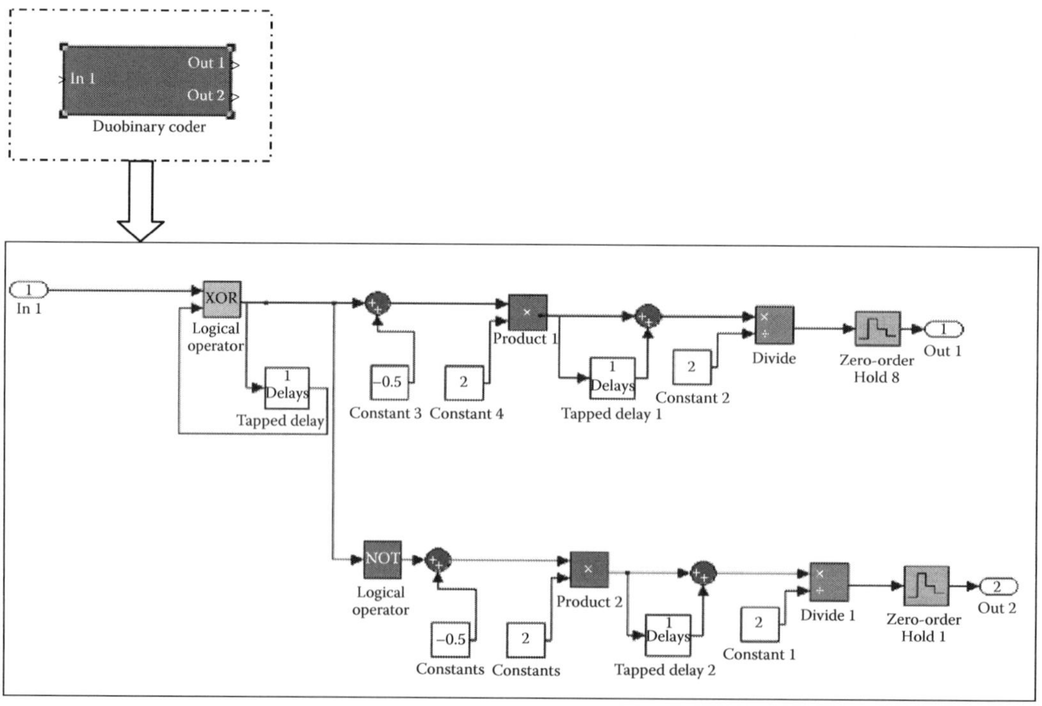

FIGURE 12.70
The DB coder of the transmitter model.

MZIM shown in Figure 12.71 is usually driven by a two-level electrical signal. In some cases, it occurred that this three-level driving signal may experience significant distortion in electrical amplifiers operating at saturation. This may lead to penalties for long word lengths. It may also occur that there will be a degradation of receiver sensitivity. Due to these uncertainties, an alternative DB transmitter, as shown in Figure 12.72, is proposed. This second type of DB transmitter has the MZIM driven by two-level electrical signals.

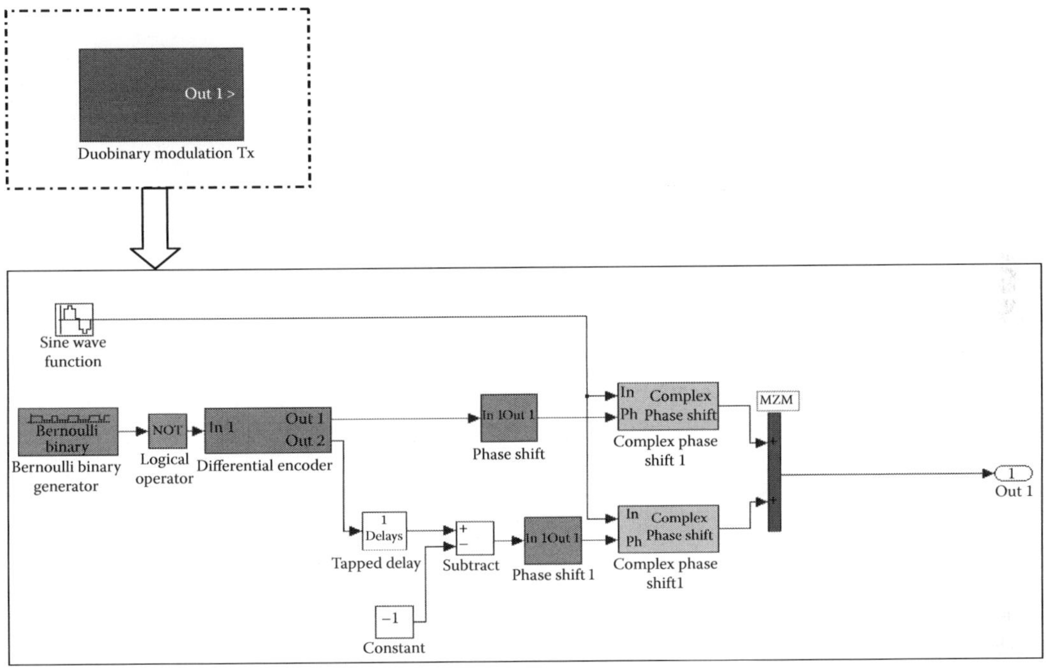

FIGURE 12.71
Simulink model of MZIM.

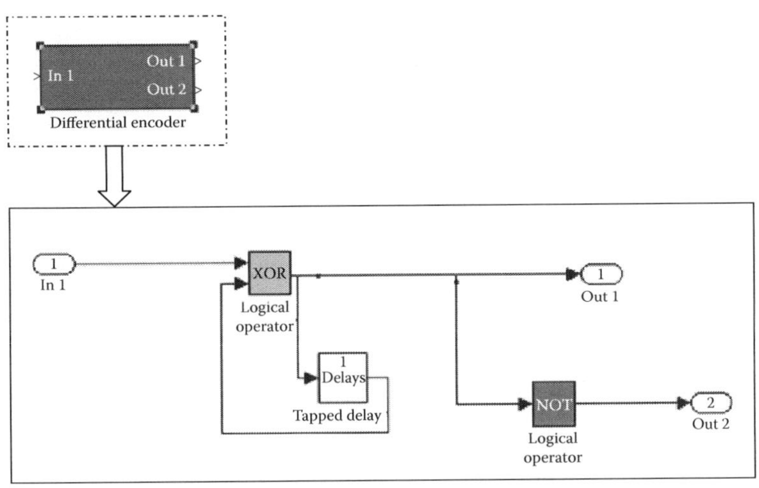

FIGURE 12.72
DB transmitter (type 2).

It consists of a differential encoder, a 1-bit period electrical time delay, and an MZM. One arm of the dual-drive MZM is driven with the DB signal, whereas the other arm is driven by the inverted DB signal, delayed by 1-bit period. Both DB transmitters produce the same result, which is constant phase in blocks of 1s.

12.8 MSK Transmission Model

12.8.1 Introduction

The generation of MSK requires a linear variation of the phase, hence a constant frequency of the OC. However, the generation of the optical phase may be preferred by driving an optical modulator using sinusoidal signal for practical implementation. Thus, a nonlinear variation of the carrier phase, hence some distortion effects is produced. In this chapter, we investigate the use of linear and nonlinear phase shaping filtering and their impacts on MSK-modulated optical signals transmission over optically amplified long-haul communication systems. The evolution of the phasor of the in-phase and quadrature components is illustrated for lightwave-modulated signal transmission. The distinct features of three different MSK modulation formats, linear MSK, weakly nonlinear MSK, and strongly nonlinear MSK, and their transmission are simulated. Transmission performance obtained indicates the resilience of the MSK signals in transmission over multi-optically amplified multi-spans.

Recent years have witnessed intensive interests in the employment of advanced modulation formats to explore their advantages and performance in high-density and long-haul transmission systems. RZ and NRZ with or without carrier suppression (CS-RZ) formats are associated with shift keying modulation schemes such as amplitude (ASK), phase (PSK), and differential phase (DPSK, DQPSK) [1,2]. Differential detection offers the best technological implementation due to the non-requirement of coherent sources and avoidance of polarization control of the mixing of the signals and a local oscillator at multi-terahertz frequency range. CPM is also another form of PSK in which the phase of the OC evolves continuously from one phase state to the other. For minimum phase shift keying (MSK), the phase change is limited to $\pi/2$. Although MSK is a well-known modulation format in radio frequency digital communications, it has only been attracting interests in optical system research in the last few years [4,5,21]. The phase continuous evolution of the MSK has many interesting features: the main lobe of the power spectrum is wider than that of QPSK and DPSK, and the side lobes of the MSK signal spectrum are much lower allowing easy optical filtering and hence less distortion due to dispersion effects. In addition, higher signal energy is concentrated in the main lobe of MSK spectrum than its side lobes leading to better signal-to-noise ratio (SNR) at the receiver.

Advanced modulation formats for 40 Gb/s and higher bit rate long-haul optical transmission systems have recently attracted intensive research including various amplitude and discrete differential phase modulation and pulse shape formats (ASK-NRZ/RZ/CSRZ, DPSK, and DQPSK-NRZ/RZ/CSRZ). However, there are only a few reports on optical CPM schemes using external EO modulators [1–3,21]. Compared to discrete phase modulation, CPM signals have very interesting and attractive characteristics including high spectral efficiency, higher energy concentration in the signal bands, and more robustness to interchannel cross talk in DWDM due to greater suppression of the side lobes, which has recently risen as a critical issue in DWDM optical systems [4,5].

In bandwidth-limited digital communication including wireless and satellite digital transmission, MSK has been proven as a very efficient modulation format due to its prominent spectral efficiency, high sideband suppression, constant envelop, and high energy concentration in the main spectral lobe. In modern high capacity and high performance optical systems, we have witnessed the acceleration of transmission bit rate approaching 40 Gb/s and even to 160 Gb/s. At these ultrahigh bit transmission speeds, the essence of efficient modulation formats for long-haul transmission is critical. There are several technical works published on advanced modulation techniques for optical transmission, which mostly focus on discrete phase modulation including binary DPSK and DQPSK [1,2,21] with various pulse carving formats including NRZ and CS-RZ [2,3]. However, there is only a limited number of works on MSK modulation formats [4–6,22]. The features of MSK, compared to other modulation formats, have prompted researchers to investigate its suitability for high-capacity long-haul transmission. The side lobes of MSK power spectrum are greatly suppressed giving it good dispersion tolerance and avoiding interchannel cross talks. Thus this modulation is of interest for further investigations.

Two different proposals of transmitter configurations for the generation of linear and nonlinear phase optical MSK modulation are reported in another paper in this conference [23]. The brief description of the operation of these two transmitter configurations are presented in Section 12.8.2 whereas the direct detection techniques for both linear and nonlinear optical MSK signals are described in Section 12.8.3. Our simulation models for the modulation and system transmission are based on MATLAB and Simulink platforms [8] with detailed discussions in Section 12.8.4. In this chapter, we have also proven that the optical MSK signal is capable of propagating over an optically amplified and fully dispersion-compensated system. Not only the performance of conventional MSK format is evaluated but also the weakly and strongly nonlinear optical MSK sequences are examined.

The MSK format exhibits dual alternating frequencies and offers orthogonal property of the two consecutive bit periods. More interestingly, MSK can be considered as either a special case of continuous-phase FSK (CPFSK) or a staggered/offset QPSK in which I and Q components are interleaved with each other [22]. These characteristics are greatly advantageous to optically amplified long-haul transmissions because such a compact spectrum potentially gives a good dispersion tolerance, making MSK a suitable candidate for DWDM systems.

This section thus investigates a number of novel structures of photonic transmitters for the generation of optical MSK signals. Theoretical background of MSK modulation formats discussed in two different approaches are presented in Section 12.8.2. Section 12.8.2 proposes two configurations of optical MSK transmitters that employ (1) two cascaded EO PMs (E-OPMs) and (2) parallel dual-drive MZMs. Different types of linear and nonlinear phase-shaped optical MSK sequences can be generated. The precoder for I–Q optical MSK structure is also derived in this section. We present in Section 12.8.3, a simple noncoherent configuration for the detection of the MSK and nonlinear MSK-modulated sequences. The optical detection of MSK and nonlinear MSK signals employs a $\pi/2$ phase shift in one arm of the MZ interferometric delay (MZIM) balanced receiver. New techniques in the measurement of the BERs with the probability density function (pdf) of received signals after decision sampling to be computed with the superposition of a number of weighted Gaussian pdfs. The following performance results and observations are obtained in Section 12.8.4: (a) spectral characteristics of linear and nonlinear MSK-modulated signals and (b) improvement on dispersion tolerance of MSK and nonlinear MSK over ASK and DPSK counterparts.

Recently, advanced modulation formats have attracted intensive research for long-haul optical transmission systems including various amplitude and discrete differential phase modulation and pulse shape formats (ASK-NRZ/RZ/CSRZ, DPSK, and DQPSK-NRZ/RZ/CSRZ). For the case of phase modulation, the phases of the OC are discretely coded with "0," "π" (DPSK) or "0," "π" and "π/2," "−π/2" (DQPSK). Although the differential phase modulation techniques offer better spectral properties, higher energy concentration in the signal bands, and more robustness to combat the nonlinearity impairments as compared to the amplitude modulation formats, phase continuity would offer even better spectral efficiency and at least 20 dB better in the suppression of the side lobes. MSK, which exhibits a dual alternating frequency between the two consecutive bit periods, is considered to offer the best scheme as this offers orthogonal property of the two frequency modulation schemes of the carriers contained in the consecutive bits. Furthermore, it offers the most simplicity in the implementation of the modulation in the photonic domain.

This section thus investigates a number of novel structures of photonic transmitters and differential noncoherent balanced receivers for the generation and detection of MSK optical signals as follows:

1. Cascaded EO PMs MSK transmitter: This structure employs two cascaded optical PMs (OPM). The first OPM plays the role of modulating the binary data logic into two carrier frequencies deviating from the OC of the laser source by a quarter of the bit rate. The second OPM enforces the phase continuity of the lightwave carrier at every bit transition. The driving voltage of this second OPM is precoded in such as a way that the phase discrepancy due to frequency modulation of the first OPM will be compensated, hence preserving the phase continuity characteristics of MSK signal.

   Utilizing this double phase modulation configuration, different types of linear and nonlinear CPM phase shaping signals including MSK, weakly MSK and linear-sinusoidal MSK can be generated. The optical spectra of the modulation scheme obtained confirm the bandwidth efficiency of this novel optical MSK transmitter.
2. Parallel dual-drive MZMs optical MSK transmitter: In this second configuration, MSK signals with small ripple of approximately 5% signal amplitude level can be generated. The optical spectrum is demonstrated equivalent to the configuration of (12.1). This configuration can be implemented using commercially available dual-drive intensity interferometric EO modulators.

We also describe a simple noncoherent configuration for the detection of the MSK and nonlinear MSK-modulated sequences. The optical detection of MSK and nonlinear MSK signals employs a π/2 OPM followed by the MZ interferometric delay (MZIM) balanced receiver.

The following performance results are observed:

1. The modeling platform based on MATLAB and Simulink developed for the transmission systems.
2. Spectral characteristics of linear and nonlinear MSK-modulated signals.
3. Improvement on dispersion tolerance of MSK and nonlinear MSK over ASK and DPSK.

4. New techniques in the measurement of the BERs with the pdf of received signals after decision sampling to be computed with the superposition of a number of weighted Gaussian pdfs. This technique has been proven to be effective, convenient but accurate for estimating the non-Gaussian pdf for the case of CPM and its transmission over long-haul optically amplified systems.

The MSK format has been demonstrated via modeling as a promising candidate in the selection of advanced modulation formats for DWDM long-haul optical transmission system.

12.8.2 Generation of Optical MSK-Modulated Signals

The two optical MSK transmitter configurations can be referred to in more detail in f [5]. The descriptions of these configurations are briefly addressed earlier.

12.8.2.1 Optical MSK Transmitter Using Two Cascaded Electro-Optic Phase Modulators

EO PMs and interferometers operating at high frequency using resonant-type electrodes have been studied and proposed in [6,22]. In addition, high-speed electronic driving circuits evolved with the ASIC technology using 0.1 μm GaAs P-HEMT or InP HEMTs [23] enables the feasibility in the realization of the proposed optical MSK transmitter structure. The baseband equivalent optical MSK signal is represented in Equation 12.4.

The first E-OPM enables the frequency modulation of data logics into upper side bands (USB) and lower side bands (LSB) of the OC with a frequency deviation of f_d. Differential phase precoding is not necessary in this configuration due to the nature of the continuity of the differential phase trellis. By alternating the driving sources $V_d(t)$ to sinusoidal waveforms for simple implementation or combination of sinusoidal and periodic ramp signals, which was first proposed by Amoroso in 1976 [8], different schemes of linear and nonlinear phase shaping MSK-transmitted sequences can be generated.

The second E-OPM enforces the phase continuity of the light wave carrier at every bit transition. The delay control between the E-OPMs is usually implemented by the phase shifter shown in Figure 12.73. The driving voltage of the second E-OPM is precoded to fully compensate the transitional phase jump at the output $E_{01}(t)$ of the first E-OPM. Phase continuity characteristic of the optical MSK signals is determined by

$$\Phi(t,k) = \frac{\pi}{2}\left(\sum_{j=0}^{k-1} a_j - a_k I_k \sum_{j=0}^{k-1} I_j\right) \tag{12.23}$$

where

$a_k = \pm 1$ are the logic levels

$I_k = \pm 1$ is a clock pulse whose duty cycle is equal to the period of the driving signal $V_d(t)$

f_d is the frequency deviation from the OC frequency

$h = 2f_d T$ is previously defined as the frequency modulation index

In case of optical MSK, $h = 1/2$ or $f_d = 1/(4T)$. The phase evolution of the continuous-phase optical MSK signals are shown with detailed discussion in Figure 12.74. In order to mitigate the effects of unstable stages of rising and falling edges of the electronic circuits, the clock pulse $V_c(t)$ is offset with the driving voltages $V_d(t)$.

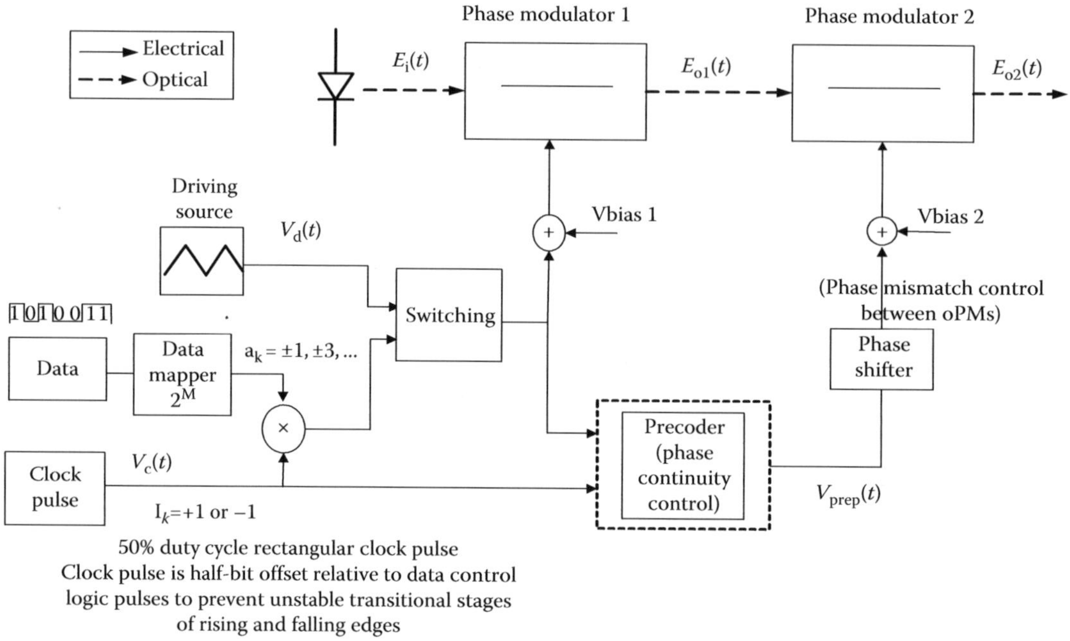

FIGURE 12.73
Block diagram of optical MSK transmitter employing two cascaded optical phase modulators.

12.8.2.2 Possibility to Generate Optical M-Ary CPFSK Format

$$\tilde{s}(t) = A\exp\left\{j\left[a_k I_k f_d \frac{2\pi t}{T} + \Phi(t,k)\right]\right\}, \quad kT \le t \le (k+1)T \tag{12.24}$$

where $a_k = \pm 1, \pm 3, \ldots$ (see Figure 12.75)

12.8.2.3 Detection of M-Ary CPFSK Modulated Optical Signal

The detection of linear and nonlinear optical M-ary CPFSK utilizes the well-known structure MZDI balanced receiver. The addition of $\pi/2$ phase on one arm of MZDI is also introduced. The time delay being a fraction of bit period enables the phase trellis detection of optical M-ary CPFSK. The detected phase trellis using the proposed technique is shown in Figure 12.76. An optimized ratio of switching frequencies results in the maximum eye open. The detection of optical M-ary CPFSK with a delay of $T_b/2$ at $t=(k+1)T_b/2$ is expressed in Equation 12.3.

$$\sin(\Delta\Phi) = \sin\left(a_{k+1}\frac{2\pi f_d (k+1)^{T_b}\!/_2}{T_b} - a_k \frac{2\pi f_d k^{T_b}\!/_2}{T_b}\right) \tag{12.25}$$

On the same slope of phase in the phase trellis, the differential phase, and hence the modulated frequency levels, can be mapped to detected amplitude levels via (12.4).

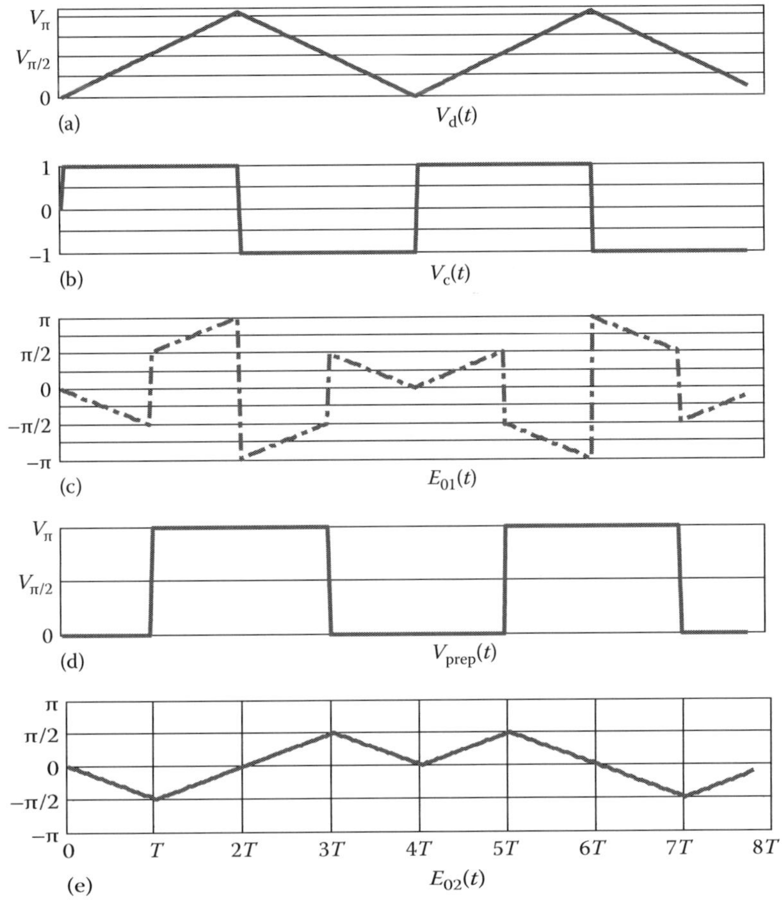

FIGURE 12.74
Evolution of time-domain phase trellis of optical MSK signal sequence [−1 1 1 −1 1 −1 1 1] as inputs and outputs at different stages of the optical MSK transmitter. The notation is denoted in Figure 12.80 accordingly: (a) $V_d(t)$: periodic triangular driving signal for optical MSK signals with a duty cycle of 4 bit period, (b) $V_c(t)$: the clock pulse with a duty cycle of 4*T*, (c) $E_{01}(t)$: phase output of oPM1, (d) $V_{prep}(t)$: precomputed phase compensation driving voltage of oPM2, and (e) $E_{02}(t)$: phase trellis of a transmitted optical MSK sequences at output of oPM2.

$$\sin(\Delta\Phi) = \sin(a_{k+1}\pi f_d) \tag{12.26}$$

12.8.2.4 Optical MSK Transmitter Using Parallel Mach–Zehnder Intensity Modulators (I–Q Approach)

The conceptual block diagram of an optical MSK transmitter is shown in Figure 12.77. The transmitter consists of two dual-drive EO MZM modulators generating chirpless I and Q components of MSK-modulated signals, which is considered as a special case of staggered or offset QPSK. The binary logic data is precoded and de-interleaved into even and odd bit streams that are interleaved with each other by one bit duration offset. Figure 12.77 shows (a) the general block diagram configuration of band-limited phase-shaped optical MSK and (b) the detailed structure of the optical modulator in which the I and Q components are generated with continuous-phase variation between the states.

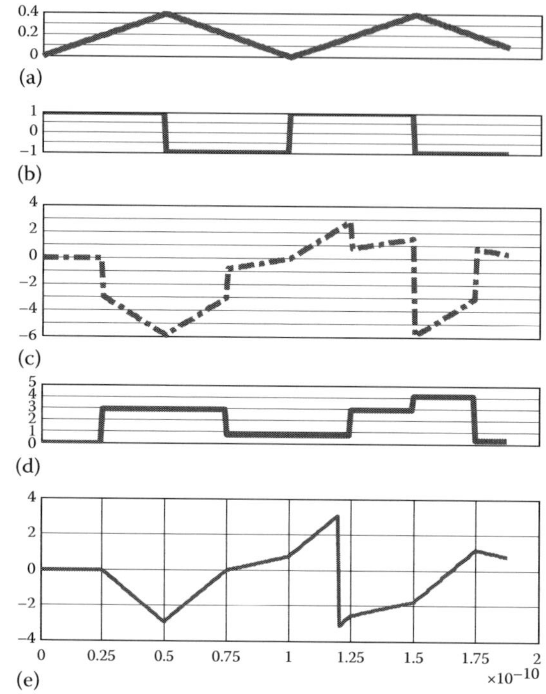

FIGURE 12.75
Demonstration of phase compensation for the enforcement of phase continuity at bit transitions: (a) shows the periodic triangular driving signal whose peak voltage is $V\pi/8$ and duty cycle of 4 bit period, (b) shows the clock pulse corresponding to the driving signal, (c) shows frequency switching with $h=\pm 1/8, \pm 7/8$, (d) shows the computed phase compensation, and (e) shows the phase trellis of an optical quaternary CPFSK.

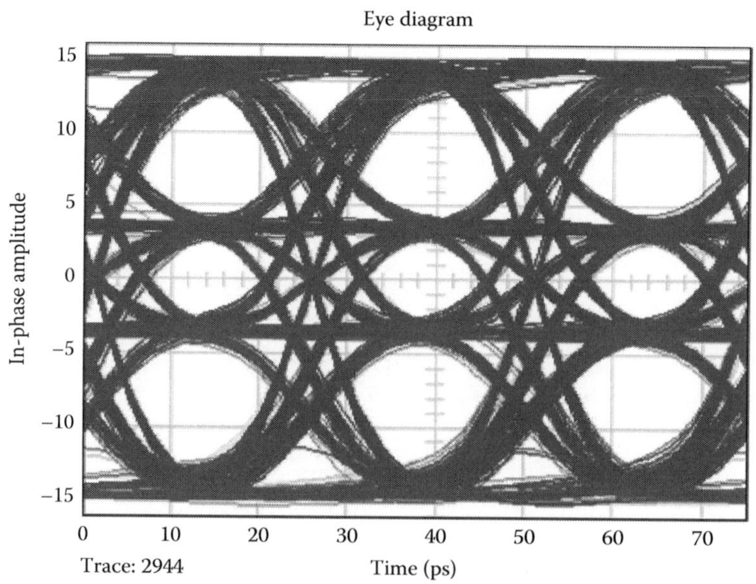

FIGURE 12.76
Eye phase trellis detection of optical M-ary CPFSK-modulated signal.

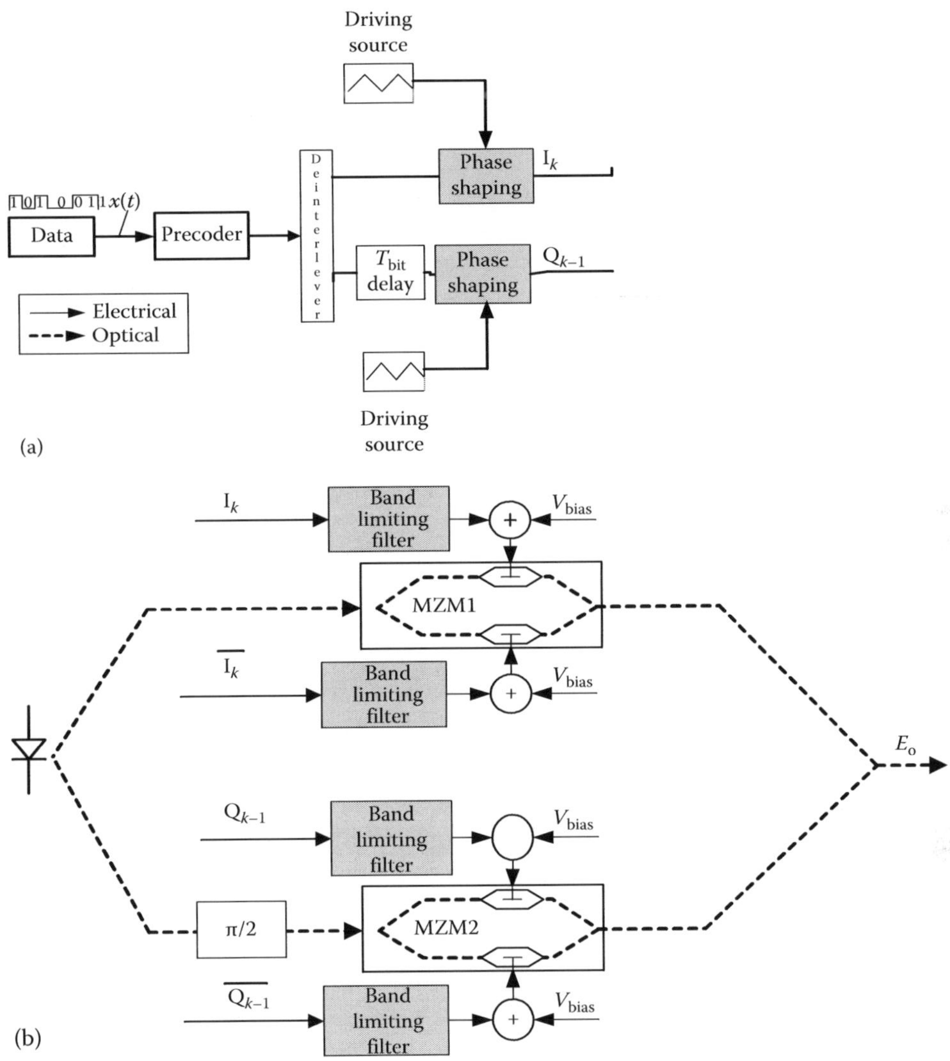

FIGURE 12.77
Block diagram configuration of band-limited phase-shaped optical MSK: (a) general structure and (b) details of the optical modulators.

Two arms of the dual-drive MZM modulator are biased at $V_{\pi/2}$ and $-V_{\pi/2}$ and driven with data and $\overline{\text{data}}$. Phase-shaping driving sources can be a periodic triangular voltage source in the case of linear MSK generation or simply a sinusoidal source for generating a nonlinear MSK-like signal, which also has linear phase trellis property but with small ripples introduced in the magnitude. The magnitude fluctuation level depends on the magnitude of the phase shaping driving source. High spectral efficiency can be achieved with tight filtering of the driving signals before modulating the EO MZMs. Three types of pulse-shaping filters are investigated including Gaussian, raised cosine, and squared-root raised cosine filters. The OC phase trellis of linear and nonlinear optical MSK signals are shown in Figure 12.78.

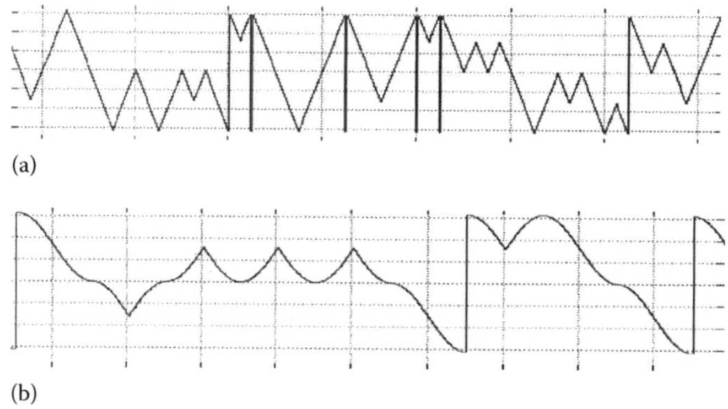

FIGURE 12.78
Phase trellis of linear (a) and nonlinear (b) MSK-transmitted signals.

The generation of linear and nonlinear optical MSK sequences can be briefly discussed in Sections 12.8.2.4.1 through 12.8.2.4.3.

12.8.2.4.1 *Linear MSK*

The pulse-shaping waveform for linear MSK is triangular waveform with a duty cycle of $4 \times T_b$. The triangular waveform for quadrature path is delayed by one bit period with respect to the in-phase path; hence they are interleaved with each other. The optical MSK signal is the superposition of both even and odd waveforms from the MZIMs. The amplitude of the signal is perfectly constant, clearly displaying the constant amplitude characteristics of CPM. The phase trellis is perfectly linear and the phase transition is continuous as shown in Figure 12.78a. The signal constellation is a perfect circle.

12.8.2.4.2 *Weakly Nonlinear MSK*

The pulse-shaping waveform for weakly nonlinear MSK is a sinusoidal waveform with an amplitude of $1/4V_\pi$ and its symbol period is equal to 2 bit period (symbol rate of 20Gb/s). The in-phase pulse shaper is a cosine waveform while the quadrature pulse shaper is a sine waveform. There is a ripple of approximately 5% signal amplitude level. Due to the sinusoidal pulse shaper, the variation of phase with time, which is represented by the phase trellis is nonlinear. Therefore, the rate of frequency change is not constant. This causes mismatch of MZIM when the modulated waveforms are added up, resulting in the ripple.

12.8.2.4.3 *Strongly Nonlinear MSK*

The pulse-shaping waveform for strongly nonlinear MSK is sinusoidal waveform, the same as for weakly nonlinear. However, the amplitude of pulse shaper is $1/2V_\pi$. The waveforms are interleaved with each other. The optical MSK signal has ripple of approximately 26%. This ripple is also caused by the mismatch of MZIM as the modulating waveform is strongly nonlinear. The effect of nonlinearity is obvious in the phase trellis in Figure 12.78b.

The signal state constellations and eye diagrams of optical MSK sequences are shown in Figure 12.79a through c for both the linear and nonlinear schemes, respectively. The magnitude ripple can be obviously observed in Figure 12.79. In the case of nonlinear configuration, MSK signals with a small ripple of approximately 5% signal amplitude level can be generated. This configuration can be implemented using commercially available dual-drive intensity interferometric EO modulators.

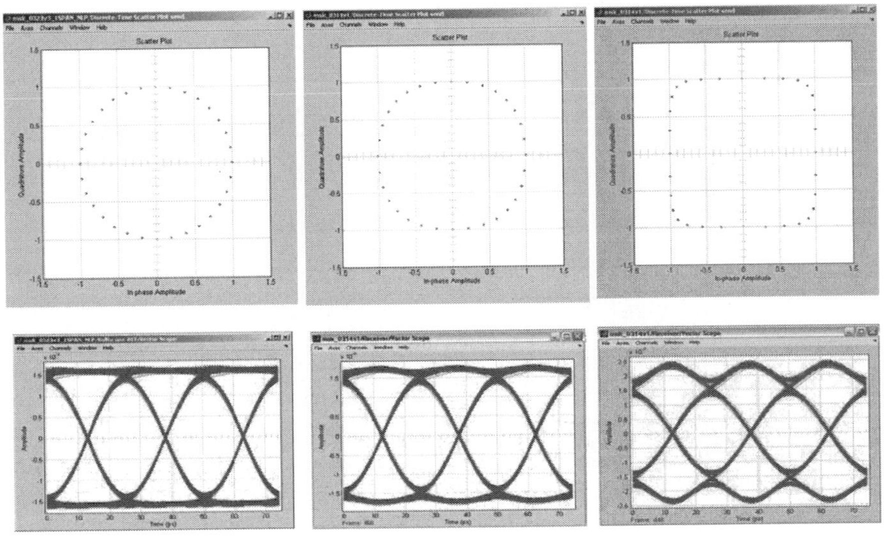

(a) (b) (c)

FIGURE 12.79
Constellation diagrams and eye diagrams of optical MSK-transmitted signals: (a) linear, (b) weakly nonlinear, and (c) strongly nonlinear transmission.

The conceptual block diagram of an optical MSK transmitter is shown in Figure 12.73. The transmitter consists of two dual-drive EO MZMs modulators generating chirpless I and Q components of MSK-modulated signals, which is considered as a special case of staggered or offset QPSK. The binary logic data is precoded and de-interleaved into even and odd bit streams that are interleaved with each other by one bit duration offset.

Two arms of the dual-drive MZM modulator are biased at $V_{\pi/2}$ and $-V_{\pi/2}$ and driven with data and $\overline{\text{data}}$. Phase-shaping driving sources can be a periodic triangular voltage source in the case of linear MSK generation or simply a sinusoidal source for generating a nonlinear MSK-like signal, which also has linear phase trellis property but with small ripples introduced in the magnitude. The magnitude fluctuation level depends on the magnitude of the phase shaping driving source. High spectral efficiency can be achieved with tight filtering of the driving signals before modulating the EO MZMs. Three types of pulse-shaping filters are investigated including Gaussian, raised cosine, and squared-root raised cosine filters. Narrow spectral width and high suppression of the side lobes can be achieved.

The logic gates of the precoder are constructed based on the state diagram. As seen from the state diagram, the current state of the signal is dependent on the previous state since the state of the signal advances corresponding to the binary data from the previous state. Therefore, memory is needed to store the previous state. The state diagram in Figure 12.80a is developed into a logic state diagram in Figure 12.80b to enable the construction of truth table. $S_0S_1=00$ or $S_0'S_1'=00$ corresponds to state 0, $S_0S_1=01$ or $S_0'S_1'=01$ corresponds to state $\pi/2$, $S_0S_1=10$ or $S_0'S_1'=10$ corresponds to state π, while $S_0S_1=11$ or $S_0'S_1'=11$ corresponds to state $-\pi/2$, with S_0S_1 as current state and $S_0'S_1'$ as previous state.

Two delay units in Figure 12.80b function as memory by delaying the current state and feedback into the precoding logic block as the previous state. The precoding logic block that consists of logic gates would compute the current state and output based on the feedback state (previous state) and binary data from the Bernoulli binary generator.

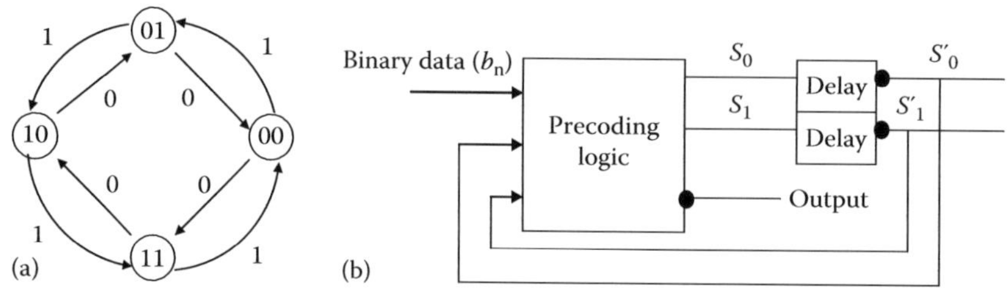

FIGURE 12.80
(a) State diagram for MSK. The arrows indicate continuous increment or decrement of the phase of the carrier and (b) combinational logic, the basis of the logic for constructing the truth table of the precoder.

The truth table is constructed based on the logic state diagram and the combinational logic diagram presented earlier. For positive half cycle cosine wave and positive half cycle sine wave, the output is 1; for negative half cycle cosine wave and negative half cycle sine wave, the output is 0. Then, Karnaugh maps are constructed to derive the logic gates within the precoding logic block, based on the truth table. The following three precoding logic equations are derived as follows:

$$S_0 = \overline{b_n}\,\overline{S_0'}\,\overline{S_1'} + b_n\overline{S_0'}\,S_1' + \overline{b_n}S_0'S_1' + b_nS_0'\overline{S_1}' \tag{12.27}$$

$$S_1 = \overline{S_1'} = \overline{b_n}\,\overline{S_1'} + b_n\overline{S_1}' \tag{12.28}$$

$$\text{Output} = \overline{S_0} \tag{12.29}$$

The final logic gates construction for the precoder is as shown in Table 12.4.

12.8.2.4.4 Optical MSK Receivers

The optical detection of MSK and nonlinear MSK signals employs a $\pi/2$ OPM followed by an optical phase comparator, an MZ interferometric delay (MZDI), and then a balanced receiver (BalRx). This receiver structure for the detection of optical MSK and nonlinear MSK signals is similar to the well-known structure employing on one arm of MZDI is introduced to detect $\pm\pi/2$ phase difference of the carrier imbedded in the two adjacent bits due to the nature of MSK modulation.

A new technique for the evaluation of the BERs is implemented. The pdf of noise-corrupted received signals after decision sampling are computed with the superposition of a number of weighted Gaussian pdfs. The technique implements the expected maximization (EM) theorem and has shown its effectiveness in determining arbitrary distributions [7,24].

TABLE 12.4
Truth Table Based on MSK State Diagram

$b_n\ S_0'\ S_1'$	$S_0\ S_1$	Output
100	01	1
001	00	1
010	01	1
101	10	0
110	11	0
111	00	1
000	11	0
011	10	0

12.8.3 Optical Binary-Amplitude MSK Format

12.8.3.1 Generation

The optical MSK transmitters from [9,10,25] can be integrated in the proposed generation of optical MAMSK signals. However, in this section, we propose a new simple-in-implementation optical MSK transmitter configuration employing high-speed cascaded EO PMs (E-OPMs) as shown in

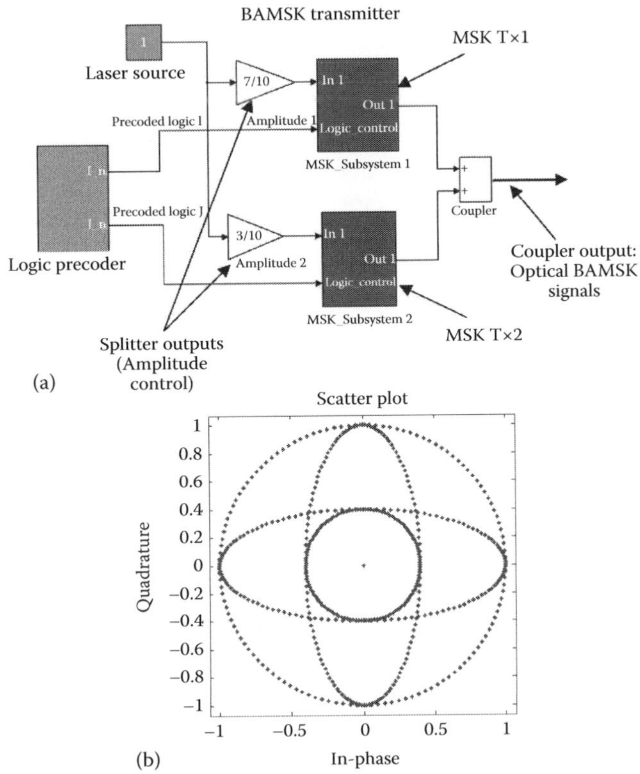

FIGURE 12.81
(a) Block diagram of optical BAMSK transmitter and (b) signal trajectories of optical BAMSK-transmitted signals.

Figure 12.81b. EO PMs and interferometers operating at high frequency using resonant-type electrodes have been studied and proposed in [6,22]. In addition, high-speed electronic driving circuits evolved with the ASIC technology using 0.1 μm GaAs P-HEMT or InP HEMTs [23] enable the feasibility in the realization of the proposed optical MSK transmitter structure. The baseband equivalent optical MSK signal is represented in (12.6).

$$\begin{aligned}\tilde{s}(t) &= A\exp\left\{j\left[a_k I_k 2\pi f_d t + \Phi(t,k)\right]\right\}, \quad kT \le t \le (k+1)T \\ &= A\exp\left\{j\left[a_k I_k \frac{\pi t}{2T} + \Phi(t,k)\right]\right\}\end{aligned} \tag{12.30}$$

where
- $a_k = \pm 1$ are the logic levels
- $I_k = \pm 1$ is a clock pulse whose duty cycle is equal to the period of the driving signal $V_d(t)$
- f_d is the frequency deviation from the OC frequency
- $h = 2f_d T$ is defined in Equations 12.2 and 12.3 as the frequency modulation index

In case of optical MSK, $h = 1/2$ or $f_d = 1/(4T)$.

The first E-OPM enables the frequency modulation of data logics into USB and LSB of the OC with frequency deviation of f_d. Differential phase precoding is not necessary in this configuration due to the nature of the continuity of the differential phase trellis. By alternating the driving sources $V_d(t)$ to sinusoidal waveforms or a combination of sinusoidal and periodic ramp signals, different schemes of linear and nonlinear phase shaping MSK-transmitted sequences can be generated [11]. The second E-OPM enforces the phase continuity of the light wave carrier at every bit transition. The delay control between the E-OPMs is usually implemented by the phase shifter shown in Figure 12.81.

The driving voltage of the second E-OPM is precoded to fully compensate the transitional phase jump at the output $E_{01}(t)$ of the first E-OPM. Phase continuity characteristics of the optical MSK signals is determined by the algorithm in Equation 12.8. In order to mitigate the effects of unstable stages of rising and falling edges of the electronic circuits, the clock pulse $V_c(t)$ is offset with the driving voltages $V_d(t)$.

$$\Phi(t,k)=\frac{\pi}{2}\left(\sum_{j=0}^{k-1}a_j-a_kI_k\sum_{j=0}^{k-1}I_j\right) \tag{12.31}$$

Binary-amplitude MSK (BAMSK) modulation format is proposed for optical communications. This section describes numerical results of 80 Gb/s 2-bit-per-symbol BAMSK optical system on spectral characteristics and residual dispersion tolerance to different types of fibers. The BER of 1e–23 is obtained for 80 Gb/s optical BAMSK transmission over 900 km Vascade-fiber systems enabling the feasibility of long-haul transmission for the modulation format.

MSK, which is well-known in radio frequency and wireless communications, has recently been adapted into optical communications. A few optical MSK transmitter configurations have recently been reported [11,26]. For optically amplified communications systems, if multilevel concepts can be incorporated in those reported schemes, the symbol rate would be reduced and hence the bandwidth efficiency can be achieved. This is the principal motivation for the proposed modulation scheme.

BAMSK is a special case of M-ary CPM format that enables binary-level (PAM- or QAM-like) transmission scheme while the bandwidth efficiency due to transitional phase continuity properties between two consecutive symbols (CPM-like signals) are preserved. The generation of M-ary CPM sequences can be expressed in Equations 12.9 through 12.11 [12]:

$$s(t)=A_n\cos\left(\omega_c t+\phi_n(t,a)\right)+\sum_{m=1}^{N-1}B_m\cos\left(\omega_c t+\phi_m(t,b_m)\right) \tag{12.32}$$

where

$$\phi_n(t,a)=\pi h a_n q(t-nT)+\pi h\sum_{k=-\infty}^{n-1}a_k \quad nT\le t\le(n+1)T \tag{12.33}$$

$$\phi_m(t,b_m)=\pi a_n\left(h+\frac{b_{mn}+1}{2}\right)q(t-nT)+\pi\sum_{k=-\infty}^{n-1}a_k\left(h+\frac{b_{mk}+1}{2}\right) \quad mT\le t\le(m+1)T \tag{12.34}$$

In a generalized M-ary CPM transmitter, values of a_n and b_{mn} are statistically independent and taken from the set of {±1, ±3, …}. A_n and B_m are the signal state amplitude levels, which are either in phase or in π phase shift with the largest level component at the end of nth symbol interval; $q(t)$ is the pulse-shaping function and h is the frequency modulation index. In the case of BAMSK, these equations that show the constraints of ϕ_m to maintain the phase continuity characteristics of CPM sequences can be simplified where $h=1/2$ and the phase shaping function $q(t-nT)$ is a periodic ramp signal with a duty cycle of $4T$.

$$\phi_n(t,a)=\pi h a_n\frac{t-nT}{T}+\pi h\sum_{k=-\infty}^{n-1}a_k \quad nT\le t\le(n+1)T \tag{12.35}$$

$$\phi_m(t,b_m)=\pi I_n\left(h+\frac{b_{mn}+1}{2}\right)\frac{t-nT}{T}+\pi\sum_{k=-\infty}^{n-1}b_k\left(h+\frac{b_{mk}+1}{2}\right) \quad mT\le t\le(m+1)T \tag{12.36}$$

Any configuration of the reported optical MSK transmitters in [9,10,26] can be utilized in the proposed generation scheme of optical BAMSK signals. Figure 12.81a shows the block structure of the optical BAMSK transmitter in which two optical MSK transmitters are integrated in parallel configuration. The amplitude levels can be determined from these simplified equations by the splitting ratio at the output of a high-precision power splitter. The logic sequences {±1,…} of a_n and b_n are precoded from the binary logic {0, 1} of d_n as $a_n=d_n-1$ and $b_n=a_n\ (1-d_n-1/h)$ [12]. The signal-space trajectories of BAMSK signals are shown in Figure 12.80b.

A simple noncoherent configuration for the detection of the optical BAMSK sequences consists of phase and amplitude detections. In the case of BAMSK, that is, $n=2$, the system effectively implements 2 bits per symbol with two amplitude levels. Phase detection is enabled with the employment of the well-known integrated optic phase comparator MZDI balanced receiver with one-bit time delay on one arm of the MZDI. An additional π/2 phase shift is introduced. Figure 12.82a and b show the eye diagrams of the amplitudes and phases of the optical BAMSK signals, respectively. In phase detection, the decision

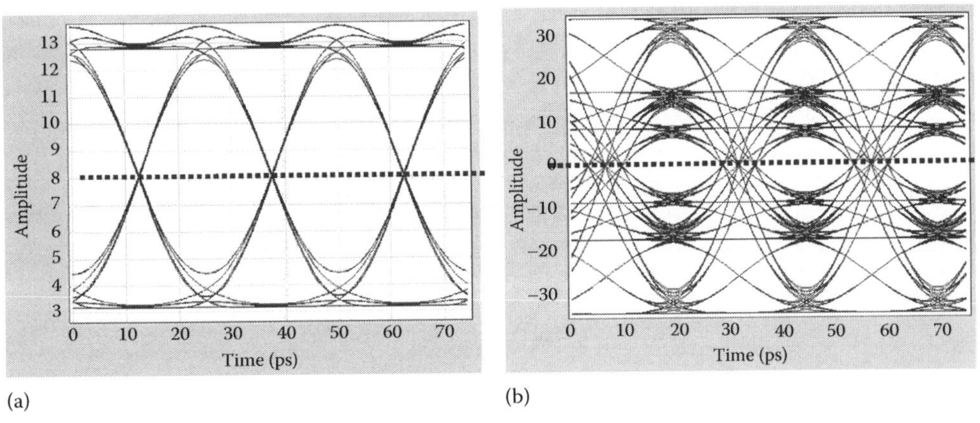

FIGURE 12.82
Eye diagrams of (a) amplitude and (b) phase detection of optical B-AMSK received signals with normalized amplitude ratio of 0.285/0.715. The decision threshold is plotted in broken-line style.

threshold that is plotted in broken-line style is at zero level whereas amplitude levels are determined by different thresholds. A new technique in the calculation of BER for dispersive and noise-corrupted received signals, which exploits the EM theorem, is implemented with the superposition of a number of weighted Gaussian probability distribution functions.

12.8.3.2 Detection

A simple noncoherent configuration for the detection of a linear and nonlinear optical M-ary MSK sequences consists of phase and amplitude detections, which are very well known in the discrete PSK schemes such as DPSK or quadrature DPSK [27]. Phase detection is enabled with the employment of the well-known integrated optic phase comparator MZDI balanced receiver with one-bit time delay on one arm of MZDI. An additional $\pi/2$ phase shift is introduced to detect the differential $\pi/2$ phase shift difference of two adjacent optical MSK pulses. Figure 12.82a and b show the eye diagrams of the amplitudes and phases of the optical B-AMSK detected signals, respectively. In the case of $N=2$ and $N=3$, the system effectively implements 2-bits-per-symbol scheme and 3 bits-per-symbol scheme with two and four amplitude levels, respectively. In phase detection, the decision threshold that is plotted in broken-line style is at zero level because only in-phase and π differential phase are of interests.

12.8.4 Typical Simulation Results

12.8.4.1 Transmission Performance of Linear and Nonlinear Optical MSK Systems

The block diagram of a simulation setup is shown in Figure 12.83. The dispersion tolerances of linear, weakly nonlinear, and strongly nonlinear optical MSK signals are numerically investigated and the results are shown in Figure 12.83. Among the three types, linear MSK is most tolerant to residual dispersion with 1 dB eye open penalty at 72 ps/nm/km. Strongly nonlinear MSK suffers a severe penalty when residual dispersion exceeds 85 ps/nm/km or equivalently 5 km SSMF.

Figure 12.84 shows the typical performance of three types of optical MSK-modulated signals versus OSNR in transmission over 540 km Vascade fibers of optically amplified

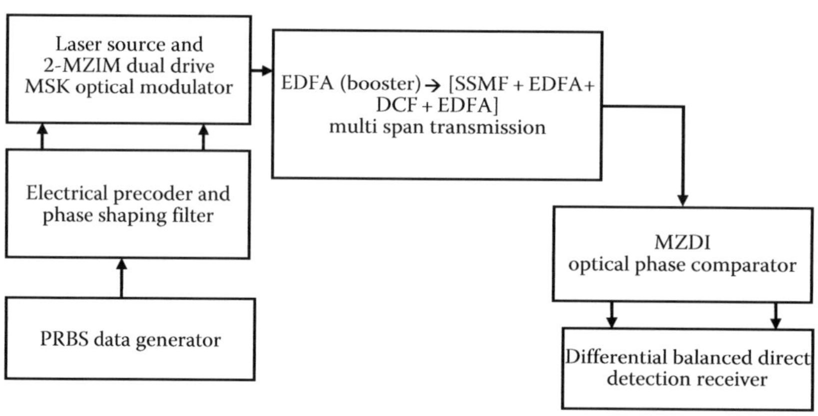

FIGURE 12.83
Schematic diagram of an optically amplified optical transmission system.

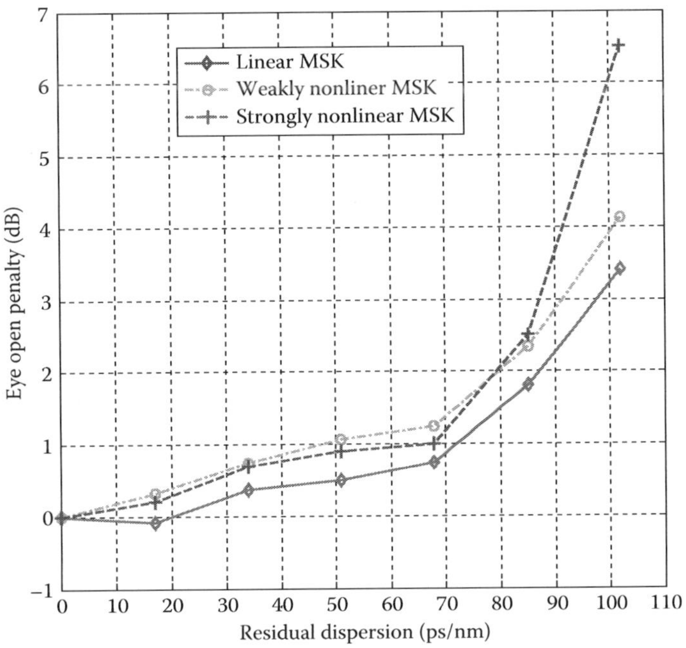

FIGURE 12.84
Dispersion tolerance of 40 Gb/s linear MSK, weakly nonlinear MSK, and strongly nonlinear MSK optical signals.

and fully compensated multi-span links (6 spans × 90 km/span). The receiver sensitivity of the differential phase comparison balanced receiver is −24.6 dBm. In Vascade fibers, the dispersion factors of the DCF is negatively opposite to that of the transmission fiber, a standard SMF, of +17.5 ps/nm/km at 1550 nm wavelength. In addition, the dispersion slopes of these fibers are also matched. Optical amplifiers of EDFA types are placed one at the end of the transmission fiber and another after the DCF so that it would boost the optical power to the right level, which is equal to that of the launched power. The EDFA optical gain of 19 and 5 dB ASE noise figure is used. The noise margin reduces severely after the propagation over 6 × 90 km spans. Effects of positive and negative dispersion mismatch and mid-link nonlinearity on phase evolution are shown in Figure 12.85.

The tolerance of these MSK modulation using transmission models to nonlinear effects is also studied and simulation results are shown in Figure 12.86. The input power into the fiber span is kept increasing whereas the length of transmission link is constant at 180 km. At BER = 1e−9, linear MSK could tolerate an increase of the input launched power up to 10.5 dBm, weakly nonlinear could tolerate up to 10.2 dBm compared to 9.2 dBm in the case of strongly nonlinear MSK. The nonlinear phase shift is proportional to the input power; therefore, increasing the input power would increase nonlinear phase shift as well. This nonlinear phase shift is observed through the asymmetries in the eye diagram and through the scatter plot, which shows that the phases of the in-phase and quadrature components have shifted from the x and y axes. The maximum phase shift that could be tolerated is approximately 15°.

Although increasing input power increases the noise margin of the eye diagram, it is paid off by the large distortion at sampling time causing the SNR to decrease. Typical eye diagrams for full compensation and after transmission over 540 km Vascade fibers

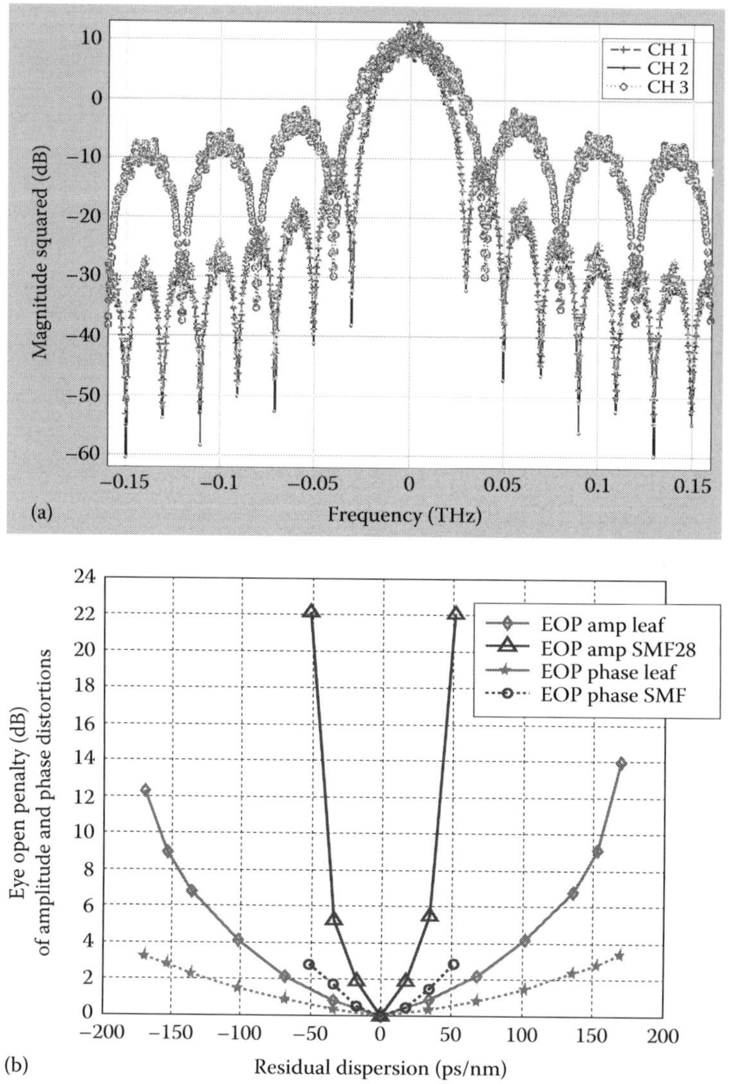

FIGURE 12.85
(a) Comparison of spectra of 80 Gb/s optical BAMSK, 40 Gb/s optical MSK, and 40 Gb/s optical binary DPSK signals; and (b) numerical results on residual dispersion tolerance of 80 Gb/s optical BAMSK systems (effectively 40 Gb/s symbol rate) with normalized amplitude ratio of 0.285/0.715 in both amplitude and phase detection.

of optically amplified and fully compensated multi-span links (6 spans × 90 km/span) are shown in Figures 12.87 and 12.88. We note that if the sampling is conducted at the center of the bit period then the error is at its minimum as the ripples of the eyes are fallen on this position. This is the principal reason why the MSK signals can suffer minimum pulse spreading due to residual and nonlinear phase dispersion.

Linear MSK and weakly and strongly nonlinear MSK phase shaping functions are investigated. It has been proven that optical MSK is a very efficient modulation that offers excellent performance. With OSNR of about 17 dB, BER is obtained to be 1e−9 and reaches 1e−17 for an optical SNR of 19 dB under linear MSK modulation. The modulation formats of linear and nonlinear phase shaping MSK is also highly resilient to

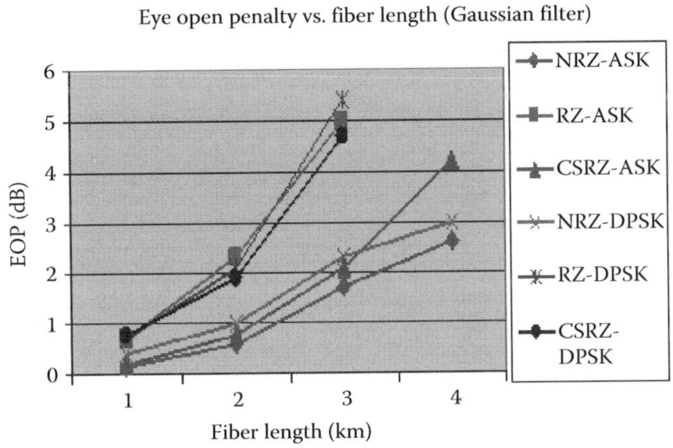

FIGURE 12.86
Simulation of EOP versus transmission distance 1–4 km of SSMF.

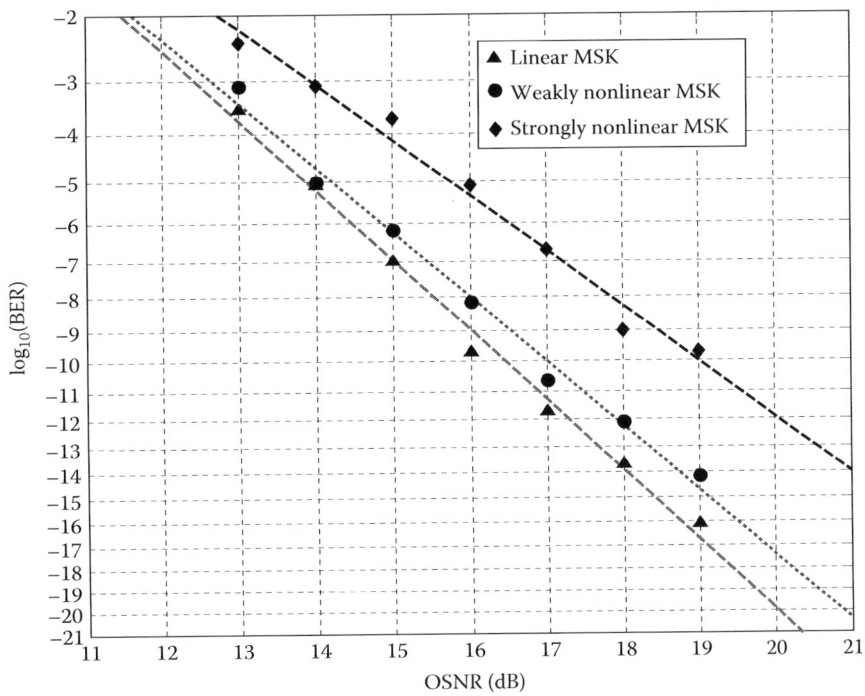

FIGURE 12.87
BER versus OSNR for transmission of three types of modulated optical MSK signals over 540 km Vascade fibers of optically amplified and fully compensated multi-span links (6 spans × 90 km/span).

nonlinear effects. Nonlinear distortion appears when the total average power reaches about 9 dBm, that is, about 3–4 dB above that of NRZ ASK format over a 50 μm diameter SSMF fiber.

The nonlinear phase shaping filters offer better implementation structures in the electronic domain for driving the dual-drive MZIMs than the linear type but suffer some

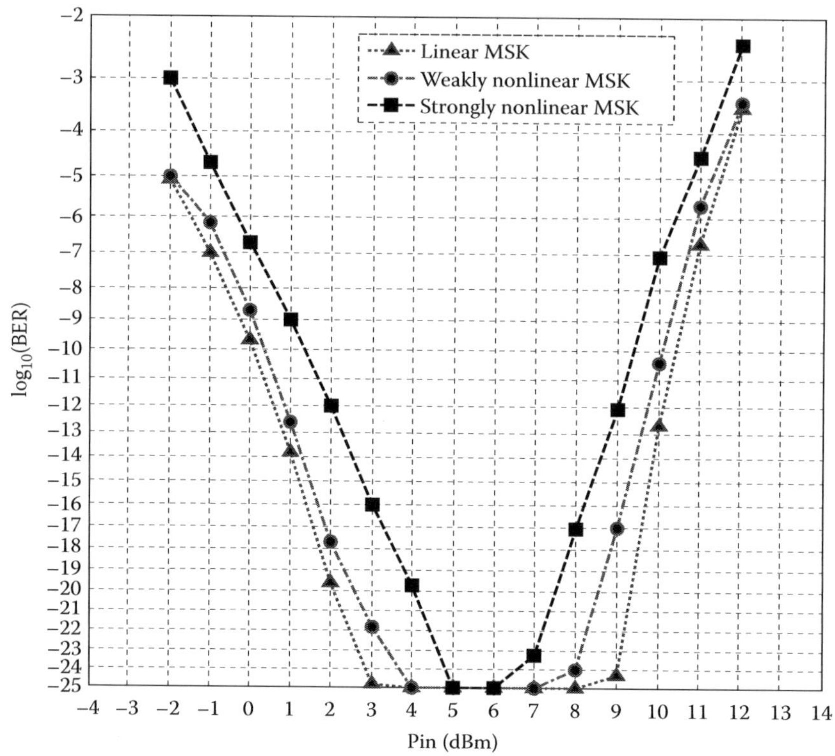

FIGURE 12.88
BER versus input power showing robustness to nonlinearity of linear, weakly nonlinear, and strongly nonlinear optical MSK signals with transmission over 180 km Vascade fibers of two optically amplified and fully compensated span links.

power penalty, still better than those candidates of other amplitude, phase or differential phase modulation formats. Weakly nonlinear MSK offers much lower power penalty and ease of implementation for precoders and phase shaping filters; thus it would be the preferred MSK format for long-haul transmission over optically amplified multi-span systems. At BER = 1e−12, linear MSK is 0.3 and 1.2 dB more resilient to nonlinear phase effects than weakly and strongly nonlinear MSK, respectively.

The transmission performance of 80 Gb/s optical BAMSK over 900 km Vascade fiber systems in two cases of normalized amplitude ratios: 0.25/0.75 (round markers and solid line) and 0.285/0.715 (diamond markers and dashed line), respectively, is shown in Figure 12.89.

12.9 Star-QAM Transmission Systems for 100 Gb/s Capacity

This section presents the simulation of the transmission performances of optical transmission systems over 10 spans of dispersion-compensated and optically amplified fiber transmission systems with a bit rate of 100 Gb/s, thus an effective symbol rate of 25 GSymbols/s, and two amplitude levels and eight phase states per amplitude level.

The modulation format employs the Star 16-QAM with two-level and eight-phase state constellations. Different QAM structures are given to determine the dual-level

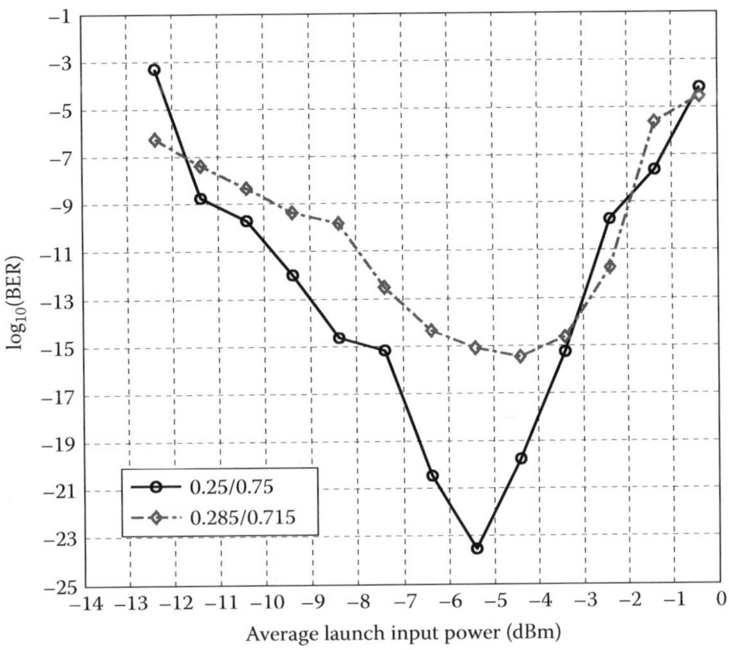

FIGURE 12.89
Transmission performance of 80 Gb/s optical BAMSK over 900 km Vascade fiber systems in two cases of normalized amplitude ratios: 0.25/0.75 (round markers and solid line) and 0.285/0.715 (diamond markers and dashed line), respectively.

and eight-PSK rings. Optical transmitters and coherent receivers are proposed, with the transmitter using only two MZIM optical modulators. The transmitted lightwave signals are coherently demodulated and detected. Simulation results have shown that it is possible to transmit and detect the data symbols for 107 Gb/s without much difficulty. The OSNR with 0.1 nm optical filters is achieved with 18 and 23 dB for back-to-back and long-haul transmission cases with a dispersion tolerance of 300 ps/nm. The multilevel modulation scheme has eight phases and two amplitude levels at a bit rate of 100 Gb/s and effectively utilizes four bits per one symbol for transmission, in which the first three bits are for coding phase information while the coding of the amplitude levels is implemented with the fourth bit. As a result, the transmission baud rate is equivalently a quarter of the bit rate from bit pattern generator. The detection of this scheme is pseudo-coherent detection in which the amplitude detection is direct detected, and the phase detection is conducted via an M-ary PSK coherent demodulator that estimates the phase state of the received symbol. Monte Carlo simulation is performed and a receiver sensitivity of −15 dBm is achieved for a BER of 10^{-5} after 1100 km of dispersion-compensated and optically amplified transmission 100 km fiber spans.

12.9.1 Introductory Remarks

To increase the channel capacity and bandwidth efficiency in optical transmission, the multilevel modulation formats like QAM formats are of interest. Especially when 100 Gb/s is expected to be the basic transmission rate for 10 Gb/s Ethernet in the near future. In digital transmission with multilevel (M-levels) modulation, *m* bits are collected and mapped onto a complex symbol from an alphabet with $M=2m$ possibilities at the transmitter side [28–32].

The symbol duration is $T_s=m\cdot T_B$ with T_B as the bit duration and the symbol rate is $f_s=f_B/m$ with $f_B=1/T_B$ as the bit rate. This shows that for a given bit rate the symbol rate decreases, if the modulation level increases. That means higher bandwidth efficiency can be achieved by a higher-order modulation format. For 16-QAM format, $m=4$ bits are collected and mapped to one symbol from an alphabet with $M=16$ possibilities. In comparison to the case of binary modulation format, there is only $m=1$ bit mapped to one symbol from an alphabet with $M=2$ possibilities. With 16-QAM format and a data source with a bit rate of $f_B=40$ Gb/s, only a symbol rate of $f_s=10$ Gbaud/s is necessary. In a commercial point of view, it means that a 40 Gb/s data rate can be transmitted with 10 Gb/s transmission devices. In the case of binary transmission, the transmitter needs a symbol rate of $f_s=40$ Gbaud/s. It means 16-QAM transmission requires four times slower transmission devices than that for the binary transmission. It is noted here that a 10.7 GSymbols/s is used as the symbol rate so as to compare the simulation results with the well-known 10.7 Gb/s modulation schemes such as DPSK and CSRZ DPSK. For a 10.7 Gb/s bit rate, the transmission performance, that is, the sensitivity and OSNR can be scaled accordingly without any difficulty.

This section gives a general approach regarding the design and simulation of Star 16-QAM using coherent and pseudo-coherent detection techniques. Section 12.9.2 gives the design of the QAM and Star-QAM with two rings and eight-phase per ring noninterleaved to determine an optimum ring ratio. The simulation of the transmission performances of a modulation Star-QAM scheme is demonstrated for the effectiveness of 100 Gb/s bit rate. The transmission format is discussed with theoretical estimates and simulation results to determine the transmission performance. The optimum Euclidean distance is defined for the design of Star 16-QAM and further analytically determined and studied with an optimum ratio of the amplitudes of the two rings of the 16-Star-QAM. Then in Section 12.9.2.1 the coherent and pseudo-coherent detection schemes for Star-QAM constellations are demonstrated. Alternatively, Section 12.9.2.2 investigates eight phase states and two amplitude levels denoted as 8-DPSK_2-ASK, which effectively utilizes four bits per symbol for transmission, in which the first three bits are for coding phase information while the coding of the amplitude levels is implemented with the fourth bit. Only one MZIM is used.

12.9.2 Design of 16-QAM Signal Constellation

There are many possibilities to design 16-QAM signal constellation. Three most popular constellations can be introduced. These 16-QAM modulation schemes are (1) Star 16-QAM; (2) square 16-QAM; and (3) shifted-square 16-QAM. The first two of these constellations are implemented. However, only the Star 16-QAM with two amplitude levels and eight phase states per amplitude level are employed as the preferred constellation in this book.

12.9.2.1 Star 16-QAM

12.9.2.1.1 Signal Constellation

The signal constellation for Star 16-QAM with gray coding is shown in Figure 12.90. As can be seen from the figure, the symbols are evenly distributed on two rings and the phase difference between the neighboring symbols on the same ring are equal ($\pi/4$). In order to detect a received symbol, its phase and amplitude must be determined. In

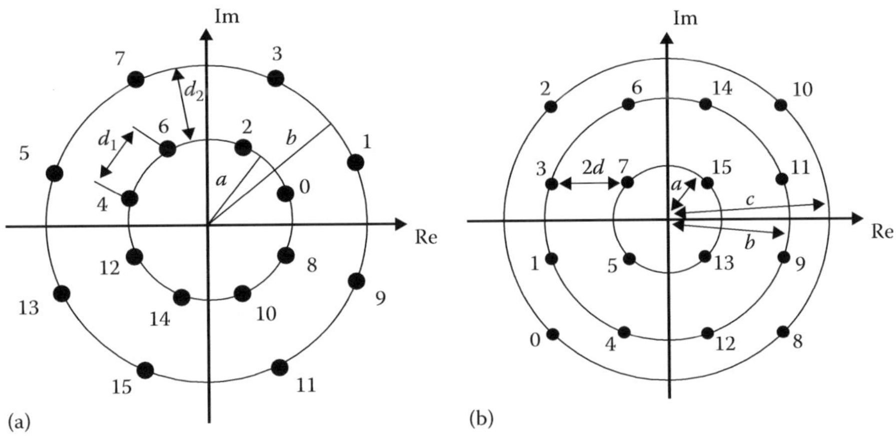

FIGURE 12.90
Theoretical arrangement of the modulation constellation for (a) Star 16-QAM and (b) square QAM.

other words, between two amplitude levels of the rings and among eight phase possibilities, there are a number of ways to form this constellation.

The ring ratio (RR) for this constellation is defined as RR = b/a where a and b are the ring radii as shown in Figure 12.90. The *RR* can be set to different values to optimize the transmission performance.

12.9.2.1.2 Optimum Ring Ratio for Star Constellation

It can be seen from Figure 12.90 that there are many possibilities to choose the RR for the Star 16-QAM constellation. Here the theoretical best RR is defined to minimize the error probability in an AWGN channel by maximizing the minimum distance d_{min} between the neighboring symbols. The results for AWGN channel can be used approximately for optically amplified fiber transmission systems. For Star 16-QAM, the minimum distance d_{min} is maximized, when

$$d_1 = d_2 = b - a = d_{min} \tag{12.37}$$

with some geometrical manipulations it can be shown that

$$d_{min} = 2a\sin\frac{\pi}{8} \tag{12.38}$$

which leads to an optimal ring ratio RR_{opt} of

$$RR_{opt} = \frac{b}{a} = \frac{d_{min} + a}{a} = \frac{2a\sin\frac{\pi}{8} + a}{a} = 1.77 \tag{12.39}$$

Under the condition of no noises, the average power of the Star 16-QAM constellation can then be determined as follows:

$$P_0 = \frac{8a^2 + 8b^2}{16} = \frac{a^2 + b^2}{2} \tag{12.40}$$

Thus we have the relationship between the average optical power and the minimum distance between the two rings of the two amplitude levels as follows:

$$d_{\min} = 0.53 P_0^{1/2} \tag{12.41}$$

The obtained $RR_{opt} = 1.77$ does not depends on P_0 and is constant for each P_0 value. For an average power of 5 dBm (3.16 mW), hence $d_{min} = 2.98 \times 10^{-2}\sqrt{W}$, $a = 3.89 \times 10^{-2}\sqrt{W}$, and $b = 6.87 \times 10^{-2}\sqrt{W}$ are obtained.

12.9.2.1.3 Square 16-QAM

The signal constellation of the square 16-QAM with gray coding is shown in Figure 12.90b. In the constellation of the square 16-QAM, the 16 symbols have equal distance with direct neighbors and totally 12 different phases, that is, three phases per quarter, distributed on three rings. The phase differences between neighboring symbols on the inner- and outer rings are equal (π/2), but the phase differences between neighboring symbols on the middle ring are different (37°or 53°). If the distance between direct neighbors in the square 16-QAM is noted as $2d$, the average symbol power (P_0) of the constellation is

$$P_0 = 10 d_2. \tag{12.42}$$

For an average power of 5 dBm (3.16 mW), $d = 1.77 \times 10^{-2}\sqrt{W}$ and $a = 2.5 \times 10^{-2}\sqrt{W}$, $b = 5.6 \times 10^{-2}\sqrt{W}$, and $c = 7.5 \times 10^{-2}\sqrt{W}$. In comparison with square 16-QAM, this case gives much smaller distances between the middle ring and the outer ring. It means that, to achieve the same BER, square 16-QAM needs a higher average power than Star 16-QAM. The decision method for the square 16-QAM is more complicated than that for Star 16-QAM.

12.9.2.1.4 Offset-Square 16-QAM

To optimize the phase detection of the middle ring, it is envisaged that the phase differences between neighboring symbols on the middle ring in square 16-QAM should be equal. Thus, the shifted-square 16-QAM is introduced by shifting (rotation) of symbols on the middle ring to obtain equal phase differences between all neighboring symbols. After shifting the symbols on the middle ring, the distances between all direct neighbors are not necessarily equal. In comparison to square 16-QAM, this constellation may offer a more robust detection against phase distortions according to our amplitude and phase detection method introduced in a later section.

12.9.2.2 8-DPSK_2-ASK 16-Star-QAM

We have examined the Star-QAM and the determination of the ring ratio between the inner and outer rings of the constellation. This part investigates a multilevel modulation scheme of Figure 12.90a, which has eight phases per ring and two amplitude levels, non-interleaved phase angles between the rings. This scheme, which is named in short as 8-DPSK_2-ASK, effectively utilizes 4 bits per one symbol for transmission, in which the first three bits are for coding phase information while the coding of the amplitude levels is implemented with the fourth bit. As a result, the transmission baud rate is equivalently a quarter of the bit rate from a bit pattern generator. This section is organized as follows: Section 12.9.2.2.1 presents a detailed description of the optical transmitter for generating

8-DPSK_2-ASK signals. In Section 12.9.2.2.2, the detailed configuration of the receiver is provided. The configuration of the optical transmitter and receiver is referred from those reported by Hoshida et al. [1]. Section 12.9.2.2.3 provides a study on dispersion tolerance and transmission performance of the 8-DPSK_2-ASK scheme.

12.9.2.2.1 Configuration of 8-DPSK_2-ASK Optical Transmitter

There have been several different configurations of an optical transmitter for generating multiphase/level optical signals with the use of amplitude or PMs arranged in either serial or parallel configurations [1–4,21]. However, the optical transmitters reported in [2–4,21] require a precoder with high complexity. On the contrary, the configuration reported in Ivan et al. [1] utilizes the gray mapping technique to differentially encode the phase information, and this significantly reduces the complexity of the optical transmitter. In addition, as elaborated in more detail in Section 12.9.2.2.2, this precoding technique enables the detection scheme using the I–Q demodulation techniques, as equivalently in coherent transmission systems.

The optical transmitter of the 8-DPSK_2-ASK scheme employs the differential coding and I–Q modulation technique with two MZIM in parallel and a $\pi/2$ optical PM, as shown in Figure 12.91.

At each kth instance, the absolute phase of transmitted lightwaves θ_k is expressed as $\theta_k = \theta_{k-1} + \Delta\theta_k$ where θ_{k-1} is the phase at $(k-1)$th instance and $\Delta\theta_k$ is the differentially coded phase information. The encoding of this $\Delta\theta_k$ for generating 8-DPSK_2-ASK modulated optical signals (4 bits per one transmitted symbol) follows the well-known gray mapping rules. This gray mapping phasor diagram is shown in Figure 12.92. The phasor can be normalized with the maximum energy on each branch, that is, $E_1/2$.

The amplitude levels are optimized in order that the Euclidean distances d_1,d_2, and d_3 are equal, that is, $d_1 = d_2 = d_3$. After derivation, we obtain $r_1 = 0.5664$. The I and Q field vector corresponding to gray mapping rules from the M-ADPSK precoder (Figure 12.92) are provided in Table 12.5.

The above-described transmitter configuration can be replaced with a dual-drive MZIM. The explanation and derivation for generating 8-DPSK_2-ASK optical signals are also based on the phasor diagram of Figure 12.92. In this case, the output field vector is the summation of two component field vectors, each of which is determined not only by the amplitude but also by initially biased phases.

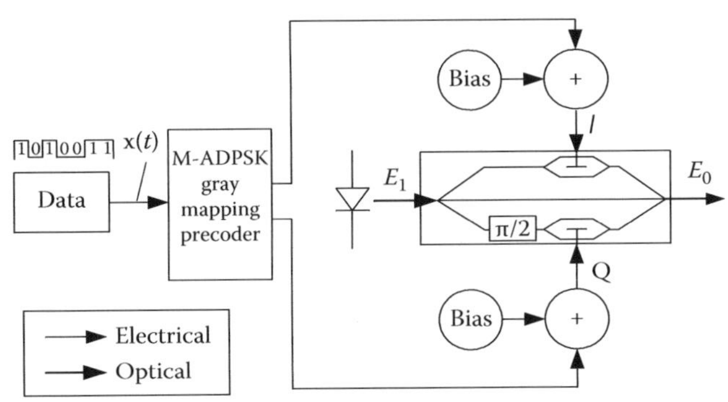

FIGURE 12.91
Optical transmitter configuration of the 8-DPSK_2-ASK modulation scheme.

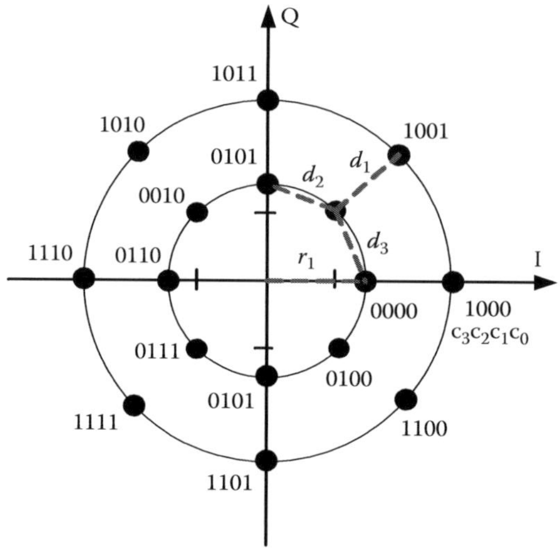

FIGURE 12.92
Gray mapping for optimal 8-DPSK_2-ASK modulation scheme.

TABLE 12.5

I and Q Field Vectors in 8-DPSK_2-ASK Modulation Scheme Using Two MZIMs in Parallel

Binary Sequence	($\Delta\theta_k$, Amplitude)	I_k	Q_k
1000	(0, 1)	1	0
1001	($\pi/4$, 1)	$\sqrt{2}/2$	$\sqrt{2}/2$
1011	($\pi/2$, 1)	0	1
1010	($3\pi/4$, 1)	$\sqrt{2}/2$	$\sqrt{2}/2$
1110	(π, 1)	−1	0
1111	($-3\pi/4$, 1)	$\sqrt{2}/2$	$\sqrt{2}/2$
1101	($-\pi/2$, 1)	0	−1
1100	($-\pi/4$, 1)	$\sqrt{2}/2$	$\sqrt{2}/2$
0000	(0, 0.5664)	1*0.5664	0
0001	($\pi/4$, 0.5664)	$\sqrt{2}/2$ *0.5664	$\sqrt{2}/2$ *0.5664
0011	($\pi/2$, 0.5664)	0	1*0.5664
0010	($3\pi/4$, 0.5664)	$\sqrt{2}/2$ *0.5664	$\sqrt{2}/2$ *0.5664
0110	(π, 0.5664)	−1*0.5664	0
0111	($-3\pi/4$, 0.5664)	$\sqrt{2}/2$ *0.5664	$\sqrt{2}/2$ *0.5664
0101	($-\pi/2$, 0.5664)	0	−1*0.5664
0100	($-\pi/4$, 0.5664)	$\sqrt{2}/2$ *0.5664	$\sqrt{2}/2$ *0.5664

12.9.2.2.2 Configuration of 8-DPSK_2-ASK Detection Scheme

The detection of 8-DPSK_2-ASK optical signals is implemented with the use of two MZ delay interferometric (MZDI) balanced receivers (see Figure 12.93).

Several key notes in this detection structure are the following: (1) the MZDI introducing a delay corresponding to the baud rate, (2) one arm of the MZDI has a $\pi/4$ optical

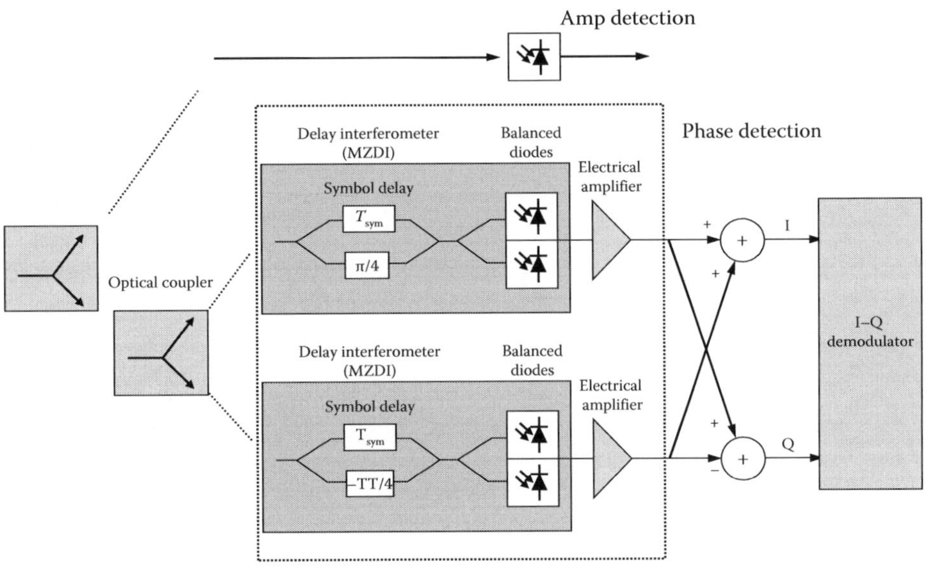

FIGURE 12.93
Detection configuration for the 8-DPSK_2-ASK modulation scheme.

phase shifter while the other arm has an optical phase shift of −π/4, (3) the outputs from two balanced receivers that are superimposed positively and negatively, which leads to I and Q detected signals, respectively. The I and Q detected components are expressed as follows: $I = \mathrm{Re}\{E_k E_{k-1}^*\}$ and $Q = \mathrm{Im}\{E_k E_{k-1}^*\}$, and (4) the I–Q detected components that are demodulated using the popular I–Q demodulator. The detected signals (after sampler) are represented in the signal constellation.

19.9.2.2.3 Transmission Performance of 100 Gb/s 8-DPSK_2-ASK Scheme

Performance characteristics of the 8-DPSK_2-ASK scheme operating at 100 Gb/s bit rate are studied in terms of receiver sensitivity, dispersion tolerance, and the feasibility for long-haul transmission. BERs are the pre-forward error correct (FEC) BERs and the pre-FEC limit is conventionally referenced at $2e^{-3}$. In addition, the BERs are evaluated by the Monte Carlo method.

19.9.2.2.4 Power Spectrum

The power spectrum of 8-DPSK_2-ASK optical signals is shown in Figure 12.94. It can be observed that the main lobe spectral width is about 25 GHz as the symbol baud rate of this modulation scheme is equal to a quarter of the bit rate from the bit pattern generator. The harmonics are not highly suppressed thus, requiring bandwidth of the optical filter to be necessarily large in order not to severely distort signals.

12.9.2.3 Receiver Sensitivity and Dispersion Tolerance

The receiver sensitivity is studied by connecting the optical transmitter of the 8-DPSK_2-ASK scheme directly to the receiver to make a back-to-back setup (see Figure 12.95). On the other hand, the dispersion tolerance is studied by varying the length of SSMF from 0 to 5 km ($|D| = 17$ ps/(nm·km)). Received powers are varied by using an optical attenuator.

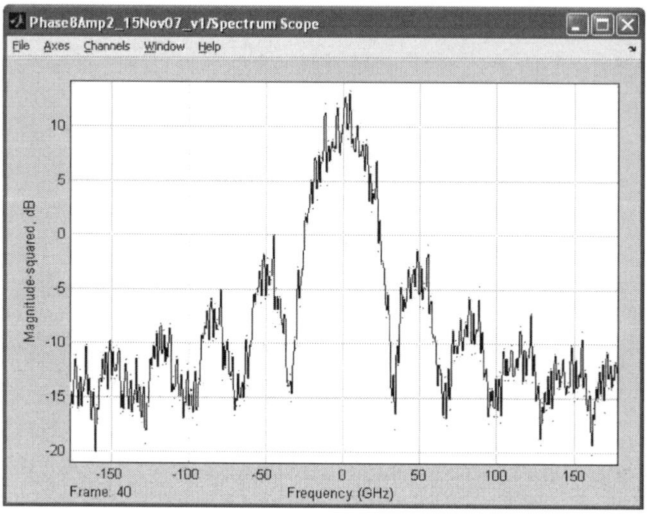

FIGURE 12.94
Power spectrum of 8-DPSK_2-ASK signals as observed by the spectrum scope of Simulink.

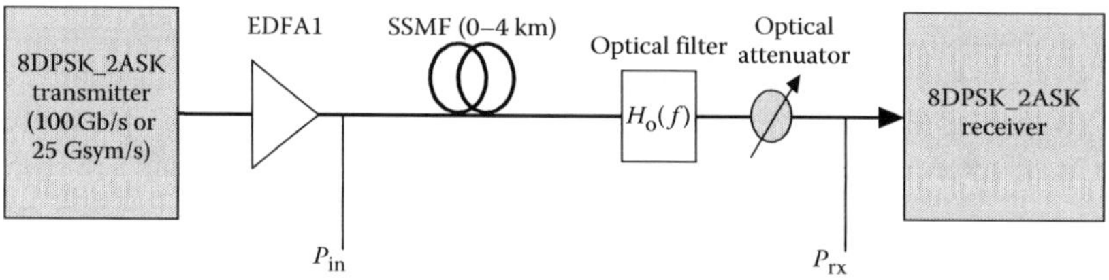

FIGURE 12.95
Setup for the study of receiver sensitivity (back to back) and dispersion tolerance (0–4 km SSMF) for the 8-DPSK_2-ASK modulation scheme.

The optical Gaussian filter has $BT=3$ (B is approximately 75 GHz). Modeling of receiver noise sources comprises of shot noise, equivalent noise current density of $20\,\text{pA}/\sqrt{\text{Hz}}$ at the input of the trans-impedance electrical amplifier, and dark current of 10 nA for each of the two photodiodes in balanced structure. A fifth-order Bessel electrical filter with a bandwidth of $BT=0.8$ is used.

The numerical BER curves of the receiver sensitivity for cases of 0–5 km SSMF are shown in Figure 12.96. The receiver sensitivity of the 8-DPSK_2-ASK scheme is approximately −18.5 dBm at BER = 1e−4. The receiver sensitivity at BER = 1e−9 can be obtained by extrapolating the BER curve of 0 km case.

The power penalty versus residual dispersion results are then obtained and plotted in Figure 12.97. It is realized that the 2 dB penalty occurs for the residual dispersion of approximately 60 ps/nm or equivalently to 3.5 km SSMF.

12.9.2.4 Long-Haul Transmission

The long-haul transmission performance of this modulation format is conducted over 10 optically amplified and fully compensated spans and each span is composed of 100 km SSMF and 10 km DCF100 (Sumitomo fiber). As a result, the length of the transmission fiber

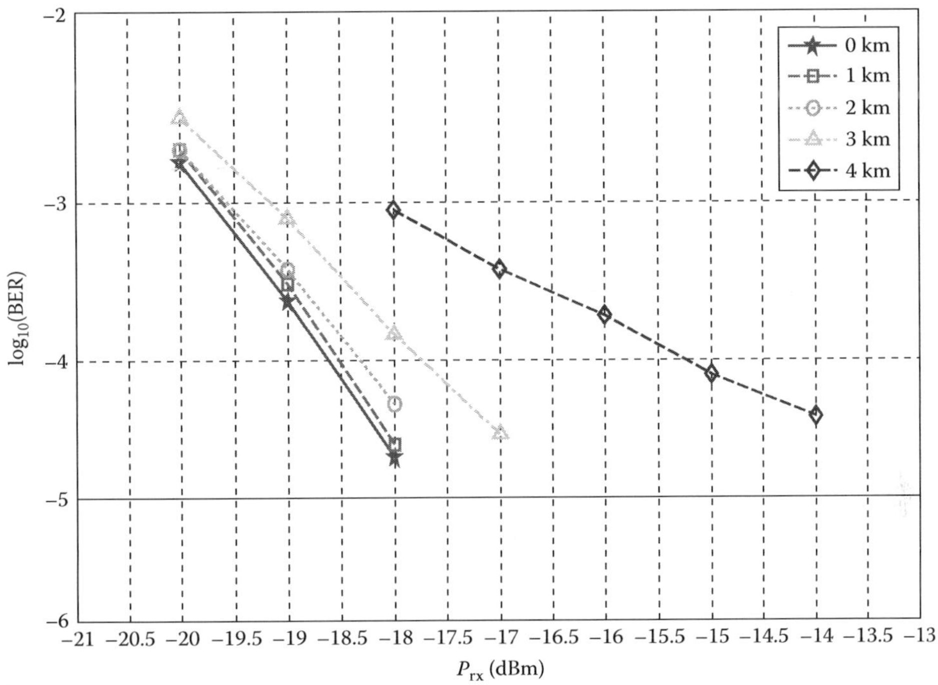

FIGURE 12.96
Receiver sensitivity (back to back) and dispersion tolerance (0–4 km SSMF) for the 8-DPSK_2-ASK modulation scheme.

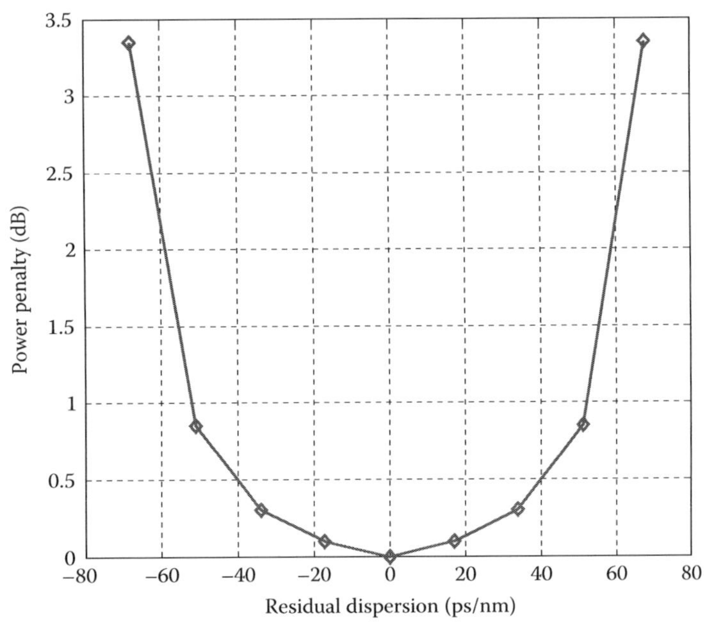

FIGURE 12.97
Power penalty due to residual dispersions for the 8-DPSK_2-ASK modulation scheme.

link is 1100 km. This long-haul range is selected to reflect the distance between Melbourne and Sydney of Australia. The wavelength of interest is 1550 nm, and the dispersion at the end of the transmission link is fully compensated. The simulation setup is shown in Figure 12.98. Additionally, the fiber attenuation due to SSMF and DCF is also fully compensated by using two EDFAs with optical gains as depicted in Figure 12.98. These EDFAs have noise figure (NF) set at 5 dB.

Numerical transmission BERs are plotted against received powers in Figure 12.99 and compared to the back-to-back BER curve. It can be observed that the BER curve of 1100 km follows a linear trend and feasibly reaches 1e−9 if extrapolated as shown in Figure 12.99. It should be noted that this transmission performance can be significantly improved with the use of high-performance FEC scheme.

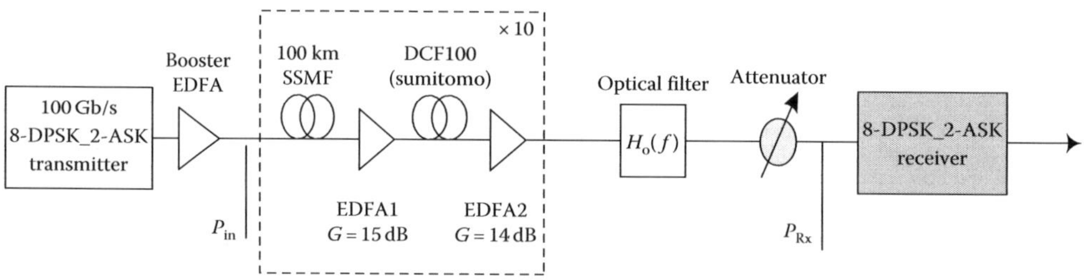

FIGURE 12.98
Transmission set up of 1100 km optically amplified and fully compensated fiber link.

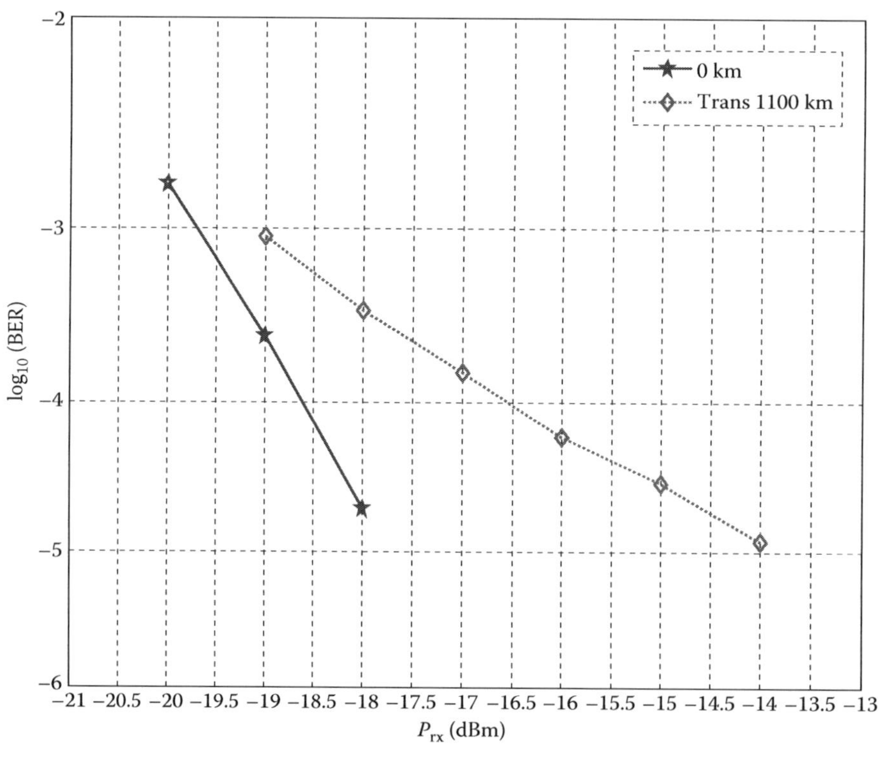

FIGURE 12.99
BER versus receiver sensitivity for 8DPSK 2ASK modulation format transmission.

This section has presented in detail the transmitter and receiver configurations and associated Simulink models for generating 8-DPSK_2-ASK optical signals. In addition, performance characteristics of this modulation format at 100 Gb/s (equivalently 25 GSymbols/s) has also been investigated in terms of the receiver sensitivity, dispersion tolerance, and the long-haul transmission performance. The simulation results show that 8-DPSK_2-ASK is a promising modulation for 100 Gb/s Ethernet and long-haul optical communications.

12.10 Concluding Remarks

This chapter presents the principles of modeling of optically amplified fiber optic transmission systems employing digital modulation techniques with different pulse-shaping methods. Generating RZ33 and CS-RZ optical pulses using optical intensity modulators are described. The phasor representation providing a clear explanation of the modulation process was verified mathematically. The generation and detection of optical DPSK signals were also described.

This chapter describes a number of digital modulation formats including ASK, and discrete and continuous PSK. Partial response technique is also outlined. We have also given examples of two MSK optical transmitters schemes. The first scheme is based on the CPFSK approach and implements two cascaded EO PMs. The second scheme consists of two dual-drive MZIMs, reflecting the I–Q configuration of the ODQPSK format. These I and Q driving signals are staggered by a one-bit period and operate at half of the bit rate. This scheme can generate either linear or nonlinear types of optical MSK pulses. Nonlinear optical MSK signals are distinguished based on the magnitude ripple levels of transmitted waveforms. Optical MSK signals with a small ripple of less than 10% of the push–pull amplitude levels are referred to as weakly nonlinear type whereas ripples of more than 10% of the signal levels are observed on strongly nonlinear optical MSK waveforms. Key differences in the implementations of these MSK optical transmitters are the waveforms and the magnitudes of the phase shaping electrical driving signals. In the case of linear MSK, periodic ramp waveforms are utilized whereas sinusoidal waveforms are used for the generation of weakly nonlinear and strongly nonlinear MSK. In order to generate weakly nonlinear MSK optical pulses, sinusoidal waveforms have its amplitude of approximately $V_\pi/4$ compared to a value of close to V_π in the case of strongly nonlinear optical MSK. The precoder for I–Q optical MSK structure was also derived. The incoherent detection of MSK-modulated optical signals is based on the MZDI balanced receiver. The main difference in this MSK receiver compared to the DPSK is the additional $\pi/2$ phase shift inserted onto one arm of the MZDI.

The dual-level MSK format has proposed for fiber-optic communications. The optical transmitter of this format is constructed from two MSK optical transmitters arranged in a parallel structure while the demodulation requires the noncoherent detection of both amplitude and phase components of the lightwaves. The phase detection for dual-level MSK format is similar to that of optical MSK whereas the amplitude detection is simply implemented with a single photodetector.

Partial modulation schemes are also described such as DB schemes with three levels; the schemes can be coded by using the PSK techniques and thus different from those in conventional three-level inverted amplitude schemes.

Multilevel amplitude and PSK schemes have also been described. Two principal schemes are presented the M-ary amplitude DPSK and Star-QAM. The transmitters including pre-coder are described; the detection is very much identical to those for differential phase keying.

Finally, spectral properties of optical 80 Gb/s dual-level MSK, 40 Gb/s MSK and 40 Gb/s NRZ/RZ33/CS-RZ DPSK, and OOK formats were discussed. There is a trade-off between RZ33/CS-RZ and NRZ pulse types. Compared to NRZ pulses, RZ33 and CS-RZ pulses are more resilient to fiber nonlinear effects but less tolerant to fiber CD. The spectral properties of 40 Gb/s MSK and 80 Gb/s dual-level MSK are similar to each other and more advantageous than 40 Gb/s optical DPSK and OOK. These advantages include (1) the spectral efficiency and filtering robustness due to the narrow spectral main lobe and (2) the mitigation of interchannel cross talks because of high suppression of side lobes. Therefore, optical MSK and dual-level MSK formats offer promising potential for high-speed and high-capacity optical long-haul and metropolitan transmission systems.

Appendix 12.A: Simulink® and Simulation Guidelines

This section gives a brief introduction to MATLAB and Simulink and guidelines for simulation of optically amplified fiber transmission employing modulation formats such as ASK, PSK, and FSK.

12.A.1 MATLAB® and Simulink®

MATLAB and Simulink are packages integrated in MATLAB (from MATLAB 7.0 onward). Simulink is a simulation platform based on blocks of functions for users to structure their mathematical development and algorithm. This means that a number of MATLAB.m extension files have been transformed into blocks.

The MATLAB and Simulink platforms are shown in Figure 12.100 in which there is a section on Simulink (Figure 12.101) that consists of several blocksets including basic Simulink blocksets, communication blocksets, and signal processing blocksets. The basic Simulink blocksets consist of commonly used blocksets for the manipulation of mathematical operations, discrete and continuous blocks for sampling and operations in the continuous and discrete domain, source and sink blocks are used as monitoring equipment and for the generation of signals. A user-defined function can be used to integrate MATLAB.m files into blocks for integration in the Simulink block diagram.

Thus Simulink can be employed as a computer experimental development system for testing and simulation of communications transmission systems. With the availability of communication blocksets and signal processing blocksets, the manipulation, modulation, and reception of optical signals can be implemented without much difficulty.

To develop a Simulink model, one needs to open a file in Simulink and then drag blocks that are required to implement an algorithm. It is noted that the advantage of Simulink is that the principles of operations of systems are important and mathematical analysis skill is minimized. Thus developing Simulink is like setting up experiments on a computer MATLAB platform without writing any .m file codes.

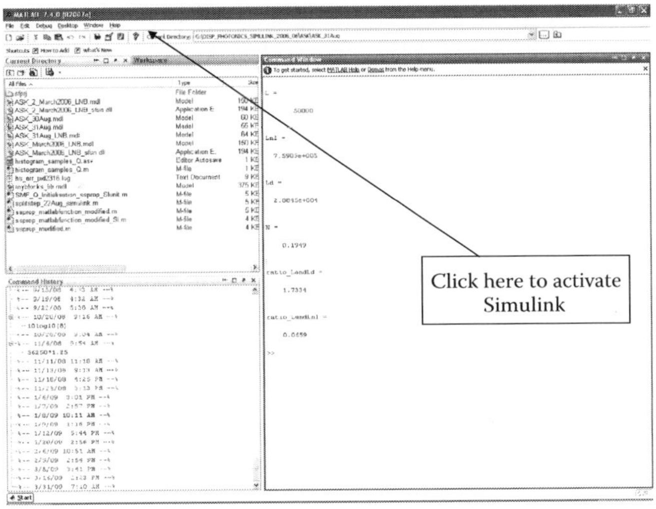

FIGURE 12.100
MATLAB platform and where to click to activate Simulink.

12.A.2 Guidelines for Using Simulink® Models

This section outlines a number of models that can be used to simulate the transmission of optical signals over the optically amplified fiber multi-span systems.

A transmission system would normally consist of the following:

- System parameters, which are defined and specified in the next section.
- A general block model given in Figure 12.102, which consists of the optical transmitters, the fiber and optical amplifier spans, and the optical receivers and the evaluation of BERs. There are also a number of blocks for monitoring signals at various points in the transmission systems.
- Optical transmitters, which accept electrical signals, data bit sequence of arbitrary distribution. That is the "0" and "1" are randomly distributed in the data sequence. This random binary sequence is essential for obtaining the eye diagram at the output of the transmitter and the optical receiver so as to evaluate the BER of the transmission system. The transmitter module is available and indicated in the model. Figure 12.102 shows the Simulink model for the generation of ASK modulation signals. The Bernoulli arbitrary signal generator is used to generate arbitrary binary data bits.
- Modules of fiber spans that consist of a length of single-mode optical fiber for transmission and that of dispersion-compensation fiber plus optical amplifiers (Figures 12.103 and 104). The propagation of optical signals can be represented by the nonlinear Schröedinger equation (only SPM is included) given as follows:

$$\frac{\partial u}{\partial z} + j\frac{\beta_2}{2}\frac{\partial^2 u}{\partial T^2} - \frac{\beta_3}{6}\frac{\partial^3 u}{\partial T^3} + \frac{\alpha}{2}u = j\gamma\left|u\right|^2 u$$

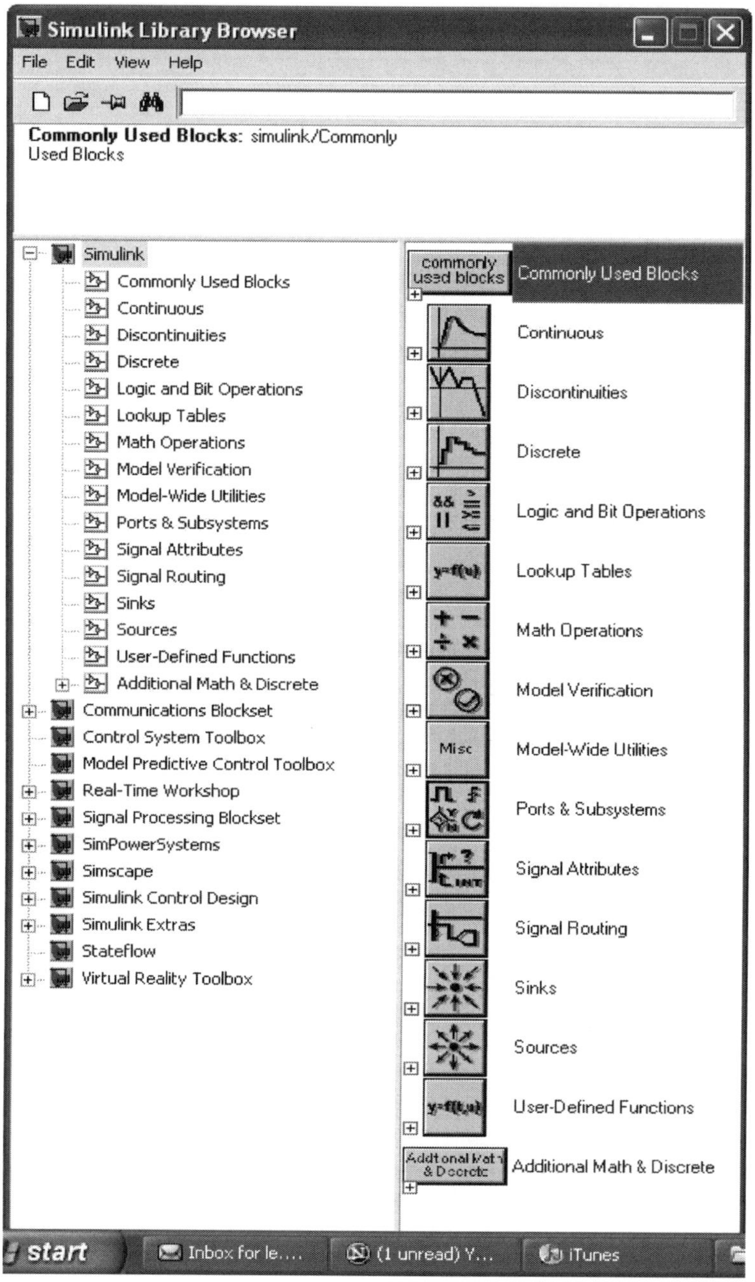

FIGURE 12.101
Blocksets of Simulink.

where the parameter u represents the normalized amplitude of the signal. Note that due to the extremely high value of the frequency of the OC, ultrahigh sampling is required, and the optical signals take complex amplitudes. This means that the amplitude is the envelope of the signals and the complex parts of the amplitude represent the phase of the OCs. The coefficient α is the attenuation factor, T is the normalized time variable, and g is the

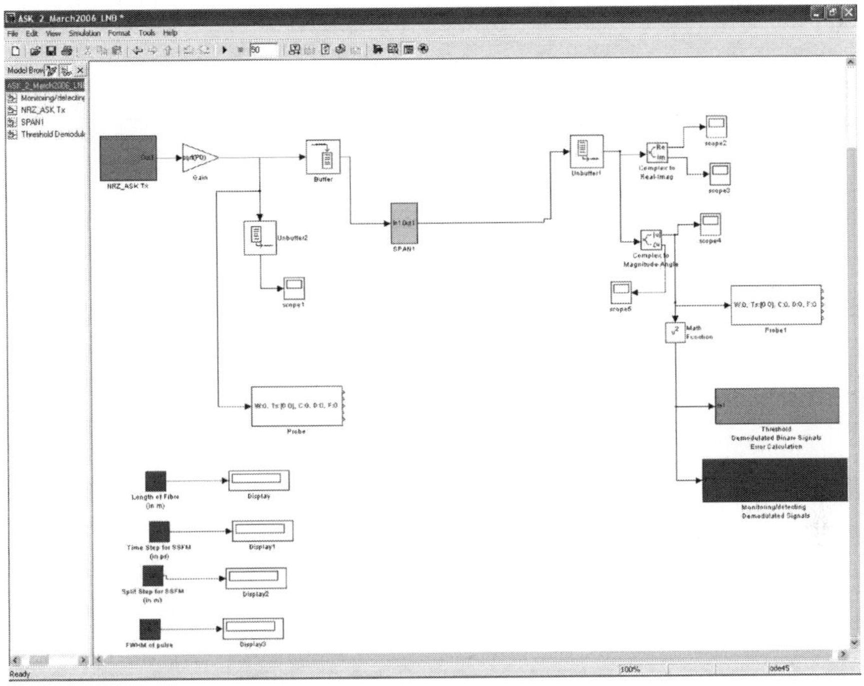

FIGURE 12.102
General structure of an ASK-modulated optically amplified transmission system.

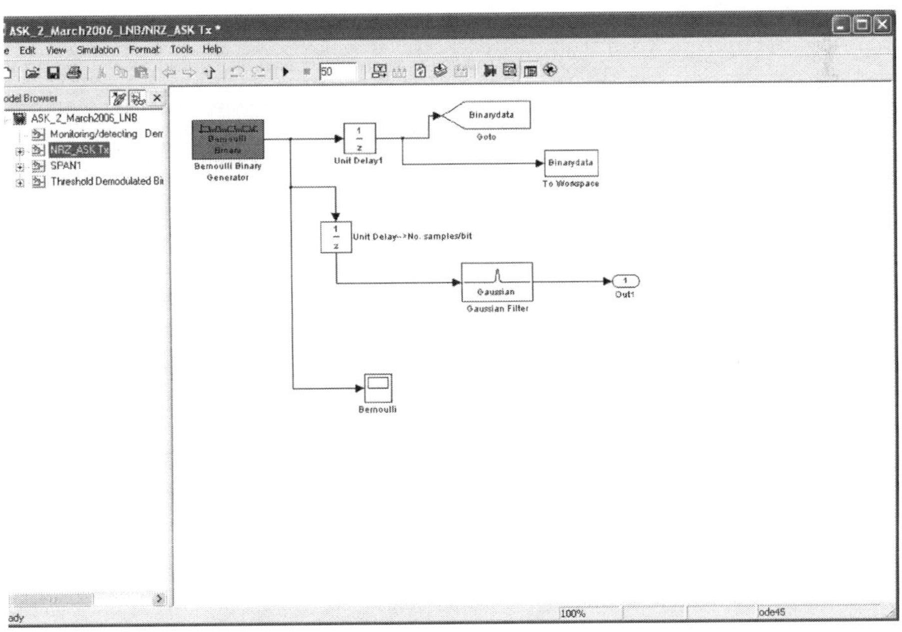

FIGURE 12.103
Optical transmitter employing ASK modulation format.

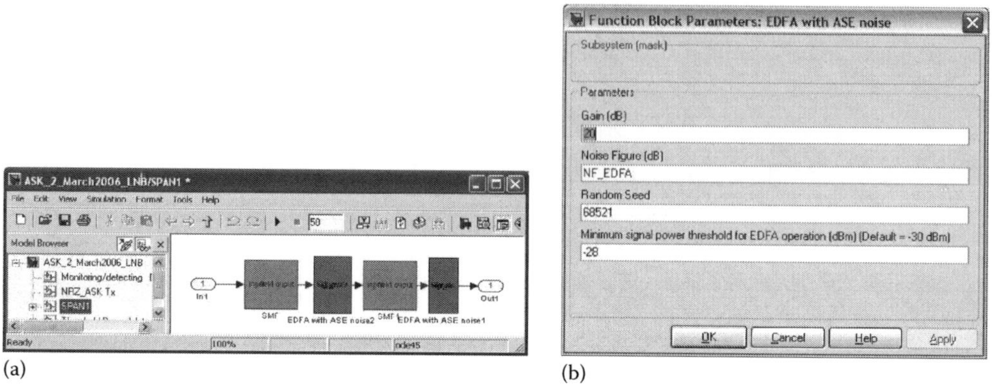

FIGURE 12.104
An optically amplified span consisting of a section of single-mode optical fiber, a section of DCF, and two optical amplifiers: (a) EDFA in transmission span and (b) under-mask model of EDFA.

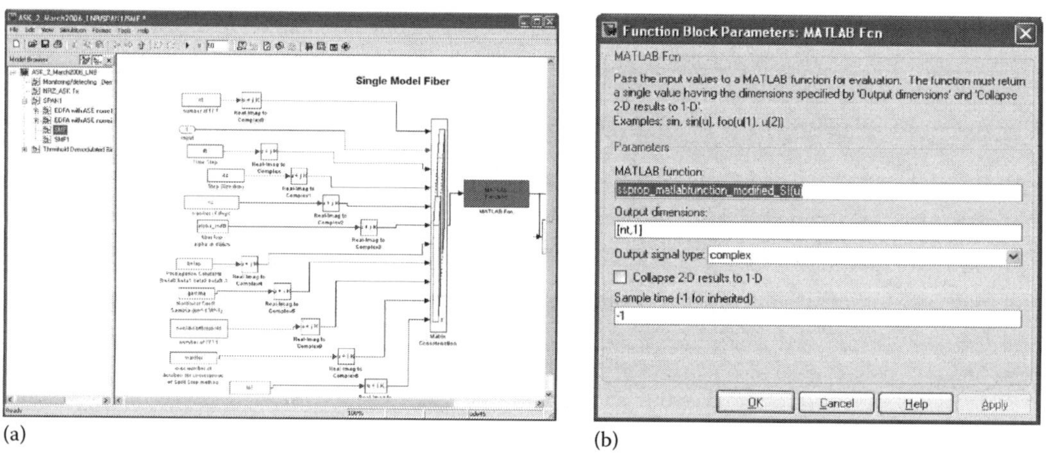

FIGURE 12.105
Simulink model of the propagation of optical signals through a section of optical fibers: (a) model with input parameters and signals and (b) under-mask model of block "MATLAB functions," which contains the m-file to calculate the propagation effects on signals by the dispersive property of the SMF.

nonlinear SPM coefficient. The .m file for solving this propagation can be found in Section 12.A.3.1. This program must be included in the "call back" of the "model property" of the Simulink model.

A block diagram for representing the fiber propagation is shown in Figure 12.105. The structures of the subsystems are given as follows:

- DCF can be structured by estimating the dispersion factor that is in the opposite sign of the transmission fiber so as to reverse the phase effects of the transmission length. This can be implemented by using the model given in Figure 12.105, but the beta parameters must be negative in sign for equal dispersion factor. For different dispersion factors between the fibers, appropriate lengths of the fibers must be modified by taking the difference in the number of small sections *nz* and *dz*, which are also specified in the initialization file given in Section 12.A.3.1.

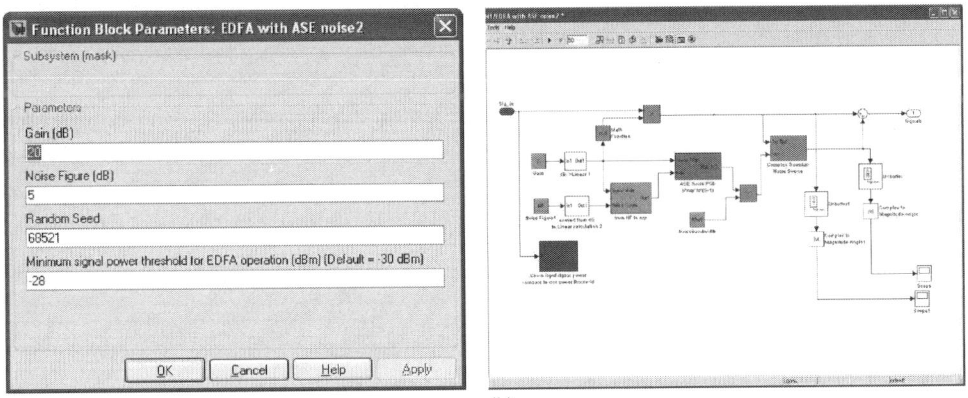

(a) (b)

FIGURE 12.106
Optical amplifier: (a) parameters for the specification of the optical amplifier, gain, and noise figure (NF) and (b) under-mask model of the amplifier.

- Optical amplifiers are inserted after the SMF transmission fiber and the dispersion-compensated fiber. The specification of the gain and NF of the amplifiers can be altered by clicking the block and entering the parameters. The under-mask model of the amplifier is shown in Figure 12.106b, and its integration in a transmission span is shown in Figure 12.106a.
- Optical receiver is represented by square-law detection. That is taking the absolute value of the amplitude and the square. This is the photodetection process. This means that the photodetector absorbs the optical power and then converts it into electronic current, which is then further amplified in the electronic domain and then displayed on an oscilloscope. At this point, the eye diagram is achieved, and the BER can be evaluated by getting the raw data generated from the Simulink model and calculation in Excel or m-files. An error detection and calculation of BER is given in the models for DPSK modulation format transmission system. Monitoring of data received at the output of the receiver is required to obtain the raw data for the evaluation of the BER (Figure 12.107).

12.A.3 MATLAB® Files

12.A.3.1 Initialization File

```
% This file solves NLSE for
% pulse propagation in an optical fiber using the split-step
% Fourier method as given in: GP. Agrawal,"Nonlinear Fiber Optics",
2nd ed.
% AP, 1995, Chapter 2
%
% The following effects are included in the model: group
% velocity dispersion (GVD), GVD-slope/third-order
% dispersion, loss, and self-phase modulation (n2). The core
% routine is implementing the split-step propagation

close all
clear all

% CONSTANTS
```

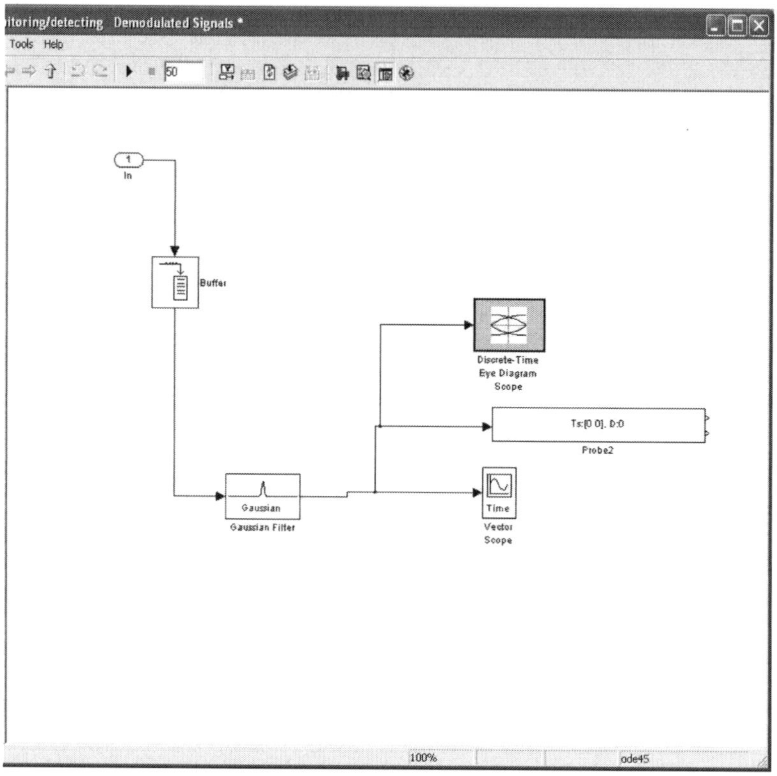

FIGURE 12.107
Block diagram of optical receiver and eye diagram monitor.

```
c=3e8;% speed of light (m/s)
% NUMERICAL PARAMETERS
numbitspersymbol=1;
P0=0.001;                          % peak power (W)
FWHM=25e-12 ;                      % pulse width FWHM (ps)
FWHMps=25e-12;
% halfwidth=FWHM/1.6651            % for Gaussian pulse
halfwidth=FWHM ;                   % for square pulse
bitrate=1/halfwidth;               % THz
bitrateG=1/FWHMps;
baudrate=bitrate/numbitspersymbol;
signalbandwidth=baudrate;
%%%%%%%%%%%%%%%%%%%%%%%%%%%%%%%%
%%%%%%%%%%%%%%%%%%%%%%%%%%%%%%%%

PRBSlength=2^8;

% Make sure : FFT time window (=nt*dt)=PRBSlength * FWHM...
% FFTlength nt=PRBSlength/block * numbersamples/bit=PRBSlength * (FWHM/dt)
% num_samplesperbit=FWHM/dt should be about 8 - 16 samples/bit

num_samplesperbit=32;              % should be 2^n
dt=FWHMps/num_samplesperbit ;      % sampling time(ps);%time step(ps)
nt=PRBSlength*num_samplesperbit;   % FFT length
dz=500;                            % distance stepsize (m)
```

```
nz=100;                              % number of z-steps
maxiter=20;                          % max # of iterations
tol=1e-5;                            % error tolerance
% OPTICAL PARAMETERS
nonlinearthreshold=0.005; % 5mW -- % Nonlinear Threshold Peak Power for
silica core fiber

lambda=1550e-9;
% lambda_carrier=14648.4375;      % wavelength (nm)with
       % Level 4 group the carrier freq is scaled down to 500G or 200GHz
optical_carrier=c/lambda; %carrier freq
num_samplesperperiod=8;
sampling_fac=16;
carrier_freq=num_samplesperperiod*num_samplesperbit*bitrateG; % artificial
freq
%dt=FWHM/(num_samplesperbit*num_samplesperperiod*sampling_fac) ; %
sampling time(ps); % time step (ps)
%nt=PRBSlength*num_samplesperbit*num_samplesperperiod*sampling_fac ; %
FFT length
%dBperkm=0.2;        % loss (dB/km)
alpha_indB=0.2;      % loss (dB/km)
D=17e-6;       % NZDSF GVD (s/m^2); if anomalous dispersion (for
compensation),D is negative
%D=17;% SSMF GVD (ps/nm.km); if anomalous dispersion (for
compensation),D is negative

ng=1.46;       % group index
n2=2.6e-20;  % nonlinear index (m^2/W)
Aeff=80e-12; % effective area (um^2)

% CALCULATED QUANTITIES

T=nt*dt;       % FFT window size in sec. (ps) -Agrawal: should be about
10-20 times of the pulse width
alpha_loss=log(10)*alpha_indB/10^4;      % alpha (1/m)
beta2=D*lambda^2/(2*pi*c);% beta2 (s^2/m);
beta3=0.3e-39;       % GVD slope (s^3/m)
%beta2 =0;
%-------------------------------------------------------------
% beta 3 can be calculated from the Slope Dispersion (S) as follows:]
% Slope Dispersion S=0.092;       % ps/(nm^2.km)
% beta31=(S - (4*pi*c./lambda.^3))./(2*pi*c./lambda.^2)
%-------------------------------------------------------------
gamma=2*pi*n2/(lambda*Aeff);      % nonlinearity coef (m^-1.W^-1)

% STARTING FIELD

L=nz*dz

Lnl=1/(P0*gamma)    % nonlinear length (m)
Ld=halfwidth^2/abs(beta2) % dispersion length (m)
N=sqrt(abs(Ld./Lnl))       % governing the which one is dominating:
dispersion or Non-linearities
ratio_LandLd=L/Ld   % if L<<Ld -> NO Dispersion Effect
ratio_LandLnl=L/Lnl % if L<<Lnl -> NO Nonlinear Effect
% Monitor the broadening of the pulse with relative the Dispersion Length
% Calculate the expected pulsewidth of the output pulse
```

```
% Eq 3.2.10 in Agrawal "Nonlinear Fiber Optics" 2001 pp67
FWHM_new=FWHM*sqrt(1+(L/Ld)^2);

% N<<1 -> GVD ; N >>1 -> SPM
Leff=(1-exp(-alpha_loss*L))/alpha_loss;
expected_normPout=exp(-alpha_loss*2*L);
NlnPhaseshiftmax=gamma*P0*Leff ;

betap=[0 0 beta2 beta3]';

% Constants for ASE of EDFA
% PSD of ASE: N(at carrier freq)=2*h*fc*nsp*(G-1) with nsp=Noise
% Figure/2 (assume saturated gain)
%*************** Standard Constant ******************************
h=6.626068e-34;     %Plank's Constant
%****************************************
```

12.A.3.1.1 Propagation of Optical Signals over a Single-Mode Optical Fiber—SSMF

```
function output=ssprop_matlabfunction_modified(input)

nt=input(1);
u0=input(2:nt+1);
dt=input(nt+2);
dz=input(nt+3);
nz=input(nt+4);
alpha_indB=input(nt+5);
betap=input(nt+6:nt+9);
gamma=input(nt+10);
P_non_thres=input(nt+11)
maxiter=input(nt+12);
tol=input(nt+13);

tic;
%tmp=cputime;

% %-----------------------------------------------------
% This function solves the nonlinear Schrodinger equation for
% pulse propagation in an optical fiber using the split-step
% Fourier method described in:
% Agrawal, Govind. Nonlinear Fiber Optics, 2nd ed. Academic
% Press, 1995, Chapter 2
%
% The following effects are included in the model: group velocity
% dispersion (GVD), higher order dispersion, loss, and self-phase
% modulation (gamma).
%
% USAGE
%
% u1=ssprop(u0,dt,dz,nz,alpha,betap,gamma);
% u1=ssprop(u0,dt,dz,nz,alpha,betap,gamma,maxiter);
% u1=ssprop(u0,dt,dz,nz,alpha,betap,gamma,maxiter,tol);
%
% INPUT
%
% u0 - starting field amplitude (vector)
% dt - time step - [in ps]
```

```
% dz - propagation stepsize - [in km]
% nz - number of steps to take, ie, ztotal=dz*nz
% alpha - power loss coefficient [in dB/km], need to convert to linear to
have P=P0*exp(-alpha*z)
% betap - dispersion polynomial coefs, [beta_0 … beta_m] [in ps^(m-1)/km]
% gamma - nonlinearity coefficient [in (km^-1.W^-1)]
% maxiter - max number of iterations (default=4)
% tol - convergence tolerance (default=1e-5)
%
% OUTPUT
%
% u1 - field at the output
%-----------------------------------------------------------
% Convert alpha_indB to alpha in linear domain
%-----------------------------------------------------------
alpha=1e-3*log(10)*alpha_indB/10;        % alpha (1/km) - see Agrawal p57
%-----------------------------------------------------------

ntt=length(u0);
w=2*pi*[(0:ntt/2-1),(-ntt/2:-1)]'/(dt*nt);
%w=2*pi*[(ntt/2:ntt-1),(1:ntt/2)]'/(dt*ntt);

clear halfstep

halfstep=-alpha/2;
for ii=0:length(betap)-1;
  halfstep=halfstep - j*betap(ii+1)*(w.^ii)/factorial(ii);
end

clear LinearOperator
% Linear Operator in Split Step method
LinearOperator=halfstep;
% pause
halfstep=exp(halfstep*dz/2);

u1=u0;
ufft=fft(u0);

% Nonlinear operator will be added if the peak power is greater than the
% Nonlinear threshold
iz=0;
while (iz<nz) & (max((abs(u1).^2+abs(u0).^2))>P_non_thres)
  iz=iz+1;

  uhalf=ifft(halfstep.*ufft);

  for ii=1:maxiter,
    uv=uhalf .* exp(-j*gamma*(abs(u1).^2+abs(u0).^2)*dz/2);
    ufft=halfstep.*fft(uv);
    uv=ifft(ufft);

    fprintf('You are using SSFMn');

    if (max(uv-u1)/max(u1)<tol)
      u1=uv;
      break;
    else
      u1=uv;
    end
```

```
  end
  if (ii == maxiter)
    warning(sprintf('Failed to converge to %f in %d iterations',…
        tol,maxiter));
  end

  u0=u1;

end
if (iz<nz) & (max((abs(u1).^2+abs(u0).^2))<P_non_thres)

% u1=u1.*rectwin(ntt);
  ufft == fft(u1);
  ufft=ufft.*exp(LinearOperator*(nz-iz)*dz);
  u1=ifft(ufft);

  fprintf('Implementing Linear Transfer Function of the Fiber
Propagation');
end

toc;
%****************
% OUTPUT
%****************
% u1=u1(1:16);
%      angle_ifft_test=angle(u1)
%      pause
  output=u1;
```

12.A.3.1.2 BER Evaluation

```
clear Q_Gaussian BER_simple varied_threshold index_bit1 index_bit0 …
    mean_bit1 mean_bit0 std_bit1 std_bit0;
%---------------------------------------------------------
Q_Gaussian=0; % Initialise Q value
% Simple Q-factor calculation with assumption:
% Gaussian Noise distribution
%---------------------------------------------------------
%**************************************************
%**************************************************
% Assign variables from SimuLink to Matlab variables
%**************************************************
  delay=32;

  % To determine received "1"
  index1_temp=find(Bernoulli==0);
  tmp1=find(index1_temp>length(demodsignal)- delay);

  index1=index1_temp(2:tmp1(1)-1);

  for i=1 : length(index1)
    var_1(i)=demodsignal(delay+index1(i));
  end

  % To determine received "0"
  index0_temp=find(Bernoulli==1);
  tmp0=find(index0_temp>length(demodsignal)-delay);

  index0=index0_temp(2:tmp0(1)-1);
```

```
  for ii=1 : length(index0)
    var_0(ii)=demodsignal(delay+index0(ii));
  end
% End
%*****************
Thres_Volt=0;
%*****************

% % BER
% % Assume bit 1 and 0 have a Gaussian normal distribution
% mean_bit1=mean(var_1); % I1
% std_bit1=std (var_1) ; % sigma1
%
% mean_bit0=mean(var_0); % I0
% std_bit0=std(var_0) ; % sigma2
% % Q-factor
% Q=(mean_bit1 - mean_bit0) / (std_bit1+std_bit0) ;
%
% BER=1/2 * (erfc(Q/sqrt(2))) ;
% Determine BER
% Determine Mean value of samples for bit 1
mean_bit1=mean(var_1);
std_bit1=std(var_1);

% Determine Mean value of samples for bit 0
mean_bit0=mean(var_0);
std_bit0=std(var_0);

% Calculation of Q-factor and BER
% Assumption: Gaussian approximation for the distributions of samples of
% bit 1 and bit 0
% Pe=P0*E(1/2 of erfc((S-X0)/sqrt(2)*sigma0)+P1*E(1/2 of …
% erfc((S-X1)/sqrt(2)*sigma0)

BER_Gaussian=length(var_0)/length(Bernoulli(delay+1:length(Bernoulli)))*
mean(1/2.*erfc(abs(Thres_Volt - var_0)./(sqrt(2)*std_bit0)))+…
length(var_1)/length(Bernoulli(delay+1:length(Bernoulli)))*
mean(1/2.*erfc(abs(var_1 - Thres_Volt)./(sqrt(2)*std_bit1)));

% clc
mean_bit1
mean_bit0
std_bit1
std_bit0
Q_Gaussian=sqrt(2)*erfcinv(BER_Gaussian*2)
BER_Gaussian
  BER_Gaussian_Agrawal=1/2*(erfc(Q_Gaussian/sqrt(2)))
  Q_Gaussian=abs(mean_bit1 - mean_bit0)/(std_bit1+std_bit0)
```

12.A.3.1.3 Linking Initialization File and Other Related Files Such as ssprop_matlab_modified.m with the Model

The initialization file is required to pass on the values of the parameters to the Simulink model by inserting these file names in the section "call back" of "model property" of the Simulink model as shown in Figure 12.108. The fiber propagation .m file "ssprop_matlab-function_modified.m" must be included in the folder that contains the Simulink model for the model to call this function for execution.

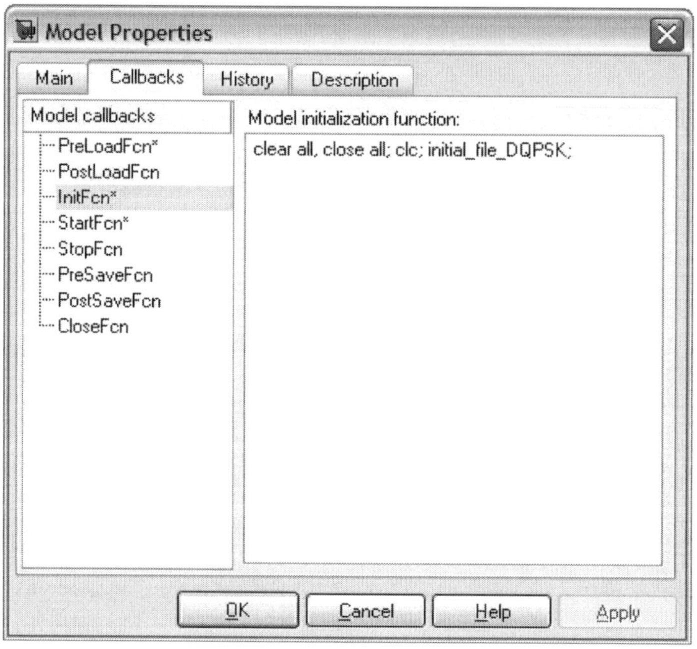

FIGURE 12.108
Model property and related m-files for model parameters.

References

1. T. Hoshida et al., Optimal 40Gb/s modulation formats for spectrally efficient long haul DWDM systems, *IEEE Journal of Lightwave Technology*, 20: 1989–1996, 2002.
2. Y. Zhu et al., 1.6 bit/s/Hz orthogonally polarized CSRZ—DQPSK transmission of 8x40Gbit/s over 320km NDSF, in *Proceedings of OFC'04*, Los Angeles, CA, Tu-F1, 2004.
3. M. Ohm and J. Spiedel, Optical minimum shift keying with direct detection, *Proceedings of SPIE on Optical Transmission, Switching and Systems*, 5281: 150–161, 2004.
4. T.L. Huynh, L.N. Binh, and K.K. Pang, Optical MSK long-haul transmission systems, to be published in *SPIE Proceedings of APOC'06*, Gwangju, Korea, Thu9a, 2006, Paper 6353–86.
5. T.L. Huynh, L.N. Binh, and K.K. Pang, Linear and weakly nonlinear optical continuous phase modulation formats for high performance DWDM long-haul transmission, presented at submitted to *ECOC'06*, Cannes, France, 2006.
6. R. Krahenbuhl, J.H. Cole, R.P. Moeller, and M.M. Howerton, High-speed optical modulator in $LiNbO_3$ with cascaded resonant-type electrodes, *IEEE Journal of Lightwave Technology*, 24(5): 2184–2189, 2006.
7. L. Ding, W.D. Zhong, C. Lu, and Y. Wang, New bit-error-rate monitoring technique based on histograms and curve fitting, *Optics Express*, 12(11): 2507–2511, 2004.
8. F. Amoroso, Pulse and spectrum manipulation in the minimum frequency shift keying (MSK) format, *IEEE Transactions on Communications*, 24: 381–384, March 1976.
9. J. Mo, D. Yi, Y. Wen, S. Takahashi, Y. Wang, and C. Lu, Optical minimum-shift keying modulator for high spectral efficiency WDM systems, in *Proceedings of ECOC '05*, Vol. 4, Glasgow, U.K.,2005, pp. 781–782.

10. M. Ohm and J. Speidel, Optical minimum-shift keying with direct detection (MSK/DD), *Proceedings of SPIE*, 5281: 150–161, 2004.
11. T.L. Huynh, L.N. Binh, K.K. Pang, and L. Chan, Photonic MSK transmitter models using linear and non-linear phase shaping for non-coherent long-haul optical transmission, in *Proceedings of APOC 2006*, Kwangju, Korea, September 3–7, 2006, Paper 6353–85.
12. J.G. Proakis, *Digital Communications*, 4th edn. New York: McGraw-Hill, 2001, pp. 199–202, Chapter 4.
13. K.P. Ho, Spectrum of externally modulated optical signals, *IEEE Journal of Lightwave Technology*, 22(2): 658–663, February 2004.
14. H. Kim, Robustness to laser frequency offset in direct-detection DPSK and DQPSK systems, *IEEE Journal of Lightwave Technology*, 21(9): 1887, September 2003.
15. S.E. Miller, Ed., *Optical Fiber Communication*. Boston, MA: Academic Press, 1988, p. 15.
16. C. Wree, J. Leibrich, J. Eick, W. Rosenkranz, and D. Mohr, Experimental investigation of receiver sensitivity of RZ-DQPSK modulation using balanced detection, in *Proceedings of OFC'03*, Vol. 2, Atlanta, GA, ThE5, 2003, pp. 456–457.
17. J.G. Proakis, *Digital Communications*, 3rd edn. New York: McGraw-Hill Publishing Company, 1995, p. 72.
18. J.C. Palais, *Fiber Optic Communications*. Englewood Cliffs, NJ: Prentice Hall, 1984, p. 57.
19. C. Wree, J. Leibrich, and W. Rosenkranz, RZ-DQPSK format with high spectral efficiency and high robustness towards fiber nonlinearities, in *28th European Conference on Optical Communication, 2002, ECOC 2002*, Vol. 4, Copenhagen, Denmark, September 2002, pp. 1–2.
20. J. Armstrong, SimuLink optical simulator, Honors Thesis, Monash University, Clayton, Victoria, Australia.
21. T. Sakamoto, T. Kawanishi, and M. Izutsu, Initial phase control method for high-speed external modulation in optical minimum-shift keying format, in *Proceedings of ECOC'05*, Vol. 4, Glasgow, U.K., 2005, pp. 853–854.
22. T. Kawanishi, S. Shinada, T. Sakamoto, S. Oikawa, K. Yoshiara, and M. Izutsu, Reciprocating optical modulator with resonant modulating electrode, *Electronics Letters*, 41(5): 271–272, 2005.
23. I.P. Kaminow and T. Li, *Optical Fiber Communications*, Vol. IVA. New York: Elsevier Science, 2002, Chapter 16.
24. I. Shake, H. Takara, and S. Kawanishi, Simple Q factor monitoring for BER estimation using opened eye diagrams captured by high-speed asynchronous. *Photonics Technology Letters*, 15: 620–622, 2003.
25. J. Mo, D. Yi, Y. Wen, S. Takahashi, Y. Wang, and C. Lu, Novel modulation scheme for optical continuous-phase frequency-shift keying, in *Proceedings of OFC'03*, Atlanta, GA, 2004, Paper OFG2.
26. T. Sakamoto, T. Kawanashi, and M. Izutsu, Optical minimum-shift keying with external modulation scheme, *Optics Express*, 13(20): 7741–7747, 2005.
27. K. Sekine, N. Kikuchi, S. Sasaki, S. Hayase, C. Hasegawa, and T. Sugawara, 40Gbit/s, 16-ary (4bit/symbol) optical modulation/demodulation scheme, *Electronics Letters*, 41(7): 430–432, March 2005.
28. I.B. Djordjevic and B. Vasic, Multilevel coding in M-ary DPSK/differential QAM high-speed optical transmission with direct detection, *IEEE Journal of Lightwave Technology*, 24: 420–428, 2006.
29. M. Serbay, C. Wree, and W. Rosenkranz, Implementation of different precoder for high-speed optical DQPSK transmission, *Electronics Letters*, 40(20): 1288–1289, September 2004.
30. M. Serbay, C. Wree, and W. Rosenkranz, Experimental investigation of RZ-8DPSK at 3 × 10.7Gb/s, in *Proceedings of LEOS'05*, Sydney, Australia, 2005.
31. M. Seimetz, M. Noelle, and E. Patzak, Optical systems with high-order DPSK and Star QAM modulation based on interferometric direct detection, *IEEE Journal of Lightwave Technology*, 25: 1515–1530, 2007.

32. H. Yoon, D. Lee, and N. Park, Performance comparison of optical 8-ary differential phase-shift keying systems with different electrical decision schemes, *Optic Express*, 13: 371–376, 2005.
33. R. Redner and H. Walker, *SIAM Review*, 26(2): 195–239, 1984.
34. J.G. Proakis, *Digital Communications*, 4th edn. New York: McGraw-Hill, 2001, pp. 185–213.
35. L.N. Binh, *Digital Optical Communications*, Boca Raton, FL: CRC Press, Taylor & Francis Group, 2009.

Appendix: A Short Glossary of Terms in Optical Fiber Communications

Absorption: Together with scattering, this forms the principal cause of attenuation of an optical fiber. It results from unwanted impurities in the fiber material and has an effect at certain wavelengths.

In the case of photodiode, absorption is the process by which the energy of incident photon raises an electron from the valence band to the conduction band. Thus, the bandgap of these two energy bands determines the operating wavelength of the photodiode.

Acceptance angle: Half the vertex angle of the cosine within which the power coupled into an optical waveguide with uniform illumination is equal to a specific fraction of the total coupled power. The sine of the acceptance angle is referred to as the numerical aperture (NA).

Analog transmission: Transmission of continuously varying signals by either direct or external modulation of the optical field or by using an electrical sub-carrier with which the optical power is then modulated.

ASK (amplitude shift keying): Digital modulation technique by switching the amplitude of the optical carrier from a "0" level to "1" level.

Attenuation: Reduction in the optical power between two cross sections of an optical fiber. The main causes are scattering and absorption, as well as optical losses in connectors and splices. The unit is decibel or the logarithm of the ratio of the input and output power of lightwaves.

Attenuation coefficient: The attenuation coefficient is the attenuation with respect to length of a uniform optical fiber with equilibrium mode distribution—normal unit dB/km.

Avalanche noise: Although the photocurrent amplification in an avalanche photodiode contributes little noise, the resulting additional noise is referred to as avalanche noise.

Avalanche photodiode: Photodiode in which the photocurrent is amplified by carrier multiplication in an electric field—avalanche effect.

Backscattering: A small fraction of lightwaves that are deflected out of the original direction of propagation by scattering, which suffers a reversal of direction.

Balanced receiver: Receiver using the detector at the front end in a push-pull manner. This is usually used to detect the differential phase between consecutive bit periods. In optical communications, this balanced receiver is coupled with an optical comparator thus detecting the "0" and "π" phase shift between consecutive bits. This type of receiver can also be integrated with a local oscillator for coherent detection.

Bandwidth: See fiber bandwidth, dispersion factor, and receiver bandwidth.

Chromatic dispersion: Combination of two related effects: material and waveguide dispersion.

Cladding: Optically transparent material comprising an optical waveguide excluding the core region.

Cladding diameter: Diameter of the circle that encloses the surface of the cladding.

Connector: See optical fiber connector.

Core: The central region provided for guiding lightwaves for transmission in an optical waveguide. It is defined by the smallest cross section enclosed by a line for which the refractive index is larger than that of the cladding region.

Core diameter: Diameter of the smallest circle that encloses the core area. The core radius is the radius of this circle.

CPM (continuous phase modulation): This is also a digital modulation scheme with the phase varying continuously from one state to the other. Thus it is equivalent to frequency shift keying except that at the transition the phase is continuous.

Cutoff wavelength: The shortest wavelength at which only the fundamental mode of an optical fiber/waveguide is capable of propagation. For single-mode optical fibers, the cutoff wavelength must be smaller than the operating wavelength to be transmitted.

Dark current noise: The noise component in a photodetector resulting from the shot noise of the current flowing in the un-illuminated photodetector.

Digital transmission: Transmission of a discrete signal (generally obtained by digitizing an analog signal by binary light pulses in a periodic time pattern).

Dispersion: Spread of signal delay in an optical fiber. It consists of various components such as modal dispersion, material dispersion, waveguide dispersion, polarization mode dispersion, group velocity dispersion, and nonlinear phase dispersion. As a result, the optical fiber acts as a low-pass optical filter.

DPSK (differential phase shift keying): Digital modulation technique by changing the phase of the optical carrier whenever there is a change in the transition of the bit period from "1" to "0" or "0" to "1." The phase change happens at the transition instant.

At the receiver, the phase difference between the "π" and "0" phase shift of the carrier can be recovered by an optical phase comparison of the consecutive bits (differential demodulation or self heterodyne detection), and then a balance photodetector pair is used to convert the phase difference to amplitude to obtain the eye diagram.

Under MATLAB® and Simulink® simulations, optical carrier is represented as 1 and its phase is included in the complex amplitude of the sampled value.

DQPSK (differential quadrature phase shift keying): This digital modulation technique uses two DPSK schemes with a $\pi/2$ phase shift between them. Thus at the transmitter there are two optical modulators with an optical phase shift of π/s. Similarly, at the optical receiver, the inphase and quadrature components are split with a $\pi/2$ phase shifter, and the detection of each DPSK can be recovered as in a DPSK receiver.

Equivalent step index profile (ESI): Description of single-mode optical fiber with arbitrary index profile by a step index fiber whose core and refractive index difference are chosen so that the mode field distributions are almost the same.

Fiber: See optical fiber, optical waveguide.

Fiber bandwidth: The frequency at which the magnitude of the transfer function of an optical fiber has fallen to half the mid-band value, that is, the frequency at which the signal attenuation has increased by 3 dB. Since the bandwidth of an optical fiber is approximately reciprocal to its wavelength, for a multimode fiber, the bandwidth–length product is the quality factor, while for single-mode optical fiber, the bit rate is increased by a factor of 2 and the length of the transmission distance is reduced by a factor of 4.

Fiber pigtail: The fiber short length connected to a photodetector or a laser or an optical fiber waveguide component.

Fundamental mode: The lowest order of an optical waveguide. See also modes. It usually has a bell-shaped (Gaussian-like) field distribution and its compass in the near field is characterized by the spot size or mode field diameter (MFD). It is the only mode capable of propagation in single-mode optical fibers.

Graded index fiber: Optical fibers with a graded index profile, that is, with a refractive index profile that varies continuously over the waveguide cross section.

Group index: The velocity of propagation of light pulses in an optical waveguide is the determined by the group index, which is wavelength dependent.

Index profile: Curve of the refractive index over the cross section of an optical waveguide. See also graded-index and step-index fibers.

Induced emission: Stimulated emission.

Injection laser diode: See laser diode.

Irradiance: Power density at a surface through which radiation passes, for example, at the radiating surface of a light source or at the section of an optical waveguide. Normal unit is watts per square meter.

Insertion loss: The attenuation caused by the insertion of an optical component, for example, a connector or an optical coupler, in an optical transmission system.

Intensity: The irradiance of the light source. It is also evaluated by the ratio of optical power over the area of acceptance of the light beam.

Intermodal dispersion: Dispersion due to the differences of the delays between the guided modes in multimode fibers.

Intermodulation effects: Modulation due to the nonlinear effects.

Laser diode: Semiconductor diode that emits coherent light driven by a current above the threshold level to activate stimulated emission. A distinction is made between gain-guided and index-guided laser diodes, whose characteristics such as spectral width of emission and noises can differ considerably. Fabry–Perot and distributed-feedback lasers are commonly employed. The latter is necessary for high-speed external modulation.

Launch angle: Angle between the propagation direction of the incident lightwave and the optical axis of the optical fiber. This angle must be less than the numerical aperture in order for a complete guiding of the lightwave into the fiber.

Launching fiber: A fiber inserted between the source and the fiber so that it can excite certain mode of the optical fiber.

Leaky modes: The mode whose propagation is not within the guided spectrum and radiative to the cladding region.

Light emitting diode: Emitting diode that emits incoherent light—normally a laser diode when biased below the threshold current would behave as an LED.

Lightwaves: Electromagnetic waves whose frequency is in the region of optical frequencies.

Macrobending: Macroscopic axial deviation of a fiber from a straight-line, in contrast to microbending.

Material dispersion: Broadening of pulse due to different group velocity of the lightwaves as a function of the refractive index of the material with respect to the wavelength.

Microbending: Curvature of the fiber that involves displacement of a few micrometers. Microbends cause loss of optical power of lightwaves.

Modal dispersion: Dispersion of modulated lightwave caused by the difference in the propagation paths of different modes of the optical fiber.

Modal noise: Disturbance in the modulated optical signals due to the distribution of energy to different modes of multimode fibers.

Mode distortion: Multimode dispersion.

Mode distribution: The distribution of power among individual modes of a multimode optical fiber.

Modes: Discrete optical waves that can propagate in optical waveguides. They are eigenvalue solutions to the differential equations (the wave equation in the transverse plane) that characterize the waveguide.

Mono-mode fiber: The optical fiber that supports one and only one mode—and only one polarized mode. Note that LP_{01}, as the fundamental mode of the single fiber, has two polarized modes.

MSK (minimum shift keying): This is a digital modulation scheme using both the inphase and quadrature phase components with a phase change from one state to the other states continuously. Thus it is equivalent to frequency shift keying (FSK) but the difference in the frequencies of the "1" and "0" equals a quarter of the bit rate, thus the name minimum shift keying.

At the receiver, the frequency difference can be detected by using narrow band optical filters or by the detection of the phase continuity by using balance optical receiver with an optical phase comparator at the front end of the receiver.

Multimode fiber: Optical fiber in which more than one mode can be guided. The number of modes can reach a few thousands for a 50 μm core diameter and refractive index difference of 0.02. The length–bandwidth product is usually specified for multimode fibers rather than the dispersion factor as for single-mode fibers.

NLSE (nonlinear Schröedinger wave equation): The Schröedinger wave equation involving the timescale, the propagation direction axis z, and the cross section as well as the nonlinear effects due to the intensity dependence of the refractive index of the core and cladding regions.

Numerical aperture: Sine of the maximum possible launched angle of an optical fiber.

Operating lifetime: Period of time in which the principal parameters of a continuously operated component remain within a prescribed range.

Optical balanced receiver: See balanced receiver.

Optical fiber connector: Component for easy connecting between two sections of optical fibers.

Optical fiber coupler: A passive optical device that couples lightwave from one port to the other output ports and vice versa.

Optical fiber splice: A slicing or "soldered" joint of glass fibers.

Optical phase comparator: Optical device used to compare the phases of two sources or one source split to two branches such as in the self heterodyne detection of DPSK and DQPSK optical signals. This device works on the principles of optical interference to give constructive and destructive outputs. A delay interferometer is arranged by using an asymmetric optical Mach–Zehnder configuration.

Optical transmitter: Subsystem for converting electrical signals into optical signals consisting of an optical source and modulated by the electrical signals by direct variation of the driving current of the laser source or by using an external optical modulator to manipulate the lightwaves in the CW mode.

Optical waveguide: Dielectric waveguide with a core and a cladding usually made of glass in pure silica or doped silica. The refractive index of the core and that of the cladding must be arranged so that guided mode or eigenvalue solution of the wave equation can be obtained.

Photodiode: A semiconductor diode that absorbs lightwave power and generating electronic current.

Photon/quantum noise: The absorption of light and its conversion into electronic current is not continuous leading to the quantization of the electrons and current and random emission of electrons in an arbitrary direction, thus quantum noise.

PIN_FET receiver: Receiver with structure consisting of a PIN photodiode followed by a field effect transistor (FET) front end small signal amplifier.

PIN photodetector: Photodiode with the depletion region is in the intrinsic layer sandwiched between the n and p regions of a semiconductor.

Power density: Irradiance.

Probability density function: The distribution of the probability of receiving a "1" or a "0" as a function of the voltage or current level, normally at the output of the receiver.

Quantum noise: The fluctuations of signals of both phase and amplitude in extremely accurate optical systems with stabilized laser and efficient detectors.

Refractive index difference: Difference in the refractive indices of the core and cladding regions.

Refractive index profile: The distribution of the refractive index of the core and the cladding across the cross section of the optical fiber.

Relative refractive index difference: The ratio of the difference between the refractive index of the core and that of the cladding region to the refractive index of the core. Normally for modern single-mode optical fibers this is about 0.3%.

Scattering: Principal cause of the attenuation of an optical waveguide. It results from the microscopic density fluctuations in the glass that deflects part of the light intensity away from its original guided direction. This effect is very weak in the 1500–1700 nm wavelength region and varies at λ^{-6}.

Single-mode fiber: See single-mode optical fiber.

Single-mode optical fiber: Optical waveguide supports only the fundamental modes and cut off all higher-order modes. The fundamental mode is normally termed as linearly polarized (LP) mode. The value of the *V*-parameter must be less than 2.405 in theory and approximately less than 3.0 in practice. The polarization of this mode can be at any direction if the core is perfectly circular and the refractive index is radially uniform. However, in practice, this may not be true; this leads to polarization mode dispersion effects due to the random variation of the core geometry and refractive index distribution across the core.

Splice: Fused junction or welding between two fiber sections.

Splitting element: See Star coupler and directional coupler.

Spontaneous emission: This occurs when there are too many electrons in the conduction band of a semiconductor. These electrons drop simultaneously into a vacant position in the valence band; a photon is emitted for every electron.

Spot size: The field distribution of the fundamental mode of a single-mode optical fiber follows a Gaussian bell shape so that it extends into the cladding; that is, all the power of the guided mode is not concentrated in the core region only. The spot size is defined as the radius of the guided field at which the intensity equals $1/e^2$ of the maximum value at the core center.

Star coupler: Optical fiber coupler or waveguide coupler from any one input port and distributed to all other output ports.

Step index fiber: Optical fiber whose refractive index profile follows a step-like distribution.

Step index optical fiber: See step index fiber.

Stimulated emission: This occurs when photons in the semiconductor simulate available excess charge carriers to radiative recombination; that is, emitted wavelength is the same for all photons and in phase with the incoming photons.

Threshold current: Current level above which the lasing power grows exponentially due to stimulated emission. Below this level, the laser behaves in spontaneous emission mode.

Transfer function: An optical fiber acts as a low-pass filter for the transmitted signals. Whereas at low frequencies only the CW attenuation affects the signal level, higher frequencies are subject to attenuation due to the dispersion of the optical waveguide. This additional attenuation is characterized by the transfer function, which has a maximum value at zero frequency and gradually approaches zero with increasing frequency. The shape of the transfer function is approximately Gaussian. Strictly speaking the transfer function of an optical fiber is complex with a parabolic phase function of square of the frequency.

Transmitter: Optical transmitter is the subsystem of the transmission system that accepts information data that is then used to modulate the optical carrier, which is then transmitted through the optical transmission line. In addition, a coder or forward error coding may be needed in order to improve the signal-to-noise ratio at the receiver.

Waveguide dispersion: The dispersion associated with a non-monochromatic light source or modulated lightwaves resulting from the fact that the propagation constant of the guided wave is dependent on the parameters of a single-mode fiber, especially the radius and the operating wavelength, that is, the V-parameter. This leads to the variation of the propagation phase velocity of lightwave components of the spectrum and hence dispersion due to different arrival times. In practice, waveguide dispersion always acts in combination with the dispersion effects due to material property, the refractive index of the guided medium is a function of the wavelength. The combined dispersion effect is called the chromatic dispersion.

Wavelength division multiplexing (WDM): Simultaneous transmission of several signals in an optical fiber at different wavelength lightwave carriers. Dense WDM is used when the spacing between the optical carriers is in the order of less than 100 GHz.

Index

E

M

P